Birkhäuser

Progress in Mathematics

Volume 257

More information about this series at http://www.springer.com/series/4848

Roger Howe • Markus Hunziker
Jeb F. Willenbring
Editors

Symmetry: Representation Theory and Its Applications

In Honor of Nolan R. Wallach

Editors
Roger Howe
Department of Mathematics
Yale University
New Haven, CT, USA

Markus Hunziker
Department of Mathematics
Baylor University
Waco, TX, USA

Jeb F. Willenbring
Department of Mathematical Sciences
University of Wisconsin-Milwaukee
Milwaukee, WI, USA

ISSN 0743-1643 ISSN 2296-505X (electronic)
ISBN 978-1-4939-1589-7 ISBN 978-1-4939-1590-3 (eBook)
DOI 10.1007/978-1-4939-1590-3
Springer New York Heidelberg Dordrecht London

Library of Congress Control Number: 2014958009

Mathematics Subject Classification (2010): 05E05, 11F70, 14L35, 17B10, 17B67, 20G05, 20G20, 22E30, 22E40, 22E41, 22E45, 22E46, 32K05, 53C40, 57T15, 58G03, 58G25, 81P15, 81P68

Printed on acid-free paper

Springer is part of Springer Science+Business Media (www.birkhauser-science.com)

Nolan R. Wallach (2013)

Nolan and Barbara Wallach, Wedding (1965)

Nolan R. Wallach, Helsinki (ICM 1978)

At the UCSD 2011 Conference, surrounding Nolan Wallach (standing in front row) are his students and postdoctoral fellows: Alvany Rocha, Roberto Miatello, Markus Hunziker, Andrew Linshaw, Mandy Cheung, Asif Shakeel, Mark Colarusso, Jon Middleton, Seung Lee, Danny Goldstein, Oded Yacobi, Karin Baur, Gilad Gour, Rino Sanchez, Jeb Willenbring, Sam Evens, Laura Barberis, Allan Keeton, Pierluigi Möseneder Frajria, Jochen Kuttler.

Preface

Nolan Wallach's contributions to mathematics, made over a period now stretching to nearly five decades, exhibit a breadth of knowledge, research and scholarship matched by few contemporary mathematicians. Although all his research is rooted in his love of geometry and symmetry, one can perhaps usefully distinguish four periods during which this underlying theme was expressed in distinctive ways.

In the first period, roughly 1965–1972, the main motivation was Riemannian geometry (spaces of positive curvature and minimal submanifolds) leading to Do Carmo–Wallach theory, Aloff–Wallach spaces, and Wallach manifolds.

In the second period (1972–1980) the focus of his research moved into the theory of infinite-dimensional representations of semisimple Lie groups and also certain infinite-dimensional groups, with emphasis on unitary representations. He discovered what is now called the Wallach set of unitary positive energy representations. He also began his research into homological methods in representation theory with an eye on applications to the theory of automorphic forms, including collaboration with Armand Borel to produce *Continuous Cohomology, Discrete Subgroups, and Representations of Reductive Groups*.

In the third period (1980–1992), Nolan's attention moved to the analytic aspects of representation theory. A notable result was the Casselman–Wallach theorem, which affirms the unity of the algebraic and analytic viewpoints toward irreducible admissible representations of semisimple Lie groups. In harmonic analysis, he gave a proof of the Whittaker Plancherel theorem. He also began his long and fruitful collaboration with Roe Goodman. This period culminated with the publication of *Real Reductive Groups II*, the second volume of his masterful synthesis of the many remarkable developments in representation theory during the 1980s.

In the fourth period, since 1992, Nolan's research shows a renewed interest in algebraic geometry and physics. Here his main emphasis involves applying mathematics related to geometric invariant theory over the real and complex numbers to representation theory, combinatorics and to physics, in particular to quantum information theory. His continuing collaboration with Roe Goodman resulted in publication of the widely influential *Representations and Invariants of the Classical Groups* (which might be thought of as a *Jugendtraum* from his early days as a differential geometer).

Nolan's early training was in differential geometry, and included the study of Lie theory. His thesis advisor was Jun-Ichi Hano, whose mathematical genealogy goes back to Takagi and Hilbert, and whose research interests included homogeneous bounded complex domains, and studies in infinite-dimensional representation theory. Nolan's thesis under Hano dealt with root systems and included a new approach to the classification of real simple Lie algebras.

While Nolan's early papers dealt primarily with differential geometry, Lie theory, and representation theory in particular, is used generously. With M. Do Carmo, he studied minimal immersions of spheres into spheres by making use of spherical harmonics. He studied homogeneous manifolds of positive curvature, including a classification of the even-dimensional ones. Perhaps the most striking contribution from this phase of his work is his construction with S. Aloff of an infinite family of 7-dimensional homogeneous spaces for SU_3 with strictly positive curvature.

His training in Lie theory made the transition from geometry to representation theory easy, and in fact, representation theory is already the focus of some of his earliest papers. When he moved to Rutgers after a postdoctoral job at Berkeley, the Institute for Advanced Study was within easy reach, and Nolan visited there frequently. He became one of the most faithful attendees at Harish-Chandra's annual lecture series on representation theory. Nolan also engaged with other faculty at IAS, notably Armand Borel. As mentioned above, they produced a major book on $(\mathfrak{g}, K)$-cohomology of representations and its application to the cohomology of locally symmetric spaces. The first edition of their book appeared as an Annals of Mathematics Study in 1980. A second edition was published as Mathematical Survey and Monograph of the American Mathematical Society.

Nolan worked at Rutgers for twenty years. While there he collaborated with a large number of colleagues both at Rutgers and elsewhere, on a great variety of topics: with many people, including J. Lepowsky, K. Johnson, T.J. Enright, and A. Knapp on representation theory; with S. Greenfield on partial differential equations; with D. DeGeorge and R. Hotta on automorphic forms; with Armand Borel, as described above. He also found time to make discoveries on his own, including his analysis of the "analytic continuation" of the holomorphic discrete series, in which he identified what is now known as the Wallach set—the set of parameters when the analytic continuation of the scalar holomorphic discrete series has a unitary irreducible quotient. This work was in some sense completed in a 1983 paper with T.J. Enright and R. Howe, classifying all holomorphic unitary representations of Hermitian symmetric groups.

When completely integrable systems became the rage in the early 1980s, Nolan quickly assimilated the ideas involved, and wrote a number of papers, with his student A. Rocha-Caridi, on his own, and with his colleague Roe Goodman. This was the beginning of a long and fruitful collaboration with Goodman, resulting in over 10 joint works, including their widely used book on invariant theory of the classical groups. The second edition of this book has been published as a Springer Graduate Text, under the title *Symmetry, Representations, and Invariants*.

The late 1970s and 1980s saw rapid progress in the representation theory of reductive groups, with the ideas of cohomological induction, D-modules, and the Kazhdan–Lusztig conjectures, as well as the classifications of irreducible admissible representations by Langlands, by Vogan, and by Beilinson–Bernstein. Nolan made key contributions, including a unification of the analytic and algebraic approaches to representation theory by proving, partly in collaboration with B. Casselman, that all Harish-Chandra modules, representations of a reductive Lie algebra which served as algebraic proxies for infinite-dimensional group representations, could in fact be globalized, in a more or less unique way, to a representation of the associated group. He also contributed a very simple argument that cohomologically induced representations could be unitarized under suitable restrictions on the parameters involved. As noted above, many of these developments are described in his remarkable two-part synthesis of the main foundational results in representation theory, *Real Reductive Groups, I* and *II*, published by Academic Press, which remain the most complete and coherent account of the striking progress due to many people throughout the 1980s; the developments in these volumes also reflect a broad picture of general ideas of representation theory.

In 1989, Nolan, who had been at Rutgers for two decades and had seemed immovable in the face of many attempts to lure him elsewhere, finally succumbed to the attractions of San Diego, and joined the Mathematics Department at UCSD. During his first years at UCSD, he completed *Representations of Reductive Groups, II*, the second volume of his account of the representation theory of reductive Lie groups. In the first volume, he had established the Langlands classification of irreducible admissible representations, the construction of the discrete series, and Harish-Chandra's character theory, including the fundamental regularity theorem, that the characters of irreducible representations, which are distributions, are in fact given by integration against locally L^1 functions (for which explicit formulas are in principle available). In the second volume, he builds on these results, plus his joint results with David A. Vogan on intertwining operators, to give a complete account of Harish-Chandra's Plancherel Theorem for reductive Lie groups.

One of the first papers Nolan wrote after completion of that massive project was *Invariant differential operators on a reductive Lie algebra and Weyl group representations* (Journal of the American Mathematical Society, 1993). This paper makes a lovely connection between Harish-Chandra's theory of invariant differential operators on a reductive Lie algebra and the theory of "Springer representations", which associates representations of the Weyl group to nilpotent orbits. It also contains a beautiful new and drastically simpler proof of Harish-Chandra's famous local L^1 theorem for invariant eigendistributions on a semisimple Lie algebra. It is a pity that these insights were not available during the writing of *Real Reductive Groups*.

Despite the distance of San Diego from other centers of mathematical research, Nolan has never lacked for visitors. Among the "regulars" have been Benedict Gross, Bertram Kostant, and Hanspeter Kraft. Nolan has coauthored several papers with each of them. With Gross, Nolan constructed a distinguished family of unitary representations for the exceptional groups of real rank $= 4$ by a continuation of "quaternionic discrete series." With Kostant, he developed Gel'fand–Zeitlin theory from the perspective of classical mechanics, and with Kraft, he studied the geometry of finite-dimensional representations.

At UCSD, Nolan also enjoyed the presence of A.M. Garsia with whom he regularly shared ideas and found mutual inspiration. Starting in the 2000s, they began writing papers together, and established several difficult results on quasi-symmetric polynomials and invariant theory.

In the late 1990s, Nolan became interested in quantum information theory, and especially, quantum entanglement. Using his extensive knowledge of invariant theory he was able to make significant contributions to the field. In a 2002 paper with D. Meyer, he defined a simple polynomial measure of multiparticle entanglement which is scalable, i.e., which applies to any number of spin 1/2 particles. A recurring theme in his work on quantum entanglement and invariant theory has been the calculation of explicit Hilbert series. In these computations, he was also able to exercise his considerable expertise in conventional computation to solve problems that not long ago seemed out of reach.

Nolan continues to mentor students, so that now approximately half of his students have degrees from Rutgers, and half from UCSD. We are delighted to recognize and celebrate Nolan's inspiring mathematical journey, which is still very much in progress.

Acknowledgments. We have very much appreciated the support of our Birkhäuser Science editors, Allen Mann and Kristin Purdy. The conference *Lie Theory and Its Applications* (supported in part by NSA grant H98230-10-1-0239 and NSF grant 1105825) provided a venue for discussions planning this volume. Without question, these discussions were led by Ann Kostant. Now, over three years later, we have the pleasure of acknowledging her continued support and guidance, which profoundly influenced this work. We thank you, Ann, for helping us honor our friend and teacher, Nolan Wallach.

New Haven, CT, USA — Roger Howe
Waco, TX, USA — Markus Hunziker
Milwaukee, WI, USA — Jeb F. Willenbring
April 2014

Publications of Nolan R. Wallach

1. Nolan Russell Wallach, *A classification of real simple Lie algebras*, ProQuest LLC, Ann Arbor, MI, 1966, Thesis (Ph.D.)–Washington University in St. Louis.
2. N. R. Wallach, *On maximal subsystems of root systems*, Canad. J. Math. **20** (1968), 555–574.
3. N. R. Wallach, *Induced representations of Lie algebras and a theorem of Borel-Weil.*, Trans. Amer. Math. Soc. **136** (1969), 181–187.
4. N. R. Wallach, *Induced representations of Lie algebras. II*, Proc. Amer. Math. Soc. **21** (1969), 161–166.
5. M. P. do Carmo and N. R. Wallach, *Minimal immersions of spheres into spheres*, Proc. Nat. Acad. Sci. U.S.A. **63** (1969), 640–642.
6. N. R. Wallach, *Homogeneous positively pinched Riemannian manifolds*, Bull. Amer. Math. Soc. **76** (1970), 783–786.
7. N. R. Wallach and F. W. Warner, *Curvature forms for 2-manifolds*, Proc. Amer. Math. Soc. **25** (1970), 712–713.
8. M. P. do Carmo and N. R. Wallach, *Representations of compact groups and minimal immersions into spheres.*, J. Differential Geometry **4** (1970), 91–104.
9. M. Cahen and N. Wallach, *Lorentzian symmetric spaces*, Bull. Amer. Math. Soc. **76** (1970), 585–591.
10. M. P. Do Carmo and N. R. Wallach, *Minimal immersions of sphere bundles over spheres*, An. Acad. Brasil. Ci. **42** (1970), 5–9.
11. N. R. Wallach, *Extension of locally defined minimal immersions into spheres*, Arch. Math. (Basel) **21** (1970), 210–213.
12. S. J. Greenfield and N. R. Wallach, *Extendibility properties of submanifolds of C2*, Proc. Carolina Conf. on Holomorphic Mappings and Minimal Surfaces (Chapel Hill, N.C., 1970), Dept. of Math., Univ. of North Carolina, Chapel Hill, N.C., 1970, pp. 77–85.
13. S. J. Greenfield and N. R. Wallach, *The Hilbert ball and bi-ball are holomorphically inequivalent*, Bull. Amer. Math. Soc. **77** (1971), 261–263.

14. M. P. do Carmo and N. R. Wallach, *Minimal immersions of spheres into spheres*, Ann. of Math. (2) **93** (1971), 43–62.
15. N. R. Wallach, *Cyclic vectors and irreducibility for principal series representations.*, Trans. Amer. Math. Soc. **158** (1971), 107–113.
16. N. R. Wallach, *Three new examples of compact manifolds admitting Riemannian structures of positive curvature*, Bull. Amer. Math. Soc. **78** (1972), 55–56.
17. S. J. Greenfield and N. R. Wallach, *Automorphism groups of bounded domains in Banach spaces*, Trans. Amer. Math. Soc. **166** (1972), 45–57.
18. S. J. Greenfield and N. R. Wallach, *Global hypoellipticity and Liouville numbers*, Proc. Amer. Math. Soc. **31** (1972), 112–114.
19. N. R. Wallach, *Compact homogeneous Riemannian manifolds with strictly positive curvature*, Ann. of Math. (2) **96** (1972), 277–295.
20. K. Johnson and N. R. Wallach, *Composition series and intertwining operators for the spherical principal series*, Bull. Amer. Math. Soc. **78** (1972), 1053–1059.
21. N. R. Wallach, *Cyclic vectors and irreducibility for principal series representations. II*, Trans. Amer. Math. Soc. **164** (1972), 389–396.
22. N. R. Wallach, *Minimal immersions of symmetric spaces into spheres*, Symmetric spaces (Short Courses, Washington Univ., St. Louis, Mo., 1969–1970), Dekker, New York, 1972, pp. 1–40. Pure and Appl. Math., Vol. 8.
23. Stephen J. Greenfield and N. R. Wallach, *Globally hypoelliptic vector fields*, Topology **12** (1973), 247–254.
24. J. Lepowsky and N. R. Wallach, *Finite- and infinite-dimensional representation of linear semisimple groups*, Trans. Amer. Math. Soc. **184** (1973), 223–246.
25. N. R. Wallach, *Kostant's P^γ and R^γ matrices and intertwining integrals*, Harmonic analysis on homogeneous spaces (Proc. Sympos. Pure Math., Vol. XXVI, Williams Coll., Williamstown, Mass., 1972), Amer. Math. Soc., Providence, R.I., 1973, pp. 269–273.
26. S. J. Greenfield and N. R. Wallach, *Remarks on global hypoellipticity*, Trans. Amer. Math. Soc. **183** (1973), 153–164.
27. N. R. Wallach, *Harmonic analysis on homogeneous spaces*, Marcel Dekker, Inc., New York, 1973, Pure and Applied Mathematics, No. 19.
28. S. Aloff and N. R. Wallach, *An infinite family of distinct* 7*-manifolds admitting positively curved Riemannian structures*, Bull. Amer. Math. Soc. **81** (1975), 93–97.
29. R. Hotta and N. R. Wallach, *On Matsushima's formula for the Betti numbers of a locally symmetric space*, Osaka J. Math. **12** (1975), no. 2, 419–431.
30. N. R. Wallach, *On Harish-Chandra's generalized C-functions*, Amer. J. Math. **97** (1975), 386–403.
31. N. R. Wallach, *On the unitarizability of representations with highest weights*, Non-commutative harmonic analysis (Actes Colloq., Marseille-Luminy, 1974), Springer, Berlin, 1975, pp. 226–231. Lecture Notes in Math., Vol. 466.

32. N. R. Wallach, *On the Selberg trace formula in the case of compact quotient*, Bull. Amer. Math. Soc. **82** (1976), no. 2, 171–195.
33. N. R. Wallach, *An asymptotic formula of Gelfand and Gangolli for the spectrum of* $G\backslash G$, J. Differential Geometry **11** (1976), no. 1, 91–101.
34. A. W. Knapp and N. R. Wallach, *Szegö kernels associated with discrete series*, Invent. Math. **34** (1976), no. 3, 163–200.
35. N. R. Wallach, *On the Enright-Varadarajan modules: a construction of the discrete series*, Ann. Sci. École Norm. Sup. (4) **9** (1976), no. 1, 81–101.
36. K. D. Johnson and N. R. Wallach, *Composition series and intertwining operators for the spherical principal series. I*, Trans. Amer. Math. Soc. **229** (1977), 137–173.
37. N. R. Wallach, *Symplectic geometry and Fourier analysis*, Math Sci Press, Brookline, Mass., 1977, With an appendix on quantum mechanics by Robert Hermann, Lie Groups: History, Frontiers and Applications, Vol. V.
38. T. J. Enright and N. R. Wallach, *The fundamental series of representations of a real semisimple Lie algebra*, Acta Math. **140** (1978), no. 1–2, 1–32.
39. D. L. de George and N. R. Wallach, *Limit formulas for multiplicities in* $L^2(\Gamma\backslash G)$, Ann. of Math. (2) **107** (1978), no. 1, 133–150.
40. N. R. Wallach, *Representations of semi-simple Lie groups and Lie algebras*, Lie theories and their applications (Proc. Ann. Sem. Canad. Math. Congr., Queen's Univ., Kingston, Ont., 1977), Academic Press, New York, 1978, pp. 154–246.
41. N. R. Wallach, *Representations of reductive Lie groups*, Automorphic forms, representations and L-functions (Proc. Sympos. Pure Math., Oregon State Univ., Corvallis, Ore., 1977), Part 1, Proc. Sympos. Pure Math., XXXIII, Amer. Math. Soc., Providence, R.I., 1979, pp. 71–86.
42. N. R. Wallach, *The analytic continuation of the discrete series. I, II*, Trans. Amer. Math. Soc. **251** (1979), 1–17, 19–37.
43. D. L. DeGeorge and N. R. Wallach, *Limit formulas for multiplicities in* $L^2(\Gamma\backslash G)$*. II. The tempered spectrum*, Ann. of Math. (2) **109** (1979), no. 3, 477–495.
44. T. J. Enright and N. R. Wallach, *Notes on homological algebra and representations of Lie algebras*, Duke Math. J. **47** (1980), no. 1, 1–15.
45. N. R. Wallach, *The spectrum of compact quotients of semisimple Lie groups*, Proceedings of the International Congress of Mathematicians (Helsinki, 1978), Acad. Sci. Fennica, Helsinki, 1980, pp. 715–719.
46. A. W. Knapp and N. R. Wallach, *Correction and addition: "Szegő kernels associated with discrete series" [Invent. Math.* **34** *(1976), no. 3, 163–200; MR* **54** *#7704]*, Invent. Math. **62** (1980/81), no. 2, 341–346.
47. R. Goodman and N. R. Wallach, *Whittaker vectors and conical vectors*, J. Funct. Anal. **39** (1980), no. 2, 199–279.
48. A. Borel and N. R. Wallach, *Continuous cohomology, discrete subgroups, and representations of reductive groups*, Annals of Mathematics Studies, Vol. 94, Princeton University Press, Princeton, N.J.; University of Tokyo Press, Tokyo, 1980.

49. N. R. Wallach, *The restriction of Whittaker modules to certain parabolic subalgebras*, Proc. Amer. Math. Soc. **81** (1981), no. 2, 181–188.
50. J. T. Stafford and N. R. Wallach, *The restriction of admissible modules to parabolic subalgebras*, Trans. Amer. Math. Soc. **272** (1982), no. 1, 333–350.
51. A. Rocha-Caridi and N. R. Wallach, *Projective modules over graded Lie algebras. I*, Math. Z. **180** (1982), no. 2, 151–177.
52. R. Goodman and N. R. Wallach, *Classical and quantum-mechanical systems of Toda lattice type. I*, Comm. Math. Phys. **83** (1982), no. 3, 355–386.
53. A. Rocha-Caridi and N. R. Wallach, *Characters of irreducible representations of the Lie algebra of vector fields on the circle*, Invent. Math. **72** (1983), no. 1, 57–75.
54. N. R. Wallach, *Asymptotic expansions of generalized matrix entries of representations of real reductive groups*, Lie group representations, I (College Park, Md., 1982/1983), Lecture Notes in Math., Vol. 1024, Springer, Berlin, 1983, pp. 287–369.
55. T. Enright, R. Howe, and N. Wallach, *A classification of unitary highest weight modules*, Representation theory of reductive groups (Park City, Utah, 1982), Progr. Math., Vol. 40, Birkhäuser Boston, Boston, MA, 1983, pp. 97–143.
56. N. R. Wallach and J. A. Wolf, *Completeness of Poincaré series for automorphic forms associated to the integrable discrete series*, Representation theory of reductive groups (Park City, Utah, 1982), Progr. Math., Vol. 40, Birkhäuser Boston, Boston, MA, 1983, pp. 265–281.
57. Alvany Rocha-Caridi and N. R. Wallach, *Highest weight modules over graded Lie algebras: resolutions, filtrations and character formulas*, Trans. Amer. Math. Soc. **277** (1983), no. 1, 133–162.
58. Alvany Rocha-Caridi and N. R. Wallach, *Characters of irreducible representations of the Virasoro algebra*, Math. Z. **185** (1984), no. 1, 1–21.
59. N. R. Wallach, *Square integrable automorphic forms and cohomology of arithmetic quotients of* $\mathrm{SU}(p, q)$, Math. Ann. **266** (1984), no. 3, 261–278.
60. Roe Goodman and N. R. Wallach, *Structure and unitary cocycle representations of loop groups and the group of diffeomorphisms of the circle*, J. Reine Angew. Math. **347** (1984), 69–133.
61. N. R. Wallach, *On the constant term of a square integrable automorphic form*, Operator algebras and group representations, Vol. II (Neptun, 1980), Monogr. Stud. Math., Vol. 18, Pitman, Boston, MA, 1984, pp. 227–237.
62. Roe Goodman and N. R. Wallach, *Erratum to the paper: "Structure and unitary cocycle representations of loop groups and the group of diffeomorphisms of the circle" [J. Reine Angew. Math.* **347** *(1984), 69–133]*, J. Reine Angew. Math. **352** (1984), 220.
63. Roe Goodman and N. R. Wallach, *Classical and quantum mechanical systems of Toda-lattice type. II. Solutions of the classical flows*, Comm. Math. Phys. **94** (1984), no. 2, 177–217.
64. N. R. Wallach, *On the unitarizability of derived functor modules*, Invent. Math. **78** (1984), no. 1, 131–141.

65. N. R. Wallach, *The asymptotic behavior of holomorphic representations*, Mém. Soc. Math. France (N.S.) (1984), no. 15, 291–305, Harmonic analysis on Lie groups and symmetric spaces (Kleebach, 1983).
66. T. J. Enright, R. Parthasarathy, N. R. Wallach, and J. A. Wolf, *Unitary derived functor modules with small spectrum*, Acta Math. **154** (1985), no. 1–2, 105–136.
67. N. R. Wallach, *Classical invariant theory and the Virasoro algebra*, Vertex operators in mathematics and physics (Berkeley, Calif., 1983), Math. Sci. Res. Inst. Publ., Vol. 3, Springer, New York, 1985, pp. 475–482.
68. Roe Goodman and N. R. Wallach, *Projective unitary positive-energy representations of* $\mathrm{Diff}(S^1)$, J. Funct. Anal. **63** (1985), no. 3, 299–321.
69. Roe Goodman and N. R. Wallach, *Positive-energy representations of the group of diffeomorphisms of the circle*, Infinite-dimensional groups with applications (Berkeley, Calif., 1984), Math. Sci. Res. Inst. Publ., Vol. 4, Springer, New York, 1985, pp. 125–135.
70. Roe Goodman and N. R. Wallach, *Classical and quantum mechanical systems of Toda-lattice type. III. Joint eigenfunctions of the quantized systems*, Comm. Math. Phys. **105** (1986), no. 3, 473–509.
71. N. R. Wallach, *On the irreducibility and inequivalence of unitary representations of gauge groups*, Compositio Math. **64** (1987), no. 1, 3–29.
72. N. R. Wallach, *A class of nonstandard modules for affine Lie algebras*, Math. Z. **196** (1987), no. 3, 303–313.
73. N. R. Wallach, *Real reductive groups. I*, Pure and Applied Mathematics, Vol. 132, Academic Press, Inc., Boston, MA, 1988.
74. N. R. Wallach, *Lie algebra cohomology and holomorphic continuation of generalized Jacquet integrals*, Representations of Lie groups, Kyoto, Hiroshima, 1986, Adv. Stud. Pure Math., Vol. 14, Academic Press, Boston, MA, 1988, pp. 123–151.
75. Roe Goodman and N. R. Wallach, *Higher-order Sugawara operators for affine Lie algebras*, Trans. Amer. Math. Soc. **315** (1989), no. 1, 1–55.
76. R. Miatello and N. R. Wallach, *Automorphic forms constructed from Whittaker vectors*, J. Funct. Anal. **86** (1989), no. 2, 411–487.
77. D. A. Vogan, Jr. and N. R. Wallach, *Intertwining operators for real reductive groups*, Adv. Math. **82** (1990), no. 2, 203–243.
78. R. Miatello and N. R. Wallach, *Kuznetsov formulas for real rank one groups*, J. Funct. Anal. **93** (1990), no. 1, 171–206.
79. N. R. Wallach, *Limit multiplicities in* $L^2(\Gamma\backslash G)$, Cohomology of arithmetic groups and automorphic forms (Luminy-Marseille, 1989), Lecture Notes in Math., Vol. 1447, Springer, Berlin, 1990, pp. 31–56.
80. Nolan Wallach, *Some applications of group representations*, AMS-MAA Joint Lecture Series, American Mathematical Society, Providence, RI, 1990, A joint AMS-MAA lecture presented in Louisville, Kentucky, January 1990.

81. N. R. Wallach and R. Miatello, *Kuznetsov formulas for products of groups of R-rank one*, Festschrift in honor of I. I. Piatetski-Shapiro on the occasion of his sixtieth birthday, Part II (Ramat Aviv, 1989), Israel Math. Conf. Proc., Vol. 3, Weizmann, Jerusalem, 1990, pp. 305–320.
82. N. R. Wallach, *The powers of the resolvent on a locally symmetric space*, Bull. Soc. Math. Belg. Sér. A **42** (1990), no. 3, 777–795, Algebra, groups and geometry.
83. N. R. Wallach, *On the distribution of eigenvalues of the Laplacian of a locally symmetric space*, Differential geometry, Pitman Monogr. Surveys Pure Appl. Math., Vol. 52, Longman Sci. Tech., Harlow, 1991, pp. 337–350.
84. N. R. Wallach, *Real reductive groups. II*, Pure and Applied Mathematics, Vol. 132, Academic Press, Inc., Boston, MA, 1992.
85. R. Miatello and N. R. Wallach, *The resolvent of the Laplacian on locally symmetric spaces*, J. Differential Geom. **36** (1992), no. 3, 663–698.
86. N. R. Wallach, *Automorphic forms*, New developments in Lie theory and their applications (Córdoba, 1989), Progr. Math., Vol. 105, Birkhäuser Boston, Boston, MA, 1992, Notes by Roberto Miatello, pp. 1–25.
87. N. R. Wallach, *Polynomial differential operators associated with Hermitian symmetric spaces*, Representation theory of Lie groups and Lie algebras (Fuji-Kawaguchiko, 1990), World Sci. Publ., River Edge, NJ, 1992, pp. 76–94.
88. N. R. Wallach, *Invariant differential operators on a reductive Lie algebra and Weyl group representations*, J. Amer. Math. Soc. **6** (1993), no. 4, 779–816.
89. N. R. Wallach, *Transfer of unitary representations between real forms*, Representation theory and analysis on homogeneous spaces (New Brunswick, NJ, 1993), Contemp. Math., Vol. 177, Amer. Math. Soc., Providence, RI, 1994, pp. 181–216.
90. Benedict H. Gross and N. R. Wallach, *A distinguished family of unitary representations for the exceptional groups of real rank* = 4, Lie theory and geometry, Progr. Math., Vol. 123, Birkhäuser Boston, Boston, MA, 1994, pp. 289–304.
91. N. R. Wallach, *C^∞ vectors*, Representations of Lie groups and quantum groups (Trento, 1993), Pitman Res. Notes Math. Ser., Vol. 311, Longman Sci. Tech., Harlow, 1994, pp. 205–270.
92. N. R. Wallach and M. Hunziker, *On the Harish-Chandra homomorphism of invariant differential operators on a reductive Lie algebra*, Representation theory and harmonic analysis (Cincinnati, OH, 1994), Contemp. Math., Vol. 191, Amer. Math. Soc., Providence, RI, 1995, pp. 223–243.
93. N. R. Wallach, *On a theorem of Milnor and Thom*, Topics in geometry, Progr. Nonlinear Differential Equations Appl., Vol. 20, Birkhäuser Boston, Boston, MA, 1996, pp. 331–348.
94. Jing-Song Huang, Toshio Oshima, and Nolan Wallach, *Dimensions of spaces of generalized spherical functions*, Amer. J. Math. **118** (1996), no. 3, 637–652.
95. Benedict H. Gross and N. R. Wallach, *On quaternionic discrete series representations, and their continuations*, J. Reine Angew. Math. **481** (1996), 73–123.

96. Thomas J. Enright and N. R. Wallach, *Embeddings of unitary highest weight representations and generalized Dirac operators*, Math. Ann. **307** (1997), no. 4, 627–646.
97. Roe Goodman and N. R. Wallach, *Representations and invariants of the classical groups*, Encyclopedia of Mathematics and its Applications, Vol. 68, Cambridge University Press, Cambridge, 1998.
98. N. R. Wallach, *A variety of solutions to the Yang-Baxter equation*, Advances in geometry, Progr. Math., Vol. 172, Birkhäuser Boston, Boston, MA, 1999, pp. 391–399.
99. Hanspeter Kraft, Lance W. Small, and N. R. Wallach, *Hereditary properties of direct summands of algebras*, Math. Res. Lett. **6** (1999), no. 3–4, 371–375.
100. R. W. Bruggeman, R. J. Miatello, and N. R. Wallach, *Resolvent and lattice points on symmetric spaces of strictly negative curvature*, Math. Ann. **315** (1999), no. 4, 617–639.
101. A. Borel and N. Wallach, *Continuous cohomology, discrete subgroups, and representations of reductive groups*, second ed., Mathematical Surveys and Monographs, Vol. 67, American Mathematical Society, Providence, RI, 2000.
102. N. R. Wallach and J. Willenbring, *On some q-analogs of a theorem of Kostant-Rallis*, Canad. J. Math. **52** (2000), no. 2, 438–448.
103. B. Gross and N. Wallach, *Restriction of small discrete series representations to symmetric subgroups*, The mathematical legacy of Harish-Chandra (Baltimore, MD, 1998), Proc. Sympos. Pure Math., Vol. 68, Amer. Math. Soc., Providence, RI, 2000, pp. 255–272.
104. Hanspeter Kraft, Lance W. Small, and N. R. Wallach, *Properties and examples of FCR-algebras*, Manuscripta Math. **104** (2001), no. 4, 443–450.
105. David A. Meyer and N. R. Wallach, *Global entanglement in multiparticle systems*, J. Math. Phys. **43** (2002), no. 9, 4273–4278, Quantum information theory.
106. N. R. Wallach, *An unentangled Gleason's theorem*, Quantum computation and information (Washington, DC, 2000), Contemp. Math., Vol. 305, Amer. Math. Soc., Providence, RI, 2002, pp. 291–298.
107. Hanspeter Kraft and N. R. Wallach, *On the separation property of orbits in representation spaces*, J. Algebra **258** (2002), no. 1, 228–254, Special issue in celebration of Claudio Procesi's 60th birthday.
108. David A. Meyer and Nolan Wallach, *Invariants for multiple qubits: the case of 3 qubits*, Mathematics of quantum computation, Comput. Math. Ser., Chapman & Hall/CRC, Boca Raton, FL, 2002, pp. 77–97.
109. N. R. Wallach, *Generalized Whittaker vectors for holomorphic and quaternionic representations*, Comment. Math. Helv. **78** (2003), no. 2, 266–307.
110. A. M. Garsia and N. Wallach, *Qsym over Sym is free*, J. Combin. Theory Ser. A **104** (2003), no. 2, 217–263.
111. A. M. Garsia and N. R. Wallach, *Some new applications of orbit harmonics*, Sém. Lothar. Combin. **50** (2003/04), Art. B50j, 47.
112. Thomas J. Enright, Markus Hunziker, and N. R. Wallach, *A Pieri rule for Hermitian symmetric pairs. I*, Pacific J. Math. **214** (2004), no. 1, 23–30.

113. J. Kuttler and N. Wallach, *Representations of* SL_2 *and the distribution of points in* $\mathbb{P}^n$, Noncommutative harmonic analysis, Progr. Math., Vol. 220, Birkhäuser Boston, Boston, MA, 2004, pp. 355–373.
114. Thomas J. Enright and N. R. Wallach, *A Pieri rule for Hermitian symmetric pairs. II*, Pacific J. Math. **216** (2004), no. 1, 51–61.
115. A. M. Garsia and N. Wallach, *Combinatorial aspects of the Baker-Akhiezer functions for* S_2, European J. Combin. **25** (2004), no. 8, 1231–1262.
116. N. R. Wallach, *Armand Borel: A reminiscence*, Asian J. Math. **8** (2004), no. 4, xxv–xxxi.
117. N. R. Wallach and Chen-Bo Zhu, *Transfer of unitary representations*, Asian J. Math. **8** (2004), no. 4, 861–879.
118. J. Bell, A. M. Garsia, and N. Wallach, *Some new methods in the theory of* m*-quasi-invariants*, Electron. J. Combin. **11** (2004/06), no. 2, Research Paper 20, 32.
119. Karin Baur and Nolan Wallach, *Nice parabolic subalgebras of reductive Lie algebras*, Represent. Theory **9** (2005), 1–29 (electronic).
120. Karin Baur and Nolan Wallach, *Erratum: "Nice parabolic subalgebras of reductive Lie algebras" [Represent. Theory* **9** *(2005), 1–29 (electronic)]*, Represent. Theory **9** (2005), 267 (electronic).
121. N. R. Wallach, *The Hilbert series of measures of entanglement for 4 qubits*, Acta Appl. Math. **86** (2005), no. 1–2, 203–220.
122. Jessica Benton, Rion Snow, and Nolan Wallach, *A combinatorial problem associated with nonograms*, Linear Algebra Appl. **412** (2006), no. 1, 30–38.
123. Bertram Kostant and Nolan Wallach, *Gelfand-Zeitlin theory from the perspective of classical mechanics. II*, The unity of mathematics, Progr. Math., Vol. 244, Birkhäuser Boston, Boston, MA, 2006, pp. 387–420.
124. Bertram Kostant and Nolan Wallach, *Gelfand-Zeitlin theory from the perspective of classical mechanics. I*, Studies in Lie theory, Progr. Math., Vol. 243, Birkhäuser Boston, Boston, MA, 2006, pp. 319–364.
125. Hanspeter Kraft and N. R. Wallach, *On the nullcone of representations of reductive groups*, Pacific J. Math. **224** (2006), no. 1, 119–139.
126. A. M. Garsia and N. Wallach, *The non-degeneracy of the bilinear form of* m*-quasi-invariants*, Adv. in Appl. Math. **37** (2006), no. 3, 309–359.
127. N. R. Wallach, *Holomorphic continuation of generalized Jacquet integrals for degenerate principal series*, Represent. Theory **10** (2006), 380–398 (electronic).
128. A. M. Garsia and N. Wallach, r*-Qsym is free over Sym*, J. Combin. Theory Ser. A **114** (2007), no. 4, 704–732.
129. Karin Baur and Nolan Wallach, *A class of gradings of simple Lie algebras*, Lie algebras, vertex operator algebras and their applications, Contemp. Math., Vol. 442, Amer. Math. Soc., Providence, RI, 2007, pp. 3–15.

130. Ron Evans and Nolan Wallach, *Pfaffians and strategies for integer choice games*, Harmonic analysis, group representations, automorphic forms and invariant theory, Lect. Notes Ser. Inst. Math. Sci. Natl. Univ. Singap., Vol. 12, World Sci. Publ., Hackensack, NJ, 2007, pp. 53–72.
131. G. Gour and N. R. Wallach, *Entanglement of subspaces and error correcting codes*, Physical Review A **76(4): 042309** (2007), (8 pp.).
132. Lei Ni and Nolan Wallach, *On four-dimensional gradient shrinking solitons*, Int. Math. Res. Not. IMRN (2008), no. 4, Art. ID rnm152, 13.
133. Michael Cowling, Edward Frenkel, Masaki Kashiwara, Alain Valette, David A. Vogan, Jr., and N. R. Wallach, *Representation theory and complex analysis*, Lecture Notes in Mathematics, Vol. 1931, Springer-Verlag, Berlin; Fondazione C.I.M.E., Florence, 2008, Lectures from the C.I.M.E. Summer School held in Venice, June 10–17, 2004, Edited by Enrico Casadio Tarabusi, Andrea D'Agnolo and Massimo Picardello.
134. N. R. Wallach, *Quantum computing and entanglement for mathematicians*, Representation theory and complex analysis, Lecture Notes in Math., Vol. 1931, Springer, Berlin, 2008, pp. 345–376.
135. Lei Ni and Nolan Wallach, *On a classification of gradient shrinking solitons*, Math. Res. Lett. **15** (2008), no. 5, 941–955.
136. Adriano Garsia, Gregg Musiker, Nolan Wallach, and Guoce Xin, *Invariants, Kronecker products, and combinatorics of some remarkable Diophantine systems*, Adv. in Appl. Math. **42** (2009), no. 3, 392–421.
137. Roe Goodman and N. R. Wallach, *Symmetry, representations, and invariants*, Graduate Texts in Mathematics, Vol. 255, Springer, Dordrecht, 2009.
138. A. Garsia, N. Wallach, G. Xin, and M. Zabrocki, *Hilbert series of invariants, constant terms and Kostka-Foulkes polynomials*, Discrete Math. **309** (2009), no. 16, 5206–5230.
139. Bertram Kostant and Nolan Wallach, *On a theorem of Ranee Brylinski*, Symmetry in mathematics and physics, Contemp. Math., Vol. 490, Amer. Math. Soc., Providence, RI, 2009, pp. 105–132.
140. Nolan Wallach and Oded Yacobi, *A multiplicity formula for tensor products of* SL_2 *modules and an explicit* Sp_{2n} *to* $\mathrm{Sp}_{2n-2} \times \mathrm{Sp}_2$ *branching formula*, Symmetry in mathematics and physics, Contemp. Math., Vol. 490, Amer. Math. Soc., Providence, RI, 2009, pp. 151–155.
141. Hanspeter Kraft and N. R. Wallach, *Polarizations and nullcone of representations of reductive groups*, Symmetry and spaces, Progr. Math., Vol. 278, Birkhäuser Boston, Inc., Boston, MA, 2010, pp. 153–167.
142. Gilad Gour and N. R. Wallach, *All maximally entangled four-qubit states*, J. Math. Phys. **51** (2010), no. 11, 112201, 24.
143. Bertram Kostant and Nolan Wallach, *On the algebraic set of singular elements in a complex simple Lie algebra*, Representation theory and mathematical physics, Contemp. Math., Vol. 557, Amer. Math. Soc., Providence, RI, 2011, pp. 215–229.

144. Benedict H. Gross and N. R. Wallach, *On the Hilbert polynomials and Hilbert series of homogeneous projective varieties*, Arithmetic geometry and automorphic forms, Adv. Lect. Math. (ALM), Vol. 19, Int. Press, Somerville, MA, 2011, pp. 253–263.
145. G. Gour and N. R. Wallach, *Necessary and sufficient conditions for local manipulation of multipartite pure quantum states*, New Journal of Physics **13(7): 073013** (2011), (28 pp.).
146. Raul Gomez and Nolan Wallach, *Bessel models for general admissible induced representations: the compact stabilizer case*, Selecta Math. (N.S.) **18** (2012), no. 1, 1–26.
147. Adriano Garsia, Nolan Wallach, Guoce Xin, and Mike Zabrocki, *Kronecker coefficients via symmetric functions and constant term identities*, Internat. J. Algebra Comput. **22** (2012), no. 3, 1250022, 44.
148. Gilad Gour and N. R. Wallach, *Classification of multipartite entanglement of all finite dimensionality*, Phys. Rev. Lett. **111, 060502** (2013).

Contents

Unitary Hecke algebra modules with nonzero Dirac cohomology

Dan Barbasch and Dan Ciubotaru

To Nolan Wallach with admiration

Abstract In this paper, we review the construction of the Dirac operator for graded affine Hecke algebras and calculate the Dirac cohomology of irreducible unitary modules for the graded Hecke algebra of $gl(n)$.

Keywords: Dirac cohomology • Unitary representations • Hecke algebra

Mathematics Subject Classification: 22, 16, 20

1 Introduction

The Dirac operator plays an important role in the representation theory of real reductive Lie groups. An account of the definition, properties and some applications can be found in [BW]. It is well known, starting with the work of [AS] and [P], that discrete series occur in the kernel of the Dirac operator. Work of Enright and Wallach [EW] generalizes these results to other types of representations. Other uses are to provide, via the *Dirac inequality*, introduced by Parthasarathy, necessary

The first author was partially supported by NSF grants DMS-0967386, DMS-0901104 and an NSA-AMS grant. The second author was partially supported by NSF DMS-0968065 and NSA-AMS 081022.

D. Barbasch
Department of Mathematics, Cornell University, Ithaca, NY 14853, USA
e-mail: barbasch@math.cornell.edu

D. Ciubotaru
Mathematics Institute, University of Oxford, Andrew Wiles Building, Radcliffe Observatory Quarter, Woodstock Road, OX 2666, Oxford, England
e-mail: dan.ciubotaru@maths.ox.ac.uk

R. Howe et al. (eds.), *Symmetry: Representation Theory and Its Applications*, Progress in Mathematics 257, DOI 10.1007/978-1-4939-1590-3_1

conditions for unitarity. One of the most striking applications is that for a regular integral infinitesimal character, the Dirac inequality gives precisely the unitary dual, and determines the unitary representations with nontrivial $(\mathfrak{g}, K)$-cohomology.

Given these properties, Vogan has introduced the notion of Dirac cohomology; this was studied extensively in [HP] and subsequent work. One can argue that Dirac cohomology is a generalization of $(\mathfrak{g}, K)$-cohomology. While a representation has nontrivial $(\mathfrak{g}, K)$-cohomology only if its infinitesimal character is regular integral, the corresponding condition necessary for Dirac cohomology to be nonzero is more general; certain representations with singular and nonintegral infinitesimal character will also have nontrivial Dirac cohomology.

In this paper, we prove new results about an analogue of the Dirac operator in the case of the graded affine Hecke algebra, introduced in [BCT]. This operator can be thought of as the analogue of the Dirac operator in the case of a p-adic group. One of our results is to determine the behaviour of the Dirac cohomology with respect to Harish-Chandra type induction. In the real case, a unitary representation has nontrivial $(\mathfrak{g}, K)$-cohomology if and only if it is (essentially) obtained from the trivial representation on a Levi component via the derived functor construction. For unitary representations with nontrivial Dirac cohomology the infinitesimal character can be nonintegral and singular. So we conjecture instead that unitary representations with nontrivial Dirac cohomology are all cohomologically induced from unipotent (in the sense of [A]) representations. To investigate this conjecture we explore the Dirac cohomology of unipotent representations for graded affine Hecke algebras. In particular, we compute part of the cohomology of spherical unipotent representations for affine Hecke algebras of all types. In the case of type A we go further; we compute the cohomology of all unitary modules.

This paper was written while we were guests of the Max Planck Institute in Bonn as part of the program *Analysis on Lie groups*. We would like to thank the institute for its hospitality, and the organizers for making the program possible, and providing the environment to do this research.

2 Dirac cohomology for graded Hecke algebras

In this section we review the construction and properties of the Dirac operator from [BCT] and the classification of spin projective Weyl group representations from [Ci].

2.1 Root systems

We fix an $\mathbb{R}$-root system $\Phi = (V, R, V^\vee, R^\vee)$. Here $V, V^\vee$ are finite-dimensional $\mathbb{R}$-vector spaces, with a perfect bilinear pairing $(\ ,\ ) : V \times V^\vee \to \mathbb{R}$, so that $R \subset V \setminus \{0\}$, $R^\vee \subset V^\vee \setminus \{0\}$ are finite subsets in bijection

$$R \longleftrightarrow R^\vee,\ \alpha \longleftrightarrow \alpha^\vee,\ \text{such that } (\alpha, \alpha^\vee) = 2. \tag{2.1.1}$$

The reflections

$$\begin{aligned} s_\alpha &: V \to V,\ s_\alpha(v) = v - (v, \alpha^\vee)\alpha, \\ s_\alpha &: V^\vee \to V^\vee,\ s_\alpha(v') = v' - (\alpha, v')\alpha^\vee, \quad \alpha \in R, \end{aligned} \tag{2.1.2}$$

leave R and $R^\vee$ invariant, respectively. Let W be the subgroup of $GL(V)$ (respectively $GL(V^\vee)$) generated by $\{s_\alpha : \alpha \in R\}$.

We will assume that the root system Φ is reduced and crystallographic. We will fix a choice of simple roots $\Pi \subset R$, and consequently, positive roots R^+ and positive coroots $R^{\vee,+}$. Often, we will write $\alpha > 0$ or $\alpha < 0$ in place of $\alpha \in R^+$ or $\alpha \in (-R^+)$, respectively.

We fix a W-invariant inner product $\langle\ ,\ \rangle$ on V. Denote also by $\langle\ ,\ \rangle$ the dual inner product on $V^\vee$. If v is a vector in V or $V^\vee$, we denote $|v| := \langle v, v\rangle^{1/2}$.

2.2 *The Clifford algebra*

A classical reference for the Clifford algebra is [Ch] (see also Section II.6 in [BW]). Denote by $C(V)$ the Clifford algebra defined by V and the inner product $\langle\ ,\ \rangle$. More precisely, $C(V)$ is the quotient of the tensor algebra of V by the ideal generated by

$$\omega \otimes \omega' + \omega' \otimes \omega + 2\langle \omega, \omega' \rangle, \quad \omega, \omega' \in V.$$

Equivalently, $C(V)$ is the associative algebra with unit generated by V with relations:

$$\omega\omega' + \omega'\omega = -2\langle \omega, \omega' \rangle. \tag{2.2.1}$$

Let $\mathsf{O}(V)$ denote the group of orthogonal transformation of V with respect to $\langle\ ,\ \rangle$. This acts by algebra automorphisms on $C(V)$, and the action of $-1 \in \mathsf{O}(V)$ induces a grading

$$C(V) = C(V)_{\mathsf{even}} + C(V)_{\mathsf{odd}}. \tag{2.2.2}$$

Let ϵ be the automorphism of $C(V)$ which is $+1$ on $C(V)_{\mathsf{even}}$ and -1 on $C(V)_{\mathsf{odd}}$. Let ^{t} be the transpose antiautomorphism of $C(V)$ characterized by

$$\omega^t = -\omega,\ \omega \in V, \quad (ab)^t = b^t a^t,\ a, b \in C(V). \tag{2.2.3}$$

The Pin group is

$$\mathsf{Pin}(V) = \{a \in C(V) : \epsilon(a) V a^{-1} \subset V,\ a^t = a^{-1}\}. \tag{2.2.4}$$

It sits in a short exact sequence

$$1 \longrightarrow \mathbb{Z}/2\mathbb{Z} \longrightarrow \mathsf{Pin}(V) \overset{p}{\longrightarrow} \mathsf{O}(V) \longrightarrow 1, \tag{2.2.5}$$

where the projection p is given by $p(a)(\omega) = \epsilon(a)\omega a^{-1}$.

If $\dim V$ is even, the Clifford algebra $C(V)$ has a unique (up to equivalence) complex simple module (γ, S) of dimension $2^{\dim V/2}$, endowed with a positive definite Hermitian form $\langle\ ,\ \rangle_S$ such that

$$\langle \gamma(a)s, s'\rangle_S = \langle s, \gamma(a^t)s'\rangle_S, \quad \text{for all } a \in C(V) \text{ and } s, s' \in S. \tag{2.2.6}$$

When $\dim V$ is odd, there are two simple inequivalent complex modules $(\gamma_+, S^+), (\gamma_-, S^-)$ of dimension $2^{[\dim V/2]}$. Analogous to (2.2.6), these modules admit an invariant positive definite Hermitian form. In order to simplify the formulation of the results, we will often refer to any one of S, S^+, S^-, as a spin module.

Via (2.2.4), a spin module S is an irreducible unitary $\mathsf{Pin}(V)$ representation.

2.3 The pin cover $\widetilde{W}$ of the Weyl group

The Weyl group W acts by orthogonal transformations on V, so one can embed W as a subgroup of $\mathsf{O}(V)$. We define the group $\widetilde{W}$ in $\mathsf{Pin}(V)$:

$$\widetilde{W} := p^{-1}(W) \subset \mathsf{Pin}(V), \text{ where } p \text{ is as in (2.2.5)}. \tag{2.3.1}$$

The group $\widetilde{W}$ has a Coxeter presentation similar to that of W. Recall that as a Coxeter group, W has a presentation:

$$W = \langle s_\alpha,\ \alpha \in \Pi \mid (s_\alpha s_\beta)^{m(\alpha,\beta)} = 1,\ \alpha, \beta \in \Pi\rangle, \tag{2.3.2}$$

for certain positive integers $m(\alpha, \beta)$. Theorem 3.2 in [Mo] exhibits $\widetilde{W}$ as

$$\widetilde{W} = \langle z, \tilde{s}_\alpha,\ \alpha \in \Pi \mid z^2 = 1,\ (\tilde{s}_\alpha \tilde{s}_\beta)^{m(\alpha,\beta)} = z,\ \alpha, \beta \in \Pi\rangle. \tag{2.3.3}$$

We call a representation $\tilde{\sigma}$ of $\widetilde{W}$ genuine (resp. non-genuine) if $\tilde{\sigma}(z) = -1$ (resp. $\tilde{\sigma}(z) = 1$). The non-genuine $\widetilde{W}$-representations are the ones that factor through W. We say that two genuine $\widetilde{W}$-types σ_1, σ_2 are associate if $\sigma_1 \cong \sigma_2 \otimes \mathsf{sign}$.

Since $\widetilde{W} \subset \mathsf{Pin}(V)$, we can regard S if $\dim V$ is even (resp. $S^\pm$ if $\dim V$ is odd) as unitary (genuine) $\widetilde{W}$-representations. If R spans V, they are irreducible representations ([Mo, Theorem 3.3]). When $\dim V$ is odd, S^+ and S^- are associate, while if $\dim V$ is even, S is self-associate.

Definition 2.3.1 ([BCT, §3.4]). Define the Casimir element of $\widetilde{W}$:

$$\Omega_{\widetilde{W}} = z \sum_{\substack{\alpha>0,\beta>0\\ s_\alpha(\beta)<0}} |\alpha^\vee||\beta^\vee|\, \tilde{s}_\alpha \tilde{s}_\beta \in \mathbb{C}[\widetilde{W}]^{\widetilde{W}}. \tag{2.3.4}$$

Every $\tilde{\sigma} \in \widehat{\widetilde{W}}$ acts on $\Omega_{\widetilde{W}}$ by a scalar, which we denote by $\tilde{\sigma}(\Omega_{\widetilde{W}})$.

Before stating Theorem 2.3.1, we need to introduce more notation. Assume that R spans V and let $\mathfrak{g}$ be the complex semisimple Lie algebra with root system Φ and Cartan subalgebra $\mathfrak{h} = V^\vee \otimes_{\mathbb{R}} \mathbb{C}$, and let G be the simply connected Lie group with Lie algebra $\mathfrak{g}$. Extend the inner product from $V^\vee$ to $\mathfrak{h}$. Let us denote by $\mathcal{T}(G)$ the set of G-conjugacy classes of Jacobson–Morozov triples (e, h, f) in $\mathfrak{g}$. We set

$$\mathcal{T}_0(G) = \big\{[(e,h,f)] \in \mathcal{T}(G) : \text{the centralizer of } (e,h,f) \text{ in } \mathfrak{g} \text{ is a toral subalgebra}\big\}. \tag{2.3.5}$$

For every class in $\mathcal{T}(G)$, we may (and will) choose a representative (e, h, f) such that $h \in \mathfrak{h}$. For every nilpotent element e, let $A(e)$ denote the A-group in G, and let $\widehat{A(e)}_0$ denote the set of representations of $A(e)$ of Springer type. For every $\phi \in \widehat{A(e)}_0$, let $\sigma_{(e,\phi)}$ be the associated Springer representation. Normalize the Springer correspondence so that $\sigma_{0,\mathrm{triv}} = \mathsf{sign}$.

Theorem 2.3.1 ([Ci]). *There is a surjective map*

$$\Psi : \widehat{\widetilde{W}}_{\mathsf{gen}} \longrightarrow \mathcal{T}_0(G), \tag{2.3.6}$$

with the following properties:

(1) *If $\Psi(\tilde{\sigma}) = [(e,h,f)]$, then we have*

$$\tilde{\sigma}(\Omega_{\widetilde{W}}) = \langle h, h\rangle, \tag{2.3.7}$$

where $\Omega_{\widetilde{W}}$ is as in (2.3.4).

(2) *Let $(e,h,f) \in \mathcal{T}_0(G)$ be given. For every Springer representation $\sigma_{(e,\phi)}$, $\phi \in \widehat{A(e)}_0$, and every spin $\widetilde{W}$-module S, there exists $\tilde{\sigma} \in \Psi^{-1}[(e,h,f)]$ such that $\tilde{\sigma}$ appears with nonzero multiplicity in the tensor product $\sigma_{(e,\phi)} \otimes S$. Conversely, for every $\tilde{\sigma} \in \Psi^{-1}[(e,h,f)]$, there exists a spin $\widetilde{W}$-module S and a Springer representation $\sigma_{(e,\phi)}$, such that $\tilde{\sigma}$ is contained in $\sigma_{(e,\phi)} \otimes S$.*

Since $\mathsf{triv}(\Omega_{\widetilde{W}}) = \mathsf{sign}(\Omega_{\widetilde{W}})$, Theorem 2.3.1(1) says in particular that any two associate genuine $\widetilde{W}$-types $\tilde{\sigma}_1, \tilde{\sigma}_2$ lie in the same fiber of Ψ.

2.4 The graded Hecke algebra

Recall the real root system $\Phi = (V, R, V^\vee, R^\vee)$. The complexifications of $V, V^\vee$ are denoted by $V_\mathbb{C}, V_\mathbb{C}^\vee$. We denote by $S(V_\mathbb{C})$ the symmetric algebra in $V_\mathbb{C}$.

Definition 2.4.1 ([Lu]). The graded affine Hecke algebra $\mathbb{H}$ (with equal parameters) is defined as follows:

(i) as a $\mathbb{C}$-vector space, it is $S(V_\mathbb{C}) \otimes \mathbb{C}[W]$;
(ii) $S(V_\mathbb{C})$ and $\mathbb{C}[W]$ have the usual algebra structures as subalgebras;
(iii) the cross relations are

$$s_\alpha \cdot \xi - s_\alpha(\xi) \cdot s_\alpha = (\xi, \alpha^\vee),$$

for every $\alpha \in \Pi$ and $\xi \in V_\mathbb{C}$.

Definition 2.4.2. Let $\{\omega_i : i = 1, n\}$ and $\{\omega^i : i = 1, n\}$ be dual bases of V with respect to $\langle\ ,\ \rangle$. Define the Casimir element of $\mathbb{H}$: $\Omega = \sum_{i=1}^n \omega_i \omega^i \in \mathbb{H}$.

It is easy to see that the element Ω is independent of the choice of bases and central in $\mathbb{H}$. Moreover, if (π, X) is an irreducible $\mathbb{H}$-module with central character χ_ν for $\nu \in V_\mathbb{C}^\vee$, then π acts on Ω by the scalar $\langle \nu, \nu \rangle$.

The algebra $\mathbb{H}$ has a natural conjugate linear anti-involution defined on generators as follows:

$$\begin{aligned} w^* &= w^{-1}, \quad w \in W, \\ \omega^* &= -\omega + \sum_{\beta>0} (\omega, \beta^\vee) s_\beta, \quad \omega \in V. \end{aligned} \tag{2.4.1}$$

An $\mathbb{H}$-module (π, X) is said to be Hermitian if there exists a Hermitian form $(\ ,\ )_X$ on X which is invariant in the sense that $(\pi(h)x, y)_X = (x, \pi(h^*)y)_X$, for all $h \in \mathbb{H}, x, y \in X$. If such a form exists which is also positive definite, then X is said to be unitary.

For every $\omega \in V$, define

$$\tilde{\omega} = \omega - \frac{1}{2} \sum_{\beta>0} (\omega, \beta^\vee) s_\beta \in \mathbb{H}. \tag{2.4.2}$$

It is immediate that $\widetilde{\omega}^* = -\widetilde{\omega}$.

Definition 2.4.3 ([BCT]). Let $\{\omega_i\}, \{\omega^i\}$ be dual bases of V. The Dirac element is defined as

$$\mathcal{D} = \sum_i \widetilde{\omega}_i \otimes \omega^i \in \mathbb{H} \otimes C(V).$$

It is elementary to verify that $\mathcal{D}$ does not depend on the choice of dual bases.

We will usually work with a fixed spin module (γ, S) for $C(V)$ and a fixed $\mathbb{H}$-module (π, X). Define the Dirac operator for X (and S) as $D = (\pi \otimes \gamma)(\mathcal{D})$.

Suppose X is a Hermitian $\mathbb{H}$-module with invariant form $(\ ,\ )_X$. Endow the tensor product $X \otimes S$ with the Hermitian form $(x \otimes s, x' \otimes s')_{X\otimes S} = (x, x')_X \langle s, s' \rangle_S$. Analogous to results of Parthasarathy in the real case, the operator D is self-adjoint with respect to $(\ ,\ )_{X\otimes S}$,

$$(D(x \otimes s), x' \otimes s')_{X\otimes S} = (x \otimes s, D(x' \otimes s'))_{X\otimes S} \tag{2.4.3}$$

Thus a Hermitian $\mathbb{H}$-module is unitary only if

$$(D^2(x \otimes s), x \otimes s)_{X\otimes S} \geq 0, \qquad \text{for all } x \otimes s \in X \otimes S. \tag{2.4.4}$$

We write $\Delta_{\widetilde{W}}$ for the diagonal embedding of $\mathbb{C}[\widetilde{W}]$ into $\mathbb{H} \otimes C(V)$ defined by extending $\Delta_{\widetilde{W}}(\widetilde{w}) = p(\widetilde{w}) \otimes \widetilde{w}$ linearly.

For $\widetilde{w} \in \widetilde{W}$, one can easily see that

$$\Delta_{\widetilde{W}}(\widetilde{w})\mathcal{D} = \mathsf{sign}(\widetilde{w})\mathcal{D}\Delta_{\widetilde{W}}(\widetilde{w}) \tag{2.4.5}$$

as elements of $\mathbb{H} \otimes C(V)$. In particular, the kernel of the Dirac operator on $X \otimes S$ is invariant under $\widetilde{W}$.

Theorem 2.4.1 ([BCT]). *The square of the Dirac element equals*

$$\mathcal{D}^2 = -\Omega \otimes 1 + \frac{1}{4}\Delta_{\widetilde{W}}(\Omega_{\widetilde{W}}), \tag{2.4.6}$$

in $\mathbb{H} \otimes C(V)$.

2.5 Dirac cohomology

To have a uniform notation, we will denote a spin module by S^ϵ. If $\dim V$ is even, then S^ϵ is S, the unique spin module, and if $\dim V$ is odd, then ϵ could be $+$ or $-$.

Definition 2.5.1. In the setting of Definition 2.4.3, define

$$H^D_\epsilon(X) := \ker D \big/ (\ker D \cap \operatorname{Im} D) \tag{2.5.1}$$

and call it the Dirac cohomology of X. (The symbol ϵ denotes the dependence on the spin module S^ϵ.) If X is unitary, the self-adjointness of D implies that $\ker(D) \cap \operatorname{Im}(D) = 0$, and so $H^D_\epsilon(X) = \ker(D)$.

Vogan's conjecture takes the following form.

Theorem 2.5.1 ([BCT, Theorem 4.8]). *Suppose (π, X) is an $\mathbb{H}$-module with central character χ_ν with $\nu \in V_{\mathbb{C}}^\vee$. Suppose that $H_\epsilon^D(X) \neq 0$ and let $(\tilde{\sigma}, \widetilde{U})$ be an irreducible representation of $\widetilde{W}$ such that $\mathrm{Hom}_{\widetilde{W}}(\widetilde{U}, H_\epsilon^D(X)) \neq 0$. If $\Psi(\tilde{\sigma}) = [(e, h, f)] \in \mathcal{T}_0(G)$, then $\nu = \frac{1}{2}h$.*

Theorem 2.5.1 has an easy weak converse, which will be useful in applications.

Proposition 2.5.1. *Assume that (π, X) is a unitary $\mathbb{H}$-module with central character χ_ν, $\nu \in V_{\mathbb{C}}^\vee$ and that there exists an irreducible $\widetilde{W}$-type $(\tilde{\sigma}, \widetilde{U})$ such that $\mathrm{Hom}_{\widetilde{W}}(\widetilde{U}, X \otimes S^\epsilon) \neq 0$ and $\langle \nu, \nu \rangle = \tilde{\sigma}(\Omega_{\widetilde{W}})$. Then $\mathrm{Hom}_{\widetilde{W}}(\widetilde{U}, H_\epsilon^D(X)) \neq 0$, and in particular $H_\epsilon^D(X) \neq 0$.*

Proof. Let $x \otimes s$ be an element of $X \otimes S^\epsilon$ in the isotypic component of $\tilde{\sigma}$. Then $D^2(x \otimes s) = -\langle \nu, \nu \rangle + \tilde{\sigma}(\Omega_{\widetilde{W}}) = 0$. Since X is assumed unitary, the operator D is self-adjoint on $X \otimes S$ and thus $\ker D^2 = \ker D$. This implies

$$x \otimes s \in \ker D (= H_\epsilon^D(X).$$

□

As a corollary, we find the following formula for $H_\epsilon^D(X)$.

Corollary 2.5.1. *Assume X is a unitary $\mathbb{H}$-module with central character $\chi_{\frac{1}{2}h}$, for some $[(e, h, f)] \in \mathcal{T}_0(G)$ (otherwise $H_\epsilon^D(X) = 0$). Then, as a $\widetilde{W}$-module*

$$H_\epsilon^D(X) = \sum_{\tilde{\sigma} \in \Psi^{-1}(e,h,f)} \sum_{\mu \in \widehat{W}} [\tilde{\sigma} : \mu \otimes S^\epsilon][X|_W : \mu]\, \tilde{\sigma}. \tag{2.5.2}$$

2.6 An induction lemma

Let $(V_M, R_M, V_M^\vee, R_M^\vee)$ be a root subsystem of $(V, R, V^\vee, R^\vee)$. Let $\Pi_M \subset \Pi$ be the corresponding simple roots and $W_M \subset W$ the reflection subgroup. Let $\mathbb{H}_M$ denote the Hecke subalgebra of $\mathbb{H}$ given by this root subsystem. Denote by V_N the orthogonal complement of V_M in V with respect to the fixed product $\langle\ ,\ \rangle$.

Recall that the graded tensor product $A \hat{\otimes} B$ of two $\mathbb{Z}/2\mathbb{Z}$-graded algebras A and B is $A \otimes B$ as a vector space, but with multiplication defined by

$$(a_1 \otimes b_1)(a_2 \otimes b_2) = (-1)^{\deg(b_1)\deg(a_2)} a_1 a_2 \otimes b_1 b_2.$$

Lemma 2.6.1. *There is an isomorphism of algebras $C(V) \cong C(V_M) \hat{\otimes} C(V_N)$.*

Proof. If an orthonormal basis of V_M is $\{\omega_1, \ldots, \omega_k\}$ and an orthonormal basis of V_n is $\{\omega_{k+1}, \ldots, \omega_n\}$, then the isomorphism is given by

$$\omega_{i_1} \ldots \omega_{i_l} \otimes \omega_{j_1} \ldots \omega_{j_r} \mapsto \omega_{i_1} \ldots \omega_{i_l} \omega_{j_1} \ldots \omega_{j_r},$$

where $i_1, \ldots, i_l \in \{1, \ldots, k\}$ and $j_1, \ldots, j_r \in \{k+1, \ldots, n\}$. □

Since W_M acts trivially on V_N, and therefore $\widetilde{W}_M$ acts trivially on every $C(V_N)$-module, we see that as $\widetilde{W}_M$-representations:

$$\begin{aligned} &S \cong \oplus_{2^{\dim V_N/2}} S_M, && \text{if } \dim V, \dim V_M \text{ are both even;} \\ &S^{\pm} \cong \oplus_{2^{\dim V_N/2}} S_M^{\pm}, && \text{if } \dim V, \dim V_M \text{ are both odd;} \\ &S^{\pm} \cong \oplus_{2^{(\dim V_N-1)/2}} S_M, && \text{if } \dim V \text{ is odd and } \dim V_M \text{ is even;} \\ &S \cong \oplus_{2^{(\dim V_N-1)/2}} (S_M^{+} + S_M^{-}), && \text{if } \dim V \text{ is even and } \dim V_M \text{ is odd.} \end{aligned} \tag{2.6.1}$$

The following lemma will be our main criterion for proving that certain induced modules have nonzero Dirac cohomology. In order to reduce the number of cases, denote $\mathcal{S} = S$ if $\dim V$ is even, and $\mathcal{S} = S^+ + S^-$ if $\dim V$ is odd, and similarly define $\mathcal{S}_M$. In particular, $\mathcal{S}$ is self-contragredient.

Lemma 2.6.2. *Let π_M be an $\mathbb{H}_M$-module, and $\pi = \mathbb{H} \otimes_{\mathbb{H}_M} \pi_M$.*

(a) $\mathrm{Hom}_{\widetilde{W}}[\tilde{\sigma}, \pi \otimes \mathcal{S}] = \frac{\dim \mathcal{S}}{\dim \mathcal{S}_M} \mathrm{Hom}_{\widetilde{W}_M}[\tilde{\sigma}|_{\widetilde{W}_M}, \pi_M \otimes \mathcal{S}_M]$.

(b) *Assume that π_M is unitary, and the $\widetilde{W}_M$-type $\tilde{\sigma}_M$ occurs in $H^D(\pi_M)$. Assume further that there exists a $\widetilde{W}$-type $\tilde{\sigma}$ such that*

  (i) *$Hom_{\widetilde{W}_M}[\tilde{\sigma}_M, \tilde{\sigma}] \neq 0$,*

  (ii) *the central character of π is $\chi_\pi = \chi_{h/2}$, where $\Psi(\tilde{\sigma}) = [(e, h, f)]$.*

  Then $\tilde{\sigma}$ occurs in $H^D(\pi)$.

Proof. Part (b) is an immediate consequence of (a) using Proposition 2.5.1. To prove (a), we use Frobenius reciprocity and the restriction of $\mathcal{S}$ to $\widetilde{W}_M$:

$$\begin{aligned} \mathrm{Hom}_{\widetilde{W}}[\tilde{\sigma}, \pi \otimes \mathcal{S}] &= \mathrm{Hom}_W[\tilde{\sigma} \otimes \mathcal{S}, \mathrm{Ind}_{W_M}^W \pi_M] = \mathrm{Hom}_{W_M}[(\tilde{\sigma} \otimes \mathcal{S})|_{W_M}, \pi_M] \\ &= \mathrm{Hom}_{\widetilde{W}_M}[\tilde{\sigma}|_{\widetilde{W}_M}, \pi_M \otimes \mathcal{S}|_{\widetilde{W}_M}] = \frac{\dim \mathcal{S}}{\dim \mathcal{S}_M} \mathrm{Hom}_{\widetilde{W}_M}[\tilde{\sigma}|_{\widetilde{W}_M}, \pi_M \otimes \mathcal{S}_M]. \end{aligned}$$

□

2.7 Spherical modules

An $\mathbb{H}$-module X is called spherical if $\mathrm{Hom}_W[\mathrm{triv}, X] \neq 0$. The (spherical) principal series modules of $\mathbb{H}$ are defined as the induced modules

$$X(\nu) = \mathbb{H} \otimes_{S(V_{\mathbb{C}})} \mathbb{C}_\nu,$$

for $\nu \in V_{\mathbb{C}}^\vee$. Since $X(\nu) \cong \mathbb{C}[W]$ as W- modules, there is a unique irreducible spherical $\mathbb{H}$-subquotient $L(\nu)$ of $X(\nu)$. It is well known that

(1) $L(\nu) \cong L(w\nu)$, for every $w \in W$;
(2) if ν is R^+-dominant, then $L(\nu)$ is the unique irreducible quotient of $X(\nu)$;
(3) every irreducible spherical $\mathbb{H}$-module is isomorphic to a quotient $L(\nu)$, ν is R^+-dominant.

Recall the Lie algebra $\mathfrak{g}$ attached to the root system Φ. The identification $\mathfrak{h} = V_{\mathbb{C}}^{\vee}$ allows us to view ν as an element of $\mathfrak{h}$. Now consider $\mathfrak{g}_1 = \{x \in \mathfrak{g} : [\nu, x] = x\}$, the ad 1-eigenspace of ν on $\mathfrak{g}$. The stabilizer $G_0 = \{g \in G : \mathrm{Ad}(g)\nu = \nu\}$ acts on $\mathfrak{g}_1$ with finitely many orbits, and let e be an element of the unique open dense G_0-orbit. Lusztig's geometric realization of $\mathbb{H}$ and classification of irreducible $\mathbb{H}$-modules implies in particular the following statement.

Proposition 2.7.1. *Let $\nu \in V_{\mathbb{C}}^{\vee}$ be given and let e be a nilpotent element of $\mathfrak{g}$ attached to ν by the procedure above. Then the spherical module $L(\nu)$ contains the Springer representation $\sigma_{(e,1)}$ with multiplicity one.*

The second result that we need is the unitarizability of the spherical unipotent $\mathbb{H}$-modules.

Proposition 2.7.2 ([BM]). *For every Lie triple (e, h, f), the spherical module $L(\frac{1}{2}h)$ is unitary.*

Now we can state the classification of spherical modules with nonzero Dirac cohomology.

Definition 2.7.1. We say that an $\mathbb{H}$-module X has nonzero Dirac cohomology if for a choice of spin module S^{ϵ}, $H_{\epsilon}^{D}(X) \neq 0$.

Let $[(e, h, f)] \in \mathcal{T}_0(G)$ be given and assume G is simple. The results of [Ci] give a concrete description in every Lie type of the map Ψ from Theorem 2.3.1. In particular, there is either only one self-associate $\widetilde{W}$-type which we denote by $\tilde{\sigma}_{(e,1)}$, or two $\widetilde{W}$-types denoted $\tilde{\sigma}_{(e,1)}^{\pm}$, which appear in the fiber $\Psi^{-1}(e, h, f)$ and can occur in the decomposition of the tensor product $\sigma_{(e,1)} \otimes S^{\epsilon}$.

Corollary 2.7.1. *An irreducible spherical module $L(\nu)$ has nonzero Dirac cohomology if and only if $\nu = w \cdot \frac{1}{2}h$ for some $[(e, h, f)] \in \mathcal{T}_0(G)$.*

Proof. Assume that $H_{\epsilon}^{D}(L(\nu)) \neq 0$. Then there exists a genuine $\widetilde{W}$-type $\tilde{\sigma}$ occurring in $H_{\epsilon}^{D}(L(\nu))$, such that $\Psi(\tilde{\sigma}) = [(e, h, f)] \in \mathcal{T}_0(G)$. By Theorem 2.5.1, $\nu = w \cdot \frac{1}{2}h$.

Conversely, fix $[(e, h, f)] \in \mathcal{T}_0(G)$. The spherical module $L(\frac{1}{2}h)$ contains $\sigma_{(e,1)}$ with multiplicity one by Proposition 2.7.1, and it is unitary by Proposition 2.7.2. From this, Proposition 2.5.1 implies immediately that one of the $\widetilde{W}$-types in $\Psi^{-1}(e, h, f)$ occurs in $H_{\epsilon}^{D}(L(\frac{1}{2}h))$, for some ϵ, and therefore $L(\frac{1}{2}h)$ has nonzero Dirac cohomology. □

In order to investigate the precise formula for $H_{\epsilon}^{D}(L(h/2))$, one uses (2.5.2) and the results of Borho–MacPherson [BMcP] about the W-structure of the Springer representations in $A(e)$-isotypic components of the full cohomology of a Springer fiber. In our setting, this says that as a W- module,

$$L(h/2) = \sigma_{(e,1)} + \sum_{e'>e} \sum_{\phi' \in \widehat{A}(e)_0} m_{e',\phi'} \sigma_{(e',\phi')}, \tag{2.7.1}$$

for some integers $m_{e',\phi'} \geq 0$. Here $e' > e$ means the closure ordering of nilpotent orbits, i.e., $e \in \overline{G \cdot e'} \setminus G \cdot e'$. We make the following conjecture.

Conjecture 2.7.1. Let $\tilde{\sigma}$ be a $\widetilde{W}$-type such that $\Psi(\tilde{\sigma}) = [(e', h', f')]$. Then

$$\mathrm{Hom}_{\widetilde{W}}[\tilde{\sigma}, \sigma_{(e,\phi)} \otimes S^{\epsilon}] \neq 0$$

only if $e' \geq e$.

If this conjecture is true, then if we tensor by S^{ϵ} in (2.7.1), every $\widetilde{W}$-type coming from a $\sigma_{(e',\phi')} \otimes S^{\epsilon}$, $e' > e$, would correspond under the map Ψ to a triple (e'', h'', f'') with $e'' \geq e' > e$. In particular, $|h''| > |h|$, so the formula for D_{ϵ}^2 (Theorem 2.4.1) implies that none of these $\widetilde{W}$- types can contribute to $H_{\epsilon}^D(L(h/2))$. Thus the only nontrivial contribution to $H_{\epsilon}^D(L(h/2))$ comes from $\sigma_{(e,1)} \otimes S^{\epsilon}$, and we would have

$$\begin{aligned} &H_{\epsilon}^D(L(h/2)) \\ &= \begin{cases} [\tilde{\sigma}_{(e,1)} : \sigma_{(e,1)} \otimes S^{\epsilon}]\, \tilde{\sigma}_{(e,1)}, & \text{if } \tilde{\sigma}_{(e,1)} \cong \tilde{\sigma}_{(e,1)} \otimes \mathsf{sign}; \\ [\tilde{\sigma}^{+}_{(e,1)} : \sigma_{(e,1)} \otimes S^{\epsilon}]\, \tilde{\sigma}^{+}_{(e,1)} + [\tilde{\sigma}^{-}_{(e,1)} : \sigma_{(e,1)} \otimes S^{\epsilon}]\, \tilde{\sigma}^{-}_{(e,1)}, & \text{otherwise.} \end{cases} \end{aligned} \tag{2.7.2}$$

In Section 3, we will show that Conjecture 2.7.1 holds when $\mathbb{H}$ is a Hecke algebra of type A, and therefore in that case (2.7.2) is true (see Lemma 3.6.2). Further evidence for this conjecture is provided by the computation of the Dirac index for tempered $\mathbb{H}$-modules in [CT, Theorem 1].

3 Nonzero Dirac cohomology for type A

In this section, we specialize to the case of the graded Hecke algebra attached to the root system $\Phi = (V, R, V^{\vee}, R^{\vee})$ of $gl(n)$. Explicitly, $V = \mathbb{R}^n$ with a basis $\{\epsilon_1, \dots, \epsilon_n\}$, $R = \{\epsilon_i - \epsilon_j : 1 \leq i \neq j \leq n\}$. To simplify notation, we will also use the coordinates $\{\epsilon_i\}$ to describe $V^{\vee} \cong \mathbb{R}^n$ and $R^{\vee}$. We choose positive roots $R^+ = \{\epsilon_i - \epsilon_j : 1 \leq i < j \leq n\}$. The simple roots are therefore $\Pi = \{\epsilon_i - \epsilon_{i+1} : 1 \leq i < n\}$. The Weyl group is the symmetric group S_n and we write $s_{i,j}$ for the reflection in the root $\epsilon_i - \epsilon_j$.

The graded Hecke algebra $\mathbb{H}_n$ for $gl(n)$ is therefore generated by S_n and the set $\{\epsilon_i : 1 \leq i \leq n\}$ subject to the commutation relations:

$$s_{i,i+1}\epsilon_k = \epsilon_k s_{i,i+1}, \qquad k \neq i, i+1;$$
$$s_{i,i+1}\epsilon_i - \epsilon_{i+1}s_{i,i+1} = 1.$$

We review the classification of the unitary dual of $\mathbb{H}_n$ and then determine which unitary $\mathbb{H}_n$-modules have nonzero Dirac cohomology.

3.1 Langlands classification

We begin by recalling the Langlands classification for $\mathbb{H}_n$.

Definition 3.1.1. The Steinberg module St is the $\mathbb{H}_n$-module whose restriction to $\mathbb{C}[S_n]$ is the sign-representation, and whose only $S(V_\mathbb{C})$ weight is $-\rho^\vee = -\frac{1}{2}\sum_{\alpha\in R^+}\alpha^\vee$.

Let $\lambda = (n_1, n_2, \dots, n_r)$ be a composition of n, i.e., $n_1 + n_2 + \dots + n_r = n$, but there is no order assumed between the n_i's. (E.g., $(2,1)$ and $(1,2)$ are different compositions of 3.) For every $1 \le i \le r$, regard the Hecke algebra $\mathbb{H}_{n_i}$ as the subalgebra of $\mathbb{H}$ generated by $\{\epsilon_j, \epsilon_{j+1}, \dots, \epsilon_{j+n_i-1}\}$ and $\{s_{j,j+1}, s_{j+1,j+2}, \dots, s_{j+n_i-1,j+n_i}\}$, where $j = n_1 + n_2 + \dots + n_{i-1} + 1$. Then

$$\mathbb{H}_\lambda = \mathbb{H}_{n_1} \times \mathbb{H}_{n_2} \times \dots \times \mathbb{H}_{n_r}$$

is a (parabolic) subalgebra of $\mathbb{H}_n$. For every r-tuple $\underline{\nu} = (\nu_1, \nu_2, \dots, \nu_r)$ of complex numbers, we may consider the induced module

$$I_\lambda(\underline{\nu}) = \mathbb{H}_n \otimes_{\mathbb{H}_\lambda} (\mathsf{St} \otimes \mathbb{C}_{\nu_1}) \boxtimes \dots \boxtimes (\mathsf{St} \otimes \mathbb{C}_{\nu_r}). \tag{3.1.1}$$

If $\underline{\nu}$ satisfies the dominance condition

$$\mathrm{Re}(\nu_1) \ge \mathrm{Re}(\nu_2) \ge \dots \ge \mathrm{Re}(\nu_r), \tag{3.1.2}$$

we call $I_\lambda(\underline{\nu})$ a standard module.

Theorem 3.1.1 ([BZ]).

(a) *Let λ be a composition of n and $I_\lambda(\underline{\nu})$ a standard module as in (3.1.1) and (3.1.2). Then $I_\lambda(\underline{\nu})$ has a unique irreducible quotient $L_\lambda(\underline{\nu})$.*

(b) *Every irreducible $\mathbb{H}_n$-module is isomorphic to an $L_\lambda(\underline{\nu})$ as in (a).*

Recall that by Young's construction, the S_n-types are in one-to-one correspondence with partitions of n. We write σ_λ for the S_n-type parameterized by the partition λ of n. For example, $\sigma_{(n)} = \mathsf{triv}$ and $\sigma_{(1^n)} = \mathsf{sign}$. If λ^t denotes the transpose partition of λ, then $\sigma_\lambda \otimes \mathsf{sign} = \sigma_{\lambda^t}$. Finally, every composition λ of n gives rise to a partition of n by reordering, and we denote the corresponding S_n-type by σ_λ again.

Theorem 3.1.2 ([Ro]). *In the notation of Theorem 3.1.1, the irreducible module $L_\lambda(\underline{\nu})$ contains the S_n-type σ_{λ^t} with multiplicity one.*

3.2 Speh modules

The building blocks of the unitary dual are the Speh modules whose construction we review now.

The following lemma is well known and elementary.

Lemma 3.2.1. *For every $c \in \mathbb{C}$, there exists a surjective algebra homomorphism $\tau_c : \mathbb{H} \to \mathbb{C}[S_n]$ given by*

$$w \mapsto w, \quad w \in S_n;$$
$$\epsilon_k \mapsto s_{k,k+1} + s_{k,k+2} + \cdots + s_{k,n} + c, \quad 1 \le k < n;$$
$$\epsilon_n \mapsto c.$$

Proof. We check the commutation relations. It is clear that if $k \neq i, i+1$, $s_{i,i+1}$ commutes with ϵ_k, since $s_{i,i+1}$ commutes with every $s_{j,n}$, $k \le j < n$, when $i+1 < k$, and it commutes with $s_{k,i} + s_{k,i+1}$ and $s_{k,j}$, $j \neq i, i+1$, when $i > k$.

Next, note that $s_{i,i+1}\epsilon_i = 1 + \sum_{j>i+1} w_{(i,j,i+1)} + cs_{i,i+1}$ and $\epsilon_{i+1}s_{i,i+1} = \sum_{j>i+1} w_{(i,j,i+1)} + cs_{i,i+1}$, where $w_{(i,j,i+1)}$ denotes the element of S_n with cycle structure $(i, j, i+1)$. The claim follows. □

For every partition λ of n and $c \in \mathbb{C}$, define the $\mathbb{H}$-module $\tau_c^*(\lambda)$ obtained by pulling back σ_λ to $\mathbb{H}$ via τ_c.

Viewing λ as a left justified Young diagram, define the c-content of the (i, j) box of λ to be $c + (j - i)$, and the c-content of λ to be the set of c-contents of boxes. This is best explained by an example. If λ is the partition of $(3, 3, 1)$ of $n = 7$, the 0-content is the Young tableau

0	1	2
−1	0	1
−2		

For the c-content, add c to the entry in every box.

Lemma 3.2.2. *The central character of $\tau_c^*(\lambda)$ is the (S_n-orbit of the) c-content of the partition λ.*

Proof. This follows from the known values of the simultaneous eigenvalues of the Jucys–Murphy elements $s_{k,k+1} + s_{k,k+2} + \cdots + s_{k,n}$ used to defined τ_c. See for example [OV, Theorem 5.8]. □

Definition 3.2.1. If λ is a box partition, i.e., $\lambda = \underbrace{(m, m, \ldots, m)}_{d}$, for some m, d such that $n = md$, and $c = 0$ when $m + d$ is even or $c = \frac{1}{2}$ when $m + d$ is odd, call the module $\tau_c^*(\lambda)$ a Speh module, and denote it by $a(m, d)$.

Lemma 3.2.3. *In the notation of Theorem 3.1.1, the Speh module $a(m,d)$ is isomorphic to $L_{\lambda^t}(\frac{m-1}{2}, \frac{m-3}{2}, \ldots, -\frac{m-1}{2})$, where $\lambda^t = \underbrace{(d, d, \ldots, d)}_{m}$.*

Proof. This is immediate from Theorem 3.1.1, Theorem 3.1.2 and Lemma 3.2.2. □

3.3 The unitary dual

The classification of irreducible $\mathbb{H}_n$-modules which admit a nondegenerate invariant Hermitian form is a particular case of the classical result of [KZ], as formulated in the Hecke algebra setting by [BM].

If $\lambda = (n_1, \ldots, n_r)$ is a composition of n, let $R_\lambda \subset R$ denote the root subsystem of the Levi subalgebra $gl(n_1) \oplus \cdots \oplus gl(n_r) \subset gl(n)$. If $w \in S_n$ has the property that $wR_\lambda^+ = R_\lambda^+$, then w gives rise to an algebra automorphism of $\mathbb{H}_\lambda$, and therefore w acts on the set of irreducible $\mathbb{H}_\lambda$-modules.

Theorem 3.3.1. *Let $\lambda = (n_1, \ldots, n_r)$ be a composition of n and let $\underline{\nu} = (\nu_1, \ldots, \nu_r)$ be a dominant r-tuple of complex numbers in the sense of (3.1.2). In the notation of Theorem 3.1.1, $L_\lambda(\underline{\nu})$ is Hermitian if and only if there exists $w \in S_n$ such that $wR_\lambda^+ = R_\lambda^+$ and*

$$w((\mathsf{St} \otimes \mathbb{C}_{\nu_1}) \boxtimes \cdots \boxtimes (\mathsf{St} \otimes \mathbb{C}_{\nu_r})) = (\mathsf{St} \otimes \mathbb{C}_{-\overline{\nu}_1}) \boxtimes \cdots \boxtimes (\mathsf{St} \otimes \mathbb{C}_{-\overline{\nu}_r}), \quad (3.3.1)$$

as $\mathbb{H}_\lambda$-modules.

Corollary 3.3.1. *Every Speh module $a(m, d)$ is a unitary $\mathbb{H}_n$-module.*

Proof. Let w_0 denote the longest Weyl group element in S_n and $w_0(\lambda)$ the longest Weyl group element in $S_{n_1} \times \cdots \times S_{n_r}$. Using Lemma 3.2.3, we see now that every Speh module $a(m, d)$ is Hermitian since the Weyl group element $w_0 w_0(\lambda^t)$ satisfies condition (3.3.1) in this case.

Since in addition $a(m, d)$ is irreducible as an S_n-module, it is in fact unitary. □

The classification of the unitary dual of $\mathbb{H}_n$ is also well known (see [Ta] for the classification of the unitary dual for $GL(n, \mathbb{Q}_p)$).

The building blocks are the Speh modules defined before. First, every Speh module $a(m, d)$ can be tensored with a unitary character $\mathbb{C}_y$, $y \in \sqrt{-1}\mathbb{R}$ by which the central element $\epsilon_1 + \cdots + \epsilon_n$ of $\mathbb{H}_n$ acts. We denote the resulting (unitary) irreducible module by $a_y(m, d)$.

Next, we consider induced complementary series representations of the form

$$\pi(a_y(m,d),\nu) = \tag{3.3.2}$$

$$= \mathbb{H}_{2k} \otimes_{\mathbb{H}_k \times \mathbb{H}_k} (a_y(m,d) \otimes \mathbb{C}_\nu) \boxtimes (a_y(m,d) \otimes \mathbb{C}_{-\nu}), \quad 0 < \nu < \frac{1}{2};$$

in this notation, it is implicit that $k = md$. An easy deformation argument shows that all $\pi(a_y(m,d))$ are irreducible unitary $\mathbb{H}_{2k}$-modules.

Theorem 3.3.2 ([Ta]).

(a) *Let $\lambda = (n_1,\dots,n_r)$ be a composition of n. If every $\pi_1,\dots,\pi_r$ is either a Speh module of the form $a_y(m,d)$ or an induced complementary series module of the form $\pi(a_y(m,d),\nu)$ as in (3.3.2), then the induced module*

$$\mathbb{H} \otimes_{\mathbb{H}_\lambda} (\pi_1 \boxtimes \cdots \boxtimes \pi_r) \tag{3.3.3}$$

is irreducible and unitary. Moreover, two such modules are isomorphic if and only if one is obtained from the other one by permuting the factors.

(b) *Every unitary $\mathbb{H}_n$-module is of the form (3.3.3).*

3.4 Nilpotent orbits in $sl(n)$

The classification of nilpotent orbits for $sl(n)$ is well known. Let $P(n)$ denote the set of all (decreasing) partitions of n and let $DP(n)$ be the set of partitions with distinct sizes. The Jordan canonical form gives a bijection between the set of nilpotent orbits of $sl(n)$ and $P(n)$. If $(e_\lambda, h_\lambda, f_\lambda)$ is a Lie triple, where the nilpotent element e_λ is the Jordan form given by the partition $\lambda = (n_1, n_2, \dots, n_r), n_1 \geq n_2 \geq \cdots \geq n_r > 0$, then, using the identification $\mathfrak{h} = \mathbb{C}^n$, the middle element h_λ can be chosen to have coordinates

$$h_\lambda = \left(\frac{n_1-1}{2},\dots,-\frac{n_1-1}{2};\dots;\frac{n_r-1}{2},\dots,-\frac{n_r-1}{2}\right). \tag{3.4.1}$$

If we write λ as $\lambda = (\underbrace{n'_1,\dots,n'_1}_{k_1},\underbrace{n'_2,\dots,n'_2}_{k_2},\dots,\underbrace{n'_l,\dots,n'_l}_{k_l})$, with

$$n'_1 > n'_2 > \cdots > n'_l > 0,$$

then the centralizer in $gl(n)$ of the triple $(e_\lambda, h_\lambda, f_\lambda)$ is $gl(k_1)\oplus gl(k_2)\oplus\cdots\oplus gl(k_l)$. In particular, the centralizer in $sl(n)$ is a toral subalgebra if and only if $\lambda \in DP(n)$. Thus, we have a natural bijection $\mathcal{T}_0(SL(n)) \leftrightarrow DP(n)$. For $\lambda \in P(n)$, ($\mathcal{T}_0$ defined in (2.3.5)) viewed as a left justified Young tableau, define

$$\text{hook}(\lambda) \tag{3.4.2}$$

to be the partition obtained by taking the hooks of λ. For example, if $\lambda = (3,3,1)$, then $\text{hook}(\lambda) = (5,2)$. It is clear that $\text{hook}(\lambda) \in DP(n)$.

We will need the following reformulation for the central character of a Speh module.

Lemma 3.4.1. *The central character of a Speh module $a(m,d)$ is the (S_n-orbit of) $h_{\lambda'}$ (see (3.4.1)), where λ' is the partition*

$$\lambda' = \text{hook}(\underbrace{m,m,\ldots,m}_{d}) = (m+d-1, m+d-3, \ldots, |m-d|+1).$$

Proof. This is immediate from Lemma 3.2.2 and (3.4.1). □

3.5 Irreducible $\widetilde{S}_n$-representations

Denote the length of a partition λ by $|\lambda|$. We say that λ is even (resp. odd) if $n - |\lambda|$ is even (resp. odd). The first part of Theorem 2.3.1 for $\widetilde{S}_n$ is a classical result of Schur.

Theorem 3.5.1 (Schur, [St]). *The irreducible $\widetilde{S}_n$-representations are parameterized by partitions in $DP(n)$ as follows:*

(i) *for every even $\lambda \in DP(n)$, there exists a unique $\tilde{\sigma}_\lambda \in \widehat{\widetilde{S}_n}$;*
(ii) *for every odd $\lambda \in DP(n)$, there exist two associate $\tilde{\sigma}_\lambda^+, \tilde{\sigma}_\lambda^- \in \widehat{\widetilde{S}_n}$.*

The dimension of $\tilde{\sigma}_\lambda$ or $\tilde{\sigma}_\lambda^\pm$, where $\lambda = (\lambda_1, \ldots, \lambda_m) \in DP(n)$, is

$$2^{[\frac{n-m}{2}]} \frac{n!}{\lambda_1! \ldots \lambda_m!} \prod_{1 \le i < j \le m} \frac{\lambda_i - \lambda_j}{\lambda_i + \lambda_j}. \tag{3.5.1}$$

In order to simplify the formulas below, we let $\tilde{\sigma}_\lambda^\epsilon$ denote any one of $\tilde{\sigma}_\lambda$, if λ is an even partition in $DP(n)$, or $\tilde{\sigma}_\lambda^\pm$, if λ is an odd partition in $DP(n)$.

The decomposition of the tensor product of an S_n-type σ_μ with a spin representation $\tilde{\sigma}_{(n)}$ is known.

Theorem 3.5.2 ([St, Theorem 9.3], [Ma, Chapter 3, (8.17)]). *If $\lambda \neq (n)$, we have*

$$\dim \text{Hom}_{\widetilde{S}_n}[\tilde{\sigma}_\lambda, \sigma_\mu \otimes \tilde{\sigma}_{(n)}] = \frac{1}{\epsilon_\lambda \epsilon_{(n)}} 2^{\frac{|\lambda|-1}{2}} g_{\lambda,\mu}, \tag{3.5.2}$$

where $\epsilon_\lambda = 1$ (resp. $\epsilon_\lambda = \sqrt{2}$) if λ is even (resp. odd), and the integer $g_{\lambda,\mu}$ is the (λ,μ) entry in the inverse matrix $K(-1)^{-1}$, where $K(t)$ is the matrix of Kostka–Foulkes polynomials. In particular:

(i) $g_{\lambda,\lambda} = 1$;
(ii) $g_{\lambda,\mu} = 0$, *unless* $\lambda \geq \mu$ *in the ordering of partitions.*

Example 3.5.1. The integers $g_{\lambda,\mu}$ have also an explicit combinatorial description in terms of "shifted tableaux" of unshifted shape μ and content λ satisfying certain admissibility conditions (see [St, Theorem 9.3]). From this description, one may see for example that if $\lambda = \text{hook}(\mu)$, then $g_{\lambda,\mu} = 1$ in (3.5.2).

3.6 *Nonzero cohomology*

We are now in position to determine the unitary modules of $\mathbb{H}_n$ with nonzero Dirac cohomology.

We remark that since $gl(n)$ is not semisimple, the spin modules S^ϵ of $C(V)$ ($V \cong \mathbb{C}^n$) are not necessarily irreducible $\widetilde{S}_n$-representations. More precisely, using (2.6.1), we see that $S^\pm|_{\widetilde{S}_n} = \tilde{\sigma}_{(n)}$, when n is odd, and $S|_{\widetilde{S}_n} = \tilde{\sigma}^+_{(n)} + \tilde{\sigma}^-_{(n)}$, when n is even.

Lemma 3.6.1. *Assume X is an irreducible $\mathbb{H}_n$-module such that $H^D(X) \neq 0$. Then the central character of X is in the set $\{h_\lambda/2 : \lambda \in DP(n)\}$, where h_λ is as in (3.4.1).*

Proof. This is just a reformulation of Theorem 2.5.1 in this particular case. □

As a consequence of (3.5.2), we obtain the following precise results for Dirac cohomology.

Lemma 3.6.2. (a) *A spherical module $L(\nu)$ has nonzero Dirac cohomology if and only if $\nu \in \{h_\lambda/2 : \lambda \in DP(n)\}$, where h_λ is as in (3.4.1), and in this case $H^D_\epsilon(L(h_{(n)}/2)) = S^\epsilon$, and if $\lambda \neq (n)$:*

$$H^D_\epsilon(L(h_\lambda/2)) = \begin{cases} 2^{[(|\lambda|-1)/2]}\tilde{\sigma}_\lambda, & \text{if } n \text{ is odd and } \lambda \text{ is even;} \\ 2^{[(|\lambda|-1)/2]}(\tilde{\sigma}^+_\lambda + \tilde{\sigma}^-_\lambda), & \text{if } n \text{ is odd and } \lambda \text{ is odd;} \\ 2^{[(|\lambda|)/2-1]}(\tilde{\sigma}^\epsilon_\lambda + \tilde{\sigma}^\epsilon_\lambda \otimes \mathsf{sign}) & \text{if } n \text{ is even.} \end{cases}$$

(b) *Every Speh module $a(m,d)$ has nonzero Dirac cohomology. More precisely,*

$$H^D_\epsilon(a(m,d)) = \begin{cases} 2^{(d-1)/2}\,(\tilde{\sigma}^+_{(m+d-1,m+d-3,\dots,|m-d|+1)} + \tilde{\sigma}^-_{(m+d-1,m+d-3,\dots,|m-d|+1)}) \\ \qquad \text{if } d \text{ is odd and } m \text{ is even,} \\ 2^{[(d-1)/2]}\,\tilde{\sigma}^\epsilon_{(m+d-1,m+d-3,\dots,|m-d|+1)} \quad \text{if } d \text{ is odd and } m \text{ is odd,} \\ 2^{[(d+1)/2]}\,\tilde{\sigma}^\epsilon_{(m+d-1,m+d-3,\dots,|m-d|+1)} \quad \text{otherwise.} \end{cases}$$

(c) *Every complementary series induced module $\pi(a_y(m,d),\nu)$ as in (3.3.2) has zero Dirac cohomology.*

Proof.

(a) This is immediate by (2.7.1) and the upper unitriangular property of the numbers $g_{\lambda,\mu}$ in Theorem 3.5.2.
(b) By Lemma 3.4.1, the central character of $a(m,d)$ is $h_{\lambda'}$ where $\lambda' = \text{hook}(\lambda) \in DP(n)$. By Example 3.5.1, the genuine $\widetilde{S}_n$-type $\tilde{\sigma}_{\lambda'}$ occurs with nonzero multiplicity in $\sigma_{(m,m,\dots,m)} \otimes S$. By construction, $a(m,d)$ is isomorphic with $\sigma_{(m,m,\dots,m)}$ as S_n-representations. This means that the hypothesis of Proposition 2.5.1 are satisfied, hence $\tilde{\sigma}_{\lambda'}$ occurs in $H^D(a(m,d))$.
(c) This is immediate from Lemma 3.6.1, since $a_y(m,d)$, $y \neq 0$ and $\pi(a_y(m,d),\nu), 0 < \nu < \frac{1}{2}$ do not have the allowable central characters. □

Theorem 3.6.1. *An irreducible unitary $\mathbb{H}_n$-module has nonzero Dirac cohomology if and only if it is isomorphic with an induced module*

$$X = \mathbb{H}_n \otimes_{\mathbb{H}_{ev} \times \mathbb{H}_{odd}} (\pi_{ev} \boxtimes \pi_{odd}), \tag{3.6.1}$$

where

$$\begin{aligned}
&\mathbb{H}_{ev} = \mathbb{H}_{k_1} \times \mathbb{H}_{k_2} \times \dots \times \mathbb{H}_{k_\ell}, \quad \mathbb{H}_{odd} = \mathbb{H}_{k_1'} \times \mathbb{H}_{k_2'} \times \dots \times \mathbb{H}_{k_t'},\\
&\pi_{ev} = a(m_1,d_1) \boxtimes a(m_2,d_2) \boxtimes \dots \boxtimes a(m_\ell,d_\ell),\\
&\pi_{odd} = a(m_1',d_1') \boxtimes a(m_2',d_2') \boxtimes \dots \boxtimes a(m_t',d_t'),\\
&m_i + d_i \equiv 0 \ (\mathrm{mod}\, 2),\ m_j' + d_j' \equiv 1 \ (\mathrm{mod}\, 2),\\
&k_1 + k_2 + \dots + k_\ell + k_1' + k_2' + \dots + k_t' = n
\end{aligned}$$

and $a(m_i,d_i)$, $a(m_j',d_j')$ are Speh modules for $\mathbb{H}_{k_i}$, $\mathbb{H}_{k_j'}$ such that the following conditions are satisfied:

$$\begin{aligned}
m_1 + d_1 - 1 \geq |m_1 - d_1| + 1 > m_2 + d_2 - 1&\\
\geq |m_2 - d_2| + 1 > \dots > m_\ell + d_\ell - 1;&\\
m_1' + d_1' - 1 \geq |m_1' - d_1'| + 1 > m_2' + d_2' - 1&\\
\geq |m_2' - d_2'| + 1 > \dots > m_t' + d_t' - 1.&
\end{aligned} \tag{3.6.2}$$

Proof. From Theorem 3.3.2, a unitary irreducible module X is induced from a combination of Speh modules and complementary series modules. It is immediate that in order for X to have one of the central characters from Lemma 3.6.1, a first restriction is that only Speh modules can appear in the induction, so X is of the form (3.6.1). Notice then that the central character of X is obtained by concatenating the central characters of $a(m_i, d_i)$. Therefore the central character of X is S_n-conjugate to h_λ, where λ is the composition

$$\lambda = \lambda^1 \sqcup \cdots \sqcup \lambda^\ell \sqcup \mu^1 \sqcup \cdots \sqcup \mu^t,$$

where $\lambda^i = (m_i + d_i - 1, m_i + d_i - 3, \ldots, |m_i - d_i| + 1)$, $1 \leq i \leq \ell$ and $\mu^j = (m'_j + d'_j - 1, m'_j + d'_j - 3, \ldots, |m'_j - d'_j| + 1)$, $1 \leq j \leq t$. The entries in the first type of strings are all even, while the entries in the second type of strings are all odd. Since we need λ to have no repetitions, condition (3.6.2) follows.

For the converse, assume X is as in (3.6.1) and (3.6.2). Then the central character of X is h_λ, where λ is as above. By Proposition 2.5.1, it remains to check that $X \otimes S$ contains the $\widetilde{S}_n$-type $\tilde{\sigma}_\lambda$. From Lemma 2.6.2, we see that $\mathrm{Hom}_{\widetilde{S}_n}[\tilde{\sigma}_\lambda, X \otimes S] = \frac{\dim \mathcal{S}}{\dim \mathcal{S}_M} \mathrm{Hom}_{\widetilde{W}_M}[\tilde{\sigma}|_{\widetilde{W}_M}, (\pi_{\mathrm{ev}} \boxtimes \pi_{\mathrm{odd}})|_{W_M} \otimes \mathcal{S}_M]$, where $\widetilde{W}_M = \widetilde{S}_{k_1} \cdot \ldots \cdot \widetilde{S}_{k_\ell} \cdot \widetilde{S}_{k'_1} \cdot \ldots \cdot \widetilde{S}_{k'_t}$, and $\mathcal{S}_M$ is the corresponding spin module. (Here $\cdot$ denotes the graded version of the direct product coming from the graded tensor product of Clifford algebras as in Section 2.6.) From Lemma 3.6.2, we know that the $\widetilde{S}_{k_i}$-representation $\tilde{\sigma}_{\lambda^i}$ occurs in $a(m_i, d_i)|_{S_{k_i}}$ tensored with the spin $\widetilde{S}_{k_i}$-module and similarly the $\widetilde{S}_{k'_j}$-representation $\tilde{\sigma}_{\mu^j}$ occurs in $a(m'_j, d'_j)|_{S_{k'_j}}$ tensored with the spin $\widetilde{S}_{k'_j}$-module. Therefore the tensor product representation $\tilde{\sigma}_{\lambda,M} := \tilde{\sigma}_{\lambda^1} \boxtimes \cdots \boxtimes \tilde{\sigma}_{\lambda^\ell} \boxtimes \tilde{\sigma}_{\mu^1} \boxtimes \cdots \boxtimes \tilde{\sigma}_{\mu^t}$ occurs in $(\pi_{\mathrm{ev}} \boxtimes \pi_{\mathrm{odd}})|_{W_M} \otimes \mathcal{S}_M$. Finally, since the composition λ is just the concatenation of the (λ^i)'s and the (μ^j)'s, one sees that $\tilde{\sigma}_{\lambda,M}$ occurs with nonzero multiplicity in $\tilde{\sigma}_\lambda|_{\widetilde{W}_M}$. □

References

[A] J. Arthur, *Unipotent automorphic representations: conjectures. Orbites unipotentes et représentations, II*, Astérisque No. 171–172 (1989), 13–71.

[AS] M. Atiyah, W. Schmid, *A geometric construction of the discrete series for semisimple Lie groups*, Invent. Math. 42 (1977), 1–62.

[BCT] D. Barbasch, D. Ciubotaru, P. Trapa, *The Dirac operator for graded affine Hecke algebras*, Acta Math. 209 (2012), 197–227.

[BM] D. Barbasch, A. Moy, *Unitary spherical spectrum for p-adic classical groups*, Acta Appl. Math. 44 (1996), no. 1–2, 3–37.

[BMcP] W. Borho, R. MacPherson, *Partial resolutions of nilpotent varieties*, Analysis and topology on singular spaces, II, III (Luminy, 1981), 23–74, Astérisque, 101–102, Soc. Math. France, Paris, 1983.

[BZ] J. Bernstein, A. Zelevinsky, *Induced representations of reductive p-adic groups. I*, Ann. Sci. École Norm. Sup. (4) 10 (1977), no. 4, 441–472.

[BW] A. Borel, N. Wallach, *Continuous cohomology, discrete subgroups, and representations of reductive groups,* Princeton University Press, Princeton, New Jersey, 1980.

[Ch] C. Chevalley, *The Algebraic Theory of Spinors*, Columbia University Press, New York, 1954. viii+131 pp.

[Ci] D. Ciubotaru, *Spin representations of Weyl groups and Springer's correspondence*, J. Reine Angew. Math. 671 (2012), 199–222.

[CT] D. Ciubotaru, P. Trapa, *Characters of Springer representations on elliptic conjugacy classes*, Duke Math. J. 162 (2013), 201–223.
[EW] T. Enright, N. Wallach, *Embeddings of unitary highest weight representations and generalized Dirac operators*, Math. Ann. 307 (1997), no. 4, 627–646.
[HP] J.-S. Huang, P. Pandzić, *Dirac cohomology, unitary representations and a proof of a conjecture of Vogan*, J. Amer. Math. Soc. 15 (2002), no. 1, 185–202.
[KZ] A. Knapp, G. Zuckerman, *Classification theorems for representations of semisimple Lie groups*, Non-commutative harmonic analysis (Actes Colloq., Marseille-Luminy, 1976), 138–159. Lecture Notes in Math., Vol. 587, Springer, Berlin, 1977.
[Lu] G. Lusztig, *Affine Hecke algebras and their graded version*, J. Amer. Math. Soc. 2 (1989), 599–635.
[Ma] I.G. MacDonald, *Symmetric functions and Hall polynomials*, Oxford Mathematical Monographs, Oxford Science Publications, The Clarendon Press, Oxford University Press, New York, 1995. x+475 pp.
[Mo] A. Morris, *Projective representations of reflection groups. II*, Proc. London Math. Soc. (3) 40 (1980), no. 3, 553–576.
[OV] A. Okounkov, A. Vershik, *A new approach to representation theory of symmetric groups*, Selecta Math. 2 (4) (1996), 581–605.
[P] R. Parthasarathy, *Dirac operator and the discrete series*, Ann. of Math. (2) 96 (1972), 1–30.
[Ro] J. Rogawski, *On modules over the Hecke algebra of a p-adic group*, Invent. Math. 79 (1985), no. 3, 443–465.
[St] J. Stembridge, *Shifted tableaux and the projective representations of symmetric groups*, Adv. Math. 74 (1989), no. 1, 87–134.
[Ta] M. Tadić, *Classification of unitary representations in irreducible representations of general linear group (non-Archimedean case)*, Ann. Sci. École Norm. Sup. (4) 19 (1986), no. 3, 335–382.

On the nilradical of a parabolic subgroup

Karin Baur

Dedicated to Nolan Wallach

Abstract We present various approaches to understanding the structure of the nilradical of parabolic subgroups in type A. In particular, we consider the complement of the open dense orbit and describe its irreducible components.

Keywords: Nilradical • Parabolic subgroups

Mathematics Subject Classification 2010: 17B45

1 Introduction

This paper is an extended version of a talk at the conference "Lie Theory and Its Applications" held at UCSD in March 2011.

Nolan Wallach had ignited my interest in parabolic subalgebras (cf. Subsection 1.4). During the time I was a post-doc at UCSD and also during later visits, I have enjoyed numerous lectures by and discussions with Nolan Wallach. I am very grateful for them. This paper allows me to display joint work with N. Wallach and to give a view on related recent progress. I will present several approaches towards understanding the nilradical of a parabolic subgroup of a reductive algebraic group.

1.1 Classical situation

Let $\mathfrak{g}$ $=\mathrm{End}(\mathbb{C}^n)$ be the Lie algebra of endomorphisms of $\mathbb{C}^n$. The nilpotent endomorphisms among them are well known. Up to conjugacy by G $=\mathrm{GL}_n(\mathbb{C})$,

---

K. Baur
Institute for Mathematics and Scientific Computing, University of Graz, Heinrichstrasse 36, A-8010 Graz, Austria
e-mail: karin.baur@uni-graz.at

R. Howe et al. (eds.), *Symmetry: Representation Theory and Its Applications*, Progress in Mathematics 257, DOI 10.1007/978-1-4939-1590-3_2

they are given by partitions, i.e., by the Jordan canonical form. In particular, these orbits are well understood. There are finitely many and we have an order on them, namely by inclusion of orbit closures.

If we put this in a more formal language, we are in the following situation: Let G be a classical algebraic group over $\mathbb{C}$ and let $\mathfrak{g}$ be its Lie algebra.[1] The group G acts on the cone $\mathcal{N} \subset \mathfrak{g}$ of nilpotent elements by conjugation. This action breaks up $\mathcal{N}$ into finitely many orbits. The nilpotent orbits are parametrized by certain partitions. The exact description can be found in [9], cf. also [15]. More generally, Jacobson–Morozov theory tells us that every nilpotent element e of $\mathfrak{g}$ can be embedded in an $\mathfrak{sl}_2$-triple (e, f, h). The action of the semisimple element h of this triple gives rise to a labeled Dynkin diagram associated to the nilpotent orbit, with labels from $\{0, 1, 2\}$. This is the so-called Dynkin–Kostant classification of nilpotent orbits, cf. [9, 15].

Since $\mathcal{N}$ is irreducible, there exists an open dense orbit, called the regular nilpotent orbit. We can give a representative of this orbits in a very nice way: If we take a generator X_α for each simple root space (with respect to a given Cartan subalgebra of $\mathfrak{g}$), we obtain a regular nilpotent element,

$$X = \sum_{\alpha \text{ simple}} X_\alpha.$$

The labeled Dynkin diagram of this orbit has a 2 at every node.

Example 1. Let G be $\mathrm{SL}_{n+1}(\mathbb{C})$. We choose the diagonal matrices in its Lie algebra as the Cartan subalgebra. The root spaces are then spanned by the elementary matrices $E_{i,j}$, $i \neq j$, where the only nonzero entry is a 1 at position (i, j). With this choice, the above representative takes the form

$$X = \sum_{i=1}^{n} E_{i,i+1} = \begin{pmatrix} 0 & 1 & & & 0 \\ & 0 & 1 & & \\ & & \ddots & \ddots & \\ & & & 0 & 1 \\ & & & & 0 \end{pmatrix}.$$

1.2 Flags in n-space

We now consider nilpotent endomorphisms of $\mathbb{C}^n$ which preserve flags of vector spaces. For this and the following section let $G = \mathrm{GL}_n(\mathbb{C})$ and let $\mathcal{F}$ be a partial flag in $\mathbb{C}^n$:

$$\mathcal{F}:\ 0 = V_0 \subsetneq V_1 \subsetneq \cdots \subsetneq V_{r-1} \subsetneq V_r = \mathbb{C}^n$$

[1] Often it would be enough to assume that G is defined over an algebraically closed field.

for some $r \geq 1$. Then we define $G \supset P$ to be the parabolic subgroup which is the stabilizer of the flag $\mathcal{F}$,

$$P := \{g \in G \mid gV_i = V_i \ \forall\, i\}.$$

We will sometimes write $P = P(\mathcal{F})$ for the parabolic subgroup corresponding to $\mathcal{F}$. We will write $\mathfrak{p}$ to denote the Lie algebra of P.

What can we say about the nilpotent endomorphisms of $\mathbb{C}^n$, which preserve the flag $\mathcal{F}$? In other words, how can we describe the $X \in \mathcal{N}$ with

$$XV_i \subset V_{i-1} \ \ \forall\, i\ ?$$

Example 2. Consider

$$\mathcal{F}:\ V_0 = 0 \subset V_1 \subset V_2 \subset V_3 = \mathbb{C}^4,$$

with $V_1 := \langle e_1 \rangle$ and $V_2 := \langle e_1, e_2 \rangle$. Then $\mathfrak{p}$ consists of the 4×4 matrices of the form

$$\begin{pmatrix} \bullet & * & * & * \\ 0 & \bullet & * & * \\ 0 & 0 & \bullet & \bullet \\ 0 & 0 & \bullet & \bullet \end{pmatrix}.$$

(with arbitrary entries at the positions of the $\bullet$'s and $*$'s). If X is a nilpotent element satisfying $XV_i \subset V_{i-1}$ for all i, X has to have the form

$$\begin{pmatrix} 0 & * & * & * \\ 0 & 0 & * & * \\ 0 & 0 & 0 & 0 \\ 0 & 0 & 0 & 0 \end{pmatrix}.$$

1.3 The Richardson orbit

The above example is an instance of the following situation. Let $G \supset P$ be a parabolic subgroup, $P = L \cdot U$ with L reductive (called the Levi factor) and U the unipotent radical of P. The Lie algebra $\mathfrak{n}$ of U is called the nilradical of P. It is convenient to assume that P contains the Borel subgroup of the upper triangular matrices and that L contains the diagonal matrices. Such P and its Lie algebra $\mathfrak{p}$ are called standard. The direct sum decomposition $\mathfrak{p} = \mathfrak{l} \oplus \mathfrak{n}$ is called the Levi decomposition of $\mathfrak{p}$. In the example above, the nilradical $\mathfrak{n}$ consists of the matrices with nonzero entries only at the positions of the $*$'s, and X belongs to $\mathfrak{n}$.

The Levi part $\mathfrak{l}$ of $\mathfrak{p}$ consists of the matrices with nonzero entries only at the $\bullet$'s. It is a general feature that $\mathfrak{l}$ consists of matrices with nonzero entries only in square blocks on the diagonal. The sizes of these square blocks are the differences $\dim V_i - \dim V_{i-1}$. We willlater denote them by $d_1, \dots, d_r$.

The parabolic subgroup P acts on its nilradical $\mathfrak{n}$ by conjugation. It is known that the nilradical can be written as the union of the intersections of $\mathfrak{n}$ with the nilpotent G-orbits in $\mathfrak{g}$, cf. [11] (Satz 4.2.8). Recall that the nilpotent G-orbits are parametrized by partitions of n. If λ is a partition of n, we will write $C(\lambda)$ to denote the corresponding nilpotent G-orbit.

Since there are only finitely many nilpotent G-orbits in $\mathfrak{g}$, one of the intersections of $\mathfrak{n}$ with the nilpotent G-orbits, say $C(\lambda) \cap \mathfrak{n}$, is open and dense in $\mathfrak{n}$. If μ is any partition of n, we get

- $C(\mu) \cap \mathfrak{n} \subset \mathfrak{n} \setminus (C(\lambda) \cap \mathfrak{n})$ with $C(\mu) \cap \mathfrak{n} \neq 0$ if and only if $\mu \leq \lambda$;
- $\dim(C(\mu) \cap \mathfrak{n}) < \dim \mathfrak{n}$ whenever $\mu \neq \lambda$.

Note that we write $\mu \leq \lambda$ if and only if the closure of the orbit $C(\mu)$ is contained in $C(\lambda)$. In particular, we see that $C(\lambda)$ is the unique nilpotent G-orbit of $\mathfrak{g}$ intersecting $\mathfrak{n}$ in an open dense set.

In fact, Richardson shows[2] that $C(\lambda) \cap \mathfrak{n}$ is a single P-orbit, [16]. We call this P-orbit the Richardson orbit of P and denote it by $\mathcal{O}_R$. Its elements are called the Richardson elements (of $\mathfrak{p}$). The G-saturation $G \cdot \mathcal{O}_R$ is also called a Richardson orbit.

Even though $\mathfrak{n}$ contains an open dense P-orbit, we cannot expect that $\mathfrak{n}$ consists of finitely many P-orbits. The Borel subgroup B of GL_6 already has infinitely many B-orbits in its nilradical.

Example 3. Let $\mathcal{F} : V_i := \mathbb{C}^i$, $0 \leq i \leq n$, be the complete flag in $\mathbb{C}^n$. The corresponding parabolic subgroup is a Borel subgroup $B \subset G$. The nilradical consists of the strictly upper triangular matrices. One can show that if $n = 6$, there is a 1-parameter family of B-orbits in $\mathfrak{n}$.

A consequence of the above example is that whenever a flag $\mathcal{F}$ consists of at least 6 nonzero vector spaces, there are infinitely many $P(\mathcal{F})$-orbits in the corresponding nilradical. For classical G, the parabolic subgroups with finitely many P-orbits are classified, cf. [12]. Roughly speaking, they are the ones with at most 5 blocks in the Levi factor.

1.4 (Very) nice parabolic subalgebras

For the moment, let G be a reductive algebraic group over $\mathbb{C}$. Let $B \subset G$ be a Borel subgroup, T a fixed maximal torus in B and $\mathfrak{b}$ the corresponding Borel subalgebra, $\mathfrak{h}$ =Lie(T) the corresponding Cartan subalgebra. This determines a basis $\{\alpha_1, \ldots, \alpha_n\}$ of simple roots of $\mathfrak{g}$ (with n the rank of $\mathfrak{g}$).

---

[2]His result holds in much greater generality for reductive algebraic groups.

We assume that $\mathfrak{p}$ is a standard parabolic subalgebra, i.e., that it contains $\mathfrak{b}$. It gives rise to a $\mathbb{Z}$-grading of $\mathfrak{g}$ as follows: The parabolic subalgebra $\mathfrak{p}$ is determined by the simple roots α_i such that $\mathfrak{p}$ does not contain the root subspace $\mathfrak{g}_{-\alpha_i}$, equivalently by the simple roots whose root space does not lie in the Levi factor. Hence $\mathfrak{p}$ gives rise to a tuple $(u_1, \dots, u_n) \in \{0,1\}^n$ with $u_i = 1$ whenever $\mathfrak{g}_{-\alpha_i}$ is not in $\mathfrak{p}$. Then we set $H \in \mathfrak{h}$ to be the element defined by $\alpha_i(H) = u_i$. The adjoint action of H on $\mathfrak{g}$ then defines the $\mathbb{Z}$-grading:

$$\mathfrak{g}_i := \{x \in \mathfrak{g} \mid [H, x] = ix\}.$$

The grading $\mathfrak{g} = \sum_{i\in\mathbb{Z}} \mathfrak{g}_i$ is such that the parabolic subalgebra is the sum of the nonnegatively graded parts and that the nilradical the sum of the positively graded part. (cf. e.g., Section 2 of [6]). In case $\mathfrak{g} = \mathrm{End}(\mathbb{C}^n)$ we can read off the graded parts from the block structure of the matrices. In particular, $\mathfrak{g}_0 = \mathfrak{l}$ is the Levi part, $\mathfrak{g}_1$ consists of the sequence of the rectangular regions to the right of the squares on the diagonal. Any element $X \in \mathfrak{g}_1$ gives rise to a character χ_X on $\mathfrak{g}_{-1}$. In case $X \in \mathfrak{g}_1$ is a Richardson element, the character χ_X is admissible in the sense of Lynch, [14]. Hence the existence of a Richardson element in $\mathfrak{g}_1$ ensures the existence of an admissible character. This is exactly Lynch's vanishing condition of certain Lie algebra cohomology spaces for a generalized Whittaker module (associated with the parabolic subalgebra). If $\mathfrak{p}$ has a Richardson element in $\mathfrak{g}_1$, we say that the parabolic subalgebra is *nice*. In joint work with N. Wallach, [6], we classified the nice parabolic subalgebras of simple Lie algebras over $\mathbb{C}$.

It is known that for X in $\mathcal{O}_R$, the identity component G_X^0 of the stabilizer subgroup in G is contained in $P_X \subset G_X$, in particular, $|G_X/P_X|$ is finite. The numbers $|G_X/P_X|$ can be found in the article [10] by Hesselink.[3] Assume that $\mathfrak{p}$ is nice. The condition $G_X = P_X$ corresponds to the birationality of the moment map from the dual of the cotangent bundle of G/P onto its image. Nice parabolic subalgebras with $G_X = P_X$ are called very nice. In [7] we continued our joint work with N. Wallach and described the very nice parabolic subalgebras. The main application of this is that under these conditions, one can prove a holomorphic continuation of Jacquet integrals for a real form of $\mathfrak{g}$, cf. [18, 19].

1.5 *P-orbit structure in* $\mathfrak{n}$

What can we say about the P-orbit structure in the nilradical $\mathfrak{n}$? This is a very difficult question. In general, it is a "wild" problem (in the language of representations of algebras).

A first approach towards understanding the nilradical is the description of the elements of the open dense orbit. One can give representatives for Richardson

[3] He attributes them to Spaltenstein, cf. [17].

elements explicitly, as has been done for the classical and exceptional groups, cf. [1,2,6] and [4]. We now go back to $G = \mathrm{GL}_n(\mathbb{C})$. Let $\mathcal{F} : 0 = V_0 \subset V_1 \subset \cdots \subset V_r = \mathbb{C}^n$ be a flag and let $P \subset G$ be the corresponding parabolic subgroup. For $i = 1, \ldots, r$, we set $d_i := \dim V_i/V_{i-1}$. The d_i are the block lengths of the Levi factor in the nilradical of P. One can show that they determine $\mathfrak{p}$ and $\mathfrak{n}$. The r-tuple $d = (d_1, \ldots, d_r)$ forms a composition of n. Let $\lambda = (\lambda_1 \geq \lambda_2 \geq \cdots \geq \lambda_s \geq 0)$ be the dual of the partition obtained by ordering the d_i by size. Then λ is the partition of the Richardson orbit.

We finish this section by illustrating how one can construct a representative of the Richardson orbit.

Example 4. If we assume $d_1 \geq d_2 \geq \cdots \geq d_r$ we obtain a representative[4] of the Richardson orbit by choosing small identity blocks of the size $d_{i+1} \times d_{i+1}$ next to the ith block in the Levi factor: For $d = (3, 2, 2)$ we get

$$\begin{pmatrix} 0 & 0 & 0 & 1 & & & \\ 0 & 0 & 0 & & 1 & & \\ 0 & 0 & 0 & & & & \\ & & & 0 & 0 & 1 & \\ & & & 0 & 0 & & 1 \\ & & & & & 0 & 0 \\ & & & & & 0 & 0 \end{pmatrix}.$$

Clearly, if $\mathcal{F}$ is the complete flag, i.e. if $d_i = 1$ for all i, then the resulting element of the nilradical is the regular nilpotent with 1's next to the diagonal.

1.6 Two approaches to $\mathfrak{n}$

As $\mathcal{O}_R$ is open dense in $\mathfrak{n}$, the knowledge about the Richardson orbit already gives a lot of information about the nilradical. However, it is very difficult to get a grasp on the remaining P-orbits, in particular if there are infinitely many of them.

So far, there exist two approaches towards understanding the structure of the nilradical. One approach is to study the complement of the open dense orbit. The other approach is an example of the process of categorification: We search for a category of representations for an algebra with the hope of finding a bijection between the P-orbits in $\mathfrak{n}$ and a class of isomorphism classes of modules in this category. Such a correspondence has been established for the general linear groups in [8]. For the orthogonal groups, we have a good candidate for the corresponding algebra, but it is not yet clear what is the class of representations corresponding to the P-orbits, [3].

---

[4]It is enough to assume that the sequence of the d_i is unimodal, i.e., that it is first increasing and then decreasing.

In this article, we explain the first approach: Consider the complement of the open dense orbit in the nilradical, i.e. the variety $Z := \mathfrak{n} \setminus \mathcal{O}_R$. We will describe the irreducible components of Z. In particular, we will see that if the flag $\mathcal{F}$ is composed of r nonzero vector spaces, then Z has at most $r-1$ components.

2 Complement of the Richardson orbit

2.1 *Notation*

In what follows we derive the description of the components of Z using rank conditions on matrices. Let $d = (d_1, \dots, d_r)$ be the sizes of the blocks in the Levi factor of the parabolic subgroup. If A is a $n \times n$-matrix, we divide A into rectangular blocks whose sizes are given by the d_i. We let A_{ij} be the $d_i \times d_j$-rectangle formed by the intersection of the d_i rows $(d_1 + \dots + d_{i-1} + 1), \dots, (d_1 + \dots + d_i)$ with the d_j columns $(d_1 + \dots + d_{j-1} + 1), \dots, (d_1 + \dots + d_j)$: A_{11} is the region formed by the intersection of the first d_1 rows and the first d_1 columns, etc. With this notation, the nilradical $\mathfrak{n}$ consists of the matrices A with $A_{ij} = 0$ whenever $i \geq j$.

Let $X = X(d)$ be a Richardson element and let λ be the partition of X. If A is any element in $\mathfrak{n} \setminus \mathcal{O}_R$, the nilpotency class μ of A is strictly smaller than λ. We can translate this as follows: X is characterized by the fact that the sequence $\operatorname{rk} X$, $\operatorname{rk} X^2$, $\operatorname{rk} X^3, \dots$ of the ranks of its powers decreases as slowly as possible. We will use this observation to characterize the elements of the complement Z.

For $A \in \mathfrak{gl}_n$ we write $A[ij]$ for the square formed by the $(j-i+1)^2$ blocks A_{lm}, $i \leq l \leq j, i \leq m \leq j$,

$$A[ij] := \begin{pmatrix} \boxed{A_{ii}} & \cdots & \boxed{A_{ij}} \\ \vdots & \ddots & \vdots \\ \boxed{A_{ji}} & \cdots & \boxed{A_{jj}} \end{pmatrix}$$

With this, we are almost ready to state our result. We first need two more definitions:

$$\kappa(i, j) := 1 + |\{l \mid i < l < j,\ d_l \geq \min(d_i, d_j)\}|$$

$$Z_{ij}^k := \{A \in \mathfrak{n} \mid \operatorname{rk}(A[ij]^k) < \operatorname{rk}(X[ij]^k)\}.$$

When $k = \kappa(i, j)$, we write Z_{ij} instead of $Z_{ij}^{\kappa(i,j)}$.

2.2 Decomposition of Z

In this subsection, we explain how to get Z as a disjoint union of irreducible components. We claim that

$$Z = \bigcup_{(i,j)\in\Lambda(d)} Z_{ij}$$

is the decomposition of Z into irreducible components, cf. [5].

It is rather unpleasant to describe the parameter set $\Lambda(d)$. We will first define a larger set $\Gamma(d)$ and then restrict to $\Lambda(d)$. Let $\Gamma(d)$ be

$$\Gamma(d) := \{(i,j) \mid d_l < \min(d_i, d_j) \text{ or } d_l > \max(d_i, d_j)\ \forall\, i < l < j\}.$$

Inside $\Gamma(d)$ we define $\Lambda(d)$. In case $d_i \neq d_j$, we put further constraints on the first $i-1$ entries $d_1, \ldots, d_{i-1}$ of d and on $d_{j+1}, \ldots, d_r$ of d:

$$\begin{aligned}\Lambda(d) := \{(i,j) \in \Gamma(i,j) \mid d_i = d_j\}& \\ \cup \{(i,j) \in \Gamma(i,j) \mid d_i \neq d_j \text{ and }& \\ &\left.\begin{array}{l}\bullet\ \forall\, k \leq r : d_k \leq \min(d_i, d_j) \text{ or } d_k \geq \max(d_i, d_j) \\ \bullet \text{ for } k < i : d_k \neq d_j \\ \bullet \text{ for } k > j : d_k \neq d_i\end{array}\right\}\end{aligned}$$

Figure 1 illustrates this: the vertical lines indicate the entries d_i and d_j of d, the $\bullet$ stand for d_i resp. d_j, the $\circ$ for $d_i - 1$, $d_i - 2$, etc. and $d_j - 1$, $d_j - 2$, etc. Assume that (i, j) belongs to $\Lambda(d)$. Figure (a): if $d_i = d_j$, the conditions on the elements of $\Lambda(d)$ tell us that there is no l between i and j such that d_l equals $d_i = d_j$. That means that the dashed line is ruled out for the d_l (with $i < l < j$). Figure (b): if $d_i \neq d_j$, the shaded area shows that among the $i < l < j$, no d_l is allowed with $\min(d_i, d_j) < d_l < \max(d_i, d_j)$. The two dashed lines to the left resp. to the right correspond to the two last conditions on elements of $\Lambda(d)$.

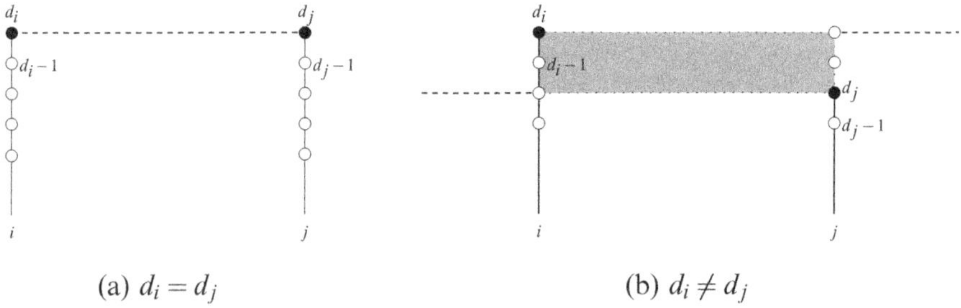

(a) $d_i = d_j$ (b) $d_i \neq d_j$

Figure 1 Dashed lines and shaded areas are not allowed in $\Lambda(d)$

Example 1. We compare $\Gamma(d)$ and $\Lambda(d)$ for several different choices of d.

(a) If d is increasing or decreasing, then

$$\Gamma(d) = \Lambda(d) = \{(1,2),(2,3),\dots,(r-1,r)\}.$$

(b) If $d = (1,1,2,1)$, then

$$\Gamma(d) = \{(1,2),(2,3),(2,4),(3,4)\}$$

and $\Lambda(d) = \{(1,2),(2,4)\}$.

(c) If $d = (7,5,2,3,5,1,2,6,5)$, then

$$\Gamma(d) = \{(i,i+1) \mid i = 1,\dots,8\} \cup$$
$$\{(1,8),(2,4),(2,5),(3,6),(3,7),(4,6),(4,7),(5,7),(5,8),(5,9),(7,9)\}$$

and $\Lambda(d) = \{(1,8),(2,5),(3,7),(5,9)\}$.

A consequence of the result above is that Z has at most $r-1$ irreducible components: when d is increasing or decreasing, it is clear that $\Lambda(d)$ has size $r-1$. If all d_i are different, the same is true. In all other cases, there is at least one pair i,j with $d_i = d_j$, $|i-j| > 1$. Then one can find an index l between i and j such that (i,l) and (l,j) do not belong to $\Lambda(d)$. Example (c) above shows that the actual number of irreducible components can be much smaller than $r-1$.

We first illustrate the decomposition of Z on an example before explaining the main ideas behind the proof.

Example 2. Let $d = (1,2,3,2)$. Then

$$\mathfrak{n} = \left\{A \in \mathfrak{gl}_8 \mid A = \begin{pmatrix} 0 & * & * & * & * & * & * & * \\ & 0 & 0 & * & * & * & * & * \\ & 0 & 0 & * & * & * & * & * \\ & & & 0 & 0 & 0 & * & * \\ & & & 0 & 0 & 0 & * & * \\ & & & 0 & 0 & 0 & * & * \\ & & & & & & 0 & 0 \\ & & & & & & 0 & 0 \end{pmatrix}\right\}.$$

Then $\Lambda(d) = \{(1,2),(2,4)\}$, with $\kappa(1,2) = 1$ and $\kappa(2,4) = 2$. The matrix $X = E_{1,2} + E_{2,4} + E_{3,5} + E_{4,7} + E_{5,8}$ is a Richardson element for the corresponding parabolic subgroup and $X^2 = E_{1,4} + E_{2,7} + E_{3,8}$. We have to compute the ranks of $X[12]$ (the matrix formed by the first 3 rows and columns) and of the second power of $X[24]$ (the matrix formed by rows and columns 2 to 8): $\operatorname{rk} X[12] = 1$ and $\operatorname{rk} X[24]^2 = 2$.

The component Z_{12} thus consists of all elements of the nilradical whose 1×2-rectangle A_{12} is zero. The component Z_{24} of the matrices $A \in \mathfrak{n}$ with $\operatorname{rk} A[24]^2 \leq 1$. Observe that the only nonzero entries of $A[24]^2$ are in the intersection of its first two rows with its last two columns. This square is just $A_{23}A_{34}$.

2.3 *Ideas of the proof*

The main steps in proving that Z is the union of the $Z_{ij} = Z_{ij}^{\kappa(i,j)}$ with $(i,j) \in \Lambda(d)$ are the following:

- Show that $Z_{ij}^k = \emptyset \iff k > j - i$, that $Z_{ij}^l \subsetneq Z_{ij}^{\kappa(i,j)}$ for $1 \leq l \leq \kappa(i,j)$ and that for l with $\kappa(i,j) < l \leq j - i$ there are $i < i_0 \leq j_0 < j$ such that $Z_{ij}^l \subset Z_{ij_0}^{\kappa(i,j_0)} \cup Z_{i_0 j}^{\kappa(i_0,j)}$.
- Argue that $Z = \cup_{1 \leq i < j \leq r} Z_{ij}^{\kappa(i,j)} = \cup_{i<j} \cup_{k \geq 1} Z_{ij}^k$.
- Prove that for any two $(i,j) \neq (k,l)$ in $\Lambda(d)$ we have $Z_{ij} \not\subset Z_{kl}$ and that the elements (i,j) in $\Lambda(d)$ are enough to get all components: For $(i,j) \notin \Gamma(d)$, we can find pairs (k_m, l_m) in $\Gamma(d)$ such that Z_{ij} lies in the union of the corresponding $Z_{(k_m,l_m)}$. If (i,j) is in $\Gamma(d) \setminus \Lambda(d)$, then there exists $k, l \in \Lambda(d)$ such that $Z_{ij} \subset Z_{kl}$.

It then remains to see that the Z_{ij} are irreducible. This can be done using Young tableaux. We first recall a result of Hille, cf. [11]: If $C(\mu)$ is a nilpotent G-orbit, then the irreducible components of $\mathfrak{n} \cap C(\mu)$ are in bijection with a set $\mathcal{T}(\mu, d)$ of Young tableaux of shape μ.

The Young tableaux of $\mathcal{T}(\mu, d)$ are all possible fillings of the Young diagram of shape μ with d_1 ones, with d_2 twos, d_3 threes, etc. If $\mu = \lambda$ is the partition of the Richardson orbit, there is exactly one way to fill the Young diagram of λ with d_1 ones, etc., i.e., $|\mathcal{T}(\lambda, d)| = 1$. We write $T(d)$ for this tableau.

Since we want to describe irreducible components of the complement, we aim for degenerations of the Young tableau $T(d)$. These degenerations should be minimal: If not, we might end up taking subsets of irreducible components. The minimal degenerations arise from $T(d)$ by moving a single box down a number of rows as follows.

Let $T(i,j)$ be the tableau obtained from $T(d)$ by removing the box containing the number j from the last row containing i and j and inserting it in the closest row in order to obtain another tableau. We call its partition $\mu(i,j)$. By construction, $\mu(i,j) \leq \lambda$. Then we set $\mathfrak{n}(T(i,j)) \subset \mathfrak{n}$ to be the irreducible component of $\mathfrak{n} \cap C(\mu(i,j))$ whose tableau is $T(i,j)$ under Hille's bijection.

We proceed by showing that the components Z_{ij} (with $(i,j) \in \Lambda(d)$) are equal to $\mathfrak{n}(T(i,j))$.

We observe that for every row a box from a Young diagram is moved down, the dimension of the GL_n-orbits of the new nilpotent orbit is decreased by two. This can be derived from the formula for the dimension of the stabilizer, [13]. The change in dimension in the nilradical is half of this. This gives us the codimension of Z_{ij} in $\mathfrak{n}$ as the number of rows the box j has been moved down to get $T(i, j)$.

2.4 Examples and remarks

(1) If $d = (1, 1, 1, 1, 1)$, $P = B$ is a Borel subgroup.

Since $\Lambda(d) = \{(1, 2), (2, 3), (3, 4), (4, 5)\}$, the complement is the union of four irreducible components. The regular nilpotent elements are the nilpotent 5×5-matrices whose 4th power is nonzero. Thus the Richardson orbit consists of strictly upper triangular matrices $A = (a_{ij})_{ij}$ with

$$A[1,5]^4 = \begin{pmatrix} 0 & 0 & 0 & 0 & x \\ & 0 & 0 & 0 & 0 \\ & & 0 & 0 & 0 \\ & & & 0 & 0 \\ & & & & 0 \end{pmatrix}$$

where $x := a_{12}a_{23}a_{34}a_{45} \neq 0$. For A to belong to the complement Z of the Richardson orbit, this product has to be zero. In other words, $A[1, 5]^4$ is the zero matrix. But this means that A belongs to $Z_{i,i+1}$ for some $i \leq 4$ as the component $Z_{i,i+1}$ is the set of matrices with $A_{i,i+1} = a_{i,i+1} = 0$. So A lies in one of the components $Z_{i,i+1}$. The components $Z_{i,i+1}$ all have codimension one. The Young tableaux in $\mathcal{T}(\mu, d)$ for $\mu = (4, 1)$ are in Figure 2.

In the case of a Borel subalgebra, the irreducible components are all orbit closures of B-orbits in $\mathfrak{n}$. So far, it is not known whether this is true in general, though we suspect that it is the case. Another question to which we do not know the answer yet is whether the Z_{ij} are reduced.

(2) The smallest interesting case is $d = (1, 1, 2, 1)$.

Here, $\Gamma(d) \supsetneq \Lambda(d) = \{(1, 2), (2, 4)\}$. From the description in Section 2.2 we expect two irreducible components. Z_{12} is the set of the matrices A with $a_{12} = 0$ and Z_{24} the set of matrices A with $A[24]^2 = 0$. We take $A = (a_{ij})_{ij} \in \mathfrak{n}$ and compute A^2, A^3. Then A^3 has $a_{12}(a_{23}a_{35} + a_{24}a_{45})$ as only nontrivial entry, it is in the upper right corner of A^3. A belongs to the Richardson orbit if and only if this product is nonzero. If it is zero, then $a_{12} = 0$ or $a_{23}a_{35} + a_{24}a_{45} = 0$. The case $a_{12} = 0$ clearly corresponds to $A \in Z_{12}$.

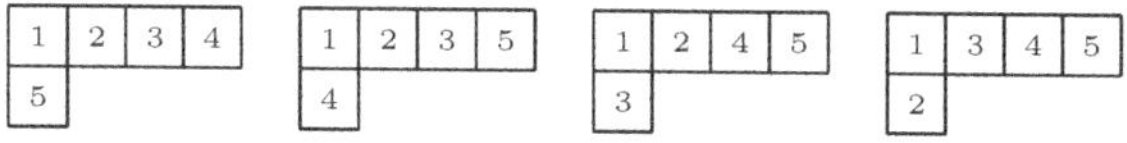

Figure 2 The four Young tableaux of $\mathcal{T}((5, 1), d)$

By definition, Z_{24} consists of the A with $A[24]^2 = 0$. We have

$$A[24]^2 = \begin{pmatrix} 0 & 0 & 0 & a_{23}a_{35} + a_{24}a_{45} \\ 0 & 0 & 0 & 0 \\ 0 & 0 & 0 & 0 \\ 0 & 0 & 0 & 0 \end{pmatrix}$$

So $A \in Z_{24}$ if and only if $a_{23}a_{35} + a_{24}a_{45} = 0$. Thus, $A \notin \mathcal{O}_R$ is equivalent to $A \in Z_{12} \cup Z_{24}$.

References

1. K. Baur, *Richardson elements for classical Lie algebras.* J. Algebra 297 (2006), no. 1, 168–185.
2. K. Baur, *A normal form for admissible characters in the sense of Lynch*, Represent. Theory 9 (2005), 30–45.
3. K. Baur, K. Erdmann, A. Parker, *Δ-filtered modules and nilpotent orbits of a parabolic subgroup in O_N*. J. Pure Appl. Algebra 215 (2011), no. 5, 885–901.
4. K. Baur, S. Goodwin, *Richardson elements for parabolic subgroups of classical groups in positive characteristic.* Algebr. Represent. Theory 11 (2008), no. 3, 275–297.
5. K. Baur, L. Hille, *On the complement of the Richardson orbit* Math. Z. 272 (2012), no. 1–2, 31–49.
6. K. Baur, N. Wallach, *Nice Parabolic Subalgebras of Reductive Lie Algebras*, Represent. Theory 9 (2005), 1–29.
7. K. Baur, N. Wallach, A class of gradings of simple Lie algebras. In *Lie algebras, vertex operator algebras and their applications*, 3–15, Contemp. Math., 442, AMS, Providence, RI, 2007.
8. T. Brüstle, L. Hille, C. M. Ringel, G. Röhrle, *The Δ-filtered modules without self-extensions for the Auslander algebra of $k[T]/\langle T^n \rangle$*, Algebr. Represent. Theory 2 (1999), no. 3, 295–312.
9. D.H. Collingwood, W.M. McGovern, *Nilpotent orbits in semisimple Lie algebras.* Van Nostrand Reinhold Mathematics Series. Van Nostrand Reinhold Co., New York, 1993. xiv+186 pp.
10. W. Hesselink, *Polarizations in the classical groups*, Math. Zeitschrift 160 (1978), 217–234.
11. L. Hille, *Aktionen algebraischer Gruppen, geometrische Quotienten und Köcher*, Habilitationsschrift, Hamburg 2003.
12. L. Hille, G. Röhrle, *A classification of parabolic subgroups of classical groups with a finite number of orbits on the unipotent radical*, Transform. Groups 4 (1) (1999), 35–52.
13. H. Kraft, C. Procesi, *Closures of conjugacy classes of matrices are normal*, Invent. Math. 53 (1979), no. 3, 227–247.
14. T. E. Lynch, *Generalized Whittaker vectors and representation theory*, Thesis, M.I.T., 1979.
15. W.M. McGovern, *The adjoint representation and the adjoint action*, in Encyclopedia of Mathematical Sciences, vol. 131, Invariant Theory and Algebraic Transformation Groups subseries, Springer, 2002, 159–238.
16. R.W. Richardson, Conjugacy classes in parabolic subgroups of semisimple algebraic groups, Bulletin London Math. Society 6 (1974), 21–24.
17. N. Spaltenstein, *Classes unipotentes et sous-groupes de Borel*, Lecture Notes in Mathematics 946, Springer-Verlag, Berlin-New York, 1982.
18. N. Wallach, *Holomorphic continuation of generalized Jacquet integrals for degenerate principal series.* Represent. Theory 10 (2006), 380–398.
19. H. Yamashita, *Multiplicity one theorems for generalized Gel'fand-Graev representations of semisimple Lie groups and Whittaker models for discrete series*, Adv. Stud. Pure Math. 14, Academic Press Boston, 1988, 31–121.

Arithmetic invariant theory

Manjul Bhargava and Benedict H. Gross

To Nolan Wallach

Abstract Let k be a field, let G be a reductive algebraic group over k, and let V be a linear representation of G. Geometric invariant theory involves the study of the k-algebra of G-invariant polynomials on V, and the relation between these invariants and the G-orbits on V, usually under the hypothesis that the base field k is algebraically closed. In favorable cases, one can determine the geometric quotient $V//G = \mathrm{Spec}(\mathrm{Sym}^*(V^\vee)^G)$ and can identify certain fibers of the morphism $V \to V//G$ with certain G-orbits on V. In this paper we study the analogous problem when k is not algebraically closed. The additional complexity that arises in the orbit picture in this scenario is what we refer to as arithmetic invariant theory. We illustrate some of the issues that arise by considering the regular semisimple orbits—i.e., the closed orbits whose stabilizers have minimal dimension—in three arithmetically rich representations of the split odd special orthogonal group $G = \mathrm{SO}_{2n+1}$.

Keywords: Invariant theory • Hyperelliptic curves

Mathematics Subject Classification 2010: 11E72, 14L24

M. Bhargava
Department of Mathematics, Princeton University, Fine Hall, Washington Road, Princeton, NJ 08544-1000, USA
e-mail: bhargava@math.princeton.edu

B.H. Gross
Department of Mathematics, Harvard University, One Oxford Street, Cambridge, MA 02138-2901, USA
e-mail: gross@math.harvard.edu

R. Howe et al. (eds.), *Symmetry: Representation Theory and Its Applications*, Progress in Mathematics 257, DOI 10.1007/978-1-4939-1590-3_3

1 Introduction

Let k be a field, let G be a reductive algebraic group over k, and let V be a linear representation of G. Geometric invariant theory involves the study of the k-algebra of G-invariant polynomials on V, and the relation between these invariants and the G-orbits on V, usually under the hypothesis that the base field k is algebraically closed. In favorable cases, one can determine the geometric quotient $V/\!\!/G = \mathrm{Spec}(\mathrm{Sym}^*(V^\vee))^G$ and can identify certain fibers of the morphism $V \to V/\!\!/G$ with certain G-orbits on V.

As an example, consider the three-dimensional adjoint representation of $G = \mathrm{SL}_2$ given by conjugation on the space V of 2×2 matrices $v = \left(\begin{smallmatrix} a & b \\ c & -a \end{smallmatrix}\right)$ of trace zero. This is irreducible when the characteristic of k is not equal to 2, which we assume here. It has the quadratic invariant $q(v) = -\det(v) = bc + a^2$, which generates the full ring of polynomial invariants. Hence $V/\!\!/G$ is isomorphic to the affine line and $q : V \to V/\!\!/G = \mathbb{G}_a$. If v and w are two vectors in V with $q(v) = q(w) \neq 0$, then they lie in the same G-orbit provided that the field k is separably closed.

For general fields the situation is more complicated. In our example, let d be a non-zero element of k and let K be the étale quadratic algebra $k[x]/(x^2 - d)$. Then the $G(k)$-orbits on the set of vectors $v \in V$ with $q(v) = d \neq 0$ can be identified with elements in the 2-group k^*/NK^*. (See §2.)

The additional complexity in the orbit picture, when k is not separably closed, is what we refer to as arithmetic invariant theory. It can be reformulated using non-abelian Galois cohomology, but that does not give a complete resolution of the problem. Indeed, when the stabilizer G_v of v is smooth, we will see that there is a bijection between the different orbits over k which lie in the orbit of v over the separable closure and the elements in the kernel of the map in Galois cohomology $\gamma : H^1(k, G_v) \to H^1(k, G)$. Since γ is only a map of pointed sets, the computation of this kernel can be non-trivial.

In this paper, we will illustrate some of the issues which remain by considering the *regular semi-simple* orbits—i.e., the closed orbits whose stabilizers have minimal dimension—in three representations of the split odd special orthogonal group $G = \mathrm{SO}_{2n+1} = \mathrm{SO}(W)$ over a field k whose characteristic is not equal to 2. Namely, we will study:

- the standard representation $V = W$;
- the adjoint representation $V = \mathfrak{so}(W) = \wedge^2(W)$; and
- the symmetric square representation $V = \mathrm{Sym}^2(W)$.

In the first case, the map γ is an injection and the arithmetic invariant theory is completely determined by the geometric invariant theory. In the second case, the stabilizer is a maximal torus and the arithmetic invariant theory is the Lie algebra version of stable conjugacy classes of regular semi-simple elements. The theory of stable conjugacy classes, introduced by Langlands [13, 14] and developed further by Shelstad [23] and Kottwitz [10], forms one of the key tools in the study of endoscopy and the trace formula. Here there are the analogous problems, involving the Galois cohomology of tori, for the adjoint representations of general reductive groups. In the third case, there are *stable* orbits in the sense of Mumford's geometric invariant theory [17], i.e., closed orbits whose stabilizers are finite. Such representations arise more generally in Vinberg's invariant theory (cf. [18, 20]), where the torsion automorphism corresponds to a regular elliptic class in the extended Weyl group. In this case, we can use the geometry of pencils of quadrics to describe an interesting subgroup of classes in the kernel of γ.

Although we have focused here primarily on the case of orbits over a general field, a complete arithmetic invariant theory would also consider the orbits of a reductive group over more general rings such as the integers. We end with some remarks on integral orbits for the three representations we have discussed.

We would like to thank Brian Conrad, for his help with étale and flat cohomology, and Mark Reeder and Jiu-Kang Yu for introducing us to Vinberg's theory. We would also like to thank Bill Casselman, Wei Ho, Alison Miller, Jean-Pierre Serre, and the anonymous referee for a number of very useful comments on an earlier draft of this paper. It is a pleasure to dedicate this paper to Nolan Wallach, who introduced one of us (BHG) to the beauties of invariant theory.

2 Galois cohomology

Let k be a field, let k^s be a separable closure of k, and let k^a denote an algebraic closure containing k^s. Let Γ be the (profinite) Galois group of k^s over k. Let G be a reductive group over k and V an algebraic representation of G on a finite-dimensional k-vector space. The problem of classifying the $G(k)$-orbits on $V(k)$ which lie in a fixed $G(k^s)$-orbit can be translated (following Serre [21, §I.5]) into the language of Galois cohomology.

Let $v \in V(k)$ be a fixed vector in this orbit, and let G_v be the stabilizer of v. We assume that G_v is a smooth algebraic group over k. If $w \in V(k)$ is another vector in the same $G(k^s)$-orbit as v, then we may write $w = g(v)$ with $g \in G(k^s)$ well-defined up to right multiplication by $G_v(k^s)$. For every $\sigma \in \Gamma$, we have $g^\sigma = ga_\sigma$ with $a_\sigma \in G_v(k^s)$. The map $\sigma \to a_\sigma$ is a continuous 1-cocycle on Γ with values in $G_v(k^s)$, whose class in the first cohomology set $H^1(\Gamma, G_v(k^s))$ is independent of the choice of g. Since $a_\sigma = g^{-1}g^\sigma$, this class is trivial when mapped to the cohomology set $H^1(\Gamma, G(k^s))$. We will use the notation $H^1(k, G_v)$ and $H^1(k, G)$ to denote these Galois cohomology sets in this paper.

Reversing the argument, one can show similarly that an element in the kernel of the map of pointed sets $H^1(k, G_v) \to H^1(k, G)$ gives rise to a $G(k)$-orbit on $V(k)$ in the $G(k^s)$-orbit of v. Hence we obtain the following.

Proposition 1. *There is a bijection between the set of $G(k)$-orbits on the vectors w in $V(k)$ that lie in the same $G(k^s)$-orbit as v and the kernel of the map*

$$\gamma : H^1(k, G_v) \to H^1(k, G) \tag{1}$$

in Galois cohomology.

When the stabilizer G_v is smooth over k, the set of all vectors $w \in V(k)$ lying in the same $G(k^s)$-orbit as v can be identified with the k-points of the quotient variety G/G_v, and the central problem of arithmetic invariant theory in this case is to understand the kernel of the map γ in Galois cohomology. This is particularly interesting when k is a finite, local, or global field, when the cohomology of the two groups G_v and G can frequently be computed.

In the example of the introduction with $G = \mathrm{SL}_2$ and V the adjoint representation (again assuming char$(k) \neq 2$), let v be a vector in $V(k)$ with $q(v) = d \neq 0$. Then the stabilizer G_v is a maximal torus in SL_2 which is split by the étale quadratic algebra K. The pointed set $H^1(k, G) = H^1(k, \mathrm{SL}_2)$ is trivial, so all classes in the abelian group $H^1(k, G_v) = k^*/NK^*$ lie in the kernel of γ. These classes index the orbits of $\mathrm{SL}_2(k)$ on the set S of non-zero vectors w with $q(w) = q(v)$, since this is precisely the set S of vectors $w \in V(k)$ which lie in the same $\mathrm{SL}_2(k^s)$-orbit as v. (This illustrates the point that one first has to solve the orbit problem over the separable closure k^s, before using Proposition 1 to descend to orbits over k.)

The vanishing of $H^1(k, G)$ occurs whenever $G = \mathrm{GL}_n$ or $G = \mathrm{SL}_n$ or $G = \mathrm{Sp}_{2n}$, and gives an elegant solution to many orbit problems. For example, when the characteristic of k is not equal to 2, the classification of the non-degenerate orbits of $\mathrm{SL}_n = \mathrm{SL}(W)$ on the symmetric square representation $V = \mathrm{Sym}^2(W^\vee)$ shows that the isomorphism classes of non-degenerate orthogonal spaces W of dimension n over k with a fixed determinant in k^*/k^{*2} correspond bijectively to classes in $H^1(k, G_v) = H^1(k, \mathrm{SO}(W))$ (cf. [11, Ch VII, §29], [21, Ch III, Appendix 2, §4]). In general, both $H^1(k, G_v)$ and $H^1(k, G)$ are non-trivial, and the determination of the kernel of γ remains a challenging problem.

Remark 2. In those cases where the stabilizer G_v is not smooth, it is at least flat of finite type over k, so one can replace the map γ in Galois (étale) cohomology with one in flat (fppf) cohomology. Indeed, the k-valued points of G/G_v can always be identified with the possibly larger set S' of vectors w' in $V(k)$ which lie in the same $G(k^a)$-orbit as v, where k^a is an algebraic closure of k^s. As an example, the group $\mathbb{G}_m$ acts on $V = \mathbb{G}_a$ by the formula $\lambda(v) = \lambda^p \cdot v$. The stabilizer of $v = 1$ is the subgroup μ_p, and $G/G_v = \mathbb{G}_m/\mu_p = \mathbb{G}_m$. The stabilizer G_v is smooth if the characteristic of k is not equal to p, in which case the set S consists of the non-zero elements of the field k, and the $G(k)$-orbits on S form a principal homogeneous space for the group $H^1(k, \mu_p) = k^*/k^{*p}$. If the characteristic of k is equal to p, the stabilizer μ_p is not smooth over k. In this case the set S consists of the pth powers in k^*. The set S' is equal to the full group of non-zero elements in k, which is strictly larger than S when the field k is imperfect. In the general case one can show that the $G(k)$ orbits on $S' = (G/G_v)(k)$ are in bijection with the kernel of the map

$\gamma_f : H^1_f(k, G_v) \to H^1_f(k, G)$ in flat (fppf) cohomology. In our example, we get a bijection of these orbits with the flat cohomology group $H^1_f(k, \mu_p) = k^*/k^{*p}$, as $H^1_f(k, \mathbb{G}_m) = 1$.

The semi-simple orbits in the three representations that we will study in this paper all have smooth stabilizers G_v. Hence we only consider the map γ in Galois cohomology.

3 Some representations of the split odd special orthogonal group

Let k be a field, with $\mathrm{char}(k) \neq 2$. Let $n \geq 1$ and let W be a fixed non-degenerate, split orthogonal space over k, of dimension $2n + 1 \geq 3$ and determinant $(-1)^n$ in k^*/k^{*2}. Such an orthogonal space is unique up to isomorphism. If $\langle v, w \rangle$ is the bilinear form on W, then we may choose an ordered basis $\{e_1, e_2, \ldots, e_n, u, f_n, \ldots, f_2, f_1\}$ of W over k with inner products given by

$$\begin{aligned} \langle e_i, e_j \rangle = \langle f_i, f_j \rangle = \langle e_i, u \rangle = \langle f_i, u \rangle = 0, \\ \langle e_i, f_j \rangle = \delta_{ij}, \\ \langle u, u \rangle = 1. \end{aligned} \tag{2}$$

The Gram matrix of the bilinear form with respect to this basis (which we will call *the standard basis*) is an anti-diagonal matrix. (A good general reference on orthogonal spaces, which gives proofs of these results, is [16].)

Let $T : W \to W$ be a k-linear transformation. We define the *adjoint transformation* T^* by the formula

$$\langle Tv, w \rangle = \langle v, T^* w \rangle.$$

The matrix of T^* in our standard basis is obtained from the matrix of T by reflection around the anti-diagonal. In particular, we have the identity $\det(T) = \det(T^*)$. We say a linear transformation $g : W \to W$ is *orthogonal* if $\langle gv, gw \rangle = \langle v, w \rangle$. Then g is invertible, with $g^{-1} = g^*$, and $\det(g) = \pm 1$ in k^*. We define the *special orthogonal group* $\mathrm{SO}(W)$ of W by

$$\mathrm{SO}(W) := \{g \in \mathrm{GL}(W) : gg^* = g^*g = 1, \ \det(g) = 1\}. \tag{3}$$

We are going to consider the arithmetic invariant theory for three representations V of the reductive group $G = \mathrm{SO}(W)$ over k.

The first is the standard representation $V = W$, which is irreducible and symmetrically self-dual (isomorphic to its dual by a symmetric bilinear pairing) of dimension $2n+1$. Here we will see that the invariant polynomial $q_2(v) := \langle v, v\rangle$ generates the ring of polynomial invariants and separates the non-zero orbits over k.

The second is the adjoint representation $V = \mathfrak{so}(W)$, which is irreducible and symmetrically self-dual of dimension $2n^2+n$. This representation is isomorphic to the exterior square $\wedge^2(W)$ of W, and can be realized as the space of skew self-adjoint operators:

$$V = \wedge^2(W) = \{T : W \to W : T = -T^*\}, \tag{4}$$

where $g \in G$ acts by conjugation: $T \mapsto gTg^{-1} = gTg^*$. The Lie bracket on V is given by the formula $[T_1, T_2] = T_1T_2 - T_2T_1$ and the duality by $\langle T_1, T_2\rangle = \mathrm{Trace}(T_1T_2)$. Here the theory of $G(k)$-orbits in a fixed $G(k^s)$-orbit is the Lie algebra version of stable conjugacy classes for the group $G = \mathrm{SO}(W)$.

The third is a representation V which arises in Vinberg's theory, from an outer involution θ of the group $\mathrm{GL}(W)$. It is isomorphic to the symmetric square $\mathrm{Sym}^2(W)$ of W, and can be realized as the space of self-adjoint operators:

$$V = \mathrm{Sym}^2(W) = \{T : W \to W : T = T^*\}, \tag{5}$$

where again $G = \mathrm{SO}(W)$ acts by conjugation. This representation has dimension $2n^2+3n+1$ and is symmetrically self-dual by the pairing $\langle T_1, T_2\rangle = \mathrm{Trace}(T_1T_2)$. We will see that there are stable orbits, and that the arithmetic invariant theory of the stable orbits involves the arithmetic of hyperelliptic curves of genus n over k, with a k-rational Weierstrass point.

We note that the third representation V is not irreducible, as it contains the trivial subspace spanned by the identity matrix, and has a non-trivial invariant linear form given by the trace. When the characteristic of k does not divide $2n+1 = \dim(W)$, the representation V is the direct sum of the trivial subspace and the kernel of the trace map, and the latter is irreducible and symmetrically self-dual of dimension $2n^2+3n$. When the characteristic of k divides $2n+1$ the trivial subspace is contained in the kernel of the trace. In this case V has two trivial factors and an irreducible factor of dimension $2n^2+3n-1$ in its composition series.

4 Invariant polynomials and the discriminant

In the standard representation $V = W$ of $G = \mathrm{SO}(W)$, the quadratic invariant $q_2(v) = \langle v, v\rangle$ generates the ring of invariant polynomials. We define $\Delta = q_2$ in this case. When $\Delta(v) \neq 0$, the stabilizer G_v is the reductive subgroup $\mathrm{SO}(U)$, where U is the hyperplane in W of vectors orthogonal to v.

In the second and third representations, the group $\mathrm{SO}(W)$ acts by conjugation on the subspace V of $\mathrm{End}(W)$. Hence the characteristic polynomial of an operator T is an invariant of the $G(k)$-orbit.

For the adjoint representation, the operator T is skew self-adjoint and its characteristic polynomial has the form

$$f(x) = \det(xI - T) = x^{2n+1} + c_2 x^{2n-1} + c_4 x^{2n-3} + \cdots + c_{2n}x = xg(x^2)$$

with coefficients $c_{2m} \in k$. The coefficients c_{2m} are polynomial invariants of the representation, with $\deg(c_{2m}) = 2m$. These polynomials are algebraically independent and generate the full ring of polynomial invariants on $V = \mathfrak{so}(W)$ over k [3, Ch 8, §8.3, §13.2, VI]. An important polynomial invariant, of degree $2n(2n+1)$, is the discriminant Δ of the characteristic polynomial of T:

$$\Delta = \Delta(c_2, c_4, \ldots, c_{2n}) = \operatorname{disc} f(x).$$

This is non-zero in k precisely when the polynomial $f(x)$ is separable, so has $2n+1$ distinct roots in the separable closure k^s of k. The condition $\Delta(T) \neq 0$ defines the regular semi-simple orbits in the Lie algebra. For such an orbit, we will see that the stabilizer G_T is a maximal torus in G, of dimension n over k.

For the third representation V on self-adjoint operators, the characteristic polynomial $f(x)$ of T can be any monic polynomial of degree $2n+1$; we write

$$f(x) = \det(xI - T) = x^{2n+1} + c_1 x^{2n} + c_2 x^{2n-1} + \cdots + c_{2n}x + c_{2n+1}$$

with coefficients $c_m \in k$. Again the c_m give algebraically independent polynomial invariants, with $\deg(c_m) = m$, which generate the full ring of polynomial invariants on V over k. The discriminant

$$\Delta = \Delta(c_1, c_2, \ldots, c_{2n+1}) = \operatorname{disc} f(x)$$

is defined as before, and is non-zero when $f(x)$ is separable. We will see that the condition $\Delta(T) \neq 0$ defines the stable orbits of G on V. For such an orbit, we will see that the stabilizer G_T is a finite commutative group scheme of order 2^{2n} over k, which embeds as a Jordan subgroup scheme of G (see [12, Ch 3]).

5 The orbits with non-zero discriminant

In this section, for each of the three representations V, we exhibit an orbit for G where the invariant polynomials described above take arbitrary values in k, subject to the single restriction that $\Delta \neq 0$. We calculate the stabilizer G_v and its cohomology $H^1(k, G_v)$ in terms of the values of the invariant polynomials on v. We also give an explicit description of the map $\gamma : H^1(k, G_v) \to H^1(k, G)$. We note that all three representations arise naturally in Vinberg's invariant theory, and the representative orbits that we will construct are in the Kostant section (cf. [18]).

When $V = W$ is the standard representation, let d be an element of k^*. The vector $v = e_1 + df_1$ has $q_2(v) = \Delta(v) = d$. The stabilizer G_v acts on the orthogonal complement U of the non-degenerate line kv in W, which is a quasi-split orthogonal space of dimension $2n$ and discriminant d in k^*/k^{*2}. (The *discriminant* of an orthogonal space of dimension $2n$ is defined as $(-1)^n$ times its determinant.) This gives an identification $G_v = \mathrm{SO}(U)$, where the special orthogonal group $\mathrm{SO}(U)$ is quasi-split over k and split by $k(\sqrt{d})$. Witt's extension theorem [16, Ch 1] shows that all vectors w with $q_2(w) = d$ lie in the $G(k)$-orbit of v, so the invariant polynomials separate the orbits over k with non-zero discriminant. One can also show that there is a single non-zero orbit with $q_2(v) = 0$, represented by the vector $v = e_1 = e_1 + 0f_1$.

The cohomology set $H^1(k, \mathrm{SO}(U))$ classifies non-degenerate orthogonal spaces U' of dimension $2n$ and discriminant d over k, and the cohomology set $H^1(k, \mathrm{SO}(W))$ classifies non-degenerate orthogonal spaces W' of dimension $2n+1$ and determinant $(-1)^n$ over k, with the trivial class corresponding to the split space W. The map

$$\gamma : H^1(k, G_v) = H^1(k, \mathrm{SO}(U)) \longrightarrow H^1(k, G) = H^1(k, \mathrm{SO}(W))$$

is given explicitly by mapping the space U' to the space $W' = U' + \langle d \rangle$. Witt's cancellation theorem [16] shows that the map γ is an injection of sets in this case, so the arithmetic invariant theory for the standard representation of any odd orthogonal group is the same as its geometric invariant theory.

For the second representation $V = \mathfrak{so}(W) = \wedge^2(W)$, let

$$f(x) = x^{2n+1} + c_2 x^{2n-1} + c_4 x^{2n-3} + \cdots + c_{2n} x$$

be a polynomial in $k[x]$ with non-zero discriminant. We will construct a skew self-adjoint operator T on W with characteristic polynomial $f(x)$. Since $f(x) = xh(x) = xg(x^2)$, we have

$$\operatorname{disc} f(x) = c_{2n}^2 \operatorname{disc} h(x) = (-4)^n c_{2n}^3 \operatorname{disc} g(x)^2.$$

Let $K = k[x]/(g(x))$, $E = k[x]/(h(x))$, and $L = k[x]/(f(x))$. By our assumption that $\Delta \neq 0$, these are étale k-algebras of ranks n, $2n$, and $2n+1$ respectively. We have $L = E \oplus k$. Furthermore the map $x \to -x$ induces an involution τ of the algebras E and of L, with fixed algebras K and $K \oplus k$ respectively.

Let β be the image of x in $L = k[x]/(f(x))$, so $f(\beta) = 0$ in L and $f'(\beta)$ is a unit in L^*. We define a symmetric bilinear form $\langle\ ,\ \rangle$ on the k-vector space $L = k + k\beta + k\beta^2 + \cdots + k\beta^{2n}$ by taking

$$\langle \lambda, \mu \rangle := \text{ the coefficient of } \beta^{2n} \text{ in the product } (-1)^n \lambda \mu^\tau. \tag{6}$$

This is non-degenerate, of determinant $(-1)^n$, and the map $t(\lambda) = \beta\lambda$ is skew self-adjoint, with characteristic polynomial $f(x)$. Finally, the subspace $M = k + k\beta + \cdots + k\beta^{n-1}$ is isotropic of dimension n, so the orthogonal space L is split and isomorphic to W over k. Choosing an isometry $\theta : L \to W$ we obtain a skew self-adjoint operator $T = \theta t \theta^{-1}$ on W with the desired separable characteristic polynomial. Since the isometry θ is unique up to composition with an orthogonal transformation of W, the orbit of T is well-defined. The stabilizer of T in $\mathrm{O}(W)$ has k-points $\{\lambda \in L^* : \lambda^{1+\tau} = 1\}$. The subgroup G_T which fixes T is a maximal torus in $G = \mathrm{SO}(W)$, isomorphic to the torus $\mathrm{Res}_{K/k}\, U_1(E/K)$ of dimension n over k.

Over the separable closure k^s of k, any skew self-adjoint operator S with (separable) characteristic polynomial $f(x)$ is in the same orbit of T. Indeed, since $f(x)$ is separable, it is also the minimal polynomial of T and S, so we can find an element g in $\mathrm{GL}(W)$ with $S = gTg^{-1}$. Since both operators are skew self-adjoint, the product g^*g is in the centralizer of T in $\mathrm{GL}(W)$. The centralizer of T in $\mathrm{End}(W)$ is the algebra $k[T] = L$. Since g^*g is self-adjoint in L^*, and its determinant is a square in k^*, we see that g^*g is an element of the subgroup $K^* \times k^{*2}$. Over the separable closure, every element of $K^* \times k^{*2}$ is a norm from L^*: $g^*g = h^{1+\tau}$. Then gh^{-1} is an orthogonal transformation of W over k^s mapping T to S. Hence S is in the $\mathrm{SO}(W)(k^s)$-orbit of T.

To understand the orbits with a fixed separable characteristic polynomial over k, we need an explicit form of the map γ in Galois cohomology. Since the stabilizer of T is abelian, the pointed set $H^1(k, G_T)$ is an abelian group, which is isomorphic to K^*/NE^* by Hilbert's Theorem 90. The map

$$\gamma : K^*/NE^* = H^1(k, G_T) \longrightarrow H^1(k, G) = H^1(k, \mathrm{SO}(W))$$

is given explicitly as follows. We first associate to an element $\kappa \in K^*$ the element $\alpha = (\kappa, 1)$ in $(L^\tau)^* = K^* \times k^*$, with square norm from L^* to k^*. We then associate to α the vector space L with symmetric bilinear form

$$\langle \lambda, \mu \rangle_\alpha := \text{the coefficient of } \beta^{2n} \text{ in the product } (-1)^n \alpha\lambda\mu^\tau. \tag{7}$$

This orthogonal space W_κ has dimension $2n+1$ and determinant $(-1)^n$ over k, and its isomorphism class depends only on the class of κ in the quotient group $K^*/NE^* = H^1(k, G_T)$.

Lemma 3. *The orthogonal space W_κ represents the class $\gamma(\kappa)$ in $H^1(k, \mathrm{SO}(W))$.*

Proof. We first recall the recipe for associating to a cocycle g_σ on the Galois group with values in $\mathrm{SO}(W)(k^s)$ a new orthogonal space W' over k. We use the inclusion $\mathrm{SO}(W) \to \mathrm{GL}(W)$ and the triviality of $H^1(k, \mathrm{GL}(W))$ to write $g_\sigma = h^{-1}h^\sigma$ for an element $h \in \mathrm{GL}(W)(k^s)$. We then define a new non-degenerate symmetric bilinear form on W by the formula

$$\langle v, w \rangle^* = \langle h^{-1}v, h^{-1}w \rangle. \tag{8}$$

This takes values in k and defines the space W', which has dimension $2n+1$ and determinant $(-1)^n$. The isomorphism class of W' over k depends only on the cohomology class of the cocycle g_σ in $H^1(k, \mathrm{SO}(W))$.

In our case, the cocycle g_σ representing $\gamma(\kappa)$ comes from a cocycle with values in the stabilizer G_v. This is a maximal torus in $\mathrm{SO}(W)$, which is a subgroup of the maximal torus $\mathrm{Res}_{L/k}\,\mathbb{G}_m$ of $\mathrm{GL}(W)$. This torus already has trivial Galois cohomology, so we can write $g_\sigma = h^\sigma/h$ with $h \in (L \otimes k^s)^*$ satisfying $h^{1+\tau} = \alpha$. Substituting this particular h into formula (8) for the new inner product on W completes the proof. □

We note that the class κ above will be in the kernel of γ precisely when the quadratic space W' with bilinear form $\langle\ ,\ \rangle_\alpha$ is split. Such classes give additional orbits of $\mathrm{SO}(W)$ on $\mathfrak{so}(W) = \wedge^2(W)$ over k with characteristic polynomial $f(x)$.

The analysis for the third representation $V = \mathrm{Sym}^2(W)$ is similar. Here we start with an arbitrary monic separable polynomial $f(x) = x^{2n+1} + c_1x^{2n} + \cdots + c_{2n+1}$ and wish to construct a self-adjoint operator T on W with characteristic polynomial $f(x)$. We let $L = k[x]/(f(x))$, which is an étale k-algebra of rank $2n+1$, and let β be the image of x in L. We define a symmetric bilinear form $\langle \lambda, \mu \rangle$ on $L = k + k\beta + \cdots + k\beta^{2n}$ by taking the coefficient of β^{2n} in the product $\lambda\mu$. This is non-degenerate of determinant $(-1)^n$, and the map $t(\lambda) = \beta\lambda$ is self-adjoint, with characteristic polynomial $f(x)$. Finally, the subspace $M = k + k\beta + \cdots + k\beta^{n-1}$ is isotropic of dimension n, so the orthogonal space L is split and isomorphic to W over k. Choosing an isometry $\theta : L \to W$, we obtain a self-adjoint operator $T = \theta t \theta^{-1}$ on W with the desired separable characteristic polynomial. Since the isometry θ is unique up to composition with an orthogonal transformation of W, the orbit of T is well-defined. The stabilizer of T in $O(W)$ has k-points $\{\lambda \in L^* : \lambda^2 = 1\}$. The subgroup G_T in $\mathrm{SO}(W)$ which fixes T is the finite étale group scheme A of order 2^{2n}, which is the kernel of the norm map $\mathrm{Res}_{L/k}(\mu_2) \to \mu_2$.

Over the separable closure k^s of k, any self-adjoint operator S with (separable) characteristic polynomial $f(x)$ is in the same orbit as T. Indeed, since $f(x)$ is separable, it is also the minimal polynomial of T and S, so we can find an element $g \in \mathrm{GL}(W)$ with $S = gTg^{-1}$. Since both operators are self-adjoint, the product g^*g is in the centralizer of T in $\mathrm{GL}(W)$. The centralizer of T in $\mathrm{End}(W)$ is the algebra $k[T] = L$, so g^*g is an element of L^*. Over the separable closure, every element of L^* is a square: $g^*g = h^2$. Then gh^{-1} is an orthogonal transformation of W over k^s mapping T to S. Hence S is in the $\mathrm{SO}(W)(k^s)$-orbit of T.

We now consider the orbits with a fixed separable characteristic polynomial over k. Since the stabilizer of T is again abelian, the pointed set $H^1(k, G_T)$ is an abelian group which is isomorphic to $(L^*/L^{*2})_{N=1}$ by Kummer theory. The map

$$\gamma : H^1(k, G_T) = (L^*/L^{*2})_{N=1} \longrightarrow H^1(k, G) = H^1(k, \mathrm{SO}(W))$$

is given explicitly as follows. We associate to an element α in $(L^*)_{N=1}$ the orthogonal space L with bilinear form $\langle \lambda, \mu \rangle_\alpha$ given by the coefficient of β^{2n} in the product $\alpha\lambda\mu$. This orthogonal space has dimension $2n+1$ and determinant

$(-1)^n$ over k. Its isomorphism class over k depends only on the image of α in the quotient group $(L^*/L^{*2})_{N=1} = H^1(k, G_T)$. This orthogonal space represents the class $\gamma(\alpha)$ in $H^1(k, \mathrm{SO}(W))$. The proof is the same as that of Lemma 3. We first observe that the map taking the cocycle g_σ from G_T to $\mathrm{SO}(W)$ to $\mathrm{GL}(W)$ can also be obtained by mapping G_T to the maximal torus $\mathrm{Res}_{L/k}\, \mathbb{G}_m$ in $\mathrm{GL}(W)$. This torus has trivial cohomology, so $\alpha = h^2$ with $h \in (L \otimes k^s)^*$, and this choice of h gives the inner product $\langle \lambda, \mu \rangle_\alpha$. The class α will be in the kernel of γ precisely when the quadratic space L with bilinear form $\langle\ ,\ \rangle_\alpha$ is split; such classes give additional orbits of $\mathrm{SO}(W)$ on $\mathrm{Sym}^2(W)$ over k with characteristic polynomial $f(x)$.

We summarize what we have established for the representations $V = \mathfrak{so}(W)$ and $V = \mathrm{Sym}^2(W)$.

Proposition 4. *For each monic separable polynomial $f(x)$ of degree $2n+1$ over k of the form $f(x) = xg(x^2)$ there is a distinguished $\mathrm{SO}(W)(k)$-orbit of skew self-adjoint operators T on W with characteristic polynomial $f(x)$. All other orbits on $\wedge^2(W)$ with this characteristic polynomial lie in the $\mathrm{SO}(W)(k^s)$-orbit of T, and correspond bijectively to the non-identity classes in the kernel of $\gamma : K^*/NE^* \to H^1(k, \mathrm{SO}(W))$, where $K = k[x]/(g(x))$ and $E = k[x]/(g(x^2))$.*

For each monic separable polynomial $f(x)$ of degree $2n+1$ over k there is a distinguished $\mathrm{SO}(W)(k)$-orbit of self-adjoint operators T on W with characteristic polynomial $f(x)$. All other orbits on $\mathrm{Sym}^2(W)$ with this characteristic polynomial lie in the $\mathrm{SO}(W)(k^s)$-orbit of T, and correspond bijectively to the non-identity classes in the kernel of $\gamma : (L^*/L^{*2})_{N=1} \to H^1(k, \mathrm{SO}(W))$, where $L = k[x]/(f(x))$.

6 Stable orbits and hyperelliptic curves

For both representations $V = \wedge^2(W)$ and $V = \mathrm{Sym}^2(W)$ of $G = \mathrm{SO}(W)$ we associated to the distinguished orbit T with separable characteristic polynomial $f(x)$ and any class α in the cohomology group $H^1(k, G_T)$ a symmetric bilinear form $\langle \lambda, \mu \rangle_\alpha$ on the k-vector space $L = k[x]/(f(x))$. The class α is in the kernel of the map $\gamma : H^1(k, G_T) \to H^1(k, G)$ precisely when this quadratic space is split over k. However, exhibiting specific classes $\alpha \neq 1$ where this space is split is a difficult general problem, so it is difficult to exhibit other orbits with this characteristic polynomial.

In the case of the third representation $V = \mathrm{Sym}^2(W)$, the orbits T with $\Delta(T) \neq 0$ are stable; namely, they are closed (defined by the values of the invariant polynomials over the separable closure) and have finite stabilizer (the commutative group scheme $A = \mathrm{Res}_{L/k}(\mu_2)_{N=1}$ of order 2^{2n}). In this case, we will use some results in algebraic geometry, on hyperelliptic curves with a Weierstrass point and the Fano variety of the complete intersection of two quadrics in $\mathbb{P}(L \oplus k)$, to produce certain classes in the kernel of the map $\gamma : H^1(k, A) \to H^1(k, \mathrm{SO}(W))$.

Let C be the smooth projective hyperelliptic curve of genus n over k with affine equation $y^2 = f(x)$ and k-rational Weierstrass point P above $x = \infty$. The functions on C which are regular outside of P form an integral domain:

$$H^0(C - P, O_{C-P}) = k[x, y]/(y^2 = f(x)) = k[x, \sqrt{f(x)}].$$

The complete curve C is covered by this affine open subset U_1, together with the affine open subset U_2 associated to the equation $w^2 = v^{2n+2}f(1/v)$ and containing the point $P = (0, 0)$. The gluing of U_1 and U_2 is by $(v, w) = (1/x, y/x^{n+1})$ and $(x, y) = (1/v, w/v^{n+1})$ wherever these maps are defined. Let J denote the Jacobian of C over k and let $J[2]$ the kernel of multiplication by 2 on J. This is a finite étale group scheme of order 2^{2n} over k.

Lemma 5. *The group scheme $J[2]$ of 2-torsion on the Jacobian of C is canonically isomorphic to the stabilizer $A = \mathrm{Res}_{L/k}(\mu_2)_{N=1}$ of the orbit T in $\mathrm{SO}(W)$.*

Proof. Write $L = k[x]/(f(x)) = k + k\beta + \cdots + k\beta^{2n}$, where $f(\beta) = 0$. The other Weierstrass points $P_\eta = (\eta(\beta), 0)$ of $C(k^s)$ correspond bijectively to algebra embeddings $\eta : L \to k^s$. Associated to such a point we have the divisor $d_\eta = (P_\eta) - (P)$ of degree zero. The divisor class of d_η lies in the 2-torsion subgroup $J[2](k^s)$ of the Jacobian, as

$$2d_\eta = \mathrm{div}(x - \eta(\beta)).$$

The Riemann-Roch theorem shows that the classes d_η generate the finite group $J[2](k^s)$, and satisfy the single relation

$$\sum(d_\eta) = \mathrm{div}(y).$$

Since the Galois group of k^s acts on these classes by permutation of the embeddings η, we have an isomorphism of group schemes: $J[2] \cong \mathrm{Res}_{L/k}(\mu_2)/\mu_2$. This quotient of $\mathrm{Res}_{L/k}(\mu_2)$ is isomorphic to the subgroup scheme $A = \mathrm{Res}_{L/k}(\mu_2)_{N=1}$, as the degree of L over k is odd. This completes the proof. □

The exact sequence of Galois modules,

$$0 \to J[2](k^s) \to J(k^s) \to J(k^s) \to 0,$$

gives an exact descent sequence

$$0 \to J(k)/2J(k) \to H^1(k, J[2]) \to H^1(k, J)[2] \to 0$$

in Galois cohomology. By Lemma 5, the middle term in this sequence can be identified with the group $H^1(k, A) = H^1(k, G_T)$, and our main result in this section is the following.

Proposition 6. *The subgroup $J(k)/2J(k)$ of $H^1(k,A) = H^1(k,G_T)$ lies in the kernel of the map $\gamma : H^1(k,G_T) \to H^1(k,G)$.*

Proof. We first make the descent map from $H^1(k,A)$ to $H^1(k,J)[2]$ more explicit. That is, we need to associate to a class α in the group

$$H^1(k,A) = (L^*/L^{*2})_{N=1}$$

a principal homogeneous space F_α of order 2 for the Jacobian J over k. The class α will be in the subgroup $J(k)/2J(k)$ precisely when the homogeneous space F_α has a k-rational point.

We have previously associated to the class α the orthogonal space L with symmetric bilinear form $\langle\lambda,\mu\rangle_\alpha :=$ the coefficient of β^{2n} in the product $\alpha\lambda\mu$. We also defined a self-adjoint operator given by multiplication by β on L, and that gives a second symmetric bilinear form on L: $\langle\beta\lambda,\mu\rangle_\alpha = \langle\lambda,\beta\mu\rangle_\alpha$.

Let $M = L \oplus k$, which has dimension $2n+2$ over k, and consider the two quadrics on M given by

$$Q(\lambda,a) = \langle\lambda,\lambda\rangle_\alpha$$
$$Q'(\lambda,a) = \langle\beta\lambda,\lambda\rangle_\alpha + a^2.$$

The pencil $uQ - vQ'$ is non-degenerate and contains exactly $2n+2$ singular elements over k^s, namely, the quadric Q at $v = 0$ and the $2n+1$ quadrics $\eta(\beta)Q - Q'$ at the points where $f(\eta(\beta)) = 0$. Hence the base locus is non-singular in $\mathbb{P}(M)$ and the Fano variety F_α of this complete intersection, consisting of the n-dimensional subspaces Z of M which are isotropic for all of the quadrics in the pencil, is a principal homogeneous space of order 2 for the Jacobian J (cf. [7]). More precisely, there is a commutative algebraic group I_α with 2 components over k, having identity component J and non-identity component F_α.

Since the discriminant of the quadric $uQ - vQ'$ in the pencil is equal to $v^{2n+2}f(x)$ with $x = u/v$, a point $c = (x,y)$ on the hyperelliptic curve $y^2 = f(x)$ determines both a quadric $Q_x = xQ - Q'$ in the pencil together with a *ruling* of Q_x, i.e., a component of the variety of $(n+1)$-dimensional Q_x-isotropic subspaces in M. Each point gives an involution of the corresponding Fano variety $\theta(c) : F_\alpha \to F_\alpha$ with 2^{2n} fixed points over a separable closure k^s of k. The involution $\theta(c)$ is defined as follows. A point of F_α consist of a common isotropic subspace Z of dimension n in $M \otimes k^s$. The point c gives a maximal isotropic subspace Y for the quadric Q_x which contains Z. If we restrict any non-singular quadric in the pencil (other than Q_x) to Y, we get a reducible quadric which is the sum of two hyperplanes: Z and another common isotropic subspace Z'. This defines the involution: $\theta(c)(Z) = Z'$. In the algebraic group I_α, we have that $Z + Z'$ is the class of the divisor $(c) - (P)$ of degree zero in J.

Now assume that the class α is in the subgroup $J(k)/2J(k)$. Then its image in $H^1(k,J)$ is trivial, and the homogenous space F_α has a k-rational point. Hence there is a k-subspace Z of $M = L \oplus k$ which is isotropic for both Q and Q'. Since it is isotropic for Q', the subspace Z does not contain the line $0 \oplus k$, so its projection to the subspace L has dimension n and is isotropic for Q. This implies that the orthogonal space L with bilinear form $(\lambda, \nu)_\alpha$ is split, so the class α is in the kernel of the map $\gamma : H^1(k, A) \to H^1(k, \mathrm{SO}(W))$. □

Note that when $c = P$, the Weierstrass point over $x = \infty$, the involution $\theta(P)$ is induced by the linear involution $(\lambda, a) \to (\lambda, -a)$ of $M = L \oplus k$. The fixed points are just the n-dimensional subspaces X over k^s which are isotropic for both quadrics

$$q(\lambda) = \langle \lambda, \lambda \rangle_\alpha ,$$
$$q'(\lambda) = \langle \beta\lambda, \lambda \rangle_\alpha$$

on the space L of dimension $2n + 1$ over k. There are 2^{2n} such isotropic subspaces over k^s, and they form a principal homogeneous space for $J[2]$. The variety F_α has a k-rational point when α lies in the subgroup $J(k)/2J(k)$, but only has a k-rational point fixed by the involution $\theta(P)$ when α is the trivial class in $H^1(k, J[2])$.

Remark 7. The finite group scheme $A = J[2]$ does not determine the hyperelliptic curve C over k. Indeed, for any class $d \in k^*/k^{*2}$, the hyperelliptic curve C_d with affine equation $dy^2 = f(x)$ has the same 2-torsion subgroup of its Jacobian. This Jacobian J_d of C_d acts on the Fano variety of the complete intersection of the two quadrics given by

$$Q(\lambda, a) = \langle \lambda, \lambda \rangle_\alpha ,$$
$$Q'(\lambda, a) = \langle \beta\lambda, \lambda \rangle_\alpha + da^2.$$

Indeed, the discriminant of the quadric $uQ - vQ'$ in the pencil is equal to $dv^{2n+2} f(x)$, where $x = u/v$. A similar argument then shows that the subgroup $J_d(k)/2J_d(k)$ is also contained in the kernel of the map γ on $H^1(k, A)$.

7 Arithmetic fields

In this section, we describe the orbits in our three representations when k is a finite, local, or global field.

7.1 *Finite fields*

First, we consider the case when k is finite, of odd order q. In this case, $H^1(k, \mathrm{SO}(W)) = 1$ by Lang's theorem, as $\mathrm{SO}(W)$ is connected. As a consequence, every quadratic space of dimension $2n + 1$ and determinant $(-1)^n$ is split, and all elements of $H^1(k, G_T)$ lie in the kernel of γ.

In the standard representation $V = W$ the stabilizer of a vector v with $q_2(v) \neq 0$ is the connected orthogonal subgroup $\mathrm{SO}(U)$, which also has trivial first cohomology. So for every non-zero element d in k^*, there is a unique orbit of vectors with $q_2(v) = d$. (We have already seen this for general fields via Witt's extension theorem.)

In the adjoint representation $V = \mathfrak{so}(W)$, the stabilizer of a vector T with $\Delta(T) \neq 0$ is the connected torus $\mathrm{Res}_{K/k}\, U_1(E/K)$, which also has trivial first cohomology. So for each separable characteristic polynomial of the form $f(x) = xg(x^2)$ there is a unique orbit of skew self-adjoint operators T with characteristic polynomial $f(x)$.

In the representation $V = \mathrm{Sym}^2(W)$ the stabilizer of T with characteristic polynomial $f(x)$ satisfying $\mathrm{disc}(f) = \Delta(T) \neq 0$ is the finite group scheme $A = (\mathrm{Res}_{L/k}\, \mu_2)_{N=1}$. In this case $H^1(k, A) = (L^*/L^{*2})_{N=1}$ is an elementary abelian 2-group of order 2^m, where $m + 1$ is the number of irreducible factors of $f(x)$ in $k[x]$. So 2^m is the number of distinct orbits with characteristic polynomial $f(x)$. But this is also the order of the stabilizer $H^0(k, A) = A(k) = (L^*[2])_{N=1}$ of any point in the orbit. Hence the number of self-adjoint operators T with any fixed separable polynomial is equal to the order of the finite group $\mathrm{SO}(W)(q)$. This is given by the formula

$$\#\,\mathrm{SO}(W)(q) = q^{n^2}(q^{2n} - 1)(q^{2n-2} - 1)\cdots(q^2 - 1).$$

By Lang's theorem, we also have $H^1(k, J) = 0$, where J is the Jacobian of the smooth hyperelliptic curve $y^2 = f(x)$ of genus n over k. Hence the homomorphism $J(k)/2J(k) \to H^1(k, A)$ is an isomorphism and every orbit with characteristic polynomial $f(x)$ comes from a k-rational point on the Jacobian.

7.2 Non-Archimedean local fields

Next, we consider the case when k is a non-Archimedean local field, with ring of integers O and finite residue field $O/\pi O$ of odd order. In this case, Kneser's theorem on the vanishing of H^1 for simply-connected groups (cf. [19, Th. 6.4], [21]) gives an isomorphism

$$H^1(k, \mathrm{SO}(W)) \cong H^2(k, \mu_2) \cong (\mathbb{Z}/2\mathbb{Z}).$$

For the standard representation $V = W$, we also have $H^1(k, G_v) = H^1(k, \mathrm{SO}(U)) \cong (\mathbb{Z}/2\mathbb{Z})$, except in the case when $\dim(V) = 3$ and $q_2(v) = 1$, when $\mathrm{SO}(U)$ is a split torus and $H^1(k, \mathrm{SO}(U)) = 1$. The map γ is a bijection except in the special case.

For the adjoint representation $V = \mathfrak{so}(W)$, Kottwitz has shown in the local case that the map

$$\gamma : H^1(k, G_v) = (K^*/NE^*) \to H^1(k, G) = (\mathbb{Z}/2\mathbb{Z})$$

is actually a homomorphism of groups [10]. Let $f(x) = xg(x^2)$, so $K = k[x]/(g(x))$ and $E = k[x]/(g(x^2))$. It follows from local class field theory that the group K^*/NE^* is elementary abelian of order 2^m, where m is the number of irreducible factors $g_i(x)$ of $g(x)$ such that $g_i(x^2)$ remains irreducible over k. Kottwitz also shows that that the map γ is surjective when $m \geq 1$. Hence the number of orbits with separable characteristic polynomial $f(x)$ is 1 when $m = 0$, and is 2^{m-1} when $m \geq 1$.

For the third representation $V = \mathrm{Sym}^2(W)$, the map

$$\gamma : H^1(k, A) = H^1(k, J[2]) \to H^2(k, \mu_2) \cong (\mathbb{Z}/2\mathbb{Z})$$

is an even quadratic form. The associated bilinear form is the cup product on $H^1(k, J[2])$ induced from the Weil pairing $J[2] \times J[2] \to \mu_2$, and $J(k)/2J(k)$ is a maximal isotropic subspace on which $\gamma = 0$. This allows us to count the number of stable orbits with a fixed characteristic polynomial.

Let $m + 1$ be the number of irreducible factors of $f(x)$ in $k[x]$, and let O_L be the integral closure of the ring O in L. Then $H^1(k, A) = (L^*/L^{*2})_{N=1}$ has order 2^{2m} and the number of stable orbits with characteristic polynomial $f(x)$ is equal $2^{m-1}(2^m + 1) = 2^{2m-1} + 2^{m-1}$. The subgroup $J(k)/2J(k)$ has order 2^m, which is also the order of the subgroup $(O_L^*/O_L^{*2})_{N=1}$ of units. These two subgroups coincide when the polynomial $f(x)$ has coefficients in O and the quotient algebra $O[x]/(f(x))$ is maximal in L.

7.3 The local field $\mathbb{R}$

We next consider the orbits in our representations when $k = \mathbb{R}$ is the local field of real numbers. Then the pointed set $H^1(k, G) = H^1(k, \mathrm{SO}(W))$ has $n+1$ elements, corresponding to the quadratic spaces W' of signature (p, q) satisfying: $p + q = 2n + 1$ and $q \equiv n \pmod 2$. The pointed set $H^1(k, G_v) = H^1(k, \mathrm{SO}(U))$ for the standard representation has $n + 1$ elements when $q_2(v)$ has sign $(-1)^n$, and has n elements when $q_2(v)$ has sign $-(-1)^n$. The map γ is a bijection in the first case and an injection in the second case, when the definite quadratic space W' does not have an orbit with $q_2(w^*) = q_2(v)$.

In the second and third representations, $H^1(k, G_T)$ is an elementary abelian 2-group, and we will consider the situations where it has maximal rank. For the adjoint representation $V = \mathfrak{so}(W)$, this occurs when all of the nonzero roots of the characteristic polynomial $f(x)$ of the skew self-adjoint transformation T are purely imaginary. Thus $f(x) = xg(x^2)$ where $g(x)$ factors completely over the real numbers and all of its roots are strictly negative. In this case, the 2-group $H^1(k, G_T) = K^*/NE^* = (\mathbb{R}^*)^n/N(\mathbb{C}^*)^n$ has rank n. The real orthogonal space W decomposes into n orthogonal T-stable planes and an orthogonal line on which $T = 0$. The signatures of these planes determine the real orbit of T. Writing $n = 2m$ or $n = 2m + 1$, we see that there are $\binom{n}{m}$ elements in the kernel of γ. One can show that γ is surjective in this case, and calculate the order of each fiber as a binomial coefficient $\binom{n}{k}$.

For the symmetric square representation $V = \mathrm{Sym}^2(W)$, the 2-group $H^1(k, G_T)$ has maximal rank when the characteristic polynomial $f(x)$ of the self-adjoint transformation T factors completely over the real numbers. In this case, $H^1(k, G_T) = ((\mathbb{R}^*)^{2n+1}/(\mathbb{R}^{*2})^{2n+1})_{N=1}$ has rank $2n$. The real orthogonal space W decomposes into $2n+1$ orthogonal eigenspaces for T, and the signatures of these lines determine the real orbit. Hence there are $\binom{2n+1}{n}$ elements in the kernel of γ. One can also show that γ is surjective in this case, and calculate the order of each fiber as a binomial coefficient $\binom{2n+1}{k}$ with $k \equiv n \pmod 2$.

7.4 *Global fields*

Finally, we consider the representation $\mathrm{Sym}^2(V)$ when k is a global field. In this case, the group $H^1(k, A) = H^1(k, J[2])$ is infinite. We will now prove that there are also infinitely many classes in the kernel of γ, so infinitely many orbits with characteristic polynomial $f(x)$.

Proposition 8. *Every class α in the 2-Selmer group $\mathrm{Sel}_2(J/k)$ of $H^1(k, J[2])$ lies in the kernel of γ, so corresponds to an orbit over k.*

Proof. By definition, the elements of the 2-Selmer group $\mathrm{Sel}_2(J/k)$ correspond to classes in $H^1(k, J[2])$ whose restriction to $H^1(k_v, J[2])$ is in the image of $J(k_v)/2J(k_v)$ for every completion k_v. Hence the orthogonal space U_v associated to the class $\gamma(\alpha_v)$ in $H^1(k_v, \mathrm{SO}(V))$ is split at every completion k_v. By the theorem of Hasse and Minkowski, a non-degenerate orthogonal space U of dimension $2n+1$ is split over k if and only if $U_v = U \otimes k_v$ is split over every completion k_v. Hence the orthogonal space U associated to $\gamma(\alpha)$ is split over k, and α lies in the kernel of γ. □

The same argument applies to the Selmer group of the Jacobian J_d of the hyperelliptic curve $dy^2 = f(x)$, for any class $d \in k^*/k^{*2}$. Since the 2-Selmer groups of the twisted curves are known to become arbitrarily large (cf. [5] for the case of genus $n = 1$), the number of k-rational orbits is infinite.

8 More general representations

The three representations V of $\mathrm{SO}(W)$ that we have studied illustrate various phenomena which occur in many other cases. For the standard representation, we have seen that the invariant polynomial q_2 distinguishes the orbits with $\Delta \neq 0$ over any field k. Here the arithmetic invariant theory is the same as the geometric invariant theory.

This pleasant situation also occurs for orbits where the stabilizer G_v is trivial! An interesting example for the odd orthogonal group $\mathrm{SO}(W)$ is the reducible representation $V = W \oplus \wedge^2(W)$. This occurs as the restriction of the adjoint representation of the split even orthogonal group of the space $W \oplus \langle -1 \rangle$. In this representation, the vector $v = (w, T)$ is stable if and only if the $2n+1$ vectors $\{w, T(w), T^2(w), \ldots, T^{2n}(w)\}$ form a basis of W, or equivalently, if the invariant polynomial $\Delta(v) = \det(\langle T^i(w), T^j(w)\rangle)$ is non-zero. In this case $G_v = 1$.

One complication in this case is that the k-orbits do not cover the k-rational points of the categorical quotient: the map on points

$$V(k)/\,\mathrm{SO}(W)(k) \to (V/\,\mathrm{SO}(W))(k)$$

is not surjective. This situation is far more typical in invariant theory than the surjectivity for the three representations we studied. Another atypical property of the three (faithful) representations we studied was that a generic vector had a nontrivial stabilizer. For a generic v in a typical faithful representation V of a reductive group G, the stabilizer G_v is trivial. For G a torus and k complex, G_v is always the kernel of the representation; meanwhile, for G simple, there are only finitely many exceptions (see [20, pp. 229–235]).

The adjoint representations $V = \mathfrak{g}$ of split reductive groups G generalize the second representation $V = \wedge^2(W) = \mathfrak{so}(W)$. Here the invariant polynomials correspond to the invariants for the Weyl group on a Cartan subalgebra, and generate a polynomial ring of dimension equal to the rank of G. The orbits where the discriminant Δ is non-zero correspond to the regular semi-simple elements in $\mathfrak{g}$, and the stabilizer G_v of such an orbit is a maximal torus in G. As an example, one can take the adjoint representation $V = \mathrm{Sym}^2(W)$ of the adjoint form $\mathrm{PGSp}(W) = \mathrm{PGSp}_{2n}$ of the symplectic group, where the degrees of the invariants are $2, 4, 6, \ldots, 2n$ ([cf. 3, Ch 8, §13.3, VI]). For some applications to knot theory, see [15].

The representations which occur in Vinberg's theory for torsion automorphisms θ generalize the third representation $V = \mathrm{Sym}^2(W)$. Here the invariants again form a polynomial ring. As an example, one can take the reducible representation $\wedge^2(W)$ of the group $\mathrm{PGSp}(W) = \mathrm{PGSp}_{2n}$, which corresponds to the pinned outer involution θ of PGL_{2n}. When θ lifts a regular elliptic class in the Weyl group, the orbits where the discriminant Δ is non-zero are stable, and the stabilizer G_v is a finite commutative group scheme over k. Several examples of this type were discussed in [8] and [1].

9 Integral orbits

In order to develop a truly complete arithmetic invariant theory, we should consider orbits in representations not just over a field, but over $\mathbb{Z}$ or a general ring. The descent from an algebraically closed field to a general field that we have discussed in Sections 2–8 gives an indication of some of the issues that arise over more general rings, and it serves as a useful guide for the more general integral theory. In particular, just as a single orbit over an algebraically closed field can split into several orbits over a subfield, an orbit over say the field $\mathbb{Q}$ of rational numbers may then split into several orbits over $\mathbb{Z}$.

Often some of the most interesting arithmetic occurs in the passage from $\mathbb{Q}$ to $\mathbb{Z}$. For example, consider the classical representation given by the action of SL_2 on binary quadratic forms $\mathrm{Sym}_2(2)$. As we have already noted, an orbit over $\mathbb{Q}$ (as over any field) is completely determined by the value of the discriminant d of the binary quadratic forms in that orbit. However, the set of primitive integral orbits inside the rational orbit of discriminant $d \in \mathbb{Z}$ does not necessarily consist of one element, but rather is in bijection with the set of (oriented) ideal classes of the quadratic order $\mathbb{Z}[(d + \sqrt{d})/2]$ in the quadratic field $\mathbb{Q}(\sqrt{d})$ (see, e.g., [6]).

In general, to discuss integral orbits we must fix an integral model of the representation being considered. We give some canonical integral models for the three representations we have studied. For the first representation, we take W to be the odd unimodular lattice of signature $(n + 1, n)$ defined by (2). Because this lattice is self-dual, we can define the adjoint of an endomorphism of W over $\mathbb{Z}$. The group G is then the subgroup of $\mathrm{GL}(W)$ consisting of those transformations g such that $gg^* = 1$ and $\det(g) = 1$. This defines a group that is smooth over $\mathbb{Z}[1/2]$ but is not smooth over $\mathbb{Z}_2$. For the other representations of G, we define $\wedge_2(W)$ as the lattice of skew self-adjoint endomorphisms of W equipped with the action of G by conjugation; we similarly define $\mathrm{Sym}_2(W)$ to be the lattice of self-adjoint endomorphisms of W. Our objective is to describe the orbits of G on each of these three G-modules, or at least those orbits where the discriminant invariant is nonzero.

For the standard representation W, we have already seen that there is a unique orbit over $\mathbb{Q}$ for each value of the discriminant $d \in \mathbb{Q}^*$. An invariant of a $\mathbb{Z}$-orbit of a vector w in the lattice W with $\langle w, w\rangle = d$ is the isomorphism class of the orthogonal complement $U = (\mathbb{Z}w)^{\perp}$, which is a lattice of rank $2n$ and discriminant d over $\mathbb{Z}$. Although W is an odd lattice, the lattice U can be either even or odd. For example, when $n = 3$, the orthogonal complement U of a primitive vector w is an even bilinear space of rank 2 and discriminant d (so corresponds to an integral binary quadratic form of discriminant d) if and only if the vector w has the form $w = ae + bv + cf$ with a and c even and b odd. In this case $d = b^2 + 2ac \equiv 1$ modulo 8, and the orbits of $G(\mathbb{Z})$ on such vectors form a principal homogeneous space for the ideal class group of the quadratic order $\mathbb{Z}[(d + \sqrt{d})/2]$ of discriminant d. We note that these are precisely the quadratic orders where the prime 2 is split. In this case the group $G(\mathbb{Z})$ is isomorphic to the normalizer $N(\Gamma_0(2))$ of $\Gamma_0(2)$ in $\mathrm{PSL}_2(\mathbb{R})$, and the orbits described above correspond to the Heegner points of odd discriminant on the modular curve $X_0(2)^+$ [9, §1].

We consider next the second representation $V = \wedge_2(W)$. Here, we find that the integral orbits of $\mathrm{SO}(M)$ on the self-adjoint transformations $T : M \to M$ with (separable) characteristic polynomial $f(x) = xg(x^2) \in \mathbb{Z}[x]$ correspond to data which generalize the notion of a "minus ideal class" for the ring $R = \mathbb{Z}[x]/(f(x))$. More precisely, the ring R in $L = \mathbb{Q}[x]/(f(x))$ has an involution τ sending β to $-\beta$, where β denotes the image of x in R. Let us consider pairs (I, α), where I is a fractional ideal for R, the element α is in the $\mathbb{Q}$-subalgebra F of L fixed by τ, the product II^τ is contained in the principal ideal (α), and $N(I)N(I^\tau) = N(\alpha)$. Such a pair (I, α) gives I the structure of an integral lattice having rank $2n + 1$ and determinant $(-1)^n$, where the symmetric bilinear form on I is defined by

$$\langle x, y \rangle := \text{coefficient of } \beta^{2n} \text{ in } (-1)^n \alpha^{-1} x y^\tau. \tag{9}$$

The pair (I', α') gives an isometric lattice if $I' = cI$ and $\alpha' = cc^\tau \alpha$ for some element $c \in L^*$. The operator $S : I \to I$ defined by $S(x) = \beta x$ is skew self-adjoint, and has characteristic polynomial $f(x)$. If the integral lattice determined by the pair (I, α) has signature $(n + 1, n)$ over $\mathbb{R}$, there is an isometry $\theta : I \to M$ (cf. [22]), which is well-defined up to composition by an element in $\mathrm{O}(M)$. We obtain an $\mathrm{SO}(M)$-orbit of skew self-adjoint operators with characteristic polynomial $f(x)$ by taking $T = \theta S \theta^{-1}$. Conversely, since a skew self-adjoint $T : W \to W$ gives W the structure of a torsion free $\mathbb{Z}[T] = R$-module of rank one, every integral orbit arises in this manner. Thus the equivalence classes of pairs (I, α) for the ring $R = \mathbb{Z}[x]/(f(x))$, as defined above, index the finite number of integral orbits on $V = \wedge_2(W)$ with characteristic polynomial $f(x)$.

Let us now consider the third representation $V = \mathrm{Sym}_2(W)$. When $\dim(W) = 3$, the kernel of the trace map gives a lattice of rank 5, closely related to the space of binary quartic forms for PGL_2. The integral orbits in this case were studied in [2] and [25]. In general, the integral orbits of $\mathrm{SO}(M)$ on the self-adjoint transformations $T : M \to M$ with (separable) characteristic polynomial $f(x)$ correspond to data which generalize the notion of an ideal class of order 2 for the order $R = \mathbb{Z}[x]/(f(x))$ in the $\mathbb{Q}$-algebra $L = \mathbb{Q}[x]/(f(x))$. More precisely, we consider pairs (I, α), where I is a fractional ideal for R, the element α lies in L^*, the square I^2 of the ideal I is contained in the principal ideal (α), and the square of the norm of I satisfies $N(I)^2 = N(\alpha)$. Then the lattice I has the integral symmetric bilinear form

$$\langle x, y \rangle := \text{coefficient of } \beta^{2n} \text{ in } \alpha^{-1} x y \tag{10}$$

of determinant $(-1)^n$, and self-adjoint operator given by multiplication by β, where β again denotes the image of x in R. The pair (I', α') gives an isometric lattice if $I' = cI$ and $\alpha' = c^2 \alpha$ for some element $c \in L^*$. When this lattice has signature $(n + 1, n)$ over $\mathbb{R}$, it is isometric to M and we obtain an integral orbit with characteristic polynomial $f(x)$. Conversely, since a self-adjoint $T : W \to W$ gives W the structure of a torsion free $\mathbb{Z}[T] = R$-module of rank one, every integral orbit arises in this way. Thus pairs (I, α) for the ring $R = \mathbb{Z}[x]/(f(x))$, up to the equivalence relation defined by c in L^*, index the finite number of integral orbits on $V = \mathrm{Sym}_2(W)$ with characteristic polynomial $f(x)$.

We summarize what we have established for the representations $V = \wedge_2(W)$ and $V = \mathrm{Sym}_2(W)$.

Proposition 9. *Let V denote either the representation $\wedge_2(W)$ or $\mathrm{Sym}_2(W)$ of G. Let $f(x)$ be a polynomial of degree $2n+1$ with coefficients in $\mathbb{Z}$ and non-zero discriminant in $\mathbb{Q}$; if $V = \wedge_2(W)$ we further assume that $f(x) = xg(x^2)$ for an integral polynomial g. Then the integral orbits of $G(\mathbb{Z})$ on $V(\mathbb{Z})$ with characteristic polynomial $f(x)$ are in bijection with the equivalence classes of pairs (I, α) for the order $R = \mathbb{Z}[x]/(f(x))$ defined above, with the property that the bilinear form $\langle\ ,\ \rangle$ on I (given by (9) or (10), respectively) is split.*

In terms of Proposition 4, the integral orbit corresponding to the pair (I, α) maps to the rational orbit of $\mathrm{SO}(W)(\mathbb{Q})$ on $V(\mathbb{Q})$ corresponding to the class of $\alpha \equiv \alpha^{-1}$. Here we view α as an element of (K^*/NE^*) when $V = \wedge_2(W)$, so $L = E + \mathbb{Q}$ and $L^\tau = K + \mathbb{Q}$. When $V = \mathrm{Sym}_2(W)$, we view α as an element of $(L^*/L^{*2})_{N\equiv 1}$.

Finally, we remark that it would be interesting and useful to develop a theory of cohomology that allows one to describe orbits over the integers as we have in the cases above. For example, let us consider again the representation V of the group $G = \mathrm{PGL}_2$ over $\mathbb{Q}$ given by conjugation on the 2×2 matrices v of trace zero. Then this is the adjoint representation, and is also the standard representation of $\mathrm{SO}_3 \cong \mathrm{PGL}_2$. The ring of invariant polynomials on V is generated by $q(v) := -\det(v)$, and the stabilizer G_v of a vector with $q(v) = d \neq 0$ is isomorphic to the one-dimensional torus over $\mathbb{Q}$ which is split by $K = \mathbb{Q}(\sqrt{d})$, and all vectors w with $q(w) = q(v) \neq 0$ lie in the same $G(\mathbb{Q})$-orbit.

A natural integral model of this representation is given by the action of the $\mathbb{Z}$-group $G = \mathrm{PGL}_2$ on the finite free $\mathbb{Z}$-module of binary quadratic forms $ax^2 + bxy + cy^2$. This is equivalent to the representation by conjugation on the matrices of trace zero in the subring $\mathbb{Z} + 2\mathcal{R}$ of the ring $\mathcal{R}$ of 2×2 integral matrices. In this model, the invariant polynomial is just the discriminant $d = b^2 - 4ac$ of the binary form. The content $e = \gcd(a, b, c)$ is also an invariant of a non-zero integral orbit.

We may calculate the $G(\mathbb{Z})$-orbits on the set S of forms with discriminant $d \in \mathbb{Z} - \{0\}$ and content $e = 1$ (so the binary quadratic form is primitive) via cohomology. Let $O = O(d)$ be the quadratic order of discriminant d. Then the stabilizer G_v of such an orbit in PGL_2 is a smooth group scheme over $\mathbb{Z}$ which lies in an exact sequence (in the étale topology)

$$1 \to \mathbb{G}_m \to \mathrm{Res}_{O/\mathbb{Z}}\, \mathbb{G}_m \to G_v \to 1.$$

Furthermore, the $\mathbb{Z}$-points of the quotient scheme G/G_v can be identified with the set S. Hence the orbits in question are in bijection with the kernel of the map $\gamma : H^1(\mathbb{Z}, G_v) \to H^1(\mathbb{Z}, \mathrm{PGL}_2)$ in étale cohomology. Since $H^1(\mathbb{Z}, \mathrm{PGL}_2) = 1$, the orbits are in bijection with the elements of $H^1(\mathbb{Z}, G_v)$. Since $H^1(\mathbb{Z}, \mathbb{G}_m) = H^2(\mathbb{Z}, \mathbb{G}_m) = 1$ the long exact sequence in cohomology gives

$$H^1(\mathbb{Z}, G_v) = H^1(\mathbb{Z}, \mathrm{Res}_{O/\mathbb{Z}}\, \mathbb{G}_m) = \mathrm{Pic}(O).$$

Hence the orbits of $\mathrm{PGL}_2(\mathbb{Z})$ on the set S of binary quadratic forms of discriminant $d \neq 0$ and content 1 form a principal homogeneous space for the finite group $\mathrm{Pic}(O(d))$ of isomorphism classes of projective $O(d)$-modules of rank one. Thus the number of primitive integral orbits contained in the rational orbit of discriminant d is given by the class number of $O(d)$.

References

1. M. Bhargava and W. Ho, Coregular spaces and genus one curves, ArXiv:1306.4424 (2013).
2. M. Bhargava and A. Shankar, Binary quartic forms having bounded invariants, and the boundedness of the average rank of elliptic curves, ArXiv: 1006.1002 (2010); to appear *Annals of Math.*
3. N. Bourbaki, *Groupes et algèbres de Lie*, Hermann, 1982.
4. B. J. Birch and H. P. F. Swinnerton-Dyer, Notes on elliptic curves I, *J. Reine Angew. Math.* **212** (1963), 7–25.
5. R. Bolling, Die Ordnung der Schafarewitsch-Tate-Gruppe kann beliebig gross werden, *Math. Nachr.* **67** (1975), 157–179.
6. D. A. Buell, *Binary Quadratic Forms: Classical Theory and Modern Computations*, Springer-Verlag, 1989.
7. R. Donagi, Group law on the intersection of two quadrics, *Annali della Scuola Normale Superiore di Pisa* **7** (1980), 217–239.
8. B. Gross, On Bhargava's representations and Vinberg's invariant theory, In: *Frontiers of Mathematical Sciences*, International Press (2011), 317–321.
9. B. Gross, W. Kohnen, and D. Zagier, Heegner points and derivatives of L-series II, *Math. Ann.* **278** (1987), 497–562.
10. R. Kottwitz, Stable trace formula: cuspidal tempered terms, *Duke Math. J.* **51** (1984), 611–650.
11. A. Knus, A. Merkurjev, M. Rost, and J.-P. Tignol, *The book of involutions*, AMS Colloquium Publications **44**, 1998.
12. A. Kostrikin and P. H. Tiep, *Orthogonal decompositions and integral lattices*, deGruyter Expositions in Mathematics **15**, Berlin, 1994.
13. R. Langlands, Stable conjugacy—definitions and lemmas, *Canadian J. Math* **31** (1979), 700–725.
14. R. Langlands, Les débuts d'une formule des traces stable, *Publ. Math. de L'Univ. Paris VII*, **13**, 1983.
15. A. Miller, Knots and arithmetic invariant theory, preprint.
16. J. Milnor and D. Husemoller, *Symmetric bilinear forms*, Springer Ergebnisse **73**, 1970.
17. D. Mumford, J. Fogarty, F. Kirwan, *Geometric invariant theory*, Springer Ergebnisse **34**, 1994.
18. D. Panyushev, On invariant theory of θ-groups, *J. Algebra* **283** (2005), 655–670.
19. V. Platonov and A. Rapinchuk, *Algebraic groups and number theory*, Translated from the 1991 Russian original by Rachel Rowen, *Pure and Applied Mathematics* **139**, Academic Press, Inc., Boston, MA, 1994.
20. V. L. Popov and E. B. Vinberg, Invariant Theory in *Algebraic Geometry IV*, Encyclopaedia of Mathematical Sciences **55**, Springer-Verlag, 1994.
21. J-P. Serre, *Galois Cohomology*, Springer Monographs in Mathematics, 2002.
22. J-P. Serre, *A Course in Arithmetic*, Springer GTM **7**, (1978).
23. D. Shelstad Orbital integrals and a family of groups attached to a real reductive group, *Ann. Sci. École Norm. Sup.* **12** (1979), 1–31.
24. M. Stoll, Implementing 2-descent for Jacobians of hyperelliptic curves, *Acta Arith* **98** (2001), 245–277.
25. M. Wood, *Moduli spaces for rings and ideals*, Ph.D. thesis, Princeton University, 2008.

Structure constants of Kac–Moody Lie algebras

Bill Casselman

To Nolan Wallach, wishing him many more years of achievement

Abstract This paper outlines an algorithm for computing structure constants of Kac–Moody Lie algebras. In contrast to the methods currently used for finite-dimensional Lie algebras, which rely on the additive structure of the roots, it reduces to computations in the extended Weyl group first defined by Jacques Tits in about 1966. The new algorithm has some theoretical interest, and its basis is a mathematical result generalizing a theorem of Tits about the finite-dimensional case. The explicit algorithm seems to be new, however, even in the finite-dimensional case. I include towards the end some remarks about repetitive patterns of structure constants, which I expect to play an important role in understanding the associated groups. That neither the idea of Tits nor the phenomenon of repetition has already been exploited I take as an indication of how little we know about Kac–Moody structures.

Keywords: Structure constants • Kac–Moody • Lie algebras

Mathematics Subject Classification: 17B05, 17B22, 17B45, 17B67

1 Introduction

This paper is based on a lecture I gave at a conference in San Diego in honor of the achievements of Nolan Wallach.

Suppose Δ to be a finite set. In this paper a *Cartan matrix* indexed by Δ will be an arbitrary integral matrix $C = (c_{\alpha,\beta})$ $(\alpha, \beta \in \Delta)$ satisfying these conditions:

B. Casselman
Mathematics Department, University of British Columbia, Vancouver, BC, Canada
e-mail: cass@math.ubc.ca

R. Howe et al. (eds.), *Symmetry: Representation Theory and Its Applications*, Progress in Mathematics 257, DOI 10.1007/978-1-4939-1590-3_4

(C1) $c_{\alpha,\alpha} = 2$;
(C2) for $\alpha \neq \beta$, $c_{\alpha,\beta} \leq 0$;
(C3) $c_{\alpha,\beta} = 0$ if and only if $c_{\beta,\alpha} = 0$.

More commonly such a matrix is called a *generalized Cartan matrix*, whereas a Cartan matrix is taken to be one with the additional requirement that

(CD) there exist a diagonal matrix D such that CD is positive definite.

But times change, and "generalized" is now closer to "usual". In this paper, if condition (CD) is satisfied I will call C positive definite. For purely technical reasons I will assume further that

(C4) C is irreducible;
(C5) C is invertible.

The first condition means that Δ cannot be expressed as $\Delta_1 \cup \Delta_2$ with $c_{\alpha,\beta} = 0$ for $\alpha \in \Delta_1$, $\beta \in \Delta_2$. This condition is no significant restriction on results, since one can work with the summands of the root systems corresponding to the Δ_i. The second condition excludes the affine Cartan matrices. This will restrict results, but the missing cases can be easily dealt with separately.

Let

$$L = \mathbb{Z}^{\Delta}, \quad L^{\vee} = \mathrm{Hom}(L, \mathbb{Z}).$$

Thus Δ embeds into L as a basis. For every subset Θ of Δ let L_{Θ} be the span of Θ in L, so that in particular $L_{\Delta} = L$. Since C is integral and invertible, there exists a unique maximal linearly independent subset $\Delta^{\vee}$ of $L^{\vee}$ and a bijection $\alpha \mapsto \alpha^{\vee}$ with

$$\langle \alpha, \beta^{\vee} \rangle = c_{\alpha,\beta}.$$

Let $L^{\vee}_{\Delta^{\vee}}$ be the submodule of $L^{\vee}$ spanned by $\Delta^{\vee}$. The linear transformation

$$s_{\alpha}\colon \lambda \mapsto \lambda - \langle \lambda, \alpha^{\vee} \rangle \alpha$$

is a (possibly skew) reflection, taking L to itself. Its contragredient $s_{\alpha^{\vee}}$ takes $L^{\vee}$ and $L^{\vee}_{\Delta^{\vee}}$ to themselves. These reflections generate the *Weyl group* W associated to the Cartan matrix. The W-orbit of Δ is the set of *real* roots $\Sigma_{\mathbb{R}}$. It is the disjoint union of two halves, positive and negative. The positive (resp. negative) roots $\Sigma^{+}_{\mathbb{R}}$ (resp. $\Sigma^{-}_{\mathbb{R}}$) are integral linear combinations of elements in Δ with non-negative (resp. non-positive) coefficients.

Let $\mathcal{D}$ be the region of all v in $L \otimes \mathbb{R}$ such that $\langle v, \alpha^{\vee} \rangle \geq 0$ for all $\alpha \in \Delta$, and define the *closed Tits cone* $\overline{\mathcal{C}}$ to be the closure of $W(\mathcal{D})$. It is convex, and has a non-empty interior, on which the group W acts discretely with fundamental domain $\mathcal{D}$. I define $\Sigma_{\mathbb{I}}$, the set of *imaginary* roots, to be the union of

$$\Sigma^{+}_{\mathbb{I}} = (L_{\Delta} \cap \overline{\mathcal{C}}) - \{0\}$$

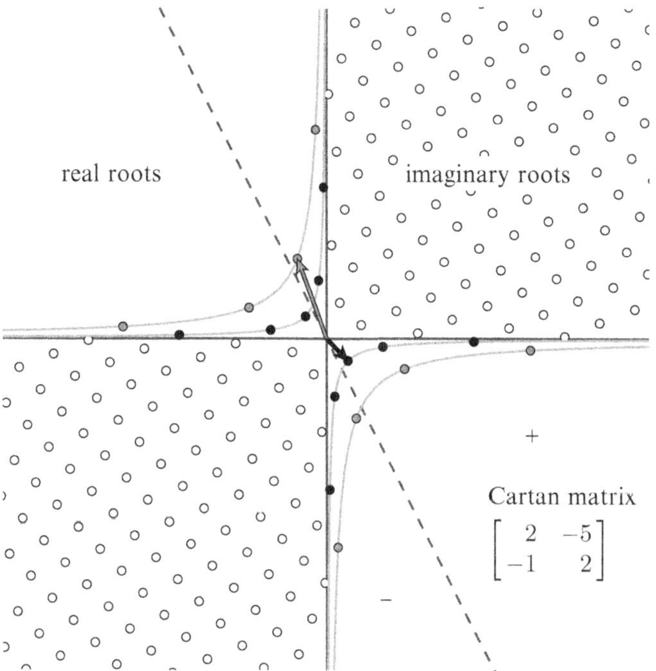

Figure 1 Real and imaginary roots for a system of rank 2

and its opposite $\Sigma_{\mathbb{I}}^{-}$. It, too, is W-stable, as is each half $\Sigma_{\mathbb{I}}^{\pm}$ and the set Σ of all roots, the union of real and imaginary ones. (For all this, refer to Proposition 5.2, Theorem 5.4, Lemma 5.8, and Exercise 5.12 in Chapter 5 of [Kac85]. Figure 1 shows how things look in a simple case.) The terminology is perhaps motivated by the fact that many root systems possess a natural quadratic form with respect to which real roots have real lengths and imaginary ones have imaginary ones.

Let

$$\mathfrak{h} = L^{\vee} \otimes \mathbb{C},$$

which may be identified with an Abelian Lie algebra, and let

$$\mathfrak{h}^{\vee} = L \otimes \mathbb{C}$$

be its complex linear dual. For each α in Δ let h_α be the image of $\alpha^\vee$ in $\mathfrak{h}$. In other words, $\alpha^\vee$ may be identified with a linear injection of $\mathbb{C}$ into $\mathfrak{h}$ taking x to xh_α. Let $\mathfrak{g}$ be the *Kac–Moody* Lie algebra associated to these data. One is given in the construction of $\mathfrak{g}$ an embedding of $\mathfrak{h}$ into it. The adjoint action of $\mathfrak{h}$ on $\mathfrak{g}$ is semi-simple and locally finite, breaking it up into the sum of $\mathfrak{h}$ and a number of *root spaces* $\mathfrak{g}_\lambda$, with $\lambda\colon \mathfrak{h} \to \mathbb{C}$ lying inside the image of L in $\mathfrak{h}^\vee$. The fundamental fact about these data is that $\mathfrak{g}_\lambda \neq 0$ *if and only if* λ *is a root in the sense spoken of earlier.* This is complemented by the assertion that *for each* α *in* Δ, $\mathfrak{g}_\alpha$ *has dimension one.* It is also true that $\mathfrak{g}_\lambda$ has dimension one for every real root λ.

The validity of these assertions depends on assumption (C5). Without it, there still exists a Kac–Moody algebra defined by C, but its description is a bit more subtle, as explained in the opening chapter of [Kac85], and elaborated on in Chapter 6. The corresponding Lie algebras turn out to be extensions of $\mathfrak{g} \otimes \mathbb{C}[t^{\pm 1}]$, with $\mathfrak{g}$ finite-dimensional, and computations in it reduce easily to computations in $\mathfrak{g}$ itself. This is why (C5) is not very restrictive.

There are two basic problems this paper will deal with:

Problem 1. *To specify a good choice of basis elements* e_λ *of* $\mathfrak{g}_\lambda$ *for all real roots* λ;

Problem 2. *To find, for every pair of real roots* λ, μ *such that* $\lambda + \mu$ *is also a real root, the structure constant* $N_{\lambda,\mu}$ *such that* $[e_\lambda, e_\mu] = N_{\lambda,\mu} e_{\lambda+\mu}$.

The method I will use to solve these problems originates in [Tits66a], which sketches what happens when $\mathfrak{g}$ has finite dimension. The extension to Kac–Moody algebras is not quite trivial, but neither is it particularly tricky. I hope that it will be useful in exploring what happens for bracket computations involving imaginary roots, which is one of the great mysteries of Kac–Moody algebras, although I say nothing about this problem here.

When $\mathfrak{g}$ is finite-dimensional, there are already in the literature two practical approaches to constructing the e_λ and computing the $N_{\lambda,\mu}$. One can be found in [Car72] and is explained in more detail in [CMT04]. The construction uses induction on roots, going from λ to $\lambda + \alpha$ ($\alpha \in \Delta$) according to a certain rule. Calculation of the structure constants amounts to transliterations of the Jacobi identity. The other method, introduced in [FK80], exhibits an interesting extension of groups to interpret the constants as related to cohomology. This technique was applied originally only to simply laced systems, but it has since been extended to the rest of the finite-dimensional $\mathfrak{g}$ in [Ryl00] by 'folding'. As far as I can tell these methods cease to be valid for arbitrary Kac–Moody algebras. In any event, in the cases for which they do work they do not seem to be a great deal faster than a program based on the method to be explained here. To the extent to which they are faster, they are 'hard-wired', incorporating for each system special short-cuts that do not apply in general.

Acknowledgments. I thank the organizers of the conference in honor of Nolan Wallach for inviting me to contribute to it. I also wish to thank my colleague Julia Gordon, who—rather casually—started me off on this project. I am extremely happy to have had this opportunity to try to understand and elaborate on Jacques Tits' work on structure constants. I first saw that volume of the *Publications de l'IHES* for sale in Schoenhof's when I was a very ignorant graduate student, and the memory of my puzzled thought, "What exactly is a structure constant?" has remained with me ever since. And finally I wish to thank the referee, who complained much about lacunae in an earlier draft but nonetheless continued to read and criticize carefully.

2 Chevalley bases

In this section I will start looking at Problem 1, that of constructing a good basis of $\mathfrak{g}$. The beginning is straightforward—for each α in Δ choose an arbitrary $e_\alpha \neq 0$ in $\mathfrak{g}_\alpha$. It doesn't make any difference what choice you make, because all choices will be conjugate with respect to an automorphism of the Lie algebra $\mathfrak{g}$. For each α in Δ, $[\mathfrak{g}_\alpha, \mathfrak{g}_{-\alpha}]$ will be the one-dimensional subspace spanned by h_α. Following [Tits66a], then choose $e_{-\alpha}$ so

$$[e_\alpha, e_{-\alpha}] = -h_\alpha .$$

The usual convention imposes a plus sign on the right hand side of this equation, but Tits' change of sign is extraordinarily convenient, in fact obligatory if the symmetries exploited later on in this paper are to remain comprehensible. In my opinion, Tits' choice should have become the conventional one long ago.

For example, if $\mathfrak{g} = \mathfrak{sl}_2$, we get

$$e_+ = \begin{bmatrix} 0 & 1 \\ 0 & 0 \end{bmatrix}, \quad e_- = \begin{bmatrix} 0 & 0 \\ -1 & 0 \end{bmatrix}, \quad h = \begin{bmatrix} 1 & 0 \\ 0 & -1 \end{bmatrix}.$$

The data of $\mathfrak{h}$ together with the e_α make up what I will call a *frame* for $\mathfrak{g}$. (It is called in French an *épinglage*, frequently translated awkwardly into English as *pinning*. The reference in French is to the way butterflies are mounted. To those of us who have been parents of young children, the proposed English term has other, less fortunate, connotations.)

From now on, assume a frame to have been fixed. With this assumption, we are given also for each α in Δ a well-defined embedding

$$\varphi_\alpha \colon \mathfrak{sl}_2 \longrightarrow \mathfrak{g} \quad \Big| \quad e_+ \longmapsto e_\alpha, \quad h \longmapsto h_\alpha, \quad e_- \longmapsto e_{-\alpha} .$$

Let $\mathfrak{sl}_2^{[\alpha]}$ be its image.

The map taking each $e_{\pm\alpha}$ to $e_{\mp\alpha}$, $h \mapsto -h$ for h in $\mathfrak{h}$ extends to an involutory automorphism θ of all of $\mathfrak{g}$ called the *canonical involution*.

For x in $\mathfrak{g}_\lambda$ and y in $\mathfrak{g}_\mu$, the bracket $[x, y]$ lies in $\mathfrak{g}_{\lambda+\mu}$ (which may be 0). If $\mu = -\lambda$, this means it will lie in $\mathfrak{h}$. For each real root λ the subspace $[\mathfrak{g}_\lambda, \mathfrak{g}_{-\lambda}]$ has dimension one in $\mathfrak{h}$, and is not contained in the kernel of λ. Hence there exists a unique h_λ in it with $\langle \lambda, h_\lambda \rangle = 2$.

Lemma 2.1. *For each real root λ there exists up to sign a unique e_λ in $\mathfrak{g}_\lambda$ such that*

$$[e_\lambda, \theta(e_\lambda)] = -h_\lambda .$$

Proof. Given $e \neq 0$ in $\mathfrak{g}_\lambda$, we know that $\theta(e)$ lies in $\mathfrak{g}_{-\lambda}$ and hence

$$[e, \theta(e)] = ah_\lambda$$

for some constant $a \neq 0$. But then

$$[ce, \theta(ce)] = c^2 ah_\lambda$$

so we choose ce with $c^2 a = -1$. □

We may thus assemble a basis of $\mathfrak{g}$ by choosing for each root λ one of the two choices this gives us. This is called a *Chevalley basis*. Part of the solution to Problem 1 above is to choose a Chevalley basis. It is unique only up to signs, and there is apparently no canonical choice.

Suppose we are given a Chevalley basis. Suppose further we are given real roots λ, μ such that $\lambda + \mu$ is also a real root. We have

$$[e_\lambda, e_\mu] = N_{\lambda,\mu} e_{\lambda+\mu}$$

in which the absolute value $|N_{\lambda,\mu}|$ is now uniquely determined. It is known never to be 0. We should not be too surprised to learn that it has a relatively simple expression. I will next explain the formula for it.

If λ and μ are roots with λ real, the λ*-string* through μ is the intersection of $\mu + \mathbb{Z}\lambda$ with Σ. It is always finite. Define $p_{\lambda,\mu}$ to be the maximum p such that $\mu - p\lambda$ is in the string, and $q_{\lambda,\mu}$ to be the maximum q such that $\mu + q\lambda$ lies in it. The string is the full segment $[\mu - p\lambda, \mu + q\lambda]$. The following identities are easy to verify:

$$\begin{aligned} p_{\mu,\lambda} &= p_{-\mu,-\lambda} \\ p_{\mu+\lambda,\lambda} &= p_{\mu,\lambda} + 1 \\ p_{\mu,-\lambda} &= n - p_{\mu,\lambda} \text{ if } n \text{ is the length of the string.} \end{aligned}$$

The following was first proved by Chevalley for finite-dimensional Lie algebras, probably by Tits more generally. See also [Mor87]. Following [Tits66a], I shall in effect reprove it later on.

Theorem 2.2 (Chevalley's Theorem). *If λ, μ, and $\lambda + \mu$ are all real roots, then*

$$|N_{\lambda,\mu}| = p_{\lambda,\mu} + 1 .$$

The constants $p_{\lambda,\mu}$ are simple to evaluate, so the whole problem of computing structure constants comes down to computing a sign.

This result makes it possible to assign a $\mathbb{Z}$-structure to $\mathfrak{g}$, and was classically the basis for Chevalley's construction of split group schemes defined over $\mathbb{Z}$. Incidentally, the usual proof of Chevalley's theorem is a case-by-case examination. This is somewhat unsatisfactory, and moreover it will not be possible when working with arbitrary Kac–Moody algebras. One is forced to come up with a proof that is illuminating even in the finite-dimensional cases.

3 The root graph

The algorithm we shall eventually see will require that we be able to list as many roots as we want. In fact, in principle we can construct even an infinite number all at the same time, as I will explain later.

Every real root is $w\alpha$ for some w in W, α in Δ. The natural way to construct real roots is therefore to start with all the α in Δ and apply the elements of W to them. How does this go? Keep in mind that the group W is generated by the elementary reflections s_α. We maintain two lists of positive roots, one the current list to be processed—the *waiting list*—and the other that of roots that have been processed—*serviced customers*. Processing a root means (a) removing it from the waiting list and (b) applying all elementary reflections to it to see whether we get new roots or roots we have already seen. When we see a new one, we put it on the waiting list, if the new root has height below some bound we have initially set. (I recall that the height of $\lambda = \sum \lambda_\alpha \alpha$ is $\sum |\lambda_\alpha|$.) Of course one may as well restrict oneself to the task of finding only positive roots.

It is not hard to use a look-up table to follow this method. It is guaranteed to give us eventually as many roots as we need, but it is not at all clear how long it will take. The problem that comes to mind is that when we apply elementary reflections we might expect *a priori* to go down in height to get a new root, and then down in height again, and so on. This is reminiscent of solving the word problem in group theory. In this case it ought not to be necessary—we would like to know that we can proceed by adding to the waiting list only reflections of greater height. This is easy to carry out, since $s_\alpha\lambda$ has height more than λ if and only if $\langle \lambda, \alpha^\vee \rangle < 0$. But seeing why this procedure will work is not quite trivial.

Define the *depth* of a positive root λ to be $\ell(w)$ for the w of least length such that $w^{-1}\lambda < 0$. Equivalently this is $\ell(w) + 1$ where w is of least length such that $\lambda = w\alpha$ with $\alpha \in \Delta$. For example, the depth of every α in Δ is 1. The fundamental fact about depth is Proposition 2.3 of [BH93], which is simple to state but not quite so simple to prove:

Proposition 3.1 (Brink and Howlett). *The depth of $s_\alpha\lambda$ is greater than that of λ if and only if $\langle \lambda, \alpha^\vee \rangle < 0$.*

This is also part of Lemma 3.3 in [Cas06].

One consequence is that the height of a root is greater than or equal to its depth, so if we have found all roots of depth $\leq n$ we have also found all roots of height $\leq n$. Another consequence is one we'll need in a later section:

Corollary 3.2. *If λ is a positive real root not in Δ, then there exists α in Δ with $\langle \lambda, \alpha^\vee \rangle > 0$.*

Here is another natural consequence:

Corollary 3.3. *If $\Theta \subseteq \Delta$, then the real roots in L_Θ are in the W_Θ-orbit of Θ.*

Here W_Θ is the subgroup of W generated by the s_α with α in Θ.

Proof. Arguing by induction on depth, it has to be shown that if λ is a positive real root in the linear span of Θ, there exists $\alpha \in \Theta$ such that $\langle \lambda, \alpha^\vee \rangle > 0$. But since λ is positive and in L_Θ, $\langle \lambda, \beta^\vee \rangle \leq 0$ for all β not in Θ, so that the claim follows from the previous Corollary. □

In finding positive roots, we are in effect building what I call the *root graph*, from the bottom up. It is an important structure. With one exception the nodes of this graph are the positive roots. The exception is that for technical convenience I add a 'dummy' node at the bottom (Figures 2 and 3).

What about edges? They are oriented. There is an edge labeled s_α from the dummy node to α, for each α in Δ. There is an edge from a root λ to another μ if $\mu = s_\alpha \lambda$ and the height of μ is greater than the height of λ. It is labeled by s_α. Since

$$s_\alpha \lambda = \lambda - \langle \lambda, \alpha^\vee \rangle \alpha$$

this will happen if and only if $\langle \lambda, \alpha^\vee \rangle < 0$.

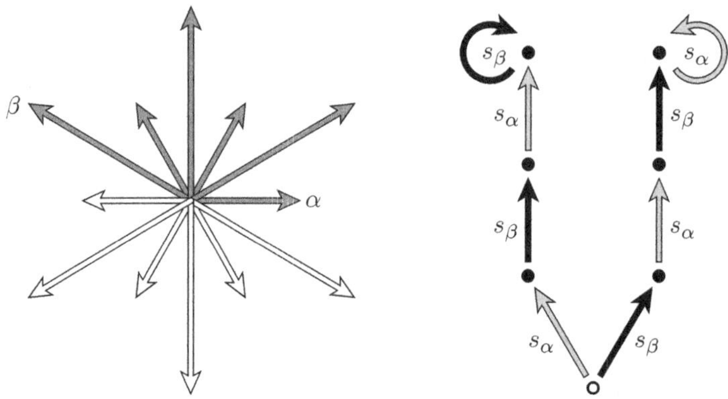

Figure 2 The roots and root graph of G_2

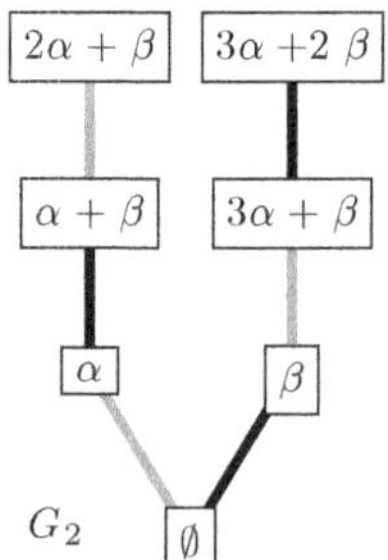

Figure 3 A different version of the root graph of G_2

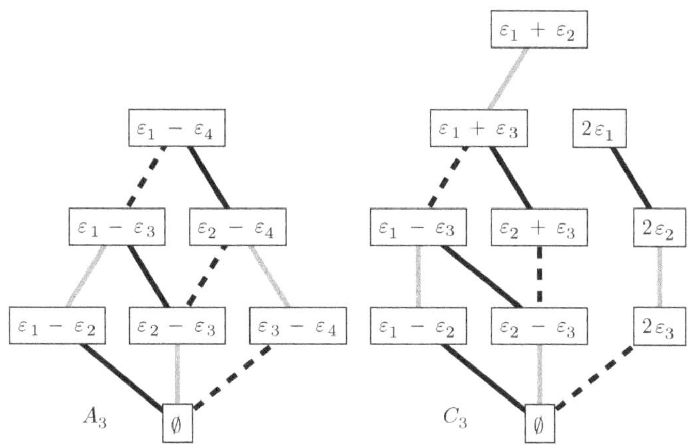

Figure 4 The root graphs for A_3 and C_3

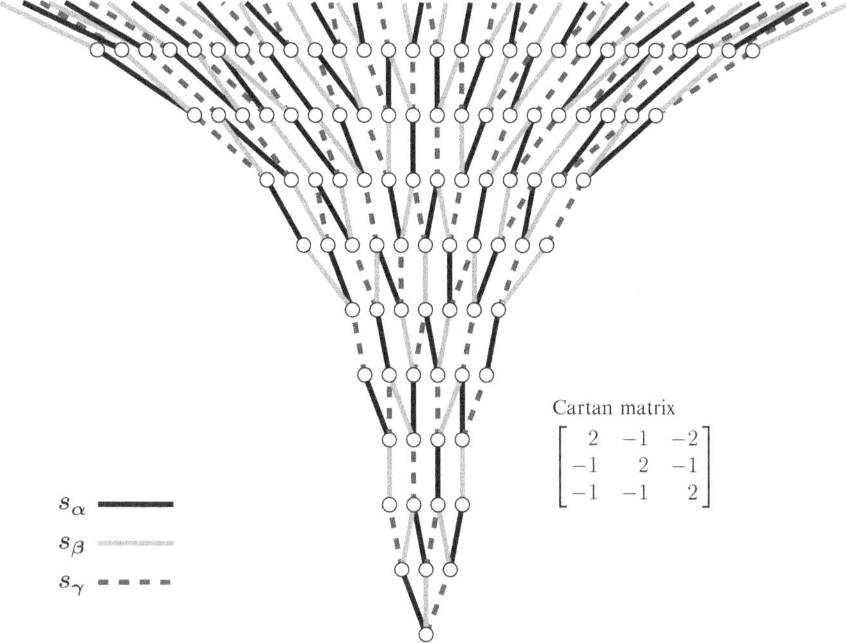

Figure 5 The bottom of an infinite root graph

In drawing the root graph, the loops are redundant, since the total number of edges coming in or out is a constant. (This is the reason for adding the 'dummy'.) Furthermore, in the diagrams that follow the arrows can be replaced by simple lines, because the orientation is always upward (Figs. 4 and 5).

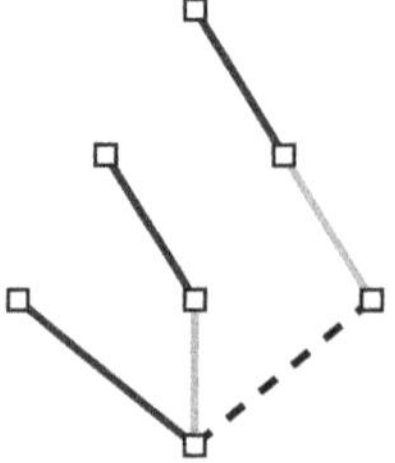

Figure 6 The ISL root graph for A_3

Each root can be reached by one or more paths in the root graph, starting at the dummy node. But if we are to construct roots by traversing paths, we want to specify a unique path to every node. I choose the *ISL* (**I**nverse **S**hort **L**ex) path, which is defined recursively (Figure 6). Assign an order to Δ. A path

$$\gamma = \lambda_0 \xrightarrow{\alpha_1} \cdots \xrightarrow{\alpha_n} \lambda_n$$

is ISL if α_n is the least label on the edges leading up to λ_n, and similarly for all its initial segments.

4 Rank two systems

Chevalley's formula for $|N_{\lambda,\mu}|$ indicates that the geometry of root configurations will likely be important in this business. In this section and the next I prove a few crucial properties.

I begin by recalling the original way of thinking about bases of roots. Suppose V to any real vector space, given a coordinate system (x_i). Impose on V the associated *lexicographic order*: $x < y$ if $x_i = y_i$ for $i < m$ but $x_m < y_m$. This is a linear order, invariant under translation, and [Sat51] points out that any translation-invariant linear order that is continuous in some sense has to be one of these. In any free $\mathbb{Z}$-module contained in V there exists a least non-zero positive vector.

Suppose Δ to be a basis of a set Σ of roots, and take V to be $L_\Delta \otimes \mathbb{R}$. If we assign an order to Δ, we get a corresponding lexicographic order on V. *The positive roots in Σ are those > 0 with respect to the lexicographic order.* This should motivate the following discussion.

Lemma 4.1. *Suppose we are given an ordered set of real and linearly independent roots λ_i, and let Λ_k be the subset of the first k. There exists a subset of roots α_i in Δ and w in W such that $w\lambda_1 = \alpha_1$, and each $w\Lambda_k$ is contained in the non-negative integral span of the α_i for $i \leq k$.*

Proof. Extending the set of λ_i if necessary, we may assume they form a basis of V. Assign V the associated lexicographic order. Let V_k be the real span of Λ_k.

Define β_1 to be λ_1, and define β_k to be the least positive root in V_k that is not in V_{k-1} (look at the linear order induced on the quotient to see that this exists). Let B be the set of all β_i, B_k the subset of the first k. The real span of B_k is V_k.

I claim that every real root λ in the real span of B_k positive with respect to this linear order (including the λ_i themselves) is in the non-negative integral span of B_k. For $k = 1$, there is nothing to prove. Proceed by induction on the smallest k such that λ is in the real span of Λ_k. Applying induction, we may assume λ is in $V_k - V_{k-1}$. If $\lambda = \beta_k$, no problem. Otherwise $\lambda > \beta_k$. According to Corollary 3.2 of § 3 there exists $i \leq k$ such that $\langle \lambda, \beta_i^\vee \rangle > 0$. Then $\lambda - \beta_i$ will again be a root, of smaller height. Repeat as required.

Now apply Theorem 2 of §5.9 in [MP95] to see that there exists w in W with $wB = \Delta$. □

Proposition 4.2. *If λ, μ are real roots, there exists w in W, α and β in Δ such that $w\lambda = \alpha$, $w\mu$ is a non-negative integral combination of α and β. The real root $w\mu$ is in $W_{\alpha,\beta}\{\alpha, \beta\}$.*

Proof. The last assertion follows from Corollary 3.3 . □

The map from Δ to $\Delta^\vee$ has a natural extension to a map $\lambda \mapsto \lambda^\vee$ from $\Sigma_{\mathbb{R}}$ to $\Sigma_{\mathbb{R}}^\vee \subset L^\vee$. It is characterized by the formula $(w\lambda)^\vee = w\lambda^\vee$ for all real roots λ. If $\lambda = w(\alpha)$, then the reflection corresponding to λ is $s_\lambda = w s_\alpha w^{-1}$:

$$s_\lambda u = u - \langle u, \lambda^\vee \rangle \lambda \,.$$

For any pair of distinct real roots λ, μ let $L_{\lambda,\mu}$ be their integral span.

Proposition 4.3. *If λ and μ are any two distinct real roots with $\langle \lambda, \mu^\vee \rangle \neq 0$, there exists on $L_{\lambda,\mu}$ an inner product with respect to which s_λ and s_μ are orthogonal reflections, and $\lambda \cdot \lambda > 0$. It is unique up to a positive scalar, and $\kappa \cdot \kappa > 0$ for all real roots κ in $L_{\lambda,\mu}$.*

Proof. The reflection s_κ is orthogonal if and only if

$$2\left(\frac{u \cdot \kappa}{\kappa \cdot \kappa}\right) = \langle u, \kappa^\vee \rangle$$

for all u in $L_{\lambda,\mu}$. This requires that

$$\langle \lambda, \mu^\vee \rangle = 2\left(\frac{\lambda \cdot \mu}{\mu \cdot \mu}\right)$$
$$\langle \mu, \lambda^\vee \rangle = 2\left(\frac{\lambda \cdot \mu}{\lambda \cdot \lambda}\right).$$

These equations tell us that $\lambda \cdot \mu$ and $\mu \cdot \mu$ are both determined by $\lambda \cdot \lambda$, so certainly there is a unique inner product defined uniquely on $L_{\lambda,\mu}$ by $\lambda \cdot \lambda$ and the requirement that reflections be orthogonal. Why are the norms of real roots then all positive?

According to Proposition 4.2, we may assume that $\lambda = \alpha$, $\mu = p\alpha + q\beta$ with α, β in Δ. In these circumstances $\langle \alpha, \beta^\vee \rangle \neq 0$, so there exists on $L_{\alpha,\beta}$ an essentially unique inner product with $\alpha \cdot \alpha > 0$. By the definition of a Cartan matrix, we also have $\beta \cdot \beta > 0$. The given inner product on $L_{\lambda,\mu}$ must be some scalar multiple of this one. But both λ and μ are Weyl transforms of α or β, so both also have positive norms. □

Corollary 4.4. *Given two real roots λ, μ we have*

$$\langle \lambda, \mu^\vee \rangle \langle \mu, \lambda^\vee \rangle \geq 0$$

and one factor is 0 *if and only if both are.*

Proof. Since

$$\langle \lambda, \mu^\vee \rangle \langle \mu, \lambda^\vee \rangle = 4 \frac{(\lambda \cdot \mu)^2}{\|\lambda\|^2 \|\mu\|^2} \,. \qquad \square$$

Since the transpose of a Cartan matrix is also a Cartan matrix, the previous result implies that there exists an essentially unique metric on $\Sigma^\vee$ as well.

Corollary 4.5. *In an irreducible root system of rank two, the product $\|\lambda\| \, \|\lambda^\vee\|$ is constant.*

Proof. It is immediate from the defining formulas that

$$\|\alpha\| \, \|\alpha^\vee\| = \|\beta\| \, \|\beta^\vee\| \,. \qquad (*)$$

Any other real root λ is of the form $w\alpha$ or $w\beta$, from which the claim follows. □

Corollary 4.6. *The map*

$$\nu \longmapsto \frac{\nu^\vee}{\|\nu^\vee\|^2}$$

is linear on $\Sigma_{\mathbb{R}} \cap L_{\lambda,\mu}$.

5 Tits triples

In this section, suppose λ, μ, and ν to be what I call a **Tits triple**—a trio of real roots with $\lambda + \mu + \nu = 0$. In these circumstances, either $\langle \mu, \lambda^\vee \rangle$ or $\langle \nu, \lambda^\vee \rangle$ does not vanish, so by Corollary 4.3 there exists a well-defined inner product on $L_{\lambda,\mu} = L_{\lambda,\nu}$, with respect to which all real roots have positive norm.

In particular, the λ-string through μ contains at least the two real roots μ and $-\nu$. It is stable under s_λ, which reflects in the point where $\langle \mu, \Lambda^\vee \rangle = 0$. By assumption, both μ and $-\nu$ are in this string.

Lemma 5.1. *Let* $n = \langle \mu, \lambda^\vee \rangle$.

(a) $n < -1$ *if and only if* $\|\mu\| > \|\nu\|$;
(b) $n = -1$ *if and only if* $\|\mu\| = \|\nu\|$;
(c) $n > -1$ *if and only if* $\|\mu\| < \|\nu\|$.

Proof. If $n = -1$, then $s_\lambda \mu = -\nu$. If $n \geq 0$, then evaluate $(\mu + \lambda) \cdot (\mu + \lambda)$ to see that $\|\nu\|^2 > \|\mu\|^2$. In the remaining case, with $n \leq -2$, consider instead $s_\lambda \lambda = -\lambda$, $s_\lambda \nu$, $s_\lambda \mu$. □

There are a limited number of configurations of the λ-string through μ. If κ is a positive (resp. negative) imaginary root in the string, then $s_\lambda \kappa$ is also one, since the positive (resp. negative) imaginary roots are stable under W. Since the positive (resp. negative) imaginary roots are the intersection of a convex set with the roots, all intervening roots must also be positive (resp. negative) and imaginary. Therefore any real roots in the string must be at its ends. Since s_λ reflects in the hyperplane $\langle \mu, \lambda^\vee \rangle = 0$ and the real roots in the string can have at most two lengths, the previous result implies:

Lemma 5.2. *There do not exist three real roots* κ *in a* λ*-string with* $\langle \kappa, \lambda^\vee \rangle \leq 0$.

In other words, the following figures, with real roots dark, show all possibilities for strings containing a real root. The lengths in the string decrease until at most the half-way point, then increase (Figure 7).

All these are in fact possible, as you can see by perusing classical root diagrams or Figure 1.

Lemma 5.3. *The following are equivalent:*

(a) $s_\lambda \mu = -\nu$;
(b) $\langle \mu, \lambda^\vee \rangle = -1$;
(c) $\|\lambda\| \geq \|\mu\|, \|\nu\|$.

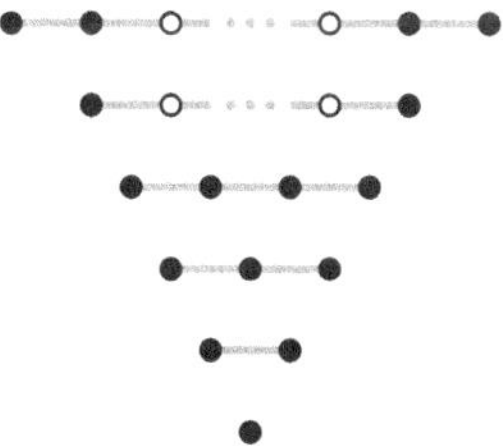

Figure 7 Possible root strings containing a real root

Figure 8 An impossible real root configuration

The point of this result is that whenever $\lambda + \mu + \nu = 0$ we can cycle to obtain $s_\lambda \mu = -\nu$, since we can certainly cycle to get the third condition by taking λ of maximum length.

Proof. The equivalence of (a) and (b) is immediate.

Assume (b). Then $s_\lambda \mu = -\nu$, so $\|\mu\| = \|\nu\|$. But also $\langle \lambda, \mu^\vee \rangle \le -1$, so by Lemma 5.1 we have $\|\lambda\| \ge \|\mu\|$. Thus (b) implies (c).

Assume (c). First of all, Lemma 5.1 implies that $\langle \lambda, \mu^\vee \rangle < 0$, for if not then $\|\nu\|^2 = \|\lambda + \mu\|^2 > \|\lambda\|^2$, a contradiction. This implies that $n = \langle \mu, \lambda^\vee \rangle < 0$ as well.

If $n = -1$ then $s_\lambda \mu = -\nu$ and $\|\mu\| = \|\nu\|$. If $n < -1$. But then by Lemma 5.1 we have $\|\mu\| \ge \|\mu + \lambda\| = \|\nu\|$. By symmetry, we also have $\|\nu\| \ge \|\mu\|$. But then $\|\mu\| = \|\nu\|$ and by Lemma 5.1 $s_\lambda \mu = -\nu$. □

One consequence is that we cannot have a trio of real roots with $\lambda + \mu + \nu = 0$, $\|\lambda\|$, $\|\mu\| > \|\nu\|$. The diagram of Figure 8 is impossible.

I shall now examine what happens when $\|\lambda\| \ge \|\mu\| = \|\nu\|$. There are two different cases.

Suppose first that $\|\lambda\| = \|\mu\| = \|\nu\|$. Then by Lemma 5.1

$$\begin{aligned} s_\lambda\colon \lambda &\longmapsto -\lambda \\ \mu &\longmapsto -\nu \\ \nu &\longmapsto -\mu \\ s_\mu\colon \lambda &\longmapsto -\nu \\ \mu &\longmapsto -\mu \\ \nu &\longmapsto -\lambda \end{aligned}$$

so $s_\lambda s_\mu$ rotates (λ, μ, ν) to (μ, ν, λ). Following this by s_μ maps

$$\begin{aligned} \lambda &\longmapsto -\mu \\ \mu &\longmapsto -\lambda \\ \nu &\longmapsto -\nu\,. \end{aligned}$$

Thus

$$p_{\lambda,\mu} = p_{\mu,\nu} \text{ etc.,} \qquad p_{\lambda,\mu} = p_{\mu,-\lambda} = p_{\mu,\lambda}\,.$$

Suppose next that $\|\lambda\| > \|\mu\| = \|\nu\|$. We may set these last to 1, $\|\lambda\|^2 = n > 1$. By Lemma 5.1 we must have $\langle \mu, \lambda^\vee \rangle = -1$. Since

$$\langle \lambda, \mu^\vee \rangle \|\mu\|^2 = \langle \mu, \lambda^\vee \rangle \|\lambda\|^2$$

we must also have $\langle \lambda, \mu^\vee \rangle = -n < -1$.

But then λ and $-\nu = \lambda + \mu$ must also lie in the initial half of the μ-string through λ, and therefore by Lemma 5.2 λ must be its beginning. Similarly, λ is at the beginning of its ν-string. Therefore

$$p_{\lambda,\mu} = p_{\lambda,\nu} = 0 .$$

If ν were not at the beginning of its λ-string, there would exist a root $\nu - \lambda$ of squared-length

$$\|\nu - \lambda\|^2 = \|\nu\|^2 - 2\nu\dot{\lambda} + \|\lambda\|^2$$

greater than that of λ. Again by Lemma 5.2 , this cannot happen. So ν is at the beginning of its λ-string. The same holds for μ. So

$$p_{\nu,\lambda} = p_{\mu,\lambda} = 0 .$$

Since λ is at the beginning of its ν-string, $s_\nu \lambda = \lambda + n\nu$ is the end of that string, which is of length n. But $\lambda + \nu = -\mu$, so

$$p_{\mu,\nu} = p_{-\mu,-\nu} = n - 1$$

and also

$$p_{\nu,\mu} = p_{-\nu,-\mu} = n - 1 .$$

These computations now imply:

Lemma 5.4 (Geometric Lemma). *We have*

$$\frac{p_{\lambda,\mu} + 1}{\|\nu\|^2} = \frac{p_{\mu,\nu} + 1}{\|\lambda\|^2} = \frac{p_{\nu,\lambda} + 1}{\|\mu\|^2} .$$

Lemma 5.5. *We have* $p_{\mu,\lambda} = p_{\lambda,\mu}$.

Proposition 5.6. *We have*

$$(p_{\lambda,\mu} + 1)\nu^\vee + (p_{\mu,\nu} + 1)\lambda^\vee + (p_{\nu,\lambda} + 1)\mu^\vee = 0 .$$

Proof. Identify λ with $2\lambda^\vee/\|\lambda^\vee\|^2$, etc. Since $\lambda + \mu + \nu = 0$ we also have by Corollary 4.6

$$\frac{\lambda^\vee}{\|\lambda^\vee\|^2} + \frac{\mu^\vee}{\|\mu^\vee\|^2} + \frac{\nu^\vee}{\|\nu^\vee\|^2} = 0$$

But the product $\|\kappa\|\,\|\kappa^\vee\|$ does not depend on κ, so

$$\|\lambda\|^2\lambda^\vee + \|\mu\|^2\mu^\vee + \|\nu\|^2\nu^\vee = 0\,.$$

Lemma 5.4 now implies that

$$(p_{\lambda,\mu} + 1)\nu^\vee + (p_{\mu,\nu} + 1)\lambda^\vee + (p_{\nu,\lambda} + 1)\mu^\vee = 0\,. \qquad \square$$

6 Representations of SL(2)

Representations of SL_2 play an important role in both Carter's and Tits' approaches to structure constants. In this section I recall briefly what is needed. Of course this is well-known material, but perhaps not in quite the form I wish to refer to.

Let

$$u = \begin{bmatrix} 1 \\ 0 \end{bmatrix}, \quad v = \begin{bmatrix} 0 \\ 1 \end{bmatrix}$$

be the standard basis of $\mathbb{C}^2$, on which $\mathrm{SL}_2(\mathbb{C})$ and $\mathfrak{sl}_2$ act. They also act on the symmetric space $S^n(\mathbb{C})$, with basis $u^k v^{n-k}$ for $0 \le k \le n$. Let π_n be this representation, which is of dimension $n+1$.

$$\begin{aligned}
\pi_n\left(\begin{bmatrix} e^x & 0 \\ 0 & e^{-x} \end{bmatrix}\right) &: u^k v^{n-k} \longmapsto e^{(2k-n)x} u^k v^{n-k} \\
\pi_n\left(\begin{bmatrix} 1 & x \\ 0 & 1 \end{bmatrix}\right) &: u^k v^{n-k} \longmapsto u^k (v + xu)^{n-k} \\
\pi_n\left(\begin{bmatrix} 1 & 0 \\ -x & 1 \end{bmatrix}\right) &: u^k v^{n-k} \longmapsto (u - xv)^k v^{n-k}\,.
\end{aligned}$$

Now to translate these formulas into those for the action of $\mathfrak{g}$. Let $w_k = u^k v^{n-k}$.

$$\begin{aligned}
\pi_n(h) &: w_k \longmapsto (2k - n)\, w_k \\
\pi_n(e_+) &: w_k \longmapsto (n - k) w_{k+1} \\
\pi_n(e_-) &: w_k \longmapsto -k w_{k-1}\,.
\end{aligned}$$

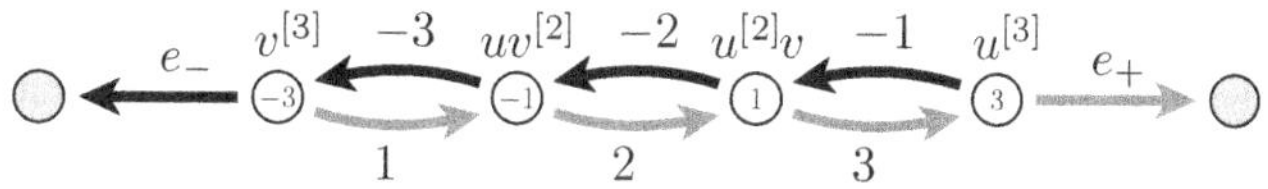

Figure 9 The representation π_3

For many reasons, it is a good idea to use a different basis of the representation. Define *divided powers*

$$u^{[k]} = \frac{u^k}{k!},$$

and set $w_{[k]} = u^{[k]} v^{[n-k]}$ (Figure 9). Then

$$\begin{aligned}\pi_n(h) &: w_{[k]} \longmapsto (2k-n)\, w_{[k]} \\ \pi_n(e_+) &: w_{[k]} \longmapsto (k+1) w_{[k+1]} \\ \pi_n(e_-) &: w_{[k]} \longmapsto -(n-k+1) w_{[k-1]}\,.\end{aligned}$$

If

$$\sigma = \begin{bmatrix} 0 & 1 \\ -1 & 0 \end{bmatrix},$$

then we also have

$$\pi_n(\sigma)\colon\ w_{[k]} \longmapsto (-1)^k w_{[n-k]}\,.$$

The π_n exhaust the irreducible finite-dimensional representations of both $\mathrm{SL}_2(\mathbb{C})$ and $\mathfrak{sl}_2$, and every finite-dimensional representation of either is a direct sum of them.

Let $d = n+1$, the dimension of π_n, assumed even. The weights of this with respect to h are

$$-d, \ldots, -1, 1, \ldots, d\,,$$

The formulas above imply immediately:

Proposition 6.1. *Suppose v to be a vector of weight -1 in this representation. Then $\pi(e_+)v$ and $\pi(\sigma)v$ are both of weight 1, and*

$$\pi(e_+)v = -(-1)^{d/2} \left(\frac{d}{2}\right) \pi(\sigma)v\,.$$

The sum of weight spaces $\mathfrak{g}_\mu$ for μ in a given λ-string is a representation of $\mathfrak{sl}_2$. The formulas laid out in this section relate closely to the numbers $p_{\lambda,\mu}$ and $q_{\lambda,\mu}$. For one thing, as the picture above suggests and is easy to verify, they tell us that we can choose basis elements e_μ for each μ in the chain so that

$$\begin{aligned}[][e_\lambda, e_\mu] &= \qquad\qquad (p_{\lambda,\mu}+1) e_{\mu+\lambda} \\ [e_{-\lambda}, e_\mu] &= (-1)^{p_{\lambda,\mu}} (q_{\lambda,\mu}+1) e_{\mu-\lambda}\,.\end{aligned}$$

The choice of basis for one chain unfortunately affects other chains as well, so this observation doesn't make the problem of structure constants trivial. But it is our first hint of a connection between structure constants and chains.

7 The extended Weyl group ...

Tits' beautiful idea is to reduce the computation of structure constants to computation in a certain extension of the Weyl group, the **extended** Weyl group, fitting into a short exact sequence

$$1 \to \{\pm 1\}^{\Delta} \to W_{\text{ext}} \to W \to 1\,.$$

If $\mathfrak{g}$ has finite dimension, the group W_{ext} can be described succinctly as the normalizer of a maximal torus in the integral form of the simply connected split group with Lie algebra $\mathfrak{g}$. For arbitrary $\mathfrak{g}$ it is a subgroup of a group G constructed in [KP85], a simply connected analogue for general Kac–Moody algebras.

I cannot resist remarking here about the literature in this field. That on Kac–Moody algebras is adequate. The primary reference is still [Kac85], but it needs updating and its exposition is dense. A useful supplement is [MP95]. For that matter, it is puzzling that there are a number of extremely basic questions about the algebras that have not yet been answered, such as whether or not they can be defined by Serre relations. This makes it very difficult to do explicit calculations with them. But the literature on the associated groups is far less satisfactory. The original paper [MT72] is very readable, but doesn't tell enough for practical purposes. There are a number of valuable expositions by Jacques Tits, such as [Tits87] and [Tits88], but many of these are difficult to obtain, and are in any event inconclusive. The paper [KP85] is exceptionally clear, but for full proofs one must refer back to earlier papers by the same authors that are not so clear. There is the book [Kum02], but it doesn't contain everything one wants. There are a number of expositions by Olivier Mathieu, but I have the impression that he stopped writing on this subject before he was through.

The good news is that we do not need to know a great deal about the Kac–Peterson group G.

(G1) There exists for each α in Δ a unique embedding

$$\varphi_\alpha\colon\ \mathrm{SL}_2(\mathbb{C}) \hookrightarrow G$$

compatible with the map φ_α of Lie algebras. *The images* $\mathrm{SL}_2^{[\alpha]}(\mathbb{C})$ *generate* G.

(G2) If π is any representation of $\mathfrak{g}$ whose restriction to $\mathfrak{sl}_2^{[\alpha]}$ is a direct sum of irreducible representations of finite dimension, it lifts to a representation I will call π_α of $SL_2(\mathbb{C})$. *There exists a unique representation of* G *that I will also call* π *such that* $\pi \circ \varphi_\alpha = \pi_\alpha$. This is true in particular if $\pi = \mathrm{ad}$.

(G3) Let ι be the embedding of $\mathbb{C}^\times$ into $\mathrm{SL}_2(\mathbb{C})$:

$$\iota\colon x \longmapsto \begin{bmatrix} x & 0 \\ 0 & 1/x \end{bmatrix}.$$

Abusing terminology slightly, for every α in Δ let $\alpha^\vee = \varphi_\alpha \circ \iota$. These give us a homomorphism

$$\varphi = \prod \alpha^\vee\colon (\mathbb{C}^\times)^\Delta \longrightarrow G\,.$$

It is an embedding, say with image H, which is its own centralizer in G. Let $H_\mathbb{Z}$ be the image of $\{\pm 1\}^\Delta$. (The notation is suggested by what happens for $\mathfrak{g}$ finite-dimensional, in which case $H_\mathbb{Z}$ is the group of integral points in a maximal torus.)

(G4) Let

$$\omega(x) = \begin{bmatrix} 0 & x \\ -1/x & 0 \end{bmatrix}$$

The normalizer $N_G(H)$ is generated by H and the elements $\omega_\alpha(x) = \varphi_\alpha\big(\omega(x)\big)$. The adjoint action of $\omega_\alpha(x)$ on $\mathfrak{h}$ is the same as that of the Weyl reflection s_α. Let

$$\sigma_\alpha = \omega_\alpha(1)\,.$$

(G5) The group W_{ext} is defined to be the subgroup of G generated by the σ_α. It fits into a short exact sequence:

$$1 \to H_\mathbb{Z} \to W_{\text{ext}} \to W \to 1\,.$$

(G6) There exists a useful cross-section $w \mapsto \widehat{w}$ of the projection from W_{ext} to W. Define $\hat{s}_\alpha$ to be σ_α. If $w = s_1 \dots s_n$ is an expression for w as a product of elementary reflections, the product

$$\widehat{w} = \hat{s}_1 \dots \hat{s}_n$$

depends only on w. Multiplication in W_{ext} is determined by the formulas

$$\widehat{s_\alpha w} = \begin{cases} \hat{s}_\alpha \widehat{w} & \text{if } \ell(s_\alpha w) = 1 + \ell(w) \\ \alpha^\vee(-I)\, \hat{s}_\alpha \widehat{w} & \text{otherwise.} \end{cases}$$

and

$$\widehat{w}\hat{s}_\alpha = \begin{cases} \widehat{w s_\alpha} & \text{if } w s_\alpha > w \\ [w\alpha^\vee](-1)\widehat{w s_\alpha} & \text{otherwise.} \end{cases}$$

The properties of the cross section were first found by Tits and proved in [Tits66b] in the case of finite-dimensional $\mathfrak{g}$. Curiously, the explicit cocycle of the extension of W by $H_{\mathbb{Z}}$ was first exhibited in [LS87].

The generalization to arbitrary Kac–Moody algebras is by Kac and Peterson.

Why is W_{ext} relevant to structure constants? The connection between $N_G(H)$ and nilpotent elements of the Lie algebra can be seen already in SL_2. In this group, the normalizer breaks up into two parts, the diagonal matrices and the inverse image M of the non-trivial element of the Weyl group, made up of the matrices $\omega(x)$ for $x \neq 0$. The matrix $\omega = \omega(x)$ satisfies

$$\omega^2 = \begin{bmatrix} 0 & x \\ -1/x & 0 \end{bmatrix}^2 = \begin{bmatrix} -1 & 0 \\ 0 & -1 \end{bmatrix}, \quad \text{hence} \quad \sigma_\alpha^{-1} = \alpha^\vee(-1)\sigma_\alpha .$$

According to the Bruhat decomposition, every element of SL_2 is either upper triangular or can be factored as $n_1 w(x) n_2$ with the n_i unipotent and upper triangular. Making this explicit:

$$\begin{bmatrix} 1 & 0 \\ x & 1 \end{bmatrix} = \begin{bmatrix} 1 & 1/x \\ 0 & 1 \end{bmatrix} \begin{bmatrix} 0 & -1/x \\ x & 0 \end{bmatrix} \begin{bmatrix} 1 & 1/x \\ 0 & 1 \end{bmatrix}$$
$$\begin{bmatrix} 0 & x \\ -1/x & 0 \end{bmatrix} = \begin{bmatrix} 1 & x \\ 0 & 1 \end{bmatrix} \begin{bmatrix} 1 & 0 \\ -1/x & 1 \end{bmatrix} \begin{bmatrix} 1 & x \\ 0 & 1 \end{bmatrix}.$$

Some easy calculating will verify further the following observation of Tits:

Lemma 7.1. *For any σ in the non-trivial coset of H in $N_G(H)$ in SL_2 there exist unique upper-triangular nilpotent e_+ and lower-triangular nilpotent e_- such that*

$$\sigma = \exp(e_+) \exp(e_-) \exp(e_+) .$$

In this equation, any one of the three determines the other two.

Applying the involution θ we see that the roles of e_+ and e_- may be reversed. In other words, *specifying an upper triangular nilpotent e_+ or a lower triangular nilpotent e_- is equivalent to specifying an element of M.*

Suppose $\lambda = w\alpha$ to be a root of $\mathfrak{g}$. If ω in W_{ext} maps to w, define $\mathrm{SL}_2^{[\lambda]}$ to be $\omega\, \mathrm{SL}_2^{[\alpha]}\, \omega^{-1}$. This group is independent of the choice of ω, although its identification with SL_2 is not. Let $H_{\mathbb{Z}}^{[\lambda]}$ be the image of the diagonal matrices in $\mathrm{SL}_2^{[\lambda]}(\mathbb{Z})$, and let $M_{\mathbb{Z}}^{[\lambda]}$ be the image of the matrices

$$\begin{bmatrix} 0 & \pm 1 \\ \mp 1 & 0 \end{bmatrix}$$

in $N_G(H)$. Each $M_{\mathbb{Z}}^{[\lambda]}$ has exactly two elements in it, corresponding to the possible choices of $e_{\pm\lambda}$. In other words, there is a well-defined map taking $\sigma \in M_{\mathbb{Z}}^{[\lambda]}$ to $e_{\lambda,\sigma}$ such that

$$\sigma = \exp(e_{\lambda,\sigma}) \exp(e_{-\lambda,\sigma}) \exp(e_{\lambda,\sigma}) .$$

Here is the whole point:

A choice of e_λ or $e_{-\lambda}$ is equivalent to a choice of σ in $M_{\mathbb{Z}}^{[\lambda]}$.

So now we can at last see the connection between structure constants and the extended Weyl group.

Lemma 7.2. *For ω in W_{ext} with image w in W we have*

$$\omega M_{\mathbb{Z}}^{[\lambda]}\omega^{-1} = M_{\mathbb{Z}}^{[w\lambda]} .$$

and furthermore for σ in $M_{\mathbb{Z}}^{[\lambda]}$

$$\mathrm{Ad}(\omega)e_{\lambda,\sigma} = e_{w\lambda,\omega\sigma\omega^{-1}} .$$

Let me now solve Problem 1. We must choose for each real root $\lambda > 0$ an element σ_λ in $M_{\mathbb{Z}}^{[\lambda]}$. We have already chosen σ_α for α in Δ. Now follow the ISL root graph to set

$$\sigma_\mu = \sigma_\alpha\sigma_\lambda\sigma_\alpha^{-1}$$

when $\lambda \xrightarrow{\alpha} \mu$ is an edge in it. This technique, of defining objects associated to nodes of a graph by finding a spanning tree in it, is common in computer algorithms.

8 ... and structure constants

How will this new scheme relate to structure constants? Suppose we are given choices of the elements σ_λ. These determine nilpotent elements $e_{\pm\lambda,\sigma_\lambda}$ etc. in $\mathfrak{sl}_2^{[\lambda]}$. Suppose $\lambda + \mu + \nu = 0$. According to our definitions,

$$[e_{\lambda,\sigma_\lambda}, e_{\mu,\sigma_\mu}] = \varepsilon(\lambda,\mu,\nu,\sigma_\lambda,\sigma_\mu,\sigma_\nu)(p_{\lambda,\mu}+1)e_{-\nu,\sigma_\nu} ,$$

for some ε factor. According to the generalization of Chevalley's theorem this factor is ± 1, but I won't assume that.

The dependence on the roots λ, μ, ν is manifest but weak, in that σ_λ determines both e_λ and $e_{-\lambda}$. In fact we don't need to take it into account at all. If $\lambda+\mu+\nu = 0$ then the only other linear combination $\pm\lambda \pm \mu \pm \nu$ that vanishes is $-\lambda-\mu-\nu$. Since $\theta(e_+) = e_-$ we know that $\varepsilon(-\lambda,-\mu,-\nu,\dots) = \varepsilon(\lambda,\mu,\nu,\dots)$, so we can just write $\varepsilon(\sigma_\lambda,\sigma_\mu,\sigma_\nu)$:

Proposition 8.1. *For real roots* λ, μ, ν *with* $\lambda + \mu + \nu = 0$ *we have*

$$[e_{\lambda,\sigma_\lambda}, e_{\mu,\sigma_\mu}] = \varepsilon(\sigma_\lambda, \sigma_\mu, \sigma_\nu)(p_{\lambda,\mu} + 1)e_{-\nu,\sigma_\nu}\,.$$

where ε *depends only on* σ_λ, σ_μ, σ_ν.

How do we compute $\varepsilon(\sigma_\lambda, \sigma_\mu, \sigma_\nu)$? Following §2.9 of [Tits66a], we get four basic rules.

Theorem 8.2. *Suppose that* λ, μ, ν *are real roots with* $\lambda + \mu + \nu = 0$. *Then*

(a) $\varepsilon(\sigma_\mu, \sigma_\lambda, \sigma_\nu) = -\varepsilon(\sigma_\lambda, \sigma_\mu, \sigma_\nu)$.
(b) $\varepsilon(\sigma_\lambda, \sigma_\mu, \sigma_\nu^{-1}) = -\varepsilon(\sigma_\lambda, \sigma_\mu, \sigma_\nu)$.
(c) *Suppose* $\|\lambda\| \ge \|\mu\|$, $\|\nu\|$. *In this case* $s_\lambda \mu = -\nu$ *and* $s_\nu = s_\lambda s_\mu s_\lambda$, *so* $\sigma_\lambda \sigma_\mu \sigma_\lambda^{-1}$ *lies in* $M_{\mathbb{Z}}^{[\nu]}$ *and satisfies*

$$\varepsilon(\sigma_\lambda, \sigma_\mu, \sigma_\lambda \sigma_\mu \sigma_\lambda^{-1}) = (-1)^{p_{\lambda,\mu}}\,.$$

(d) *The function* ε *is invariant under cyclic permutations:*

$$\varepsilon(\sigma_\lambda, \sigma_\mu, \sigma_\nu) = \varepsilon(\sigma_\mu, \sigma_\nu, \sigma_\lambda)\,.$$

Proof. (a) follows from Lemma 5.5, since $[x, y] = -[y, x]$. (b) is elementary, since $e_{\nu,\sigma_\nu^{-1}} = -e_{\nu,\sigma_\nu}$.

It is the last two results that are significant. The first tells us that there is one case in which we can calculate the constant explicitly, and the second tells us that we can manipulate any case so as to fall in this one.

For (c): Proposition 6.1 tells that in this case

$$\begin{aligned}[e_{\lambda,\sigma_\lambda}, e_{\mu,\sigma_\mu}] &= (-1)^{p_{\lambda,\mu}}(p_{\lambda,\mu} + 1)\,\mathrm{Ad}(\sigma_\lambda)e_{\mu,\sigma_\mu} \\ &= (-1)^{p_{\lambda,\mu}}(p_{\lambda,\mu} + 1)e_{s_\lambda\mu,\sigma_\lambda\sigma_\mu\sigma_\lambda^{-1}}\,.\end{aligned}$$

For (d): This reduces to the fact that the Jacobi identity has cyclic symmetry. For e_λ, e_μ, and e_ν it tells us that

$$\begin{aligned}0 &= [[e_\lambda, e_\mu], e_\nu] + [[e_\mu, e_\nu], e_\lambda] + [[e_\nu, e_\lambda], e_\mu] \\ &= \varepsilon(\sigma_\lambda, \sigma_\mu, \sigma_\nu)(p_{\lambda,\mu} + 1)[e_{-\nu}, e_\nu] \\ &\quad + \varepsilon(\sigma_\mu, \sigma_\nu, \sigma_\lambda)(p_{\mu,\nu} + 1)[e_{-\lambda}, e_\lambda] \\ &\qquad + \varepsilon(\sigma_\nu, \sigma_\lambda, \sigma_\mu)(p_{\nu,\lambda} + 1)[e_{-\mu}, e_\mu] \\ &= \varepsilon(\sigma_\lambda, \sigma_\mu, \sigma_\nu)(p_{\lambda,\mu} + 1)h_\nu \\ &\quad + \varepsilon(\sigma_\mu, \sigma_\nu, \sigma_\lambda)(p_{\mu,\nu} + 1)h_\lambda \\ &\qquad + \varepsilon(\sigma_\nu, \sigma_\lambda, \sigma_\mu)(p_{\nu,\lambda} + 1)h_\mu\,.\end{aligned}$$

But according to Proposition 5.6,

$$(p_{\lambda,\mu} + 1)h_\nu + (p_{\mu,\nu} + 1)h_\lambda + (p_{\nu,\lambda} + 1)h_\mu = 0\,.$$

The vectors h_λ and h_μ are linearly independent, so the conclusion may be deduced from the following trivial observation: if u, v, w are vectors, two of them linearly independent, and

$$\begin{aligned} au + bv + cw &= 0 \\ u + \ v + \ w &= 0, \end{aligned}$$

then $a = b = c$. □

The relevance to computation should be evident. If $\lambda + \mu + \nu = 0$, then one of the three roots will at least weakly dominate in length, and by Theorem 8.2(d) we can cycle to get the condition assumed in Lemma 5.3.

Incidentally, from these rules follows Chevalley's theorem:

Proposition 8.3. *The function ε always takes values ± 1.*

At any rate, we are now faced with the algorithmic problem: *Suppose $\lambda + \mu + \nu = 0$ with $\|\lambda\| \geq \|\mu\| = \|\nu\|$. Then $-\nu = s_\lambda \mu$ and $\sigma_\lambda \sigma_\mu \sigma_\lambda^{-1}$ lies in $M_{\mathbb{Z}}^{[\nu]}$. Is it equal to σ_ν or $\sigma_\nu^{-1} = \nu^\vee(-1)\sigma_\nu$?* If $\sigma_\lambda \sigma_\mu \sigma_\lambda^{-1} = \nu^\vee(\pm 1)\sigma_\nu$ we can deduce now:

$$[e_{\lambda,\sigma_\lambda}, e_{\mu,\sigma_\mu}] = \pm(-1)^{p_{\lambda,\mu}}(p_{\lambda,\mu} + 1)e_{\nu,\sigma_\nu}\,.$$

9 The extended root graph

To each real root λ are associated two elements $\sigma_\lambda^{\pm 1}$ in $M_{\mathbb{Z}}^{[\lambda]}$. These are the nodes of the *extended root graph*, which is a two-fold covering of the root graph itself. Make edges from each node σ to $\sigma_\alpha \sigma \sigma_\alpha^{-1}$, and from σ to $\sigma_\alpha^{-1} \sigma \sigma_\alpha$ (for $\alpha \in \Delta$). Compute these as we assign the σ_λ. This can all be calculated by using the formulas for multiplication in W_{ext} (Figure 10).

This is what we need in order to compute the edges in the extended root graph, which we use to compute all $\sigma_\lambda \sigma_\mu \sigma_\lambda^{-1}/\sigma_\nu$ for Tits triples (λ, μ, ν).

Let's look at an example, the Lie algebra $\mathfrak{sl}_3$.

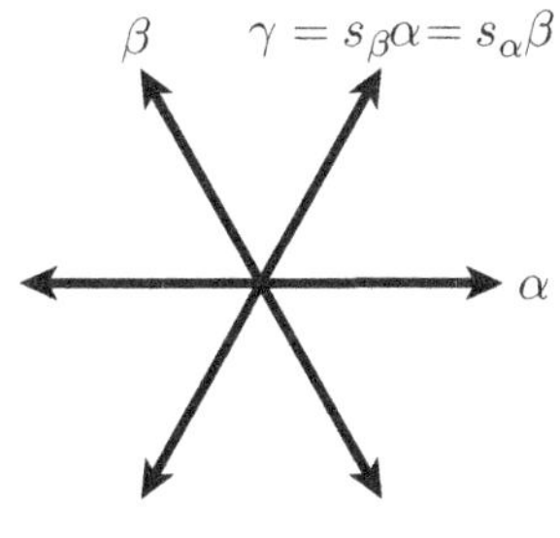

Figure 10 The root system A_2

Set

$$\sigma_\gamma = \sigma_\alpha\sigma_\beta\sigma_\alpha^{-1} = s_\alpha s_\beta \alpha^\vee(-1)\hat{s}_\alpha\hat{s}_\beta\hat{s}_\alpha = \beta^\vee(-1)\hat{s}_\gamma .$$

Thus by definition $[e_\alpha, e_\beta] = e_\gamma$.

We can also use Tits' trick to calculate $[e_\beta, e_\alpha]$. We know

$$[e_\beta, e_\alpha] = e_{\gamma,\sigma_\beta\sigma_\alpha\sigma_\beta^{-1}} .$$

But

$$\sigma_\beta\sigma_\alpha\sigma_\beta^{-1} = s_\beta s_\alpha \beta^\vee(-1)\hat{s}_\beta\hat{s}_\alpha\hat{s}_\beta = \alpha^\vee(-1)\hat{s}_\gamma = \gamma^\vee(-1)\sigma_\gamma,$$

so $[e_\beta, e_\alpha] = -e_\gamma$. Of course this just confirms what we already knew.

10 Admissible triples

At this point I have explained how to construct the σ_λ and computed as much as we want of the extended root graph.

According to Proposition 8.2, we can reduce the computation of the factors ε to the case in which the arguments come from a Tits triple (λ, μ, ν), and in this case it reduces more precisely to a calculation of a comparison of $\sigma_\lambda\sigma_\mu\sigma_\lambda^{-1}/\sigma_\nu$.

We therefore start by making a list of Tits triples, and the ISL tree can be used to do this. We start by dealing directly with all cases in which $\lambda = \alpha$ lies in Δ and $\mu > 0$, following from λ up the ISL tree. The cases where $\mu < 0$ can be dealt with at the same time, since if $\alpha + \mu = \nu$, then $-\nu + \alpha = -\mu$. Then we add the ones where λ is not in Δ. If (λ, μ, ν) is a Tits triple with $s_\alpha\lambda > 0$, then so is

$$s_\alpha(\lambda, \mu, \nu) = (s_\alpha\lambda, s_\alpha\mu, s_\alpha\nu) .$$

Thus we can compile a complete list of Tits triples by going up the ISL tree.

We must then compare $\sigma_\lambda\sigma_\mu\sigma_\lambda^{-1}$ to σ_ν for all Tits triples (λ, μ, ν). This computation may also be done inductively in the ISL tree, since

$$\sigma_\alpha\sigma_\lambda\sigma_\alpha^{-1} \cdot \sigma_\alpha\sigma_\mu\sigma_\alpha^{-1} \cdot \sigma_\alpha\sigma_\lambda^{-1}\sigma_\alpha^{-1} = \sigma_\alpha \cdot \sigma_\lambda\sigma_\mu\sigma_\lambda^{-1} \cdot \sigma_\alpha^{-1} .$$

To see exactly how this goes, fix for the moment assignments $\lambda \mapsto \sigma_\lambda$. For every triple (λ, μ, ν) with $s_\lambda\mu = \nu$ (not just Tits triples) define the factor $\tau^\vee_{\lambda,\mu,\nu}$ by the formula

$$\sigma_\lambda\sigma_\mu\sigma_\lambda^{-1} = \tau^\vee_{\lambda,\mu,\nu} \cdot \sigma_\nu .$$

The factor $\tau^\vee_{\lambda,\mu,\nu}$ will lie in $H_\nu(\mathbb{Z})$, hence will be either 1 or $\nu^\vee(-1)$. I will show in the next section how to compute these factors when λ lies in Δ. This will depend on something we haven't seen yet. But for the moment assume that the $\tau^\vee_{\alpha,\mu,\nu}$ have been calculated for all α in Δ and μ an arbitrary positive root. We can then proceed to calculate the τ-factors for all Tits triples by induction. Let

$$(\lambda_\bullet, \mu_\bullet, \nu_\bullet) = (s_\alpha\lambda, s_\alpha\mu, s_\alpha\nu) .$$

Then

$$
\begin{aligned}
\sigma_\lambda \sigma_\mu \sigma_\lambda^{-1} &= \tau^\vee_{\lambda,\mu,\nu} \cdot \sigma_\nu \\
\sigma_\alpha \sigma_\lambda \sigma_\alpha^{-1} \cdot \sigma_\alpha \sigma_\mu \sigma_\alpha^{-1} \cdot \sigma_\alpha \sigma_\lambda^{-1} \sigma_\alpha^{-1} &= \sigma_\alpha \cdot \sigma_\lambda \sigma_\mu \sigma_\lambda^{-1} \cdot \sigma_\alpha^{-1} \\
(\tau^\vee_{\alpha,\lambda,\lambda_\bullet} \cdot \sigma_{\lambda_\bullet})(\tau^\vee_{\alpha,\mu,\mu_\bullet} \cdot \sigma_{\mu_\bullet})(\tau^\vee_{\alpha,\lambda,\lambda_\bullet} \cdot \sigma_{\lambda_\bullet})^{-1} &= \sigma_\alpha \cdot \tau^\vee_{\lambda,\mu,\nu} \sigma_\nu \cdot \sigma_\alpha^{-1} \\
(\tau^\vee_{\alpha,\lambda,\lambda_\bullet} + s_{\lambda_\bullet} \tau^\vee_{\alpha,\mu,\mu_\bullet} + s_{\lambda_\bullet} s_{\mu_\bullet} s_{\lambda_\bullet} \tau^\vee_{\alpha,\lambda,\lambda_\bullet}) \cdot \sigma_{\lambda_\bullet} \sigma_{\mu_\bullet} \sigma_{\lambda_\bullet}^{-1} &= s_\alpha \tau^\vee_{\lambda,\mu,\nu} \cdot \sigma_\alpha \sigma_\nu \sigma_\alpha^{-1} \\
(\tau^\vee_{\alpha,\lambda,\lambda_\bullet} + s_{\lambda_\bullet} \tau^\vee_{\alpha,\mu,\mu_\bullet} + s_{\nu_\bullet} \tau^\vee_{\alpha,\lambda,\lambda_\bullet}) \cdot \sigma_{\lambda_\bullet} \sigma_{\mu_\bullet} \sigma_{\lambda_\bullet}^{-1} &= (s_\alpha \tau^\vee_{\lambda,\mu,\nu} + \tau^\vee_{\alpha,\nu,\nu_\bullet}) \cdot \sigma_{\nu_\bullet}
\end{aligned}
$$

leading to:

Lemma 10.1. *If (λ, μ, ν) is a Tits triple and $\lambda \neq \alpha \in \Delta$, then so is*

$$
(\lambda_\bullet, \mu_\bullet, \nu_\bullet) = (s_\alpha \lambda, s_\alpha \mu, s_\alpha \nu),
$$

and

$$
\tau^\vee_{\lambda_\bullet,\mu_\bullet,\nu_\bullet} = s_\alpha \tau^\vee_{\lambda,\mu,\nu} + \tau^\vee_{\alpha,\lambda,\lambda_\bullet} + s_{\lambda_\bullet} \tau^\vee_{\alpha,\mu,\mu_\bullet} + s_{\nu_\bullet} \tau^\vee_{\alpha,\lambda,\lambda_\bullet} + \tau^\vee_{\alpha,\nu,\nu_\bullet}.
$$

I wish this formula were more enlightening. One must conclude that the relationship between the groups $\mathrm{SL}_2^{[\lambda]}$ and Tits' cross section is complicated.

The principal conclusion of these preliminary formulas is that for both the specification of the σ_λ and the calculation of the $\sigma_\lambda \sigma_\mu \sigma_\lambda^{-1}$ we are reduced to the single calculation: *for α in Δ and $\lambda > 0$, given σ_λ how do we calculate $\sigma_\alpha \sigma_\lambda \sigma_\alpha^{-1}$?* I must explain in detail not only how calculations are made, but also how elements of $N_H(\mathbb{Z})$ are interpreted in a computer program. I have already explained the basis of calculation in the extended Weyl group.

In understanding how efficient the computation of structure constants will be, we have to know roughly how many admissible triples there are. Following [Car72] and [CMT04], I assume the positive roots to be ordered, and I define a trio of roots λ, μ, ν to be *special* if $0 < \lambda < \mu$ and $\lambda + \mu = -\nu$ is again a root. If λ, μ, ν is any triple of roots with $\lambda + \mu + \nu = 0$, then (as Carter points out) exactly one of the following twelve triples is special:

$$
\begin{gathered}
(\lambda, \mu, \nu),\ (-\lambda, -\mu, -\nu),\ (\mu, \lambda, \nu),\ (-\mu, -\lambda, -\nu) \\
(\nu, \lambda, \mu),\ (-\nu, -\lambda, -\mu),\ (\lambda, \nu, \mu),\ (-\lambda, -\nu, -\mu) \\
(\mu, \nu, \lambda),\ (-\mu, -\nu, -\lambda),\ (\nu, \mu, \lambda),\ (-\nu, -\mu, -\lambda).
\end{gathered}
$$

Hence there are at most 12 times as many Tits triples as special triples. How many special triples are there? This is independent of the ordering of Σ^+, since it is one-half the number of pairs of positive roots λ, μ with $\lambda + \mu$ also a root. In [CMT04] it is asserted that for all classical systems the number is $O(n^3)$, where n is the rank of the system. Don Taylor has given me the following more precise table:

System	Number of special triples
A_n	$n(n^2-1)/6$
B_n	$n(n-1)(2n-1)/3$
C_n	$n(n-1)(2n-1)/3$
D_n	$2n(n-1)(n-2)/3$
E_6	120
E_7	336
E_8	1120
F_4	68
G_2	5

11 Some details of computation

For this section I am going to simplify notation. Every element of the extended Weyl group may be represented uniquely as $t^\vee(-1)\cdot\widehat{w}$, where $t^\vee$ is in $X_*(H) = L_{\Delta^\vee}$ and w is in W. I will express it as just $t^\vee\cdot\widehat{w}$, and of course it is only $t^\vee$ modulo $2X_*(H)$ that counts. Also, I will refer to the group operation in $L_{\Delta^\vee}$ additively.

Proposition 11.1. *Suppose α to be in Δ, $\lambda \neq \alpha > 0$. Then*

$$\hat{s}_\alpha\hat{s}_\lambda\hat{s}_\alpha = \begin{cases} \widehat{s_{s_\alpha\lambda}} & \textit{if } \langle\alpha,\lambda^\vee\rangle < 0 \\ \alpha^\vee\cdot\hat{s}_\lambda & \langle\alpha,\lambda^\vee\rangle = 0 \\ (\alpha^\vee + s_\alpha s_\lambda\alpha^\vee)\cdot\widehat{s_{s_\alpha\lambda}} & \langle\alpha,\lambda^\vee\rangle > 0. \end{cases}$$

Proof. A preliminary calculation:

$$\begin{aligned} s_\lambda\alpha &= \alpha - \langle\alpha,\lambda^\vee\rangle\lambda \\ s_\alpha s_\lambda\alpha &= -\alpha - \langle\alpha,\lambda^\vee\rangle s_\alpha\lambda \\ &= -\alpha - \langle\alpha,\lambda^\vee\rangle(\lambda - \langle\lambda,\alpha^\vee\rangle\alpha) \\ &= -\alpha - \langle\alpha,\lambda^\vee\rangle\lambda + \langle\alpha,\lambda^\vee\rangle\langle\lambda,\alpha^\vee\rangle\alpha \\ &= -\langle\alpha,\lambda^\vee\rangle\lambda + \big(\langle\alpha,\lambda^\vee\rangle\langle\lambda,\alpha^\vee\rangle - 1\big)\alpha. \end{aligned}$$

Recall that by Corollary 4.4 the product $\langle\alpha,\lambda^\vee\rangle\langle\lambda,\alpha^\vee\rangle \geq 0$. Recall also that $ws_\alpha > w$ if and only if $w\alpha > 0$.

(a) $\langle\alpha,\lambda^\vee\rangle < 0$. Here $s_\lambda\alpha > 0$ and $s_\alpha s_\lambda\alpha > 0$ so $s_\lambda < s_\alpha s_\lambda < s_\alpha s_\lambda s_\alpha$, and $\ell(s_\alpha s_\lambda s_\alpha) = \ell(s_\lambda) + 2$, and

$$\hat{s}_\alpha\hat{s}_\lambda\hat{s}_\alpha = \widehat{s_\alpha s_\lambda s_\alpha} = \widehat{s_{s_\alpha\lambda}}\,.$$

(b) $\langle \alpha, \lambda^\vee \rangle = 0$. Here $s_\alpha \lambda = \lambda$, and $\hat{s}_\lambda \hat{s}_\alpha = \widehat{s_\lambda s_\alpha}$, but $s_\alpha(s_\lambda s_\alpha) = s_\lambda$ so

$$\hat{s}_\alpha \hat{s}_\lambda \hat{s}_\alpha = \hat{s}_\alpha \widehat{s_\lambda s_\alpha} = \alpha^\vee \cdot \hat{s}_\lambda \,.$$

(c) $\langle \alpha, \lambda^\vee \rangle > 0$. Here one sees easily that $s_\lambda \alpha < 0$. But since $\lambda \neq \alpha$ we also have $s_\alpha s_\lambda \alpha < 0$ also. So $\ell(s_{s_\alpha \lambda}) = \ell(s_\lambda) - 2$.

$$\begin{aligned}
\hat{s}_\alpha \hat{s}_\lambda &= \alpha^\vee \cdot \widehat{s_\alpha s_\lambda} \\
\hat{s}_\alpha \hat{s}_\lambda \hat{s}_\alpha &= \alpha^\vee \cdot \widehat{s_\alpha s_\lambda} \hat{s}_\alpha \\
&= \alpha^\vee \cdot \widehat{s_{s_\alpha \lambda}} \cdot \alpha^\vee \cdot 1 \\
&= (\alpha^\vee + s_\alpha s_\lambda \alpha^\vee) \cdot \widehat{s_{s_\alpha \lambda}} \,.
\end{aligned}$$

□

Corollary 11.2. *Suppose $\sigma_\lambda = t_\lambda^\vee \cdot \hat{s}_\lambda$. Then for $\alpha \neq \lambda$*

$$\sigma_\alpha \sigma_\lambda \sigma_\alpha^{-1} = \begin{cases} (s_\alpha t_\lambda^\vee + s_\alpha s_\lambda \alpha^\vee) \cdot \widehat{s_{s_\alpha \lambda}} & \langle \alpha, \lambda^\vee \rangle < 0 \\ (s_\alpha t_\lambda^\vee) \cdot \widehat{s_{s_\alpha \lambda}} & \langle \alpha, \lambda^\vee \rangle = 0 \\ (s_\alpha t_\lambda^\vee + \alpha^\vee) \widehat{s_{s_\alpha \lambda}} & \langle \alpha, \lambda^\vee \rangle > 0. \end{cases}$$

Keep in mind that the reflection associated to $s_\alpha \lambda$ is $s_\alpha s_\lambda s_\alpha$.

Proof. We start with

$$\sigma_\alpha \sigma_\lambda \sigma_\alpha^{-1} = \hat{s}_\alpha t_\lambda^\vee \hat{s}_\lambda \alpha^\vee \hat{s}_\alpha = (s_\alpha t_\lambda^\vee + s_\alpha s_\lambda \alpha^\vee) \cdot \hat{s}_\alpha \hat{s}_\lambda \hat{s}_\alpha \,.$$

Apply the Proposition. □

There are three cases, according to whether $\langle \alpha, \lambda^\vee \rangle$ is $<$, $=$, or $>$ 0. These correspond to how the length $\ell(s_\alpha s_\lambda s_\alpha)$ relates to $\ell(s_\lambda)$. So now finally we can compute the factors $\tau^\vee_{\alpha,\mu,s_\alpha \mu}$, comparing $\sigma_\alpha \sigma_\mu \sigma_\alpha^{-1}$ to $\sigma_{s_\alpha \mu}$.

Lemma 11.3. *Suppose $s_\alpha \mu = \nu$. If $\sigma_\alpha \sigma_\mu \sigma_\alpha^{-1} = t^\vee \hat{s}_\nu$ and $\sigma_\nu = t_\nu^\vee \hat{s}_\nu$, then*

$$\tau^\vee_{\alpha,\mu,\nu} = t^\vee + t_\nu^\vee \,.$$

12 Patterns in the computation

As a consequence of the main theorem of [BH93], Bob Howlett proved that the root graph of an arbitrary Coxeter group is described by a finite automaton. What this means is that paths in the root graph are the same as paths in a certain finite state machine. The machine, although finite, can be quite large. As Figure 11 indicates, there are 31 states in the machine producing paths in the root graph of the root system we have seen earlier in Figure 1. (In this diagram, nodes give rise to equivalent states if and only if the subsequent paths out of them are equivalent.

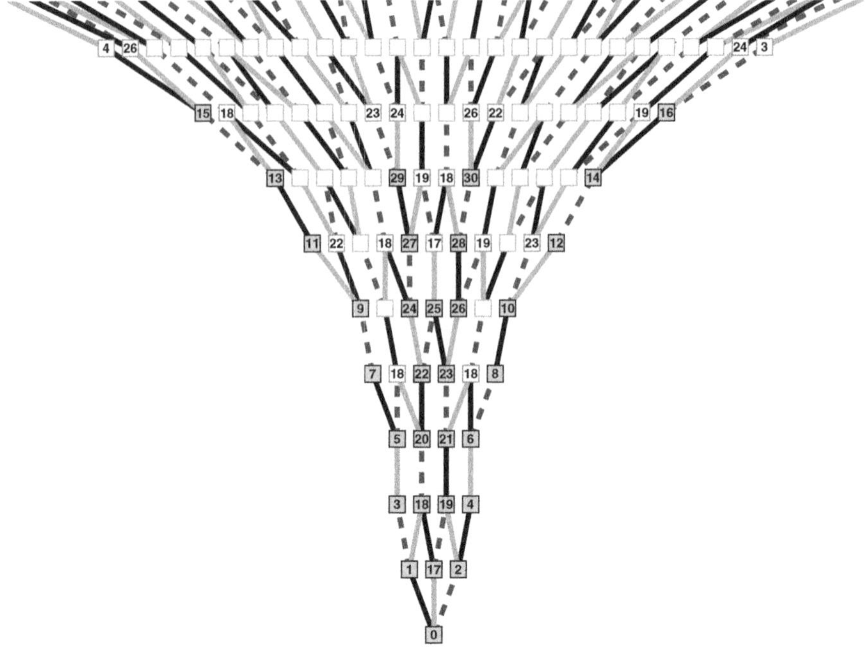

Figure 11 The finite state machine for the root graph of an infinite root system

Each state is noted by a unique shaded box. An unshaded one with the same label signifies another occurrence. The labeled boxes are all one needs to specify the finite state machine.)

One thing evident in this picture is the repetition of the weak Bruhat order of the finite groups W_Θ. (Lemma 5.2 of [Cas06] explains this.) This as well as regularity should have some significance for the extended root graph as well, but I don't yet completely understand what is going on, and I will not discuss this topic here.

References

[BH93] Brigitte Brink and Robert Howlett, A finiteness property and an automatic structure for Coxeter groups, *Mathematische Annalen* **296** (1993), 179–190.

[Cas06] Bill Casselman, Computations in Coxeter groups II. Constructing minimal roots, *Representation Theory* **12** (2008), 260–293.

[Car72] Roger Carter, *Simple groups of Lie type*, Wiley, 1972.

[Che55] Claude Chevalley, Sur certains groupes simples, *Tohoku Mathematics Journal* **48** (1955), 14–66.

[CMT04] Arjeh Cohen, Scott Murray, and Don Taylor, Computing in groups of Lie type, *Mathematics of Computation* **73** (2004), 1477–1498.

[FK80] Igor Frenkel and Victor Kac, Affine Lie algebras and dual resonance models, *Inventiones Mathematicae* **62** (1980), 23–66.

[Kac85] Victor Kac, *Infinite Dimensional Lie Algebras* (second edition). Cambridge University Press, Cambridge, New York, 1985.

[KP85] V. G. Kac et D. H. Peterson, 'Defining relations of certain infinite dimensional groups', in *Élie Cartan et les mathématiques d'aujourd'hui*, *Astérisque*, volume hors série (1985), 165–208.

[Kum02] Shrawan Kumar, *Kac–Moody Groups, Their Flag Varieties, and Representation Theory*, Birkhäuser, Boston, 2002.

[LS87] R. P. Langlands and D. Shelstad, On the definition of transfer factors, *Mathematische Annalen* **278** (1987), 219–271.

[MP95] Robert Moody and Arturo Pianzola, *Lie Algebras with Triangular Decompositions*, J. Wiley, New York, 1995.

[MT72] Robert Moody and K. L. Teo, Tits' systems with crystallographic Weyl groups, *Journal of Algebra* **21** (1972), 178–190.

[Mor87] Jun Morita, Commutator relations in Kac–Moody groups, *Proceedings of the Japanese Academy* **63** (1987), 21–22.

[Ryl00] L. J. Rylands, Fast calculation of structure constants, preprint, 2000.

[Sat51] Ichiro Satake, On a theorem of Cartan, *Journal of the Mathematical Society of Japan* **2** (1951), 284–305.

[Tits66a] Jacques Tits, Sur les constants de structure et le théorème d'existence des algèbres de Lie semi-simple, *Publications de l'I. H. E. S.* **31** (1966), 21–58.

[Tits66b] ——, Normalisateurs de tores I. Groupes de Coxeter étendus, *Journal of Algebra* **4** (1966), 96–116.

[Tits68] ——, Le problème de mots dans les groupes de Coxeter, *Symposia Math.* **1** (1968), 175–185.

[Tits87] ——, Uniqueness and presentation of Kac–Moody groups over fields, *Journal of Algebra* **105** (1987), 542–573.

[Tits88] ——, Groupes associés aux algèbres de Kac–Moody, *Séminaire Bourbaki*, exposé 700 (1988).

The Gelfand–Zeitlin integrable system and K-orbits on the flag variety

Mark Colarusso and Sam Evens

To Nolan Wallach, on the occasion of his 70th birthday, with gratitude and admiration

Abstract In this paper, we provide an overview of the Gelfand–Zeitlin integrable system on the Lie algebra of $n \times n$ complex matrices $\mathfrak{gl}(n, \mathbb{C})$ introduced by Kostant and Wallach in 2006. We discuss results concerning the geometry of the set of strongly regular elements, which consists of the points where the Gelfand–Zeitlin flow is Lagrangian. We use the theory of $K_n = GL(n-1, \mathbb{C}) \times GL(1, \mathbb{C})$-orbits on the flag variety $\mathcal{B}_n$ of $GL(n, \mathbb{C})$ to describe the strongly regular elements in the nilfiber of the moment map of the system. We give an overview of the general theory of orbits of a symmetric subgroup of a reductive algebraic group acting on its flag variety, and illustrate how the general theory can be applied to understand the specific example of K_n and $GL(n, \mathbb{C})$.

Keywords: Flag variety • Symmetric subgroup • Nilpotent matrices • Integrable systems • Gelfand–Zeitlin theory

Mathematics Subject Classification: 20G20, 14M15, 14L30, 70H06, 17B08, 37J35

The work by the second author was partially supported by NSA grants H98230-08-0023 and H98230-11-1-0151.

M. Colarusso
Department Mathematical Sciences, University of Wisconsin-Milwaukee, Milwaukee, WI 53201-0413, USA
e-mail: colaruss@uwm.edu

S. Evens
Department of Mathematics, University of Notre Dame, Notre Dame, IN 46556, USA
e-mail: sevens@nd.edu

R. Howe et al. (eds.), *Symmetry: Representation Theory and Its Applications*, Progress in Mathematics 257, DOI 10.1007/978-1-4939-1590-3_5

1 Introduction

In a series of papers [24, 25], Kostant and Wallach study the action of an abelian Lie group $A \cong \mathbb{C}^{\frac{n(n-1)}{2}}$ on $\mathfrak{g} = \mathfrak{gl}(n, \mathbb{C})$. The Lie algebra $\mathfrak{a}$ of A is the abelian Lie algebra of Hamiltonian vector fields of the Gelfand–Zeitlin[1] collection of functions $J_{\mathrm{GZ}} := \{f_{i,j} : i = 1, \ldots, n,\ j = 1, \ldots, i\}$ (see Section 2 for precise notation). The set of functions J_{GZ} is Poisson commutative, and its restriction to each regular adjoint orbit in $\mathfrak{g}$ forms an integrable system. For each function in the collection, the corresponding Hamiltonian vector field on $\mathfrak{g}$ is complete, and the action of A on $\mathfrak{g}$ is given by integrating the action of $\mathfrak{a}$.

Kostant and Wallach consider a Zariski open subset $\mathfrak{g}_{\mathrm{sreg}}$ of $\mathfrak{g}$, which consists of all elements $x \in \mathfrak{g}$ such that the differentials of the functions J_{GZ} are linearly independent at x. Elements of $\mathfrak{g}_{\mathrm{sreg}}$ are called strongly regular, and Kostant and Wallach show that $\mathfrak{g}_{\mathrm{sreg}}$ is exactly the set of regular elements x of $\mathfrak{g}$ such that the orbit $A \cdot x$ is Lagrangian in the adjoint orbit of x. In [7, 9], the first author determined the A-orbits in $\mathfrak{g}_{\mathrm{sreg}}$ through explicit computations. We denote by $\Phi : \mathfrak{g} \to \mathbb{C}^{\frac{n(n+1)}{2}}$ the map given by $\Phi(x) = (f_{i,j}(x))$, and note that in [7, 9], the most subtle and interesting case is the nilfiber $\Phi^{-1}(0)$.

The Gelfand–Zeitlin functions are defined using a sequence of projections $\pi_i : \mathfrak{gl}(i, \mathbb{C}) \to \mathfrak{gl}(i-1, \mathbb{C})$ given by mapping an $i \times i$ matrix y to its $(i-1) \times (i-1)$ submatrix in the upper left hand corner. Our paper [11] exploits the fact that each projection π_i is equivariant with respect to the action of $GL(i-1, \mathbb{C})$ on $\mathfrak{gl}(i, \mathbb{C})$ by conjugation, where $GL(i-1, \mathbb{C})$ is embedded in the top left hand corner of $GL(i, \mathbb{C})$ in the natural way. In particular, we use the theory of $GL(i-1, \mathbb{C})$-orbits on the flag variety $\mathcal{B}_i$ of $\mathfrak{gl}(i, \mathbb{C})$ for $i = 1, \ldots, n$, to provide a more conceptual understanding of the A-orbits in the nilfiber. In addition, we prove that every Borel subalgebra contains strongly regular elements, and hope to develop these methods in order to better understand the topology of $\mathfrak{g}_{\mathrm{sreg}}$.

In this paper, we review results of Kostant, Wallach, and the first author, and then explain how to use the theory of $GL(i-1, \mathbb{C})$-orbits on $\mathcal{B}_i$ in order to derive the results from [11]. In Section 2, we recall the basic symplectic and Poisson geometry needed to construct the Gelfand–Zeitlin integrable system. We then discuss the work of Kostant and Wallach in constructing the system and the action of the group A, and the work of the first author in describing the A-orbit structure of $\mathfrak{g}_{\mathrm{sreg}}$. In Section 3, we give an overview of our results from [11] and sketch some of the proofs. In Section 4, we review the rich theory of orbits of a symmetric subgroup K on the flag variety $\mathcal{B}$ of a reductive group G, as developed by Richardson, Springer, and others. In particular, we show explicitly how the theory applies if $K = GL(n-1, \mathbb{C}) \times GL(1, \mathbb{C})$ and $G = GL(n, \mathbb{C})$, and we hope this section will make the general theory of K-orbits more accessible to researchers interested in applying this theory.

---

[1]Alternate spellings of Zeitlin include Cetlin, Tsetlin, Tzetlin, and Zetlin. In this paper, we follow the convention from our earlier work.

It would be difficult to overstate the influence of Nolan Wallach on the work discussed in this paper. We look forward to further stimulating interactions with Nolan in the future, and note that our plans for developing this work may well depend on utilizing completely different work of Nolan than that discussed here.

2 The Gelfand–Zeitlin integrable system on $\mathfrak{gl}(n, \mathbb{C})$

2.1 Integrable systems

In this section, we give a brief discussion of integrable systems. For further details, we refer the reader to [1, 2]. We denote by M an analytic (respectively smooth) manifold with holomorphic (resp. smooth) functions $\mathcal{H}(M)$.

Let (M, ω) be a $2n$-dimensional symplectic manifold with symplectic form $\omega \in \wedge^2 T^* M$. For $f \in \mathcal{H}(M)$, we let ξ_f be the unique vector field such that

$$df(Y) = \omega(Y, \xi_f), \tag{1}$$

for all vector fields Y on M. The vector field ξ_f is called the *Hamiltonian* vector field of f. We can use these vector fields to give $\mathcal{H}(M)$ the structure of a Poisson algebra with Poisson bracket:

$$\{f, g\} := \omega(\xi_f, \xi_g), \tag{2}$$

for f, $g \in \mathcal{H}(M)$. That is $\{\cdot, \cdot\}$ makes $\mathcal{H}(M)$ into a Lie algebra and $\{\cdot, \cdot\}$ satisfies a Leibniz rule with respect to the associative multiplication of $\mathcal{H}(M)$.

To define an integrable system on (M, ω), we need the following notion.

Definition 2.1. We say the functions $\{F_1, \ldots, F_r\} \subset \mathcal{H}(M)$ are *independent* if the open set $U = \{m \in M : (dF_1)_m \wedge \cdots \wedge (dF_r)_m \neq 0\}$ is dense in M.

Definition 2.2. Let (M, ω) be a $2n$-dimensional symplectic manifold. An *integrable system* on M is a collection of n independent functions $\{F_1, \ldots, F_n\} \subset \mathcal{H}(M)$ such that $\{F_i, F_j\} = 0$ for all i, j.

Remark 2.3. This terminology originates in Hamiltonian mechanics. In that context, (M, ω, H) is a phase space of a classical Hamiltonian system with n degrees of freedom and Hamiltonian function $H \in \mathcal{H}(M)$ (the total energy of the system). The trajectory of the Hamiltonian vector field ξ_H describes the time evolution of the system. If we are given an integrable system $\{F_1 = H, \ldots, F_n\}$, then this trajectory can be found using only the operations of function integration and function inversion ([1], Section 4.2). Such a Hamiltonian system is said to be integrable by quadratures.

Integrable systems are important in Lie theory, because they are useful in geometric constructions of representations through the theory of quantization

[14, 19] (see Remark 2.12 below). For example, integrable systems provide a way to construct polarizations of symplectic manifolds (M, ω). By a polarization, we mean an integrable subbundle of the tangent bundle $P \subset TM$ such that each of the fibers $P_m \subset T_m(M)$ is Lagrangian, i.e., $P_m = P_m^{\perp}$, where $P_m^{\perp}$ is the annihilator of P_m with respect to the symplectic form ω_m on $T_m(M)$. A submanifold $S \subset (M, \omega)$ is said to be Lagrangian if $T_m(S)$ is Lagrangian for each $m \in S$, so that the leaves of a polarization are Lagrangian submanifolds of M. The existence of a polarization is a crucial ingredient in constructing a geometric quantization of M (for M a real manifold) (see for example [39]), and Lagrangian submanifolds are also important in the study of deformation quantization (see for example [28]).

To see how an integrable system on (M, ω) gives rise to a polarization, we consider the moment map of the system $\{F_1, \dots, F_n\}$:

$$\mathbf{F} : M \to K^n,\ \mathbf{F}(m) = (F_1(m), \dots, F_n(m)) \text{ for } m \in M, \tag{3}$$

where $K = \mathbb{R}$ or $\mathbb{C}$. Let $U = \{m \in M : (dF_1)_m \wedge \cdots \wedge (dF_n)_m \neq 0\}$ and let $P \subset TU$ be $P = \text{span}\{\xi_{F_i} : i = 1, \dots, n\}$. Then P is a polarization of the symplectic manifold $(U, \omega|_U)$ whose leaves are the connected components of the level sets of $\mathbf{F}|_U$, i.e., the regular level sets of $\mathbf{F}$. Indeed, if $S \subset U$ is a regular level set of $\mathbf{F}$, then $\dim S = \dim M - n = n$. It then follows that for $m \in S$, $T_m(S) = \text{span}\{(\xi_{F_i})_m : i = 1, \dots, n\}$, since the vector fields $\xi_{F_1}, \dots, \xi_{F_n}$ are tangent to S and independent on U. Thus, $T_m(S)$ is isotropic by Equation (2) and of dimension $\dim S = n = \frac{1}{2} \dim U$, so that $T_m(S)$ is Lagrangian.

2.2 *Poisson manifolds and the Lie–Poisson structure*

To study integrable systems in Lie theory, we need to consider not only symplectic manifolds, but Poisson manifolds. We briefly review some of the basic elements of Poisson geometry here. For more detail, we refer the reader to [37] and [1].

A Poisson manifold $(M, \{\cdot,\cdot\})$ is an analytic (resp. smooth) manifold where the functions $\mathcal{H}(M)$ have the structure of a Poisson algebra with Poisson bracket $\{\cdot,\cdot\}$. For example, any symplectic manifold is a Poisson manifold where the Poisson bracket of functions is given by Equation (2). For a Poisson manifold $(M, \{\cdot,\cdot\})$, the Hamiltonian vector field for $f \in \mathcal{H}(M)$ is given by

$$\xi_f(g) := \{f, g\}, \tag{4}$$

where $g \in \mathcal{H}(M)$. In the case where (M, ω) is symplectic, it is easy to see that this definition of the Hamiltonian vector field of f agrees with the definition given in Equation (1).

If we have two Poisson manifolds $(M_1, \{\cdot,\cdot\}_1)$ and $(M_2, \{\cdot,\cdot\}_2)$, an analytic (resp. smooth) map $\Phi : M_1 \to M_2$ is said to be *Poisson* if

$$\{f \circ \Phi, g \circ \Phi\}_1 = \{f, g\}_2 \circ \Phi, \tag{5}$$

for $f, g \in \mathcal{H}(M_2)$. That is, $\Phi^* : \mathcal{H}(M_2) \to \mathcal{H}(M_1)$ is a homomorphism of Poisson algebras. Equivalently, for $f \in \mathcal{H}(M_2)$,

$$\Phi_* \xi_{\Phi^* f} = \xi_f. \tag{6}$$

In particular, a submanifold $(S, \{\cdot,\cdot\}_S) \subset (M, \{\cdot,\cdot\}_M)$ with Poisson structure $\{\cdot,\cdot\}_S$ is said to be a *Poisson submanifold* of $(M, \{\cdot,\cdot\}_M)$ if the inclusion $i : S \hookrightarrow M$ is Poisson.

In general, Poisson manifolds $(M, \{\cdot,\cdot\})$ are not symplectic, but they are foliated by symplectic submanifolds called *symplectic leaves*. Consider the (singular) distribution on M given by

$$\chi(M) := \operatorname{span}\{\xi_f : f \in \mathcal{H}(M)\}. \tag{7}$$

The distribution $\chi(M)$ is called the *characteristic distribution* of $(M, \{\cdot,\cdot\})$. Using the Jacobi identity for the Poisson bracket $\{\cdot,\cdot\}$, one computes that

$$[\xi_f, \xi_g] = \xi_{\{f,g\}}, \tag{8}$$

so that the distribution $\chi(M)$ is involutive. Using a general version of the Frobenius theorem, one can then show that $\chi(M)$ is integrable and the leaves $(S, \{\cdot,\cdot\}_S)$ are Poisson submanifolds of $(M, \{\cdot,\cdot\})$, where the Poisson bracket $\{\cdot,\cdot\}_S$ is induced by a symplectic form ω_S on S as in Equation (2). For further details, see [37], Chapter 2.

Let $\mathfrak{g}$ be a reductive Lie algebra over $\mathbb{R}$ or $\mathbb{C}$ and let G be any connected Lie group with Lie algebra $\mathfrak{g}$. Let $\beta(\cdot,\cdot)$ be a nondegenerate, G-invariant bilinear form on $\mathfrak{g}$. Then $\mathfrak{g}$ has the structure of a Poisson manifold, which we call the Lie–Poisson structure. If $f \in \mathcal{H}(\mathfrak{g})$, we can use the form β to identify the differential $df_x \in T_x^*(\mathfrak{g}) = \mathfrak{g}^*$ at $x \in \mathfrak{g}$ with an element $\nabla f(x) \in \mathfrak{g}$. The element $\nabla f(x)$ is determined by its pairing against $z \in \mathfrak{g} \cong T_x(\mathfrak{g})$ by the formula,

$$\beta(\nabla f(x), z) = \frac{d}{dt}\bigg|_{t=0} f(x + tz) = df_x(z). \tag{9}$$

We then define a Poisson bracket on $\mathcal{H}(\mathfrak{g})$ by

$$\{f, g\}(x) = \beta(x, [\nabla f(x), \nabla g(x)]). \tag{10}$$

It can be shown that this definition of the Poisson structure on $\mathfrak{g}$ is independent of the choice of form β in the sense that a different form gives rise to an isomorphic Poisson manifold structure on $\mathfrak{g}$.

From (10) it follows that

$$(\xi_f)_x = [x, \nabla f(x)] \in T_x(\mathfrak{g}) = \mathfrak{g}. \tag{11}$$

For $x \in \mathfrak{g}$, let $G \cdot x$ denote its adjoint orbit. From Equation (11), it follows that the fiber of the characteristic distribution of $(\mathfrak{g}, \{\cdot,\cdot\})$ at x is

$$\chi(\mathfrak{g})_x = \{[x, y] : y \in \mathfrak{g}\} = T_x(G \cdot x).$$

One can then show that the symplectic leaves of $(\mathfrak{g}, \{\cdot,\cdot\})$ are the adjoint orbits of G on $\mathfrak{g}$ with the canonical Kostant–Kirillov–Souriau (KKS) symplectic structure (see for example [6], Proposition 1.3.21). Since $G \cdot x \subset \mathfrak{g}$ is a Poisson submanifold, it follows from Equations (5) and (6) that

$$\{f, g\}_{\mathrm{LP}}|_{G\cdot x} = \{f|_{G\cdot x}, g|_{G\cdot x}\}_{\mathrm{KKS}} \text{ and } \xi_f^{\mathrm{LP}}|_{G\cdot x} = \xi_{f|_{G\cdot x}}^{\mathrm{KKS}} \tag{12}$$

for $f, g \in \mathcal{H}(\mathfrak{g})$, where the Poisson bracket and Hamiltonian field on the left side of the equations are defined using the Lie–Poisson structure, and on the right side they are defined using the KKS symplectic structure as in Section 2.1.

This description of the symplectic leaves allows us to easily identify the Poisson central functions of $(\mathfrak{g}, \{\cdot,\cdot\})$. We call a function $f \in \mathcal{H}(\mathfrak{g})$ a *Casimir* function if $\{f, g\} = 0$ for all $g \in \mathcal{H}(\mathfrak{g})$. Clearly, f is a Casimir function if and only if $\xi_f = 0$. Equation (12) implies this occurs if and only if $df|_{G\cdot x} = 0$ for every $x \in \mathfrak{g}$, since each $G \cdot x$ is symplectic. Thus, the Casimir functions for the Lie–Poisson structure on $\mathfrak{g}$ are precisely the $\mathrm{Ad}(G)$-invariant functions, $\mathcal{H}(\mathfrak{g})^G$.

The symplectic leaves of $(\mathfrak{g}, \{\cdot,\cdot\})$ of maximal dimension play an important role in our discussion. For $x \in \mathfrak{g}$, let $\mathfrak{z}_{\mathfrak{g}}(x)$ denote the centralizer of x. We call an element $x \in \mathfrak{g}$ *regular* if $\dim \mathfrak{z}_{\mathfrak{g}}(x) = \mathrm{rank}(\mathfrak{g})$ is minimal [23]. The orbit $G \cdot x$ then has maximum possible dimension, i.e., $\dim(G \cdot x) = \dim \mathfrak{g} - \mathrm{rank}(\mathfrak{g})$.

2.3 *Construction of the Gelfand–Zeitlin integrable system on* $\mathfrak{gl}(n, \mathbb{C})$

Let $\mathfrak{g} = \mathfrak{gl}(n, \mathbb{C})$ and let $G = GL(n, \mathbb{C})$. Then $\mathfrak{g}$ is reductive with nondegenerate, invariant form $\beta(x, y) = \mathrm{tr}(xy)$, where $\mathrm{tr}(xy)$ denotes the trace of the matrix xy for $x, y \in \mathfrak{g}$. Thus, $\mathfrak{g}$ is a Poisson manifold with the Lie–Poisson structure. In this section, we construct an independent, Poisson commuting family of functions on $\mathfrak{g}$, whose restriction to each regular adjoint orbit $G \cdot x$ forms an integrable system in the sense of Definition 2.2. We refer to this family of functions as the Gelfand–Zeitlin integrable system on $\mathfrak{g}$. The family is constructed using Casimir functions for certain Lie subalgebras of $\mathfrak{g}$ and extending these functions to Poisson commuting functions on all of $\mathfrak{g}$.

We consider the following Lie subalgebras of $\mathfrak{g}$. For $i = 1, \ldots, n-1$, we embed $\mathfrak{gl}(i, \mathbb{C})$ into $\mathfrak{g}$ in the upper left corner and denote its image by $\mathfrak{g}_i$. That is to say, $\mathfrak{g}_i = \{x \in \mathfrak{g} : x_{k,j} = 0, \text{ if } k > i \text{ or } j > i\}$. Let $G_i \subset GL(n, \mathbb{C})$ be the corresponding closed subgroup. If $\mathfrak{g}_i^{\perp}$ denotes the orthogonal complement of $\mathfrak{g}_i$ with respect to the form β, then $\mathfrak{g} = \mathfrak{g}_i \oplus \mathfrak{g}_i^{\perp}$. Thus, the restriction of the form β to $\mathfrak{g}_i$ is nondegenerate, so we can use it to define the Lie–Poisson structure of $\mathfrak{g}_i$ via Equation (10). We have a natural projection $\pi_i : \mathfrak{g} \to \mathfrak{g}_i$ given by $\pi_i(x) = x_i$, where x_i is the upper left $i \times i$ corner of x, that is, $(x_i)_{k,j} = x_{k,j}$ for $1 \leq k, j \leq i$ and is zero otherwise. The following lemma is the key ingredient in the construction of the Gelfand–Zeitlin integrable system on $\mathfrak{g}$.

Lemma 2.4. *The projection $\pi_i : \mathfrak{g} \to \mathfrak{g}_i$ is Poisson with respect to the Lie–Poisson structures on $\mathfrak{g}$ and $\mathfrak{g}_i$.*

Proof. Since the Poisson brackets on $\mathcal{H}(\mathfrak{g})$ and $\mathcal{H}(\mathfrak{g}_i)$ satisfy the Leibniz rule, it suffices to show Equation (5) for linear functions $\lambda_x, \mu_y \in \mathcal{H}(\mathfrak{g}_i)$, where $\lambda_x(z) = \beta(x,z)$ and $\mu_y(z) = \beta(y,z)$ for $x, y, z \in \mathfrak{g}_i$. This is an easy computation using the definition of the Lie–Poisson structure in Equation (10) and the decomposition $\mathfrak{g} = \mathfrak{g}_i \oplus \mathfrak{g}_i^{\perp}$. □

Let $\mathbb{C}[\mathfrak{g}]$ denote the algebra of polynomial functions on $\mathfrak{g}$. Let

$$J(n) := \langle \pi_1^*(\mathbb{C}[\mathfrak{g}]^{G_1}), \ldots, \pi_{n-1}^*(\mathbb{C}[\mathfrak{g}_{n-1}]^{G_{n-1}}), \mathbb{C}[\mathfrak{g}]^G \rangle \tag{13}$$

be the associative subalgebra of $\mathbb{C}[\mathfrak{g}]$ generated by $\pi_i^*(\mathbb{C}[\mathfrak{g}_i]^{G_i})$ for $i \leq n-1$ and $\mathbb{C}[\mathfrak{g}]^G$.

Proposition 2.5. *The algebra $J(n)$ is a Poisson commutative subalgebra of $\mathbb{C}[\mathfrak{g}]$.*

Proof. The proof proceeds by induction on n, the case $n = 1$ being trivial. Suppose that $J(n-1)$ is Poisson commutative. Then

$$J(n) = \langle \pi_{n-1}^*(J(n-1)), \mathbb{C}[\mathfrak{g}]^G \rangle$$

is the associative algebra generated by $\pi_{n-1}^*(J(n-1))$ and $\mathbb{C}[\mathfrak{g}]^G$. By Lemma 2.4, $\pi_{n-1}^*(J(n-1))$ is Poisson commutative, and the elements of $\mathbb{C}[\mathfrak{g}]^G$ are Casimir functions, so that $J(n)$ is Poisson commutative. □

Remark 2.6. It can be shown that the algebra $J(n)$ is a maximal Poisson commutative subalgebra of $\mathbb{C}[\mathfrak{g}]$ ([24], Theorem 3.25).

The Gelfand–Zeitlin integrable system is obtained by choosing a set of generators for the algebra $J(n)$. We note that the map $\pi_i : \mathfrak{g} \to \mathfrak{g}_i$ is surjective, so that we can identify $\mathbb{C}[\mathfrak{g}_i]^{G_i}$ with its image $\pi_i^*(\mathbb{C}[\mathfrak{g}_i]^{G_i})$. Let $\mathbb{C}[\mathfrak{g}_i]^{G_i} = \mathbb{C}[f_{i,1}, \ldots, f_{i,i}]$, where $f_{i,j}(x) = \operatorname{tr}(x_i^j)$ for $j = 1, \ldots, i$. Then the functions

$$J_{\mathrm{GZ}} := \{f_{i,j} : i = 1, \ldots, n,\ j = 1, \ldots, i\} \tag{14}$$

generate the algebra $J(n)$ as an associative algebra. We claim that J_{GZ} is an algebraically independent, Poisson commuting set of functions whose restriction to each regular $G \cdot x$ forms an integrable system.

By Proposition 2.5, the functions J_{GZ} Poisson commute. To see that the functions J_{GZ} are algebraically independent, we study the following morphisms:

$$\Phi_i : \mathfrak{g}_i \to \mathbb{C}^i, \; \Phi_i(y) = (f_{i,1}(y), \ldots, f_{i,i}(y)),$$

for $i = 1, \ldots, n$. We define the Kostant–Wallach map to be the morphism

$$\Phi : \mathfrak{g} \to \mathbb{C}^{\binom{n+1}{2}} \text{ given by } \Phi(x) = (\Phi_1(x_1), \ldots, \Phi_i(x_i), \ldots, \Phi_n(x_n)). \tag{15}$$

For $z \in \mathfrak{g}_i$, let $\sigma_i(z)$ equal the collection of i eigenvalues of z counted with repetitions, where here we regard z as an $i \times i$ matrix.

Remark 2.7. If $x, y \in \mathfrak{g}$, then $\Phi(x) = \Phi(y)$ if and only if $\sigma_i(x_i) = \sigma_i(y_i)$ for $i = 1, \ldots, n$. This follows from the fact that $\mathbb{C}[\mathfrak{g}_i]^{G_i} = \mathbb{C}[f_{i,1}, \ldots, f_{i,i}] = \mathbb{C}[p_{i,1}, \ldots, p_{i,i}]$, where $p_{i,j}$ is the coefficient of t^{j-1} in the characteristic polynomial of x_i thought of as an $i \times i$ matrix. In particular, $\Phi(x) = (0, \ldots, 0)$ if and only if x_i is nilpotent for $i = 1, \ldots, n$.

Kostant and Wallach produce a cross-section to the map Φ using the (upper) Hessenberg matrices. For $1 \leq i, j \leq n$, let $E_{i,j} \in \mathfrak{g}$ denote the elementary matrix with 1 in the (i, j)-th entry and zero elsewhere. Let $\mathfrak{b}_+ \subset \mathfrak{g}$ be the standard Borel subalgebra of upper triangular matrices and let $e = \sum_{i=2}^{n} E_{i,i-1}$ be the sum of the negative simple root vectors. We call elements of the affine variety $e + \mathfrak{b}$ (upper) Hessenberg matrices:

$$e + \mathfrak{b} = \begin{bmatrix} a_{11} & a_{12} & \cdots & a_{1n-1} & a_{1n} \\ 1 & a_{22} & \cdots & a_{2n-1} & a_{2n} \\ 0 & 1 & \cdots & a_{3n-1} & a_{3n} \\ \vdots & \vdots & \ddots & \vdots & \vdots \\ 0 & 0 & \cdots & 1 & a_{nn} \end{bmatrix}.$$

Kostant and Wallach prove the following remarkable fact ([24], Theorem 2.3).

Theorem 2.8. *The restriction of the Kostant–Wallach map $\Phi|_{e+\mathfrak{b}} : e+\mathfrak{b} \to \mathbb{C}^{\binom{n+1}{2}}$ to the Hessenberg matrices $e + \mathfrak{b}$ is an isomorphism of algebraic varieties.*

Remark 2.9. For $x \in \mathfrak{g}$, let $\mathcal{R}(x) = \{\sigma_1(x_1), \ldots, \sigma_i(x_i), \ldots, \sigma_n(x)\}$ be the collection of $\binom{n+1}{2}$-eigenvalues of $x_1, \ldots, x_i, \ldots, x$ counted with repetitions. The numbers $\mathcal{R}(x)$ are called the Ritz values of x and play an important role in numerical linear algebra (see for example [29, 30]). In this language, Theorem 2.8 says that any $\binom{n+1}{2}$-tuple of complex numbers can be the Ritz values of an $x \in \mathfrak{g}$ and that there is a unique Hessenberg matrix having those numbers as Ritz values. Contrast this with the Hermitian case in which the necessarily real eigenvalues of x_i must interlace those of x_{i-1} (see for example [21]). This discovery has led to some new work on Ritz values by linear algebraists [29, 34].

Theorem 2.8 suggests the following definition from [24].

Definition 2.10. We say that $x \in \mathfrak{g}$ is strongly regular if the differentials $\{(df_{i,j})_x : i = 1,\ldots,n,\ j = 1,\ldots,i\}$ are linearly independent. We denote the set of strongly regular elements of $\mathfrak{g}$ by $\mathfrak{g}_{\rm sreg}$.

By Theorem 2.8, $e + \mathfrak{b} \subset \mathfrak{g}_{\rm sreg}$, and since $\mathfrak{g}_{\rm sreg}$ is Zariski open, it is dense in both the Zariski topology and the Hausdorff topology on $\mathfrak{g}$ [27]. Hence, the polynomials J_{GZ} in (14) are independent. For $c \in \mathbb{C}^{\binom{n+1}{2}}$, let $\Phi^{-1}(c)_{\rm sreg} := \Phi^{-1}(c) \cap \mathfrak{g}_{\rm sreg}$ denote the strongly regular elements of the fiber $\Phi^{-1}(c)$. It follows from Theorem 2.8 that $\Phi^{-1}(c)_{\rm sreg}$ is nonempty for any $c \in \mathbb{C}^{\binom{n+1}{2}}$.

By a well-known result of Kostant [23], if x is strongly regular, then $x_i \in \mathfrak{g}_i$ is regular for all i. We state several equivalent characterizations of strong regularity.

Proposition 2.11 ([24], Proposition 2.7 and Theorem 2.14). *The following statements are equivalent.*

(i) *x is strongly regular.*
(ii) *The tangent vectors $\{(\xi_{f_{i,j}})_x;\ i = 1,\ldots,n-1,\ j = 1,\ldots,i\}$ are linearly independent.*
(iii) *The elements $x_i \in \mathfrak{g}_i$ are regular for all $i = 1,\ldots,n$ and $\mathfrak{z}_{\mathfrak{g}_i}(x_i) \cap \mathfrak{z}_{\mathfrak{g}_{i+1}}(x_{i+1}) = 0$ for $i = 1,\ldots,n-1$, where $\mathfrak{z}_{\mathfrak{g}_i}(x_i)$ denotes the centralizer of x_i in $\mathfrak{g}_i$.*

To see that the restriction of the functions J_{GZ} to a regular adjoint orbit $G \cdot x$ forms an integrable system, we first observe that $G \cdot x \cap \mathfrak{g}_{\rm sreg} \neq \emptyset$ for any regular x. This follows from the fact that any regular matrix is conjugate to a companion matrix, which is Hessenberg and therefore strongly regular. Note that the functions $f_{n,1},\ldots,f_{n,n}$ restrict to constant functions on $G \cdot x$, so we only consider the restrictions of $\{f_{i,j} : i = 1,\ldots,n-1,\ j = 1,\ldots,i\}$. Let $q_{i,j} = f_{i,j}|_{G\cdot x}$ for $i = 1,\ldots,n-1$, $j = 1,\ldots,i$ and let $U = G \cdot x \cap \mathfrak{g}_{\rm sreg}$. Then U is open and dense in $G \cdot x$. By Equation (12), part (ii) of Proposition 2.11 and Proposition 2.5 imply respectively that the functions $\{q_{i,j} : i = 1,\ldots,n-1,\ j = 1,\ldots i\}$ are independent and Poisson commute on U. Observe that there are

$$\sum_{i=1}^{n-1} i = \frac{n(n-1)}{2} = \frac{\dim(G \cdot x)}{2}$$

such functions. Hence, they form an integrable system on regular $G \cdot x$.

It follows from our work in Section 2.1 that the connected components of the regular level sets of the moment map

$$y \to (q_{1,1}(y),\ldots,q_{i,j}(y),\ldots,q_{n-1,n-1}(y))$$

are the leaves of a polarization of $G \cdot x \cap \mathfrak{g}_{\rm sreg}$. It is easy to see that such regular level sets coincide with certain strongly regular fibers of the Kostant–Wallach map, namely the fibers $\Phi^{-1}(c)_{\rm sreg}$ where $c = (c_1,\ldots,c_n)$, $c_i \in \mathbb{C}^i$ with $c_n = \Phi_n(x)$ (see Equation (15)). This follows from Proposition 2.11 and the fact that regular matrices which have the same characteristic polynomial are conjugate (see Remark 2.7).

We therefore turn our attention to studying the geometry of the strongly regular set $\mathfrak{g}_{\text{sreg}}$ and Lagrangian submanifolds $\Phi^{-1}(c)_{\text{sreg}}$ of regular $G \cdot x$.

Remark 2.12. The Gelfand–Zeitlin system described here can be viewed as a complexification of the one introduced by Guillemin and Sternberg [19] on the dual to the Lie algebra of the unitary group. They show that the Gelfand–Zeitlin integrable system on $\mathfrak{u}(n)^*$ is a geometric version of the classical Gelfand–Zeitlin basis for irreducible representations of $U(n)$ [18]. More precisely, they construct a geometric quantization of a regular, integral coadjoint orbit of $U(n)$ on $\mathfrak{u}(n)^*$ using the polarization from the Gelfand–Zeitlin integrable system and show that the resulting quantization is isomorphic to the corresponding highest weight module for $U(n)$ using the Gelfand–Zeitlin basis for the module.

There is strong empirical evidence (see [17]) that the quantum version of the complexified Gelfand–Zeitlin system is the category of Gelfand–Zeitlin modules studied by Drozd, Futorny, and Ovsienko [13]. These are Harish-Chandra modules for the pair $(U(\mathfrak{g}), \Gamma)$, where $\Gamma \subset U(\mathfrak{g})$ is the Gelfand–Zeitlin subalgebra of the universal enveloping algebra $U(\mathfrak{g})$ [16]. It would be interesting to produce such modules geometrically using the geometry of the complex Gelfand–Zeitlin system developed below and deformation quantization.

2.4 *Integration of the Gelfand–Zeitlin system and the group A*

We can study the Gelfand–Zeitlin integrable system on $\mathfrak{gl}(n, \mathbb{C})$ and the structure of the fibers $\Phi^{-1}(c)_{\text{sreg}}$ by integrating the corresponding Hamiltonian vector fields to a holomorphic action of $\mathbb{C}^{\binom{n}{2}}$ on $\mathfrak{g}$. The first step is the following observation.

Theorem 2.13. *Let $f_{i,j}(x) = \operatorname{tr}(x_i^j)$ for $i = 1, \ldots, n-1$, $j = 1, \ldots, i$. Then the Hamiltonian vector field $\xi_{f_{i,j}}$ is complete on $\mathfrak{g}$ and integrates to a holomorphic action of $\mathbb{C}$ on $\mathfrak{g}$ whose orbits are given by*

$$t_{i,j} \cdot x := \operatorname{Ad}(\exp(t_{i,j}\, j x_i^{j-1})) \cdot x, \tag{16}$$

for $x \in \mathfrak{g}$, $t_{i,j} \in \mathbb{C}$.

Proof. Denote the right side of Equation (16) by $\theta(t_{i,j}, x)$. We show that $\theta'(t_{i,j}, x) = (-\xi_{f_{i,j}})_{\theta(t_{i,j},x)}$ for any $t_{i,j} \in \mathbb{C}$, so that $\theta(-t_{i,j}, x)$ is an integral curve of the vector field $\xi_{f_{i,j}}$. For the purposes of this computation, replace the variable $t_{i,j}$ by the variable t. Then

$$\begin{aligned} \left.\frac{d}{dt}\right|_{t=t_0} \operatorname{Ad}(\exp(t\, j x_i^{j-1})) \cdot x &= \operatorname{ad}(j x_i^{j-1}) \cdot \operatorname{Ad}(\exp(t_0\, j x_i^{j-1})) \cdot x \\ &= \operatorname{ad}(j x_i^{j-1}) \cdot \theta(t_0, x). \end{aligned}$$

Clearly, $\exp(t_0 j x_i^{j-1})$ centralizes x_i, so that $\theta(t_0, x)_i = x_i$. This implies

$$\mathrm{ad}(jx_i^{j-1}) \cdot \theta(t_0, x) = \mathrm{ad}(j(\theta(t_0, x)_i)^{j-1}) \cdot \theta(t_0, x).$$

Now it is easily computed that $\nabla f_{i,j}(y) = jy_i^{j-1}$ for any $y \in \mathfrak{g}$. Thus, Equation (11) implies that

$$\mathrm{ad}(j(\theta(t_0, x)_i)^{j-1}) \cdot \theta(t_0, x) = -(\xi_{f_{i,j}})_{\theta(t_0,x)}. \qquad \square$$

We now consider the Lie algebra of Gelfand–Zeitlin vector fields

$$\mathfrak{a} := \mathrm{span}\{\xi_{f_{i,j}} : i = 1, \ldots, n-1,\ j = 1, \ldots, i\}. \tag{17}$$

By Equation (8), $\mathfrak{a}$ is an abelian Lie algebra, and since $\mathfrak{g}_{\mathrm{sreg}}$ is nonempty, $\dim \mathfrak{a} = \binom{n}{2}$, by (ii) of Proposition 2.11. Let A be the corresponding simply connected Lie group, so that $A \cong \mathbb{C}^{\binom{n}{2}}$. We take as coordinates on A,

$$\underline{t} = (\underline{t}_1, \ldots, \underline{t}_i, \ldots, \underline{t}_{n-1}) \in \mathbb{C} \times \cdots \times \mathbb{C}^i \times \cdots \times \mathbb{C}^{n-1} = \mathbb{C}^{\binom{n}{2}},$$

where $\underline{t}_i \in \mathbb{C}^i$ with $\underline{t}_i = (t_{i,1}, \ldots, t_{i,i})$, with $t_{i,j} \in \mathbb{C}$ for $i = 1, \ldots, n-1$, $j = 1, \ldots, i$. Since $\mathfrak{a}$ is abelian the actions of the various $t_{i,j}$ given in Equation (16) commute. Thus, we can define an action of A on $\mathfrak{g}$ by composing the actions of the various $t_{i,j}$ in any order. For $a = (\underline{t}_1, \ldots, \underline{t}_{n-1}) \in A$, $a \cdot x$ is given by the formula

$$\begin{aligned} a \cdot x = \mathrm{Ad}(\exp(t_{1,1})) \cdot \ldots \cdot \mathrm{Ad}(\exp(jt_{i,j} x_i^{j-1})) \cdot \ldots \\ \cdot \mathrm{Ad}(\exp((n-1)t_{n-1,n-1} x_{n-1}^{n-2})) \cdot x. \end{aligned} \tag{18}$$

Theorem 2.13 shows that this action integrates the action of $\mathfrak{a}$ on $\mathfrak{g}$, so that

$$T_x(A \cdot x) = \mathrm{span}\{(\xi_{f_{i,j}})_x : i = 1, \ldots, n-1,\ j = 1, \ldots, i\}. \tag{19}$$

Since the functions J_{GZ} Poisson commute, it follows from Equation (12) that $A \cdot x \subset G \cdot x$ is isotropic with respect to the KKS symplectic structure on $G \cdot x$. Note also that Equation (4) implies that $\xi_{f_{i,j}} f_{k,l} = 0$ for any i, j and k, l. It follows that $f_{k,l}$ is invariant under the flow of $\xi_{f_{i,j}}$ for any i, j and therefore is invariant under the action of A given in Equation (18). Thus, the action of A preserves the fibers of the Kostant–Wallach map Φ defined in Equation (15).

It follows from Equation (19) and Part (ii) of Proposition 2.11 that $x \in \mathfrak{g}_{\mathrm{sreg}}$ if and only if $\dim(A \cdot x) = \binom{n}{2}$, which holds if and only if $A \cdot x \subset G \cdot x$ is Lagrangian in regular $G \cdot x$. Thus, the group A acts on the strongly regular fibers $\Phi^{-1}(c)_{\mathrm{sreg}}$ and its orbits form the connected components of the Lagrangian submanifold $\Phi^{-1}(c)_{\mathrm{sreg}} \subset G \cdot x$. Hence, the leaves of the polarization of a regular adjoint orbit $G \cdot x$ constructed from the Gelfand–Zeitlin integrable system are exactly the A-orbits on $G \cdot x \cap \mathfrak{g}_{\mathrm{sreg}}$. Moreover, there are only finitely many A-orbits in $\Phi^{-1}(c)_{\mathrm{sreg}}$.

Theorem 2.14 ([24], Theorem 3.12). *Let $c \in \mathbb{C}^{\binom{n+1}{2}}$ and let $\Phi^{-1}(c)_{\mathrm{sreg}}$ be a strongly regular fiber of the Kostant–Wallach map. Then $\Phi^{-1}(c)_{\mathrm{sreg}}$ is a smooth algebraic variety of dimension $\binom{n}{2}$ whose irreducible components in the Zariski topology coincide with the orbits of A on $\Phi^{-1}(c)_{\mathrm{sreg}}$.*

Remark 2.15. Our definition of the Gelfand–Zeitlin integrable system involved choosing the specific set of algebraically independent generators J_{GZ} for the algebra $J(n)$ in Equation (13). However, it can be shown that if we choose another algebraically independent set of generators, J'_{GZ}, then their restriction to each regular adjoint orbit $G \cdot x$ forms an integrable system, and the corresponding Hamiltonian vector fields are complete and integrate to an action of a holomorphic Lie group A' whose orbits coincide with those of A, [24], Theorem 3.5. Our particular choice of generators J_{GZ} is to facilitate the easy integration of the Hamiltonian vector fields ξ_f, $f \in J_{\mathrm{GZ}}$ in Theorem 2.13.

2.5 *Analysis of the A-action on* $\Phi^{-1}(c)_{\mathrm{sreg}}$

Kostant and Wallach [24] studied the action of A on a special set of regular semisimple elements in $\mathfrak{g}$ defined by:

$$\mathfrak{g}_{\Omega} = \{x \in \mathfrak{g} : x_i \text{ is regular semisimple and } \sigma_i(x_i) \cap \sigma_{i+1}(x_{i+1}) = \emptyset \text{ for all } i\}. \tag{20}$$

Let $\Omega = \Phi(\mathfrak{g}_{\Omega}) \subset \mathbb{C}^{\binom{n+1}{2}}$. By Remark 2.7, we have $\mathfrak{g}_{\Omega} = \Phi^{-1}(\Omega)$. In [24], the authors show that the action of A is transitive on the fibers $\Phi^{-1}(c)$ for $c \in \Omega$ and that these fibers are $\binom{n}{2}$-dimensional tori.

Theorem 2.16 ([24], Theorems 3.23 and 3.28). *The elements of $\mathfrak{g}_{\Omega}$ are strongly regular, so that $\Phi^{-1}(c) = \Phi^{-1}(c)_{\mathrm{sreg}}$ for $c \in \Omega$. Moreover, $\Phi^{-1}(c)$ is a homogeneous space for a free algebraic action of the torus $(\mathbb{C}^{\times})^{\binom{n}{2}}$ and therefore is precisely one A-orbit.*

Remark 2.17. An analogous Gelfand–Zeitlin integrable system exists for complex orthogonal Lie algebras $\mathfrak{so}(n, \mathbb{C})$. One can also show that this system integrates to a holomorphic action of $\mathbb{C}^d$ on $\mathfrak{so}(n, \mathbb{C})$, where d is half the dimension of a regular adjoint orbit in $\mathfrak{so}(n, \mathbb{C})$. One can then prove the analogue of Theorem 2.16 for $\mathfrak{so}(n, \mathbb{C})$. We refer the reader to [8] for details.

The thesis of the first author generalizes Theorem 2.16 to an arbitrary fiber $\Phi^{-1}(c)_{\mathrm{sreg}}$ for $c \in \mathbb{C}^{\binom{n+1}{2}}$ (see [7]). The methods used differ from those used to prove Theorem 2.16, but the idea originates in some unpublished work of Wallach, who used a similar strategy to describe the A-orbit structure of the set $\mathfrak{g}_\Omega$. We briefly outline this strategy, which can be found in detail in [9], Section 4. The key observation is that the vector field $\xi_{f_{i,j}}$ acts via Equation (16) by the centralizer $Z_{G_i}(x_i)$ of x_i in G_i. The problem is that the group $Z_{G_i}(x_i)$ is difficult to describe for arbitrary x_i, so that the formula for the A-action in Equation (18) is too difficult to use directly. However, if $x \in \mathfrak{g}_{\text{sreg}}$ and J_i is the Jordan canonical form of x_i, then the group $Z_i := Z_{G_i}(J_i)$ is easy to describe, since $x_i \in \mathfrak{g}_i$ is regular for $i = 1, \ldots, n$ by (iii) of Proposition 2.11. Further, for $x \in \Phi^{-1}(c)_{\text{sreg}}$, x_i is in a fixed regular conjugacy class for $i = 1, \ldots, n$. This allows us to construct morphisms $\Phi^{-1}(c)_{\text{sreg}} \to G_i$, given by $x \to g_i(x)$, where $\text{Ad}(g_i(x)^{-1})\cdot x = J_i$, with J_i a fixed Jordan matrix (depending only on $\Phi^{-1}(c)_{\text{sreg}}$). We can then use these morphisms to define a free algebraic action of the group $Z := Z_1 \times \cdots \times Z_{n-1}$ on $\Phi^{-1}(c)_{\text{sreg}}$ such that the Z-orbits coincide with the A-orbits. The action of Z is given by

$$
\begin{aligned}
(z_1, \ldots, z_{n-1}) \cdot x = \text{Ad}(g_1(x) z_1 g_1(x)^{-1}) \cdot \ldots \cdot \text{Ad}(g_i(x) z_i g_i(x)^{-1}) \cdot \ldots \\
\cdot \text{Ad}(g_{n-1}(x) z_{n-1} g_{n-1}(x)^{-1}) \cdot x,
\end{aligned}
\tag{21}
$$

where $z_i \in Z_i$ for $i = 1, \ldots, n-1$ and $x \in \Phi^{-1}(c)_{\text{sreg}}$, (cf. Equation (18)).

The action of the group Z in Equation (21) is much easier to work with than the action of A in Equation (18) and allows us to understand the structure of an arbitrary fiber $\Phi^{-1}(c)_{\text{sreg}}$. The first observation is that we can enlarge the set of elements on which the action of A is transitive on the fibers of the Kostant–Wallach map from the set $\mathfrak{g}_\Omega$ to the set $\mathfrak{g}_\Theta$ defined by

$$\mathfrak{g}_\Theta = \{x \in \mathfrak{g} : \sigma_i(x_i) \cap \sigma_{i+1}(x_{i+1}) = \emptyset\}.$$

Let $\Theta = \Phi(\mathfrak{g}_\Theta)$. Note that by Remark 2.7, $\Phi^{-1}(\Theta) = \mathfrak{g}_\Theta$.

Theorem 2.18 ([9], Theorem 5.15). *The elements of $\mathfrak{g}_\Theta$ are strongly regular. If $c \in \Theta$, then $\Phi^{-1}(c) = \Phi^{-1}(c)_{\text{sreg}}$ is a homogeneous space for a free algebraic action of the group $Z = Z_1 \times \cdots \times Z_{n-1}$ given in Equation (21), and thus is exactly one A-orbit. Moreover, $\mathfrak{g}_\Theta$ is the maximal subset of $\mathfrak{g}$ for which the action of A is transitive on the fibers of Φ.*

For general fibers the situation becomes more complicated.

Theorem 2.19 ([9], Theorem 5.11). *Let $x \in \mathfrak{g}_{\text{sreg}}$ be such that there are j_i distinct eigenvalues in common between x_i and x_{i+1} for $1 \le i \le n-1$, and let $c = \Phi(x)$. Then there are exactly 2^j A-orbits in $\Phi^{-1}(c)_{\text{sreg}}$, where $j = \sum_{i=1}^{n-1} j_i$. The orbits of A on $\Phi^{-1}(c)_{\text{sreg}}$ coincide with the orbits of a free algebraic action of the group $Z = Z_1 \times \cdots \times Z_{n-1}$ defined on $\Phi^{-1}(c)_{\text{sreg}}$ in Equation (21).*

Remark 2.20. After the proof of Theorem 2.19 was established in [7], a similar result appeared in an interesting paper of Bielwaski and Pidstrygach [3]. Their arguments are independent and completely different from ours. It would be interesting to study the relation between the two different approaches to establishing the result of Theorem 2.19.

We highlight a special case of Theorem 2.19, which we will investigate in much greater detail below in Section 3.

Corollary 2.21. *Consider the strongly regular nilfiber*

$$\Phi^{-1}(0)_{\mathrm{sreg}} := \Phi^{-1}(0, \ldots, 0)_{\mathrm{sreg}}.$$

Then there are exactly 2^{n-1} *A-orbits in* $\Phi^{-1}(0)_{\mathrm{sreg}}$. *These orbits coincide with the orbits of a free algebraic action of* $(\mathbb{C}^{\times})^{n-1} \times \mathbb{C}^{\binom{n}{2}-n+1}$ *on* $\Phi^{-1}(0)_{\mathrm{sreg}}$.

Proof. The first statement follows immediately from Remark 2.7 and Theorem 2.19. For the second statement, we observe that in this case the group

$$Z = Z_{G_1}(e_1) \times \cdots \times Z_{G_{n-1}}(e_{n-1}),$$

where $e_i \in \mathfrak{g}_i$ is the principal nilpotent Jordan matrix. It follows that $Z = (\mathbb{C}^{\times})^{n-1} \times \mathbb{C}^{\binom{n}{2}-n+1}$. □

Theorem 2.19 gives a complete description of the local structure of the Lagrangian foliation of regular adjoint orbits of $\mathfrak{g}$ by the Gelfand–Zeitlin integrable system and shows the system is locally algebraically integrable, giving natural algebraic "angle coordinates" coming from the action of the group $Z = Z_1 \times \cdots \times Z_{n-1}$. However, Theorem 2.19 does not say anything about the global nature of the foliation. Motivated by Theorem 2.19, we would like to extend the local Z-action on $\Phi^{-1}(c)_{\mathrm{sreg}}$ given in (21) to larger subvarieties of $\mathfrak{g}$. However, this is not possible, except in certain special cases. The definition of the Z-action uses the fact that the Jordan form of each x_i for $i = 1, \ldots, n-1$ is fixed on the fiber $\Phi^{-1}(c)_{\mathrm{sreg}}$. The problem with trying to extend this action is that there is in general no morphism on a larger variety which assigns to x_i its Jordan form. The issue is that the ordered eigenvalues of a matrix are not in general algebraic functions of the matrix entries.

For the set $\mathfrak{g}_{\Omega}$, Kostant and Wallach resolve this issue by producing an étale covering $\mathfrak{g}_{\Omega}(\mathfrak{e})$ of $\mathfrak{g}_{\Omega}$ on which the eigenvalues are algebraic functions [25]. They then lift the Lie algebra $\mathfrak{a}$ of Gelfand–Zeitlin vector fields in Equation (17) to the covering where they integrate to an algebraic action of the torus $(\mathbb{C}^{\times})^{\binom{n}{2}}$. In our paper [10], we extend this to the full strongly regular set using the theory of decomposition classes [4] and Poisson reduction [15].

3 The geometry of the strongly regular nilfiber

In recent work [11], we take a very different approach to describing the geometry of $\mathfrak{g}_{\mathrm{sreg}}$ by studying the Borel subalgebras that contain elements of $\mathfrak{g}_{\mathrm{sreg}}$. We develop a new connection between the orbits of certain symmetric subgroups K_i on the flag varieties of $\mathfrak{g}_i$ for $i = 2, \ldots, n$ and the Gelfand–Zeitlin integrable system on $\mathfrak{g}$.

We use this connection to prove that every Borel subalgebra of $\mathfrak{g}$ contains strongly regular elements, and we determine explicitly the Borel subalgebras which contain elements of the strongly regular nilfiber $\Phi^{-1}(0)_{\mathrm{sreg}} = \Phi^{-1}(0,\ldots,0)_{\mathrm{sreg}}$. We show that there are 2^{n-1} such Borel subalgebras, and that the subvarieties of regular nilpotent elements of these Borel subalgebras are the 2^{n-1} irreducible components of $\Phi^{-1}(0)_{\mathrm{sreg}}$ given in Corollary 2.21. This description of the nilfiber is much more explicit than the one given in Corollary 2.21, since the $Z = (\mathbb{C}^{\times})^{n-1} \times \mathbb{C}^{\binom{n}{2}-n+1}$-action of Equation (21) is not easy to compute explicitly. We refer the reader to our paper [11] for proofs of the results of this section.

3.1 K-orbits and $\Phi^{-1}(0)_{\mathrm{sreg}}$

We begin by considering the strongly regular nilfiber $\Phi^{-1}(0)_{\mathrm{sreg}}$ of the Kostant–Wallach map. By Remark 2.7 and (iii) of Proposition 2.11, we note that $x \in \Phi^{-1}(0)_{\mathrm{sreg}}$ if and only if the following two conditions are satisfied for every $i = 2,\ldots,n$:

$$\begin{aligned} &\text{(a)} \quad x_{i-1}, x_i \text{ are regular nilpotent.} \\ &\text{(b)} \quad \mathfrak{z}_{\mathfrak{g}_{i-1}}(x_{i-1}) \cap \mathfrak{z}_{\mathfrak{g}_i}(x_i) = 0. \end{aligned} \tag{22}$$

We proceed by finding the Borel subalgebras in $\mathfrak{g}_i$ which contain elements satisfying (a) and (b), and we then use these Borel subalgebras to construct the Borel subalgebras of $\mathfrak{g}$ which contain elements of $\Phi^{-1}(0)_{\mathrm{sreg}}$.

Let $K_i := GL(i-1,\mathbb{C}) \times GL(1,\mathbb{C}) \subset GL(i,\mathbb{C})$ be the group of invertible block diagonal matrices with an $(i-1)\times(i-1)$ block in the upper left corner and a 1×1 block in the lower right corner. Let $\mathcal{B}_i$ be the flag variety of $\mathfrak{g}_i$. Then K_i acts on $\mathcal{B}_i$ by conjugation with finitely many orbits (see for example [35]). We observe that the conditions (a) and (b) in (22) are $\mathrm{Ad}(K_i)$-equivariant. Thus, the problem of finding the Borel subalgebras of $\mathfrak{g}_i$ containing elements satisfying these conditions reduces to the problem of studying the conditions for a representative in each K_i-orbit. In this section, we find all K_i-orbits Q_i through Borel subalgebras containing such elements, and in the process reveal some new facts about the geometry of K_i-orbits on $\mathcal{B}_i$. In the following sections, we explain how to link the orbits Q_i together for $i = 2,\ldots,n$ to produce the Borel subalgebras of $\mathfrak{g}$ that contain elements of $\Phi^{-1}(0)_{\mathrm{sreg}}$ and use these Borel subalgebras to study the geometry of the fiber $\Phi^{-1}(0)_{\mathrm{sreg}}$.

For concreteness, let us fix $i = n$, so that $K_n = GL(n-1,\mathbb{C}) \times GL(1,\mathbb{C})$ and $\mathcal{B}_n$ is the flag variety of $\mathfrak{gl}(n,\mathbb{C})$. For $\mathfrak{b} \in \mathcal{B}_n$, let $K_n \cdot \mathfrak{b}$ denote the K_n-orbit through $\mathfrak{b}$. We analyze each of the conditions in (22) in turn.

Theorem 3.1 ([11], Proposition 3.6). *Suppose $x \in \mathfrak{g}$ satisfies condition* (a) *in* (22) *and that $x \in \mathfrak{b}$, with $\mathfrak{b} \subset \mathfrak{g}$ a Borel subalgebra of $\mathfrak{g}$. Then $\mathfrak{b} \in Q$, where Q is a closed K_n-orbit.*

Theorem 3.1 follows from a stronger result. The group K_n is the group of fixed points of the involution θ on G, where $\theta(g) = cgc^{-1}$ with $c = \mathrm{diag}[1, \ldots, 1, -1]$. Let $\mathfrak{k}_n = \mathrm{Lie}(K_n)$, so that $\mathfrak{k}_n$ is the Lie algebra of block diagonal matrices $\mathfrak{k}_n = \mathfrak{gl}(n-1, \mathbb{C}) \oplus \mathfrak{gl}(1, \mathbb{C})$. Then $\mathfrak{g} = \mathfrak{k}_n \oplus \mathfrak{p}_n$, where $\mathfrak{p}_n$ is the -1-eigenspace for the involution θ on $\mathfrak{g}$. Let $\pi_{\mathfrak{k}_n} : \mathfrak{g} \to \mathfrak{k}_n$ be the projection of $\mathfrak{g}$ onto $\mathfrak{k}_n$ along $\mathfrak{p}_n$, and let $\mathcal{N}_{\mathfrak{k}_n}$ be the nilpotent cone in $\mathfrak{k}_n$.

Theorem 3.2 ([11], Theorem 3.7). *Let $\mathfrak{b} \subset \mathfrak{g}$ be a Borel subalgebra and let $\mathfrak{n} = [\mathfrak{b}, \mathfrak{b}]$, with $\mathfrak{n}^{\mathrm{reg}}$ the regular nilpotent elements in $\mathfrak{b}$. Suppose that $\mathfrak{b} \in Q$ with Q a K_n-orbit in $\mathcal{B}_n$ which is not closed. Then $\pi_{\mathfrak{k}_n}(\mathfrak{n}^{\mathrm{reg}}) \cap \mathcal{N}_{\mathfrak{k}_n} = \emptyset$.*

Remark 3.3. By the K_n-equivariance of the projection $\pi_{\mathfrak{k}_n} : \mathfrak{g} \to \mathfrak{k}_n$, it suffices to prove Theorem 3.2 for a representative of the K_n-orbit Q. Standard representatives are given by the Borel subalgebras $\mathfrak{b}_{i,j}$ discussed later in Notation 4.23 and Example 4.30. Let $\mathfrak{b} = \mathfrak{b}_{i,j}$ be such a representative. To compute $\pi_{\mathfrak{k}_n}(\mathfrak{n}^{\mathrm{reg}})$, one needs to understand the action of θ on the roots of $\mathfrak{b}$ with respect to a θ-stable Cartan subalgebra $\mathfrak{h}' \subset \mathfrak{b}$. In general, this action is difficult to compute. It is easier to replace the pair $(\mathfrak{b}, \theta)$ with an equivalent pair $(\mathfrak{b}_+, \theta')$ where $\mathfrak{b}_+ \subset \mathfrak{g}$ is the standard Borel subalgebra of upper triangular matrices and θ' is an involution of $\mathfrak{g}$ which stabilizes the standard Cartan subalgebra of diagonal matrices $\mathfrak{h} \subset \mathfrak{b}_+$. We then prove the statement of the theorem for the pair $(\mathfrak{b}_+, \theta')$. The construction and computation of the involution θ' is explained in detail in Equation (31) and Example 4.30, where it is denoted by $\theta_{\hat{v}}$ and $\theta_{\widehat{v_{i,j}}}$ respectively.

Theorem 3.1 permits us to focus only on closed K_n-orbits. There are n such orbits in $\mathcal{B}_n$, two of which are $Q_{+,n} = K_n \cdot \mathfrak{b}_+$, the orbit of the $n \times n$ upper triangular matrices, and $Q_{-,n} = K_n \cdot \mathfrak{b}_-$, the orbit of the $n \times n$ lower triangular matrices (see Example 4.16). We now study the second condition in (22).

Proposition 3.4. *Let $Q = K_n \cdot \mathfrak{b}$ be a closed K_n-orbit and let $x \in \mathfrak{n} = [\mathfrak{b}, \mathfrak{b}]$ satisfy condition* (b) *in* (22)*. Then $Q = Q_{+,n}$ or $Q = Q_{-,n}$.*

This is an immediate consequence of the following result. Recall the projection $\pi_{n-1} : \mathfrak{g} \to \mathfrak{g}_{n-1}$ defined by $\pi_{n-1}(x) = x_{n-1}$.

Proposition 3.5 ([11], Proposition 3.8). *Let $\mathfrak{b} \subset \mathfrak{g}$ be a Borel subalgebra that generates a closed K_n-orbit Q, which is neither the orbit of the upper nor the lower triangular matrices. Let $\mathfrak{n} = [\mathfrak{b}, \mathfrak{b}]$ and let $\mathfrak{n}_{n-1} := \pi_{n-1}(\mathfrak{n})$. Let $\mathfrak{z}_{\mathfrak{g}}(\mathfrak{n})$ denote the centralizer of $\mathfrak{n}$ in $\mathfrak{g}$ and let $\mathfrak{z}_{\mathfrak{g}_{n-1}}(\mathfrak{n}_{n-1})$ denote the centralizer of $\mathfrak{n}_{n-1}$ in $\mathfrak{g}_{n-1}$. Then*

$$\mathfrak{z}_{\mathfrak{g}_{n-1}}(\mathfrak{n}_{n-1}) \cap \mathfrak{z}_{\mathfrak{g}}(\mathfrak{n}) \neq 0. \tag{23}$$

Remark 3.6. We note that the projection $\pi_{n-1} : \mathfrak{g} \to \mathfrak{g}_{n-1}$ is K_n-equivariant, so that it suffices to prove Equation (23) for a representative $\mathfrak{b}$ of the closed K_n-orbit Q. We can take $\mathfrak{b}$ to be one of the representatives given below in Example 4.16.

For any $i = 2, \ldots, n$, let $Q_{+,i}$ denote the K_i-orbit of the $i \times i$ upper triangular matrices in $\mathcal{B}_i$ and let $Q_{-,i}$ denote the K_i-orbit of the $i \times i$ lower triangular matrices in $\mathcal{B}_i$. Combining the results of Theorem 3.1 and Proposition 3.4, we obtain:

Theorem 3.7. *Let $x \in \mathfrak{g}_i$ satisfy the two conditions in* (22) *and suppose that $x \in \mathfrak{b}$, with $\mathfrak{b} \subset \mathfrak{g}_i$ a Borel subalgebra. Then $K_i \cdot \mathfrak{b} = Q_{+,i}$ or $K_i \cdot \mathfrak{b} = Q_{-,i}$.*

3.2 Constructing Borel subalgebras out of K_i-orbits

In this section we explain how to link together the K_i-orbits $Q_{+,i}$ and $Q_{-,i}$ for $i = 2, \ldots, n$ to construct all the Borel subalgebras containing elements of $\Phi^{-1}(0)_{\text{sreg}}$. The key to the construction is the following lemma.

Lemma 3.8 ([11], Proposition 4.1). *Let Q be a closed K_n-orbit in $\mathcal{B}_n$ and let $\mathfrak{b} \in Q$. Then $\pi_{n-1}(\mathfrak{b}) \subset \mathfrak{g}_{n-1}$ is a Borel subalgebra.*

We can use Lemma 3.8 to give an inductive construction of special subvarieties of $\mathcal{B}_n$ by linking together closed K_i-orbits Q_i for $i = 2, \ldots, n$. For this construction, we view $K_i \subset K_{i+1}$ by embedding K_i in the upper left corner of K_{i+1}. We also make use of the following notation. If $\mathfrak{m} \subset \mathfrak{g}$ is a subalgebra, we denote by $\mathfrak{m}_i$ the image of $\mathfrak{m}$ under the projection $\pi_i : \mathfrak{g} \to \mathfrak{g}_i$.

Suppose we are given a sequence $\mathcal{Q} = (Q_2, \ldots, Q_n)$ with Q_i a closed K_i-orbit in $\mathcal{B}_i$. We call $\mathcal{Q}$ a sequence of closed K_i-orbits. For $\mathfrak{b} \in Q_n$, $\mathfrak{b}_{n-1}$ is a Borel subalgebra by Lemma 3.8. Since K_n acts transitively on $\mathcal{B}_{n-1}$, there is $k \in K_n$ such that $\text{Ad}(k)\mathfrak{b}_{n-1} \in Q_{n-1}$ and the variety

$$X_{Q_{n-1},Q_n} := \{\mathfrak{b} \in \mathcal{B}_n : \mathfrak{b} \in Q_n,\ \mathfrak{b}_{n-1} \in Q_{n-1}\}$$

is nonempty. Lemma 3.8 again implies that $(\text{Ad}(k)\mathfrak{b}_{n-1})_{n-2} = (\text{Ad}(k)\mathfrak{b})_{n-2}$ is a Borel subalgebra in $\mathfrak{g}_{n-2}$, so that there exists an $l \in K_{n-1}$ such that $\text{Ad}(l)(\text{Ad}(k)\mathfrak{b})_{n-2} \in Q_{n-2}$. Since $K_{n-1} \subset K_n$, the variety

$$X_{Q_{n-2},Q_{n-1},Q_n} := \{\mathfrak{b} \in \mathcal{B}_n : \mathfrak{b} \in Q_n,\ \mathfrak{b}_{n-1} \in Q_{n-1},\ \mathfrak{b}_{n-2} \in Q_{n-2}\}$$

is nonempty. Proceeding in this fashion, we can define a nonempty closed subvariety of $\mathcal{B}_n$ by

$$X_{\mathcal{Q}} := \{\mathfrak{b} \in \mathcal{B}_n : \mathfrak{b}_i \in Q_i,\ 2 \le i \le n\}. \tag{24}$$

Theorem 3.9 ([11], Theorem 4.2). *Let $\mathcal{Q} = (Q_2, \ldots, Q_n)$ be a sequence of closed K_i-orbits. Then the variety $X_{\mathcal{Q}}$ is a single Borel subalgebra of $\mathfrak{g}$ that contains the standard Cartan subalgebra of diagonal matrices. Moreover, if $\mathfrak{b} \subset \mathfrak{g}$ is a Borel* *subalgebra which contains the diagonal matrices, then* $\mathfrak{b} = X_{\mathcal{Q}}$ *for some sequence of closed* K_i*-orbits* $\mathcal{Q}$.

Notation 3.10. In light of Theorem 3.9, we refer to the Borel subalgebras $X_{\mathcal{Q}}$ as $\mathfrak{b}_{\mathcal{Q}}$ for the remainder of the discussion.

3.3 Borel subalgebras containing elements of $\Phi^{-1}(0)_{\rm sreg}$

Now we can at last describe the Borel subalgebras of $\mathfrak{g}$ that contain elements of $\Phi^{-1}(0)_{\rm sreg}$ and use these to determine the irreducible component decomposition of $\Phi^{-1}(0)_{\rm sreg}$ explicitly. Since $x \in \Phi^{-1}(0)_{\rm sreg}$ if and only if $x_i \in \mathfrak{g}_i$ satisfies the two conditions in (22) for all $i = 2, \ldots, n$, Theorem 3.7 implies:

Proposition 3.11 ([11], Theorem 4.5). *Let* $x \in \Phi^{-1}(0)_{\rm sreg}$. *Then* $x \in \mathfrak{b}_{\mathcal{Q}}$, *where the sequence of closed* K_i*-orbits* $\mathcal{Q} = (Q_2, \ldots, Q_n)$ *has* $Q_i = Q_{+,i}$ *or* $Q_i = Q_{-,i}$ *for each* $i = 2, \ldots, n$.

Example 3.12. It is easy to describe explicitly these Borel subalgebras. For example, for $\mathfrak{g} = \mathfrak{gl}(3, \mathbb{C})$ there are four such Borel subalgebras:

$$\mathfrak{b}_{Q_-,Q_-} = \begin{bmatrix} h_1 & 0 & 0 \\ a_1 & h_2 & 0 \\ a_2 & a_3 & h_3 \end{bmatrix} \quad \mathfrak{b}_{Q_+,Q_+} = \begin{bmatrix} h_1 & a_1 & a_2 \\ 0 & h_2 & a_3 \\ 0 & 0 & h_3 \end{bmatrix}$$

$$\mathfrak{b}_{Q_+,Q_-} = \begin{bmatrix} h_1 & a_1 & 0 \\ 0 & h_2 & 0 \\ a_2 & a_3 & h_3 \end{bmatrix} \quad \mathfrak{b}_{Q_-,Q_+} = \begin{bmatrix} h_1 & 0 & a_1 \\ a_2 & h_2 & a_3 \\ 0 & 0 & h_3 \end{bmatrix},$$

where $a_i, h_i \in \mathbb{C}$.

We can use these Borel subalgebras to describe the fiber $\Phi^{-1}(0)_{\rm sreg}$. Let $\mathfrak{n}_{\mathcal{Q}}^{\rm reg}$ be the subvariety of regular nilpotent elements of $\mathfrak{b}_{\mathcal{Q}}$. Proposition 3.11 implies that

$$\Phi^{-1}(0)_{\rm sreg} \subseteq \bigsqcup_{\mathcal{Q}} \mathfrak{n}_{\mathcal{Q}}^{\rm reg}, \tag{25}$$

where $\mathcal{Q} = (Q_2, \ldots, Q_n)$ ranges over all 2^{n-1} sequences where $Q_i = Q_{+,i}$ or $Q_{-,i}$. We note that the union on the right side of (25) is disjoint, since a regular nilpotent element is contained in a unique Borel subalgebra (see for example [6], Proposition 3.2.14). We claim that the inclusion in (25) is an equality and that the right side of (25) is an irreducible component decomposition of the variety $\Phi^{-1}(0)_{\rm sreg}$. The key observation is the converse to Proposition 3.11.

Proposition 3.13 ([11], Prop. 3.11, Thm. 4.5). *Let* $\mathcal{Q} = (Q_2, \ldots, Q_n)$ *be a sequence of closed* K_i*-orbits with* $Q_i = Q_{+,i}$ *or* $Q_{-,i}$. *Then* $\mathfrak{n}_{\mathcal{Q}}^{\rm reg} \subset \Phi^{-1}(0)_{\rm sreg}$.

Thus, the variety $\mathfrak{n}_{\mathcal{Q}}^{\mathrm{reg}}$ is an irreducible subvariety of $\Phi^{-1}(0)_{\mathrm{sreg}}$ of dimension $\dim \mathfrak{n}_{\mathcal{Q}} = \binom{n}{2}$. It follows from Theorem 2.14 that $\mathfrak{n}_{\mathcal{Q}}^{\mathrm{reg}}$ is an open subvariety of a unique irreducible component $\mathcal{Y}$ of $\Phi^{-1}(0)_{\mathrm{sreg}}$. But then by (25), we have

$$\mathcal{Y} = \bigsqcup_{\mathcal{Q}'} \mathfrak{n}_{\mathcal{Q}'}^{\mathrm{reg}},$$

where the disjoint union is taken over a subset of the set of all sequences $(\mathcal{Q}_2', \ldots, \mathcal{Q}_n')$ with $\mathcal{Q}_i' = \mathcal{Q}_{+,i}$ or $\mathcal{Q}_{-,i}$. Since $\mathcal{Y}$ is irreducible, we must have $\mathfrak{n}_{\mathcal{Q}}^{\mathrm{reg}} = \mathcal{Y}$. This yields the main theorem of [11].

Theorem 3.14 ([11], Theorem 4.5). *The irreducible component decomposition of the variety $\Phi^{-1}(0)_{\mathrm{sreg}}$ is*

$$\Phi^{-1}(0)_{\mathrm{sreg}} = \bigsqcup_{\mathcal{Q}} \mathfrak{n}_{\mathcal{Q}}^{\mathrm{reg}}, \tag{26}$$

where $\mathcal{Q} = (\mathcal{Q}_2, \ldots, \mathcal{Q}_n)$ ranges over all 2^{n-1} sequences where $\mathcal{Q}_i = \mathcal{Q}_{+,i}$ or $\mathcal{Q}_{-,i}$. The A-orbits in $\Phi^{-1}(0)_{\mathrm{sreg}}$ are exactly the varieties $\mathfrak{n}_{\mathcal{Q}}^{\mathrm{reg}}$, for $\mathcal{Q}$ as above.

The description of $\Phi^{-1}(0)_{\mathrm{sreg}}$ in Equation (26) is much more explicit than the one given in Corollary 2.21, where the components are described as orbits of the group $Z = (\mathbb{C}^{\times})^{n-1} \times \mathbb{C}^{\binom{n}{2}-n+1}$ where Z acts via the formula in Equation (21). In fact, we can describe easily the varieties $\mathfrak{n}_{\mathcal{Q}}^{\mathrm{reg}} \cong (\mathbb{C}^{\times})^{n-1} \times \mathbb{C}^{\binom{n}{2}-n+1}$.

Example 3.15. For $\mathfrak{g} = \mathfrak{gl}(3, \mathbb{C})$, Theorem 3.14 implies that the four A-orbits in $\Phi^{-1}(0)_{\mathrm{sreg}}$ are the regular nilpotent elements of the four Borel subalgebras given in Example 3.12.

$$\mathfrak{n}_{\mathcal{Q}_-,\mathcal{Q}_-}^{\mathrm{reg}} = \begin{bmatrix} 0 & 0 & 0 \\ a_1 & 0 & 0 \\ a_3 & a_2 & 0 \end{bmatrix} \quad \mathfrak{n}_{\mathcal{Q}_+,\mathcal{Q}_+}^{\mathrm{reg}} = \begin{bmatrix} 0 & a_1 & a_3 \\ 0 & 0 & a_2 \\ 0 & 0 & 0 \end{bmatrix}$$

$$\mathfrak{n}_{\mathcal{Q}_+,\mathcal{Q}_-}^{\mathrm{reg}} = \begin{bmatrix} 0 & a_1 & 0 \\ 0 & 0 & 0 \\ a_2 & a_3 & 0 \end{bmatrix} \quad \mathfrak{n}_{\mathcal{Q}_-,\mathcal{Q}_+}^{\mathrm{reg}} = \begin{bmatrix} 0 & 0 & a_1 \\ a_2 & 0 & a_3 \\ 0 & 0 & 0 \end{bmatrix},$$

where $a_1, a_2 \in \mathbb{C}^{\times}$ and $a_3 \in \mathbb{C}$.

Remark 3.16. We note that the 2^{n-1} Borel subalgebras appearing in Theorem 3.14 are exactly the Borel subalgebras $\mathfrak{b}$ with the property that each projection of $\mathfrak{b}$ to $\mathfrak{g}_i$ for $i = 2, \ldots, n$ is a Borel subalgebra of $\mathfrak{g}_i$ whose K_i-orbit in $\mathcal{B}_i$ is related via the Beilinson–Bernstein correspondence to Harish-Chandra modules for the pair $(\mathfrak{g}_i, K_i)$ coming from holomorphic and anti-holomorphic discrete series. It would be interesting to relate our results to representation theory, especially to work of

Kobayashi [22]. For more on the relation between geometry of orbits of a symmetric subgroup and Harish-Chandra modules, see [12, 20, 38].

3.4 Strongly regular elements and Borel subalgebras

It would be interesting to study strongly regular fibers $\Phi^{-1}(c)_{\text{sreg}}$ for arbitrary $c \in \mathbb{C}^{\binom{n+1}{2}}$ using the geometry of K_i-orbits on $\mathcal{B}_i$. The following result is a step in this direction.

Theorem 3.17 ([11], Theorem 5.3). *Every Borel subalgebra* $\mathfrak{b} \subset \mathfrak{g}$ *contains strongly regular elements.*

We briefly outline the proof of Theorem 3.17. For complete details see [11], Section 5. For ease of notation, we denote the flag variety $\mathcal{B}_n$ of $\mathfrak{gl}(n,\mathbb{C})$ by $\mathcal{B}$. Let $\mathfrak{h} \subset \mathfrak{g}$ denote the standard Cartan subalgebra of diagonal matrices and let H be the corresponding Cartan subgroup. Define

$$\mathcal{B}_{\text{sreg}} = \{\mathfrak{b} \in \mathcal{B} : \mathfrak{b} \cap \mathfrak{g}_{\text{sreg}} \neq \emptyset\}.$$

We want to show that $\mathcal{B}_{\text{sreg}} = \mathcal{B}$. Consider the variety $Y = \mathcal{B} \setminus \mathcal{B}_{\text{sreg}}$. We show that Y is closed and H-invariant. Let $\mathfrak{b} \in Y$ and consider its H-orbit, $H \cdot \mathfrak{b}$. Since Y is closed, $\overline{H \cdot \mathfrak{b}} \subset Y$. We know that $\overline{H \cdot \mathfrak{b}}$ contains a closed H-orbit. But the closed H-orbits on $\mathcal{B}$ are precisely the Borel subalgebras $\mathfrak{b}$ which contain the Cartan subalgebra $\mathfrak{h}$ ([6], Lemma 3.1.10). Thus, it suffices to show that no Borel subalgebra $\mathfrak{b}$ with $\mathfrak{h} \subset \mathfrak{b}$ can be contained in Y. This can be shown using the characterization of such Borel subalgebras as $\mathfrak{b}_{\mathcal{Q}}$, with $\mathcal{Q} = (Q_2, \ldots, Q_n)$ a sequence of closed K_i-orbits (see Theorem 3.9) and properties of closed K_i-orbits (see [11], Proposition 5.2).

4 The geometry of K-orbits on the flag variety

Proofs of the results discussed in Section 3 require an understanding of aspects of the geometry and parametrization of K_n-orbits on the flag variety $\mathcal{B}_n$ of $\mathfrak{gl}(n,\mathbb{C})$. In this section, we develop the theory of orbits of a symmetric subgroup K of an algebraic group G acting on the flag variety $\mathcal{B}$ of G, as developed by Richardson, Springer, and others. Our aim is to apply this theory in the specific example of $G = GL(n,\mathbb{C})$ and $K = GL(n-1,\mathbb{C}) \times GL(1,\mathbb{C})$, which provides the details behind the computations of [11], Section 3.1. We hope our exposition will make this important theory more accessible. See the papers [32, 33], and [38] for results concerning orbits of a general symmetric subgroup on the flag variety.

4.1 Parameterization of K-orbits on G/B

Let G be a reductive group over $\mathbb{C}$ such that $[G, G]$ is simply connected. Let $\theta : G \to G$ be a holomorphic involution, and we also refer to the differential of θ as $\theta : \mathfrak{g} \to \mathfrak{g}$. Since $\theta : \mathfrak{g} \to \mathfrak{g}$ is a Lie algebra homomorphism, it preserves $[\mathfrak{g}, \mathfrak{g}]$ and the Killing form $< \cdot, \cdot >$ of $\mathfrak{g}$. Let $K = G^{\theta}$ and assume that the fixed set $(Z(G)^{0})^{\theta}$ is connected, where $Z(G)^{0}$ is the identity component of the center of G. Then by a theorem of Steinberg ([36], Corollary 9.7), K is connected.

Let $\mathcal{B}$ be the flag variety of $\mathfrak{g}$, and recall that if B is a Borel subgroup of G, the morphism $G/B \to \mathcal{B}$, $gB \mapsto \mathrm{Ad}(g)\mathfrak{b}$, where $\mathfrak{b} = \mathrm{Lie}(B)$, is a G-equivariant isomorphism $G/B \cong \mathcal{B}$. The involution θ acts on the variety $\mathcal{T}$ of Cartan subalgebras of $\mathfrak{g}$ by $\mathfrak{t} \mapsto \theta(\mathfrak{t})$ for $\mathfrak{t} \in \mathcal{T}$, and the fixed set $\mathcal{T}^{\theta}$ is the variety of θ-stable Cartan subalgebras. We consider the variety

$$\mathcal{C} = \{(\mathfrak{b}, \mathfrak{t}) \in \mathcal{B} \times \mathcal{T} : \mathfrak{t} \subset \mathfrak{b}\}.$$

Then G acts on $\mathcal{C}$ through the adjoint action, and the subvariety $\mathcal{C}_{\theta} = \mathcal{C} \cap (\mathcal{B} \times \mathcal{T}^{\theta})$ is K-stable. Consider the G-equivariant map $\pi : \mathcal{C} \to \mathcal{B}$ given by projection onto the first coordinate, $\pi(\mathfrak{b}, \mathfrak{t}) = \mathfrak{b}$. It induces a map

$$\gamma : K\backslash\mathcal{C}_{\theta} \to K\backslash\mathcal{B},\ \gamma(K \cdot (\mathfrak{b}, \mathfrak{t})) = K \cdot \mathfrak{b} \tag{27}$$

from the set of K-orbits on $\mathcal{C}_{\theta}$ to the set of K-orbits on $\mathcal{B}$.

Proposition 4.1. *The map γ is a bijection.*

For a proof of this proposition, we refer the reader to [33], Proposition 1.2.1. We summarize the main ideas. To show the map γ is surjective, it suffices to show that every Borel subalgebra contains a θ-stable Cartan subalgebra. This follows from [36], Theorem 7.5. To show that the map is injective, it suffices to show that if $\mathfrak{t}, \mathfrak{t}'$ are θ-stable Cartan subalgebras of a Borel subalgebra $\mathfrak{b}$, then $\mathfrak{t}$ and $\mathfrak{t}'$ are $K \cap B$-conjugate, which is verified in [33].

Throughout the discussion, we will fix a θ-stable Borel subalgebra $\mathfrak{b}_0$ and θ-stable Cartan subalgebra $\mathfrak{t}_0 \subset \mathfrak{b}_0$. Such a pair exists by [36], Theorem 7.5, and is called a *standard pair*. Let $N = N_G(T_0)$ be the normalizer of T_0, where T_0 is the Cartan subgroup with Lie algebra $\mathfrak{t}_0$. We consider the map $\zeta_0 : G \to \mathcal{C}$ given by $\zeta_0(g) = (\mathrm{Ad}(g)\mathfrak{b}_0, \mathrm{Ad}(g)\mathfrak{t}_0)$, which is clearly G-equivariant with respect to the left translation action on G and the adjoint action on $\mathcal{C}$. It is easy to see that ζ_0 is constant on left T_0-cosets, and induces an isomorphism of varieties

$$\zeta : G/T_0 \to \mathcal{C}. \tag{28}$$

To parameterize the K-orbits on $\mathcal{B}$ using Proposition 4.1, we introduce the variety $\mathcal{V} = \zeta_0^{-1}(\mathcal{C}_{\theta})$. It is easy to show that $\mathcal{V}$ is the set

$$\mathcal{V} = \{g \in G : g^{-1}\theta(g) \in N\}. \tag{29}$$

By Equation (28) and the G-equivariance of the map ζ_0, it follows that the morphism ζ induces a bijection,

$$\zeta : K\backslash\mathcal{V}/T_0 \to K\backslash\mathcal{C}_\theta, \tag{30}$$

which we also denote by ζ. Combining Equation (30) with Proposition 4.1, we obtain the following useful parametrization of K-orbits on $\mathcal{B}$ (cf. [33], Proposition 1.2.2).

Proposition 4.2. *There are natural bijections*

$$K\backslash\mathcal{V}/T_0 \leftrightarrow K\backslash\mathcal{C}_\theta \leftrightarrow K\backslash\mathcal{B} \leftrightarrow K\backslash G/B_0.$$

Let V denote the set of (K, T_0)-double cosets in $\mathcal{V}$. By [35], Corollary 4.3, V is a finite set and hence

The number of K-orbits on $\mathcal{B}$ is finite.

Notation 4.3. For $v \in V$, let $\hat{v} \in \mathcal{V}$ denote a representative, so that $v = K\hat{v}T$. Denote the corresponding K-orbit in $\mathcal{B}$ by $K \cdot \mathfrak{b}_{\hat{v}}$, where $\mathfrak{b}_{\hat{v}} = \mathrm{Ad}(\hat{v})\mathfrak{b}_0$.

We end this section with a discussion of how θ acts on the root decomposition of $\mathfrak{g}$ with respect to a θ-stable Cartan subalgebra $\mathfrak{t}$.

Definition 4.4. For $(\mathfrak{b}, \mathfrak{t}) \in \mathcal{C}_\theta$ and $\alpha \in \Phi = \Phi(\mathfrak{g}, \mathfrak{t})$, let $e_\alpha \in \mathfrak{g}_\alpha$ be a root vector in the corresponding root space. We say that α is positive for $(\mathfrak{b}, \mathfrak{t})$ if $\mathfrak{g}_\alpha \subset \mathfrak{b}$. We define the *type* of α for the pair $(\mathfrak{b}, \mathfrak{t})$ with respect to θ as follows.

(1) If $\theta(\alpha) = -\alpha$, then α is said to be *real*.
(2) If $\theta(\alpha) = \alpha$, then α is said to be *imaginary*. In this case, there are two subcases:
   (a) If $\theta(e_\alpha) = e_\alpha$, then α is said to be *compact imaginary*.
   (b) If $\theta(e_\alpha) = -e_\alpha$, then α is said to be *noncompact imaginary*.
(3) If $\theta(\alpha) \neq \pm\alpha$, then α is said to *complex*. If also α and $\theta(\alpha)$ are both positive, we say α is complex θ-stable.

Remark 4.5. Let α be a positive root. Then $\theta(\alpha)$ is positive if and only if α is imaginary or complex θ-stable.

For $v \in V$ with representative $\hat{v} \in \mathcal{V}$, we define a new involution by the formula,

$$\theta_{\hat{v}} = \mathrm{Ad}(\hat{v}^{-1}) \circ \theta \circ \mathrm{Ad}(\hat{v}) = \mathrm{Ad}(\hat{v}^{-1}\theta(\hat{v})) \circ \theta. \tag{31}$$

Note that $\theta_{\hat{v}}(\mathfrak{t}_0) = \mathfrak{t}_0$, and consider the induced action of $\theta_{\hat{v}}$ on $\Phi(\mathfrak{g}, \mathfrak{t}_0)$.

Definition 4.6. Let $\alpha \in \Phi(\mathfrak{g}, \mathfrak{t}_0)$, $v \in V$, and $\hat{v} \in \mathcal{V}$ be a representative for v. We define *the type of the root α for v* to be the type of the root α for the pair $(\mathfrak{b}_0, \mathfrak{t}_0)$ with respect to the involution $\theta_{\hat{v}}$.

For example, a root α is imaginary for v if and only if $\theta_{\hat{v}}(\alpha) = \alpha$. Note that if $k\hat{v}t$ is a different representative for v, then $\theta_{k\hat{v}t} = \mathrm{Ad}(t^{-1}) \circ \theta_{\hat{v}} \circ \mathrm{Ad}(t)$. It follows easily that the type of α for v does not depend on the choice of a representative $\hat{v}$. Further, the involution $\theta_{\hat{v}}$ of $\Phi(\mathfrak{g}, \mathfrak{t}_0)$ does not depend on the choice of $\hat{v}$, and we refer to $\theta_{\hat{v}}$ as the involution associated to the orbit v.

For $v \in V$ and $\mathfrak{b}_{\hat{v}} = \mathrm{Ad}(\hat{v})\mathfrak{b}_0$, consider the θ-stable Cartan subalgebra $\mathfrak{t}' = \mathrm{Ad}(\hat{v})\mathfrak{t}_0 \subset \mathfrak{b}_{\hat{v}}$. For $\alpha \in \Phi(\mathfrak{g}, \mathfrak{t}_0)$, we define $\mathrm{Ad}(\hat{v})\alpha := \alpha \circ \mathrm{Ad}(\hat{v}^{-1}) \in \Phi(\mathfrak{g}, \mathfrak{t}')$.

Proposition 4.7. *For $\alpha \in \Phi(\mathfrak{g}, \mathfrak{t}_0)$, the type of α for v is the same as the type of $\mathrm{Ad}(\hat{v})\alpha$ for the pair $(\mathfrak{b}_{\hat{v}}, \mathfrak{t}')$ with respect to θ.*

Proof. This follows easily from the identity $\theta \circ \mathrm{Ad}(\hat{v}) = \mathrm{Ad}(\hat{v}) \circ \theta_{\hat{v}}$. □

By Proposition 4.7, we may compute the action of θ on the positive roots in $\Phi(\mathfrak{g}, \mathfrak{t}')$ for the pair $(\mathfrak{b}_{\hat{v}}, \mathfrak{t}')$ using the involution $\theta_{\hat{v}}$ on our standard positive system $\Phi^+(\mathfrak{g}, \mathfrak{t}_0)$ in $\Phi(\mathfrak{g}, \mathfrak{t}_0)$.

Remark 4.8. We also denote the corresponding involution on G by $\theta_{\hat{v}}$. By abuse of notation, we denote conjugation on G by Ad, i.e., for $g,\ h \in G$; $\mathrm{Ad}(g)h = ghg^{-1}$. Thus $\theta_{\hat{v}} : G \to G$ is also given by the formula in Equation (31). Its differential at the identity is $\theta_{\hat{v}} : \mathfrak{g} \to \mathfrak{g}$.

4.2 *The W-action on V*

The fact that K-orbits on the flag variety have representatives coming from $\mathcal{V}$ was used by Springer [35] to associate a Weyl group element $\phi(v)$ to the K-orbit indexed by $v \in V$. The element $\phi(v)$ plays a crucial role in understanding the action of the involution $\theta_{\hat{v}}$ associated to v on the roots for the standard pair $\Phi(\mathfrak{g}, \mathfrak{t}_0)$.

We first consider the map $\tau : G \to G$ given by $\tau(g) = g^{-1}\theta(g)$. Note that $\tau^{-1}(N) = \mathcal{V}$. Then following [35], Section 4.5, we define for $v = K\hat{v}T_0$

$$\phi(v) = \tau(\hat{v})T_0 \in N/T_0 = W. \tag{32}$$

We refer to the map ϕ as the Springer map and $\phi(v)$ as the Springer invariant of $v \in V$. It is easy to check that $\phi(v)$ is independent of the choice of representative $\hat{v}$.

The Springer map is not injective, but we can study its fibers using an action of W on V, which we now describe. The group N acts on $\mathcal{V}$ on the left by $n \cdot \hat{v} = \hat{v}n^{-1}$ for $\hat{v} \in \mathcal{V}$ and $n \in N$. This action induces a W-action on V given by

$$w \times v := K\hat{v}\dot{w}^{-1}T_0, \tag{33}$$

where $\hat{v} \in \mathcal{V}$ is a representative of $v \in V$ and $\dot{w} \in N$ is a representative of $w \in W$. It is easy to check that the formula in Equation (33) does not depend on the choice of representatives $\dot{w}$ or $\hat{v}$. We refer to this action as the *cross action* of W on V. The Springer map intertwines the cross action of W on V with a certain twisted action of W on itself. We note that since T_0 is θ-stable, θ acts on N and hence on W. We define the twisted conjugation action of W on itself by:

$$w' * w = w'w\theta((w')^{-1}), \text{ for } w, w' \in W. \tag{34}$$

Proposition 4.9. (1) *The Springer map $\phi : V \to W$ is W-equivariant with respect the cross action on V and the twisted W-action on W.*
(2) **([32], Proposition 2.5)** *Suppose for v, $v' \in V$, we have $\phi(v) = \phi(v')$. Then $v' = w \times v$ for some $w \in W$.*

Part (1) is an easy calculation using the definition of ϕ. Part (2) is nontrivial and relies on many of the results of [32], Section 2.

4.3 Closed K-orbits on $\mathcal{B}$

In this section we use the properties of the Springer map developed in the previous section to find representatives for the closed K-orbits on $\mathcal{B}$ and describe the involution $\theta_{\hat{v}}$ associated to such orbits.

Since θ acts on W, we can consider the W-fixed point subgroup, W^θ. By [31], Lemma 5.1, $T_0 \cap K$ is a maximal torus of K, and by [31], Lemma 5.3, the subgroup $N_K(T_0 \cap K) \subset N_G(T_0)$. It follows that the group homomorphism $N_K(T_0 \cap K)/(T_0 \cap K) \to N_G(T_0)/T_0$ is injective. Hence, we may regard W_K as a subgroup of W, and it is easy to see that it has image in W^θ.

Theorem 4.10. *There is a one-to-one correspondence between the set of closed K-orbits on $\mathcal{B}$ and the coset space W^θ / W_K. The correspondence is given by*

$$w W_K \to K\dot{w}^{-1}T_0, \tag{35}$$

for $\dot{w} \in N$ a representative of $w \in W^\theta$.

To prove Theorem 4.10, we describe equivalent conditions for a K-orbit on $\mathcal{B}$ to be closed. We begin with the following lemma (see [5], Lemma 3).

Lemma 4.11. *Let $B \subset G$ be a Borel subgroup. Then the following statements are equivalent.*

(i) *The Borel subgroup B is θ-stable.*
(ii) *The subgroup $(B \cap K)^0$ is a Borel subgroup of K, where $(B \cap K)^0$ denotes the identity component of $B \cap K$.*

Let $v_0 \in V$ correspond to the K-orbit $K \cdot \mathfrak{b}_0$ so that $v_0 = KT_0$, and we choose the representative $\widehat{v_0} = 1$. Define $V_0 := \{v \in V : K \cdot \mathfrak{b}_{\hat{v}} \text{ is closed}\}$.

Proposition 4.12. *The following statements are equivalent.*

(i) $v \in V_0$.
(ii) *For any representative $\hat{v} \in \mathcal{V}$ of $v \in V$, the Borel subalgebra $\mathfrak{b}_{\hat{v}} = \mathrm{Ad}(\hat{v})\mathfrak{b}_0$ is θ-stable.*

(iii) $\phi(v) = 1$.
(iv) $v \in W^{\theta} \times v_0$.

Proof. We first show that (i) implies (ii). Let $v \in V_0$, and let $B_{\hat{v}} \subset G$ be the Borel subgroup of G corresponding to the Borel subalgebra $\mathfrak{b}_{\hat{v}}$. Then $K \cdot \mathfrak{b}_{\hat{v}} \subset \mathcal{B}$ is projective, so that the homogeneous space $K/(K \cap B_{\hat{v}}) \cong K \cdot \mathfrak{b}_{\hat{v}}$ is projective, and hence $K \cap B_{\hat{v}}$ is parabolic. Since $K \cap B_{\hat{v}}$ is solvable, it follows that $K \cap B_{\hat{v}}$ is a Borel subgroup of K. Part (ii) now follows from Lemma 4.11.

We now prove that (ii) implies (iii). Suppose that $v \in V$ and that $\mathfrak{b}_{\hat{v}} = \mathrm{Ad}(\hat{v})\mathfrak{b}_0$ is θ-stable. Thus, $\mathrm{Ad}(\theta(\hat{v}))\theta(\mathfrak{b}_0) = \mathrm{Ad}(\hat{v})\mathfrak{b}_0$. But $\mathfrak{b}_0$ is itself θ-stable, implying that $\hat{v}^{-1}\theta(\hat{v}) \in B_0$. But then $\hat{v}^{-1}\theta(\hat{v}) = \tau(\hat{v}) \in B_0 \cap N = T_0$ by definition of $\mathcal{V}$. Thus, $\phi(v) = \tau(\hat{v})T_0 = 1$.

We next show that (iii) implies (iv). Suppose that $\phi(v) = 1$. Clearly, $\phi(v_0) = 1$. It then follows from part (2) of Proposition 4.9 that $v = w \times v_0$ for some $w \in W$. But then part (1) of Proposition 4.9 implies

$$1 = \phi(v) = \phi(w \times v_0) = w\phi(v_0)\theta(w^{-1}) = w\theta(w^{-1}),$$

whence $w \in W^{\theta}$ and $v \in W^{\theta} \times v_0$.

Lastly, we show that (iv) implies (i). If $v \in W^{\theta} \times v_0$, then $v = K\dot{w}T_0$, where $\dot{w} \in N$ is a representative of $w \in W^{\theta}$. We note that since $w \in W^{\theta}$, $\theta(\dot{w}) = \dot{w}t$ for some $t \in T_0$. It follows that $\mathfrak{b}_{\hat{v}} = \mathrm{Ad}(\dot{w})\mathfrak{b}_0$ is θ-stable, since $\mathfrak{t}_0 \subset \mathfrak{b}_0$. Let $B_{\hat{v}}$ be the Borel subgroup corresponding to $\mathfrak{b}_{\hat{v}}$, so that $B_{\hat{v}}$ is θ-stable. It follows from [31], Lemma 5.1 that $B_{\hat{v}} \cap K$ is connected and therefore is a Borel subgroup by Lemma 4.11. Since $(B_{\hat{v}} \cap K)$ is a Borel subgroup, the variety $K/(B_{\hat{v}} \cap K)$ is complete, and the orbit $K \cdot \mathfrak{b}_{\hat{v}} \cong K/(B_{\hat{v}} \cap K)$ is a complete subvariety of $\mathcal{B}$ and is therefore closed. □

We now prove Theorem 4.10.

Proof (of Theorem 4.10). It follows from Proposition 4.12 that

$$V_0 = W^{\theta} \times v_0. \tag{36}$$

By [32], Proposition 2.8, the stabilizer of v_0 in W is precisely $W_K \subset W^{\theta}$. Thus, the elements of the orbit $W^{\theta} \times v_0$ are in bijection with the coset space W^{θ}/W_K. Equation (35) then follows from the definition of the cross action of V on W. □

Recall the notion of the type of a root $\alpha \in \Phi(\mathfrak{g}, \mathfrak{t}_0)$ for v from Definition 4.6, and note that by Equation (31),

$$\theta_{\hat{v}} = \mathrm{Ad}(\hat{v}^{-1}\theta(\hat{v})) \circ \theta = \mathrm{Ad}(\tau(\hat{v})) \circ \theta. \tag{37}$$

Proposition 4.13. *For $v \in V_0$, every positive root $\alpha \in \Phi^+(\mathfrak{g}, \mathfrak{t}_0)$ is imaginary or complex θ-stable for v. Moreover, a positive root $\alpha \in \Phi^+(\mathfrak{g}, \mathfrak{t}_0)$ is imaginary (resp. complex) for v if and only if it is imaginary (resp. complex) for v_0.*

Proof. By Equation (37), for $v \in V$, $\theta_{\hat{v}}(\alpha) = \phi(v)(\theta(\alpha))$ for $\alpha \in \Phi(\mathfrak{g}, \mathfrak{t}_0)$. Since $v \in V_0$, then $\phi(v) = 1$ by Proposition 4.12, so

$$\theta_{\hat{v}}(\alpha) = \theta(\alpha) \tag{38}$$

for any $\alpha \in \Phi(\mathfrak{g}, \mathfrak{t}_0)$. Since $\mathfrak{b}_0 \subset \mathfrak{g}$ is θ-stable, Remark 4.5 implies that any $\alpha \in \Phi^+(\mathfrak{g}, \mathfrak{t}_0)$ is complex θ-stable or imaginary for v_0. Both statements of the proposition now follow immediately from Equation (38). □

Remark 4.14. Let $v \in V_0$ and let $\theta_{\hat{v}}$ be the involution associated to the orbit v. To determine the action of $\theta_{\hat{v}}$ on $\Phi(\mathfrak{g}, \mathfrak{t}_0)$, Proposition 4.13 implies that it suffices to find which roots are compact (resp. noncompact) imaginary for v. By Theorem 4.10, we may take $\hat{v} = \dot{w}^{-1}$, where $\dot{w}$ is a representative for $w \in W^{\theta}$. By Proposition 4.7, it follows that a root $\alpha \in \Phi(\mathfrak{g}, \mathfrak{t}_0)$ is compact (resp. noncompact) imaginary for v if and only if $w^{-1}(\alpha)$ is compact (resp. noncompact) for the pair $(\mathrm{Ad}(w^{-1})\mathfrak{b}_0, \mathfrak{t}_0)$ with respect to θ.

Notation 4.15. We will make use of the following notation for flags in $\mathbb{C}^n$. Let

$$\mathcal{F} = (V_0 = \{0\} \subset V_1 \subset \cdots \subset V_i \subset \cdots \subset V_n = \mathbb{C}^n)$$

be a flag in $\mathbb{C}^n$, with $\dim V_i = i$ and $V_i = \mathrm{span}\{v_1, \ldots, v_i\}$, with each $v_j \in \mathbb{C}^n$. We will denote this flag $\mathcal{F}$ by

$$\mathcal{F} = (v_1 \subset v_2 \subset \cdots \subset v_i \subset v_{i+1} \subset \cdots \subset v_n).$$

We denote the standard ordered basis of $\mathbb{C}^n$ by $\{e_1, \ldots, e_n\}$. For $1 \leq i, j \leq n$, let $E_{i,j}$ be the matrix with 1 in the (i, j)-entry and 0 elsewhere.

Example 4.16. Let $G = GL(n, \mathbb{C})$ and let θ be conjugation by the diagonal matrix $c = \mathrm{diag}[1, 1, \ldots, 1, -1]$. Then $K = GL(n-1, \mathbb{C}) \times G(1, \mathbb{C})$ and $\mathfrak{k} = \mathfrak{gl}(n-1, \mathbb{C}) \oplus \mathfrak{gl}(1, \mathbb{C})$. Since this involution is inner, $W^{\theta} = W = \mathcal{S}_n$, the symmetric group on n letters and $W_K = \mathcal{S}_{n-1}$. We can take $\mathfrak{b}_0$ to be the standard Borel subalgebra of $n \times n$ upper triangular matrices and $\mathfrak{t}_0 \subset \mathfrak{b}_0$ to be the diagonal matrices. By Theorem 4.10, the n closed orbits are then parameterized by the identity permutation and the $n-1$ cycles $\{(n-1\, n), (n-2\, n-1\, n), \ldots, (i \ldots n), \ldots, (1 \ldots n)\}$. We consider the closed K-orbit $v \in V_0$ corresponding to the cycle $w = (i \ldots n)$. By Equation (35), it is generated by the Borel subalgebra $\mathfrak{b}_i := \mathrm{Ad}(w^{-1})\mathfrak{b}_0$, which is the stabilizer of the flag:

$$\mathcal{F}_i := (e_1 \subset \cdots \subset e_{i-1} \subset \underbrace{e_n}_{i} \subset e_i \subset \cdots \subset e_{n-1}). \tag{39}$$

Notice that $\mathcal{F}_n$ is the standard flag in $\mathbb{C}^n$ and $\mathcal{F}_1$ is K-conjugate to the opposite flag. We denote $Q_i := K \cdot \mathfrak{b}_i$, so $Q_1, \ldots, Q_n$ are the n closed orbits.

Let $\epsilon_i \in \mathfrak{t}_0^*$ be the linear functional $\epsilon_i(t) = t_i$ for $t \in \mathfrak{t}_0$, where $t = \operatorname{diag}[t_1, \ldots, t_i, \ldots, t_n]$ with each $t_i \in \mathbb{C}$. According to [26], any root of the form $\epsilon_i - \epsilon_k$ or $\epsilon_k - \epsilon_i$ is noncompact imaginary for v while all other roots are compact imaginary, and the involution $\theta_{\hat{v}}$ associated to v acts on the functionals by $\theta_{\hat{v}}(\epsilon_i) = \epsilon_i$ for all i. The second assertion follows easily from Equation (38). By Remark 4.14, $\alpha = \epsilon_k - \epsilon_j$ is compact (resp. noncompact) imaginary for v if and only if $w^{-1}(\alpha)$ is compact (resp. noncompact) imaginary with respect to θ. The first assertion then follows from the observation that roots of the form $\epsilon_n - \epsilon_k$ and $\epsilon_k - \epsilon_n$ are noncompact imaginary with respect to θ and all other roots are compact imaginary.

4.4 The case of general K-orbits in $\mathcal{B}$

In this section we compute $\tau(\hat{v})$ and $\phi(v)$ inductively based on the closed orbit case in Section 4.3. We thus obtain a formula for $\theta_{\hat{v}}$ for any K-orbit in $\mathcal{B}$.

For the first step, we take a K-orbit Q and a simple root α and construct a K-orbit denoted $m(s_\alpha) \cdot Q$ which either coincides with Q or contains Q in its closure as a divisor. Let $Q = K \cdot \mathfrak{b}_{\hat{v}} \subset \mathcal{B}$ for $v \in V$, let $\alpha \in \Phi(\mathfrak{g}, \mathfrak{t}_0)$ be a simple root, and let $\mathfrak{p}_\alpha$ be the minimal parabolic subalgebra generated by α. Let P_α denote the corresponding parabolic subgroup, and let $\pi_\alpha : G/B_0 \to G/P_\alpha$ denote the canonical projection, which is a $P_\alpha/B_0 = \mathbb{P}^1$-bundle.

Lemma-Definition 4.17. *The variety $\pi_\alpha^{-1}\pi_\alpha(Q)$ is irreducible and K acts on $\pi_\alpha^{-1}\pi_\alpha(Q)$ with finitely many orbits. The unique open K-orbit in $\pi_\alpha^{-1}\pi_\alpha(Q)$ is denoted by $m(s_\alpha) \cdot Q$.*

Proof. Note that $\pi_\alpha^{-1}\pi_\alpha(Q) = K\hat{v}P_\alpha/B_0$, and it follows easily that $\pi_\alpha^{-1}\pi_\alpha(Q)$ is irreducible, since it is the image of the double coset KvP_α under the projection $p : G \to G/B_0$. The variety $K\hat{v}P_\alpha/B_0$ is clearly K-stable. It follows that it has a unique open orbit, since the set of K-orbits in $K\hat{v}P_\alpha/B_0$ is a subset of the set of K-orbits on $\mathcal{B}$, and hence is finite. □

The orbit $m(s_\alpha) \cdot Q$ may be equal to Q itself. However, in the case where $m(s_\alpha) \cdot Q \neq Q$, then $\dim m(s_\alpha) \cdot Q = \dim Q + 1$, since the map $\pi_\alpha : G/B_0 \to G/P_\alpha$ is a $\mathbb{P}^1$-bundle. To compute $m(s_\alpha) \cdot Q$ explicitly (following [38], Lemma 5.1), we recall first some facts about involutions for $SL(2, \mathbb{C})$.

Let $\Pi \subset \Phi^+(\mathfrak{g}, \mathfrak{t}_0)$ denote the set of simple roots and let $\alpha \in \Pi$. Let $h_\alpha = \frac{2H_\alpha}{<\alpha,\alpha>}$ with $H_\alpha \in \mathfrak{t}_0$ such that $\langle H_\alpha, x \rangle = \alpha(x)$ for $x \in \mathfrak{t}_0$, and let $e_\alpha \in \mathfrak{g}_\alpha$, $f_\alpha \in \mathfrak{g}_{-\alpha}$ be chosen so that $[e_\alpha, f_\alpha] = h_\alpha$. Hence, the subalgebra $\mathfrak{s}(\alpha) = \operatorname{span}\{e_\alpha, f_\alpha, h_\alpha\}$ forms a Lie algebra isomorphic to $\mathfrak{sl}(2, \mathbb{C})$. Let $\phi_\alpha : \mathfrak{sl}(2, \mathbb{C}) \to \mathfrak{s}(\alpha)$ be the map

$$\phi_\alpha : \begin{bmatrix} 0 & 1 \\ 0 & 0 \end{bmatrix} \to e_\alpha, \; \phi_\alpha : \begin{bmatrix} 0 & 0 \\ 1 & 0 \end{bmatrix} \to f_\alpha, \; \phi_\alpha : \begin{bmatrix} 1 & 0 \\ 0 & -1 \end{bmatrix} \to h_\alpha. \tag{40}$$

Then $\phi_\alpha : \mathfrak{sl}(2,\mathbb{C}) \to \mathfrak{s}(\alpha)$ is a Lie algebra isomorphism, which integrates to an injective homomorphism of Lie groups $\phi_\alpha : SL(2,\mathbb{C}) \to G$, which we will also denote by ϕ_α. We let $S(\alpha)$ be its image.

To perform computations, it is convenient for us to choose specific representatives for the Cayley transform u_α with respect to α and the simple reflection s_α. Let

$$u_\alpha = \phi_\alpha\left(\frac{1}{\sqrt{2}}\begin{bmatrix} 1 & \imath \\ \imath & 1 \end{bmatrix}\right). \tag{41}$$

Note that $g = \frac{1}{\sqrt{2}}\begin{bmatrix} 1 & \imath \\ \imath & 1 \end{bmatrix} \in SL(2,\mathbb{C})$ is the Cayley transform which conjugates the torus in $SL(2,\mathbb{C})$ containing the diagonal split maximal torus of $SL(2,\mathbb{R})$ to a torus of $SL(2,\mathbb{C})$ containing a compact maximal torus of $SL(2,\mathbb{R})$. Let

$$\dot{s}_\alpha = \phi_\alpha\left(\begin{bmatrix} 0 & \imath \\ \imath & 0 \end{bmatrix}\right). \tag{42}$$

Then $\dot{s}_\alpha$ is a representative for $s_\alpha \in W$. Note that $u_\alpha^2 = \dot{s}_\alpha$.

Let $\theta_{1,1} : SL(2,\mathbb{C}) \to SL(2,\mathbb{C})$ be the involution on $SL(2,\mathbb{C})$ given by

$$\theta_{1,1}(g) = \begin{bmatrix} 1 & 0 \\ 0 & -1 \end{bmatrix} g \begin{bmatrix} 1 & 0 \\ 0 & -1 \end{bmatrix}$$

for $g \in SL(2,\mathbb{C})$.

Lemma 4.18. *Suppose $\alpha \in \Pi$ is compact (resp. noncompact) imaginary for v. Then $-\alpha$ is compact (resp. noncompact) imaginary for v.*

Proof. Since $\theta_{\hat{v}}(\mathfrak{g}_\alpha) = \mathfrak{g}_\alpha$, it follows easily that $\theta_{\hat{v}}(\mathfrak{g}_{-\alpha}) = \mathfrak{g}_{-\alpha}$. The rest of the proof follows since $\theta_{\hat{v}}$ preserves the Killing form. □

Lemma 4.19. *If α is noncompact imaginary for v, then*

$$\theta_{\hat{v}} \circ \phi_\alpha = \phi_\alpha \circ \theta_{1,1}. \tag{43}$$

Proof. It suffices to verify Equation (43) on the Lie algebra $\mathfrak{sl}(2,\mathbb{C})$. On $\mathfrak{sl}(2,\mathbb{C})$ the maps in Equation (43) are linear, and we need only check the equation on a basis for $\mathfrak{sl}(2,\mathbb{C})$. Since α is noncompact imaginary for v, we have $\theta_{\hat{v}}(e_\alpha) = -e_\alpha$, $\theta_{\hat{v}}(f_\alpha) = -f_\alpha$, and $\theta_{\hat{v}}(h_\alpha) = h_\alpha$ by Lemma 4.18, and the result follows. □

Remark 4.20. It follows from the proof of Lemma 4.19 that $\mathfrak{s}(\alpha)^{\theta_{\hat{v}}} = \mathbb{C}h_\alpha$.

Proposition 4.21. *Let $Q = K \cdot \mathfrak{b}_{\hat{v}}$ with $v \in V$ and let $\alpha \in \Phi^+(\mathfrak{g},\mathfrak{t}_0)$ be a simple root. Then $m(s_\alpha) \cdot Q \neq Q$ if and only if α is noncompact imaginary for v or α is complex θ-stable for v. If α is noncompact imaginary, then $m(s_\alpha) \cdot Q = K \cdot \mathfrak{b}'$, with $\mathfrak{b}' = \mathrm{Ad}(\hat{v}u_\alpha)\mathfrak{b}_0$, where u_α is the Cayley transform with respect to α. If α is complex θ-stable, then $m(s_\alpha) \cdot Q = K \cdot \mathfrak{b}'$, with $\mathfrak{b}' = \mathrm{Ad}(\hat{v}s_\alpha)\mathfrak{b}_0$.*

Proof. Let $K_{\hat{v}} = K \cap \mathrm{Ad}(\hat{v})P_\alpha$ be the stabilizer in K of $\pi_\alpha(\hat{v}B_0/B_0)$. Let $L_{\hat{v}} = \pi_\alpha^{-1}\pi_\alpha(\hat{v}B_0/B_0)$, which is identified with $\mathrm{Ad}(\hat{v})P_\alpha/\mathrm{Ad}(\hat{v})B_0 \cong \mathbb{P}^1$. We claim that the map χ from the set of $K_{\hat{v}}$-orbits in $L_{\hat{v}}$ to the set of K-orbits in $K\hat{v}P_\alpha/B_0$ given by $\chi(\hat{Q}) = K \cdot \hat{Q}$ is bijective. Indeed, if $Q_1 \subset K\hat{v}P_\alpha/B_0$ is a K-orbit, then for $z_1, z_2 \in Q_1 \cap L_{\hat{v}}$, we have $z_2 = k \cdot z_1$ for some $k \in K$, and $\pi_\alpha(z_1) = \pi_\alpha(z_2)$. It follows that k stabilizes $\pi_\alpha(\hat{v}B_0/B_0)$, so $k \in K_{\hat{v}}$. Hence, $Q_1 \cap L_{\hat{v}}$ is a $K_{\hat{v}}$-orbit, and it is routine to check that $Q_1 \mapsto Q_1 \cap L_{\hat{v}}$ is inverse to χ, giving the claim. Let U^α be the unipotent radical of P_α, and let $Z(M_\alpha)^0$ be the identity component of the center of a Levi subgroup of P_α. Then $\mathrm{Ad}(\hat{v})P_\alpha$ acts on the fiber $L_{\hat{v}}$ through its quotient $\widetilde{S}_{\hat{v}} := \mathrm{Ad}(\hat{v})P_\alpha/\mathrm{Ad}(\hat{v})(Z(M_\alpha)^0U^\alpha)$, which is locally isomorphic to $\mathrm{Ad}(\hat{v})S(\alpha)$. Hence $K_{\hat{v}}$ acts on $L_{\hat{v}}$ through its image $\widetilde{K}_{\hat{v}}$ in $\widetilde{S}_{\hat{v}}$. For α noncompact imaginary for v, it follows from Remark 4.20 that $\widetilde{K}_{\hat{v}}$ has Lie algebra $\mathrm{Ad}(\hat{v})(\mathbb{C}h_\alpha)$, and hence $\widetilde{K}_{\hat{v}}$ is either a torus of $\widetilde{S}_{\hat{v}}$ normalizing $\hat{v}B_0/B_0$ or the normalizer of such a torus. Hence, the points $\hat{v}B_0/B_0$ and $\hat{v}s_\alpha B_0/B_0$ are in zero-dimensional $\widetilde{K}_{\hat{v}}$-orbits, and the complement $L_{\hat{v}} - (\hat{v}B_0/B_0 \cup \hat{v}s_\alpha B_0/B_0)$ is a single $\widetilde{K}_{\hat{v}}$-orbit containing $\hat{v}u_\alpha B_0/B_0$. From the definition of the bijection χ, it follows that $K\hat{v}B_0/B_0$ is a proper subset of $\overline{K\hat{v}u_\alpha B_0/B_0}$, where the closure is taken in the variety $K\hat{v}P_\alpha/B_0$. Since $\dim(K\hat{v}P_\alpha/B_0) = \dim(K\hat{v}B_0/B_0) + 1$, we conclude that $m(s_\alpha) \cdot Q = K\hat{v}u_\alpha B_0/B_0$. This verifies the proposition in the case of noncompact imaginary roots, and the other cases are similar, and discussed in detail in Section 2 of [33]. □

Remark 4.22. In [38], the author discriminates between two types of noncompact roots. For $G = GL(n,\mathbb{C})$ and $K = GL(p,\mathbb{C}) \times GL(n-p,\mathbb{C})$, all noncompact roots for all orbits are type I.

Notation 4.23. We let $G = GL(n,\mathbb{C})$ and $K = GL(n-1,\mathbb{C}) \times G(1,\mathbb{C})$ as in Example 4.16. We let $\mathfrak{b}_{i,j}$ be the Borel subalgebra stabilizing the flag

$$\mathcal{F}_{i,j} = (e_1 \subset \cdots \subset \underbrace{e_i + e_n}_{i} \subset e_{i+1} \subset \cdots \subset e_{j-1} \subset \underbrace{e_i}_{j} \subset e_j \subset \cdots \subset e_{n-1}),$$

and we let $Q_{i,j} = K \cdot \mathfrak{b}_{i,j}$.

Example 4.24. We let G and K be as in Example 4.16 and compute $m(s_\alpha) \cdot Q_c$ for each closed K-orbit Q_c. By Example 4.16, $Q_c = Q_i = K \cdot \mathfrak{b}_i$, where $\mathfrak{b}_i$ is the stabilizer of the flag $\mathcal{F}_i$ from Equation (39). Let v_i be the corresponding element of V. By Example 4.16, the simple roots $\alpha_{i-1} = \epsilon_{i-1} - \epsilon_i$ and $\alpha_i = \epsilon_i - \epsilon_{i+1}$ are the only noncompact imaginary simple roots for v_i, and all other simple roots are compact (for $i = 1$ and $i = n$, one of these two roots does not exist). Since $Q_i = K \cdot \mathrm{Ad}(\dot{w})\mathfrak{b}_0$, where $\dot{w}$ is a representative for the element $(n\ \ldots\ i)$ of W, it follows from Proposition 4.21 that $m(s_{\alpha_{i-1}}) \cdot Q_i = K \cdot \mathrm{Ad}(\dot{w}u_{\alpha_{i-1}})\mathfrak{b}_0$. A routine computation shows that the K-orbit $K \cdot \mathrm{Ad}(\dot{w}u_{\alpha_{i-1}})\mathfrak{b}_0$ contains the stabilizer of the flag

$$\mathcal{F}_{i-1,i} = (e_1 \subset \cdots \subset \underbrace{e_{i-1} + e_n}_{i-1} \subset e_{i-1} \subset \cdots \subset e_{n-1}).$$

Hence,

$$m(s_{\alpha_{i-1}}) \cdot Q_i = Q_{i-1,i}. \tag{44}$$

A similar calculation shows that

$$m(s_{\alpha_i}) \cdot Q_i = Q_{i,i+1}. \tag{45}$$

Let $Q_c = K \cdot \mathfrak{b}_{\hat{v}}$ be a closed K-orbit and let $B_{\hat{v}} \subset G$ be the Borel subgroup with $\mathrm{Lie}(B_{\hat{v}}) = \mathfrak{b}_{\hat{v}}$. We observed in the proof of Proposition 4.12 that $K \cap B_{\hat{v}}$ is a Borel subgroup of K so that $Q_c \cong K/(K \cap B_{\hat{v}})$ is isomorphic to the flag variety $\mathcal{B}_K$ of K.

Definition-Notation 4.25. For a K-orbit Q on $\mathcal{B}$, we let $l(Q) := \dim(Q) - \dim(\mathcal{B}_K)$. The number $l(Q)$ is called the *length* of the K-orbit Q.

Proposition 4.26. *Let Q be any K-orbit in $\mathcal{B}$. Then there exists a sequence of simple roots $\alpha_{i_1}, \ldots, \alpha_{i_k} \in \Phi^+(\mathfrak{g}, \mathfrak{t}_0)$ and a closed orbit Q_c such that $Q = m(s_{\alpha_{i_k}}) \cdot \ldots \cdot m(s_{\alpha_{i_1}}) \cdot Q_c$. We let $Q_j = m(s_{\alpha_{i_j}}) \cdot \ldots \cdot m(s_{\alpha_{i_1}}) \cdot Q_c$. If for $j = 1, \ldots, k$, the root α_{i_j} is complex θ-stable or noncompact imaginary for Q_{j-1}, then $l(Q) = k$.*

Proof. This follows easily from [32], Theorem 4.6 . □

Let Q_v be the K-orbit corresponding to $v \in V$. We now compute the involution associated to the orbit $m(s_\alpha) \cdot Q_v$ when α is complex θ-stable or noncompact imaginary for v from the involution for the orbit Q_v. We denote the parameter $v' \in V$ for $m(s_\alpha) \cdot Q_v$ by $v' = m(s_\alpha) \cdot v$. By results from Section 4.3 and Proposition 4.26, we can then determine $\theta_{\widehat{v'}}$ for any v' in V.

There are two different cases we need to consider.

Case 1: α is noncompact imaginary for v. Let $v' = m(s_\alpha) \cdot v$. Then by Proposition 4.21, $K \cdot \mathfrak{b}_{\widehat{v'}} = K \cdot \mathrm{Ad}(\hat{v} u_\alpha)\mathfrak{b}_0$, where u_α is the representative for the Cayley transform with respect to α given in Equation (41).

We can now compute $\theta_{\widehat{v'}}$ in terms of $\theta_{\hat{v}}$.

Proposition 4.27. *Let $v' = m(s_\alpha) \cdot v$, where α is noncompact imaginary for v.*

(1) *Then $\hat{v} u_\alpha \in \mathcal{V}$ is a representative of v', and*

$$\tau(\widehat{v'}) = \tau(\hat{v} u_\alpha) = \dot{s_\alpha}^{-1} \tau(\hat{v}),$$

and

$$\phi(v') = s_\alpha \phi(v).$$

(2) *The involution for v' is given by*

$$\theta_{\widehat{v'}} = \mathrm{Ad}(\tau(\widehat{v'})) \circ \theta = \mathrm{Ad}(\dot{s_\alpha}^{-1})\mathrm{Ad}(\tau(\hat{v})) \circ \theta = \mathrm{Ad}(\dot{s_\alpha}^{-1}) \circ \theta_{\hat{v}},$$

and $\theta_{\widehat{v'}}$ *acts on the roots* $\Phi(\mathfrak{g}, \mathfrak{t}_0)$ *by:*

$$\theta_{\widehat{v'}} = s_\alpha \theta_{\hat{v}}.$$

Proof. It is easy to verify that if $g = \frac{1}{\sqrt{2}} \begin{bmatrix} 1 & \imath \\ \imath & 1 \end{bmatrix}$, then $\theta_{1,1}(g) = g^{-1}$. Hence, by Lemma 4.19, it follows that $\theta_{\hat{v}}(u_\alpha) = u_\alpha^{-1}$. Thus, by Equation (37), $\theta(u_\alpha) = \mathrm{Ad}(\tau(\hat{v})^{-1})(u_\alpha^{-1})$. It follows that

$$\tau(\hat{v}u_\alpha) = u_\alpha^{-1}\tau(\hat{v})\theta(u_\alpha) = u_\alpha^{-1}\tau(\hat{v})\tau(\hat{v})^{-1}u_\alpha^{-1}\tau(\hat{v}) = u_\alpha^{-2}\tau(\hat{v}).$$

Since $u_\alpha^{-2} = \dot{s_\alpha}^{-1}$, it follows that $\tau(\hat{v}u_\alpha) = \dot{s_\alpha}^{-1}\tau(\hat{v})$. By Equation (29) and Proposition 4.21, it follows that $\hat{v}u_\alpha \in \mathcal{V}$ is a representative of $m(s_\alpha) \cdot v$. By Equation (32), we have $\phi(m(s_\alpha) \cdot v) = s_\alpha \phi(v)$. Part (2) of the proposition now follows from part (1) and Equation (37). □

Case 2: α is complex θ-stable for v.

Proposition 4.28. *Let α be complex θ-stable for v.*

(1) Let $v' = m(s_\alpha) \cdot v$. Then v' has representative $\widehat{v'} = \hat{v}\dot{s_\alpha}$, so that $v' = s_\alpha \times v \in V$ and

$$\tau(\hat{v}\dot{s_\alpha}) = \dot{s_\alpha}^{-1}\tau(\hat{v})\theta(\dot{s_\alpha}),$$

whence

$$\phi(v') = s_\alpha \phi(v)\theta(s_\alpha).$$

(2) The involution $\theta_{\widehat{v'}}$ on $\mathfrak{g}$ associated to v' is given by

$$\theta_{\widehat{v'}} = \mathrm{Ad}(\dot{s_\alpha}^{-1}\tau(\hat{v})\theta(\dot{s_\alpha})) \circ \theta = \mathrm{Ad}(\dot{s_\alpha}^{-1}) \circ \theta_{\hat{v}} \circ \mathrm{Ad}(\dot{s_\alpha}),$$

so that the action of $\theta_{\widehat{v'}}$ on the roots $\Phi(\mathfrak{g}, \mathfrak{t}_0)$ is given by

$$\theta_{\widehat{v'}} = s_\alpha \phi(v)\theta(s_\alpha)\theta = s_\alpha \theta_{\hat{v}} s_\alpha.$$

Proof. By Proposition 4.21, we have $\mathfrak{b}_{\widehat{v'}} = \mathrm{Ad}(\hat{v}\dot{s_\alpha})\mathfrak{b}_0$ so that $v' = s_\alpha \times v$ by Equation (33). The rest of the proof follows by definitions. □

Lemma 4.29. *Let Q_v be the K-orbit corresponding to $v \in V$, and let α be a complex θ-stable simple root for v. Let β be a root of $\Phi^+(\mathfrak{g}, \mathfrak{t}_0)$. Then β is noncompact imaginary for v if and only if $s_\alpha(\beta)$ is noncompact imaginary for $m(s_\alpha) \cdot v$.*

Proof. Let $v' = m(s_\alpha) \cdot v$. Then by Proposition 4.28 (2), $\theta_{\widehat{v'}}(s_\alpha(\beta)) = s_\alpha(\theta_{\hat{v}}(\beta))$. Hence, β is imaginary for v if and only if $s_\alpha(\beta)$ is imaginary for v'. To prove the noncompactness assertion, it suffices to apply Proposition 4.28 (2) to a root vector $\mathrm{Ad}(\dot{s_\alpha}^{-1})(x_\beta)$, where x_β is a nonzero root vector in $\mathfrak{g}_\beta$. □

Example 4.30. We show how this theory helps describe the K-orbits $Q_{i,j}$ in the case when $G = GL(n, \mathbb{C})$ and $K = GL(n-1, \mathbb{C}) \times G(1, \mathbb{C})$. We let $v_{i,i+1} \in V$ parametrize the orbit $Q_{i,i+1}$. By Equation (45) and Propositions 4.12 and 4.27 (1), the Springer invariant $\phi(v_{i,i+1}) = (i \; i+1) = s_{\alpha_i}$, and using also Example 4.16, $v_{i,i+1}$ has representative $\widehat{v_{i,i+1}} = (n \; n-1 \ldots i)u_{\alpha_i}$, where u_{α_i} is the Cayley transform from Equation (41). Hence, α_i is real for $v_{i,i+1}$, while α_{i-1} and α_{i+1} are the only θ-stable complex simple roots (as before, in the case $i = 1$ or $n-1$, only one of these complex roots exists). Further, the imaginary roots for $v_{i,i+1}$ are the roots $\epsilon_j - \epsilon_k$ with $j, k \notin \{i, i+1\}$ and have root vectors $E_{j,k}$. Then by Proposition 4.27 (2), $\theta_{\widehat{v_{i,i+1}}}(E_{j,k}) = \mathrm{Ad}(\dot{s_{\alpha_i}}^{-1})\theta_{\widehat{v_i}}(E_{j,k})$, where $\dot{s_{\alpha_i}}$ is the representative for $s_{\alpha_i} \in W$ given in Equation (42). But by Example 4.16, $\theta_{\widehat{v_i}}(E_{j,k}) = E_{j,k}$, so the roots $\epsilon_j - \epsilon_k$ are compact. Hence, there are no noncompact imaginary roots for $Q_{i,i+1}$.

We now consider all orbits $Q_{i,j}$ with $i < j$. We let $v_{i,j} \in V$ denote the corresponding parameter, and we let $s_i = (i \; i+1)$ with representative $\dot{s_i}$ given by the corresponding permutation matrix.

Claim. (1) $Q_{i,j} = m(s_{j-1}) \cdot \ldots \cdot m(s_i) \cdot Q_i$ and $l(Q_{i,j}) = j - i$.
(2) $\phi(v_{i,j})$ is the transposition $(i \; j)$, $\theta_{\widehat{v_{i,j}}} = (i \; j)$ on roots, and $Q_{i,j}$ has representative given by the element $\widehat{v_{i,j}} = (n \; n-1 \; \ldots \; i)u_{\alpha_i}\dot{s}_{i+1} \ldots \dot{s}_{j-1}$.
(3) The simple roots $\alpha_{i-1} = \epsilon_{i-1} - \epsilon_i$ and $\alpha_j = \epsilon_j - \epsilon_{j+1}$ are the only complex θ-stable simple roots for $v_{i,j}$, and there are no noncompact imaginary roots for $v_{i,j}$.

We prove these claims by induction on $j-i$. Example 4.24 and our discussion in the first paragraph proves the claim when $j-i = 1$. It suffices to show that (1)–(3) of the claim for $Q_{i,j}$ imply the claim for $Q_{i,j+1}$. By Proposition 4.21 and Claims (2) and (3) for $Q_{i,j}$, it follows that $m(s_j) \cdot Q_{i,j} \neq Q_{i,j}$ and $m(s_j) \cdot Q_{i,j}$ has representative $\widehat{v_{i,j+1}}$. A routine computation with flags then shows that $K \cdot \mathrm{Ad}(\widehat{v_{i,j+1}})\mathfrak{b}_0 = Q_{i,j+1}$. Hence,

$$m(s_j) \cdot Q_{i,j} = Q_{i,j+1}. \tag{46}$$

Claim (1) for $Q_{i,j+1}$ then follows by induction. Claim (2) for $Q_{i,j}$ and Proposition 4.28 (1) imply that $\phi(v_{i,j+1})$ is the transposition $(i \; j+1)$. The formula for $\theta_{\widehat{v_{i,j+1}}}$ in Claim (2) follows from Proposition 4.28 part (2). Claim (3) now follows by Lemma 4.29 and an easy computation. This verifies Claims (1)–(3) for the orbit $Q_{i,j+1}$.

We remark that a computation similar to the one above verifies that

$$m(s_{i-1}) \cdot Q_{i,j} = Q_{i-1,j}. \tag{47}$$

Example 4.31. We retain the notation from the last example. We assert that every K-orbit Q in $\mathcal{B}$ is either of the form Q_i or $Q_{i,j}$ with $i < j$ and that these orbits are all distinct. We prove the first assertion by induction on $l(Q)$. If $l(Q) = 0$, then Q is closed, so $Q = Q_i$ by Example 4.16. If $l(Q) = 1$, then by Proposition 4.26, $Q = m(s_i) \cdot Q_c$ for some closed orbit Q_c, so by Example 4.24 and Equations (44) and (45), it follows that $Q = Q_{i,i+1}$ for some i. If $l(Q) = k > 1$, then Proposition 4.26 implies $Q = m(s_i) \cdot \widetilde{Q}$, where $l(\widetilde{Q}) = k - 1$, so by induction $\widetilde{Q} = Q_{j,j+k-1}$ for some j, and by Claim (3) of Example 4.30, the simple root α_i is either α_{j-1} or α_{j+k-1}. The first assertion now follows by Equations (46) and (47). By Example 4.16, the orbits Q_i are distinct. By Claim (2) of Example 4.30, the Springer invariant for $Q_{i,j}$ is $(i\ j)$, so that $Q_{i,j} = Q_{i',j'}$ if and only if $i = i'$ and $j = j'$. We now have a complete classification of the K-orbits on $\mathcal{B}$.

Example 4.32. We claim that $Q_{1,n}$ is the unique open orbit of K on $\mathcal{B}$, where we retain notation from the previous two examples. Indeed, by Claim (1) from Example 4.30, $l(Q_{1,n}) = n - 1 = \dim Q_{1,n} - \dim(\mathcal{B}_K)$, so that $\dim Q_{1,n} = n - 1 + \dim(\mathcal{B}_K) = \dim(\mathcal{B})$. It follows that $Q_{1,n}$ is open in $\mathcal{B}$.

Remark 4.33. The last three examples verify the assertions of [40], Section 2, and [26] for the case $G = GL(n, \mathbb{C})$ and $K = GL(n-1, \mathbb{C}) \times GL(1, \mathbb{C})$. In particular, they justify the statements made in [11], Section 3.1. Example 4.30 explains the definition of the element v in Equation (3.3) of [11] and the construction of the involution θ' in [11], Section 3.1, which is the critical ingredient in the proof of Theorem 3.2 above (see Remark 3.3).

References

1. Mark Adler, Pierre van Moerbeke, and Pol Vanhaecke. *Algebraic integrability, Painlevé geometry and Lie algebras*, Vol. 47, *Ergebnisse der Mathematik und ihrer Grenzgebiete. 3. Folge. A Series of Modern Surveys in Mathematics [Results in Mathematics and Related Areas. 3rd Series. A Series of Modern Surveys in Mathematics]*. Springer-Verlag, Berlin, 2004.
2. Michèle Audin. *Hamiltonian systems and their integrability*, Vol. 15, *SMF/AMS Texts and Monographs*. American Mathematical Society, Providence, RI, 2008. Translated from the 2001 French original by Anna Pierrehumbert, Translation edited by Donald Babbitt.
3. Roger Bielawski and Victor Pidstrygach. Gelfand-Zeitlin actions and rational maps. *Math. Z.*, 260(4):779–803, 2008.
4. Walter Borho and Hanspeter Kraft. Über Bahnen und deren Deformationen bei linearen Aktionen reduktiver Gruppen. *Comment. Math. Helv.*, 54(1):61–104, 1979.
5. Michel Brion and Aloysius G. Helminck. On orbit closures of symmetric subgroups in flag varieties. *Canad. J. Math.*, 52(2):265–292, 2000.
6. Neil Chriss and Victor Ginzburg. *Representation Theory and Complex Geometry*. Birkhäuser Boston Inc., Boston, MA, 1997.
7. Mark Colarusso. *The Gelfand-Zeitlin algebra and polarizations of regular adjoint orbits for classical groups*. PhD thesis, University of California, San Diego, 2007.
8. Mark Colarusso. The Gelfand-Zeitlin integrable system and its action on generic elements of $\mathfrak{gl}(n)$ and $\mathfrak{so}(n)$. In *New Developments in Lie Theory and Geometry (Cruz Chica, Córdoba, Argentina, 2007)*, Vol. 491, *Contemp. Math.*, pp. 255–281. Amer. Math. Soc., Providence, RI, 2009.

9. Mark Colarusso. The orbit structure of the Gelfand-Zeitlin group on $n \times n$ matrices. *Pacific J. Math.*, 250(1):109–138, 2011.
10. Mark Colarusso and Sam Evens. On algebraic integrability of Gelfand-Zeitlin fields. *Transform. Groups*, 15(1):46–71, 2010.
11. Mark Colarusso and Sam Evens. K-orbits on the flag variety and strongly regular nilpotent matrices. *Selecta Math. (N.S.)*, 18(1):159–177, 2012.
12. David H. Collingwood. *Representations of rank one Lie groups*, Vol. 137, *Research Notes in Mathematics*. Pitman (Advanced Publishing Program), Boston, MA, 1985.
13. Yu. A. Drozd, V. M. Futorny, and S. A. Ovsienko. Harish-Chandra subalgebras and Gel′fand-Zetlin modules. In *Finite-dimensional algebras and related topics (Ottawa, ON, 1992)*, Vol. 424, *NATO Adv. Sci. Inst. Ser. C Math. Phys. Sci.*, pp. 79–93. Kluwer Acad. Publ., Dordrecht, 1994.
14. Pavel Etingof. *Calogero-Moser systems and representation theory*. Zurich Lectures in Advanced Mathematics. European Mathematical Society (EMS), Zürich, 2007.
15. Sam Evens and Jiang-Hua Lu. Poisson geometry of the Grothendieck resolution of a complex semisimple group. *Mosc. Math. J.*, 7(4):613–642, 2007.
16. Vyacheslav Futorny. Harish-Chandra categories and Kostant's theorem. *Resenhas*, 6(2–3):177–186, 2004.
17. Vyacheslav Futorny. personal communication, 2008.
18. I. M. Gel′fand and M. L. Cetlin. Finite-dimensional representations of the group of unimodular matrices. *Doklady Akad. Nauk SSSR (N.S.)*, 71:825–828, 1950.
19. V. Guillemin and S. Sternberg. The Gel′fand-Cetlin system and quantization of the complex flag manifolds. *J. Funct. Anal.*, 52(1):106–128, 1983.
20. Henryk Hecht, Dragan Miličić, Wilfried Schmid, and Joseph A. Wolf. Localization and standard modules for real semisimple Lie groups. I. The duality theorem. *Invent. Math.*, 90(2):297–332, 1987.
21. Roger A. Horn and Charles R. Johnson. *Matrix analysis*. Cambridge University Press, Cambridge, 1985.
22. Toshiyuki Kobayashi. Restrictions of unitary representations of real reductive groups. In *Lie Theory*, Vol. 229, *Prog. Math.*, Birkhäuser, Boston, MA, 2005, pp. 139–207.
23. Bertram Kostant. Lie group representations on polynomial rings. *Amer. J. Math.*, 85:327–404, 1963.
24. Bertram Kostant and Nolan Wallach. Gelfand-Zeitlin theory from the perspective of classical mechanics. I. In *Studies in Lie Theory*, Vol. 243, *Prog. Math.*, Birkhäuser, Boston, MA, 2006, pp. 319–364.
25. Bertram Kostant and Nolan Wallach. Gelfand-Zeitlin theory from the perspective of classical mechanics. II. In *The unity of mathematics*, Vol. 244, *Prog. Math.*, Birkhäuser, Boston, MA, 2006, pp. 387–420.
26. Toshihiko Matsuki and Toshio Ōshima. Embeddings of discrete series into principal series. In *The orbit method in representation theory (Copenhagen, 1988)*, Vol. 82, *Prog. Math.*, Birkhäuser, Boston, MA, 1990, pp. 147–175.
27. David Mumford. *The red book of varieties and schemes*, Vol. 1358, *Lecture Notes in Mathematics*. Springer-Verlag, Berlin, expanded edition, 1999. Includes the Michigan lectures (1974) on curves and their Jacobians, With contributions by Enrico Arbarello.
28. Ryszard Nest and Boris Tsygan. Remarks on modules over deformation quantization algebras. *Mosc. Math. J.*, 4(4):911–940, 982, 2004.
29. Beresford Parlett and Gilbert Strang. Matrices with prescribed Ritz values. *Linear Algebra Appl.*, 428(7):1725–1739, 2008.
30. Beresford N. Parlett. *The symmetric eigenvalue problem*, Vol. 20, *Classics in Applied Mathematics*. Society for Industrial and Applied Mathematics (SIAM), Philadelphia, PA, 1998. Corrected reprint of the 1980 original.
31. R. W. Richardson. Orbits, invariants, and representations associated to involutions of reductive groups. *Invent. Math.*, 66(2):287–312, 1982.

32. R. W. Richardson and T. A. Springer. The Bruhat order on symmetric varieties. *Geom. Dedicata*, 35(1–3):389–436, 1990.
33. R. W. Richardson and T. A. Springer. Combinatorics and geometry of K-orbits on the flag manifold. In *Linear algebraic groups and their representations (Los Angeles, CA, 1992)*, Vol. 153, *Contemp. Math.*, Amer. Math. Soc., Providence, RI, 1993, pp. 109–142.
34. Noam Shomron and Beresford N. Parlett. Linear algebra meets Lie algebra: the Kostant-Wallach theory. *Linear Algebra Appl.*, 431(10):1745–1767, 2009.
35. T. A. Springer. Some results on algebraic groups with involutions. In *Algebraic groups and related topics (Kyoto/Nagoya, 1983)*, Vol. 6, *Adv. Stud. Pure Math.*, North-Holland, Amsterdam, 1985, pp. 525–543.
36. Robert Steinberg. *Endomorphisms of linear algebraic groups.* Memoirs of the American Mathematical Society, No. 80. American Mathematical Society, Providence, R.I., 1968.
37. Izu Vaisman. *Lectures on the geometry of Poisson manifolds*, Vol. 118, *Prog. in Math.*, Birkhäuser Verlag, Basel, 1994.
38. David A. Vogan. Irreducible characters of semisimple Lie groups. III. Proof of Kazhdan-Lusztig conjecture in the integral case. *Invent. Math.*, 71(2):381–417, 1983.
39. N. M. J. Woodhouse. *Geometric quantization.* Oxford Mathematical Monographs. The Clarendon Press, Oxford University Press, New York, second edition, 1992. Oxford Science Publications.
40. Atsuko Yamamoto. Orbits in the flag variety and images of the moment map for classical groups. I. *Represent. Theory*, 1:329–404 (electronic), 1997.

Diagrams of Hermitian type, highest weight modules, and syzygies of determinantal varieties

Thomas J. Enright, Markus Hunziker, and W. Andrew Pruett

Dedicated to Nolan Wallach, with gratitude and admiration

Abstract In this mostly expository paper, a natural generalization of Young diagrams for Hermitian symmetric spaces is used to give a concrete and uniform approach to a wide variety of interconnected topics including posets of noncompact roots, canonical reduced expressions, rational smoothness of Schubert varieties, parabolic Kazhdan–Lusztig polynomials, equivalences of categories of highest weight modules, BGG resolutions of unitary highest weight modules, and finally, syzygies and Hilbert series of determinantal varieties.

Keywords: Hermitian symmetric spaces • Kazhdan–Lusztig polynomials • Category $\mathcal{O}$ • Determinantal varieties • Syzygies

Mathematics Subject Classification: 22E47, 17B10, 13D02

T.J. Enright
Department of Mathematics, University of California San Diego, 9500 Gilman Drive #0112, La Jolla, CA 92093-0112, USA
e-mail: tenright@ucsd.edu

M. Hunziker
Department of Mathematics, Baylor University, One Bear Place #97328, Waco, TX 76798-7328, USA
e-mail: Markus_Hunziker@baylor.edu

W.A. Pruett
The University of Mississippi Medical Center, 2500 North State Street, Jackson, MS 39216, USA
e-mail: wpruett@umc.edu

R. Howe et al. (eds.), *Symmetry: Representation Theory and Its Applications*, Progress in Mathematics 257, DOI 10.1007/978-1-4939-1590-3_6

1 Introduction

In [44], Schubert gave explicit cell decompositions of the complex Grassmannians. The cells in these decompositions, the *Schubert cells*, are defined by certain incidence conditions and are naturally parametrized by integer partitions, or equivalently, Young diagrams. More precisely, the Schubert cells in the Grassmannian $\mathrm{Gr}(p, p+q)$ of complex p-planes in $\mathbb{C}^{p+q}$ can be parametrized by the Young diagrams that fit inside a $p \times q$-rectangle, such that the (complex) dimension of each Schubert cell is equal to the number of boxes of the corresponding Young diagram and the partial ordering on the set of the Schubert cells given by inclusion of their closures, the *Schubert varieties*, corresponds to the partial ordering given by inclusion of Young diagrams.

In [35], Kostant generalized Schubert's cell decompositions of the Grassmannians to flag varieties G/P, where G is a complex (semi-)simple algebraic group and P is any parabolic subgroup of G. These cell decompositions are obtained from the Bruhat decomposition (due to Harish-Chandra [24]) of G and the cells are parametrized by a certain subset W^{I} of the Weyl group W of G. The (complex) dimension of each Schubert cell is equal to the length of the corresponding Weyl group element and the partial ordering on the set of Schubert varieties is given by the Bruhat ordering on W restricted to W^{I}. In the special case when G/P is a Hermitian symmetric space (or, equivalently, a cominuscule flag variety), it follows from Proctor's thesis [41] that the poset W^{I} is a distributive lattice that can be identified with the lattice of lower-order ideals of the poset of positive noncompact roots. It turns out that the poset of positive noncompact roots can always be embedded into a two-dimensional square lattice and its lower-order ideals are represented by generalized Young diagrams. These diagrams have recently been (re)discovered by several authors [36,42,47].

The main purpose of this paper is to show how these generalized Young diagrams can be used to study categories of highest weight modules for Hermitian symmetric spaces and related topics. These categories were studied extensively by Enright and Shelton in [19,20] and this paper may be viewed as an illustrated guide to the results in [19, 20]. This paper also extends work by Lascoux and Schützenberger [38] and Boe [6] on the calculation of Kazhdan–Lusztig polynomials for Hermitian symmetric spaces and work by Enright and Willenbring [22] and Enright and Hunziker [16, 17] on Bernstein–Gelfand–Gelfand resolutions and Hilbert series of unitary highest weight representations.

To assuage the reader's curiosity how all of this is related to syzygies of determinantal varieties, let us consider the determinantal variety $Y_{p-2} \subset \mathrm{M}_p$ of complex $p \times p$ matrices of rank $\leq p-2$. Via Howe duality, the coordinate ring $\mathbb{C}[Y_{p-2}]$ carries the structure of a unitary highest weight module of the real Lie algebra $\mathfrak{u}(p, p)$. This highest weight module—and in fact every unitary highest weight module—admits a Bernstein–Gelfand–Gelfand resolution in terms of parabolic Verma modules. Each parabolic Verma module, as a $\mathrm{GL}_p \times \mathrm{GL}_p$-module, is of the form $S \otimes F$, where $S := \mathbb{C}[\mathrm{M}_p]$ is the coordinate ring of M_p

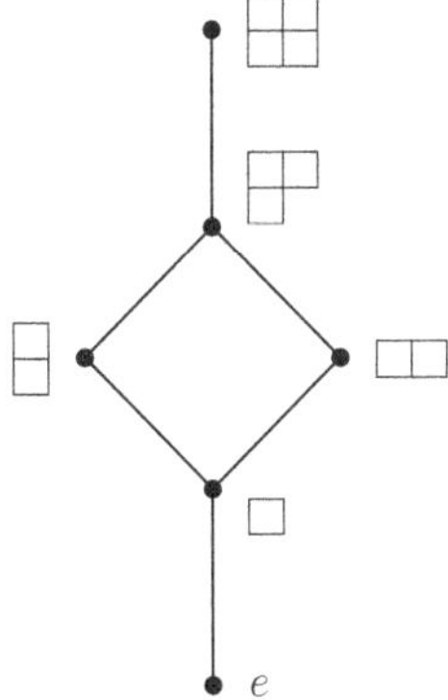

Figure 1 Young lattice associated to the Grassmannian Gr(2, 4)

with the usual $\mathrm{GL}_p \times \mathrm{GL}_p$-action and F is a finite-dimensional simple $\mathrm{GL}_p \times \mathrm{GL}_p$-module. As we will explain in Section 7, the parabolic Verma modules that appear in the resolution of $\mathbb{C}[Y_{p-2}]$ are in a natural bijection with the Young diagrams that were shown in Figure 1 and we have a minimal free resolution of $\mathbb{C}[Y_{p-2}]$ as a graded S-module of the form

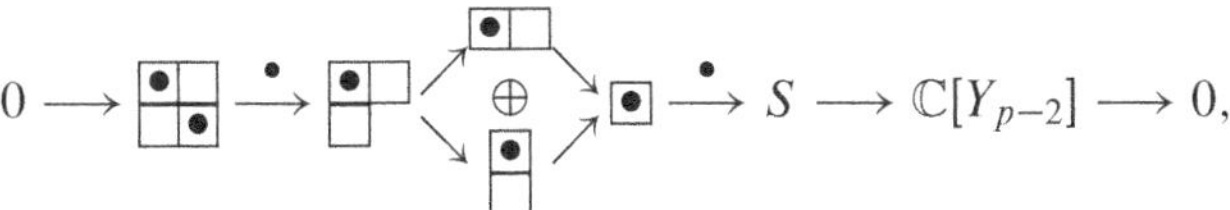

where the two arrows that are marked with a dot are homogenous maps of degree $p+1$ and the other maps are of degree 1. The mystery of what precisely the diagrams stand for is revealed in Example 7.11.

We remark that much of this work should be considered expository. In particular, we added background material for students and experts who are familiar with flag varieties and determinantal varieties, but who have not had much exposure to category $\mathcal{O}$ (and vice versa).

Acknowledgments. This paper grew out of a series of lectures that the second author gave at the University of Georgia in Athens during the VIGRE 2010 Summer School on Geometry and Representation Theory. The second author would like to thank the organizers for their hospitality as well as the participants for their contribution to the wonderful atmosphere of the school. Special thanks go to B. Boe, S. Evens, W. Graham, S. Kumar, D. Nakano, P. Trapa, R. Varley, and R. Zierau. We also thank J. Alexander for his careful proofreading.

2 Parabolic subalgebras of Hermitian type

Throughout the paper let $\mathfrak{g}$ be a complex simple Lie algebra of rank n. We fix a Cartan subalgebra $\mathfrak{h}$ of $\mathfrak{g}$ and denote by $\Phi \subset \mathfrak{h}^*$ the root system of $\mathfrak{g}$ relative to $\mathfrak{h}$. For every $\alpha \in \Phi$ we have a 1-dimensional root space

$$\mathfrak{g}_\alpha := \{x \in \mathfrak{g} \mid [h, x] = \alpha(h)x \text{ for all } h \in \mathfrak{h}\}.$$

We fix a simple system $\Pi \subset \Phi$ and write $\Pi = \{\alpha_1, \ldots, \alpha_n\}$, where the simple roots α_i are labelled as in Bourbaki [9]. The simple system determines a positive system $\Phi^+ \subset \Phi$. Let

$$\mathfrak{b} := \mathfrak{h} \oplus \bigoplus_{\alpha \in \Phi^+} \mathfrak{g}_\alpha$$

be the corresponding standard Borel subalgebra of $\mathfrak{g}$. Any Borel subalgebra of $\mathfrak{g}$ is conjugate to $\mathfrak{b}$ under the action of the adjoint group of $\mathfrak{g}$.

A *parabolic subalgebra* $\mathfrak{p}$ of $\mathfrak{g}$ is a subalgebra containing some Borel subalgebra of $\mathfrak{g}$. If $\mathfrak{p}$ contains the fixed Borel subalgebra $\mathfrak{b}$, we say that $\mathfrak{p}$ is a *standard* parabolic subalgebra. By the remark above, any parabolic subalgebra $\mathfrak{p}$ of $\mathfrak{g}$ is conjugate to a standard parabolic subalgebra under the action of the adjoint group of $\mathfrak{g}$. There is a one-to-one correspondence between subsets of Π and standard parabolic subalgebras of $\mathfrak{g}$ as follows. For $\mathrm{I} \subset \Pi$, define a root system Φ_I by

$$\Phi_\mathrm{I} := \Phi \cap \sum_{\alpha \in \mathrm{I}} \mathbb{Z}\alpha.$$

Next define a reductive subalgebra $\mathfrak{m}_\mathrm{I}$ of $\mathfrak{g}$ and a nilpotent subalgebra $\mathfrak{u}_\mathrm{I}$ of $\mathfrak{g}$ by

$$\mathfrak{m}_\mathrm{I} := \mathfrak{h} \oplus \bigoplus_{\alpha \in \Phi_\mathrm{I}} \mathfrak{g}_\alpha \quad \text{and} \quad \mathfrak{u}_\mathrm{I} := \bigoplus_{\alpha \in \Phi^+ \setminus \Phi_\mathrm{I}} \mathfrak{g}_\alpha.$$

Then $\mathfrak{p}_\mathrm{I} := \mathfrak{m}_\mathrm{I} \oplus \mathfrak{u}_\mathrm{I}$ is a standard parabolic subalgebra of $\mathfrak{g}$ with Levi subalgebra $\mathfrak{m}_\mathrm{I}$ and nilradical $\mathfrak{u}_\mathrm{I}$. If $|\Pi \setminus \mathrm{I}| = 1$ we say that $\mathfrak{p}_\mathrm{I}$ is a maximal parabolic subalgebra. The reductive Lie algebra $\mathfrak{m}_\mathrm{I}$ can be further decomposed as a direct sum $\mathfrak{m}_\mathrm{I} = \mathfrak{z}_\mathrm{I} \oplus [\mathfrak{m}_\mathrm{I}, \mathfrak{m}_\mathrm{I}]$, where $\mathfrak{z}_\mathrm{I} := \bigcap_{\alpha \in \mathrm{I}} \ker \alpha$ is the center of $\mathfrak{m}_\mathrm{I}$ and $[\mathfrak{m}_\mathrm{I}, \mathfrak{m}_\mathrm{I}]$ is the semisimple part of $\mathfrak{m}_\mathrm{I}$. Note that $\dim \mathfrak{z}_\mathrm{I} = |\Pi \setminus \mathrm{I}|$; in particular, $\mathfrak{p}_\mathrm{I}$ is a maximal parabolic subalgebra if and only if the center of $\mathfrak{m}_\mathrm{I}$ is one-dimensional.

To simplify notation, we often omit the subscript I. Thus, we would simply write $\mathfrak{p} = \mathfrak{m} \oplus \mathfrak{u}$ instead of $\mathfrak{p}_\mathrm{I} = \mathfrak{m}_\mathrm{I} \oplus \mathfrak{u}_\mathrm{I}$. Of particular importance is the set

$$\Phi(\mathfrak{u}) := \Phi^+ \setminus \Phi_\mathrm{I},$$

which we will later study as a partially ordered subset of Φ^+.

2.1 Parabolic subalgebras of Hermitian type

Suppose that $\mathfrak{g}_{\mathbb{R}}$ is a noncompact real form of $\mathfrak{g}$ with Cartan decomposition $\mathfrak{g}_{\mathbb{R}} = \mathfrak{m}_{\mathbb{R}} \oplus \mathfrak{s}_{\mathbb{R}}$ and assume that $\mathfrak{m}_{\mathbb{R}}$ has a nontrivial center $\mathfrak{z}_{\mathbb{R}}$. This latter condition is equivalent to $(\mathfrak{g}_{\mathbb{R}}, \mathfrak{m}_{\mathbb{R}})$ being the pair of Lie algebras of an irreducible Hermitian symmetric pair $(G_{\mathbb{R}}, M_{\mathbb{R}})$ of noncompact type, which means that $G_{\mathbb{R}}/M_{\mathbb{R}}$ is a noncompact symmetric space that admits a complex structure such that $G_{\mathbb{R}}$ acts as biholomorphic transformations.

Example 2.1. The simple Lie algebra $\mathfrak{g}_{\mathbb{R}} = \mathfrak{sl}(n, \mathbb{R})$ has a Cartan decomposition $\mathfrak{g}_{\mathbb{R}} = \mathfrak{m}_{\mathbb{R}} \oplus \mathfrak{s}_{\mathbb{R}}$, where $\mathfrak{m}_{\mathbb{R}} = \mathfrak{so}(n)$ and $\mathfrak{s}_{\mathbb{R}}$ is the space of real symmetric matrices of trace 0. The center of $\mathfrak{m}_{\mathbb{R}}$ is nontrivial if and only if $n = 2$ (in which case $\mathfrak{m}_{\mathbb{R}}$ is equal to its center). The Hermitian symmetric space $\mathrm{SL}(2, \mathbb{R})/\,\mathrm{SO}(2)$ can be identified with the upper halfplane $\mathfrak{H} := \{z \in \mathbb{C} \mid \operatorname{Im} z > 0\}$ on which $\mathrm{SL}(2, \mathbb{R})$ acts as linear fractional transformations as usual.

The general theory implies that $\mathfrak{z}_{\mathbb{R}} = \mathbb{R}h_0$, where h_0 can be chosen such that $\operatorname{ad}(h_0) : \mathfrak{g} \to \mathfrak{g}$ has eigenvalues 0 and $\pm i$ with the 0-eigenspace being the complexification $\mathfrak{m}$ of $\mathfrak{m}_{\mathbb{R}}$. Define

$$\mathfrak{s}^{\pm} := \{x \in \mathfrak{g} \mid [h_0, x] = \pm i x\}.$$

Then the space $\mathfrak{s}^+$ is the holomorphic tangent space to the Hermitian symmetric space $G_{\mathbb{R}}/M_{\mathbb{R}}$ at $eM_{\mathbb{R}}$. Furthermore, $[\mathfrak{m}, \mathfrak{s}^+] = \mathfrak{s}^+$, $[\mathfrak{s}^+, \mathfrak{s}^+] = 0$, and $\mathfrak{p} := \mathfrak{m} \oplus \mathfrak{s}^+$ is a maximal parabolic subalgebra of $\mathfrak{g}$ with Levi component $\mathfrak{m}$ and nilradical $\mathfrak{s}^+$. If a (standard) parabolic subalgebra $\mathfrak{p} \subset \mathfrak{g}$ can be obtained in this way, we say that $\mathfrak{p}$ is of *Hermitian type*. The distinguished class of parabolic subalgebras of Hermitian type (which are also called parabolic subalgebras of *cominuscule type* in the literature) can be characterized in a variety of equivalent ways.

Lemma 2.2. *Let $\mathfrak{p} = \mathfrak{p}_{\mathrm{I}}$ be a maximal parabolic subalgebra of $\mathfrak{g}$ with Levi decomposition $\mathfrak{p} = \mathfrak{m} \oplus \mathfrak{u}$. Then the following are equivalent:*

(i) *The parabolic subalgebra $\mathfrak{p}$ is of Hermitian type.*
(ii) *There exists an element $h \in \mathfrak{z}(\mathfrak{m})$ such that $\mathfrak{u} = \{x \in \mathfrak{g} \mid [h, x] = x\}$.*
(iii) *The nilradical $\mathfrak{u}$ is abelian.*
(iv) *If $\alpha, \beta \in \Phi(\mathfrak{u})$, then the sum $\alpha + \beta$ is not a root.*
(v) *If $\alpha, \beta \in \Phi(\mathfrak{u})$ and $\beta - \alpha \in \Pi$, then $\beta - \alpha \in \mathrm{I}$.*
(vi) *The simple root in $\Pi \setminus \mathrm{I}$ occurs with coefficient 1 in the highest root θ.*
(vii) *The pair $(\mathfrak{g}, \mathfrak{m})$ is of type $(\mathsf{A}_n, \mathsf{A}_{p-1} \times \mathsf{A}_{n-p})$ with $1 \leq p \leq n$, $(\mathsf{B}_n, \mathsf{B}_{n-1})$, $(\mathsf{C}_n, \mathsf{A}_{n-1})$, $(\mathsf{D}_n, \mathsf{A}_{n-1})$, $(\mathsf{D}_n, \mathsf{D}_{n-1})$, $(\mathsf{E}_6, \mathsf{D}_5)$, or $(\mathsf{E}_7, \mathsf{E}_6)$.*

Proof. We prove the implications (i) $\Rightarrow$ (ii) $\Rightarrow \cdots \Rightarrow$ (vii) $\Rightarrow$ (i).

(i) $\Rightarrow$ (ii): Take $h := -ih_0$, where h_0 is defined as above.

(ii) $\Rightarrow$ (iii): Assume that $\mathfrak{u} = \{x \in \mathfrak{g} \mid [h, x] = x\}$ for some $h \in \mathfrak{g}$. Then, by the Jacobi identity, for every $x, y \in \mathfrak{u}$ we have

$$0 = [h, [x, y]] + [x, [y, h]] + [y, [h, x]] = [x, y] - [x, y] - [x, y] = -[x, y].$$

Thus $\mathfrak{u}$ is abelian.

(iii) $\Rightarrow$ (iv): Suppose that there exist $\alpha, \beta \in \Phi(\mathfrak{u})$ such that $\alpha + \beta \in \Phi$. Then $[\mathfrak{g}_\alpha, \mathfrak{g}_\beta] = \mathfrak{g}_{\alpha+\beta} \neq 0$ and hence $\mathfrak{u}$ is not abelian.

(iv) $\Rightarrow$ (v): Let $\alpha, \beta \in \Phi(\mathfrak{u})$ such that $\beta - \alpha \in \Pi$. Suppose that $\beta - \alpha = \alpha_p \in \Pi \setminus \mathrm{I}$. Then $\beta - \alpha \in \Phi(\mathfrak{u})$ and $\beta = \alpha + (\beta - \alpha)$ is a sum of two roots in $\Phi(\mathfrak{u})$.

(v) $\Rightarrow$ (vi): There exists a sequence of positive roots $\beta_1, \dots, \beta_k \in \Phi^+$ such that β_1 is the simple root in $\Pi \setminus \mathrm{I}$, β_k is the highest root θ, and $\beta_{i+1} - \beta_i \in \Pi$ for $1 \leq i \leq k-1$. Since $\Phi(\mathfrak{u}) = \{\sum_j c_j \alpha_j \in \Phi^+ \mid c_p \geq 1\}$, the roots β_i are in fact in $\Phi(\mathfrak{u})$. Assuming (v) it follows that $\beta_{i+1} - \beta_i \in \mathrm{I}$ for $1 \leq i \leq k-1$ and hence $\theta = \sum_j c_j \alpha_j$ with $c_p = 1$.

(vi) $\Rightarrow$ (vii): By inspection.

(vii) $\Rightarrow$ (i): For each of the types listed there exists a pair of real Lie algebras $(\mathfrak{g}_\mathbb{R}, \mathfrak{m}_\mathbb{R})$ such that (i) is satisfied. Explicitly, these pairs are given in Table 1. □

If $\mathfrak{p} = \mathfrak{m} \oplus \mathfrak{u}$ is of Hermitian type, the roots in Φ_I are called the *compact roots* and the roots in $\Phi \setminus \Phi_\mathrm{I}$ are called the *noncompact roots*. Note that there is a unique noncompact simple root, namely the root $\alpha \in \Pi \setminus \mathrm{I}$. The set of positive noncompact roots, $\Phi(\mathfrak{u}) = \Phi^+ \setminus \Phi_\mathrm{I}$, will play a distinguished role in this paper.

Table 1. Hermitian symmetric pairs of noncompact type

$\mathfrak{g}_\mathbb{R}$	$\mathfrak{m}_\mathbb{R}$	Dynkin diagram
$\mathfrak{su}(p, q)$	$\mathfrak{s}(\mathfrak{u}(p) \oplus \mathfrak{u}(q))$	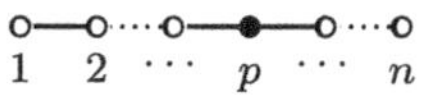 1 2 ⋯ p ⋯ n
$\mathfrak{so}(2n-1, 2)$	$\mathfrak{so}(2n-1) \oplus \mathfrak{so}(2)$	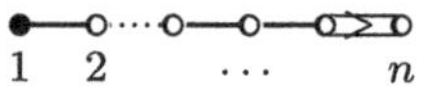 1 2 ⋯ n
$\mathfrak{sp}(n, \mathbb{R})$	$\mathfrak{u}(n)$	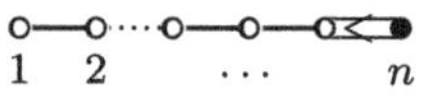 1 2 ⋯ n
$\mathfrak{so}(2n-2, 2)$	$\mathfrak{so}(2n-2) \oplus \mathfrak{so}(2)$	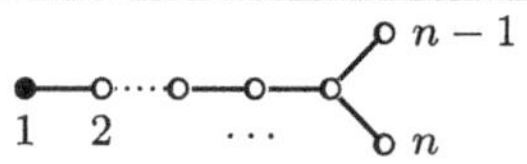 $n-1$ 1 2 ⋯ n
$\mathfrak{so}^*(2n)$	$\mathfrak{u}(n)$	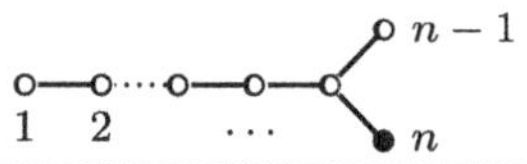 $n-1$ 1 2 ⋯ n
$\mathfrak{e}_{6(-14)}$	$\mathfrak{so}(10) \oplus \mathbb{R}$	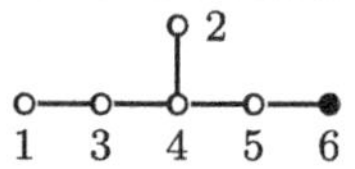 2 1 3 4 5 6
$\mathfrak{e}_{7(-25)}$	$\mathfrak{e}_6 \oplus \mathbb{R}$	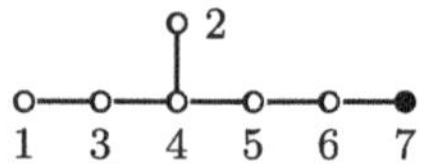 2 1 3 4 5 6 7

3 Generalized Young diagrams of Hermitian type

In the following, let (,) denote the nondegenerate bilinear form on $\mathfrak{h}^*$ which is induced from the Killing form of $\mathfrak{g}$. For $\alpha \in \Phi$, set $\alpha^\vee := 2\alpha/(\alpha,\alpha)$ and define the reflection $s_\alpha : \mathfrak{h}^* \to \mathfrak{h}^*$ by $s_\alpha(\lambda) = \lambda - (\lambda, \alpha^\vee)\alpha$. The reflections are elements of the Weyl group W which is generated by the simple reflections s_α, $\alpha \in \Pi$. The length of an element $w \in W$, denoted $\ell(w)$, is the length of the shortest word representing w as a product of the simple reflections.

3.1 Some basic observations due to Kostant

For $w \in W$, define

$$\Phi_w := \Phi^+ \cap w\Phi^-.$$

It is well known that $|\Phi_w| = \ell(w)$, the length of w. Furthermore, there is a unique element $w_\circ \in W$ such that $\Phi_{w_\circ} = \Phi^+$. For any subset $\Psi \subset \Phi^+$ define $\langle \Psi \rangle := \sum_{\alpha \in \Psi} \alpha$.

Lemma 3.1 (Kostant [34, (5.10.1)]). *For every $w \in W$,*

$$w\rho = \rho - \langle \Phi_w \rangle,$$

where $\rho := \frac{1}{2}\langle \Phi^+ \rangle$ as usual.

Proof. It follows immediately from the definitions that $\langle \Phi^+ \rangle + \langle w\Phi^- \rangle = 2\langle \Phi_w \rangle$. Since $\langle w\Phi^- \rangle = -\langle w\Phi^+ \rangle$, this implies $\rho - w\rho = \langle \Phi_w \rangle$. □

For subsets $\Psi_1, \Psi_2 \subset \Phi^+$ define $\Psi_1 \dot{+} \Psi_2 := (\Psi_1 + \Psi_2) \cap \Phi^+$.

Proposition 3.2 (Kostant [34, Prop. 5.10]). *The mapping $w \mapsto \Phi_w$ is a bijection of W onto the family of subsets $\Psi \subset \Phi^+$ such that both Ψ and $\Psi^c := \Phi^+ \setminus \Psi$ are closed under $\dot{+}$.* □

Proof. Since the W-orbit of ρ consists of $|W|$ elements, the lemma above implies that the mapping $w \mapsto \Phi_w$ is injective. It follows immediately from the definition that Φ_w is closed under $\dot{+}$. Since $(\Phi_w)^c = \Phi^+ \cap w\Phi^+ = \Phi^+ \cap ww_\circ\Phi^- = \Phi_{ww_\circ}$, it also follows that $(\Phi_w)^c$ is closed under $\dot{+}$. Now consider a subset $\Psi \subset \Phi^+$ such that both Ψ and Ψ^c are closed under $\dot{+}$. Define $\widetilde{\Psi} := \Psi \cup -(\Psi^c)$. Then $\widetilde{\Psi}$ is closed under $\dot{+}$ and $\Phi = \widetilde{\Psi} \cup -\widetilde{\Psi}$ is a disjoint union. This means that $\widetilde{\Psi}$ is a positive system of roots. Since the Weyl group acts transitively on positive systems, there exists a (unique) $w \in W$ such that $w\Phi^- = \widetilde{\Psi}$. Obviously, $\Psi = \Phi_w$ □

3.2 Minimal length coset representatives

Let $\mathfrak{p} = \mathfrak{p}_{\mathrm{I}}$ be a standard parabolic subalgebra with Levi decomposition $\mathfrak{p} = \mathfrak{m} \oplus \mathfrak{u}$ and let $W_{\mathrm{I}} := \langle s_\alpha \mid \alpha \in \mathrm{I} \rangle$, which we identify with the Weyl group of the reductive Lie algebra $\mathfrak{m}$. Note that since the Levi decomposition $\mathfrak{p} = \mathfrak{m} \oplus \mathfrak{u}$ is ad $\mathfrak{m}$-invariant, Φ_{I} and $\Phi(\mathfrak{u})$ are stable under the action of W_{I}.

Definition 3.3. Following Kostant [34, (5.13.1)], define

$$^{\mathrm{I}}W := \{w \in W \mid \Phi_w \subset \Phi(\mathfrak{u})\}.$$

For later reference, we also define $W^{\mathrm{I}} := \{w^{-1} \mid w \in {}^{\mathrm{I}}W\}$.

Proposition 3.4 (Kostant [34, Prop. 5.13]). *Every element $w \in W$ can be uniquely written in the form $w = uv$, where $u \in W_{\mathrm{I}}$ and $v \in {}^{\mathrm{I}}W$. Furthermore, if $w = uv$ is such a decomposition, then $\ell(w) = \ell(u) + \ell(v)$.*

Proof. Let $v_1, v_2 \in {}^{\mathrm{I}}W$ and suppose that $v_1 v_2^{-1} \in W_{\mathrm{I}}$. Since $\Phi(\mathfrak{u})$ is stable under W_{I}, it follows that $\Phi_{v_2^{-1}} \subset \Phi_{v_1^{-1}}$. Since $v_2 v_1^{-1} = (v_1 v_2^{-1})^{-1} \in W_{\mathrm{I}}$ it also follows that $\Phi_{v_1^{-1}} \subset \Phi_{v_2^{-1}}$ and hence $\Phi_{v_2^{-1}} = \Phi_{v_1^{-1}}$. By Proposition 3.2, $v_1 = v_2$. Now let $w \in W$ be arbitrary and consider $\Psi := \Phi_{\mathrm{I}}^+ \cap w\Phi^-$. Then Ψ and $\Psi^c := \Phi_{\mathrm{I}}^+ \setminus \Psi$ are both closed under $\dot{+}$. By Proposition 3.2, with Φ^+ replaced by Φ_{I}^+ and W replaced by W_{I}, there exists $u \in W_{\mathrm{I}}$ such that $\Psi = \Phi_{\mathrm{I}}^+ \cap u\Phi_{\mathrm{I}}^-$. Since the decomposition $\Phi^- = \Phi_{\mathrm{I}}^- \cup -\Phi(\mathfrak{u})$ is stable under W_{I}, it follows that $\Psi = \Phi_{\mathrm{I}}^+ \cap u\Phi^-$ and hence $\Phi_w \cap \Phi_{\mathrm{I}}^+ = \Phi_u$. Set $v := u^{-1}w$. Again using that $\Phi(\mathfrak{u})$ is stable under W_{I}, it is straightforward to show that $\Phi_w \cap \Phi(\mathfrak{u}) = u\Phi_v$. In particular, $\Phi_v \subset u^{-1}\Phi(\mathfrak{u}) = \Phi(\mathfrak{u})$ and hence $v \in {}^{\mathrm{I}}W$. Since $\Phi_w \cap \Phi_{\mathrm{I}}^+ = \Phi_u$ and $\Phi_w \cap \Phi(\mathfrak{u}) = u\Phi_v$, it follows that $\Phi_w = \Phi_u \cup u(\Phi_v)$ is a disjoint union. This shows that $\ell(w) = \ell(u) + \ell(v)$. □

Corollary 3.5. *The set ${}^{\mathrm{I}}W$ is the set of minimal length coset representatives of $W_{\mathrm{I}}\backslash W$ (i.e., the set of right cosets of W_{I}).* □

Remark 3.6 (cf. [34, Rem. 5.13]). In the literature, the set ${}^{\mathrm{I}}W$ of minimal length coset representatives of $W_{\mathrm{I}}\backslash W$ is often characterized in different equivalent ways. In fact, for $w \in W$, the following are equivalent:

(i) $\Phi_w \subset \Phi(\mathfrak{u})$.
(ii) $w^{-1}\Phi_{\mathrm{I}}^+ \subset \Phi^+$.
(iii) $w\rho$ is Φ_{I}^+-dominant integral.
(iv) $\ell(s_\alpha w) = \ell(w) + 1$ for all $\alpha \in \mathrm{I}$.

Proof. Let $\alpha \in \Phi^+$. Then $\alpha \in \Phi_w$ if and only if $w^{-1}\alpha \in \Phi^-$. Since $\Phi(\mathfrak{u}) = \Phi^+ \setminus \Phi_{\mathrm{I}}$, this immediately gives the equivalence of (i) and (ii). The equivalence of (ii) and (iii) is also obvious since $(w\rho, \alpha^\vee) = (\rho, w^{-1}\alpha^\vee) = (\rho, (w^{-1}\alpha)^\vee)$. Finally, (i) and (iv) are equivalent precisely because ${}^{\mathrm{I}}W$ is the set of minimal length coset representatives of $W_{\mathrm{I}}\backslash W$. □

3.3 Ideals

If $(\mathcal{P}, \leq)$ is any partially ordered set, we say that a subset $\mathcal{I} \subset \mathcal{P}$ is a *lower-order ideal* of $\mathcal{P}$ if $x \in \mathcal{I}$ and $x \geq y \in \mathcal{P}$ implies $y \in \mathcal{I}$. Similarly, a subset $\mathcal{J} \subset \mathcal{P}$ is called an *upper order ideal* of $\mathcal{P}$ if $x \in \mathcal{J}$ and $x \leq y \in \mathcal{P}$ implies $y \in \mathcal{J}$. In the following, we will view the sets Φ^+ and $\Phi(\mathfrak{u})$ as partially ordered sets with the usual ordering induced from the ordering on $\mathfrak{h}^*$, i.e., $\mu \leq \lambda$ if and only if $\lambda - \mu$ is a sum of positive roots or zero.

Lemma 3.7. *Suppose $\mathfrak{p}$ is of Hermitian type. Let $w \in {}^{\mathrm{I}}W$. Then Φ_w is a lower-order ideal of $\Phi(\mathfrak{u})$ and $\Phi(\mathfrak{u}) \setminus \Phi_w$ is an upper order ideal of Φ^+.*

Proof. Let $w \in {}^{\mathrm{I}}W$, $\alpha \in \Phi_w$, and $\beta \in \Phi(\mathfrak{u})$ such that $\beta < \alpha$. We have to show that $w^{-1}\beta < 0$. We may assume $\delta := \alpha - \beta \in \Phi^+$. First, note that $\delta \in \Phi_{\mathrm{I}}^+$ since $\delta \in \Phi(\mathfrak{u})$ would imply (by Lemma 2.2 (iv)) that $\alpha = \beta + \delta$ is not a root which is absurd. By Remark 3.6 (ii), it then follows that $w^{-1}\delta > 0$ and hence $w^{-1}\beta = w^{-1}\alpha - w^{-1}\delta < 0$. This proves the first statement of the lemma. Now let $w \in {}^{\mathrm{I}}W$, $\alpha \in \Phi(\mathfrak{u}) \setminus \Phi_w$ and $\beta \in \Phi^+$ such that $\beta > \alpha$. This time we have to show that $w^{-1}\beta > 0$. By a similar argument as above we may assume $\delta := \beta - \alpha \in \Phi^+$. Again we must have $\delta \in \Phi_{\mathrm{I}}^+$ since otherwise $\beta = \alpha + \delta$ is not a root. It then follows that $w^{-1}\beta = w^{-1}\alpha + w^{-1}\delta > 0$. □

Proposition 3.8. *Suppose that $\mathfrak{p}$ is of Hermitian type. Then we have a bijection*

$$ {}^{\mathrm{I}}W \to \{\text{ideals of } \mathfrak{b} \text{ contained in } \mathfrak{u}\} $$

given by $w \mapsto \mathfrak{a}_w := \sum_{\alpha \in \Phi(\mathfrak{u}) \setminus \Phi_w} \mathfrak{g}_\alpha$.

Proof. By Lemma 3.7, $\Phi(\mathfrak{u}) \setminus \Phi_w$ is an upper order ideal of Φ^+. This implies that $\sum_{\alpha \in \Phi(\mathfrak{u}) \setminus \Phi_w} \mathfrak{g}_\alpha$ is an ideal of $\mathfrak{b}$. Conversely, if $\mathfrak{a}$ is an ideal of $\mathfrak{b}$ contained in $\mathfrak{u}$, then $\mathfrak{a} = \sum_{\alpha \in \Phi(\mathfrak{u}) \setminus \Psi} \mathfrak{g}_\alpha$ for some subset $\Psi \subset \Phi(\mathfrak{u})$ such that $\Phi(\mathfrak{u}) \setminus \Psi$ is an upper order ideal of Φ^+. By Proposition 3.2, it follows that $\Psi = \Phi_w$ for some $w \in {}^{\mathrm{I}}W$ if we show that both Ψ and $\Psi^c = \Phi^+ \setminus \Psi$ are closed under $\dot{+}$. Since $\Psi \subset \Phi(\mathfrak{u})$ and since $\mathfrak{u}$ is abelian we have $\Psi \dot{+} \Psi = \varnothing$. Now let $\alpha, \beta \in \Phi^c$. Since $\Psi^c = \Phi_{\mathrm{I}}^+ \cup (\Phi(\mathfrak{u}) \setminus \Psi)$ and since Φ_{I}^+ is closed under $\dot{+}$, we may assume that $\alpha \in \Phi(\mathfrak{u}) \setminus \Psi$. Then, since $\Phi(\mathfrak{u}) \setminus \Psi$ is an ideal of Φ^+, $\alpha + \beta \in \Phi(\mathfrak{u}) \setminus \Psi$. □

3.4 Bruhat ordering

For $v, w \in W$, write $v \to w$ if $\ell(w) > \ell(v)$ and $w = vs_\alpha$ for some $\alpha \in \Phi^+$. Then define $v < w$ if there exists a sequence $v = w_1 \to w_2 \to \ldots \to w_m = w$. The resulting partial ordering "$\leq$" on W is called the *Bruhat ordering*. If in the definition of $v \to w$ we insist that $w = vs_\alpha$ for some $\alpha \in \Pi$, then the resulting ordering is called the *(right) weak Bruhat ordering*. For the weak ordering, the following terminology is useful. For $\alpha \in \Pi$, we say that the simple reflection s_α is a *right ascent* for $w \in W$ if $\ell(ws_\alpha) = \ell(w) + 1$ and hence $w \to ws_\alpha$; similarly, s_α is called a *right descent* for w if $\ell(ws_\alpha) = \ell(w) - 1$ and hence $ws_\alpha \to w$.

Lemma 3.9. *Let $w \in W$ and $\alpha \in \Pi$. Then s_α is an ascent for w if and only if $w\alpha > 0$. Furthermore, if $w\alpha > 0$, then $\Phi_{ws_\alpha} = \Phi_w \cup \{w\alpha\}$.* □

Proposition 3.10. *Let $v, w \in W$. Then $v \leq_{\text{weak}} w$ if and only if $\Phi_v \subset \Phi_w$.*

Proof. By Lemma 3.9, if $w = vs_\alpha$ with $\alpha \in \Pi$, then $\Phi_v \subset \Phi_w = \Phi_v \cup \{v\alpha\}$. For the converse suppose that $\Phi_v \subset \Phi_w$ and $\ell(w) = \ell(v) + 1$. We claim that $w = vs_\alpha$ for some $\alpha \in \Pi$. Suppose not. Then, by the lemma, for all right ascents s_α for v we have must $v\alpha \notin \Phi_w$ since otherwise $\Phi_w = \Phi_{vs_\alpha}$ which would imply $w = vs_\alpha$ by Proposition 3.2. Furthermore, for all right descents s_α for v we have $-v\alpha \in \Phi_v \subset \Phi_w$ since $-v\alpha > 0$ and $v^{-1}(-v\alpha) = -\alpha < 0$. It follows that $w^{-1}v\alpha > 0$ for all $\alpha \in \Pi$, which means that $w^{-1}v$ has no right descents. This implies $w^{-1}v = e$ and hence $w = v$, which contradicts $\ell(w) = \ell(v) + 1$. □

Corollary 3.11. *Suppose $\mathfrak{p}$ is of Hermitian type. Let $w \in {}^{\mathrm{I}}W$ and let $s_i \neq s_j$ be two simple reflections such that $ws_i, ws_j \in {}^{\mathrm{I}}W$ and $\ell(ws_i) = \ell(ws_j) = \ell(w) - 1$. Then $s_i s_j = s_j s_i$, $ws_i s_j \in {}^{\mathrm{I}}W$, and $\ell(ws_i s_j) = \ell(w) - 2$.*

Proof. By Lemma 3.9 and Proposition 3.8, we have $\mathfrak{a}_{ws_i} = \mathfrak{a}_w \oplus \mathfrak{g}_{w\alpha_i}$ and $\mathfrak{a}_{ws_j} = \mathfrak{a}_w \oplus \mathfrak{g}_{w\alpha_j}$. Since $\mathfrak{a}_{ws_i}, \mathfrak{a}_{ws_j} \subset \mathfrak{u}$ and $\mathfrak{u}$ is abelian(!), the sum $\mathfrak{a}_{ws_i} + \mathfrak{a}_{ws_j} = \mathfrak{a}_w \oplus \mathfrak{g}_{w\alpha_j} \oplus \mathfrak{g}_{w\alpha_j}$ is also an abelian ideal of $\mathfrak{b}$ contained in $\mathfrak{u}$. By Proposition 3.8, $\mathfrak{a}_w \oplus \mathfrak{g}_{w\alpha_j} \oplus \mathfrak{g}_{w\alpha_j} = \mathfrak{a}_v$ for some $v \in {}^{\mathrm{I}}W$ such that $\ell(v) = \ell(w) - 2$. Since $\Psi_v \subset \Psi_{ws_i}$, $\Psi_v \subset \Psi_{ws_j}$, and $\ell(ws_i) = \ell(ws_j) = \ell(v) + 1$, by Proposition 3.10 there exist simple reflections s_k, s_l such that $ws_i s_k = ws_j s_l = v$. Therefore, $s_i s_k = s_j s_l$ and it follows from the exchange condition for Coxeter groups that this is only possible if $i = l$ and $j = k$. Thus, $s_i s_j = s_j s_i$ and $ws_i s_j = v \in {}^{\mathrm{I}}W$. □

Corollary 3.12. *Suppose $\mathfrak{p}$ is of Hermitian type. Then the Bruhat ordering and the weak Bruhat ordering of ${}^{\mathrm{I}}W$ coincide, i.e., for any $v, w \in {}^{\mathrm{I}}W$, $v \leq w$ if and only if $v \leq_{\text{weak}} w$.*

Proof. Suppose the Bruhat ordering and the weak Bruhat ordering on ${}^{\mathrm{I}}W$ do not coincide. Then there would exist an element $w \in {}^{\mathrm{I}}W$ and simple reflections $s_i \neq s_j$ such that $ws_i, ws_j \in {}^{\mathrm{I}}W$, $\ell(ws_i) = \ell(ws_j) = \ell(w) - 1$, and $s_i s_j \neq s_j s_i$. By Corollary 3.11, this is impossible if $\mathfrak{u}$ is abelian. □

Remark 3.13. The fact that the Bruhat ordering and the weak Bruhat ordering of a (co-)minuscule quotient coincide is well known and is already implicit in Proctor's thesis [41] (see Stembridge [46] for a nice discussion).

3.5 Involutions

There is a canonical, order-inverting, length complementary involution on ${}^{\mathrm{I}}W$. This involution will play an important role later on when we study categories of highest weight modules in §5.

Lemma 3.14. *Let $w_\circ$ and w_{I} be the longest elements in W and W_{I}, respectively. Then the mapping $\widetilde{\ }: {}^{\mathrm{I}}W \to {}^{\mathrm{I}}W$ given by $x \mapsto \tilde{x} := w_{\mathrm{I}} x w_\circ$ is an order-inverting, length complementary involution.*

Proof. Since $w_{\mathrm{I}}(\Phi_{\mathrm{I}}^+) = -\Phi_{\mathrm{I}}^+$ and $w_\circ(\Phi^+) = -\Phi^+$, if $x \in {}^{\mathrm{I}}W$, then $(w_{\mathrm{I}} x w_\circ)^{-1}(\Phi_{\mathrm{I}}^+) = w_\circ x^{-1} w_{\mathrm{I}}(\Phi_{\mathrm{I}}^+) \subset \Phi^+$ and hence $w_{\mathrm{I}} x w_\circ \in {}^{\mathrm{I}}W$. (Here we used that $w_\circ^{-1} = w_\circ$ and $w_{\mathrm{I}}^{-1} = w_{\mathrm{I}}$.) Now let $x, y \in {}^{\mathrm{I}}W$ such that $\ell(y) = \ell(x) + 1$ and $y = x s_\alpha$ for some $\alpha \in \Pi$. By Lemma 3.9 we have $x\alpha \in \Phi_y$ and hence $x\alpha \in \Phi(\mathfrak{u})$. Since $w_{\mathrm{I}}\Phi(\mathfrak{u}) = \Phi(\mathfrak{u})$, this implies that $\widetilde{x}(w_\circ\alpha) = (w_{\mathrm{I}} x w_\circ)(w_\circ\alpha) = w_{\mathrm{I}} x\alpha > 0$. Next observe that

$$\widetilde{y} = w_{\mathrm{I}} y w_\circ = w_{\mathrm{I}} x s_\alpha w_\circ = w_{\mathrm{I}} x w_\circ w_\circ s_\alpha w_\circ = w_{\mathrm{I}} x w_\circ s_{w_\circ\alpha} = \widetilde{x} s_{w_\circ\alpha}.$$

The reflection $s_{w_\circ\alpha} = s_{-w_\circ\alpha}$ is a simple reflection since $w_\circ\Phi^+ = -\Phi$ and hence $w_\circ\Pi = -\Pi$. Since $\widetilde{x}(-w_\circ\alpha) < 0$ it follows that $\ell(\widetilde{y}) = \ell(\tilde{x}) - 1$ by Lemma 3.9. □

Corollary 3.15. *If $w \in {}^{\mathrm{I}}W$, then $\Phi(\mathfrak{u}) \setminus \Phi_w = w_{\mathrm{I}}\Phi_{\tilde{w}}$.*

Proof. Let $w \in {}^{\mathrm{I}}W$. By the definition of $\widetilde{w}$, $\Phi_{\widetilde{w}} = \Phi^+ \cap w_{\mathrm{I}} w \Phi^-$ and hence $w_{\mathrm{I}}\Phi_{\widetilde{w}} = w_{\mathrm{I}}\Phi^+ \cap w\Phi^-$. Now let $\alpha \in \Phi(\mathfrak{u})$ and $w \in {}^{\mathrm{I}}W$. Then $\alpha \in w_{\mathrm{I}}\Phi(\mathfrak{u}) \subset w_{\mathrm{I}}\Phi^+$ by Lemma 3.14. Thus, if $\alpha \in \Phi(\mathfrak{u}) \setminus \Phi_w$, then $w^{-1}\alpha < 0$ and $\alpha \in w_{\mathrm{I}}\Phi_{\widetilde{w}} = w_{\mathrm{I}}\Phi^+ \cap w\Phi^-$. Conversely, if $\alpha \in w_{\mathrm{I}}\Phi_{\widetilde{w}} = w_{\mathrm{I}}\Phi^+ \cap w\Phi^-$, then $w^{-1}\alpha < 0$ which implies that $\alpha \in \Phi(\mathfrak{u}) \setminus \Phi_w$. □

3.6 Diagrams of Hermitian type

The Hasse diagrams of the posets $\Phi(\mathfrak{u})$ for $\mathfrak{p}$ of Hermitian type appeared in the paper [29] by Jakobsen, who called them *diagrams of Hermitian type*. They played a key role in his approach to the classification of unitary highest weight representations. The same diagrams appeared already in Proctor's thesis [41], but Proctor did not interpret them explicitly in terms of roots. Here we follow Jakobsen's approach.

Lemma 3.16 (Jakobsen [29, Lemma 4.1]). *Suppose $\mathfrak{p}$ is of Hermitian type. Let $\beta \in \Phi(\mathfrak{u})$ and let $\alpha_{i_1}, \ldots, \alpha_{i_k}$ be distinct elements of I such that $\beta + \alpha_{i_j} \in \Phi(\mathfrak{u})$ for $j = 1, \ldots, k$. Then $k \leq 2$. Furthermore, if $k = 2$, then $\alpha_{i_1} \perp \alpha_{i_2}$ and $\beta + \alpha_{i_1} + \alpha_{i_2} \in \Phi(\mathfrak{u})$.*

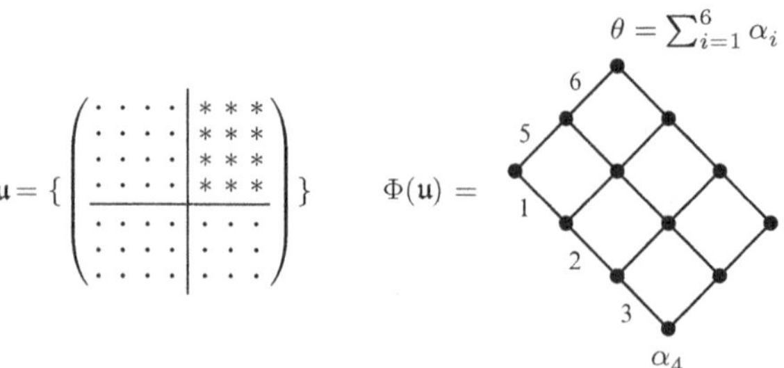

Figure 2 The poset $\Phi(\mathfrak{u})$ for $(\mathsf{A}_6, \mathsf{A}_3 \times \mathsf{A}_2)$

Proof. Jakobsen gave a nice proof which does not use the classification of Hermitian symmetric pairs, but since it is quite long we do not include it here. If we allow ourselves to use the classification, it is straightforward to verify the lemma case-by-case. □

By the lemma, if $\mathfrak{p}$ is of Hermitian type, the Hasse diagram of the poset $\Phi(\mathfrak{u})$ is two-dimensional and can be drawn on a square lattice. These diagrams are shown in the appendix for each Hermitian symmetric pair. The following example shows how to intuitively understand these diagrams in type A.

Example 3.17. Consider the Hermitian symmetric pair $(\mathsf{A}_6, \mathsf{A}_3 \times \mathsf{A}_2)$ corresponding to $\mathfrak{g}_{\mathbb{R}} = \mathfrak{su}(4,3)$ and let $\mathfrak{p} = \mathfrak{m} \oplus \mathfrak{u}$ be the corresponding parabolic subgroup of $\mathfrak{g} = \mathfrak{sl}(7, \mathbb{C})$. Then $\mathfrak{u}$ can be identified with the space of complex 4×3 matrices, $\mathrm{M}_{4,3}(\mathbb{C})$, which is embedded in $\mathfrak{g}$ as a block in the top right corner. Figure 2 shows the nilradical $\mathfrak{u}$ and Hasse diagram of $\Phi(\mathfrak{u})$.

Note that the unique simple root in $\Phi(\mathfrak{u})$ is $\alpha_4 = \epsilon_4 - \epsilon_5$ and the corresponding root subspace is spanned by the elementary matrix $E_{4,5}$. The highest root is $\theta = \epsilon_1 - \epsilon_7$ and the corresponding root subspace is spanned by $E_{1,7}$. For two roots $\alpha, \beta \in \Phi(\mathfrak{u})$ such that β covers α, the edge connecting the nodes of α and β is labelled by the subscript of the simple root $\beta - \alpha \in \mathrm{I}$. For example, the neighboring node in the NW direction of the node of α_4 corresponds to the root $\alpha_4 + \alpha_3 = \epsilon_3 - \epsilon_5$. Note that not all edge labels are shown with the understanding that for every diamond in the diagram, opposite sides (edges) have the same label. For example, the neighboring node in the NE direction of the node of α_4 corresponds to the root $\alpha_4 + \alpha_5 = \epsilon_4 - \epsilon_6$.

Suppose $w \in {}^{\mathrm{I}}W$. Then $\Phi_w \subset \Phi(\mathfrak{u})$ is a lower-order ideal and $\Phi(\mathfrak{u}) \setminus \Phi_w$ is an upper-order ideal corresponding to an abelian ideal $\mathfrak{a}_w$ of $\mathfrak{b}$ contained in $\mathfrak{u}$. We can view the Hasse diagram of Φ_w as a sub diagram of the Hasse diagram of the poset $\Phi(\mathfrak{u})$. The fact that Φ_w is a lower-order ideal of $\Phi(\mathfrak{u})$ means that whenever a node belongs to the Hasse diagram of Φ_w, its neighboring nodes in the SE or SW direction also belong to the Hasse diagram of Φ_w. Figure 3 shows an abelian ideal $\mathfrak{a}_w$ of $\mathfrak{b}$ contained in $\mathfrak{u}$ and the Hasse diagram of the lower-order ideal $\Phi_w \subset \Phi(\mathfrak{u})$ for a typical element $w \in {}^{\mathrm{I}}W$. (The Hasse diagram of Φ_w is shown in bold.)

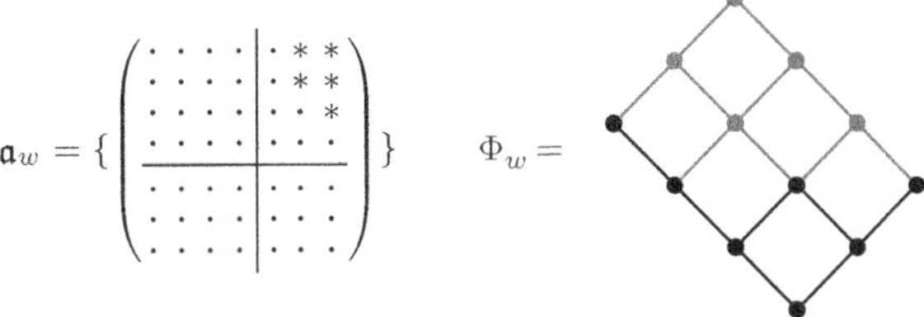

Figure 3 An ideal $\mathfrak{a}_w \subset \mathfrak{u}$ for $(\mathsf{A}_6, \mathsf{A}_3 \times \mathsf{A}_2)$

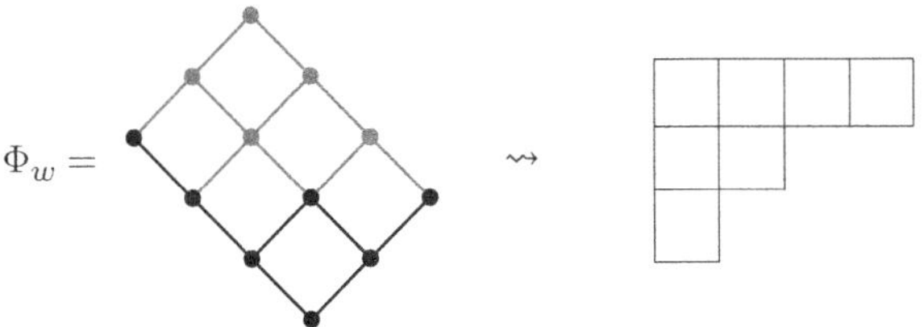

Figure 4 A Young diagram for $(\mathsf{A}_6, \mathsf{A}_3 \times \mathsf{A}_2)$

Finally, we will associate a *generalized Young diagram* to each lower-order ideal $\Phi_w \subset \Phi(\mathfrak{u})$ as follows. Start with the poset Φ_w viewed as a subdiagram of the Hasse diagram of $\Phi(\mathfrak{u})$. Then replace the nodes of the diagram of Φ_w with boxes (squares) so that two boxes share a side if and only if the two corresponding nodes are connected. Finally rotate the resulting diagram 135° clockwise so that the node corresponding to the simple root in $\Pi \setminus \mathrm{I}$ now corresponds to the top left box. Figure 4 shows the Young diagram corresponding to Φ_w for the same element $w \in {}^{\mathrm{I}}W$ as in Figure 3. (The intermediate diagram, i.e., the diagram before the 135° clockwise rotation, is not shown in Figure 3, but it is shown in Figure 6 on the left.)

For each Hermitian symmetric pair the generalized Young diagram corresponding to the longest element $w \in {}^{\mathrm{I}}W$, i.e., the element such that $\Phi_w = \Phi(\mathfrak{u})$, is shown in the appendix. (The numbers that are filled into the boxes will be explained shortly.) For the classical pairs of type $(\mathsf{D}_n, \mathsf{A}_{n-1})$ and $(\mathsf{C}_n, \mathsf{A}_{n-1})$, the generalized Young diagrams corresponding to a general element $w \in {}^{\mathrm{I}}W$ are also called *shifted Young diagrams*. An example of a shifted Young diagram for type $(\mathsf{C}_6, \mathsf{A}_5)$ is shown in Figure 7 on the right.

Using Corollary 3.12, it is easy to obtain the Hasse diagram of the poset ${}^{\mathrm{I}}W$ by starting with the generalized Young diagram of the longest element in ${}^{\mathrm{I}}W$. Several examples of the resulting Hasse diagrams for Hermitian symmetric pairs of small rank can be found at various places in this paper: for the pair $(\mathsf{A}_3, \mathsf{A}_1 \times \mathsf{A}_1)$ see Figure 1 on page 123; for $(\mathsf{A}_4, \mathsf{A}_2 \times \mathsf{A}_1)$ see Figure 8 on page 140; for $(\mathsf{C}_3, \mathsf{A}_2)$ see Figure 9 on page 141; for $(\mathsf{E}_6, \mathsf{D}_5)$ see Figure 13 on page 154, diagram on the left; for $(\mathsf{E}_7, \mathsf{E}_6)$ and $(\mathsf{D}_6, \mathsf{D}_5)$ see Figure 14 on page 177.

Remark 3.18. The Hasse diagrams of ${}^{\mathrm{I}}W$ should not be confused with the Hasse diagrams of $\Phi(\mathfrak{u})$. For starters, the Hasse diagram of $\Phi(\mathfrak{u})$ can always be drawn on a two-dimensional lattice, whereas the Hasse diagram of ${}^{\mathrm{I}}W$ cannot if the rank of the Hermitian symmetric pair is > 2.

However, if the Hermitian symmetric pair is of rank 2, then the poset ${}^{\mathrm{I}}W$ is isomorphic to a poset of the form $\Phi(\mathfrak{u})$ for some other Hermitian symmetric pair. For example, the poset ${}^{\mathrm{I}}W$ for $(\mathsf{E}_6, \mathsf{D}_5)$ is isomorphic to the poset $\Phi(\mathfrak{u})$ for $(\mathsf{E}_7, \mathsf{E}_6)$. Here is the explanation. For the Hermitian symmetric pair $(\mathfrak{g}, \mathfrak{m})$ of type $(\mathsf{E}_7, \mathsf{E}_6)$, the 27-dimensional $\mathfrak{m}$-module $\mathfrak{u}$ is a minuscule representation of E_6. Thus, the Weyl group W_{I} of type E_6 acts transitively on $\Phi(\mathfrak{u})$. The stabilizer $(W_{\mathrm{I}})_{\alpha_7} := \{w \in W_{\mathrm{I}} \mid w\alpha_7 = \alpha_7\}$ is generated by the simple reflections s_1, s_2, s_3, s_4, s_5 and hence $(W_{\mathrm{I}})_{\alpha_7}$ is a Weyl group of type D_5. The resulting bijection $W_{\mathrm{I}}/(W_{\mathrm{I}})_{\alpha_7} \cong \Phi(\mathfrak{u})$ is in fact an isomorphism of posets.

3.7 *Canonical reduced expressions*

Definition 3.19. Suppose $\mathfrak{p}$ is of Hermitian type. Define a map

$$f : \Phi(\mathfrak{u}) \to \Pi$$

as follows. For $\beta \in \Phi(\mathfrak{u})$, the sets $\Phi(\mathfrak{u})_{<\beta} := \{\gamma \in \Phi(\mathfrak{u}) \mid \gamma < \beta\}$ and $\Phi(\mathfrak{u})_{\leq\beta} := \{\gamma \in \Phi(\mathfrak{u}) \mid \gamma \leq \beta\}$ are lower-order ideals of $\Phi(\mathfrak{u})$. Therefore, by Lemma 3.7 and Proposition 3.8, there exist $v, w \in {}^{\mathrm{I}}W$ such that $\Phi(\mathfrak{u})_{<\beta} = \Phi_v$ and $\Phi(\mathfrak{u})_{\leq\beta} = \Phi_w$. Since $\Phi_w = \Phi_v \dot{\cup} \{\beta\}$ there exists a unique $f(\beta) \in \Pi$ such that $w = v s_{f(\beta)}$, namely $f(\beta) := v^{-1}\beta$.

Lemma 3.20. *Suppose $\mathfrak{p}$ is of Hermitian type. If $v', w' \in {}^{\mathrm{I}}W$ are two elements such that $\Phi_{w'} = \Phi_{v'} \dot{\cup} \{\beta\}$, then $v'^{-1}\beta = f(\beta)$.*

Proof. Let $v, w \in {}^{\mathrm{I}}W$ be as above, i.e., $\Phi_v = \Phi(\mathfrak{u})_{<\beta}$ and $\Phi_w = \Phi(\mathfrak{u})_{\leq\beta}$. Then $\Phi_v \subset \Phi_{v'}$ and $\Phi_w \subset \Phi_{w'}$ and hence $v \leq v'$ and $w \leq w'$. We will prove that $v'^{-1}\beta = f(\beta)$ by induction on $\ell(w') - \ell(w)$. If $\ell(w') = \ell(w)$, there is nothing to prove since $w = w'$ and $v = v'$ in this case. So assume $\ell(w') > \ell(w)$ and let $\alpha \in \Pi$ such that $w \leq w's_\alpha < w'$. Note that $\Phi_w \subset \Phi_{w's_\alpha} \subset \Phi_{w'}$ and hence $\beta \in \Phi_{w's_\alpha}$. Set $\alpha' := v'^{-1}\beta$. Then $s_\alpha \neq s_{\alpha'}$ are two right descents for w. By Corollary 3.11, $s_\alpha s_{\alpha'} = s_{\alpha'} s_\alpha$ and $w' s_{\alpha'} s_\alpha = v' s_\alpha < v'$ in ${}^{\mathrm{I}}W$. In particular, $\Phi_{w's_\alpha} = \Phi_{v's_\alpha} \cup \{\beta\}$ and we may assume that $(v's_\alpha)^{-1}\beta = f(\beta)$ (induction hypothesis). Since s_α and $s_{\alpha'}$ commute, $\alpha \perp \alpha'$ and hence $s_\alpha\alpha' = \alpha'$. Thus, $v'^{-1}\beta = s_\alpha v'^{-1}\beta = (v's_\alpha)^{-1}\beta = f(\beta)$. □

Proposition 3.21. *Suppose $\mathfrak{p}$ is of Hermitian type. Let $w \in {}^{\mathrm{I}}W$ and write*

$$\Phi_w = \{\beta_1, \beta_2, \ldots, \beta_l\}$$

such that for every $1 \leq j \leq l$, *the set* $\{\beta_1, \beta_2, \dots, \beta_j\}$ *is a lower-order ideal of* $\Phi(\mathfrak{u})$. *Then*

$$w = s_{f(\beta_1)} s_{f(\beta_2)} \cdots s_{f(\beta_l)}$$

is a reduced expression for w. *Furthermore,* $w = s_{\beta_l} s_{\beta_{l-1}} \cdots s_{\beta_1}$.

Proof. Set $w_0 := 0$ and for $1 \leq j \leq l$, let $w_j \in {}^{\mathrm{I}}W$ be such that $\Phi_{w_j} = \{\beta_1, \beta_2, \dots, \beta_j\}$. Then $\Phi_j = \Phi_{j-1} \dot{\cup} \{\beta_j\}$ and hence $w_j = w_{j-1} s_{f(\beta_j)}$ and $f(\beta_j) = (w_{j-1})^{-1} \beta_j$ by Lemma 3.20. It follows that $w = s_{f(\beta_1)} s_{f(\beta_2)} \cdots s_{f(\beta_l)}$ which is a reduced expression for w since $l = |\Phi_w| = \ell(w)$. For the second statement, recall that for $v \in W$ and $\alpha \in \Phi$, we have $v s_\alpha v^{-1} = s_{v\alpha}$. With $v = w_{j-1}$ and $\alpha = f(\beta_j) = (w_{j-1})^{-1} \beta_j$ we obtain $w_{j-1} s_{f(\beta_j)} (w_{j-1})^{-1} = s_{\beta_j}$ and hence $w_j = w_{j-1} s_{f(\beta_j)} = s_{\beta_j} w_{j-1}$. Thus, $w = s_{\beta_l} s_{\beta_{l-1}} \cdots s_{\beta_1}$. □

This proposition combined with the generalized Young diagrams from the previous section give a very concrete way to write canonical reduced expressions for the elements of ${}^{\mathrm{I}}W$. First fill the boxes of the diagram of the longest element of ${}^{\mathrm{I}}W$ with the numbers $1, 2, \dots, n$ so that the box corresponding to $\beta \in \Phi(\mathfrak{u})$ is assigned the number i such that $f(\beta) = \alpha_i$. The top-left box is always assigned the label of the unique simple root in $\Pi \setminus \mathrm{I}$.

Example 3.22. Consider the Hermitian symmetric pair of type $(\mathsf{A}_3, \mathsf{A}_1 \times \mathsf{A}_1)$ corresponding to $\mathfrak{g}_{\mathbb{R}} = \mathfrak{su}(2,2)$. Then the canonical reduced expressions of the elements $w \in {}^{\mathrm{I}}W$ are as shown in Figure 5.

Example 3.23. Recall that an element $w \in W$ is called a *Coxeter element* if a reduced expression of w contains every simple reflection exactly once. It is a nice

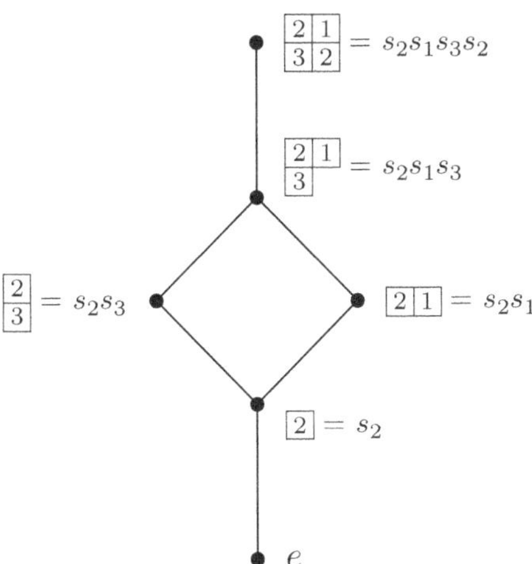

Figure 5 Diagrams associated to the Grassmannian $\mathrm{Gr}_2(\mathbb{C}^4)$

little fact that if $\mathfrak{p}$ is of Hermitian type, then ${}^{\mathrm{I}}W$ contains a unique Coxeter element and its generalized Young diagram is a fattened version of the Dynkin diagram of $\mathfrak{g}$. For example, if $(\Phi, \Phi_{\mathrm{I}}) = (\mathsf{E}_7, \mathsf{E}_6)$, the Coxeter element in ${}^{\mathrm{I}}W$ is the following element:

$$\begin{array}{|c|c|c|c|c|c|}\hline 7 & 6 & 5 & 4 & 3 & 1 \\ \hline \multicolumn{3}{c|}{} & 2 & \multicolumn{2}{c}{} \\ \cline{4-4} \end{array} = s_7 s_6 s_5 s_4 s_3 s_1 s_2.$$

The reader is invited to verify this fact for the other Hermitian symmetric pairs by consulting the appendix.

3.8 Lascoux–Schützenberger notation

For the classical Hermitian symmetric pairs corresponding to the real Lie algebras $\mathfrak{g}_{\mathbb{R}} = \mathfrak{su}(p, q)$, $\mathfrak{sp}(n, \mathbb{R})$ and $\mathfrak{so}^*(2n)$ there is another description of the elements in ${}^{\mathrm{I}}W$ in terms of binary sequences. This description was used for $\mathfrak{su}(p, q)$ by Lascoux–Schützenberger in [38] and later for $\mathfrak{sp}(n, \mathbb{R})$ and $\mathfrak{so}^*(2n)$ by Boe in [6] to calculate Kazhdan–Lusztig polynomials for Hermitian symmetric spaces. This description will be particularly useful when we describe certain equivalences of categories of highest weight categories in Section 5.

For $w \in {}^{\mathrm{I}}W$, draw the generalized Young diagram of w rotated 135° counterclockwise. The upper boundary of this diagram is a zigzag path consisting of (unit) line segments and the generalized Young diagram of w is uniquely determined by this path. To describe the path (from left to right) we construct a binary sequence (also from left to right) by writing a 1 for each line segment of positive slope and a 0 for each line segment of negative slope. The length of the binary we create is $p + q$ for $\mathfrak{g}_{\mathbb{R}} = \mathfrak{su}(p, q)$, n for $\mathfrak{g}_{\mathbb{R}} = \mathfrak{sp}(n, \mathbb{R})$, and $n - 1$ for $\mathfrak{g}_{\mathbb{R}} = \mathfrak{so}^*(2n)$. By abuse of notation, we will write the binary sequence to denote the corresponding element $w \in {}^{\mathrm{I}}W$. Figures 6 and 7 show the binary sequence of a typical element $w \in {}^{\mathrm{I}}W$ for $\mathfrak{g}_{\mathbb{R}} = \mathfrak{su}(4, 3)$ and $\mathfrak{g}_{\mathbb{R}} = \mathfrak{sp}(6, \mathbb{R})$, respectively.

The Bruhat ordering of ${}^{\mathrm{I}}W$ can be described in terms of the binary sequences as follows. Suppose $\mathfrak{g}_{\mathbb{R}} = \mathfrak{sp}(n, \mathbb{R})$. If $v = a_1 a_2 \cdots a_n$ and $w = b_1 b_2 \cdots b_n$, then $v \leq w$ if and only if

$$\sum_{i=1}^{k} a_i \leq \sum_{i=1}^{k} b_i$$

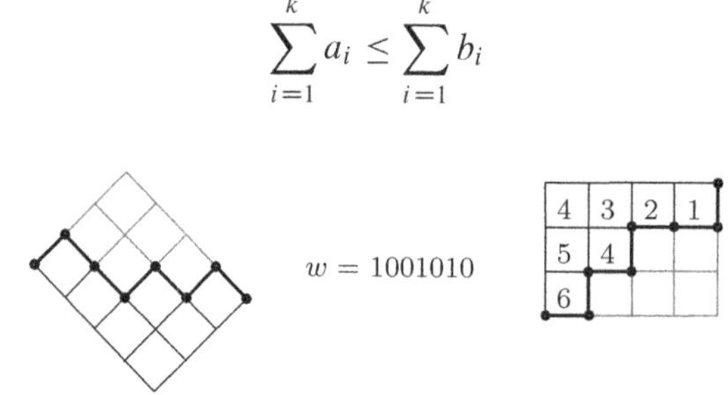

Figure 6 Lascoux–Schützenberger notation for type $(\mathsf{A}_6, \mathsf{A}_3 \times \mathsf{A}_2)$

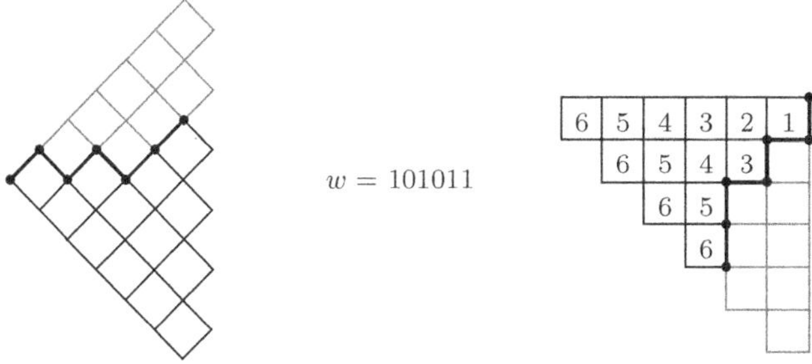

Figure 7 Lascoux–Schützenberger notation for type $(\mathsf{C}_6, \mathsf{A}_5)$

for $1 \leq k \leq n$. For $1 \leq i \leq n-1$, the simple reflection s_i is a right descent of $w = b_1 b_2 \cdots b_n$ if $b_i b_{i+1} = 10$. The simple reflection s_n is a right descent if $b_n = 1$. For example, the right descents for $w = 101011$ are s_1, s_3, and s_6. The descriptions of the Bruhat ordering for the Hermitian symmetric pairs corresponding to $\mathfrak{su}(p,q)$ and $\mathfrak{so}^*(2n)$ are similar and left to the reader.

4 Schubert varieties

4.1 Generalized flag varieties and Hermitian symmetric spaces

Let G be a connected, simple, complex linear algebraic group with Lie algebra $\mathfrak{g}$ and let $B \subset G$ be the closed algebraic subgroup corresponding to the Borel subalgebra $\mathfrak{b} \subset \mathfrak{g}$. The variety G/B is called the full flag variety of G. For $\mathrm{I} \subset \Pi$, let $P = P_{\mathrm{I}}$ be the parabolic subgroup of G with Lie algebra $\mathfrak{p} = \mathfrak{p}_{\mathrm{I}}$. The variety G/P is called a *generalized flag variety*. It is a smooth complex projective variety of dimension $|\Phi(\mathfrak{u})|$. If $\mathfrak{p}$ is a maximal parabolic of Hermitian type, we also say that G/P is a *cominuscule* flag variety. The connection to Hermitian symmetric spaces is as follows. Let $\mathfrak{p}$ be a maximal parabolic subalgebra of Hermitian type and let $\mathfrak{g}_{\mathbb{R}} = \mathfrak{m}_{\mathbb{R}} \oplus \mathfrak{s}_{\mathbb{R}}$ be the corresponding real form of $\mathfrak{g}$. Define

$$\mathfrak{k}_{\mathbb{R}} := \mathfrak{m}_{\mathbb{R}} \oplus i\mathfrak{s}_{\mathbb{R}}.$$

Then the Lie algebra $\mathfrak{k}_{\mathbb{R}}$ is a compact real form of $\mathfrak{g}$. Let $K_{\mathbb{R}}$ and $M_{\mathbb{R}}$ be the analytic Lie subgroups of G (viewed as a real Lie group) corresponding to the Lie subalgebras $\mathfrak{k}_{\mathbb{R}}$ and $\mathfrak{m}_{\mathbb{R}}$ of $\mathfrak{g}$. Then $K_{\mathbb{R}}$ is a maximal compact subgroup of G and

$$K_{\mathbb{R}}/M_{\mathbb{R}} \cong G/P.$$

Thus, $K_{\mathbb{R}}/M_{\mathbb{R}}$ is a complex manifold on which $K_{\mathbb{R}}$ acts by biholomorphic transformations. It follows that $K_{\mathbb{R}}/M_{\mathbb{R}}$ is a compact Hermitian symmetric space. Conversely, it follows from the classification of compact Hermitian symmetric spaces that every compact Hermitian symmetric spaces every arises that way (up to isomorphism).

Proposition 4.1 (Kostant [35]). *A generalized flag variety G/P is a compact Hermitian symmetric space if and only if P is of Hermitian type.* □

There is also a Hermitian symmetric space of noncompact type in the picture. Let $G_{\mathbb{R}}$ be the analytic Lie subgroup of G corresponding to the noncompact real form $\mathfrak{g}_{\mathbb{R}}$ of $\mathfrak{g}$. Harish-Chandra proved that

$$G_{\mathbb{R}}/M_{\mathbb{R}} \hookrightarrow G/P$$

is a diffeomorphism onto an open subset of G/P. Thus, $G_{\mathbb{R}}/M_{\mathbb{R}}$ is a complex manifold on which $G_{\mathbb{R}}$ acts by biholomorphic transformations and it follows that $G_{\mathbb{R}}/M_{\mathbb{R}}$ is a Hermitian symmetric space of noncompact type. The corresponding pair of Lie algebras $(\mathfrak{g}_{\mathbb{R}}, \mathfrak{m}_{\mathbb{R}})$ is one of the pairs in Table 1.

Example 4.2. Let $G = \mathrm{SL}(2,\mathbb{C})$ and P the parabolic subgroup of upper triangular matrices. This parabolic subgroup is of Hermitian type corresponding to the real form $G_{\mathbb{R}} = \mathrm{SU}(1,1)$. Then $G/P = \mathbb{CP}^1 = \mathbb{C} \cup \{\infty\}$ and $G_{\mathbb{R}}/M_{\mathbb{R}} = \{z \in \mathbb{C} \mid |z| < 1\}$, where the actions of G and $G_{\mathbb{R}}$ are given by linear fractional transformations.

4.2 Schubert cells and Schubert varieties

Let $T \subset B$ be the maximal torus of G corresponding to the Cartan subalgebra $\mathfrak{h} \subset \mathfrak{b}$. Then the Weyl group can be identified by $W = N_G(T)/T$ and we have the Bruhat decomposition (cf. Harish-Chandra [24])

$$G = \bigcup_{w \in W} BwB,$$

where, by abuse of notation, the w in BwB stands for any of the representatives of the coset $w \in N_G(T)/T$. If P is a parabolic subgroup of G containing B, then the torus T acts with finitely many fixed points on the generalized flag variety G/P and these fixed points are naturally parametrized by the elements of the poset $W^{\mathrm{I}} := \{w \in W \mid w^{-1} \in {}^{\mathrm{I}}W\}$. Explicitly, the T-fixed point in G/P that corresponds to $w \in W^{\mathrm{I}}$ is the coset wP. The B-orbit of wP is the *generalized Schubert cell*

$$C(w) = C_{\mathrm{I}}(w) := BwP.$$

Let $X(w) := \overline{C(w)}$ be the Zariski closure of $C(w)$ in G/P. Then $X(w)$ is a projective variety of dimension $\ell(w)$. Furthermore,

$$X(w) = \bigcup_{v \le w} C(v)$$

is a cell decomposition. This in fact gives a CW complex, where all the cells $C(v) \cong \mathbb{C}^{\ell(v)} \cong \mathbb{R}^{2\ell(v)}$ have even real dimension. Therefore, the differential of the CW complex are all zero and it follows that the $2i$-th homology group of $X(w)$ with integer coefficients is

$$H_{2i}(X(w), \mathbb{Z}) \cong \mathbb{Z}^{\#\{v \in W^1 | v \le w \text{ and } \ell(v)=i\}}.$$

For $w \in W^1$ define the *Poincaré polynomial* of $X(w)$ by

$$p_w(t) := \sum t^i \dim_{\mathbb{Q}} H_{2i}((X(w), \mathbb{Q}) = \sum_{v \le w} t^{\ell(v)}.$$

If $X(w)$ is smooth, then $X(w)$ is a compact manifold and hence by Poincaré duality, the Poincaré polynomial $p_w(t)$ is palindromic. It is perhaps surprising that the converse is almost true (see Proposition 4.3 below).

4.3 Rational smooth Schubert varieties

A complex algebraic variety Z of complex dimension d is said to be *rationally smooth* if for every point $z \in Z$, the relative cohomology group $H_{2i}(Z, Z \setminus \{z\}; \mathbb{Q}) = 0$ if $i \neq d$. If the variety Z is smooth (and hence a complex manifold of complex dimension d), then Z is rationally smooth. By a result of D. Peterson, if the group G is of type A-D-E, then every rationally smooth Schubert variety in G/P is in fact smooth (cf. Carrell–Kuttler [12] for a proof). Furthermore, we have the following useful criterion for rational smoothness of Schubert varieties in G/P (Figure 8).

Proposition 4.3 (Carrell–Peterson [11]). *For $w \in W^1$ the following are equivalent:*

(i) *The generalized Schubert variety $X(w)$ in G/P is rationally smooth.*
(ii) *The Poincaré polynomial $p_w(t)$ is palindromic.* □

Now suppose that G/P is a Hermitian symmetric space. Let $w \in W^1$. Then $w^{-1} \in {}^1W$ and hence $\Phi_{w^{-1}} \subset \Phi(\mathfrak{u})$ is a lower-order ideal which we can represent by a generalized Young diagram as in Section 6. (In other words, the generalized Young diagram of w is the generalized Young diagram of $w^{-1} \in {}^1W$ that was previously defined.) If $w \neq e$, then the Poincaré polynomial of w is of the form $p_w(t) =$

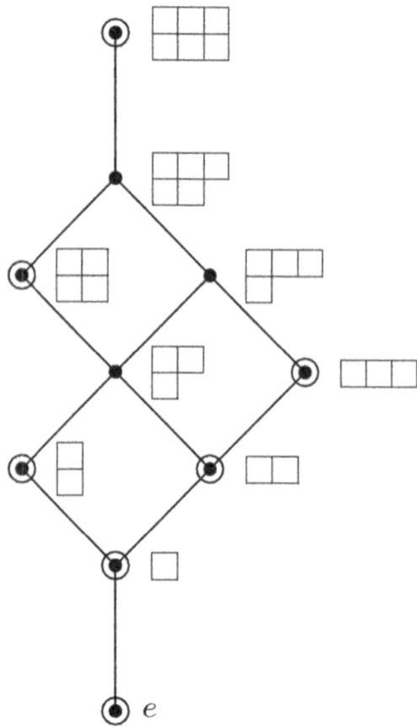

Figure 8 Smooth Schubert varieties for $(\mathsf{A}_4, \mathsf{A}_2 \times \mathsf{A}_1)$

$1+t+\cdots$ since there exists a unique element of length 1 in W^{I}, namely the simple reflection s_α with $\alpha \in \Pi \setminus \mathrm{I}$. Thus, a necessary condition for the polynomial $p_w(t)$ to be palindromic is that w can have at most one (left) descent. This means that the generalized Young diagram of $w \in W^{\mathrm{I}}$ can have at most one SE corner. In type A, where the generalized Young diagrams are ordinary Young diagrams, this condition is equivalent to the diagram being a rectangle (see Figure 8).

For the other Hermitian symmetric spaces G/P the situation is similar. Suppose that G is not of type A. Then the poset W^{I} also contains a unique element of length 2. This has to do with fact that there exists a unique compact simple root that is connected to the simple noncompact root $\alpha \in \Pi$ in the Dynkin diagram. Thus, for $w \in W^{\mathrm{I}}$ with $\ell(w) \geq 2$, the Poincaré polynomial of w is of the form $p_w(t) = 1+t+t^2+\cdots$. This means that if $p_w(t)$ is palindromic, the generalized Young diagram of w has a unique SE corner and the diagram with this corner removed also has a unique SE corner. For the Hermitian symmetric pairs of type $(\mathsf{C}_n, \mathsf{A}_{n-1})$ and $(\mathsf{D}_n, \mathsf{A}_{n-1})$, where the generalized Young diagrams are shifted Young diagrams, the diagrams that satisfy this condition are the diagrams that have a single row and the shifted diagrams corresponding to strict partitions of the form $(k, k-1, \ldots, 3, 2, 1)$.

4.4 Smooth Schubert varieties

As we observed above, if G is of simply laced type, then every rationally smooth Schubert variety in G/P is also smooth (Figure 9). It turns out that for any Hermitian symmetric space G/P the smooth Schubert varieties are also easy to

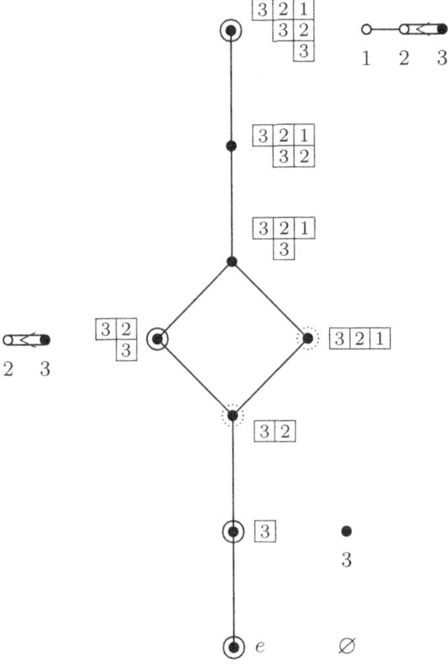

Figure 9 Smooth and rationally smooth Schubert varieties for $(\mathsf{C}_3, \mathsf{A}_2)$

describe. We start with a general observation. Let $Q \subset G$ be another parabolic subgroup containing B and with Levi decomposition $Q = LN$. Then $L \cap B$ is a Borel subalgebra and $L \cap P$ is a parabolic subgroup of the reductive group L. Furthermore, $B = (L \cap B)N$. Now let L act on G/P by left multiplication. Since the stabilizer of eP is the parabolic subgroup $L \cap P$, the orbit map $\varphi : L \to G/P$, $x \mapsto x(eP) = xP$, induces a closed embedding of the generalized flag variety $L/(L \cap P)$ into G/P.

Lemma 4.4. *Under the closed embedding $L/(L \cap P) \hookrightarrow G/P$, the image of any $(L \cap B)$-orbit in $L/(L \cap P)$ is a B-orbit in G/P. In particular, the image of $L/\ (L \cap P)$ is a smooth Schubert variety in G/P. Furthermore, this Schubert variety is the Q-orbit of the coset eP.*

Proof. Let $x \in L$ and $b \in B$. Since $B = (L \cap B)N$, we may write $b = b_L n$ with $b_L \in L \cap B$ and $n \in N$. Then since N is normalized by L, we have $bx = b_L nx = b_L xn'$ for some $n' \in N$. Since $N \subset P$ (in fact, $N \subset B \subset P$), it follows that $(L \cap B)xP = BxP$. This shows that the image of any $(L \cap B)$-orbit in $L/(L \cap P)$ is a B-orbit in G/P. Since $L/(L \cap P)$ is the closure of an $(L \cap B)$-orbit and since $L/(L \cap P) \hookrightarrow G/P$ is a closed embedding, it follows that $L/(L \cap P)$ is isomorphically mapped onto the closure of a B-orbit. Hence the image of $L/(L \cap P)$ is a smooth Schubert variety. Since $Q = LN$ and $N \subset P$, it follows that the L-orbit of eP is in fact a Q-orbit. □

Suppose P is any (standard) maximal parabolic subgroup of G, not necessarily of Hermitian type. For $\mathrm{J} \subset \Pi$, let $Q_{\mathrm{J}} = L_{\mathrm{J}}N_{\mathrm{J}}$ be the standard parabolic subgroup corresponding to J. We say that J is *connected* if the corresponding subgraph of the Dynkin diagram of G is connected. Then we can define an injective map from the set of connected subsets of Π containing α to the set of smooth Schubert varieties in G/P by $\mathrm{J} \mapsto L_{\mathrm{J}}/(L_{\mathrm{J}} \cap P) \hookrightarrow G/P$. Using results of Brion and Polo [10], J. Hong, in her paper [26], showed that this map is also surjective if G/P is a Hermitian symmetric space.

Proposition 4.5 (cf. Hong [26, Prop. 2.11]). *Suppose that G/P is a Hermitian symmetric space. Let Γ be the Dynkin diagram of G. Then we have a bijection*

$$\left\{ \begin{array}{c} \textit{connected subdiagrams of } \Gamma \\ \textit{containing the marked node} \end{array} \right\} \to \{\textit{smooth Schubert varieties in } G/P\}$$

given as above by $\mathrm{J} \mapsto L_{\mathrm{J}}/(L_{\mathrm{J}} \cap P) \hookrightarrow G/P$. □

Example 4.6. Consider the cominuscule flag variety G/P corresponding to the Hermitian symmetric pair of type $(\mathsf{C}_3, \mathsf{A}_2)$. In this case there are 8 Schubert varieties, of which 6 are rationally smooth. Four of the rationally smooth Schubert varieties are in fact smooth (see Figure 9).

Remark 4.7. In [7], Boe and Hunziker proved that the conclusion of Proposition 4.5 holds for arbitrary maximal parabolic subgroups $P \subset G$ as long as the Dynkin diagram Γ is simply laced.

4.5 *Rational smoothness and Kazhdan–Lusztig polynomials*

For $x \leq w \in {}^{\mathrm{I}}W$ define

$$\ell(x, w) := \ell(w) - \ell(x).$$

Proposition-Definition 4.8 (Deodhar [14]). *First, there exists a unique family $\{R^{\mathrm{I}}_{x,w}(q) \mid x \leq w \in W^{\mathrm{I}}\}$ of polynomials in $\mathbb{Z}[q]$ such that*

(1) *$\deg R^{\mathrm{I}}_{x,w}(q) = \ell(x, w)$ and $R^{\mathrm{I}}_{w,w}(q) = 1$;*
(2) *if s_α is a simple reflection in W such that $s_\alpha x \in W^{\mathrm{I}}$ and $\ell(s_\alpha w) < \ell(w)$, then*

$$R^{\mathrm{I}}_{x,w}(q) = \begin{cases} R^{\mathrm{I}}_{s_\alpha x, s_\alpha w}(q) & \textit{if } \ell(s_\alpha x) < \ell(x) \\ (q-1)R^{\mathrm{I}}_{x, s_\alpha w}(q) + qR^{\mathrm{I}}_{s_\alpha x, s_\alpha w}(q) & \textit{if } \ell(s_\alpha x) > \ell(x) \\ & \textit{and } s_\alpha x \in W^{\mathrm{I}} \\ qR^{\mathrm{I}}_{x, s_\alpha w}(q) & \textit{if } \ell(s_\alpha w) > \ell(x) \\ & \textit{and } s_\alpha x \notin W^{\mathrm{I}}. \end{cases}$$

Second, there exists a unique family $\{P^{\mathrm{I}}_{x,w}(q) \mid x \leq w \in W^{\mathrm{I}}\}$ *of polynomials in* $\mathbb{Z}[q]$ *such that*

(3) $P^{\mathrm{I}}_{w,w}(q) = 1$ *and* $\deg P_{x,w}(q) \leq (\ell(x,w) - 1)/2$ *if* $x < w$;
(4) $q^{\ell(x,w)} P^{\mathrm{I}}_{x,w}(q^{-1}) = \sum\limits_{x \leq y \leq w} R^{\mathrm{I}}_{x,y}(q) \cdot P^{\mathrm{I}}_{y,w}(q)$

The polynomials $P^{\mathrm{I}}_{x,w}(q)$ are called *parabolic Kazhdan–Lusztig polynomials.* In the special case when $\mathrm{I} = \varnothing$, the parabolic Kazhdan–Lusztig polynomials are the ordinary Kazhdan–Lusztig polynomials that were in introduced by Kazhdan and Lusztig in [33]. The following well known result is a straightforward generalization of a result of Kazhdan and Lusztig.

Proposition 4.9. *For* $w \in W^{\mathrm{I}}$ *the following are equivalent.*

(i) *The generalized Schubert variety* $X(w)$ *in* G/P *is rationally smooth.*
(ii) $P^{\mathrm{I}}_{x,w}(q) = 1$ *for all* $x \leq w$.

□

If G/P is a Hermitian symmetric space, the polynomials $P^{\mathrm{I}}_{x,w}(q)$ are easy to calculate. This was done by Lascoux and Schützenberger for Grassmannians in [38] and by Boe for the other classical Hermitian symmetric spaces in [6]. In Section 5 we will extend these results to all Hermitian symmetric spaces by using generalized Young diagrams.

5 Parabolic category $\mathcal{O}^{\mathfrak{p}}$

Let $\mathfrak{g}$ be a complex simple Lie algebra and let $\mathfrak{p} = \mathfrak{p}_{\mathrm{I}}$ be a standard parabolic subalgebra of $\mathfrak{g}$ with Levi decomposition $\mathfrak{p} = \mathfrak{m} \oplus \mathfrak{u}$. As usual, let $U(\mathfrak{g})$ denote the universal enveloping algebra of $\mathfrak{g}$. Following Rocha–Caridi, the category $\mathcal{O}^{\mathfrak{p}}$ ("parabolic category $\mathcal{O}$") is defined as the full subcategory of the category of $U(\mathfrak{g})$-modules whose objects V satisfy the following conditions:

(1) V is a finitely generated $U(\mathfrak{g})$-module;
(2) V is a semisimple $U(\mathfrak{m})$-module, i.e., V is a direct sum of finite-dimensional simple modules;
(3) V is locally $\mathfrak{u}$-finite, i.e., $\dim U(\mathfrak{u})v < \infty$ for all $v \in V$.

The key objects in $\mathcal{O}^{\mathfrak{p}}$ are the parabolic Verma modules. Define

$$\Lambda^{+}_{\mathrm{I}} := \{\lambda \in \mathfrak{h}^* \mid (\lambda + \rho, \alpha^{\vee}) \in \mathbb{Z}_{>0} \quad \forall \alpha \in \Phi^{+}_{\mathrm{I}}\}.$$

For $\lambda \in \Lambda^{+}_{\mathrm{I}}$, let F_λ be the finite-dimensional simple $\mathfrak{m}$-module with highest weight λ. As usual, F_λ may be viewed as a $\mathfrak{p}$-module by letting the nilradical $\mathfrak{u}$ act trivially.

Then the *parabolic Verma module* (PVM) with highest weight λ is the induced module

$$M_{\mathrm{I}}(\lambda) := U(\mathfrak{g}) \otimes_{U(\mathfrak{p})} F_\lambda.$$

The module $M_{\mathrm{I}}(\lambda)$ is a quotient of the ordinary Verma module $M(\lambda)$ with highest weight λ. Let $L(\lambda)$ denote the simple quotient of both $M_{\mathrm{I}}(\lambda)$ and $M(\lambda)$. The Verma module $M(\lambda)$ and all of its subquotients, in particular $M_{\mathrm{I}}(\lambda)$ and $L(\lambda)$, admit an *infinitesimal character* which we denote χ_λ. Recall that this means that $\chi_\lambda : Z(U(\mathfrak{g})) \to \mathbb{C}$ is a character of the center of $U(\mathfrak{g})$ such that $z \cdot v = \chi_\lambda(z)v$ for all $z \in Z(U(\mathfrak{g}))$ and all v in the module. For a character $\chi : Z(U(\mathfrak{g})) \to \mathbb{C}$, let $\mathcal{O}^{\mathfrak{p}}_\chi$ denote the full subcategory of $\mathcal{O}^{\mathfrak{p}}$ consisting of modules V such that $z - \chi(z)$ acts locally nilpotently on V for all $z \in Z$. Then the category $\mathcal{O}^{\mathfrak{p}}$ decomposes as a direct sum

$$\mathcal{O}^{\mathfrak{p}} = \bigoplus_\chi \mathcal{O}^{\mathfrak{p}}_\chi.$$

The categories $\mathcal{O}^{\mathfrak{p}}_\chi$ are called the *infinitesimal blocks* of $\mathcal{O}^{\mathfrak{p}}$. In general, the infinitesimal blocks may be decomposable.

The Weyl group W acts on $\mathfrak{h}^*$ via the "dot action", i.e., the action given by

$$w \cdot \lambda = w(\lambda + \rho) - \rho.$$

For $\lambda, \mu \in \mathfrak{h}^*$, $\chi_\lambda = \chi_\mu$ if and only if $\lambda \in W \cdot \mu$ by Harish-Chandra's theorem. An element $\mu \in \mathfrak{h}^*$ is called *antidominant* if $(\mu + \rho, \alpha^\vee) \notin \mathbb{Z}_{>0}$ for all $\alpha \in \Phi^+$. In the following we will restrict our attention to integral weights. If $\lambda \in \mathfrak{h}^*$ is integral, i.e., $(\lambda, \alpha^\vee) \in \mathbb{Z}$ for all $\alpha \in \Phi^+$, then there exists a unique antidominant $\mu \in \mathfrak{h}^*$ such that $\lambda \in W \cdot \mu$. From now until the remainder of this section, μ will always denote an antidominant integral element of $\mathfrak{h}^*$, i.e.,

$$(\mu + \rho, \alpha^\vee) \in \mathbb{Z}_{\leq 0} \text{ for all } \alpha \in \Phi^+.$$

If $\chi = \chi_\mu$, we will also write $\mathcal{O}^{\mathfrak{p}}_\mu$ instead of $\mathcal{O}^{\mathfrak{p}}_\chi$. We will say that μ is *regular* if $|W \cdot \mu| = |W|$. Equivalently, μ is regular if and only if $(\mu + \rho, \alpha^\vee) \neq 0$ for all $\alpha \in \Phi$. By the Jantzen–Zuckerman translation principle, the categories $\mathcal{O}^{\mathfrak{p}}_\mu$ for μ regular are all Morita equivalent and we will write $\mathcal{O}^{\mathfrak{p}}_{\mathrm{reg}}$ for any such category.

5.1 *Parameterization of the simple modules in $\mathcal{O}^{\mathfrak{p}}_\mu$*

Let w_{I} be the longest element of W_{I}.

Lemma 5.1. *If $w \in {}^{\mathrm{I}}W$, then $w_{\mathrm{I}} w \cdot \mu \in \Lambda^+_{\mathrm{I}}$.*

Proof. Let $w \in {}^{\mathrm{I}}W$. By Remark 3.6, $w^{-1}\Phi_{\mathrm{I}}^{+} \subset \Phi^{+}$. Then, since we have that $(w_{\mathrm{I}}w(\mu+\rho), \alpha^{\vee}) = (\mu+\rho, w^{-1}w_{\mathrm{I}}\alpha^{\vee})$ and $w_{\mathrm{I}}\Phi_{\mathrm{I}}^{+} = -\Phi_{\mathrm{I}}^{+}$, we find that $(w_{\mathrm{I}}w(\mu + \rho), \alpha^{\vee}) > 0$ for all $\alpha \in \Phi_{\mathrm{I}}^{+}$. Thus, $w_{\mathrm{I}}w \cdot \mu \in \Lambda_{\mathrm{I}}^{+}$. □

Definition 5.2. For $w \in {}^{\mathrm{I}}W$, define

$$L(w) := L(w_{\mathrm{I}}w \cdot \mu).$$

Then, $w \mapsto L(w)$ defines a surjective map from ${}^{\mathrm{I}}W$ to the set of simple modules in $\mathcal{O}_{\mu}^{\mathfrak{p}}$. If μ is regular, this map is a bijection. However, if μ is singular, i.e., if $|W \cdot \mu| < |W|$, the map is not injective. To see this, define the set of *singular simple roots* by

$$\Sigma = \Sigma_{\mu} := \{\alpha \in \Pi \mid (\mu + \rho, \alpha^{\vee}) = 0\}.$$

Note that $W_{\Sigma} := \langle s_{\alpha} \mid \alpha \in \Sigma \rangle$ is equal to the stabilizer group

$$\{w \in W \mid w \cdot \mu = \mu\}$$

and μ is regular if and only if $\Sigma = \varnothing$. If $w' \in wW_{\Sigma}$ then $L(w) = L(w')$.

Definition 5.3. Define

$${}^{\mathrm{I}}W^{\Sigma} := \{w \in {}^{\mathrm{I}}W \mid w < ws_{\alpha} \text{ and } ws_{\alpha} \in {}^{\mathrm{I}}W \quad \forall \alpha \in \Sigma\}.$$

If $\Sigma = \varnothing$ then ${}^{\mathrm{I}}W^{\Sigma} = {}^{\mathrm{I}}W$. If $\mathrm{I} = \varnothing$ then ${}^{\mathrm{I}}W^{\Sigma} = W^{\Sigma}$ by Remark 3.6 (iv).

Proposition 5.4 (cf. Boe–Nakano [8, Prop. 2.3]). *We have a bijection*

$${}^{\mathrm{I}}W^{\Sigma} \xrightarrow{\sim} \{\textit{simple modules in } \mathcal{O}_{\mu}^{\mathfrak{p}}\}$$

given by $w \mapsto L(w)$. □

Remark 5.5. Suppose that μ is regular and let w be the longest element in ${}^{\mathrm{I}}W^{\Sigma} = {}^{\mathrm{I}}W$, i.e., the element such that $w_{\circ} = w_{\mathrm{I}}w$. Then $L(w)$ is the unique finite-dimensional simple module in $\mathcal{O}_{\mu}^{\mathfrak{p}}$.

Remark 5.6. Given $\mathrm{I}, \Sigma \subset \Pi$, it is in general not obvious whether ${}^{\mathrm{I}}W^{\Sigma} \neq \varnothing$, i.e., whether the corresponding block $\mathcal{O}_{\mu}^{\mathfrak{p}}$ is nonzero. In [40], K. Platt gave a beautiful characterization of the nonzero blocks using the theory of nilpotent orbits.

In the Hasse diagram of ${}^{\mathrm{I}}W$, a node that corresponds to an element of ${}^{\mathrm{I}}W^{\Sigma}$ is a node having an edge with label s_{α} going up from it for every $\alpha \in \Sigma$.

Example 5.7. Consider $(\Phi, \mathrm{I}) = (\mathsf{A}_4, \{\alpha_1, \alpha_2, \alpha_4\})$. Then $|{}^{\mathrm{I}}W| = 8$ and hence $\mathcal{O}_{\mathrm{reg}}^{\mathfrak{p}}$ contains eight simple modules. If $\Sigma = \{\alpha_1\}$, then ${}^{\mathrm{I}}W^{\Sigma} = \{\yng(2,2), \yng(2,1), \yng(2)\}$. If $\Sigma = \{\alpha_2\}$, then ${}^{\mathrm{I}}W^{\Sigma} = \{\yng(3,2), \yng(1,1), \yng(1)\}$. If $\Sigma = \{\alpha_1, \alpha_3\}$, then ${}^{\mathrm{I}}W^{\Sigma} = \{\yng(2,1)\}$. If $\Sigma = \{\alpha_1, \alpha_2\}$, then ${}^{\mathrm{I}}W^{\Sigma} = \varnothing$. All of this follows by looking at Figure 10.

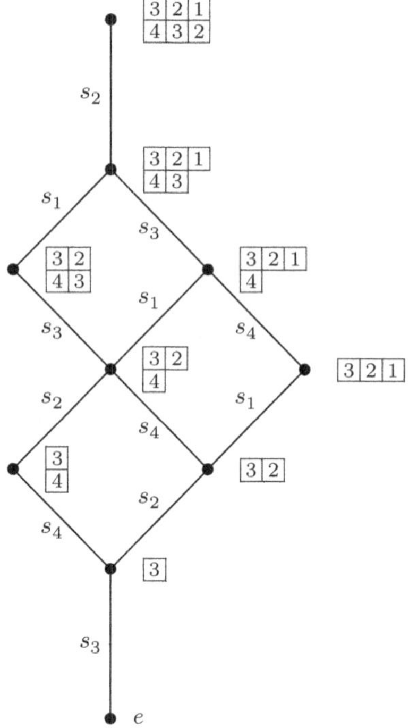

Figure 10 The poset ${}^{\mathrm{I}}W$ for $(\mathsf{A}_4, \mathsf{A}_2 \times \mathsf{A}_1)$

Example 5.8. Consider $(\Phi, \Phi_{\mathrm{I}}) = (\mathsf{E}_7, \mathsf{E}_6)$. Then $\Sigma = \{\alpha_2, \alpha_5, \alpha_7\}$ is the only subset $\Sigma \subset \Pi$ such that $|\Sigma| = 3$ and ${}^{\mathrm{I}}W^{\Sigma} \neq \varnothing$. More precisely, if $\Sigma = \{\alpha_2, \alpha_5, \alpha_7\}$, then ${}^{\mathrm{I}}W^{\Sigma} = \{w\}$, where w is the element

7	6	5	4	3	1
			2	4	3
				5	4
				6	

$$w = (s_7 s_6 s_5 s_4 s_3 s_1)(s_2 s_4 s_3)(s_5 s_4)(s_6)$$

Remark 5.9. Suppose $\mathfrak{p}$ is a parabolic subalgebra of Hermitian type. Then $\max\{s \mid s = |\Sigma| \text{ for some } \Sigma \subset \Pi \text{ with } {}^{\mathrm{I}}W^{\Sigma} \neq \varnothing\}$ is equal to the real rank of $\mathfrak{g}_{\mathbb{R}}$. Furthermore, for any $\Sigma \subset \Pi$ with ${}^{\mathrm{I}}W^{\Sigma} \neq \varnothing$, the roots in Σ are orthogonal to one another. In particular, Φ_{Σ} is of type $(\mathsf{A}_1)^s = \mathsf{A}_1 \times \cdots \times \mathsf{A}_1$, where $s = |\Sigma|$.

5.2 *Kostant modules in $\mathcal{O}^{\mathfrak{p}}_{\text{reg}}$*

Theorem 5.10 (cf. Kostant [34, Thm. 5.14]). *Let $L(\lambda)$ be a finite-dimensional irreducible $\mathfrak{g}$-module of highest weight λ, viewed as a module in $\mathcal{O}^{\mathfrak{p}}_{\text{reg}}$. Then, as an $\mathfrak{m}$-module,*

$$H^i(\mathfrak{u}, L(\lambda)) \cong \bigoplus_{\substack{y \in {}^{\mathrm{I}}W \\ l(y)=i}} F_{y\cdot\lambda}$$

for $i \leq 0$. □

To extend this result to infinite-dimensional simple modules in $\mathcal{O}^{\mathfrak{p}}_{\text{reg}}$ it is convenient to use the involution on ${}^{\mathrm{I}}W$ from Lemma 3.14 to rewrite the sum in the theorem. Let $L(\lambda)$ be a finite-dimensional $\mathfrak{g}$ module with highest weight λ, i.e., $\lambda + \rho = w_\circ \cdot \mu$, where μ is antidominant integral and regular. Then $L(\lambda) = L(w)$, where w is the longest element in ${}^{\mathrm{I}}W$, i.e., the element $w \in {}^{\mathrm{I}}W$ such that $w_\circ = w_{\mathrm{I}} w$. If $y \in {}^{\mathrm{I}}W$ with $\ell(y) = i$, then $y \cdot \lambda = w_{\mathrm{I}} \widetilde{y} \cdot \mu$ and $\ell(\widetilde{y}, w) = i$. Thus, $H^i(\mathfrak{u}, L(w)) \cong \bigoplus_x F_x$, where the sum is over all $x \in {}^{\mathrm{I}}W$ with $\ell(x, w) = i$.

Definition 5.11. A simple module $L(w)$ in $\mathcal{O}^{\mathfrak{p}}_{\text{reg}}$ is called a *Kostant module* if, as an $\mathfrak{m}$-module,

$$H^i(\mathfrak{u}, L(w)) \cong \bigoplus_{\substack{x \leq w \\ l(x,w)=i}} F_x.$$

for $i \geq 0$.

Thus, if $w \in {}^{\mathrm{I}}W$ is the longest element, then $L(w)$ is a Kostant module by Kostant's theorem (Theorem 5.10). The remaining Kostant modules in $\mathcal{O}^{\mathfrak{p}}_{\text{reg}}$ can be characterized in terms of Kazhdan–Lusztig polynomials.

Definition 5.12. For $x, w \in {}^{\mathrm{I}}W$ define the parabolic *Kazhdan–Lusztig–Vogan polynomial*

$${}^{\mathrm{I}}P_{x,w}(q) := \sum_{i \geq 0} q^{\frac{l(x,w)-i}{2}} \dim \operatorname{Ext}^i_{\mathcal{O}^{\mathfrak{p}}}(M_{\mathrm{I}}(x), L(w)).$$

It follows from the Kazhdan–Lusztig conjectures (which were proved by Casian and Collingwood in [13] for category $\mathcal{O}^{\mathfrak{p}}$) that ${}^{\mathrm{I}}P_{x,w}(q) = P^{\mathrm{I}}_{x^{-1},w^{-1}}(q)$, where the polynomial on the right hand side is the parabolic Kazhdan–Lusztig polynomial that was defined by Deodhar [14].

Theorem 5.13. *For $w \in {}^{\mathrm{I}}W$ the following are equivalent.*

(i) *The simple module $L(w)$ in $\mathcal{O}^{\mathfrak{p}}_{\mathrm{reg}}$ is a Kostant module.*
(ii) ${}^{\mathrm{I}}P_{x,w}(q) = 1$ *for all $x \leq w \in {}^{\mathrm{I}}W$.*
(iii) ${}^{\mathrm{I}}P_{e,w}(q) = 1$.

Proof. It is well known (and follows, for example, by a straightforward extension of the proof of Schmid [43, Lemma 5.13], the corresponding statement in ordinary category $\mathcal{O}$), that

$$\mathrm{Hom}_{\mathfrak{m}}(F_x, H^i(\mathfrak{u}, L(w))) \cong \mathrm{Ext}^i_{\mathcal{O}^{\mathfrak{p}}}(M_{\mathrm{I}}(x), L(w)) \quad \text{as vector spaces.}$$

It then follows immediately from the Definition 5.11 of a Kostant module that the simple module $L(w)$ in $\mathcal{O}^{\mathfrak{p}}_{\mathrm{reg}}$ is a Kostant module if and only if ${}^{\mathrm{I}}P_{x,w}(q) = 1$ for every $x \leq w$. The equivalence of (ii) and (iii) follows from the fact that ${}^{\mathrm{I}}P_{x,w}(q) - {}^{\mathrm{I}}P_{e,w}(q)$ is a polynomial with nonnegative integers which was first observed by Irving. □

5.3 BGG resolutions

The construction of the resolution of a Kostant module $L(w)$ in $\mathcal{O}^{\mathfrak{p}}_{\mathrm{reg}}$ is combinatorial and completely analogous to the construction of the resolution of a finite-dimensional simple module that was given by Lepowsky [39] (and originally by Bernstein–Gelfand–Gelfand [5] for $\mathfrak{p} = \mathfrak{b}$). For the convenience of the reader, we repeat some of the details. By [39, Prop. 3.7], for every arrow $x \to y$ in ${}^{\mathrm{I}}W$ there exists a nonzero $\mathfrak{g}$-module map $f_{x,y} : M_{\mathrm{I}}(x) \to M_{\mathrm{I}}(y)$, which lifts to the standard map between the (ordinary) Verma modules having the same highest weights as $M_{\mathrm{I}}(x)$ and $M_{\mathrm{I}}(y)$, respectively. Recall that a quadruple (w_1, w_2, w_3, w_4) of elements in W is called a square if $w_2 \neq w_3$ and $w_1 \to w_2 \to w_4$ and $w_1 \to w_3 \to w_4$. By [5, Lemma 10.4], it is possible to assign to each arrow $x \to y$ in W a number $\varepsilon_{x,y} = \pm 1$ such that for every square, the product of the four numbers assigned to the sides of the square is -1. Now for any $x, y \in {}^{\mathrm{I}}W$ with $\ell(y) = \ell(x) + 1$, define a $\mathfrak{g}$-module map $h_{x,y} : M_{\mathrm{I}}(x) \to M_{\mathrm{I}}(y)$ by

$$h_{x,y} := \begin{cases} \varepsilon_{x,y} f_{x,y} & \text{if } x < y; \\ 0 & \text{otherwise.} \end{cases}$$

Let $x, z \in {}^{\mathrm{I}}W$ with $\ell(z) = \ell(x) + 2$. It was proved in [5, Lemma 10.3] that the number of elements $y \in W$ such that $x \to y \to z$ is either zero or two. It follows immediately that the number of such $y \in {}^{\mathrm{I}}W$ is either zero, one, or two. Lepowsky showed [39, Thm. 4.3] that if there is one, then $f_{y,z} \circ f_{x,y} = 0$ and hence $h_{y,z} \circ h_{x,y} = 0$. And if there are two, i.e., if $x \to y \to z$ and $x \to y' \to z$ with $y \neq y'$, then $h_{z,y} \circ h_{x,y} = -h_{z,y'} \circ h_{x,y'}$.

These observations can now be used to construct a complex for *any* simple module $L(w)$ in $\mathcal{O}^{\mathfrak{p}}_{\text{reg}}$. For $0 \leq i \leq \ell(w)$, define

$$C_i := \bigoplus_{\substack{x \leq w \\ l(x,w)=i}} M_{\mathrm{I}}(x)$$

and for $1 \leq i \leq \ell(w)$, define $d_i : C_i \to C_{i-1}$ as the matrix of maps $d_i = (h_{x,y})$, where $x \in [e,w]_{\ell(w)-i}$ and $y \in [e,w]_{\ell(w)-(i-1)}$. Furthermore, let $d_0 : C_0 = M_{\mathrm{I}}(w) \to L(w)$ be the canonical quotient map.

Lemma 5.14. *The sequence* $0 \to C_{\ell(w)} \to \cdots \to C_1 \to C_0 \to L(w) \to 0$ *is a complex, i.e.,* $d_{i-1} \circ d_i = 0$ *for* $1 \leq i \leq l(w)$. *Furthermore, for every* $x \in [e,w]_{\ell(w)-i}$, *the restriction of* d_i *to* $M_{\mathrm{I}}(x)$ *is nonzero.*

Proof. Note that if $x, z \in [e,w]$ and $y \in {}^{\mathrm{I}}W$ with $x \to y \to z$, then $y \in [e,w]$. The observations above then immediately imply that $d_{i-1} \circ d_i = 0$ for $2 \leq i \leq \ell(w)$. Thus it remains to show that $d_0 \circ d_1 = 0$. This follows since for every $x \to w$ in ${}^{\mathrm{I}}W$, the image of $f_{x,y} : M_{\mathrm{I}}(x) \to M_{\mathrm{I}}(w)$ is contained in the radical of $M_{\mathrm{I}}(w)$, which is equal to the kernel of the quotient map $M_{\mathrm{I}}(w) \to L(w)$. □

Theorem 5.15 (cf. Boe–Hunziker [7]). *Let* $L(w)$ *be a simple module in* $\mathcal{O}^{\mathfrak{p}}_{\text{reg}}$. *Then for* $w \in {}^{\mathrm{I}}W$ *the following are equivalent:*

(i) *The truncated BGG–Lepowsky complex for* $L(w)$ *is exact.*
(ii) $L(w)$ *is a Kostant module.*
(iii) *The Schubert variety* $X(w^{-1})$ *in* G/P *is rationally smooth.*

Proof. (i) $\Rightarrow$ (ii): Identifying $\overline{\mathfrak{u}}$ with $\mathfrak{u}^*$ via the Killing form, we obtain an isomorphism $H^i(\mathfrak{u}, L(w)) \cong H_i(\overline{\mathfrak{u}}, L(w))$ of $\mathfrak{m}$-modules. Since $H_i(\overline{\mathfrak{u}}, L(w))$ is also isomorphic to $\mathrm{Tor}_i^{U(\overline{\mathfrak{u}})}(\mathbb{C}, L(w))$, we can compute $H_i(\overline{\mathfrak{u}}, L(w))$ by any projective $U(\overline{\mathfrak{u}})$-module resolution of $L(w)$. Now suppose that the truncated BGG–Lepowsky complex for $L(w)$ is exact. Then this complex is a free $U(\overline{\mathfrak{u}})$-module resolution of $L(w)$ since by the PBW Theorem every parabolic Verma module $M_{\mathrm{I}}(x)$, as a $U(\overline{\mathfrak{u}})$-module, is isomorphic to the free module $U(\overline{\mathfrak{u}}) \otimes F_x$. By applying the functor $\mathbb{C} \otimes_{U(\overline{\mathfrak{u}})} \_\_$ to this resolution and by taking the i-th homology of the resulting complex we obtain

$$H_i(\overline{\mathfrak{u}}, L(w)) \cong \bigoplus_{\substack{x \leq w \\ l(x,w)=i}} F_x$$

and hence $L(w)$ is a Kostant module.

(ii) $\Rightarrow$ (i): Suppose that $L(w)$ is a Kostant module. By the lemma above, $L(w)$ is a generalized Kostant module in the sense of [17, 2.7]. By [17, Thm. 2.8], the truncated BGG complex is exact.

The equivalence of (ii) and (iii) follows from Proposition 4.9 and Theorem 5.13 (and the comments just before Theorem 5.13). □

Example 5.16. Let $\mathfrak{p}$ be the parabolic subalgebra of $\mathfrak{g} = \mathfrak{sl}(4, \mathbb{C})$ corresponding to the Hermitian symmetric pair $(\mathsf{A}_3, \mathsf{A}_1 \times \mathsf{A}_1)$. Fix a regular block $\mathcal{O}^{\mathfrak{p}}_{\mathrm{reg}}$. Then the BGG resolution of the finite-dimensional simple module in $\mathcal{O}^{\mathfrak{p}}_{\mathrm{reg}}$ is of the form

$$0 \to M_{\mathrm{I}}(e) \to M_{\mathrm{I}}(\square) \to M_{\mathrm{I}}(\ydiagram{2}) \oplus M_{\mathrm{I}}(\ydiagram{1,1}) \to M_{\mathrm{I}}(\ydiagram{2,1}) \to M_{\mathrm{I}}(\ydiagram{2,2}) \to L(\ydiagram{2,2}) \to 0.$$

As explained above, by truncating the BGG resolution we obtain complexes for the other simple modules in $\mathcal{O}^{\mathfrak{p}}_{\mathrm{reg}}$. For example, the BGG complex

$$0 \to M_{\mathrm{I}}(e) \to M_{\mathrm{I}}(\square) \to M_{\mathrm{I}}(\ydiagram{2}) \to L(\ydiagram{2}) \to 0$$

is exact, whereas the BGG complex

$$0 \to M_{\mathrm{I}}(e) \to M_{\mathrm{I}}(\square) \to M_{\mathrm{I}}(\ydiagram{2}) \oplus M_{\mathrm{I}}(\ydiagram{1,1}) \to M_{\mathrm{I}}(\ydiagram{2,1}) \to L(\ydiagram{2,1}) \to 0$$

is not exact (by Proposition 4.3 and Theorem 5.15).

5.4 *A modified ordering for singular blocks*

For regular blocks, it is possible to compute Ext-groups, and hence $\mathfrak{u}$-cohomology, using Kazhdan–Lusztig polynomials. It turns out that this is also possible for singular blocks. Fix a singular block $\mathcal{O}^{\mathfrak{p}}_{\mu}$ and, as before, let ${}^{\mathrm{I}}W^{\Sigma}$ denote the set that parameterizes the simple modules in $\mathcal{O}^{\mathfrak{p}}_{\mu}$.

Definition 5.17. For $x, w \in {}^{\mathrm{I}}W^{\Sigma}$ define the polynomial ${}^{\mathrm{I}}P^{\Sigma}_{x,w}(q)$ by

$${}^{\mathrm{I}}P^{\Sigma}_{x,w}(q) := \sum_{i \geq 0} q^{\frac{\ell(x,w)-i}{2}} \dim \operatorname{Ext}^{i}_{\mathcal{O}^{\mathfrak{p}}}(M_{\mathrm{I}}(x), L(w)).$$

There is a beautiful formula due to Soergel [45] and Irving [28] that allows one to compute the ${}^{\mathrm{I}}P^{\Sigma}_{x,w}(q)$ in terms of (regular parabolic) KLV polynomials. It says that for $x, w \in {}^{\mathrm{I}}W^{\Sigma}$,

$${}^{\mathrm{I}}P^{\Sigma}_{x,w}(q) = \sum_{z \in W_{\Sigma}} (-1)^{\ell(z)}\, {}^{\mathrm{I}}P_{xz,w}(q)\,.$$

Definition 5.18. For $x < w$ in ${}^{\mathrm{I}}W$ define $\mu_{\mathrm{I}}(x, w)$ by

$$\mu_{\mathrm{I}}(x, w) := \dim \operatorname{Ext}^{1}_{\mathcal{O}^{\mathfrak{p}}}(M_{\mathrm{I}}(x), L(w)).$$

For $x, w \in {}^{\mathrm{I}}W^{\Sigma}$, write $x \to_{\mu} w$ if $x < w$ in the Bruhat ordering, $\mu_{\mathrm{I}}(x, w) \neq 0$ and there is no $x < z < w$ in ${}^{\mathrm{I}}W^{\Sigma}$ with $\mu_{\mathrm{I}}(z, w) \neq 0$. Then define $\leq_{\mu}$ as the ordering on ${}^{\mathrm{I}}W^{\Sigma}$ generated by the covering relations $x \to_{\mu} w$. We call this ordering on ${}^{\mathrm{I}}W^{\Sigma}$ the *μ-ordering* or the Ext^1*-ordering*.

Remark 5.19. Note that by definition, $x \leq_\mu w$ implies $x \leq w$ in the Bruhat ordering. If $\Sigma = \varnothing$ or $\mathrm{I} = \varnothing$, then the Ext^1-ordering and the Bruhat ordering coincide. The proof is easy in the case when $\Sigma = \varnothing$. The case when $\mathrm{I} = \varnothing$ follows via Koszul duality for parabolic and singular category $\mathcal{O}$ (cf. Beilinson–Ginzburg–Soergel [4] and Backelin [3]).

In general, the posets ${}^{\mathrm{I}}W^{\Sigma}$ are very complicated. For example, they need not be graded, i.e., two chains connecting two elements in ${}^{\mathrm{I}}W^{\Sigma}$ can have different length (see Boe–Hunziker [7] for examples). However, in the Hermitian symmetric cases, the posets ${}^{\mathrm{I}}W^{\Sigma}$ are always nice. In fact, ${}^{\mathrm{I}}W^{\Sigma}$ is either isomorphic to a poset of the form ${}^{\mathrm{I}'}W'$ corresponding to another Hermitian symmetric pair (of smaller rank) or isomorphic to a disjoint union of two such posets.

Example 5.20. Consider the Hermitian symmetric pair of type $(\mathsf{C}_n, \mathsf{A}_{n-1})$ and suppose that Σ contains the long simple root α_n. If $w \in {}^{\mathrm{I}}W^{\Sigma}$ is written in Lascoux–Schützenberger notation (see Figure 7) as a binary sequence of length n, we say that w is *even* if the binary sequence has an even number of 1's and *odd* if the binary sequence has an odd number of 1's. Suppose that $x, w \in {}^{\mathrm{I}}W^{\Sigma}$ such that $x < w$ in the Bruhat ordering. In [19] it was proved that

$$\mu_{\mathrm{I}}(x, w) = \begin{cases} 1 & \text{if } x \text{ and } w \text{ have the same parity,} \\ 0 & \text{if } x \text{ and } w \text{ have opposite parity.} \end{cases}$$

It follows that the poset ${}^{\mathrm{I}}W^{\Sigma}$ with respect to the μ-ordering can be written as a disjoint union

$${}^{\mathrm{I}}W^{\Sigma} = {}^{\mathrm{I}}W^{\Sigma}_{\mathrm{even}} \cup {}^{\mathrm{I}}W^{\Sigma}_{\mathrm{odd}},$$

where the elements from ${}^{\mathrm{I}}W^{\Sigma}_{\mathrm{even}}$ and ${}^{\mathrm{I}}W^{\Sigma}_{\mathrm{odd}}$ are not comparable (see Figure 11). Furthermore, if $|\Sigma| = m$, then the posets ${}^{\mathrm{I}}W^{\Sigma}_{\mathrm{even}}$ and ${}^{\mathrm{I}}W^{\Sigma}_{\mathrm{odd}}$ are both isomorphic to the poset ${}^{\mathrm{I}'}W'$ corresponding to the Hermitian symmetric pair $(\mathsf{C}_{n-2m}, \mathsf{A}_{n-2m-1})$. The isomorphisms ${}^{\mathrm{I}}W^{\Sigma}_{\mathrm{even}} \cong {}^{\mathrm{I}'}W'$ and ${}^{\mathrm{I}}W^{\Sigma}_{\mathrm{odd}} \cong {}^{\mathrm{I}'}W'$ are described explicitly as follows. Suppose that $\Sigma = \{\alpha_{i_1}, \alpha_{i_2}, \ldots, \alpha_{i_m}\}$ with $i_1 < i_2 < \cdots i_{m-1} < i_m = n$. We note that in fact $|i_j - i_{j+1}| \geq 2$ for $1 \leq j \leq m-1$. Let $w \in {}^{\mathrm{I}}W^{\Sigma}_{\mathrm{even}}$ or $w \in {}^{\mathrm{I}}W^{\Sigma}_{\mathrm{odd}}$ and write w in Lascoux–Schützenberger notation as a binary sequence $w = b_1 b_2 \cdots b_n$. Delete all the segments $b_{i_j} b_{i_j+1}$ for $j < m$ and the last digit b_n. The resulting binary sequence has length $n - 2m + 1$. If we remove its right most digit we obtain a binary sequence of length $n - 2m$ that corresponds to an element $w' \in {}^{\mathrm{I}'}W'$.

For a representative example, consider $n = 9$ and $\Sigma = \{\alpha_2, \alpha_7, \alpha_9\}$. Then every $w = {}^{\mathrm{I}}W^{\Sigma}$ is of the form $w = b_1 01 b_4 b_5 b_6 010$. The mappings ${}^{\mathrm{I}}W^{\Sigma}_{\mathrm{even}} \to {}^{\mathrm{I}'}W'$ and ${}^{\mathrm{I}}W^{\Sigma}_{\mathrm{odd}} \to {}^{\mathrm{I}'}W'$ are given by $w = b_1 01 b_4 b_5 b_6 010 \mapsto w' = b_1 b_4 b_5$. These maps are invertible since for a given element $w' = b_1 b_4 b_5 \in {}^{\mathrm{I}'}W'$, there is a unique choice for $b_6 \in \{0, 1\}$ such that $w = b_1 01 b_4 b_5 b_6 010$ is even or odd, respectively.

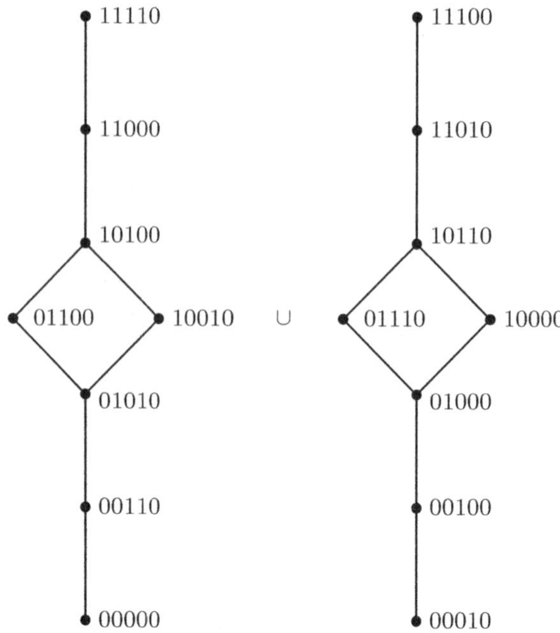

Figure 11 The μ-ordering of ${}^{\mathrm{I}}W^{\Sigma}$ for $(\Phi, \mathrm{I}, \Sigma) = (\mathsf{C}_5, \{\alpha_1, \ldots, \alpha_4\}, \{\alpha_5\})$

5.5 *Equivalences of categories*

Theorem 5.21 (cf. Enright–Shelton [19,20]). *Suppose $\mathfrak{p}$ is of Hermitian type.*

(a) *Suppose that either Φ has one root length, or Φ has two root lengths and all the roots in Σ are short. Then there exists an equivalence of categories*

$$\mathcal{E} : \mathcal{O}^{\mathfrak{p}}_{\mu} \to \mathcal{O}^{\mathfrak{p}'}_{\mathrm{reg}},$$

where $\mathfrak{p}'$ is a parabolic subalgebra of Hermitian type of a complex simple Lie algebra $\mathfrak{g}'$ of rank $\leq n$. More precisely, there exists an isomorphism of posets ${}^{\mathrm{I}}W^{\Sigma} \to {}^{\mathrm{I}'}W'$ such that $\mathcal{E} : L(w) \mapsto L(w')$ and $\mathcal{E} : M_{\mathrm{I}}(w) \mapsto M_{\mathrm{I}'}(w')$. Furthermore, if $x \leq w \in {}^{\mathrm{I}}W^{\Sigma}$ then ${}^{\mathrm{I}}P^{\Sigma}_{x,w}(q) = {}^{\mathrm{I}'}P_{x',w'}(q)$.

(b) *Suppose that Φ has two root lengths and Σ contains a long root. Then there exists an equivalence of categories*

$$\mathcal{E} : \mathcal{O}^{\mathfrak{p}}_{\mu} \to \mathcal{O}^{\mathfrak{p}'}_{\mathrm{reg}} \oplus \mathcal{O}^{\mathfrak{p}'}_{\mathrm{reg}},$$

where $\mathfrak{p}'$ is a parabolic subalgebra of Hermitian type of a complex simple Lie algebra $\mathfrak{g}'$ of rank $\leq n$. More precisely, the poset ${}^{\mathrm{I}}W^{\Sigma}$ with respect to the μ-ordering is a disjoint union ${}^{\mathrm{I}}W^{\Sigma} = {}^{\mathrm{I}}W^{\Sigma}_{\mathrm{even}} \cup {}^{\mathrm{I}}W^{\Sigma}_{\mathrm{odd}}$ and the category $\mathcal{O}^{\mathfrak{p}}_{\mu}$ has a *decomposition into a direct sum* $\mathcal{O}^{\mathfrak{p}}_{\mu} = \mathcal{O}^{\mathfrak{p}}_{\mu,\mathrm{even}} \oplus \mathcal{O}^{\mathfrak{p}}_{\mu,\mathrm{odd}}$ *such that all extensions between modules in different summands are zero. There exist isomorphisms of posets* ${}^{\mathrm{I}}W^{\Sigma}_{\mathrm{even}} \to {}^{\mathrm{I}'}W'$ *and* ${}^{\mathrm{I}}W^{\Sigma}_{\mathrm{odd}} \to {}^{\mathrm{I}'}W'$ *and corresponding equivalences of categories* $\mathcal{E}_{\mathrm{even}} : \mathcal{O}^{\mathfrak{p}}_{\mu,\mathrm{even}} \to \mathcal{O}^{\mathfrak{p}'}_{\mathrm{reg}}$ *and* $\mathcal{E}_{\mathrm{even}} : \mathcal{O}^{\mathfrak{p}}_{\mu,\mathrm{odd}} \to \mathcal{O}^{\mathfrak{p}'}_{\mathrm{reg}}$ *such that* $L(w) \mapsto L(w')$ *and* $M_{\mathrm{I}}(w) \mapsto M_{\mathrm{I}'}(w')$*. Furthermore, if* $x \leq w \in {}^{\mathrm{I}}W^{\Sigma}$ *are either both even or both odd, then* ${}^{\mathrm{I}}P^{\Sigma}_{x,w}(q) = {}^{\mathrm{I}'}P_{x',w'}(q)$*.*

Proof. The construction of the functors $\mathcal{E}$, $\mathcal{E}_{\mathrm{even}}$ and $\mathcal{E}_{\mathrm{odd}}$ is complicated and beyond the scope of this paper. We refer the reader to [19] for details. Here we content ourselves with understanding the isomorphisms of the underlying posets. For the Hermitian pairs corresponding to $\mathfrak{g}_{\mathbb{R}} = \mathfrak{su}(p,q)$, $\mathfrak{sp}(n,\mathbb{R})$, and $\mathfrak{so}^*(2n)$ these isomorphisms are given by using Lascoux–Schützenberger notation as in Example 5.20. For the remaining Hermitian symmetric pairs, the poset isomorphisms are easy to check using Hasse diagram of ${}^{\mathrm{I}}W$. Two examples are shown in Figures 12 and 13. We leave the details to the reader. □

Because of Theorem 5.21, if $\mathfrak{p}$ is of Hermitian type it makes sense to define Kostant modules and Bernstein–Gelfand–Gelfand resolutions in singular infinitesimal blocks $\mathcal{O}^{\mathfrak{p}}_{\mu}$.

Example 5.22. Suppose $\mathfrak{g}_{\mathbb{R}} = \mathfrak{su}(3,2)$ and consider the simple module $L(\lambda)$ of highest weight $\lambda = -\omega_3$. Then $\lambda = w_{\mathrm{I}}w \cdot \mu$, where $\mu + \rho = -\omega_1 - \omega_3 - \omega_4$ and $w = s_3 s_2 s_1 s_4 s_3$. Note that $\Sigma = \{\alpha_2\}$ and ${}^{\mathrm{I}}W^{\Sigma} = \{⊞\!\!\square, \boxminus, \square)\}$. The poset ${}^{\mathrm{I}}W^{\Sigma}$ is isomorphic to the poset ${}^{\mathrm{I}'}W' = \{\square\!\square, \square, e\}$ corresponding to $\mathfrak{g}'_{\mathbb{R}} = \mathfrak{su}(2,1)$. By applying the exact functor $\mathcal{E}^{-1}$ to the BGG resolution

$$0 \to M'_{\mathrm{I}'}(e) \to M'_{\mathrm{I}'}(\square) \to M'_{\mathrm{I}'}(\square\square) \to L'(\square\square) \to 0$$

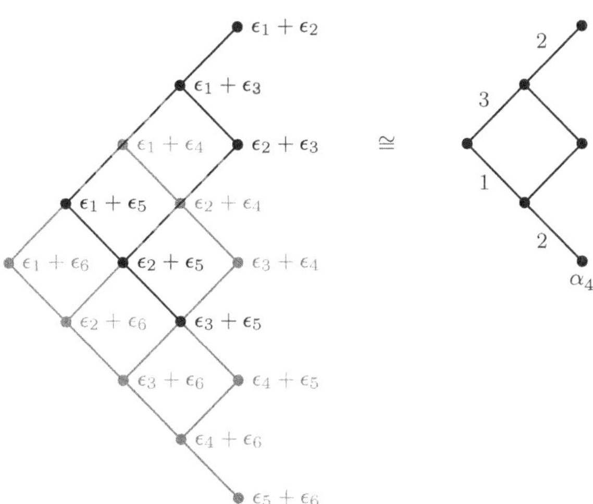

Figure 12 The poset $\Phi(\mathfrak{u})_{[\lambda]}$ for $\mathfrak{g}_{\mathbb{R}} = \mathfrak{so}^*(12)$ and $\lambda = -\omega_6$

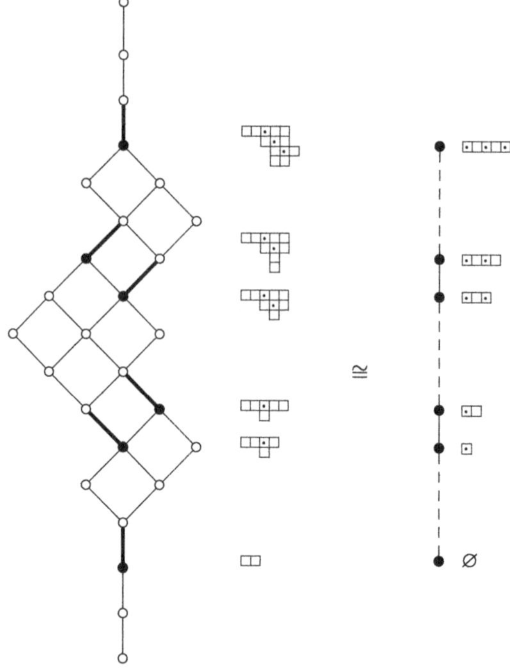

Figure 13 The isomorphism ${}^{\mathrm{I}}W^{\{\alpha_4\}} \cong {}^{\mathrm{I}'}W'$ for $(\Phi, \Phi_{\mathrm{I}}) = (\mathsf{E}_6, \mathsf{D}_5)$, $(\Phi', \Phi'_{\mathrm{I}'}) = (\mathsf{A}_5, \mathsf{A}_4)$

of the finite-dimensional module $L'(\square\square)$ in $\mathcal{O}^{\mathfrak{p}'}_{\mathrm{reg}}$, we obtain a resolution

$$0 \to M_{\mathrm{I}}(\square) \to M_{\mathrm{I}}(\boxminus) \to M_{\mathrm{I}}(\square\square\square) \to L(\square\square\square) \to 0$$

of the simple module $L(\lambda) = L(\square\square\square)$ in the singular block $\mathcal{O}^{\mathfrak{p}}_{\mu}$.

5.6 *A recursion for parabolic Kazhdan–Lusztig polynomials*

In [38], Lascoux and Schützenberger gave an efficient combinatorial algorithm to calculate Kazhdan–Lusztig polynomials for Grassmannians. In [6], Boe extended the techniques of Lascoux and Schützenberger to obtain a similar algorithm for the Hermitian symmetric spaces of type $(\mathsf{C}_n, \mathsf{A}_{n-1})$ and $(\mathsf{D}_n, \mathsf{A}_{n-1})$. Here we will use diagrams of Hermitian type to give an algorithm that works for all Hermitian symmetric spaces. The key is the notion of capacity which we will shortly define. Consider a diagram of Hermitian type, i.e., the Hasse diagram of a poset $\Phi(\mathfrak{u})$ as shown in the appendix. Note how all these Hasse diagrams are drawn on a square lattice that is tilted 45°. We will assume in the following that the square lattice is normalized such that the diagonals of each square (diamond) have unit length. Recall the map $f : \Phi(\mathfrak{u}) \to \Pi$ from Definition 3.19. We observe that for any simple root $\alpha \in \Pi$ except α_2 in the two exceptional cases, the fiber $f^{-1}(\alpha)$ consists of roots in $\Phi(\mathfrak{u})$ whose corresponding nodes in the Hasse diagram of $\Phi(\mathfrak{u})$ lie on a vertical line. (The latter is equivalent to the statement that in the generalized Young diagram of the longest element of ${}^{\mathrm{I}}W$, the boxes that are filled with the same number all lie on the same diagonal.) In particular, the vertical distance between two nodes in the Hasse diagram corresponding to two roots in the fiber $f^{-1}(\alpha)$ is a nonnegative integer. This last statement is also true if $\alpha = \alpha_2$ in the two exceptional cases. For example, in type $(\mathsf{E}_6, \mathsf{D}_5)$, the fiber $f^{-1}(\alpha_2)$ contains two roots and the vertical distance between the two corresponding nodes is 2.

Definition 5.23. For $x, w \in {}^{\mathrm{I}}W^{\{\alpha\}}$ such that $x \leq w$ in ${}^{\mathrm{I}}W$, define the *capacity* of w with respect to x as the nonnegative integer

$$c(x, w) := \text{vertical distance between the nodes of } \beta_x \text{ and } \beta_w,$$

where $\beta_x, \beta_w \in \Phi(\mathfrak{u})$ are the roots such that $\Phi_{xs_\alpha} \setminus \Phi_x = \{\beta_x\}$, $\Phi_{ws_\alpha} \setminus \Phi_w = \{\beta_w\}$, respectively.

Theorem 5.24. *Let $x < w \in {}^{\mathrm{I}}W$ such that $x \in {}^{\mathrm{I}}W^{\{\alpha\}}$ and $xs_\alpha \leq w$ in ${}^{\mathrm{I}}W$.*

(a) *Suppose that either $w \notin {}^{\mathrm{I}}W^{\{\alpha\}}$ or $w \in {}^{\mathrm{I}}W^{\{\alpha\}}$, Φ has two root lengths, α is long, and $c(x, w)$ is odd. Then*

$${}^{\mathrm{I}}P_{x,w}(q) = {}^{\mathrm{I}}P_{xs_\alpha,w}(q).$$

(b) *Suppose that $w \in {}^{\mathrm{I}}W^{\{\alpha\}}$. Also, if Φ has two root lengths and α is long, assume that $c(x, w)$ is even. Then*

$${}^{\mathrm{I}}P_{x,w}(q) = {}^{\mathrm{I}}P_{xs_\alpha,w}(q) + q^{c(x,w)} \cdot {}^{\mathrm{I}'}P_{x',w'}(q).$$

Proof. If $c(x, w)$ is replaced by $(\ell(x, w) - \ell(x', w'))/2$, then the proposition is Theorem 15.4 in [19]. This theorem also holds in the exceptional cases as follows from the work in [20]. Thus, it remains to show that

$$c(x, w) = \frac{\ell(x, w) - \ell'(x', w')}{2}.$$

For the Hermitian symmetric pairs corresponding to $\mathfrak{su}(p, q)$, $\mathfrak{sp}(n, \mathbb{R})$, and $\mathfrak{so}^*(2n)$, this follows from the results of Lascoux–Schützenberger and Boe. For example, for $\mathfrak{sp}(n, \mathbb{R})$ it was shown by Boe in the proof of [6, Propostion 3.14].

For the remaining Hermitian symmetric pairs we proceed case-by-case. First note that if $x \leq y \leq w \in {}^{\mathrm{I}}W^{\{\alpha\}}$, then $c(x, w) = c(x, y) + c(y, w)$, $\ell(x, w) = \ell(x, y) + \ell(y, w)$, and $\ell'(x', w') = \ell(x', y') + \ell'(y', w')$. This means that we may assume that w' covers x' in ${}^{\mathrm{I}'}W'$. Now, as a representative case, consider the Hermitian symmetric pair of type $(\mathsf{E}_7, \mathsf{E}_6)$ and $\alpha = \alpha_4$. The poset isomorphism ${}^{\mathrm{I}}W^{\{\alpha\}} \cong {}^{\mathrm{I}'}W'$ is shown in Figure 13. The edges in the Hasse diagram of ${}^{\mathrm{I}}W$ corresponding to the simple reflection s_4 are shown in bold. The bold nodes at the top of the bold edges correspond to the elements of ${}^{\mathrm{I}}W^{\{\alpha\}}$ multiplied by s_4. Suppose that $x < w \in {}^{\mathrm{I}}W^{\{\alpha\}}$ are the elements with

$$xs_4 = \begin{array}{|c|c|c|c|}\hline 7&6&5&4\\ \hline\end{array} \quad \text{and} \quad ws_4 = \begin{array}{|c|c|c|c|c|}\hline 7&6&5&4&3\\ \hline \multicolumn{3}{c|}{}&2&4\\ \cline{4-5}\end{array}.$$

Then $c(x,w) = 1$ and $(\ell(x,w) - \ell(x',w'))/2 = (3-1)/2 = 1$. The identity $c(x,w) = (\ell(x,w) - \ell(x',w'))/2$ is quickly verified in the same way for the other 11 pairs of elements $x < w \in {}^{\mathrm{I}}W^{\{\alpha\}}$ such that w' covers x'. □

Example 5.25. Consider the Hermitian symmetric pair $(\mathsf{D}_6, \mathsf{D}_5)$. Suppose that $x < w \in {}^{\mathrm{I}}W^{\{\alpha_2\}}$are the elements with

$$xs_2 = \begin{array}{|c|c|}\hline 1&2\\ \hline\end{array} \qquad ws_2 = \begin{array}{|c|c|c|c|c|}\hline 1&2&3&4&5\\ \hline \multicolumn{3}{c|}{}&6&4\\ \cline{4-5} \multicolumn{4}{c|}{}&3\\ \cline{5-5} \multicolumn{4}{c|}{}&2\\ \cline{5-5}\end{array}.$$

Then the capacity $c(x,w) = 3$ and hence, by Theorem 5.24 (b),

$${}^{\mathrm{I}}P_{x,w}(q) = {}^{\mathrm{I}}P_{xs_2,w}(q) + q^3 \cdot {}^{\mathrm{I}'}P_{x',w'}(q)$$

The poset ${}^{\mathrm{I}'}W'$ contains two elements in linear order which implies that ${}^{\mathrm{I}'}P_{x',w'}(q) = 1$. Therefore, ${}^{\mathrm{I}}P_{x,w}(q) = 1 + q^3$.

Example 5.26. Consider the Hermitian symmetric pair $(\mathsf{E}_7, \mathsf{E}_6)$. Let $x = e$ and let $w \in {}^{\mathrm{I}}W$ be the submaximal element. Then $x, w \in {}^{\mathrm{I}}W^{\{\alpha_7\}}$. The capacity $c(x,w) = 8$ and hence, by Theorem 5.24 (b),

$${}^{\mathrm{I}}P_{e,w}(q) = {}^{\mathrm{I}}P_{s_7,w}(q) + q^8 \cdot {}^{\mathrm{I}'}P_{e',w'}(q).$$

The poset ${}^{\mathrm{I}'}W'$ contains three elements in linear order which implies that ${}^{\mathrm{I}'}P_{x',w'}(q) = 1$. Note that ${}^{\mathrm{I}}W^{\{\alpha_7\}}$ also contains three elements. Let $u \in {}^{\mathrm{I}}W^{\{\alpha_7\}}$ be element with $e < u < w$. Then, by using Theorem 5.24 (a) repeatedly, ${}^{\mathrm{I}}P_{s_7,w}(q) = {}^{\mathrm{I}}P_{u,w}(q)$. The capacity $c(u,w) = 4$ and hence, by Theorem 5.24 (b),

$${}^{\mathrm{I}}P_{u,w}(q) = {}^{\mathrm{I}}P_{us_7,w}(q) + q^4 \cdot {}^{\mathrm{I}'}P_{u',w'}(q).$$

By the same arguments as before, ${}^{\mathrm{I}'}P_{u',w'}(q) = 1$ and ${}^{\mathrm{I}}P_{us_7,w}(q) = {}^{\mathrm{I}}P_{w,w}(q) = 1$. Therefore, ${}^{\mathrm{I}}P_{x,w}(q) = 1 + q^4 + q^8$.

6 BGG resolutions of unitary highest weight modules

Suppose that $\mathfrak{g}_{\mathbb{R}}$ is a noncompact real form of $\mathfrak{g}$ with Cartan decomposition $\mathfrak{g}_{\mathbb{R}} = \mathfrak{m}_{\mathbb{R}} \oplus \mathfrak{s}_{\mathbb{R}}$ and assume that $\mathfrak{m}_{\mathbb{R}}$ has a nontrivial center $\mathfrak{z}_{\mathbb{R}}$. Let $\mathfrak{p}$ be the corresponding parabolic subalgebra of $\mathfrak{g}$ of Hermitian type. A simple module $L(\lambda)$ in the category $\mathcal{O}^{\mathfrak{p}}$ is called *unitarizable* if there exists a $\mathfrak{g}_{\mathbb{R}}$-invariant Hermitian scalar product on $L(\lambda)$. In [15], Enright showed that every unitary highest weight module is a Kostant module. Hence every unitary highest weight module has a generalized BGG resolution.

6.1 Reduced Hermitian symmetric pairs

Definition 6.1 (cf. Enright [15]). For $\lambda \in \Lambda_{\mathrm{I}}^{+}$, define $\Phi(\mathfrak{u})_{[\lambda]}$ as the set of all roots $\beta \in \Phi(\mathfrak{u})$ such that:

(1) $(\lambda + \rho, \beta^{\vee}) \in \mathbb{Z}_{>0}$;
(2) β is orthogonal to $\Xi_{\lambda} := \{\alpha \in \Phi(\mathfrak{u}) \mid (\lambda + \rho, \alpha^{\vee}) = 0\}$;
(3) β is short if Φ has two root lengths and Ξ_{λ} contains a long root .

Next define $W_{[\lambda]} := \langle s_{\beta} \mid \beta \in \Phi(\mathfrak{u})_{[\lambda]} \rangle$ and $\Phi_{[\lambda]} := \{\alpha \in \Phi \mid s_{\alpha} \in W_{[\lambda]}\}$. Then $\Phi_{[\lambda]}$ is an (abstract) irreducible root system. Let $\mathfrak{g}_{[\lambda]}$ be the semisimple part of the reductive Lie algebra with Cartan subalgebra $\mathfrak{h}$ and root system $\Phi_{[\lambda]}$ and let $\mathfrak{b}_{[\lambda]}$ be the Borel subalgebra of $\mathfrak{g}_{[\lambda]}$ corresponding to the positive system $\Phi_{[\lambda]} \cap \Phi^{+} \subset \Phi_{[\lambda]}$. Let $\mathfrak{m}_{[\lambda]}$ be the reductive Lie subalgebra of $\mathfrak{g}_{[\lambda]}$ with Cartan subalgebra $\mathfrak{g}_{[\lambda]} \cap \mathfrak{h}$ and root system $\Phi_{[\lambda]} \cap \Phi_{\mathrm{I}}$ and let $\mathfrak{u}_{[\lambda]}$ be the sum of the root spaces of $\mathfrak{g}_{[\lambda]}$ corresponding to the roots in $(\Phi_{[\lambda]} \cap \Phi^{+}) \setminus \Phi_{\mathrm{I}}$. Then $\mathfrak{p}_{[\lambda]} := \mathfrak{m}_{[\lambda]} \oplus \mathfrak{u}_{[\lambda]}$ is a parabolic subalgebra of $\mathfrak{g}_{[\lambda]}$ containing $\mathfrak{b}_{[\lambda]}$. Note that $\mathfrak{u}_{[\lambda]}$ is abelian since $\Phi(\mathfrak{u}_{[\lambda]}) := (\Phi_{[\lambda]} \cap \Phi^{+}) \setminus \Phi_{\mathrm{I}}$ is a subset of $\Phi(\mathfrak{u})$ and hence the sum of two roots in $\Phi(\mathfrak{u}_{[\lambda]})$ is never a root. A parabolic subalgebra with abelian nilradical is always a maximal parabolic subalgebra. Therefore, by Lemma 2.2, $\mathfrak{p}_{[\lambda]}$ is of Hermitian type and $(\mathfrak{g}_{[\lambda]}, \mathfrak{m}_{[\lambda]})$ is a complexified Hermitian symmetric pair, called the *reduced Hermitian symmetric pair* associated to $L(\lambda)$.

The standard parabolic subalgebra $\mathfrak{p}_{[\lambda]}$ of $\mathfrak{g}_{[\lambda]}$ corresponds to a subset $\mathrm{E} \subset \Pi_{[\lambda]}$, where $\Pi_{[\lambda]}$ denotes the set of simple roots of the positive system $\Phi_{[\lambda]} \cap \Phi^{+}$. The set $\Pi_{[\lambda]} \setminus \mathrm{E}$ contains exactly one element, namely the unique noncompact root in $\Pi_{[\lambda]}$. We caution the reader that $\Pi_{[\lambda]}$ is *not* a subset of Π in general. In particular, if $w \in W_{[\lambda]}$ and $\ell_{[\lambda]}(w)$ denotes the length of w with respect to the simple system $\Pi_{[\lambda]}$, then $\ell_{[\lambda]}(w) \neq \ell(w)$ in general. Furthermore, the poset ${}^{\mathrm{E}}(W_{[\lambda]})$ is *not* a subset of ${}^{\mathrm{I}}W$ in general. However, by definition, ${}^{\mathrm{E}}(W_{[\lambda]}) \subset W_{[\lambda]} \subset W$.

Remark 6.2. Note that $\Phi(\mathfrak{u})_{[\lambda]} \subset \Phi(\mathfrak{u}_{[\lambda]})$, but in general the two sets do not have to be equal. However, in all the examples that are discussed in this paper, we always have equality.

Before we continue, we will discuss three examples in detail. A good understanding of these examples will be essential in the proof of the main theorems in Section 7. In all these examples, we use coordinate descriptions for roots and weights with respect to the standard bases given by the ϵ_i's. For more general examples (and rigorous proofs), we refer the reader to the paper [22] by Enright and Willenbring.

Example 6.3. Consider $\mathfrak{g}_{\mathbb{R}} = \mathfrak{u}(p,q)$[1] and $\lambda = -k\omega_p$, where ω_p is the p-th fundamental weight and $1 \leq k \leq \min\{p,q\}$. Let $n = p+q$. Then $\lambda \sim (-k,\dots,-k;0,\dots,0)$ and

$$\lambda + \rho \sim (n-k, n-k-1, \dots, n-k-p+1\ ;\ q, q-1, \dots, 1),$$

where $x \sim y$ is the equivalence relation on $\mathfrak{h}^* = \mathbb{C}^n$ given by $x-y \in \mathbb{C}(1,\dots,1) \subset \mathfrak{h}^* = \mathbb{C}^n$. Furthermore,

$$\Phi(\mathfrak{u}) = \{\epsilon_i - \epsilon_j \mid 1 \leq i \leq p\ ,\ p+1 \leq j \leq n\}.$$

We claim (and the reader is invited to verify) that

$$\Xi_\lambda = \{\epsilon_i - \epsilon_{i+k} \mid p-k+1 \leq i \leq p\}$$

and

$$\Phi(\mathfrak{u})_{[\lambda]} = \{\epsilon_i - \epsilon_j \mid 1 \leq i \leq p-k\ ,\ p+k+1 \leq j \leq p+q\}.$$

This poset is isomorphic to the poset of noncompact roots of a Hermitian symmetric pair corresponding to $\mathfrak{u}(p-k, q-k)$. The minimal element in $\Phi(\mathfrak{u})_{[\lambda]}$ is the root $\epsilon_{p-k} - \epsilon_{p+k+1}$, which is the noncompact root in $\Pi_{[\lambda]}$. By calculating the differences of neighboring roots in $\Phi(\mathfrak{u})_{[\lambda]}$, we find that the compact roots in $\Pi_{[\lambda]}$ are the roots of the form $\epsilon_i - \epsilon_{i+1}$ with $1 \leq i \leq p-k-1$ or $p+k+1 \leq i \leq p+q-1$. We do have $\Phi(\mathfrak{u})_{[\lambda]} = \Phi(\mathfrak{u}_{[\lambda]})$ in this case.

As a representative example, consider $(p,q) = (6,4)$ and $k = 2$. In this case, $\lambda + \rho \sim (8,7,6,5,\mathbf{4},\mathbf{3};\mathbf{4},\mathbf{3},2,1)$. Note that the segments shown in bold to the left and to the right of the semicolon are identical and of length $k = 2$. From this we see that $\Xi_\lambda = \{\alpha \in \Phi(\mathfrak{u}) \mid (\lambda+\rho, \alpha^\vee) = 0\}$ consists of the two roots $\epsilon_5 - \epsilon_7$ and $\epsilon_6 - \epsilon_8$. (In the general case we would find $\Xi_\lambda = \{\epsilon_i - \epsilon_{i+k} \mid p-k+1 \leq i \leq p\}$ as claimed above.) The roots in $\Phi(\mathfrak{u})$ that are orthogonal to $\epsilon_5 - \epsilon_7$ and $\epsilon_6 - \epsilon_8$ are of the form $\beta = \epsilon_i - \epsilon_j$ with $1 \leq i \leq p-k = 4$ and $p+k+1 = 9 \leq j \leq p+q = 10$. For all of these roots β we have $(\lambda+\rho, \beta^\vee) \in \mathbb{Z}_{>0}$. This shows that $\Phi(\mathfrak{u})_{[\lambda]}$ is as claimed.

[1] We work with the nonsimple Lie algebra $\mathfrak{u}(p,q)$ rather than $\mathfrak{su}(p,q)$ for convenience.

It will be important later to know how to evaluate $y(\lambda+\rho)$ for $y \in {}^{\mathrm{E}}(W_{[\lambda]})$. As a representative example, consider again $(p,q)=(6,4)$ and $k=2$ as above. Since $\Phi(\mathfrak{u})_{[\lambda]}$ is the poset of noncompact roots of a Hermitian symmetric pair corresponding to $\mathfrak{u}(p-k,q-k)=\mathfrak{u}(4,2)$, the elements of ${}^{\mathrm{E}}(W_{[\lambda]})$ are represented by Young diagrams contained in $\yng(4,4)$. Suppose that $y \in {}^{\mathrm{E}}(W_{[\lambda]})$ is represented by the Young diagram $\yng(3,2)$. The boxes of the top row (from left to right) correspond to the roots $\beta_1 := \epsilon_4-\epsilon_9$, $\beta_2 := \epsilon_3-\epsilon_9$, and $\beta_3 := \epsilon_2-\epsilon_9$. The boxes of the second row (from left to right) correspond to the roots $\beta_4 := \epsilon_4-\epsilon_{10}$ and $\beta_5 := \epsilon_3-\epsilon_{10}$. By Proposition 3.21, $y = s_{\beta_5}s_{\beta_4}s_{\beta_3}s_{\beta_2}s_{\beta_1}$. To evaluate $y(\lambda+\rho)$, we proceed row by row. Recall that $\lambda+\rho \sim (8,7,6,5,4,3;4,3,2,1)$. First we find $s_{\beta_3}s_{\beta_2}s_{\beta_1}(\lambda+\rho) \sim (8,6,5,\mathbf{2},4,3;4,3,\mathbf{7},1)$. Then we apply $s_{\beta_5}s_{\beta_4}$ to find $y(\lambda+\rho) \sim (8,6,\mathbf{2},\mathbf{1},4,3;4,3,\mathbf{7},\mathbf{5})$. Here the numbers printed in bold are the numbers that moved to a different sides of the semicolon. In the proof of Theorem 7.9, we will reveal the general pattern.

Example 6.4. Consider $\mathfrak{g}_{\mathbb{R}} = \mathfrak{sp}(n,\mathbb{R})$ and $\lambda = -\frac{k}{2}\omega_n$, where ω_n is the n-th fundamental weight and $1 \le k \le n$. Then $\lambda = (-\frac{k}{2},-\frac{k}{2},\dots,-\frac{k}{2})$ and $\lambda+\rho = (n-\frac{k}{2}, n-1-\frac{k}{2},\dots,1-\frac{k}{2})$. Furthermore, $\Phi(\mathfrak{u}) = \{\epsilon_i+\epsilon_j \mid 1 \le i \le j \le n\}$. We claim that

$$\Phi(\mathfrak{u})_{[\lambda]} = \{\epsilon_i+\epsilon_j \mid 1 \le i < j \le n-k+1\},$$

which is the poset of noncompact roots of a Hermitian symmetric pair of type $(\mathsf{D}_{n-k+1},\mathsf{A}_{n-k})$. The simple system $\Pi_{[\lambda]}$ is the usual simple system of D_{n-k+1}.

As a first representative example, consider $n=5$, $k=1$. In this case, $\lambda = (\frac{9}{2},\frac{7}{2},\frac{5}{2},\frac{3}{2},\frac{1}{2})$ and $\Xi_\lambda = \varnothing$. Since $\Xi_\lambda = \varnothing$, the set $\Phi(\mathfrak{u})_{[\lambda]}$ consists of all roots $\beta \in \Phi(\mathfrak{u})$ satisfying condition (1). If $\beta \in \Phi(\mathfrak{u})$ is short, i.e., of the form $\epsilon_i+\epsilon_j$ with $i<j$, then $(\lambda+\rho,\beta^\vee) \in \mathbb{Z}_{>0}$. However, if $\beta \in \Phi(\mathfrak{u})$ is long, i.e., of the form $2\epsilon_i$, then $(\lambda+\rho,\beta^\vee)$ is not an integer. Thus, it follows that

$$\Phi(\mathfrak{u})_{[\lambda]} = \{\epsilon_i+\epsilon_j \mid 1 \le i < j \le 6\}.$$

As a second representative example, consider $n=6$, $k=2$. In this case, $\lambda = (5,4,3,2,1,0)$ and $\Xi_\lambda = \{2\epsilon_5\}$. We start with condition (2). The set of $\beta \in \Phi(\mathfrak{u})$ that are orthogonal to $\Xi_\lambda = \{2\epsilon_5\}$ is the set $\{\epsilon_i+\epsilon_j \mid 1 \le i \le j \le 4\}$. Every root in this set also satisfies condition (1). Since Ξ_λ contains a long root, condition (3) requires that the roots in $\Phi(\mathfrak{u})_{[\lambda]}$ must be short, i.e., of the form $\epsilon_i+\epsilon_j$ with $i<j$. Thus, $\Phi(\mathfrak{u})_{[\lambda]} = \{\epsilon_i+\epsilon_j \mid 1 \le i < j \le 4\}$.

Example 6.5. Suppose that $\mathfrak{g}_{\mathbb{R}} = \mathfrak{so}^*(2n)$ and $\lambda = -2k\omega_n$, where ω_n is the n-th fundamental weight and $1 \le k \le n/2$. Then $\lambda = (-k,-k,\dots,-k)$ and $\lambda+\rho = (n-1-k, n-2-k,\dots,-k)$. Furthermore,

$$\Phi(\mathfrak{u}) = \{\epsilon_i+\epsilon_j \mid 1 \le i < j \le n\}.$$

We claim that

$$\Phi(\mathfrak{u})_{[\lambda]} = \{\varepsilon_i + \varepsilon_j \mid 1 \le i < j \text{ with } j \le n - 2k - 1 \text{ or } j = n - k\},$$

which is the poset of noncompact roots of a Hermitian symmetric pair of type $(\mathsf{D}_{n-2k}, \mathsf{A}_{n-2k-1})$. The noncompact root in $\Pi_{[\lambda]}$ is the root $\epsilon_{n-2k-1} + \epsilon_{n-k}$. The compact roots in $\Pi_{[\lambda]}$ are the roots $\{\epsilon_i - \epsilon_{i+1} \mid 1 \le i \le n-2k-2\} \cup \{\epsilon_{n-2k-1} - \epsilon_{n-k}\}$ (see Figure 12).

As a representative example, consider $n = 6$, $k = 1$. In this case, $\lambda + \rho = (4, 3, 2, 1, 0, -1)$ and $\Xi_\lambda = \{\epsilon_4 + \epsilon_6\}$. All roots in $\Phi(\mathfrak{u})$ except $\epsilon_4 + \epsilon_6$ and $\epsilon_5 + \epsilon_6$ satisfy condition (1). The roots in $\Phi(\mathfrak{u})$ that are orthogonal to Ξ_λ are the roots of the form $\epsilon_i + \epsilon_j$ with $i < j$ and $i, j \notin \{4, 6\}$. Thus,

$$\Phi(\mathfrak{u})_{[\lambda]} = \{\epsilon_i + \epsilon_j \mid 1 \le i < j \text{ with } j \le 3 \text{ or } j = 5\}.$$

The poset $\Phi(\mathfrak{u})_{[\lambda]}$, viewed as a subposet of $\Phi(\mathfrak{u})$, is shown in Figure 12 on the left. Note that there is a unique poset isomorphism from the poset of noncompact roots of $(\mathsf{D}_4, \mathsf{A}_3)$, which is shown in Figure 12 on the right, to the poset $\Phi(\mathfrak{u})_{[\lambda]}$. This induces a bijection from the set of simple roots of D_4 to $\Pi_{[\lambda]}$ given as follows:

$$\begin{aligned}
\alpha_1 &\mapsto (\epsilon_1 + \epsilon_5) - (\epsilon_2 + \epsilon_5) = \epsilon_1 - \epsilon_2, \\
\alpha_2 &\mapsto (\epsilon_2 + \epsilon_5) - (\epsilon_3 + \epsilon_5) = \epsilon_2 - \epsilon_3, \\
\alpha_3 &\mapsto (\epsilon_1 + \epsilon_3) - (\epsilon_1 + \epsilon_5) = \epsilon_3 - \epsilon_5, \\
\alpha_4 &\mapsto \epsilon_3 + \epsilon_5.
\end{aligned}$$

Note that the roots $\epsilon_3 \pm \epsilon_5 \in \Pi_{[\lambda]}$ are not in Π.

6.2 *Unitary highest weight modules are Kostant modules*

For $y \in W$, let $\overline{y} \in {}^{\mathrm{I}}W$ denote the minimal length coset representative of $W_{\mathrm{I}}\, y$. We note that by Remark 3.6 (iii), if $\lambda \in \Lambda_{\mathrm{I}}^{+}$ and $y \in W$, then $\overline{y} \cdot \lambda \in \Lambda_{\mathrm{I}}^{+}$.

Theorem 6.6 (Enright [15]). *Let $L(\lambda)$ be a unitary highest weight module of highest weight λ. Then, as an $\mathfrak{m}$-module,*

$$H^i(\mathfrak{u}, L(\lambda)) \cong \bigoplus_y F_{\overline{y}\cdot\lambda},$$

where the sum is over all $y \in {}^{\mathrm{E}}(W_{[\lambda]}) \subset W$ such that $\ell_{[\lambda]}(y) = i$. □

Remark 6.7. The unitary highest weight modules are not the only Kostant modules in category $\mathcal{O}^{\mathfrak{p}}$ for which the formula for the $\mathfrak{u}$-cohomology groups can be written in the form above. However, we did find some counterexamples of Kostant modules not satisfying this formula in the case when $\mathfrak{p}$ corresponds to $\mathfrak{sp}(n,\mathbb{R})$. In regular blocks $\mathcal{O}^{\mathfrak{p}}_{\mathrm{reg}}$, the formula seems to hold for the Kostant module $L(w)$ if the Schubert variety $X(w^{-1})$ in G/P is smooth (see Proposition 4.5).

Corollary 6.8 (cf. Enright–Willenbring [21] and Enright–Hunziker [16]). *Let $L(\lambda)$ be a unitary highest weight module of highest weight λ. Then there exists an exact sequence of $\mathfrak{g}$-modules, $0 \to C_r \to \cdots \to C_1 \to C_0 \to L(\lambda) \to 0$, with*

$$C_i := \bigoplus_y M_{\mathrm{I}}(\overline{y} \cdot \lambda),$$

where the sum is over all $y \in {}^{\mathrm{E}}(W_{[\lambda]}) \subset W$ such that $\ell_{[\lambda]}(y) = i$. □

6.3 A curious observation

By Lemma 3.1, if $\lambda = 0$ is the highest weight of the trivial representation, then

$$y \cdot \lambda = \lambda - \langle \Phi_y \rangle.$$

This means that if $M_{\mathrm{I}}(y \cdot \lambda) \to M_{\mathrm{I}}(y' \cdot \lambda)$ is a map in the BGG resolution of $L(0)$ in $\mathcal{O}^{\mathfrak{p}}_{\mathrm{reg}}$, then the highest weight of $y \cdot \lambda = y' \cdot \lambda - \beta$, where β is the unique root in $\Phi_{y'} \backslash \Phi_y$. This nice property characterizes the trivial representations among all finite-dimensional representations. Incidentally, if $\mathfrak{p}$ is of Hermitian type corresponding to the noncompact simple Lie algebra $\mathfrak{g}_{\mathbb{R}}$, then the trivial representation is also the only unitarizable finite-dimensional representation of $\mathfrak{g}_{\mathbb{R}}$.

Remarkably, it appears that the BGG resolution of every unitary highest weight module satisfies an analogous property as long as Φ has only one root length (i.e., the Dynkin diagram is of simply laced type). This is based on the following conjecture.

Conjecture 6.9. Suppose that $L(w)$ is a unitarizable highest weight module in $\mathcal{O}^{\mathfrak{p}}_{\mu}$ such that $L(w) \neq M_{\mathrm{I}}(w)$. If Φ has only one root length, then

$$-\langle \mu + \rho, \alpha^{\vee} \rangle \in \{0, 1\} \quad \text{for all } \alpha \in \mathrm{supp}(w).$$

Lemma 6.10. *Let $x < w \in {}^{\mathrm{I}}W$. If $\Phi_w = \Phi_x \cup \{\beta_1, \ldots, \beta_r\}$ such that $\Phi_x \cup \{\beta_1, \ldots, \beta_k\}$ is a lower-order ideal of $\Phi(\mathfrak{u})$ for every $1 \leq k \leq r$, then*

$$x \cdot \mu = w \cdot \mu + \sum_{i=1}^{r} \langle \mu + \rho, f(\beta_i)^{\vee} \rangle \beta_i.$$

Proof. By Proposition 3.21, $w = xs_{f(\beta_1)}s_{f(\beta_2)}\cdots s_{f(\beta_r)} = s_{\beta_r}\cdots s_{\beta_1}x$. We will show by induction that

$$s_{\beta_k}\cdots s_{\beta_1}x = x\cdot\mu - \sum_{i=1}^{k}\langle \mu+\rho, f(\beta_i)^\vee\rangle\beta_i. \qquad (*)$$

For $k = 1$, we have

$$s_{\beta_1}x\cdot\mu = x\cdot\mu - \langle x\cdot\mu, \beta_1^\vee\rangle\beta_1 = x\cdot\mu - \langle\mu+\rho, x^{-1}\beta_1^\vee\rangle\beta_1.$$

By Lemma 3.20, $x^{-1}\beta_1^\vee = f(\beta_1)^\vee$. Thus, (*) holds when $k = 1$. Using the same argument in the induction step proves (*) in general. □

Example 6.11. Consider $\mathfrak{g}_\mathbb{R} = \mathfrak{su}(3,2)$. Let $\lambda = -\omega_3$, which is the highest weight of a Wallach representation. Then $\lambda = w_\mathrm{I}w\cdot\mu$, where $\mu+\rho = -\omega_1-\omega_3-\omega_4$ and

$$w = \begin{ytableau} 3 & 2 & 1 \\ 4 & 3 \end{ytableau} = s_3s_2s_1s_4s_3.$$

The following shows $w_\mathrm{I}\beta$ for $\beta\in\Phi(\mathfrak{u})$.

$$w_\mathrm{I}\begin{pmatrix} \epsilon_3-\epsilon_4\ , \epsilon_2-\epsilon_4\ , \epsilon_1-\epsilon_4 \\ \epsilon_3-\epsilon_5\ , \epsilon_2-\epsilon_5\ , \epsilon_1-\epsilon_5 \end{pmatrix} = \begin{pmatrix} \epsilon_1-\epsilon_5\ , \epsilon_2-\epsilon_5\ , \epsilon_3-\epsilon_5 \\ \epsilon_1-\epsilon_4\ , \epsilon_2-\epsilon_4\ , \epsilon_3-\epsilon_4 \end{pmatrix}.$$

The BGG resolution of $L(\ydiagram{3,2}) = L(\lambda)$ in $\mathcal{O}_\mu^\mathfrak{p}$ is of the form

$$0\to M_\mathrm{I}(\ydiagram{1})\to M_\mathrm{I}(\ydiagram{1,1})\to M_\mathrm{I}(\ydiagram{3,2})\to L(\ydiagram{3,2})\to 0.$$

By Lemma 6.10, the highest weights of the terms in the resolution are

$$\begin{aligned} M_\mathrm{I}(\ydiagram{3,2}) &= M_\mathrm{I}(\lambda),\\ M_\mathrm{I}(\ydiagram{1,1}) &= M_\mathrm{I}(\lambda-(\epsilon_2-\epsilon_4)-(\epsilon_3-\epsilon_5)),\\ M_\mathrm{I}(\ydiagram{1}) &= M_\mathrm{I}(\lambda-(\epsilon_2-\epsilon_4)-(\epsilon_3-\epsilon_5)-(\epsilon_1-\epsilon_4)). \end{aligned}$$

7 Syzygies of determinantal ideals

In [37], Lascoux (using geometric methods originally developed by Kempf in his thesis) gave an explicit construction of minimal free resolutions of the determinantal ideals of general $p\times q$-matrices (in characteristic 0). In [31], Józefiak, Pragacz, and Weyman gave a similar construction of minimal free resolutions of the determinantal ideals of symmetric and antisymmetric matrices. By Weyl's fundamental theorems

Table 2. Data for Howe duality

$\mathfrak{g}_{\mathbb{R}}$	$M_{\mathbb{R}}$	$\mathfrak{u}$	$K_{\mathbb{R}}$	Z
$\mathfrak{u}(p,q)$	$\mathrm{U}(p)\times\mathrm{U}(q)$	$\mathrm{M}_{p,q}$	$\mathrm{U}(k)$	$\mathrm{M}_{p,k}\oplus\mathrm{M}_{k,q}$
$\mathfrak{sp}(n,\mathbb{R})$	$\widetilde{\mathrm{U}}(n)$	$\mathrm{Sym}_n := \{x\in\mathrm{M}_n \mid x = x^t\}$	$\mathrm{O}(k)$	$\mathrm{M}_{n,k}$
$\mathfrak{so}^*(2n)$	$\mathrm{U}(n)$	$\mathrm{Alt}_n := \{x\in\mathrm{M}_n \mid x = -x^t\}$	$\mathrm{Sp}(k)$	$\mathrm{M}_{n,2k}$

of classical invariant theory, these resolutions also give minimal free resolutions of rings of invariants of the classical groups. As was first observed by Howe, the latter carry the structure of a unitary highest weight module for a dual group.

7.1 Howe duality

Consider Table 2. In the case when $\mathfrak{g}_{\mathbb{R}} = \mathfrak{sp}(n,\mathbb{R})$, the group $M_{\mathbb{R}}$ is the double cover of $\mathrm{U}(n)$ given by

$$\widetilde{\mathrm{U}}(n) := \{(u,s)\in\mathrm{U}(n)\times\mathbb{C} \mid \det u = s^2\}.$$

The group $M_{\mathbb{R}}$ acts on $\mathfrak{u}$ by the usual actions:

$$\mathrm{U}(p)\times\mathrm{U}(q) \circlearrowright \mathrm{M}_{p,q} : ((u_1,u_2),x)\mapsto u_1 x\, u_2^{-1}$$

$$\widetilde{\mathrm{U}}(n) \circlearrowright \mathrm{Sym}_n : ((u,s),x)\mapsto uxu^t$$

$$\mathrm{U}(n) \circlearrowright \mathrm{Alt}_n : (u,x)\mapsto uxu^t.$$

The group $K_{\mathbb{R}}$ acts on Z as follows:

$$\mathrm{U}(k) \circlearrowright \mathrm{M}_{p,k}\oplus\mathrm{M}_{k,q} : (h,(z_1,z_2))\mapsto (z_1h^{-1}, hz_2)$$

$$\mathrm{O}(k) \circlearrowright \mathrm{M}_{n,k} : (h,z)\mapsto zh^{-1}$$

$$\mathrm{Sp}(k) \circlearrowright \mathrm{M}_{n,2k} : (h,z)\mapsto zh^{-1}.$$

Finally, the group $M_{\mathbb{R}}$ also acts on Z:

$$\mathrm{U}(p)\times\mathrm{U}(q) \circlearrowright \mathrm{M}_{p,k}\oplus\mathrm{M}_{k,q} : ((u_1,u_2),(z_1,z_2))\mapsto (u_1z_1, z_2u_2^{-1})$$

$$\widetilde{\mathrm{U}}(n) \circlearrowright \mathrm{M}_{n,k} : ((u,s),z)\mapsto uz$$

$$\mathrm{U}(n) \circlearrowright \mathrm{M}_{n,2k} : (u,z)\mapsto uz.$$

In each case, define a polynomial map $\pi : Z \to \mathfrak{u}$ of degree 2 by:

$$\pi : \mathrm{M}_{p,k} \oplus \mathrm{M}_{k,q} \to \mathrm{M}_{p,q}, \ (z_1, z_2) \mapsto z_1 z_2$$

$$\pi : \mathrm{M}_{n,k} \to \mathrm{Sym}_n, \ z \mapsto zz^t$$

$$\pi : \mathrm{M}_{n,2k} \to \mathrm{Alt}_n, \ z \mapsto zJz^t.$$

Then π is $M_{\mathbb{R}}$-equivariant and constant on $K_{\mathbb{R}}$-orbits. Define

$$Y_k := \pi(Z) \subset \mathfrak{u}.$$

In each case, Y_k is a *determinantal variety* or $Y_k = \mathfrak{u}$. More precisely, for $K = \mathrm{GL}_k$ we have $Y_k = \{x \in \mathrm{M}_{p,q} \mid \mathrm{rk}\, x \leq k\}$, for $K = \mathrm{O}_k$ we have $Y_k = \{x \in \mathrm{Sym}_n \mid \mathrm{rk}\, x \leq k\}$, and for $K = \mathrm{Sp}_{2k}$ we have $Y_k = \{x \in \mathrm{Alt}_n \mid \mathrm{rk}\, x \leq 2k\}$. Let $\mathbb{C}[Y_k]$ denote the coordinate ring of Y_k and let $\mathbb{C}[Z]^K$ denote the ring of K-invariant polynomials on Z.

Theorem 7.1 (cf. Weyl [48]). *The map $\pi : Z \to Y_k \subset \mathfrak{u}$ induces an isomorphism $\pi^* : \mathbb{C}[Y_k] \to \mathbb{C}[Z]^K$ of algebras given by $\pi^*(f) = f \circ \pi$.*

Proof. Since $\pi : Z \to Y_k$ is surjective, $\pi^* : \mathbb{C}[Y_k] \to \mathbb{C}[Z]$ is injective. Since π is constant on K-orbits, $\pi^*\mathbb{C}[Y_k] \subset \mathbb{C}[Z]^K$. By Weyl's first fundamental theorems for the classical groups $K = \mathrm{GL}_k$, O_k, and Sp_{2n}, it follows that $\mathbb{C}[Z]^K$ is generated by the functions $\pi^*(x_{ij})$ and hence $\pi^*\mathbb{C}[Y_k] = \mathbb{C}[Z]^K$. For example, if $K = \mathrm{O}_k$, then the algebra of invariants $\mathbb{C}[Z]^K$ is generated by the quadratic functions $\pi^*(x_{ij}) = \sum_{l=1}^{k} z_{il} z_{lj}$. The other two cases are similar. □

Since $\pi : Z \to Y_k \subset \mathfrak{u}$ is $M_{\mathbb{R}}$-equivariant, the isomorphism $\pi^* : \mathbb{C}[Y_k] \to \mathbb{C}[Z]^K$ is also $M_{\mathbb{R}}$-equivariant, i.e., an intertwining map. To be able to extend the $M_{\mathbb{R}}$-action on $\mathbb{C}[Y_k]$ and $\mathbb{C}[Z]$ to a $(\mathfrak{g}, M_{\mathbb{R}})$-action, we need to twist the usual $M_{\mathbb{R}}$-action on $\mathbb{C}[Z]$ as follows:

$$\mathrm{U}(p) \times \mathrm{U}(q) \circlearrowright \mathbb{C}[\mathrm{M}_{p,k} \oplus \mathrm{M}_{q,k}] : \ ((u_1, u_2).f)(z) = (\det u_1)^{-k} f(u_1^{-1} z_1, z_2 u_2)$$

$$\widetilde{\mathrm{U}}(n) \circlearrowright \mathbb{C}[\mathrm{M}_{n,k}] : \ ((u, s).f)(z) = s^{-k} f(u^{-1} z)$$

$$\mathrm{U}(n) \circlearrowright \mathbb{C}[\mathrm{M}_{n,2k}] : \ (u.f)(z) = (\det u)^{-k} f(u^{-1} z).$$

Let $\mathcal{D}(Z)^K$ denote the algebra of K-invariant polynomial differential operators.

Lemma 7.2 (cf. Howe [27]). *There exists an injective Lie algebra homomorphism $\phi : \mathfrak{g} \to \mathcal{D}(Z)^K$ such that*

(a) *the $\mathfrak{g}$-action on $\mathbb{C}[Z]$ that is defined via ϕ is compatible with the $M_{\mathbb{R}}$-action given above, i.e., $\mathbb{C}[Z]$ is a $(\mathfrak{g}, M_{\mathbb{R}})$-module;*
(b) *the action of $S(\overline{\mathfrak{u}}) = \mathbb{C}[\mathfrak{u}]$ on $\mathbb{C}[Z]$ that is defined via ϕ is given by left multiplication via $\pi^* : \mathbb{C}[\mathfrak{u}] \to \mathbb{C}[Z]$;*
(c) *$\phi(\mathfrak{g})$ generates the algebra $\mathcal{D}(Z)^K$.*

Proof. Consider the case when $K = \mathrm{O}(k, \mathbb{C})$. For $1 \le i, j \le n$, define

$$q_{ij} := \pi^*(x_{ij}) = \sum_{l=1}^{k} z_{il} z_{jl}, \quad E_{ij} := \sum_{l=1}^{k} z_{jl} \frac{\partial}{\partial z_{il}}, \quad \Delta_{ij} := \sum_{l=1}^{k} \frac{\partial^2}{\partial z_{il} \partial z_{jl}}.$$

All these operators are in $\mathcal{D}(Z)^K$. For $a \in \mathbb{C}^\times$, define $\phi_a : \mathfrak{g} = \mathfrak{sp}(n, \mathbb{C}) \to \mathcal{D}(Z)^K$ by

$$e_{n+i,j} + e_{n+j,i} \mapsto a q_{ij}, \ e_{ij} - e_{n+j,n+i} \mapsto -E_{ij} - \tfrac{k}{2}\delta_{ij}, \ e_{i,n+j} + e_{j,n+i} \mapsto -\tfrac{1}{a}\Delta_{ij}.$$

Then one can check that ϕ_a is a Lie algebra homomorphism. Furthermore, the action of $\mathfrak{m} = \mathfrak{gl}(n, \mathbb{C})$ on $\mathbb{C}[Z]$, which is found by differentiating the action of $M_{\mathbb{R}} = \widetilde{\mathrm{U}}(n)$, is given by

$$(e_{ij} . f)(z) = (-E_{ij} - \tfrac{k}{2}\delta_{ij}) f(z)$$

and coincides with the action of $\mathfrak{m}$ via ϕ_a since the embedding $\mathfrak{gl}(n, \mathbb{C}) \to \mathfrak{sp}(n, \mathbb{C})$ is given by $e_{ij} \mapsto e_{ij} - e_{n+j,n+i}$. Thus, ϕ_a satisfies (b).

If we identify $\mathfrak{u}^*$ with $\overline{\mathfrak{u}}$ via the trace form $\operatorname{tr} AB$ on $\mathfrak{sp}(n, \mathbb{C})$, then the linear form on $\mathfrak{u}$ that we called x_{ij} is identified with the element $\frac{1}{2}(e_{n+i,j} + e_{n+j,i})$. Thus, if we choose $a = 2$, then ϕ_a satisfies (c). If we identify $\mathfrak{u}^*$ with $\overline{\mathfrak{u}}$ by using a different invariant form on $\mathfrak{sp}(n, \mathbb{C})$, say the Killing form, then another choice for a is necessary.

To show that $\phi_a(\mathfrak{g})$ generates the algebra $\mathcal{D}(Z)^K$, we note that the symbols (with respect to the Bernstein filtration of the Weyl algebra $\mathcal{D}(Z)$) of the operators $a q_{ij}$, $-E_{ij} - \frac{k}{2}$, and Δ_{ij} generate the associated graded algebra $\operatorname{gr} \mathcal{D}(Z)^K = \mathbb{C}[Z \oplus Z^*]^K$ by Weyl's fundamental theorem for $K = \mathrm{O}(n, \mathbb{C})$.

The proofs in the other two cases are similar. □

Let $\hat{K}$ denote the set of isomorphism classes of (finite-dimensional) simple K-modules. For $\sigma \in \hat{K}$, let V_σ be a simple K-module in the class σ.

Theorem 7.3 (Howe duality). *Let* $\Sigma := \{\sigma \in \hat{K} \mid \operatorname{Hom}_K(V_\sigma, \mathbb{C}[Z]) \neq 0\}$. *Then, as a* $K \times (\mathfrak{g}, M_{\mathbb{R}})$*-module,*

$$\mathbb{C}[Z] = \bigoplus_{\sigma \in \Sigma} V_\sigma \otimes V^\sigma,$$

where $V^\sigma := \operatorname{Hom}_K(V_\sigma, \mathbb{C}[Z])$ *is a simple highest weight module of* $\mathfrak{g}$ *with* $V^\sigma \not\cong V^{\sigma'}$ *whenever* $\sigma \neq \sigma'$. □

Proof. Since $\phi(\mathfrak{g})$ generates the algebra $\mathcal{D}(Z)^K$ and since $\mathbb{C}[Z]$ is a simple $\mathcal{D}(Z)$-module, the result follows from a general duality theorem of Goodman and Wallach [23]. For explicit formulas for the highest weights of the $\mathfrak{g}$-modules V^σ see Kashiwara–Vergne [32]. □

Remark 7.4. Every $\mathfrak{g}$-module V^{σ} is in fact unitarizable. Furthermore, if $\mathfrak{g}_{\mathbb{R}} = \mathfrak{su}(p,q)$ or $\mathfrak{sp}(n,\mathbb{R})$, every unitary highest weight module $L(\lambda)$ such that $L(\lambda) \neq M_{\mathrm{I}}(\lambda)$, i.e., for which the highest weight λ is a reduction point, arises in that way.

Corollary 7.5. *The coordinate rings $\mathbb{C}[Y_k]$ carry the structure of a unitary highest weight module of $\mathfrak{g}$. For $\mathfrak{g}_{\mathbb{R}} = \mathfrak{u}(p,q)$, $\mathfrak{sp}(n,\mathbb{R})$, and $\mathfrak{so}^*(2n)$ these $\mathfrak{g}$-modules are the Wallach representations $L(-k\omega_p)$, $L(-\frac{k}{2}\omega_n)$, and $L(-2k\omega_n)$, respectively.* □

Proof. By Theorem 7.3, the invariant ring $V^{\mathrm{triv}} = \mathbb{C}[Z]^K$ is a highest weight module. The highest weight vector is given by the constant function $1 \in \mathbb{C}[Z]^K$. We can then use Theorem 7.1 to define a $\mathfrak{g}$-action on $\mathbb{C}[Y_k]$. Again, the highest weight vector is given by the constant function $1 \in \mathbb{C}[Y_k]$. By the definition of the twisted $M_{\mathbb{R}}$-action on $\mathbb{C}[Z]^K$, it follows that the highest weights are $\lambda = -k\omega_p$, $-\frac{k}{2}\omega_n$, and $-2k\omega_n$, respectively. □

Remark 7.6. Note that as an $M_{\mathbb{R}}$-module with respect to the usual action on $\mathbb{C}[Y_k]$ we have

$$\mathbb{C}[Y_k] \cong L(\lambda) \otimes F_{-\lambda}.$$

7.2 *Minimal free resolutions of determinantal ideals*

Let $\mathfrak{g} = \overline{\mathfrak{u}} \oplus \mathfrak{m} \oplus \mathfrak{u}$ as above. Since $\overline{\mathfrak{u}}$ is abelian, we have $U(\mathfrak{g}) = S(\overline{\mathfrak{u}}) \otimes U(\mathfrak{p})$ by the PBW Theorem. It follows that the parabolic Verma module $M_{\mathrm{I}}(\nu)$ of highest weight ν, as an $S(\overline{\mathfrak{u}})$- and as an $\mathfrak{m}$-module, is isomorphic to

$$M_{\mathrm{I}}(\nu) = S(\overline{\mathfrak{u}}) \otimes F_\nu,$$

where the actions are the obvious actions. In particular $M_{\mathrm{I}}(\nu)$ is a free $S(\overline{\mathfrak{u}})$-module of rank $\dim F_\nu$. If we identify $\overline{\mathfrak{u}}$ with the dual $\mathfrak{u}^*$ via the Killing form, then $S(\overline{\mathfrak{u}})$ is identified with $\mathbb{C}[\mathfrak{u}]$, the coordinate ring of $\mathfrak{u}$. In the following, to simplify notation, we will write $S := S(\overline{\mathfrak{u}})$.

Theorem 7.7. *Let $\mathbb{C}[Y_k]$ be the coordinate ring of the determinantal variety $Y_k \subset \mathfrak{u}$ as above, viewed as an S-module and as an $M_{\mathbb{R}}$-module with respect to the usual (untwisted) action. Let $L(\lambda)$ be the Wallach representation of $\mathfrak{g}$ of highest weight λ such that $\mathbb{C}[Y_k] = L(\lambda) \otimes F_{-\lambda}$ as an $\mathfrak{m}$-module. Then the BGG resolution of $L(\lambda)$, tensored by the 1-dimensional $\mathfrak{m}$-module $F_{-\lambda}$, gives an $M_{\mathbb{R}}$-equivariant, minimal free resolution of the coordinate ring $\mathbb{C}[Y_k]$ as an S-module.*

Proof. Note that the functor $\_ \otimes F_{-\lambda}$ is an exact since $F_{-\lambda}$ is 1-dimensional. The functor also commutes with the S-action which is given by multiplication from the left. Furthermore, note that the S-action on $\mathbb{C}[Y_k]$ is the usual S-action in light of Lemma 7.2 (b). Thus we do obtain an $M_{\mathbb{R}}$-equivariant, free resolution of the coordinate ring $\mathbb{C}[Y_k]$ as an S-module. The minimality of the resolution follows by the argument given in the proof of Theorem 5.15. □

The resolution given by Theorem 7.7 is in fact a minimal free resolution of graded S-modules. Recall from Lemma 2.2 that there exists and element $h \in \mathfrak{h}$ in the center of $\mathfrak{m}$ such that $\mathfrak{u} = \{x \in \mathfrak{g} \mid [h, x] = x\}$. If V is any semisimple $\mathfrak{m}$-module with integral weights, then h has integer eigenvalues on V and we obtain a $\mathbb{Z}$-grading on V having the property that each simple $\mathfrak{m}$-submodule of V is contained in a homogenous component. If we view the coordinate ring $\mathbb{C}[Y_k]$ as an $\mathfrak{m}$-module (via the untwisted action), the resulting grading on $\mathbb{C}[Y_k]$ is the usual grading. Similarly, each term in the resolution of $\mathbb{C}[Y_k]$ is a graded $\mathfrak{m}$-module. Furthermore, if $M_{\mathrm{I}}(\nu) \otimes F_{-\lambda} \to M_{\mathrm{I}}(\nu') \otimes F_{-\lambda}$ is a map in the resolution, then it is homogenous of degree $(\nu' - \lambda)(h) - (\nu - \lambda)(h) = (\nu' - \nu)(h)$. In algebraic geometry it is customary to shift the degrees of the terms in a graded resolution such that all maps end up being homogenous of degree 0. For a given integer $d \geq 0$, define $S(-d)$ as the free S-module of rank one with the i-th homogenous component given by

$$S(-d)_i := S_{d+i}.$$

Similarly, we can shift the grading of any graded S-module V. We now shift the degree of each term in our resolution as follows

$$(M_{\mathrm{I}}(\nu) \otimes F_{-\lambda})(-d) = S(-d) \otimes F_{\nu-\lambda},$$

where $d = (\nu - \lambda)(h)$. Then all maps of the resulting graded resolution are homogenous of degree 0.

Our goal is now to compute the minimal free resolution of $\mathbb{C}[Y_k]$ explicitly for each of the determinantal varieties $Y_k \subset \mathfrak{u}$ and describe its terms by using generalized Young diagrams. To do this efficiently, we need more notation.

7.3 *Frobenius notation*

It will be convenient to use Frobenius notation for integer partitions. Let $\lambda = (\lambda_1, \ldots, \lambda_l)$ be a partition and let $\lambda' = (\lambda'_1, \ldots, \lambda'_m)$ be its dual partition, whose Young diagram is the transpose of the Young diagram of λ. The length r of the diagonal of the Young diagram of λ is called the *Frobenius rank* (or *Durfee rank*) of λ. For $1 \leq i \leq r$, define $a_i := \lambda_i - i + 1$ and $b_i := \lambda'_i - i + 1$. Then $(a_1, \ldots, a_r \mid b_1, \ldots, b_r)$ is called the *Frobenius symbol* of λ. The partition λ is uniquely determined by its Frobenius symbol and we will, by abuse of notation, write $\lambda = (a_1, \ldots, a_r \mid b_1, \ldots, b_r)$. For example,

[Young diagram with rows of 4, 3, 1 boxes; diagonal boxes marked •] $= (4, 3, 1) = (4, 2 \mid 3, 1).$

Here we filled the diagonal boxes with dots for emphasis. We can also use Frobenius notation for the shifted partitions (Young diagrams) that arise for $\mathfrak{g}_{\mathbb{R}} = \mathfrak{sp}(n,\mathbb{R})$ and $\mathfrak{so}^*(2n)$. Given positive integers $a_1 > \cdots > a_r$, we write $(a_1,\ldots,a_r \mid 1^r)$ for the shifted partition (Young diagram), whose i-th row has length a_i. For example,

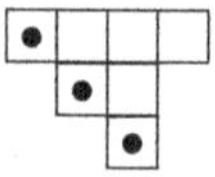

$$= (4,2,1 \mid 1^3).$$

7.4 *Schur functors*

To be able to compare our resolutions of the determinantal ideals directly to the resolutions as they are presented in [49], we will use the Schur functors that were defined by Akin–Buchsbaum–Weyman in [1]. Let E be a finite-dimensional complex vector space and let $\lambda = (\lambda_1,\ldots,\lambda_l)$ be a partition with its dual partition $\lambda' = (\lambda'_1,\ldots,\lambda'_m)$.

Definition 7.8. Following [1], define the Schur functor $L_\lambda E$ as the image of the composite map

$$\wedge^{\lambda_1} E \otimes \cdots \otimes \wedge^{\lambda_l} E \to \otimes^{|\lambda|} E \to S^{\lambda'_1} E \otimes \cdots \otimes S^{\lambda'_m} E,$$

where $\wedge^p E$ and $S^p E$ denote exterior and symmetric powers, respectively, and $|\lambda| = \sum_{i=1}^{l} \lambda_i = \sum_{j=1}^{m} \lambda'_j$.

For example, $L_{(p)} E = \wedge^p E$ and $L_{(1^p)} E = S^p E$. Note that this differs from the more common definition of the Schur functor $S^\lambda E$ via Young symmetrizers,[2] where $S^{(p)} E = S^p E$ and $S^{(1^p)} E = \wedge^p E$.

The Schur functors are useful in constructing the irreducible representations of the general linear group $\mathrm{GL}_n(\mathbb{C})$, or equivalently, the unitary group $\mathrm{U}(n)$. Let $E = \mathbb{C}^n$ and identify $\mathrm{GL}(E) = \mathrm{GL}_n(\mathbb{C})$. Then $L_\lambda E$ is the simple $\mathrm{GL}_n(\mathbb{C})$-module of highest weight $\sum_{i=j}^{m} \lambda'_j \epsilon_j$ and $L_\lambda E^*$ is the simple $\mathrm{GL}_n(\mathbb{C})$-module of highest weight $-\sum_{j=1}^{m} \lambda'_{n-j+1} \epsilon_j$, where the ϵ_j are defined as usual for the standard maximal torus consisting of diagonal matrices and highest weights are defined with respect to the standard Borel subgroup consisting of upper triangluar matrices.

[2]The Schur functor $L_\lambda E$ has the advantage that it can be defined for a free module E over a commutative ring or a field of positive characteristics.

7.5 The Lascoux resolution

We identify $\mathrm{M}_{p,q} = \mathrm{Hom}(F, E)$, where $E = \mathbb{C}^p$ and $F = \mathbb{C}^q$ are the natural representations of $\mathrm{U}(p)$ and $\mathrm{U}(q)$, respectively. Note that as a $\mathrm{U}(p)\times\mathrm{U}(q)$-module, $\mathrm{Hom}(F, E) = E \otimes F^*$, where $\otimes$ denotes the (external) tensor product.

Theorem 7.9. *Let $\mathfrak{g}_{\mathbb{R}} = \mathfrak{u}(p,q)$ and let $\lambda = -k\omega_p$ be the highest weight of the k-th Wallach representation, where $k \leq \min\{p,q\}$. If $y \in {}^{\mathrm{E}}(W_{[\lambda]})$ is the element represented by the partition $(a_1,\ldots,a_r \mid b_1,\ldots,b_r)$, then as an $M_{\mathbb{R}} = \mathrm{U}(p)\times\mathrm{U}(q)$-module,*

$$F_{\overline{y}\cdot\lambda-\lambda} = L_{(a_1+k,\ldots,a_r+k|b_1,\ldots,b_r)}E^* \otimes L_{(b_1+k,\ldots,b_r+k|a_1,\ldots,a_r)}F.$$

Furthermore, $(\overline{y}\cdot\lambda-\lambda)(h) = |y|+rk$, where $|y|$ is the content of the corresponding partition.

Proof. Using the same notation as in Example 6.3, we have $\lambda+\rho \sim (n-k, n-k-1,\ldots,n-k-p+1\,;\,q,q-1,\ldots,1)$, where $n = p+q$. By Example 6.3,

$$\Phi(\mathfrak{u}_{[\lambda]}) = \Phi(\mathfrak{u})_{[\lambda]} = \{\epsilon_i - \epsilon_j \mid i < j \text{ and } i,j \in [1,p-k]\cup[p+k,n]\}.$$

If $y \in {}^{\mathrm{E}}(W_{[\lambda]})$ is the element represented by the partition $(a_1,\ldots,a_r \mid b_1,\ldots,b_r)$, then

$$\overline{y}(\lambda+\rho) \sim (*,\ldots,*,q-k-b_r+1,\ldots,q-k-b_1+1;q+a_1,\ldots,q+a_r,*,\ldots,*),$$

where the coordinates $*$ are a permutation of the coordinates of $(n-k,n-k-1,\ldots,n-k-p+1\,;\,q,q-1,\ldots,1)$ not contained in $\{q-k-b_r+1,\ldots,q-k-b_1+1\}\cup\{q+a_1,\ldots,q+a_r\}$. It follows that $\overline{y}\cdot\lambda-\lambda = \overline{y}(\lambda+\rho)-(\lambda+\rho)$ is the highest weight of $L_{(a_1+k,\ldots,a_r+k|b_1,\ldots,b_r)}E^* \otimes L_{(b_1+k,\ldots,b_r+k|a_1,\ldots,a_r)}F$.

As a representative example, consider the case when $(p,q) = (6,4)$ and $k = 2$. Then $\lambda+\rho \sim (8,7,6,5,4,3;4,3,2,1)$. If y is represented by the partition $(3,2) = (3,1 \mid 2,1)$, then $y(\lambda+\rho) \sim (8,6,\mathbf{2},\mathbf{1},4,3;4,3,\mathbf{7},\mathbf{5})$ by Example 6.3. Note that numbers 7 and 5 moved from left to right (relative to the semicolon), and the numbers 2 and 1 moved from right to left (relative to the semicolon). In general, the numbers that move from left to right are the numbers $q+a_1,\ldots,q+a_r$ and the number that move from right to left are the numbers $q-k-b_r+1,\ldots,q-k-b_1+1$. (This can be proved rigorously by an induction on the number of rows of the diagram of y.) By sorting the segments to the left and right of the semicolon we obtain $\overline{y}(\lambda+\rho) \sim (8,6,4,3,\mathbf{2},\mathbf{1};\mathbf{7},\mathbf{5},4,3)$ and hence $\overline{y}\cdot\lambda-\lambda = \overline{y}(\lambda+\rho)-(\lambda+\rho) = (0,-1,-2,-2,-2,-2;3,2,2,2)$. This is the highest weight of $L_{(5,3|2,1)}E^* \otimes L_{(4,3|3,1)}F$. $\square$

The result of Theorem 7.9 can be visualized as follows:

$$y = \begin{smallmatrix} \boxdot & \square & \square & \square \\ \square & \boxdot & & \\ \square & & & \end{smallmatrix} \quad \rightsquigarrow \quad S(-d) \otimes \left(\begin{smallmatrix} \boxdot & \square & \square & \square & \boxtimes & \boxtimes & \boxtimes \\ \square & \boxdot & \boxtimes & \boxtimes & \boxtimes & & \\ \square & & & & & & \end{smallmatrix}^{\,*} \otimes \begin{smallmatrix} \boxdot & \square & \square & \boxtimes & \boxtimes & \boxtimes \\ \square & \boxdot & \boxtimes & \boxtimes & \boxtimes & \\ \square & & & & & \\ \square & & & & & \end{smallmatrix} \right).$$

The diagram on the left is the Young diagram representing $y \in {}^{\mathrm{E}}(W_{[\lambda]})$ with the diagonal boxes filled with a dot. The integer d is the number of boxes of this diagram plus k times the number of dots. In the example shown, $k = 3$ and $d = 7 + 2 \cdot 3 = 13$. The two diagrams on the right correspond to the partitions $(a_1 + k, \ldots, a_r + k \mid b_1, \ldots, b_r)$ and $(b_1 + k, \ldots, b_r + k \mid a_1, \ldots, a_r)$, respectively. The diagram of the partition $(a_1 + k, \ldots, a_r + k \mid b_1, \ldots, b_r)$ is obtained from the diagram of y by adding k boxes to each of its first r rows. The added boxes are filled with an $\times$. Similarly, the diagram of the partition $(b_1 + k, \ldots, b_r + k \mid a_1, \ldots, a_r)$ is obtained from the transposed diagram of y by adding k boxes to each of its first r rows.

In the following examples, we simply write the Young diagram of y for the term $S \otimes F_{\overline{y} \cdot \lambda - \lambda}$. The arrows that are marked with a dot correspond to a homogenous map of degree $k + 1$ and the other maps are homogenous of degree 1. The dotted arrows arise where the diagram at the tail has one more dotted box than the diagram at the head.

Example 7.10 (The Eagon–Northcott Complex). Consider the case when $p \geq q$ and $k = q - 1$.

$$0 \longrightarrow \underbrace{\boxdot\square\square\cdots\square}_{p-q+1 \text{ boxes}} \longrightarrow \cdots \longrightarrow \boxdot\square \longrightarrow \boxdot \overset{\bullet}{\longrightarrow} S \longrightarrow \mathbb{C}[Y_{q-1}] \longrightarrow 0,$$

where the diagrams are shorthand for

$$\underbrace{\boxdot\square\square\cdots\square}_{i \text{ boxes}} = S(-q-i+1) \otimes \left(L_{(q+i-1)}E^* \otimes L_{(q,1^{i-1})}F \right).$$

In the special case when $q = 1$ this is the Koszul complex.

Example 7.11 (The Gulliksen–Negard Complex). Consider the case when $p = q$ and $k = p - 2$.

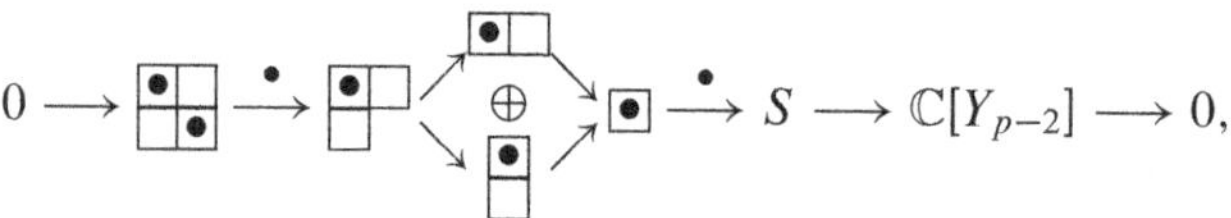

where the diagrams are shorthand for

$$= S(-2p) \otimes (L_{(p,p)}E^* \otimes L_{(p,p)}F)$$

$$= S(-p-1) \otimes (L_{(p,1)}E^* \otimes L_{(p,1)}F)$$

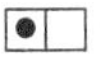

$$= S(-p) \otimes (L_{(p,1)}E^* \otimes L_{(p)}F)$$

$$= S(-p) \otimes (L_{(p)}E^* \otimes L_{(p,1)}F)$$

$$= S(-p+1) \otimes (L_{(p-1)}E^* \otimes L_{(p-1)}F).$$

For an explicit description of the differentials, see Weyman [49, (6.1.8)].

7.6 *The determinantal ideals for symmetric matrices*

We identify $\mathrm{Sym}_n = S^2E$, where $E = \mathbb{C}^n$ is the natural representation of $\mathrm{U}(n)$.

Theorem 7.12. *Let $\mathfrak{g}_{\mathbb{R}} = \mathfrak{sp}(n,\mathbb{R})$ and let $\lambda = -\frac{k}{2}\omega_n$ be the highest weight of the k-th Wallach representation, where $1 \leq k \leq n$. If $\underset{\sim}{y} \in {}^{\mathrm{E}}(W_{[\lambda]})$ is the element represented by the partition $(a_1, \ldots, a_r \mid 1^r)$, then as a $\mathrm{U}(n)$-representation,*

$$F_{\overline{y}\cdot\lambda} \otimes F_{-\lambda} = \begin{cases} L_{(a_1+k,\ldots,a_r+k|a_1+1,\ldots,a_r+1)}E^* & \text{if } r \text{ is even} \\ L_{(a_1+k,\ldots,a_r+k,k|a_1+1,\ldots,a_r+1,1)}E^* & \text{if } r \text{ is odd,} \end{cases}$$

where E^* is the dual of E. Furthermore, $(\overline{y}\cdot\lambda - \lambda)(h) = |y| + rk/2$ if r is even and $(\overline{y}\cdot\lambda - \lambda)(h) = |y| + (r+1)k/2$ if r is odd.

Proof. We have $\lambda = (-\frac{k}{2}, -\frac{k}{2}, \ldots, -\frac{k}{2})$, $\rho = (n, n-1, \ldots, 1)$, and hence $\lambda + \rho = (n - \frac{k}{2}, n-1-\frac{k}{2}, \ldots, 1-\frac{k}{2})$. By Example 6.4,

$$\Phi(\mathfrak{u}_{[\lambda]}) = \Phi(\mathfrak{u})_{[\lambda]} = \{\epsilon_i + \epsilon_j \mid 1 \leq i < j \leq n-k+1\}.$$

If $y \in {}^{\mathrm{E}}(W_{[\lambda]})$ is the element represented by the partition $(a_1, \ldots, a_r \mid 1^r)$, then

$$\overline{y}(\lambda+\rho) = \begin{cases} (*, \ldots, *, -a_r - \frac{k}{2}, \cdots, -a_1 - \frac{k}{2}) & \text{if } r \text{ is even,} \\ (*, \ldots, *, -\frac{k}{2}, -a_r - \frac{k}{2}, \cdots, -a_1 - \frac{k}{2}) & \text{if } r \text{ is odd,} \end{cases}$$

where the coordinates $*$ are permutation of the coordinates of $\lambda + \rho$ not contained in $\{a_1 + \frac{k}{2}, \ldots, a_r + \frac{k}{2}\}$ or $\{a_1 + \frac{k}{2}, \ldots, a_r + \frac{k}{2}, \frac{k}{2}\}$, respectively. It follows that $\overline{y}\cdot\lambda - \lambda$ is the highest weight of the Schur module

$$L_{(a_1+k,\ldots,a_r+k|a_1+1,\ldots,a_r+1)}E^*$$

or

$$L_{(a_1+k,\ldots,a_r+k,k|a_1+1,\ldots,a_r+1,1)}E^*,$$

respectively. As a representative example, consider the case when $n = 6$ and $k = 1$. Then $\lambda + \rho = (\frac{11}{2}, \frac{9}{2}, \frac{7}{2}, \frac{5}{2}, \frac{3}{2}, \frac{1}{2})$ and $n - k + 1 = 6$. If y is represented by the partition $(4, 1 \mid 1^2)$, then $\overline{y}(\lambda + \rho) = (\frac{11}{2}, \frac{7}{2}, \frac{5}{2}, \frac{1}{2}, -\frac{3}{2}, -\frac{9}{2})$ and hence $\overline{y} \cdot \lambda - \lambda = \overline{y}(\lambda + \rho) - (\lambda + \rho) = (0, -1, -1, -2, -3, -5)$. This is the highest weight of the Schur module $L_{(5,2|5,2)}E$. If y is represented by the partition $(4, 2, 1 \mid 1^3)$, then $\overline{y}(\lambda + \rho) = (\frac{11}{2}, \frac{7}{2}, -\frac{1}{2}, -\frac{3}{2}, -\frac{5}{2}, -\frac{9}{2})$ and hence $\overline{y} \cdot \lambda - \lambda = \overline{y}(\lambda + \rho) - (\lambda + \rho) = (0, -1, -4, -4, -4, -5)$. This is the highest weight of the Schur module $L_{(5,3,2,1|5,3,2,1)}E$. □

The result of Theorem 7.12 can be visualized as follows:

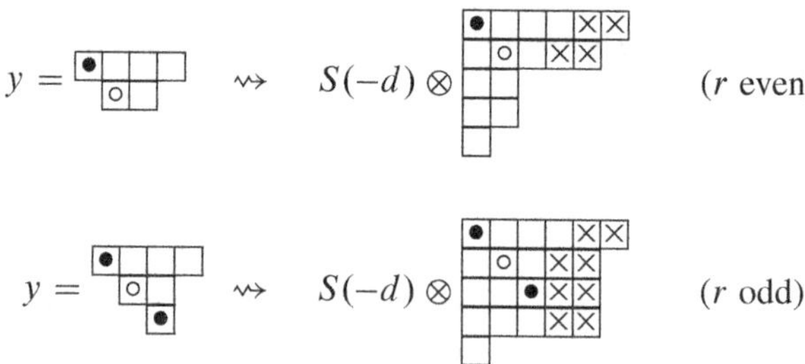

Here the diagrams on the left are the Young diagrams representing $y \in {}^{\mathrm{E}}(W_{[\lambda]})$ with every other diagonal box filled with a solid dot. The integer d equals the number of boxes plus k times the number of solid dots. Depending whether r is even or odd, we add k boxes to the first r rows or to the first $r + 1$ rows to obtain the diagrams on the right.

In the following examples, we follow the same conventions as in the case of the Lascoux resolution.

Example 7.13 (The Goto–Józefiak–Tachibana Complex). Consider the case when $k = n - 2$. Then we have a resolution

$$0 \longrightarrow \quad \longrightarrow \quad \longrightarrow \quad \overset{\bullet}{\longrightarrow} S \longrightarrow \mathbb{C}[Y_{n-2}] \longrightarrow 0,$$

where the diagrams are shorthand for

$$= S(-n-1) \otimes L_{(n,n,2)}E^*$$

$$= S(-n) \otimes L_{(n,n-1,1)}E^*$$

$$= S(-n+1) \otimes L_{(n-1,n-1)}E^*.$$

Example 7.14 (The Complex of Length 6). Consider the case when $k = n - 3$. Then we have a resolution

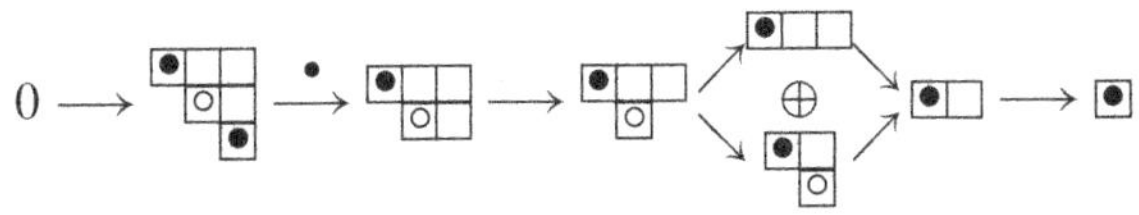

$$0 \longrightarrow \cdots \overset{\bullet}{\longrightarrow} \cdots \longrightarrow \cdots \longrightarrow \oplus \longrightarrow \cdots \longrightarrow \cdots \overset{\bullet}{\longrightarrow} S \longrightarrow \mathbb{C}[Y_{n-3}] \longrightarrow 0,$$

where the diagrams are shorthand for

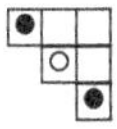

$$= S(-2n) \otimes L_{(n,n,n,n)}E^*$$

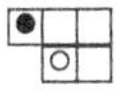

$$= S(-n-2) \otimes L_{(n,n,2,2)}E^*$$

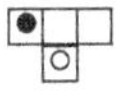

$$= S(-n-1) \otimes L_{(n,n-1,2,1)}E^*$$

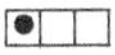

$$= S(-n) \otimes L_{(n-1,n-1,2)}E^*$$

$$= S(-n) \otimes L_{(n,n-2,1,1)}E^*$$

$$= S(-n+1) \otimes L_{(n-1,n-2,1)}E^*$$

$$= S(-n+2) \otimes L_{(n-2,n-2)}E^*.$$

7.7 *The determinantal ideals for skew symmetric matrices*

We identify $\mathrm{Alt}_n = \wedge^2 E$, where $E = \mathbb{C}^n$ is the natural representation of $\mathrm{U}(n)$.

Theorem 7.15. *Let $\mathfrak{g}_{\mathbb{R}} = \mathfrak{so}^*(2n)$ and let $\lambda = -2k\omega_n$ be the highest weight of the k-th Wallach representation, where $1 \le k \le n/2$. If $y \in {}^{\mathrm{E}}(W_{[\lambda]})$ is the element represented by the partition $(a_1, \dots, a_r \mid 1^r)$, then as a $\widetilde{\mathrm{U}}(n)$-representation,*

$$F_{\overline{y}\cdot\lambda} \otimes F_{-\lambda} = L_{(a_1+2k+1,\dots,a_r+2k+1|a_1,\dots,a_r)}E^*.$$

Proof. We have $\lambda = -2k\omega_n = (-k, -k, \dots, -k)$, $\rho = (n-1, n-2, \dots, 0)$, and hence $\lambda + \rho = (n-k-1, n-k-2, \dots, -k)$. Let $m = n - 2k - 1$. By Enright–Willenbring, $\Phi^+_{[\lambda]} = \{\varepsilon_i + \varepsilon_j \mid 1 \le i < j \text{ with } j \le m \text{ or } j = n-k\}$. If $y \in {}^{\mathrm{E}}(W_{[\lambda]})$ is the element represented by the partition $(a_1, \dots, a_r \mid 1^r)$, then

$$\overline{y}(\lambda + \rho) = (*, \dots, *, -a_r - k, \dots, -a_1 - k),$$

where the coordinates $*$ are permutation of the coordinates of $\lambda + \rho$ not contained in $\{a_1 + k, \dots, a_r + k\}$. It follows that $\overline{y}\cdot\lambda - \lambda$ is the highest weight of the Schur module

$L_{(a_1+2k+1,\dots,a_r+2k+1|a_1,\dots,a_r)}E^*$. To illustrate this consider the case when $n = 6$ and $k = 1$. Then $\lambda + \rho = (4, 3, 2, 1, 0, -1)$ and $m = 3$. If y is represented by the partition $(3, 1 \mid 1^2)$, then $\overline{y}(\lambda + \rho) = (3, 1, 0, \mathbf{-1}, \mathbf{-2}, \mathbf{-4})$ and hence $\overline{y} \cdot \lambda - \lambda = \overline{y}(\lambda + \rho) - (\lambda + \rho) = (-1, -2, -2, -2, -2, -3)$. This is the highest weight of the Schur module $L_{(6,4|3,1)}E$. □

Example 7.16 (The Buchsbaum–Eisenbud Complex). Consider the case when $n = 2t + 1, k = t - 1$. Then we have a resolution

$$0 \longrightarrow \quad \overset{\bullet}{\longrightarrow} \quad \longrightarrow \quad \overset{\bullet}{\longrightarrow} S \longrightarrow \mathbb{C}[Y_{t-1}] \longrightarrow 0$$

where the diagrams are shorthand for

$$= S(-2t-1) \otimes L_{(2t+1,2t+1)}E^*$$

$$= S(-t-1) \otimes L_{(2t+1,1)}E^*$$

$$= S(-t) \otimes L_{(2t)}E^*.$$

For an explicit description of the differentials see Weyman [49, (6.18)].

Example 7.17 (The Józefiak–Pragacz Complex). Consider the case when $n = 2t + 2$ and $k = t - 1$. Then $Y_k = Y_{t-1} = \{A \in \mathrm{Alt}_{2t+2} \mid \mathrm{rk}\, A \leq 2t - 1\}$ and we have a resolution

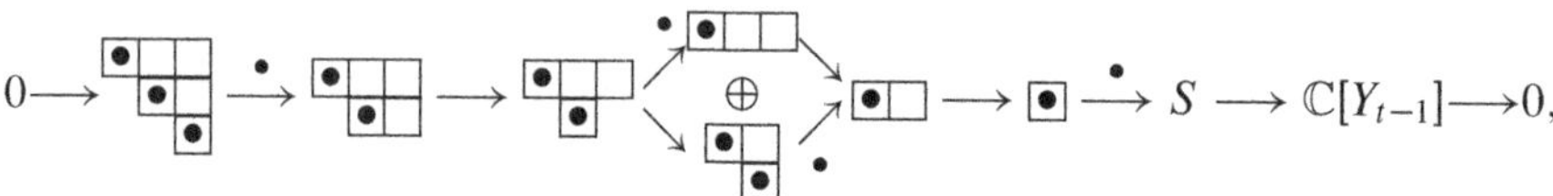

where the diagrams are shorthand for

$$= S(-3t-3) \otimes L_{(2t+2,2t+2,2t+2)}E^*$$

$$= S(-2t-3) \otimes L_{(2t+2,2t+2,2)}E^*$$

$$= S(-2t-2) \otimes L_{(2t+2,2t+1,1)}E^*$$

$$= S(-2t-1) \otimes L_{(2t+1,2t+1)}E^*$$

$$= S(-t-2) \otimes L_{(2t+2,1,1)}E^*$$

$$= S(-t-1) \otimes L_{(2t+1,1)}E^*$$

$$= S(-t) \otimes L_{(2t)}E^*.$$

7.8 Determinantal varieties for the exceptional groups

Let now $(G_{\mathbb{R}}, M_{\mathbb{R}})$ be any irreducible Hermitian symmetric pair of noncompact type and let M be the complexification of the (compact) group $M_{\mathbb{R}}$. Then M acts with finitely many orbits on $\mathfrak{u}$ and the orbit closures form a linear chain of affine varieties

$$\{0\} = Y_0 \subset Y_1 \subset \cdots \subset Y_{r-1} \subset Y_r = \mathfrak{u},$$

where r is the rank of the Hermitian symmetric pair. (For the Hermitian symmetric pairs in the dual pair setting that we discussed above, the varieties Y_k are the classical determinantal varieties.) For $1 \leq k \leq r-1$, it follows from work by Joseph [30] that the coordinate ring $\mathbb{C}[Y_k]$ carries the structure of a unitary highest weight module, namely the k-th Wallach representation of $\mathfrak{g}_{\mathbb{R}}$. If λ is the highest weight of the k-th Wallach representation, then as an $M_{\mathbb{R}}$-module,[3]

$$\mathbb{C}[Y_k] \cong L(\lambda) \otimes F_{-\lambda},$$

where the $M_{\mathbb{R}}$-action on $\mathbb{C}[Y_k]$ is the natural action obtained from the $M_{\mathbb{R}}$-action on Y_k. As in Theorem 7.7, the BGG resolution of $L(\lambda)$ gives rise to a minimal free $M_{\mathbb{R}}$-equivariant resolution of $\mathbb{C}[Y_k]$ as a graded S-module, where $S = S(\overline{\mathfrak{u}}) = \mathbb{C}[\mathfrak{u}]$.

Example 7.18. Consider the case when the Hermitian symmetric pair is of type $(\mathsf{E}_6, \mathsf{D}_5)$. Then $\mathfrak{u}$ is a 16-dimensional spin representation of $\mathfrak{m}$ and Y_1 is an 11-dimensional closed subvariety of $\mathfrak{u}$ whose ideal is generated by homogenous polynomials of degree 2. The corresponding Wallach representation has highest weight $\lambda = -3\omega_6$ and lies in a semiregular block with $\Sigma = \{\alpha_4\}$. The diagram on the left in Figure 13 shows the Hasse diagram of ${}^{\mathrm{I}}W^{\Sigma}$ as a subposet of ${}^{\mathrm{I}}W$. More precisely, the solid nodes are the nodes of the Hasse diagram of ${}^{\mathrm{I}}W$ corresponding to the elements of ${}^{\mathrm{I}}W^{\Sigma}$. The node corresponding to the Wallach representation $L(\lambda)$ is the maximal element of ${}^{\mathrm{I}}W^{\Sigma}$. (The bold edges of the Hasse diagram of ${}^{\mathrm{I}}W$ are the edges with label s_4.) Also shown are the generalized Young diagrams corresponding to the elements of ${}^{\mathrm{I}}W^{\Sigma}$. The boxes with label s_4 (in the sense of Definition 3.19) are filled with a dot. The poset ${}^{\mathrm{I}}W^{\Sigma}$ is isomorphic to a regular parabolic poset ${}^{\mathrm{I}'}W' \cong {}^{\mathrm{E}}(W_{[\lambda]})$ corresponding to a Hermitian symmetric pair of type $(\mathsf{A}_5, \mathsf{A}_4)$. The Hasse diagram of the poset ${}^{\mathrm{I}'}W'$ is the diagram on the right in Figure 13.

Note that some of the boxes of the Young diagrams corresponding to the elements of ${}^{\mathrm{I}'}W'$ have been filled with a dot according to the following rule. Let $x, y \in {}^{\mathrm{I}}W^{\Sigma}$ such that $x' \to y'$ in ${}^{\mathrm{I}'}W'$. Then the (unique) box of the diagram of y' that is not in the diagram of x' is left blank if $x \to y$ in ${}^{\mathrm{I}}W$ and filled with a dot if $x \not\to y$ in ${}^{\mathrm{I}}W$.

With this notation in place we can now describe the resolution of $\mathbb{C}[Y_1]$. As before in our classical examples, we simply write the generalized Young diagram of $y \in$

[3]It may be necessary to replace $G_{\mathbb{R}}$ and $M_{\mathbb{R}}$ by suitable finite covers.

${}^{\mathrm{E}}(W_{[\lambda]})$ for the term $S \otimes F_{\overline{y}\cdot\lambda-\lambda}$. The arrows that are marked with a dot correspond to a homogenous map of degree 2 and the other maps are homogenous of degree 1:

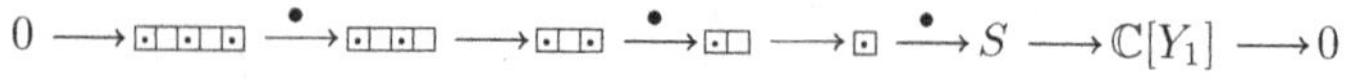

$$0 \longrightarrow \boxdot\square\boxdot\square\boxdot \overset{\bullet}{\longrightarrow} \boxdot\square\boxdot\square \longrightarrow \boxdot\square\boxdot \overset{\bullet}{\longrightarrow} \boxdot\square \longrightarrow \boxdot \overset{\bullet}{\longrightarrow} S \longrightarrow \mathbb{C}[Y_1] \longrightarrow 0$$

By calculating the dimensions of the $\mathfrak{m}$-modules $F_{\overline{y}\cdot\lambda-\lambda}$ for $y \in {}^{\mathrm{E}}(W_{[\lambda]})$ we can write the minimal free resolution of $\mathbb{C}[Y_1]$ more explicitly as follows:

$$0 \to S(-8) \to S(-6)^{10} \to S(-5)^{16} \to S(-3)^{16} \to S(-2)^{10} \to S \to \mathbb{C}[Y_1] \to 0.$$

Example 7.19. Consider a Hermitian symmetric pair of type $(\mathsf{E}_7, \mathsf{E}_6)$. Then $\mathfrak{u}$ is a 27-dimensional minuscule representation of $\mathfrak{m}$ and Y_1 is a 17-dimensional closed subvariety of $\mathfrak{u}$ whose ideal is generated by homogenous polynomials of degree 2. The corresponding Wallach representation has highest weight $\lambda = -4\omega_7$ and lies in a semiregular block with $\Sigma = \{\alpha_4\}$. The diagram on the left in Figure 14 shows the Hasse diagram of ${}^{\mathrm{I}}W^{\Sigma}$ as a subposet of ${}^{\mathrm{I}}W$. The poset ${}^{\mathrm{I}}W^{\Sigma}$ is isomorphic to a regular parabolic poset ${}^{\mathrm{I}'}W' \cong {}^{\mathrm{E}}(W_{[\lambda]})$ corresponding to a Hermitian symmetric pair of type $(\mathsf{D}_6, \mathsf{D}_5)$. The Hasse diagram of the poset ${}^{\mathrm{I}'}W'$ is the diagram on the right in Figure 14. Some of the boxes of the generalized Young diagrams have been filled with a dot according to the rule that was described in Example 7.18.

Using the same conventions in Example 7.18, we can describe the minimal free resolution of $\mathbb{C}[Y_1]$ as follows:

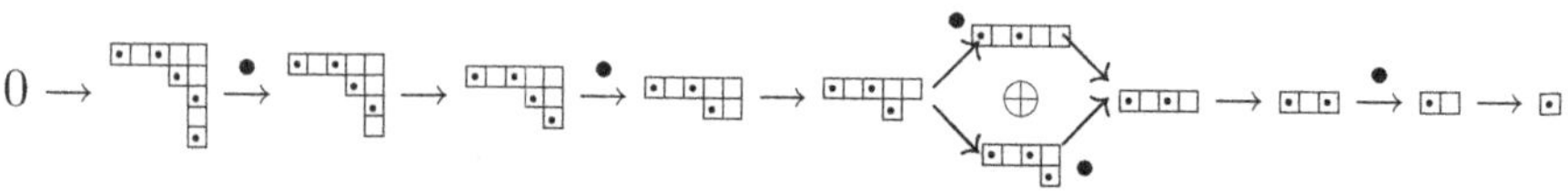

$$0 \to [\text{diagram}] \overset{\bullet}{\to} [\text{diagram}] \to [\text{diagram}] \overset{\bullet}{\to} [\text{diagram}] \to [\text{diagram}] \to \left([\text{diagram}] \oplus [\text{diagram}] \right) \to \boxdot\square\boxdot\square \to \boxdot\square\boxdot \overset{\bullet}{\to} \boxdot\square \to \boxdot$$

$$\overset{\bullet}{\to} S \to \mathbb{C}[Y_1] \to 0$$

By calculating the dimensions of the $\mathfrak{m}$-modules $F_{\overline{y}\cdot\lambda-\lambda}$ for $y \in {}^{\mathrm{E}}(W_{[\lambda]})$ we can write the minimal free resolution of $\mathbb{C}[Y_1]$ more explicitly as follows:

$$\begin{aligned} 0 &\to S(-15) \to S(-13)^{27} \to S(-12)^{78} \to S(-10)^{351} \to S(-9)^{650} \\ &\to S(-8)^{351} \oplus S(-7)^{351} \to S(-6)^{650} \to S(-5)^{351} \to S(-3)^{78} \to S(-2)^{27} \\ &\to S \to \mathbb{C}[Y_1] \to 0. \end{aligned}$$

8 Hilbert series of determinantal varieties

Let S be a polynomial ring over $\mathbb{C}$ in finitely many variables and let $V = \bigoplus_{i\geq 0} V_i$ be a finitely generated, graded S-module. Then the Hilbert series of V is defined as the formal power series

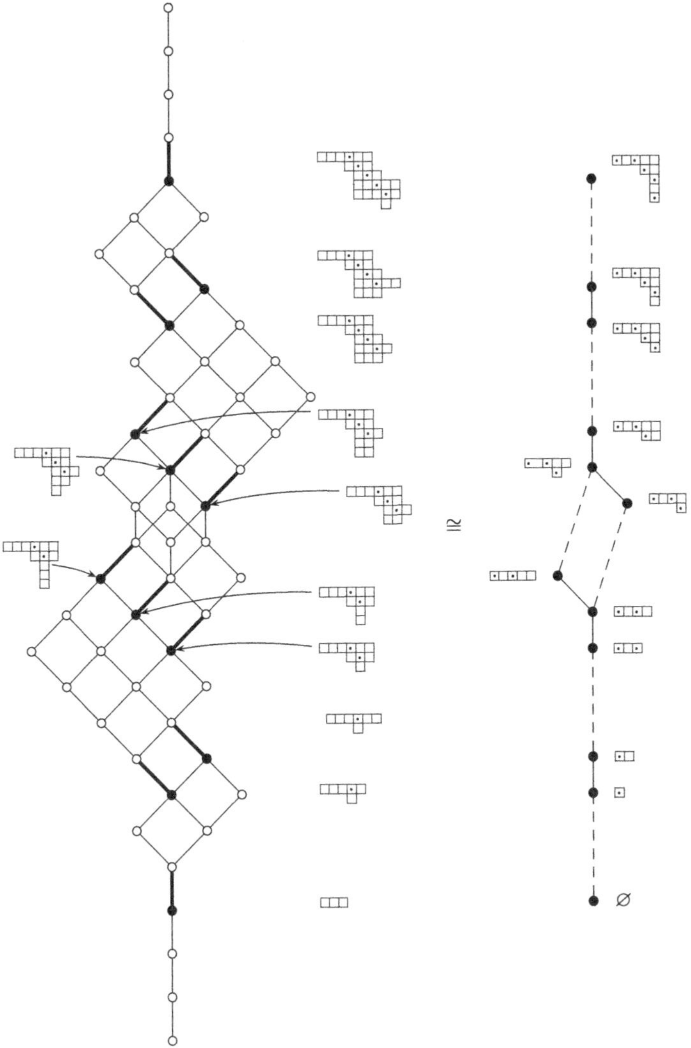

Figure 14 The isomorphism ${}^{\mathrm{I}}W^{\{\alpha_4\}} \cong {}^{\mathrm{I}'}W'$ for $(\Phi, \Phi_{\mathrm{I}}) = (\mathsf{E}_7, \mathsf{E}_6)$, $(\Phi', \Phi'_{\mathrm{I}'}) = (\mathsf{D}_6, \mathsf{D}_5)$

$$H_V(t) := \sum_{i \geq 0} (\dim V_i) t^i.$$

Suppose that V is generated by the elements in degree 1. Then it follows from Hilbert's syzygy theorem that $H_V(t)$ is a rational function of the form

$$H_V(t) = \frac{f(t)}{(1-t)^D},$$

where $f(t) \in \mathbb{Z}[t]$ such that $f(1) \neq 0$. The order of the pole at $t = 1$ is equal to the Krull dimension $D := \mathrm{Dim}\, V$ of the module V which is also equal to the Gelfand–Kirillov dimension of V in this case. The positive integer $f(1)$ is the Bernstein degree of V. Furthermore, if V is a Cohen–Macaulay module, then the coefficients of the numerator polynomial $f(t)$ are nonnegative integers.

8.1 A wonderful correspondence

Let $S = S(\overline{\mathfrak{u}}) = \mathbb{C}[\mathfrak{u}]$, where $\mathfrak{u}$ is the nilradical of a parabolic subalgebra of Hermitian type corresponding to $\mathfrak{g}_{\mathbb{R}} = \mathfrak{su}(p,q)$, $\mathfrak{sp}(n,\mathbb{R})$, or $\mathfrak{so}^*(2n)$, and let $Y_k \subset \mathfrak{u}$ be a determinantal variety as in §7. Recall that $\mathbb{C}[Y_k] \cong \mathbb{C}[Z]^K$ by Theorem 7.1, i.e., the coordinate ring $\mathbb{C}[Y_k]$ is isomorphic to an invariant ring of a reductive group. By a result of Hochster and Eagon [25], $\mathbb{C}[Y_k]$ is a Cohen-Macaulay ring and hence the Hilbert series of $\mathbb{C}[Y_k]$ is of the form $H_{\mathbb{C}[Y_k]}(t) = f(t)/(1-t)^D$, where the numerator polynomial $f(t) \in \mathbb{Z}[t]$ has nonnegative coefficients. It turns out that the coefficients of $f(t)$ can be interpreted as the Hilbert series of a certain finite-dimensional representation that corresponds to the Wallach representation on $\mathbb{C}[Y_k]$. Special cases of this correspondence were discovered by Enright and Willenbring in [22] and extended by Enright and Hunziker in [16]. The picture has recently been completed by Alexander, Hunziker, and Willenbring in [2].

To any of the Wallach representations $L(-kc\zeta)$ of $\mathfrak{g}_{\mathbb{R}} = \mathfrak{su}(p,q)$, $\mathfrak{sp}(n,\mathbb{R})$, and $\mathfrak{so}^*(2n)$, associate a real Lie algebra $\mathfrak{g}'_{\mathbb{R}}$ and a distinguished dominant integral weight $k\zeta'$ as in Table 3. Then, in the language of §5, the poset ${}^{I}W^{\Sigma}$ is isomorphic to the poset ${}^{I'}W'$ and the simple $\mathfrak{g}$-module $L(-kc\zeta)$ corresponds to the finite-dimensional $\mathfrak{g}'$-module of highest weight $k\zeta'$.

Theorem 8.1 (Alexander–Hunziker–Willenbring [2]). *Let L be the k-th Wallach representation of highest weight $-kc\zeta$, $1 \leq k \leq r-1$, and let E be the finite-dimensional $\mathfrak{g}'$-representation of highest weight $k\zeta'$ as in the Table 3. Then*

$$H_L(q) = \frac{H_E(q)}{(1-q)^D},$$

where $D = \dim \mathfrak{u} - \dim \mathfrak{u}'$. □

Table 3. The wonderful correspondence

$\mathfrak{g}_{\mathbb{R}}$	$-kc\zeta$	$\mathfrak{g}'_{\mathbb{R}}$	$k\zeta'$
$\mathfrak{su}(p,q)$	$-k\omega_p$	$\mathfrak{su}(p-k,q-k)$	$k\omega'_{p-k}$
$\mathfrak{sp}(n,\mathbb{R})$	$-\frac{k}{2}\omega_n$	$\mathfrak{so}^*(2n-2k+2)$	$k\omega'_{n-k+1}$
$\mathfrak{so}^*(2n)$	$-2k\omega_n$	$\mathfrak{sp}(n-2k-1,\mathbb{R})$	$k\omega'_{n-2k-1}$

Proof. We sketch the idea of the proof. Details will appear in [2].

Let $S = \mathbb{C}[\mathfrak{u}]$ and $S' = \mathbb{C}[\mathfrak{u}']$. Let $h \in \mathfrak{h}$ such that $\mathfrak{u} = \{x \in \mathfrak{g} \mid [h, x] = x\}$ and $h' \in \mathfrak{h}'$ such that $\mathfrak{u}' = \{x' \in \mathfrak{g} \mid [h', x'] = x'\}$. Furthermore, set $\lambda := -kc\zeta$ and $\lambda' := k\zeta'$. If $M_{\mathrm{I}}(\nu) = S \otimes F_\nu$ and $M'_{\mathrm{I}'}(\nu) = S' \otimes F'_{\nu'}$ are any two corresponding terms (via the poset isomorphism ${}^{\mathrm{I}}W^{\Sigma} \cong {}^{\mathrm{I}'}W'$) in the BGG resolutions of L and E, respectively, then miraculously

$$\dim F_\nu = \dim F'_{\nu'} \quad \text{and} \quad (\nu - \lambda)(h) = (\nu' - \lambda')(h').$$

By using the Euler-Poincaré principle to calculate the Hilbert series from the graded resolutions, the theorem follows. □

Theorem 8.2 (Enright–Hunziker–Wallach [18, Thm. 3.1]). *Let* $(\mathfrak{g}, \mathfrak{m})$ *be an irreducible complexified Hermitian symmetric pair and let* $E_{k\zeta}$ *be the finite-dimensional* $\mathfrak{g}$*-module with highest weight* $k\zeta$*, where* ζ *is the fundamental weight that is orthogonal to the compact roots. Then, as an* $\mathfrak{m}$*-module,*

$$E_{k\zeta} \otimes F_{-k\zeta} = \bigoplus_{k \geq m_1 \geq \cdots \geq m_r \geq 0} F_{-(m_1\gamma_1 + \cdots + m_r\gamma_r)},$$

where $\gamma_1 < \gamma_2 < \cdots < \gamma_r$ *are Harish-Chandra's strongly orthogonal noncompact roots with* γ_1 *being the noncompact simple root.* □

For $m \geq 1$ and $\nu = (\nu_1, \ldots, \nu_m)$ an integer partition with at most m parts, let $F_\nu^{(m)}$ denote the simple $\mathrm{GL}(m, \mathbb{C})$-module with highest weight $\nu_1\epsilon_1 + \cdots + \nu_m\epsilon_m$. Combining Theorem 8.1 and Theorem 8.2 we obtain the following Hilbert series.

Corollary 8.3. *Let* $Y_k = \{x \in \mathrm{M}_{p,q} \mid \mathrm{rk}\, x \leq k\}$ *with* $1 \leq k \leq \min\{p, q\} - 1$*. Then*

$$H_{\mathbb{C}[Y_k]}(t) = \frac{1}{(1-t)^{k(p+q-k)}} \sum_\nu (\dim F_\nu^{(p-k)})(\dim F_\nu^{(q-k)}) t^{|\nu|},$$

where the sum is over all partitions ν *whose Young diagram fit inside a rectangle of size* $\min\{p-k, q-k\} \times k$*.* □

Corollary 8.4. *Let* $Y_k = \{x \in \mathrm{Sym}_n \mid \mathrm{rk}\, x \leq k\}$ *with* $1 \leq k \leq n-1$*. Then*

$$H_{\mathbb{C}[Y_k]}(t) = \frac{1}{(1-t)^{k(2n-k+1)/2}} \sum_\nu (\dim F_\nu^{(n-k+1)}) t^{|\nu|/2},$$

where the sum is over all partitions ν *whose Young diagram has only columns of even length and fit inside a rectangle of size* $(n-k+1) \times k$*.* □

Corollary 8.5. *Let* $Y_k = \{x \in \mathrm{Alt}_n \mid \mathrm{rk}\, x \leq 2k\}$ *with* $1 \leq k \leq [n/2] - 1$. *Then*

$$H_{\mathbb{C}[Y_k]}(t) = \frac{1}{(1-t)^{k(2n-2k-1)}} \sum_{\nu} (\dim F_{\nu}^{(n-2k-1)}) t^{|\nu|/2},$$

where the sum is over all partitions ν *whose Young diagram has only rows of even length and fit inside a rectangle of size* $(n-2k-1) \times 2k$. □

Appendix: diagrams of Hermitian type

Below, for each Hermitian symmetric pair, the Hasse diagram of the poset $\Phi(\mathfrak{u})$ and the generalized Young diagram of the longest element of ${}^{I}W$ are shown. For $\beta, \beta' \in \Phi(\mathfrak{u})$ such that $\beta' - \beta = \alpha_i \in \Pi$, the edge connecting the nodes corresponding to β and β' is labelled by the integer i. (Not all labels are shown. The omitted labels are easily obtained by noting that opposite sides of every diamond have the same label.)

Type $(\mathsf{A}_n, \mathsf{A}_{p-1} \times \mathsf{A}_{n-p})$, here shown for $n = 6$ and $p = 4$:

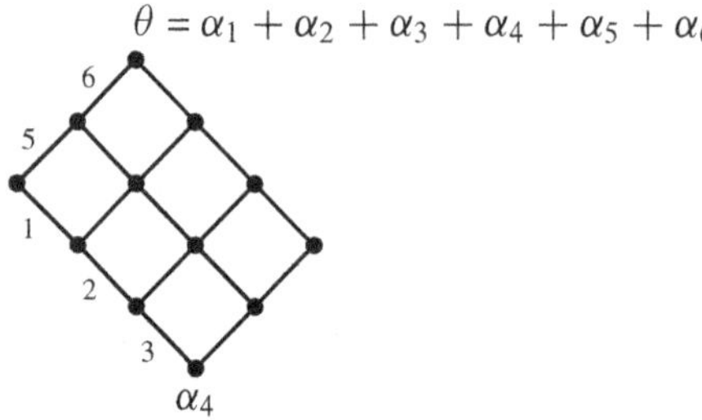

4	3	2	1
5	4	3	2
6	5	4	3

Type $(\mathsf{B}_n, \mathsf{B}_{n-1})$, here shown for $n = 5$:

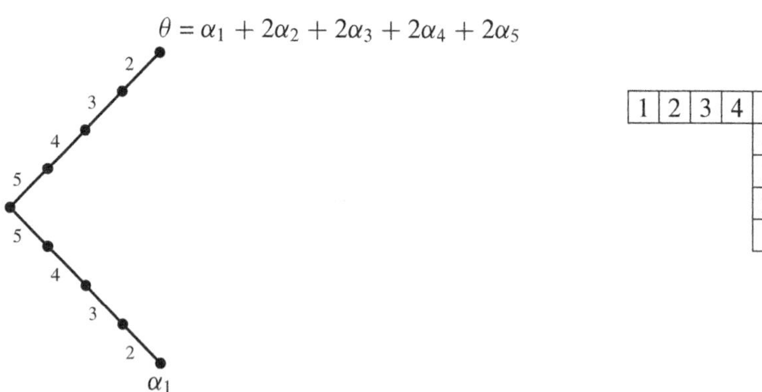

Type $(\mathsf{C}_n, \mathsf{A}_{n-1})$, here shown for $n = 5$:

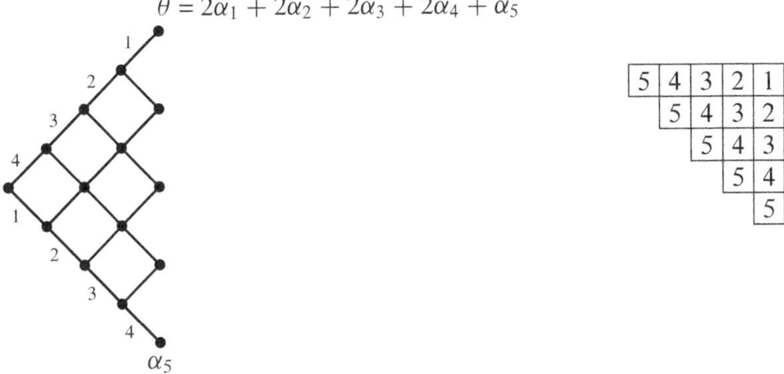

Type $(\mathsf{D}_n, \mathsf{D}_{n-1})$, here shown for $n = 6$:

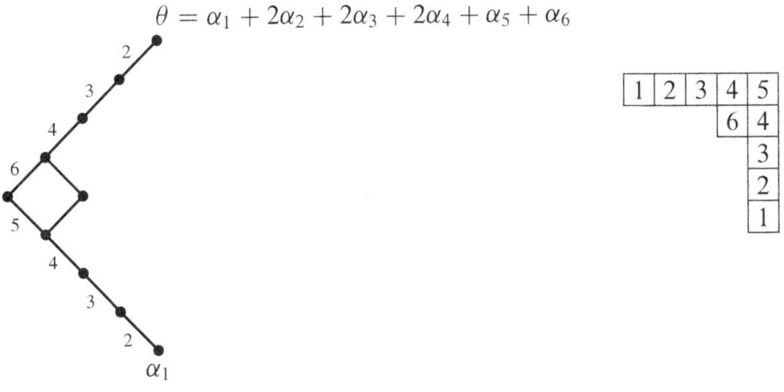

Type $(\mathsf{D}_n, \mathsf{A}_{n-1})$, here shown for $n = 6$:

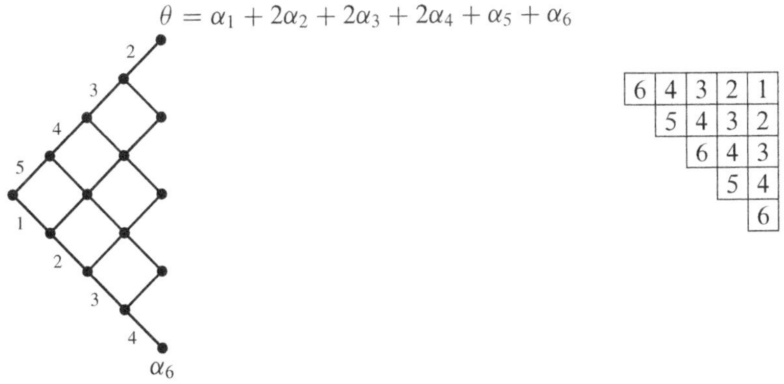

Type $(\mathsf{E}_6, \mathsf{D}_5)$:

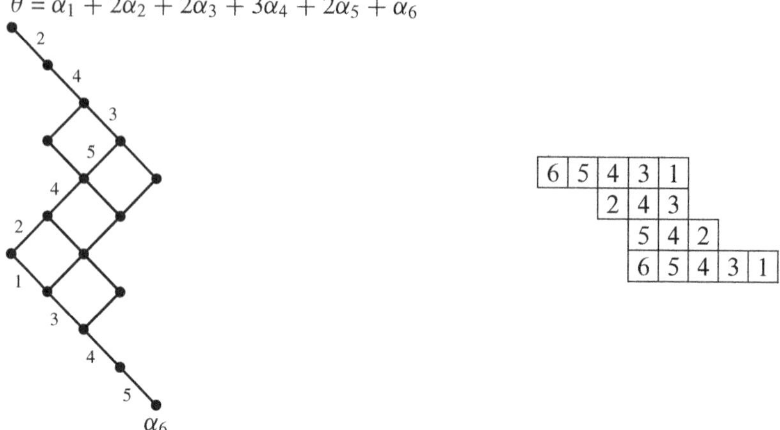

6	5	4	3	1			
		2	4	3			
			5	4	2		
			6	5	4	3	1

Type $(\mathsf{E}_7, \mathsf{E}_6)$:

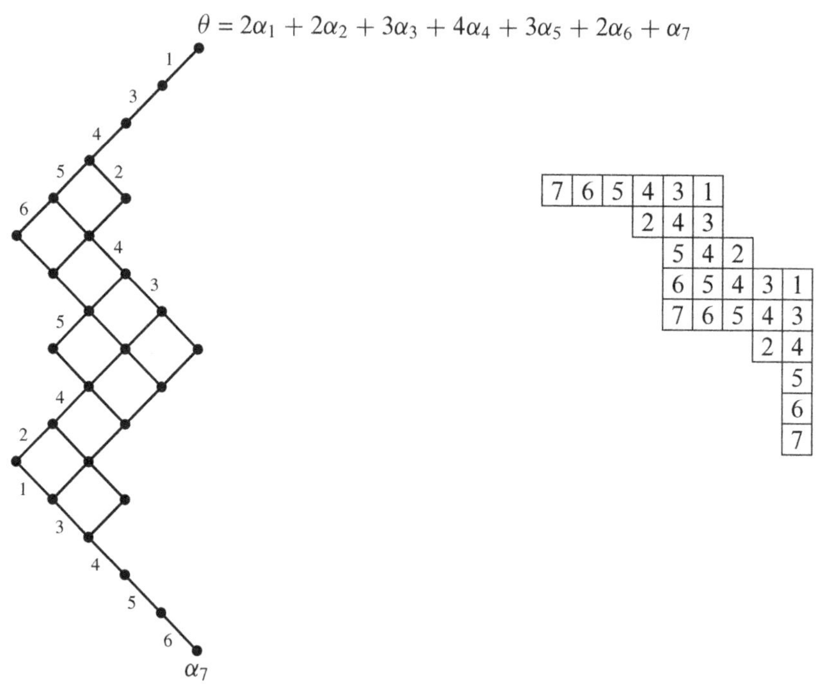

7	6	5	4	3	1			
			2	4	3			
				5	4	2		
				6	5	4	3	1
				7	6	5	4	3
							2	4
								5
								6
								7

References

1. Kaan Akin, David A. Buchsbaum, and Jerzy Weyman, *Schur functors and Schur complexes*, Adv. in Math. **44** (1982), no. 3, 207–278.
2. Jordan Alexander, Markus Hunziker, and Jeb F. Willenbring, *Hilbert series of determinantal varieties and strongly orthogonal roots*, in preparation.
3. Erik Backelin, *Koszul duality for parabolic and singular category $\mathcal{O}$*, Represent. Theory **3** (1999), 139–152 (electronic).
4. Alexander Beilinson, Victor Ginzburg, and Wolfgang Soergel, *Koszul duality patterns in representation theory*, J. Amer. Math. Soc. **9** (1996), no. 2, 473–527.
5. I. N. Bernstein, I. M. Gelfand, and S. I. Gelfand, *Differential operators on the base affine space and a study of $\mathfrak{g}$-modules*, Lie groups and their representations (Proc. Summer School, Bolyai János Math. Soc., Budapest, 1971), Halsted, New York, 1975, pp. 21–64.
6. Brian D. Boe, *Kazhdan–Lusztig polynomials for Hermitian symmetric spaces*, Trans. Amer. Math. Soc. **309** (1988), no. 1, 279–294.
7. Brian D. Boe and Markus Hunziker, *Kostant modules in blocks of category $\mathcal{O}_S$*, Comm. Algebra **37** (2009), no. 1, 323–356.
8. Brian D. Boe and Daniel K. Nakano, *Representation type of the blocks of category $\mathcal{O}_S$*, Adv. Math. **196** (2005), no. 1, 193–256.
9. N. Bourbaki, *Éléments de mathématique. Fasc. XXXIV. Groupes et algèbres de Lie. Chapitre IV: Groupes de Coxeter et systèmes de Tits. Chapitre V: Groupes engendrés par des réflexions. Chapitre VI: systèmes de racines*, Actualités Scientifiques et Industrielles, No. 1337, Hermann, Paris, 1968.
10. Michel Brion and Patrick Polo, *Generic singularities of certain Schubert varieties*, Math. Z. **231** (1999), no. 2, 301–324.
11. James B. Carrell, *The Bruhat graph of a Coxeter group, a conjecture of Deodhar, and rational smoothness of Schubert varieties*, Algebraic groups and their generalizations: classical methods (University Park, PA, 1991), Proc. Sympos. Pure Math., vol. 56, Amer. Math. Soc., Providence, RI, 1994, pp. 53–61.
12. James B. Carrell and Jochen Kuttler, *Smooth points of T-stable varieties in G/B and the Peterson map*, Invent. Math. **151** (2003), no. 2, 353–379.
13. Luis G. Casian and David H. Collingwood, *The Kazhdan–Lusztig conjecture for generalized Verma modules*, Math. Z. **195** (1987), no. 4, 581–600.
14. Vinay V. Deodhar, *On some geometric aspects of Bruhat orderings. II. The parabolic analogue of Kazhdan-Lusztig polynomials*, J. Algebra **111** (1987), no. 2, 483–506.
15. Thomas J. Enright, *Analogues of Kostant's $\mathfrak{u}$-cohomology formulas for unitary highest weight modules*, J. Reine Angew. Math. **392** (1988), 27–36.
16. Thomas J. Enright and Markus Hunziker, *Resolutions and Hilbert series of determinantal varieties and unitary highest weight modules*, J. Algebra **273** (2004), no. 2, 608–639.
17. — , *Resolutions and Hilbert series of the unitary highest weight modules of the exceptional groups*, Represent. Theory **8** (2004), 15–51 (electronic).
18. Thomas J. Enright, Markus Hunziker, and Nolan R. Wallach, *A Pieri rule for Hermitian symmetric pairs. I*, Pacific J. Math. **214** (2004), no. 1, 23–30.
19. Thomas J. Enright and Brad Shelton, *Categories of highest weight modules: applications to classical Hermitian symmetric pairs*, Mem. Amer. Math. Soc. **67** (1987), no. 367, iv+94.
20. ———, *Highest weight modules for Hermitian symmetric pairs of exceptional type*, Proc. Amer. Math. Soc. **106** (1989), no. 3, 807–819.
21. Thomas J. Enright and Jeb F. Willenbring, *Hilbert series, Howe duality, and branching rules*, Proc. Natl. Acad. Sci. USA **100** (2003), no. 2, 434–437 (electronic).
22. ———, *Hilbert series, Howe duality and branching for classical groups*, Ann. of Math. (2) **159** (2004), no. 1, 337–375.
23. Roe Goodman and Nolan R. Wallach, *Symmetry, Representations, and Invariants*, Graduate Texts in Mathematics, Vol. 255, Springer, New York, 2009.

24. Harish-Chandra, *On a lemma of F. Bruhat*, J. Math. Pures Appl. (9) **35** (1956), 203–210.
25. M. Hochster and John A. Eagon, *Cohen-Macaulay rings, invariant theory, and the generic perfection of determinantal loci*, Amer. J. Math. **93** (1971), 1020–1058.
26. Jaehyun Hong, *Rigidity of smooth Schubert varieties in Hermitian symmetric spaces*, Trans. Amer. Math. Soc. **359** (2007), no. 5, 2361–2381 (electronic).
27. Roger Howe, *Remarks on classical invariant theory*, Trans. Amer. Math. Soc. **313** (1989), no. 2, 539–570.
28. Ronald S. Irving, *Singular blocks of the category $\mathcal{O}$*, Math. Z. **204** (1990), no. 2, 209–224.
29. Hans Plesner Jakobsen, *Hermitian symmetric spaces and their unitary highest weight modules*, J. Funct. Anal. **52** (1983), no. 3, 385–412.
30. Anthony Joseph, *Annihilators and associated varieties of unitary highest weight modules*, Ann. Sci. École Norm. Sup. (4) **25** (1992), no. 1, 1–45.
31. T. Józefiak, P. Pragacz, and J. Weyman, *Resolutions of determinantal varieties and tensor complexes associated with symmetric and antisymmetric matrices*, Young tableaux and Schur functors in algebra and geometry (Toruń, 1980), Astérisque, vol. 87, Soc. Math. France, Paris, 1981, pp. 109–189.
32. Masaki Kashiwara and Michèle Vergne, *On the Segal–Shale–Weil representations and harmonic polynomials*, Invent. Math. **44** (1978), no. 1, 1–47.
33. David Kazhdan and George Lusztig, *Representations of Coxeter groups and Hecke algebras*, Invent. Math. **53** (1979), no. 2, 165–184.
34. Bertram Kostant, *Lie algebra cohomology and the generalized Borel–Weil theorem*, Ann. of Math. (2) **74** (1961), 329–387.
35. ———, *Lie algebra cohomology and generalized Schubert cells*, Ann. of Math. (2) **77** (1963), 72–144.
36. Thomas Lam and Lauren Williams, *Total positivity for cominuscule Grassmannians*, New York J. Math. **14** (2008), 53–99.
37. Alain Lascoux, *Syzygies des variétés déterminantales*, Adv. in Math. **30** (1978), no. 3, 202–237.
38. Alain Lascoux and Marcel-Paul Schützenberger, *Polynômes de Kazhdan & Lusztig pour les grassmanniennes*, Young tableaux and Schur functors in algebra and geometry (Toruń, 1980), Astérisque, Vol. 87, Soc. Math. France, Paris, 1981, pp. 249–266.
39. J. Lepowsky, *A generalization of the Bernstein–Gelfand–Gelfand resolution*, J. Algebra **49** (1977), no. 2, 496–511.
40. Kenyon J. Platt, *Nonzero infinitesimal blocks of category $\mathcal{O}_S$*, Algebr. Represent. Theory **14** (2011), no. 4, 665–689.
41. Robert A. Proctor, *Interactions between combinatorics, Lie theory and algebraic geometry via the Bruhat orders*, Ph.D. thesis, MIT, 1981.
42. W. Andrew Pruett, *Diagrams and reduced decompositions for cominuscule flag varieties and affine Grassmannians*, Ph.D. thesis, Baylor University, 2010.
43. Wilfried Schmid, *Vanishing theorems for Lie algebra cohomology and the cohomology of discrete subgroups of semisimple Lie groups*, Adv. in Math. **41** (1981), no. 1, 78–113.
44. H. Schubert, *Beiträge zur abzählenden Geometrie*, Math. Ann. **10** (1876), no. 1, 1–116.
45. W. Soergel, *$\mathfrak{n}$-cohomology of simple highest weight modules on walls and purity*, Invent. Math. **98** (1989), no. 3, 565–580.
46. John R. Stembridge, *On the fully commutative elements of Coxeter groups*, J. Algebraic Combin. **5** (1996), no. 4, 353–385.
47. Hugh Thomas and Alexander Yong, *A combinatorial rule for (co)minuscule Schubert calculus*, Adv. Math. **222** (2009), no. 2, 596–620.
48. Hermann Weyl, *The Classical Groups. Their Invariants and Representations*, Princeton University Press, Princeton, N.J., 1939.
49. Jerzy Weyman, *Cohomology of vector bundles and syzygies*, Cambridge Tracts in Mathematics, Vol. 149, Cambridge University Press, Cambridge, 2003.

A conjecture of Sakellaridis–Venkatesh on the unitary spectrum of spherical varieties

Wee Teck Gan and Raul Gomez

To Nolan Wallach, with admiration and appreciation

Abstract We describe the spectral decomposition of certain spherical varieties of low rank, verifying a recent conjecture of Sakellaridis and Venkatesh in these cases.

Keywords: Spherical varieties • Unitary spectrum • Theta correspondence • Sakellaridis–Venkatesh conjecture

Mathematics Subject Classification: 22E50

1 Introduction

The spectral decomposition of the unitary representation $L^2(H\backslash G)$, when $X = H\backslash G$ is a symmetric space, has been studied extensively, especially in the case when G is a real Lie group. In particular, through the work of many authors (such as [1, 3, 7, 20, 26]), one now has the full Plancherel theorem in this setting.

In a recent preprint [23], Sakellaridis and Venkatesh considered the more general setting where $X = H\backslash G$ is a spherical variety and G is a real or p-adic group. Motivated by the study of periods in the theory of automorphic forms and the comparison of relative trace formulas, they formulated an approach to this

The research of the first author is partially supported by NSF grant 0801071 and a startup grant from the National University of Singapore. The second author was supported by a Postdoctoral Research Fellowship at the National University of Singapore during the course of this work.

W.T. Gan
Department of Mathematics, National University of Singapore, Block S17, 10 Lower Kent Ridge Road, Singapore 587628, Singapore
e-mail: matgwt@nus.edu.sg; matrgm@nus.edu.sg

R. Gomez
Department of Mathematics, 310 Malott Hall, Cornell University, Ithaca, NY 14853, USA
e-mail: rg558@cornell.edu

R. Howe et al. (eds.), *Symmetry: Representation Theory and Its Applications*, Progress in Mathematics 257, DOI 10.1007/978-1-4939-1590-3_7

problem in the framework of Langlands functoriality. More precisely, led by and refining the work of Gaitsgory–Nadler [9] in the geometric Langlands program, they associated to a spherical variety $X = H\backslash G$ (satisfying some additional technical hypotheses)

- a dual group $\check{G}_X$;
- a natural map $\iota : \check{G}_X \times SL_2(\mathbb{C}) \longrightarrow \check{G}$.

The map ι induces a map from the set of tempered L-parameters of G_X to the set of Arthur parameters of G, and if one is very optimistic, it may even give rise to a map

$$\iota_* : \widehat{G}_X \longrightarrow \widehat{G}$$

where G_X is a (split) group with dual group $\check{G}_X$ and $\widehat{G}_X$ and $\widehat{G}$ refer to the unitary dual of the relevant groups. Assuming for simplicity that this is the case, one has the following conjecture:

Sakellaridis–Venkatesh Conjecture. One has a spectral decomposition

$$L^2(H\backslash G) \cong \int_{\widehat{G}_X} W(\pi) \otimes \iota_*(\pi)\, d\mu(\pi)$$

where μ is the Plancherel measure of $\widehat{G}_X$ and $W(\pi)$ is some (finite-dimensional) multiplicity space. The multiplicity space $W(\pi)$ should be related to the space of continuous H-invariant functionals on the representation $\iota_*(\pi)$.

In particular, the class of the spectral measure of $L^2(H\backslash G)$ is absolutely continuous with respect to that of the pushforward by ι_* of the Plancherel measure on $\widehat{G}_X$, and its support is contained in the set of those Arthur parameters of G which factor through ι. Moreover, from the point of view of Arthur parameters, the multiplicity space should be related to the number of inequivalent ways an Arthur parameter valued in $\check{G}$ can be lifted to $\check{G}_X$ (i.e., factored through ι).

The main purpose of this paper is to verify the above conjecture in many cases when $H\backslash G$, or equivalently G_X, has low rank, and to specify the multiplicity space $W(\pi)$. In particular, we demonstrate this conjecture for many cases when G_X has rank 1, and also some cases when G_X has rank 2 or 3 (see the tables in [23, §15 and §16]). More precisely, our main result is:

Theorem 1. *The conjecture of Sakellaridis–Venkatesh holds for the spherical varieties $X = H\backslash G$ listed in Tables 1 and 2.*

Table 1. Classical cases

$X = H\backslash G$	G_X
$GL_{n-1}\backslash GL_n$	GL_2
$SO_{n-1}\backslash SO_n$	$\widetilde{SL}_2$
$Sp_{2n-2}\backslash Sp_{2n}$	$SO(4)$

Table 2. Exceptional cases

$X = H\backslash G$	G_X
$SO_3\backslash SL_3$	$\widetilde{SL}_3$
$Sp_6\backslash SL_6$	SL_3
$SL_3\backslash G_2$	$\widetilde{SL}_2$
$(J,\psi)\backslash G_2$	PGL_3
$SU_3\backslash Spin_7$	$(Spin_3 \times Spin_5)/\Delta\mu_2$
$G_2\backslash Spin_7$	SL_2
$G_2\backslash Spin_8$	$SL_2^3/\Delta\mu_2$
$Spin_9\backslash F_4$	PGL_2
$F_4\backslash E_6$	SL_3

For the classical cases, the precise results are contained in Theorem 3 in §3.8 and the ensuing discussion in §3.9. We note that Theorem 3 gives the spectral decomposition, in the spirit of the Sakellaridis–Venkatesh conjecture, of the so-called generalized Stiefel manifolds [17], which are homogeneous but not necessarily spherical varieties. The exceptional cases are covered in §8.4 (Theorem 4), §8.5 and §9. We note that over $\mathbb{R}$, Kobayashi has given in [15] an explicit description of the discrete spectrum of the generalized Stiefel manifolds in terms of $A_{\mathfrak{q}}(\lambda)$ modules; he has also described in [16] the spectrum of certain special spherical varieties such as $SL_3\backslash G_2$ which can be related to symmetric spaces (as we explain in §4.4). His viewpoint is quite disjoint from that of this paper.

Theorem 1 is proved using the technique of theta correspondence. More precisely, it turns out that for the groups listed in the above table, one has a reductive dual pair

$$G_X \times G \subset S$$

for some larger group S. One then studies the restriction of the minimal representation of S to the subgroup $G_X \times G$. In the context of theta correspondence in smooth representation theory, one can typically show the following rough statement: For a representation π of G,

π has ψ-generic (and hence nonzero) theta lift to G_X

π has nonzero H-period.

Our main theorem is thus the L^2-manifestation of this phenomenon, giving a description of $L^2(H\backslash G)$ in terms of $L^2(G_X)$. This idea is not really new: a well-known example of this kind of result is the correspondence between the irreducible components of the spherical harmonics on $\mathbb{R}^n$ under the action of $O(n,\mathbb{R})$, and holomorphic discrete series of the group $\widetilde{SL}(2,\mathbb{R})$, the double cover of $SL(2,\mathbb{R})$.

Another example is given by the classical paper of Rallis and Schiffmann [21], where they used the oscillator representation to relate the discrete spectrum of

$$L^2(O(p,q-1)\backslash O(p,q))$$

with the discrete series representations of $\widetilde{SL}(2,\mathbb{R})$. Later, Howe [12] showed how these results can be inferred from his general theory of reductive dual pairs, and essentially provided a description of the Plancherel measure of $L^2(O(p,q-1)\backslash O(p,q))$ in terms of the representation theory of $\widetilde{SL}(2,\mathbb{R})$. Then Ørsted and Zhang [29, 30] proved a similar result for the space $L^2(U(p,q-1)\backslash U(p,q))$ in terms of the representation theory of $U(1,1)$. We give a more steamlined treatment of these classical cases in Section 2, which accounts for Table 1. The rest of the paper is then devoted to the exceptional cases listed in Table 2.

Acknowledgement. Both authors would like to pay tribute to Nolan Wallach for his guidance, encouragement and friendship over the past few years. It is an honor to be his colleague and student respectively. We wish him all the best in his retirement from UCSD, and hope to continue to interact with him mathematically and personally for many years to come.

We thank T. Kobayashi for a number of helpful and illuminating conversations during his visit to Singapore in March 2012, and the referee for his/her careful reading of the manuscript and his/her many pertinent comments which helped improve the exposition of the paper.

2 Classical dual pairs

We begin by introducing the classical dual pairs.

2.1 Division algebra D

Let $\mathfrak{k}$ be a local field, and let $|\cdot|$ denote its absolute value. Let $D = \mathfrak{k}$, a quadratic field extension of $\mathfrak{k}$ or the quaternion division $\mathfrak{k}$-algebra, and let $x \mapsto \overline{x}$ be its canonical involution. The case when D is the split quadratic algebra or quaternion algebra can also be included in the discussion, but for simplicity, we shall stick with division algebras. We have the *reduced trace* map $\mathrm{Tr} : D \to \mathfrak{k}$ and the *reduced norm* map $Q : D \to \mathfrak{k}$. If $D \neq \mathfrak{k}$, one has $\mathrm{Tr}(x) = x + \overline{x} \in \mathfrak{k}$ and $Q(x) = x \cdot \overline{x} \in \mathfrak{k}$.

2.2 Hermitian D-modules

Let V and W be two right D-modules. We will denote the set of right D-module morphisms between V and W by

$$\mathrm{Hom}_D(V, W) = \{T : V \longrightarrow W \mid T(v_1 a + v_2 b) = T(v_1)a + T(v_2)b \quad \text{for all } v_1, v_2 \in V, a, b \in D\}.$$

In the same way, if V and W are two left D-modules, we set

$$\mathrm{Hom}_D(V, W) = \{T : V \longrightarrow W \mid (av_1 + bv_2)T = a(v_1)T + b(v_2)T \quad \text{for all } v_1, v_2 \in V, a, b \in D\}.$$

If $V = W$, we will denote this set by $\mathrm{End}_D(V)$. Notice that for right D-module morphisms we are putting the argument on the right, while for left D-module morphisms we are putting it on the left.

In general, for every statement involving *right* D-modules one can make an analogous one involving *left* D-modules. From now on, we will focus on right D-modules, and we will leave the reader with the task of making the corresponding definitions and statements involving left D-modules. Set

$$GL(V, D) = \{T \in \mathrm{End}_D(V) \mid T \text{ is invertible}\}.$$

When it is clear from the context what the division algebra is, we will just denote this group by $GL(V)$.

Let V' be the set of right D-linear functionals on V. There is a natural left D-module structure on V' given by setting

$$(a\lambda)(v) = a\lambda(v), \quad \text{for all } a \in D, v \in V, \text{and } \lambda \in V'.$$

Observe that with this structure, $W \otimes_D V'$ is naturally isomorphic to $\mathrm{Hom}_D(V, W)$ as a $\Bbbk$-vector space. Given $T \in \mathrm{Hom}_D(V, W)$, we will define an element in $\mathrm{Hom}_D(W', V')$, which we will also denote T, by setting $(\lambda T)(v) := \lambda(Tv)$. This correspondence gives rise to natural isomorphisms between $\mathrm{End}_D(V)$ and $\mathrm{End}_D(V')$ and between $GL(V)$ and $GL(V')$.

Definition 1. Let $\varepsilon = \pm 1$. We say that (V, B) is a right ε-Hermitian D-module, if V is a right D-module and B is an ε-Hermitian form, i.e., $B : V \times V \longrightarrow D$ is a map such that

1. B is *sesquilinear*. That is, for all $v_1, v_2, v_3 \in V, a, b \in D$,

$$B(v_1, v_2 a + v_3 b) = B(v_1, v_2)a + B(v_1, v_3)b \quad \text{and}$$

$$B(v_1 a + v_2 b, v_3) = \overline{a}B(v_1, v_3) + \overline{b}B(v_2, v_3).$$

2. B is *ε-Hermitian*. That is,

$$B(v, w) = \varepsilon \overline{B(w, v)} \qquad \text{for all } v, w \in V.$$

3. B is *nondegenerate*.

Usually, 1-Hermitian D-modules are simply called Hermitian, while -1-Hermitian D-modules are called skew-Hermitian. To define left ε-Hermitian D-modules (V, B), we just have to replace the sesquilinear condition by

$$B(av_1 + bv_2, v_3) = aB(v_1, v_3) + bB(v_2, v_3) \quad \text{and}$$

$$B(v_1, av_2 + bv_3) = B(v_1, v_2)\overline{a} + B(v_1, v_3)\overline{b},$$

for all v_1, v_2, $v_3 \in V$, $a, b \in D$.

Given a right ε-Hermitian D-module (V, B), we will define

$$G(V, B) = \{g \in GL(V) \mid B(gv, gw) = B(v, w) \text{ for all } v, w \in V\},$$

to be the subgroup of $GL(V)$ preserving the ε-Hermitian form B. When there is no risk of confusion regarding B, we will denote this group by $G(V)$. Later on, we shall sometimes need to use the same notation to denote a covering group of $G(V, B)$; see §2.4.

Given a right ε-Hermitian D-module (V, B), we can construct a left ε-Hermitian D-module (V^*, B^*) in the following way: as a set, V^* will be the set of symbols $\{v^* \mid v \in V\}$. Then we give V^* a left D-module structure by setting, for all $v, w \in V, a \in D$,

$$v^* + w^* = (v + w)^* \text{ and } av^* = (v\overline{a})^*.$$

Finally, we set

$$B^*(v^*, w^*) = \overline{B(w, v)} \qquad \text{for all } v, w \in V.$$

In an analogous way, if V is a left D-module, we can define a right D-module V^*, and V^{**} is naturally isomorphic with V. Given $T \in \mathrm{End}_D(V)$, we can define $T^* \in \mathrm{End}_D(V^*)$ by setting $v^*T^* := (Tv)^*$. With this definition, it is easily seen that $(TS)^* = S^*T^*$, for all $S, T \in \mathrm{End}_D(V)$. Therefore the map $g \mapsto (g^*)^{-1}$ defines an algebraic group isomorphism between $GL(V)$ and $GL(V^*)$.

Now observe that the form B induces a left D-module isomorphism $B^\flat : V^* \longrightarrow V'$ given by $B^\flat(v^*)(w) = B(v, w)$ for $v, w \in V$. In what follows, we will make implicit use of this map to identify these two spaces. With this identification we can think of T^* as a map in $\mathrm{End}_D(V)$ defined by $v^*(T^*w) := (v^*T^*)(w)$, i.e., T^* is defined by the condition that

$$B(v, T^*w) = B(Tv, w) \qquad \text{for all } v, w \in V.$$

Observe that this agrees with the usual definition of T^*.

A D-submodule $X \subset V$ is said to be *totally isotropic* if $B|_{X \times X} = 0$. If X is a totally isotropic submodule, then there exists a totally isotropic submodule $Y \subset V$ such that $B|_{X \oplus Y \times X \oplus Y}$ is nondegenerate. If we set

$$U = (X \oplus Y)^\perp := \{u \in V \mid B(u, w) = 0 \text{ for all } w \in X \oplus Y\},$$

then $V = X \oplus Y \oplus U$, and $B|_{U\times U}$ is nondegenerate. In this case we say that X and Y are totally isotropic, *complementary* submodules. Observe that then $B^\flat|_{Y^*} : Y^* \longrightarrow X'$ is an isomorphism. As before we will make implicit use of this isomorphism to identify Y^* with X'.

2.3 Reductive dual pairs

Let (V, B_V) be a right ε_V-Hermitian D-module and (W, B_W) a right ε_W-Hermitian D-module such that $\varepsilon_V\varepsilon_W = -1$. On the $\mathfrak{k}$-vector space $V \otimes_D W^*$ we can define a symplectic form B by setting

$$B(v_1 \otimes_D \lambda_1, v_2 \otimes_D \lambda_2) = \mathrm{Tr}(B_W(w_1, w_2)B_V^*(\lambda_2, \lambda_1))$$

for all $v_1, v_2 \in V$ and $\lambda_1, \lambda_2 \in V^*$. Let

$$Sp(V \otimes_D W^*) = \{g \in GL(V \otimes_D W^*, \mathfrak{k}) \mid B(gv, gw) = B(v, w) \quad \text{for all } v, w \in V \otimes_D W^*\}.$$

Observe that

$$Sp(V \otimes_D W^*) = G(V \otimes_D W^*, B) = G(V \otimes_D W^*).$$

Moreover, there is a natural map $G(V) \times G(W) \longrightarrow Sp(V \otimes_D W^*)$ given by

$$(g_1, g_2) \cdot v \otimes_D \lambda = g_1 v \otimes \lambda g_2^*.$$

We will use this map to identify $G(V)$ and $G(W)$ with subgroups of $Sp(V \otimes_D W^*)$. These two subgroups are mutual commutants of each other, and is an example of a *reductive dual pair*.

2.4 Metaplectic cover

The group $Sp(V \otimes_D W^*)$ has an S^1- cover $Mp(V \otimes_D W^*)$ which is called a metaplectic group. It is known that this S^1-cover splits over the subgroups $G(V)$ and $G(W)$, except when V is an odd-dimensional quadratic space, in which case it does not split over $G(W)$. In this exceptional case, we shall simply redefine $G(W)$ to be the induced double cover, so as to simplify notation. We remark also that though the splittings (when they exist) are not necessarily unique, the precise choice of the splittings is of secondary importance in this paper.

2.5 Siegel parabolic

Assume in addition that there is a complete polarization $W = E \oplus F$, where E, F, are complementary totally isotropic subspaces of W. We will use the ε_W-Hermitian form B_W to identify F^* with E' by setting $f^*(e) = B_W(f, e)$. Observe that this induces an identification between E^* and F' given by

$$e^*(f) = \overline{f^*(e)} = \overline{B_W(f, e)} = \varepsilon_W B_W(e, f).$$

In what follows, we will use this identifications between F^* and E', and between E^* and F'.

Let

$$P = \{p \in G(W) \mid p \cdot E = E\}$$

be the Siegel parabolic subgroup of $G(W)$, and let $P = MN$ be its Langlands decomposition. To give a description of the groups M and N, we introduce some more notation.

Let $A \in \mathrm{End}_D(E)$. We will define $A^* \in \mathrm{End}_D(F)$, by setting, for all $e \in E$, $f \in F$,

$$B_W(e, A^* f) = B_W(Ae, f). \tag{1}$$

Now given $T \in \mathrm{Hom}_D(F, E)$, define $T^* \in \mathrm{Hom}_D(F, E)$ by setting, for all $f_1, f_2 \in F$,

$$B_W(f_1, T^* f_2) = \varepsilon_W B_W(Tf_1, f_2). \tag{2}$$

Given $\varepsilon = \pm 1$, set

$$\mathrm{Hom}_D(F, E)_\varepsilon = \{T \in \mathrm{Hom}_D(F, E) \mid T^* = \varepsilon T\}.$$

It is then clear that $\mathrm{Hom}_D(F, E) = \mathrm{Hom}_D(F, E)_1 \oplus \mathrm{Hom}_D(F, E)_{-1}$.

Now we have

$$M = \left\{ \begin{bmatrix} A & \\ & (A^*)^{-1} \end{bmatrix} \middle| \, A \in GL(E) \right\} \cong GL(E)$$

and

$$N = \left\{ \begin{bmatrix} 1 & X \\ & 1 \end{bmatrix} \middle| \, X^* = -\varepsilon_W X \right\} \cong \mathrm{Hom}_D(F, E)_{-\varepsilon_W}.$$

2.6 Characters of N

Given a nontrivial character $\chi : \mathfrak{k} \to \mathbb{C}^\times$ and $Y \in \mathrm{Hom}_D(E, F)_{-\varepsilon_W}$, define a character

$$\chi_Y\left(\begin{bmatrix} 1 & X \\ & 1 \end{bmatrix}\right) = \chi(\mathrm{Tr}_F(YX)).$$

Here Tr_F is the trace of $YX : F \longrightarrow F$ seen as a map between $\mathfrak{k}$-vector spaces. The map $Y \mapsto \chi_Y$ defines a group isomorphism between $\mathrm{Hom}_D(E, F)_{-\varepsilon_W}$ and $\widehat{N}$.

Observe that the adjoint action of M on N induces an action of M on $\widehat{N}$. Using the isomorphisms of $M \cong GL(E)$ and $\widehat{N} \cong \mathrm{Hom}_D(E, F)_{-\varepsilon_W}$, we can describe the action of M on $\widehat{N}$ by the formula

$$A \cdot Y = (A^*)^{-1} Y A^{-1} \quad \text{for all } A \in GL(E),\, Y \in \mathrm{Hom}_D(E, F)_{-\varepsilon_W}.$$

Given $Y \in \mathrm{Hom}_D(E, F)_{-\varepsilon_W}$ we can define a $-\epsilon_W$-Hermitian form on E, that we will also denote by Y, by setting

$$Y(e_1, e_2) = e_1^*(Ye_2) = \varepsilon_W B_W(e_1, Ye_2).$$

Hence the action of M on $\widehat{N}$ is equivalent to the action of $GL(E)$ on sesquilinear, $-\varepsilon_W$-Hermitian forms on E.

Let Ω be the set of orbits for the action of M on $\widehat{N}$. Given $Y \in \mathrm{Hom}_D(E, F)_{-\varepsilon_W}$, let $\mathcal{O} = \mathcal{O}_Y$ be its orbit under the action of $GL(E)$ and set

$$M_{\chi_Y} = \{m \in M \mid \chi_Y(m^{-1}nm) = \chi_Y(n) \text{ for all } n \in N\}.$$

Using the identification of M with $GL(E)$, and of $\widehat{N}$ with $\mathrm{Hom}_D(E, F)_{-\varepsilon_W}$, we see that

$$M_{\chi_Y} \cong \{A \in GL(E) \mid (A^*)^{-1} Y A^{-1} = Y\} = \{A \in GL(E) \mid Y = A^* Y A\}.$$

3 Oscillator representation and theta correspondence

After the preparation of the previous section, we can now consider the theta correspondence associated to the dual pair $G(V) \times G(W)$ and use it to establish certain cases of the Sakellaridis–Venkatesh conjecture for classical groups.

3.1 Oscillator representation

Fix a nontrivial unitary character χ of $\mathfrak{k}$. Associated to this character, there exists a very special representation of the metaplectic group, called the oscillator representation Π of $Mp(V \otimes_D W^*)$. On restricting this representation to $G(V) \times G(W)$, one may write

$$\Pi|_{G(W)\times G(V)} = \int_{G(W)^\wedge} \pi \otimes \Theta(\pi)\, d\mu_\theta(\pi), \tag{3}$$

as a $G(W) \times G(V)$-module, for some measure μ_θ on $G(W)^\wedge$ and where $\Theta(\pi)$ is a (possibly zero, possibly reducible) unitary representation of $G(V)$. We shall call the map Θ the L^2-theta correspondence.

3.2 Smooth vs. L^2-theta correspondence

One may consider the above restriction of the oscillator representation in the category of smooth representations (the so-called smooth theta correspondence). Namely, for $\pi \in G(W)^\wedge$, let π^∞ denote the subspace of smooth vectors of π. Then one may consider the maximal π^∞-isotypic quotient of Π^∞ (the smooth representation underlying Π), which has the form $\pi^\infty \otimes \Theta^\infty(\pi^\infty)$ for some smooth representation $\Theta^\infty(\pi^\infty)$ of $G(V)$, known as the (big) smooth theta lift of π^∞. It is known that $\Theta^\infty(\pi^\infty)$ is an admissible representation of finite length. Moreover, unless $\mathfrak{k}$ is a 2-adic field, one knows further that $\Theta^\infty(\pi^\infty)$ has a unique irreducible quotient $\theta^\infty(\pi^\infty)$ (the small smooth theta lift of π^∞); this is the so-called Howe duality conjecture. In any case, we may define $\theta^\infty(\pi^\infty)$ to be the maximal semisimple quotient of $\Theta^\infty(\pi^\infty)$.

It is natural to wonder how the L^2-theta correspondence and the smooth theta correspondence are related. One can show using the machinery developed in Bernstein's paper [2] that, in the context of (3), for μ_θ-almost all π, there is a nonzero surjective equivariant map

$$\Pi^\infty \longrightarrow \pi^\infty \otimes \Theta(\pi)^\infty.$$

Such a map necessarily factors through

$$\Pi^\infty \twoheadrightarrow \pi^\infty \otimes \Theta^\infty(\pi^\infty) \twoheadrightarrow \pi^\infty \otimes \Theta(\pi)^\infty,$$

so that one has a surjection

$$\Theta^\infty(\pi^\infty) \twoheadrightarrow \Theta(\pi)^\infty.$$

Thus, we see that $\Theta(\pi)^\infty$ is of finite length and unitarizable, so that $\Theta(\pi)^\infty$ is semisimple. Hence, we have a surjection

$$\theta^\infty(\pi^\infty) \twoheadrightarrow \Theta(\pi)^\infty.$$

Since $\theta^\infty(\pi^\infty)$ is semisimple, we deduce that

$$\Theta(\pi)^\infty \subseteq \theta^\infty(\pi^\infty),$$

so that $\Theta(\pi)$ is a direct sum of finitely many irreducible unitary representations for μ_θ-almost all π. Indeed, if $\mathfrak{k}$ is not a 2-adic field, $\Theta(\pi)$ is irreducible with

$$\Theta(\pi)^\infty = \theta^\infty(\pi^\infty)$$

for μ_θ-almost all π.

Hence, the L^2-theta correspondence gives a map

$$\Theta : G(W)^\wedge \longrightarrow R_{\geq 0}(G(V)^\wedge)$$

where $R_{\geq 0}(G(V)^\wedge)$ is the Grothendieck semigroup of unitary representations of $G(V)$ of finite length. If $\mathfrak{k}$ is not 2-adic, Θ takes value in $G(V)^\wedge \cup \{0\}$. Moreover, one has the compatibility of L^2-theta lifts (considered in this paper) with the smooth theta lifts.

3.3 Restriction to $P \times G(V)$

We may restrict Π further to $P \times G(V)$. By Mackey theory, for a unitary representation π of $G(W)$,

$$\pi|_P = \bigoplus_{\mathcal{O}_Y \in \Omega} \mathrm{Ind}_{M_{\chi_Y} N}^P W_{\chi_Y}(\pi), \tag{4}$$

where $W_{\chi_Y}(\pi)$ is an $M_{\chi_Y} N$-module such that $n \cdot \lambda = \chi_Y(n)\lambda$, for all $n \in N$, $\lambda \in W_{\chi_Y}(\pi)$.

Therefore, from (3) and (4), we have

$$\Pi = \bigoplus_{\mathcal{O}_Y \in \Omega} \int_{\widehat{G(W)}} \mathrm{Ind}_{M_{\chi_Y} N}^P W_{\chi_Y}(\pi) \otimes \Theta(\pi)\, d\mu_\theta(\pi). \tag{5}$$

3.4 The Schrödinger model

On the other hand, we may compute the restriction of Π to $P \times G(V)$ using an explicit model of Π. The complete polarization $W = E \oplus F$ induces a complete polarization

$$V \otimes_D W^* = V \otimes_D E^* \oplus V \otimes_D F^*.$$

With the identifications introduced above, $V \otimes_D F^* = \mathrm{Hom}_D(E, V)$, and the representation Π can be realized on the Hilbert space $L^2(\mathrm{Hom}_D(E, V))$; this realization of Π is called the Schrödinger model. The action of $P \times G(V)$ in this model can be described as follows.

Let $B_V^\flat : V \longrightarrow (V^*)'$ be given by

$$(w^*)(B_V^\flat v) = B_V(w, v).$$

Then the action of $P \times G(V)$ on $L^2(\mathrm{Hom}_D(E, V))$ is given by the formulas

$$\begin{bmatrix} 1 & X \\ & 1 \end{bmatrix} \cdot \phi(T) = \chi(\mathrm{Tr}_F(XT^* B_V^\flat T))\phi(T) \quad \forall X \in \mathrm{Hom}_D(F, E)_{-\varepsilon_W}, \tag{6}$$

$$\begin{bmatrix} A & \\ & (A^*)^{-1} \end{bmatrix} \cdot \phi(T) = |\det_F(A)|^{-\dim_D(V)/2} \phi(TA) \quad \forall A \in GL(E), \tag{7}$$

$$g \cdot \phi(T) = \phi(g^{-1}T) \quad \forall g \in G(V). \tag{8}$$

Let

$$\begin{aligned} \Omega_V = \{\mathcal{O}_Y \mid \mathcal{O}_Y & \text{ is open in } \mathrm{Hom}_D(E, F)_{-\varepsilon_W} \\ & \text{and } Y = T^* B_V^\flat T \text{ for some } T \in \mathrm{Hom}_D(E, V)\}. \end{aligned}$$

Given $\mathcal{O}_Y \in \Omega_V$, we will set

$$\Upsilon_Y = \{T \in \mathrm{Hom}_D(E, V) \mid T^* B_V^\flat T \in \mathcal{O}_Y\}.$$

Then

$$\bigcup_{\mathcal{O}_Y \in \Omega_V} \Upsilon_Y \subset \mathrm{Hom}_D(E, V)$$

is a dense open subset, and its complement in $\mathrm{Hom}_D(E, V)$ has measure 0. Therefore

$$L^2(\mathrm{Hom}_D(E, V)) \cong \bigoplus_{\mathcal{O}_Y \in \Omega_V} L^2(\Upsilon_Y) \tag{9}$$

and each of these spaces is clearly $P \times G(V)$-invariant, according to the formulas given in equations (6)–(8).

We want to show that the spaces $L^2(\Upsilon_Y)$ are equivalent to some induced representation for $P \times G(V)$. To do this, observe that the "geometric" part of the action of $P \times G(V)$ on $L^2(\Upsilon_Y)$ is transitive on Υ_Y. In other words, under the action of $P \times G(V)$ on $\mathrm{Hom}_D(E, V)$ given by

$$\left(\begin{bmatrix} A & X \\ & (A^*)^{-1} \end{bmatrix}, g\right) \cdot T = gTA^{-1} \quad \text{for all} \begin{bmatrix} A & X \\ & (A^*)^{-1} \end{bmatrix} \in P,\ g \in G(V),$$

and $T \in \mathrm{Hom}_D(E, V)$, each of the Υ_Y's is a single orbit. Fix $T_Y \in \Upsilon_Y$ such that $T_Y^* B_V^\flat T_Y = Y$. The stabilizer of T_Y in $P \times G(V)$ is the subgroup

$$(P \times G(V))_{T_Y} = \left\{ \left(\begin{bmatrix} A & X \\ & (A^*)^{-1} \end{bmatrix}, g \right) \in P \times G(V) \,\middle|\, gT_Y = T_Y A \right\}.$$

Let $g \in G(V)$ be such that $gT_Y = T_Y A$ for some $A \in GL(E)$. Then by the definition of $G(V)$

$$Y = T_Y^* B_V^\flat T_Y = T_Y^* g^* B_V^\flat g T_Y = A^* Y A,$$

that is, A is an element in M_{χ_Y}.

Define an equivalence relation in $\mathrm{Hom}_D(E, V)$ by setting $T \sim S$ if $T = SA$ for some $A \in M_{\chi_Y}$. Given $T \in \mathrm{Hom}_D(E, V)$ we will denote its equivalence class, under this equivalence relation, by $[T]$. Let

$$P_{M_{\chi_Y}}(\mathrm{Hom}_D(E, V)) = \{[T] \mid T \in \mathrm{Hom}_D(E, V)\}.$$

Since $G(V)$ acts by left multiplication on $\mathrm{Hom}_D(E, V)$, there is natural action of $G(V)$ on the space $P_{M_{\chi_Y}}(\mathrm{Hom}_D(E, V))$. Set

$$G(V)_{T_Y} = \{g \in G(V) \mid gT_Y = T_Y\} \text{ and}$$
$$G(V)_{[T_Y]} = \{g \in G(V) \mid g[T_Y] = [T_Y]\}.$$

Then $(P \times G(V))_{T_Y} \subset M_{\chi_Y} \times G(V)_{[T_Y]}$, and according to equations (6)–(8),

$$L^2(\Upsilon_Y) \cong \mathrm{Ind}_{(P \times G(V))_{T_Y}}^{P \times G(V)} \chi_Y \tag{10}$$

$$\cong \mathrm{Ind}_{M_{\chi_Y} N \times G(V)_{[T_Y]}}^{P \times G(V)} \mathrm{Ind}_{(P \times G(V))_{T_Y}}^{M_{\chi_Y} N \times G(V)_{[T_Y]}} \chi_Y. \tag{11}$$

Now consider the short exact sequence

$$1 \longrightarrow 1 \times G(V)_{T_Y} \longrightarrow (P \times G(V))_{T_Y} \xrightarrow{q} M_{\chi_Y} N \longrightarrow 1,$$

where q is the projection into the first component. Observe that the map q induces an isomorphism $G(V)_{T_Y} \backslash G(V)_{[T_Y]} \cong M_{\chi_Y}$. From this exact sequence and equation (11), we get that

$$\begin{aligned} L^2(\Upsilon_Y) &\cong \mathrm{Ind}_{M_{\chi_Y} N \times G(V)_{[T_Y]}}^{P \times G(V)} L^2(G(V)_{T_Y} \backslash G(V)_{[T_Y]})_{\chi_Y} \\ &\cong \mathrm{Ind}_{M_{\chi_Y} N}^{P} L^2(G(V)_{T_Y} \backslash G(V))_{\chi_Y}. \end{aligned} \tag{12}$$

The action of $M_{\chi_Y}N$ on $L^2(G(V)_{T_Y}\backslash G(V)_{[T_Y]})_{\chi_Y}$ is given (by definition) as follows: N acts by the character χ_Y, and M_{χ_Y} acts on $L^2(G(V)_{T_Y}\backslash G(V)_{[T_Y]})_{\chi_Y}$ on the left using the isomorphism $G(V)_{T_Y}\backslash G(V)_{[T_Y]} \cong M_{\chi_Y}$. Then according to equations (9) and (12)

$$L^2(\mathrm{Hom}_D(E,V)) \cong \bigoplus_{\mathcal{O}_Y \in \Omega_V} \mathrm{Ind}_{M_{\chi_Y}N}^{P} L^2(G(V)_{T_Y}\backslash G(V))_{\chi_Y}. \tag{13}$$

But now, from equations (5), (13) and the uniqueness of the decomposition of the N-spectrum, we obtain

Proposition 1. *As an $M_{\chi_Y}N \times G(V)$-module,*

$$L^2(G(V)_{T_Y}\backslash G(V))_{\chi_Y} \cong \int_{\widehat{G}(W)} W_{\chi_Y}(\pi) \otimes \Theta(\pi)\, d\mu_\theta(\pi), \tag{14}$$

Our goal now is to give a more explicit characterization of the spaces $W_{\chi_Y}(\pi)$ and the measure μ_θ appearing in this formula.

3.5 Stable range

Let (V, B_V) and (W, B_W) be as before. Assume now that there is a totally isotropic D-submodule $X \subset V$ such that $\dim_D(X) = \dim_D(W)$; in other words, the dual pair $(G(V), G(W))$ is in the *stable range*. In this case, the map

$$\Theta : \widehat{G}(W) \longrightarrow \widehat{G}(V)$$

can be understood in terms of the results of J. S. Li [18]. The measure μ_θ appearing in equation (3) is also known in this case: it is precisely the Plancherel measure of the group $G(W)$. In order to make this paper more self-contained, we will include an alternative calculation of the measure μ_θ using the so-called *mixed model* of the oscillator representation.

3.6 Mixed model

Let X, Y be totally isotropic, complementary subspaces of V such that $\dim_D(X) = \dim_D(W)$, and let $U = (X \oplus Y)^\perp$. We will use B_V to identify Y with $(X^*)'$ by setting

$$(x^*)y = B_V(x, y), \qquad \text{for all } x \in X,\ y \in Y.$$

Given $A \in GL(X)$, we can use the above identification to define an element $A^* \in GL(Y)$ in the following way: given $x \in X$ and $y \in Y$, we will set $(x^*)(A^*y) := (x^*A^*)y$, i.e., we will define $A^* \in GL(Y)$ by requiring that

$$B_V(x, A^*y) = B_V(Ax, y), \qquad \text{for all } x \in X, y \in Y.$$

Observe that the map $A \mapsto (A^*)^{-1}$ defines an isomorphism between $GL(X)$ and $GL(Y)$. Furthermore if $x \in X$, $y \in Y$ and $A \in GL(X)$, then

$$B_V(Ax, (A^*)^{-1}y) = B_V(x, y).$$

Therefore, we can define a map $GL(X) \times G(U) \hookrightarrow G(V)$ that identifies $GL(X) \times G(U)$ with the subgroup of $G(V)$ which preserves the direct sum decomposition $V = X \oplus Y \oplus U$.

Consider the polarization

$$V \otimes_D W^* = (X \otimes W^* \oplus U \otimes F^*) \bigoplus (Y \otimes W^* \oplus U \otimes E^*).$$

Then as a vector space

$$L^2(X \otimes W^* \oplus U \otimes F^*) \cong L^2(\mathrm{Hom}_D(W, X)) \otimes L^2(\mathrm{Hom}_D(E, U)). \tag{15}$$

Let $(\omega_U, L^2(\mathrm{Hom}_D(E, U)))$ be the Schrödinger model of the oscillator representation associated to the metaplectic group $\widetilde{Sp}(U \otimes_D W^*)$. We will identify the space appearing on the right-hand side of equation (15) with the space of L^2 functions from $\mathrm{Hom}_D(W, X)$ to $L^2(\mathrm{Hom}_D(E, U))$. This is the so-called *mixed model* of the oscillator representation.

The action of $G(W) \times GL(X) \times G(U)$ on this model can be described in the following way: If $T \in \mathrm{Hom}_D(W, X)$ and $S \in \mathrm{Hom}_D(E, U)$, then

$$g \cdot \phi(T)(S) = [\omega_U(g)\phi(Tg)](S) \qquad \forall g \in G(W) \tag{16}$$

$$h \cdot \phi(T)(S) = \phi(T)(h^{-1}S) \qquad \forall h \in G(U) \tag{17}$$

$$A \cdot \phi(T)(S) = |\det_X(A)|^{\dim W/2}\phi(A^{-1}T)(S) \qquad \forall A \in GL(X). \tag{18}$$

We now want to describe this space as an induced representation. To do this, observe that the set of invertible elements in $\mathrm{Hom}_D(W, X)$ forms a single orbit under the natural action of $G(W) \times GL(X)$. Furthermore this orbit is open and dense, and its complement has measure 0. Fix $T_0 \in \mathrm{Hom}_D(W, X)$ invertible, and define an ε_W-Hermitian form B_{T_0} on X, by setting

$$B_{T_0}(x_1, x_2) = B_W(T_0^{-1}x_1, T_0^{-1}x_2).$$

The group that preserves this form is precisely

$$G(X, B_{T_0}) = \{T_0 g T_0^{-1} \mid g \in G(W)\} \subset GL(X).$$

Let

$$(G(W) \times GL(X))_{T_0} = \{(g, T_0 g T_0^{-1}) \mid g \in G(W)\} \cong G(W)$$

be the stabilizer of T_0 in $G(W) \times GL(X)$. Then, according to equations (16)–(18),

$$L^2(W \otimes X) \otimes L^2(\mathrm{Hom}_D(E,U)) \cong \mathrm{Ind}_{(G(W)\times GL(X))_{T_0}}^{G(W)\times GL(X)} L^2(\mathrm{Hom}_D(E,U))$$

$$\cong \mathrm{Ind}_{G(W)\times G(X,B_{T_0})}^{G(W)\times GL(X)} \mathrm{Ind}_{(G(W)\times GL(X))_{T_0}}^{G(W)\times G(X,B_{T_0})} L^2(\mathrm{Hom}_D(E,U)).$$

Here $(G(W) \times GL(X))_{T_0}$ is acting on $L^2(\mathrm{Hom}_D(E,U))$ by taking projection into the first component, and then using the oscillator representation to define an action of $G(W)$ on $L^2(\mathrm{Hom}_D(E,U))$. But this representation is equivalent to taking projection into the second component and using the Schrödinger model of the oscillator representation of $\widetilde{Sp}(U \otimes X^*)$ (where X is equipped with the form B_{T_0}) to define an action of $G(X, B_{T_0})$ on $L^2(\mathrm{Hom}_D(T_0(E),U))$. Therefore

$$L^2(W^* \otimes X) \otimes L^2(\mathrm{Hom}_D(E,U))$$

$$\cong \mathrm{Ind}_{G(W)\times G(X,B_{T_0})}^{G(W)\times GL(X)} \mathrm{Ind}_{(G(W)\times GL(X))_{T_0}}^{G(W)\times G(X,B_{T_0})} L^2(\mathrm{Hom}_D(T_0(E),U))$$

$$\cong \mathrm{Ind}_{G(W)\times G(X,B_{T_0})}^{G(W)\times GL(X)} (\mathrm{Ind}_{(G(W)\times GL(X))_{T_0}}^{G(W)\times G(X,B_{T_0})} 1) \otimes L^2(\mathrm{Hom}_D(T_0(E),U))$$

$$\cong \mathrm{Ind}_{G(W)\times G(X,B_{T_0})}^{G(W)\times GL(X)} \int_{\widehat{G(W)}} \pi^* \otimes (\pi^{T_0} \otimes L^2(\mathrm{Hom}_D(T_0(E),U)))\, d\mu_{G(W)}(\pi)$$

$$\cong \int_{\widehat{G(W)}} \pi^* \otimes \mathrm{Ind}_{G(X,B_{T_0})}^{GL(X)} \pi^{T_0} \otimes L^2(\mathrm{Hom}_D(T_0(E),U))\, d\mu_{G(W)}(\pi). \qquad (19)$$

Here π^* is the contragredient representation of π, π^{T_0} is the representation of $G(X, B_{T_0})$ given by $\pi^{T_0}(g) = \pi(T_0^{-1} g T_0)$, for all $g \in G(X, B_{T_0})$, and $\mu_{G(W)}$ is the Plancherel measure of $G(W)$. Note that the multiplicity space of π^* in (19) is nonzero for each π in the support of $\mu_{G(W)}$, i.e., as a representation of $G(W)$, Π is weakly equivalent to the regular representation $L^2(G(W))$.

Comparing (3) with (19), we obtain

Proposition 2. *If $(G(W), G(V))$ is in the stable range, with $G(W)$ the smaller group, then in equations (3) and (14), $\mu_\theta = \mu_{G(W)}$ is the Plancherel measure of $\widehat{G}(W)$.*

3.7 *The Bessel–Plancherel theorem*

Finally, we want to identify the multiplicity space $W_{\chi_Y}(\pi)$ in (14). Note that this is purely an issue about representations of $G(W)$; a priori, it has nothing to do with theta correspondence. What we know is summarized in the following theorem.

Theorem 2 (Bessel–Plancherel Theorem). *Let (W, B_W) be an ε_W-Hermitian D-module, and assume that W has a complete polarization*

$$W = E \oplus F,$$

where E, F are totally isotropic complementary subspaces. Let

$$P = \{p \in G(W) \mid p \cdot E = E\}$$

be a Siegel parabolic subgroup of G, and let $P = MN$ be its Langlands decomposition. Given $\chi \in \widehat{N}$, let $\mathcal{O}_\chi$ be its orbit under the action of M, and let M_χ be the stabilizer of χ in M. Then

1. *For $\mu_{G(W)}$-almost all tempered representations π of $G(W)$,*

$$\pi|_P \cong \bigoplus_{\mathcal{O}_\chi \in \Omega_W} \mathrm{Ind}_{M_\chi N}^P V_\chi(\pi).$$

   Here $\mu_{G(W)}$ is the Plancherel measure of $G(W)$,

$$\Omega_W = \{\mathcal{O}_\chi \in \Omega \mid \mathcal{O}_\chi \text{ is open in } \widehat{N}\},$$

   and $V_\chi(\pi)$ is some $M_\chi N$-module such that the action of N is given by the character χ.

2. *If $\mathcal{O}_\chi \in \Omega_W$, then there is an isomorphism of $M_\chi \times G(W)$-modules:*

$$L^2(N\backslash G(W); \chi) \cong \int_{\widehat{G(W)}} V_\chi(\pi) \otimes \pi \, d\mu_{G(W)}(\pi). \tag{20}$$

   where $V_\chi(\pi)$ is the same space appearing in (1).

3. *If $\dim_D(W) = 2$, then for $\mathcal{O}_\chi \in \Omega_W$, $\dim V_\chi(\pi) < \infty$ and*

$$V_\chi(\pi) \cong Wh_\chi(\pi) = \{\lambda : \pi^\infty \longrightarrow \mathbb{C} \mid \lambda(\pi(n)v) = \chi(n)\lambda(v) \text{ for all } n \in N\}$$

   as an $M_\chi N$-module. Here π^∞ stands for the set of smooth vectors of π and the space on the RHS is the space of continuous χ-Whittaker functionals on π^∞.

4. *If $\mathfrak{k}$ is Archimedean, and M_χ is compact, then*

$$V_\chi(\pi) \subset Wh_\chi(\pi)$$

   as a dense subspace, and for any irreducible representation τ of M_χ, one has an equality of τ-isotypic parts:

$$V_\chi(\pi)[\tau] = Wh_\chi(\pi)[\tau].$$

   Moreover, this space is finite-dimensional.

Proof. Part 2 follows from an argument analogous to the proof of the Whittaker–Plancherel theorem given by Sakellaridis–Venkatesh [23, §6.3]. For the proof of part 1 observe that by the Harish-Chandra Plancherel theorem

$$L^2(G(W))|_{P\times G(W)} = \int_{\widehat{G(W)}} \pi^*|_P \otimes \pi \, d\mu_{G(W)}(\pi).$$

On the other hand,

$$\begin{aligned} L^2(G(W))|_{P\times G(W)} &= \bigoplus_{\mathcal{O}_\chi\in\Omega_W} \mathrm{Ind}_{M_\chi N}^{P} L^2(N\backslash G(W);\chi) \\ &= \bigoplus_{\mathcal{O}_\chi\in\Omega_W} \mathrm{Ind}_{M_\chi N}^{P} \int_{\widehat{G(W)}} V_\chi(\pi)\otimes\pi \, d\mu_{G(W)}(\pi) \\ &= \int_{\widehat{G(W)}} \left[\bigoplus_{\mathcal{O}_\chi\in\Omega_W} \mathrm{Ind}_{M_\chi N}^{P} V_\chi(\pi) \right] \otimes \pi \, d\mu_{G(W)}(\pi). \end{aligned}$$

Therefore

$$\pi^*|_P \cong \bigoplus_{\mathcal{O}_\chi\in\Omega_W} \mathrm{Ind}_{M_\chi N}^{P} V_\chi(\pi)$$

for $\mu_{G(W)}$-almost all π. In the Archimedean case, this result has also been proved in the thesis [10] of the second named author without the $\mu_{G(W)}$-almost all restriction, yielding an alternative proof of part 2 for the Archimedean case.

Part 3 is part of the Whittaker–Plancherel theorem, which was proved by Wallach in the Archimedean case [27], and independently by Delorme, Sakellaridis–Venkatesh and U-Liang Tang in the p-adic case [4, 5, 23, 25].

Finally, Part 4 was shown by Wallach and the second named author in [11].

We note that Theorem 2(1) is a refinement of equation (4): it implies that in (4), only the open orbits $\mathcal{O}_\chi$ in Ω contribute. Moreover, for $\chi \in \Omega_W$, the space $V_\chi(\pi)$ in Theorem 2 is the same as the space $W_\chi(\pi)$ in (4) and (14).

3.8 Spectral decomposition of generalized Stiefel manifolds

We may now assemble all the previous results together. For $\mathcal{O}_Y \in \Omega_V$, the space $G(V)_{T_Y}\backslash G(V)$ is known as a *generalized Stiefel manifold*. From equations (14) and (19), we deduce the following.

Theorem 3. *Suppose that $G(V)_{T_Y}\backslash G(V)$ is a generalized Stiefel manifold. If, in the notation of equation (20)*

$$L^2(N\backslash G(W);\chi_Y) \cong \int_{\widehat{G(W)}} W_{\chi_Y}(\pi)\otimes\pi \, d\mu_{G(W)}(\pi),$$

then

$$L^2(G(V)_{T_Y}\backslash G(V)) \cong \int_{\widehat{G(W)}} W_{\chi_Y}(\pi) \otimes \Theta(\pi)\, d\mu_{G(W)}(\pi).$$

In a certain sense, the last pair of equations says that the Plancherel measure of the generalized Stiefel manifold $G(V)_{T_Y}\backslash G(V)$ is the pushforward of the Bessel–Plancherel measure of $G(W)$ under the θ-correspondence. We note that in [15], Kobayashi has obtained an explicit description of the discrete spectrum of these generalized Stiefel manifold in the real case, in terms of $A_{\mathfrak{q}}(\lambda)$ modules.

3.9 The Sakellaridis–Venkatesh conjecture

Using the previous theorem, we can obtain certain examples of the Sakellaridis–Venkatesh conjecture:

- Taking $D = \mathfrak{k}, \mathfrak{k} \times \mathfrak{k}$ or $M_2(\mathfrak{k})$ to be a split $\mathfrak{k}$-algebra and W to be skew-hermitian with $\dim_D W = 2$, we obtain the spectral decomposition of $H\backslash G := G(V)_{T_Y}\backslash G(V)$ in terms of the Bessel–Plancherel (essentially the Whittaker–Plancherel) decomposition for G_X, where $H\backslash G$ and G_X are listed in the following table.

$X = H\backslash G$	G_X
$GL_{n-1}\backslash GL_n$	GL_2
$SO_{n-1}\backslash SO_n$	$\widetilde{SL}_2$ or SL_2
$Sp_{2n-2}\backslash Sp_{2n}$	$SO(4)$

  This establishes the cases listed in Table 1 in Theorem 1.

- Taking D to be a quadratic field extension of $\mathfrak{k}$ or the quaternion division $\mathfrak{k}$-algebra, and W to be skew-hermitian, we obtain the spectral decomposition of

$$H\backslash G = U_{n-1}\backslash U_n, \quad Sp_{n-1}(D)\backslash Sp_n(D)$$

  in terms of the Bessel–Plancherel decomposition of U_2 and $O_2(D)$. This gives non-split versions of the examples above.

3.10 Multiplicity space

In addition, the multiplicity space $W_\chi(\pi) = Wh_\chi(\pi)$ can be described in terms of the space of H-invariant (continuous) functionals on $\Theta(\pi)^\infty$. Indeed, by the smooth analog of our computation with the Schrödinger model in §3.4, one can show:

Lemma 1. *For any irreducible smooth representation σ^∞ of $G(V)$, let $\Theta^\infty(\sigma^\infty)$ denote the big (smooth) theta lift of σ^∞ to $G(W)$. Then for $\chi \in \mathcal{O}_Y \in \Omega_W \subset \widehat{N}$, there is a natural isomorphism of M_χ-modules:*

$$Wh_\chi(\Theta^\infty(\sigma^\infty)) \cong \mathrm{Hom}_{G(V)_{T_Y}}(\sigma^\infty, \mathbb{C}).$$

In the cases we are considering above, one can show that if π is an irreducible tempered representation of $G(W)$, then the small (smooth) theta lift $\sigma = \theta^\infty(\pi^\infty)$ is irreducible (even when k is 2-adic), and moreover, the big (smooth) theta lift $\Theta^\infty(\sigma^\infty)$ of σ^∞ back to $G(W)$ is irreducible and thus isomorphic to π^∞. By our discussion in §3.2, we see that for $\mu_{G(W)}$-almost all π, one has

$$\Theta(\pi)^\infty \cong \theta^\infty(\pi^\infty) = \sigma.$$

Thus the above lemma implies that

$$W_\chi(\pi) = Wh_\chi(\pi) \cong \mathrm{Hom}_H(\Theta(\pi)^\infty, \mathbb{C}).$$

This concludes the proof of the classical cases of Theorem 1.

3.11 Unstable range

Though we have assumed that $(G(W), G(V))$ is in the stable range from §3.5, it is possible to say something when one is not in the stable range as well. Namely, in §3.6, one would take X to be a maximal isotropic space in V (so $\dim X < \dim W$ here), and consider the mixed model defined on $L^2(\mathrm{Hom}_D(W, X))$ $\otimes L^2(\mathrm{Hom}_D(E, U))$. As an illustration, we note the result for the case when W is a symplectic space of dimension 2 and V is a split quadratic space of dimension 3, so that

$$G(W) \times G(V) \cong \widetilde{SL}_2 \times SO_3 \cong \widetilde{SL}_2 \times PGL_2.$$

For a nonzero $Y \in \widehat{N}$, the subgroup $G(V)_{T_Y}$ of $G(V)$ is simply a maximal torus A_Y of PGL_2.

Proposition 3. *We have*

$$L^2(G(V)_{T_Y}\backslash G(V)) = L^2(A_Y\backslash PGL_2)$$

$$\cong \int_{\widehat{G(W)}} (W_\chi(\sigma) \otimes W_{\chi_Y}(\sigma)) \otimes \Theta_\chi(\pi)\, d\mu_{G(W)}(\pi).$$

We record the following corollary which is needed in the second half of this paper:

Corollary 1. *The unitary representation $L^2(\mathfrak{sl}_2)$ associated to the adjoint action of PGL_2 on its Lie algebra $\mathfrak{sl}_2$ is weakly equivalent to the regular representation $L^2(PGL_2)$.*

Proof. Since the union of strongly regular semisimple classes are open dense in $\mathfrak{sl}_2$, we see that $L^2(\mathfrak{sl}_2)$ is weakly equivalent to $\bigoplus_A L^2(A\backslash PGL_2)$, where the sum runs over conjugacy classes of maximal tori A in PGL_2. Applying Proposition 3, one deduces that

$$\bigoplus_A L^2(A\backslash PGL_2) \cong \int_{\widehat{G(W)}} M_\chi(\pi) \otimes \Theta_\chi(\pi)\, d\mu_{G(W)}(\pi)$$

with

$$M_\chi(\pi) = W_\chi(\pi) \otimes \left(\bigoplus_A W_{\chi_A}(\pi) \right).$$

One can show that the theta correspondence with respect to χ induces a bijection

$$\Theta_\chi : \{\pi \in \widehat{G(W)} : W_\chi(\pi) \neq 0\} \longleftrightarrow \widehat{G(V)}.$$

Moreover, one can write down this bijection explicitly (in terms of the usual coordinates on the unitary duals of $\widetilde{SL}_2$ and PGL_2). From this description, one sees that

$$(\Theta_\chi)_*(\mu_{G(W)}) = \mu_{G(V)}.$$

This shows that

$$\int_{\widehat{G(W)}} M_\chi(\pi) \otimes \Theta_\chi(\pi)\, d\mu_{G(W)}(\pi) \cong \int_{\widehat{G(V)}} M_\chi(\Theta_\chi^{-1}(\sigma)) \otimes \sigma\, d\mu_{G(V)}(\sigma),$$

with $M_\chi(\Theta_\chi^{-1}(\sigma)) \neq 0$. This proves the corollary.

4 Exceptional structures and groups

The argument of the previous section can be adapted to various dual pairs in exceptional groups, thus giving rise to more exotic examples of the Sakellaridis–Venkatesh conjecture. In particular, we shall show that the spectral decomposition of $L^2(X) = L^2(H\backslash G)$ can obtained from that of $L^2(G_X)$, with X and G_X given in Table 3.

Table 3.

$X = H\backslash G$	G_X
$SO_3\backslash SL_3$	$\widetilde{SL}_3$
$Sp_6\backslash SL_6$	SL_3
$SL_3\backslash G_2$	$\widetilde{SL}_2$
$(J,\psi)\backslash G_2$	PGL_3
$SU_3\backslash Spin_7$	$(Spin_3 \times Spin_5)/\Delta\mu_2$
$G_2\backslash Spin_7$	SL_2
$G_2\backslash Spin_8$	$SL_2^3/\Delta\mu_2$
$Spin_9\backslash F_4$	PGL_2
$F_4\backslash E_6$	SL_3

The unexplained notation will be explained in due course. Comparing with the tables in [23, §15 and §16], we see that these exceptional examples, together with the classical examples treated earlier, verify the conjecture of Sakallaridis–Venkatesh for almost all the rank-1 spherical varieties (with certain desirable properties), and also some rank-2 or rank-3 ones. Indeed, they also include low rank examples of several infinite families of spherical varieties, such as $Sp_{2n}\backslash SL_{2n}$, $SO_n\backslash SL_n$ and $SU_n\backslash SO_{2n+1}$.

Though the proof will be similar in spirit to that of the previous section, we shall need to deal with the geometry of various exceptional groups, and this is ultimately based on the geometry of the (split) octonion algebra $\mathbb{O}$ and the exceptional Jordan algebra $J(\mathbb{O})$. Thus we need to recall some basic properties of $\mathbb{O}$ and its automorphism group. A good reference for the material in this section is the book [13]. One may also consult [19] and [28].

4.1 Octonions and G_2

Let $\mathfrak{k}$ be a local field of characteristic zero and let $\mathbb{O}$ denote the (8-dimensional) split octonion algebra over $\mathfrak{k}$. The octonion algebra $\mathbb{O}$ is non-commutative and non-associative. Like the quaternion algebra, it is endowed with a conjugation $x \mapsto \bar{x}$ with an associated trace map $\mathrm{Tr}(x) = x + \bar{x}$ and an associated norm map $N(x) = x \cdot \bar{x}$. It is a composition algebra, in the sense that $N(x \cdot y) = N(x) \cdot N(y)$.

A useful model for $\mathbb{O}$ is the so-called Zorn's model, which consists of 2×2-matrices

$$\begin{pmatrix} a & v \\ v' & b \end{pmatrix}, \quad \text{with } a, b \in \mathfrak{k},\ v \in V \cong k^3 \text{ and } v' \in V',$$

with V a 3-dimensional $\mathfrak{k}$-vector space with dual V'. By fixing an isomorphism $\wedge^3 V \cong \mathfrak{k}$, one deduces natural isomorphisms

$$\wedge^2 V \cong V' \quad \text{and} \quad \wedge^2 V' \cong V,$$

and let $\langle -, - \rangle$ denote the natural pairing on $V' \times V$. The multiplication on $\mathbb{O}$ is then defined by

$$\begin{pmatrix} a & v \\ v' & b \end{pmatrix} \cdot \begin{pmatrix} c & w \\ w' & d \end{pmatrix} = \begin{pmatrix} ac + \langle w', v \rangle & aw + dv + v' \wedge w' \\ cv' + bw' + v \wedge w & bd + \langle v', w \rangle \end{pmatrix}.$$

The conjugation map is

$$\begin{pmatrix} a & v \\ v' & b \end{pmatrix} \mapsto \begin{pmatrix} b & -v \\ -v' & a \end{pmatrix}$$

so that

$$Tr \begin{pmatrix} a & v \\ v' & b \end{pmatrix} = a + b \quad \text{and} \quad N \begin{pmatrix} a & v \\ v' & b \end{pmatrix} = ab - \langle v', v \rangle.$$

Any non-central element $x \in \mathbb{O}$ satisfies the quadratic polynomial $x^2 - \mathrm{Tr}(x) \cdot x + N(x) = 0$. Thus, a non-central element $x \in \mathbb{O}$ generates a quadratic $\mathfrak{k}$-subalgebra described by this quadratic polynomial. A nonzero element x has rank 2 if $N(x) \neq 0$; otherwise it has rank 1.

The automorphism group of the algebra $\mathbb{O}$ is the split exceptional group of type G_2. The group G_2 contains the subgroup $SL(V) \cong SL_3$ which fixes the diagonal elements in Zorn's model, and acts on V and V' naturally. Clearly, G_2 fixes the identity element $1 \in \mathbb{O}$, so that it acts on the subspace $\mathbb{O}_0$ of trace zero elements. The following proposition summarizes various properties of the action of G_2 on $\mathbb{O}_0$.

Proposition 4. *(i) Fix $a \in \mathfrak{k}^{\times}$, and let Ω_a denote the subset of $x \in \mathbb{O}_0$ with $N(x) = a$, then Ω_a is nonempty and G_2 acts transitively on Ω_a with stabilizer isomorphic to $SU_3(E_a)$, where $E_a = \mathfrak{k}[x]/(x^2 - a)$.*

(ii) The automorphism group G_2 acts transitively on the set Ω_0 of trace zero, rank 1 elements. For $x \in \Omega_0$, the stabilizer of the line $\mathfrak{k} \cdot x$ is a maximal parabolic subgroup $Q = L \cdot U$ with Levi factor $L \cong GL_2$ and unipotent radical U a 3-step unipotent group.

Now we note:

- When $a \in (\mathfrak{k}^{\times})^2$ in (i), the stabilizer of an element in Ω_a is isomorphic to SL_3; this explains the third entry in Table 3.
- In (ii), the 3-step filtration of U is given by

$$U \supset [U, U] \supset Z(U) \supset \{1\}$$

where $[U, U]$ is the commutator subgroup and $Z(U)$ is the center of U. Moreover,

$$\dim Z(U) = 2 \quad \text{and} \quad \dim [U, U] = 3,$$

so that $[U,U]/Z(U) \cong k$. If ψ is a nontrivial character of k, then ψ gives rise to a nontrivial character of $[U,U]$ which is fixed by the subgroup $[L,L] \cong SL_2$. Setting $J = [L,L]\cdot[U,U]$, we may extend ψ to a character of J trivially across $[L,L]$. This explains the fourth entry of Table 3.

Though the octonionic multiplication is neither commutative or associative, the trace form satisfies

$$\mathrm{Tr}((x\cdot y)\cdot z) = \mathrm{Tr}(x\cdot(y\cdot z)),$$

(so there is no ambiguity in denoting this element of k by $\mathrm{Tr}(x\cdot y\cdot z)$) and G_2 is precisely the subgroup of $SO(\mathbb{O},N)$ satisfying

$$\mathrm{Tr}((gx)\cdot(gy)\cdot(gy)) = \mathrm{Tr}(x\cdot y\cdot z) \quad \text{for all } x,y,z\in\mathbb{O}.$$

4.2 Exceptional Jordan algebra and F_4

Let $J = J(\mathbb{O})$ denote the 27-dimensional vector space consisting of all 3×3 Hermitian matrices with entries in $\mathbb{O}$. Then a typical element in J has the form

$$\alpha = \begin{pmatrix} a & z & \bar{y} \\ \bar{z} & b & x \\ y & \bar{x} & c \end{pmatrix}, \quad \text{with } a,b,c\in k \text{ and } x,y,z\in\mathbb{O}.$$

The set J is endowed with a multiplication

$$\alpha\circ\beta = \frac{1}{2}\cdot(\alpha\beta+\beta\alpha)$$

where the multiplication on the RHS refers to usual matrix multiplication. With this multiplication, J is the exceptional Jordan algebra.

The algebra J carries a natural cubic form $d = \det$ given by the determinant map on J, and a natural linear form tr given by the trace map. Moreover, every element $\alpha\in J$ satisfies a cubic polynomial $X^3 - \mathrm{tr}(\alpha)X^2 + s(\alpha)X - d(\alpha)$. One says that $\alpha\neq 0$ has rank 3 if $d(\alpha)\neq 0$, rank 2 if $d(\alpha) = 0$ but $s(\alpha)\neq 0$, and rank 1 if $d(\alpha) = s(\alpha) = 0$. For example, $\alpha\in J$ has rank 1 if and only if its entries satisfy

$$N(x) = bc,\ N(y) = ca,\ N(z) = ab,\ xy = c\bar{z},\ yz = a\bar{x},\ zx = b\bar{y}.$$

More generally, the above discussion holds if one uses any composition k-algebra in place of $\mathbb{O}$. Thus, if $B = k$, a quadratic algebra K, a quaternion algebra D or the octonion algebra $\mathbb{O}$, one has the Jordan algebra $J(B)$. One may consider the group $\mathrm{Aut}(J(B),\det)$ of invertible linear maps on $J(B)$ which fixes the cubic form det, and its subgroup $\mathrm{Aut}(J,\det,e)$ which fixes an element e with $\det(e)\neq 0$. For the various B's, these groups are listed in Table 4.

Table 4.

B	$\mathrm{Aut}(J(B), \det)$	$\mathrm{Aut}(J(B), \det, e)$
$\Bbbk$	SL_3	SO_3
K	$SL_3(K)/\Delta\mu_3$	SL_3
D	$SL_3(D)/\mu_2 = SL_6/\mu_2$	$PGSp_6$
$\mathbb{O}$	E_6	F_4

Proposition 5. *(i) For any $a \in \Bbbk^\times$, the group $\mathrm{Aut}(J(B), \det)$ acts transitively on the set of all $e \in J$ with $\det(e) = a$, with stabilizer group $\mathrm{Aut}(J(B), \det, e)$ described in Table 4. If e is the unit element of $J(B)$, then $\mathrm{Aut}(J(B), \det, e)$ is the automorphism group of the Jordan algebra $J(B)$.*

(ii) The group $F_4 = \mathrm{Aut}(J(\mathbb{O}))$ acts transitively on the set of rank 1 elements in $J(\mathbb{O})$ of trace $a \neq 0$. The stabilizer of a point is isomorphic to the group $Spin_9$ of type B_4.

In particular, the proposition explains the first, second, eighth and nineth entries of Table 3.

4.3 *Triality and $Spin_8$*

An element $\alpha \in J = J(\mathbb{O})$ of rank 3 generates a commutative separable cubic subalgebra $\Bbbk(\alpha) \subset J$. For any such cubic F-algebra E, one may consider the set Ω_E of algebra embeddings $E \hookrightarrow J$. Then one has

Proposition 6. *(i) The set Ω_E is nonempty and the group F_4 acts transitively on Ω_E.*

(ii) The stabilizer of a point in Ω_E is isomorphic to the quasi-split simply-connected group $Spin_8^E$ of absolute type D_4.

(iii) Fix an embedding $j : E \hookrightarrow J$ and let $E^\perp$ denote the orthogonal complement of the image of E with respect to the symmetric bilinear form $(\alpha, \beta) = \mathrm{Tr}(\alpha \circ \beta)$. The action of the stabilizer $Spin_8^E$ of j on $E^\perp$ is the 24-dimensional spin representation, which on extending scalars to $\overline{\Bbbk}$, is the direct sum of the three 8-dimensional irreducible representations of $Spin_8(\overline{\Bbbk})$ whose highest weights correspond to the 3 satellite vertices in the Dynkin diagram of type D_4.

As an example, suppose that $E = \Bbbk \times \Bbbk \times \Bbbk$, and we fix the natural embedding $E \hookrightarrow J$ whose image is the subspace of diagonal elements in J. Then $E^\perp$ is naturally $\mathbb{O} \oplus \mathbb{O} \oplus \mathbb{O}$, and the split group $Spin_8$ acts on this, preserving each copy of $\mathbb{O}$. This gives an injective homomorphism

$$\rho : Spin_8 \longrightarrow SO(\mathbb{O}, N) \times SO(\mathbb{O}, N) \times SO(\mathbb{O}, N)$$

whose image is given by

$$Spin_8 \cong \{g = (g_1, g_2, g_3) \mid \mathrm{Tr}((g_1x)\cdot(g_2y)\cdot(g_3z)) = \mathrm{Tr}(x\cdot y\cdot z) \text{ for all } x, y, z \in \mathbb{O}\}.$$

From this description, one sees that there is an action of $\mathbb{Z}/3\mathbb{Z}$ on $Spin_8$ given by the cyclic permutation of the components of g, and the subgroup fixed by this action is precisely

$$G_2 = Spin_8^{\mathbb{Z}/3\mathbb{Z}}.$$

This explains the 7th entry of Table 3.

More generally , the stabilizer of a triple $(x, y, z) \in \mathbb{O}^3$ with $(x\cdot y)\cdot z \in \Bbbk^\times$ is a subgroup of $Spin_8$ isomorphic to G_2 (see [28]). For example, the stabilizer in $Spin_8$ of the vector $(1, 0, 0) \in \mathbb{O}^3$ is isomorphic to the group $Spin_7$ which acts naturally on $\mathbb{O}_0 \oplus \mathbb{O} \oplus \mathbb{O}$. The action of $Spin_7$ on $\mathbb{O}_0$ is via the standard representation of SO_7, whereas its action on the other two copies of $\mathbb{O}$ is via the Spin representation. From the discussion above, we see that the stabilizer in $Spin_7$ of $(x, \bar{x}) \in \mathbb{O}^2$, with $N(x) \neq 0$, is isomorphic to the group G_2. In particular, this explains the 6th entry of Table 3.

On the other hand, the stabilizer in $Spin_8$ of a triple $(x, y, z) \in \mathbb{O}^3$ with $(x\cdot y)\cdot z \notin \Bbbk^\times$ is isomorphic to $SU_3 \subset G_2 \subset Spin_7 \subset Spin_8$ (see [28]). For example, if one takes $x = y = 1 \in \mathbb{O}$ and $z \notin \Bbbk$, then $K = \Bbbk[z]$ is an étale quadratic subalgebra of $\mathbb{O}$ and it follows by Proposition 4 that the stabilizer of $(1, 1, z)$ is isomorphic to $SU_3(K) \subset Spin_7$. This explains the 5th entry in Table 3.

By the above discussion, it is not difficult to show the following.

Proposition 7. *(i) The group $Spin_8$ acts transitively on the set of rank* 1 *elements*

$$\alpha = \begin{pmatrix} a & z & \bar{y} \\ \bar{z} & b & x \\ y & \bar{x} & c \end{pmatrix} \in J(\mathbb{O})$$

with diagonal part $(a, b, c) \in \Bbbk^\times \times \Bbbk^\times \times \Bbbk^\times$ fixed. Moreover, the stabilizer of a point is isomorphic to G_2.

(ii) The group $Spin_7$ acts transitively on the set of rank one elements $\alpha \in J(\mathbb{O})$ as in (i) above, with $\mathrm{Tr}(x) = 0$.

4.4 $SL_3\backslash G_2$ and $G_2\backslash Spin_7$

From the discussion above, we see that there are isomorphisms of homogeneous varieties

$$SL_3\backslash G_2 \cong SO_6\backslash SO_7 \quad \text{and} \quad G_2\backslash Spin_7 \cong Spin_7\backslash Spin_8 \cong SO_7\backslash SO_8.$$

Since we have already determined the spectral decomposition of $L^2(SO_6\backslash SO_7)$ and $L^2(SO_7\backslash SO_8)$ in terms of the spectral decomposition of $L^2(\widetilde{SL}_2)$ and $L^2(SL_2)$ respectively, we obtain the desired description for $SL_3\backslash G_2$ and $G_2\backslash Spin_7$. We note that in [16], Kobayashi used the same observation to deduce the Plancherel theorem of these spherical varieties from the Plancherel theorem for the corresponding symmetric spaces given above in the real case. He also gave an explicit description of the branching from SO_7 to G_2 (for representations of SO_7 occurring in $L^2(SO_6\backslash SO_7)$) in the real case; the p-adic case of this branching is shown in [8].

The rest of the paper is devoted to the remaining cases in Table 3.

5 Exceptional dual pairs

In this section, we introduce some exceptional dual pairs contained in the adjoint groups of type F_4, E_6, E_7 and E_8. We begin with a uniform construction of the exceptional Lie algebras of the various exceptional groups introduced above. This construction can be found in [22] and will be useful for exhibiting various reductive dual pairs. The reader may consult [19, 22, 24] and [28] for the material of this section.

5.1 Exceptional Lie algebras

Consider the chain of Jordan algebras

$$\mathfrak{k} \subset \mathfrak{k} \times \mathfrak{k} \subset E \subset J(\mathfrak{k}) \subset J(K) \subset J(D) \subset J(\mathbb{O})$$

where E is a cubic $\mathfrak{k}$-algebra, K a quadratic $\mathfrak{k}$-algebra and D a quaternion $\mathfrak{k}$-algebra, and one has the containment $\mathfrak{k}\times\mathfrak{k} \subset E$ only when $E = \mathfrak{k}\times K$ is not a field. Denoting such an algebra by $\mathcal{R}$, the determinant map det of $J(\mathbb{O})$ restricts to give a cubic form on $\mathcal{R}$. Now set

$$\mathfrak{s}_{\mathcal{R}} = \mathfrak{sl}_3 \oplus \mathfrak{m}_{\mathcal{R}} \oplus (\mathfrak{k}^3 \otimes \mathcal{R}) \oplus (\mathfrak{k}^3 \otimes \mathcal{R})', \tag{21}$$

with

$$\mathfrak{m}_{\mathcal{R}} = \mathrm{Lie}(\mathrm{Aut}(\mathcal{R}, \det)).$$

One can define a Lie algebra structure on $\mathfrak{s}_{\mathcal{R}}$ [22] whose type is given by the following table.

$\mathcal{R}$	$\mathfrak{m}_{\mathcal{R}}$	$\mathfrak{s}_{\mathcal{R}}$
k	0	$\mathfrak{g}_2$
$k \times k$	k	$\mathfrak{b}_3$
E	E_0	$\mathfrak{d}_4$
$J(k)$	$\mathfrak{sl}_3$	$\mathfrak{f}_4$
$J(K)$	$\mathfrak{sl}_3(K)$	$\mathfrak{e}_6$
$J(D)$	$\mathfrak{sl}_6$	$\mathfrak{e}_7$
$J(\mathbb{O})$	$\mathfrak{e}_6$	$\mathfrak{e}_8$

We denote the corresponding adjoint group with Lie algebra $\mathfrak{s}_{\mathcal{R}}$ by $S_{\mathcal{R}}$, or simply by S if $\mathcal{R}$ is fixed and understood.

Let $\{e_1, e_2, e_3\}$ be the standard basis of k^3 with dual basis $\{e_i'\}$. The subalgebra of $\mathfrak{sl}_3$ stabilizing the lines ke_i is the diagonal torus $\mathfrak{t}$. The nonzero weights under the adjoint action of $\mathfrak{t}$ on $\mathfrak{s}_{\mathcal{R}}$ form a root system of type G_2. The long root spaces are of dimension-1 and are precisely the root spaces of $\mathfrak{sl}_3$, i.e., the spaces spanned by $e_i' \otimes e_j$. We shall label these long roots by β, β_0 and $\beta_0 - \beta$, with corresponding 1-parameter subgroups

$$u_\beta(x) = \begin{pmatrix} 1 & x & 0 \\ & 1 & 0 \\ & & 1 \end{pmatrix}, \quad u_{\beta_0}(x) = \begin{pmatrix} 1 & 0 & x \\ & 1 & 0 \\ & & 1 \end{pmatrix}, \quad u_{\beta_0-\beta}(x) = \begin{pmatrix} 1 & 0 & 0 \\ & 1 & x \\ & & 1 \end{pmatrix}.$$

We also let

$$w_\beta = \begin{pmatrix} 0 & 1 & 0 \\ -1 & 0 & 0 \\ 0 & 0 & 1 \end{pmatrix}$$

denote the Weyl group element associated to β. The short root spaces, on the other hand, are $e_i \otimes \mathcal{R}$ and $e_i' \otimes \mathcal{R}'$ and are thus identifiable with $\mathcal{R}$.

5.2 Exceptional dual pairs

We can now exhibit two families of dual pairs in $S_{\mathcal{R}}$.

- From (21), one has

$$\mathfrak{sl}_3 \oplus \mathfrak{m}_{\mathcal{R}} \subset \mathfrak{s}_{\mathcal{R}}.$$

This gives a family of dual pairs

$$SL_3 \times \mathrm{Aut}(\mathcal{R}, \det) \longrightarrow S_{\mathcal{R}}. \tag{22}$$

We shall only be interested in these dual pairs when $\mathcal{R} = J(B)$.

- For a pair of Jordan algebras $\mathcal{R}_0 \subset \mathcal{R}$, we have $\mathfrak{s}_{\mathcal{R}_0} \subset \mathfrak{s}_{\mathcal{R}}$ which gives a subgroup $G_{\mathcal{R}_0} \subset S_{\mathcal{R}}$, where $G_{\mathcal{R}_0}$ is isogenous to $S_{\mathcal{R}_0}$. If $G'_{\mathcal{R}_0,\mathcal{R}} = \mathrm{Aut}(\mathcal{R}, \mathcal{R}_0)$, then one has a second family of dual pairs

$$G_{\mathcal{R}_0} \times G'_{\mathcal{R}_0,\mathcal{R}} \longrightarrow S_{\mathcal{R}}. \tag{23}$$

With $\mathcal{R}_0 \subset \mathcal{R}$ fixed, we shall simply write $G \times G'$ for this dual pair. For the various pairs $\mathcal{R}_0 \subset \mathcal{R}$ of interest here, we tabulate the associated dual pairs in the table below.

$\mathcal{R}_0 \subset \mathcal{R}$	$G \times G'$
$\mathfrak{k} \subset J(K)$	$G_2 \times PGL_3$
$\mathfrak{k} \times \mathfrak{k} \subset J(D)$	$Spin_7 \times (Spin_3 \times Spin_5)/\Delta\mu_2$
$E \subset J(D)$	$Spin_8 \times SL_2(E)/\Delta\mu_2$
$J(\mathfrak{k}) \subset J(D)$	$F_4 \times PGL_2$

Observe that in the language of Table 3, with $X = H\backslash G$, the dual pairs described above are precisely $G_X \times G$.

5.3 *Heisenberg parabolic*

The presentation (21) also allows one to describe certain parabolic subalgebras of $\mathfrak{s}_{\mathcal{R}}$. If we consider the adjoint action of

$$t = \mathrm{diag}(1, 0, -1) \in \mathfrak{sl}_3$$

on $\mathfrak{s}$, we obtain a grading $\mathfrak{s} = \oplus_i \mathfrak{s}[i]$ by the eigenvalues of t. Then

$$\begin{cases} \mathfrak{s}[0] = \mathfrak{t} \oplus \mathfrak{m} \oplus (e_2 \otimes \mathcal{R}) \oplus (e'_2 \otimes \mathcal{R}') \\ \mathfrak{s}[1] = \mathfrak{k}e'_2 \otimes e_1 \oplus (e_1 \otimes \mathcal{R}) \oplus (e'_3 \otimes \mathcal{R}') \oplus \mathfrak{k}e'_3 \otimes e_2 \\ \mathfrak{s}[2] = \mathfrak{k}e'_3 \otimes e_1, \end{cases}$$

and $\mathfrak{p} = \oplus_{i \geq 0} \mathfrak{s}[i]$ is a Heisenberg parabolic subalgebra.

We denote the corresponding Heisenberg parabolic subgroup by $P_S = M_S \cdot N_S$. In particular, its unipotent radical is a Heisenberg group with 1-dimensional center $Z_S \cong u_{\beta_0}(\mathfrak{k}) \cong \mathfrak{s}[2]$ and

$$N_S/Z_S \cong \mathfrak{s}[1] = \mathfrak{k} \oplus \mathcal{R} \oplus \mathcal{R}' \oplus \mathfrak{k}.$$

The semisimple type of its Levi factor M_S is given in the table below.

S	M_S
F_4	C_3
E_6	A_5
E_7	D_6
E_8	E_7

The Lie bracket defines an alternating form on N_S/Z_S which is fixed by $P_S^1 = [P_S, P_S]$. This gives an embedding

$$P_S^1 = M_S^1 \cdot N_s \hookrightarrow Sp(N_S/Z_S) \ltimes N_S.$$

5.4 Intersection with dual pairs

For a pair $\mathcal{R}_0 \subset \mathcal{R}$, with associated dual pair given in (23), it follows by construction that

$$(G_{\mathcal{R}_0} \times G'_{\mathcal{R}_0,\mathcal{R}}) \cap P_S = P \times G'_{\mathcal{R}_0,\mathcal{R}},$$

where P is the Heisenberg parabolic subgroup of $G_{\mathcal{R}_0}$. On the other hand, for the family of dual pairs given in (22),

$$(SL_3 \times \mathrm{Aut}(\mathcal{R}, \det)) \cap P_S = B \times \mathrm{Aut}(\mathcal{R}, \det)$$

where B is a Borel subgroup of SL_3.

5.5 Siegel parabolic

The group S of type E_6 or E_7 has a Siegel parabolic subgroup $Q_S = L_S \cdot U_S$ whose unipotent radical U_S is abelian; we call this a Siegel parabolic subgroup. The semisimple type of L_S and the structure of U_S as an L_S-module is summarized in the following table.

S	L_S	U_S	U_S as L_S-module
E_6	D_5	$\mathbb{O} \oplus \mathbb{O}$	half-spin representation of dimension 16
E_7	E_6	$J(\mathbb{O})$	minuscule representation of dimension 27

Let $\Omega_Q \subset \overline{U}_S$ be the orbit of a highest weight vector in $\overline{U}_S$. The following proposition describes the set Ω_Q:

Proposition 8. *(i) If S is of type E_6, then*

$$\Omega_Q = \{(x, y) \in \mathbb{O}^2 : N(x) = N(y) = 0 = x \cdot \bar{y}\}.$$

(ii) If S is of type E_7, then

$$\Omega_Q = \{\alpha \in J : \operatorname{rank}(\alpha) = 1\}.$$

5.6 Intersection with dual pairs

With $\mathcal{R}_0 \subset \mathcal{R}$ fixed, with associated dual pair $G \times G'$ as given in (23), one may choose Q_S so that

$$(G \times G') \cap Q_S = G \times Q_0$$

with $Q_0 = L_0 \cdot U_0$ a Siegel parabolic subgroup of G', so that U_0 is abelian. The group Q_0 and the embedding $U_0 \subset U_S$ can be described by the following table.

G'	Q_0	$U_0 \subset U_S$
PGL_3	maximal parabolic	$\mathfrak{k}^2 \subset \mathbb{O}^2$
$(Spin_3 \times Spin_5)/\Delta\mu_2$	(Borel)×(Siegel parabolic)	$\mathfrak{k} \oplus Sym^2(\mathfrak{k}) \subset J(\mathbb{O})$
$SL_2(E)/\Delta\mu_2$	Borel	$E \subset J(\mathbb{O})$
PGL_2	Borel	$\mathfrak{k} \subset J(\mathbb{O})$

Identifying the opposite unipotent radical $\bar{U}_0$ with the dual space of U_0 using the Killing form, one has a natural projection

$$\tau : \overline{U}_S \longrightarrow \overline{U}_0.$$

This is simply given by the projection from $\overline{U}_S$ to $\overline{U}_0$ along $U_0^{\perp}$.

6 Generic orbits

In this section, we consider an orbit problem which will be important for our applications. Namely, with the notation at the end of the last section, we have an action of $L_0 \times G$ on the set $\Omega_Q \subset \overline{U}_S$. We would like to determine the generic orbits of this action. For simplicity, we shall consider the case when $S = E_6$ and E_7 separately.

6.1 Dual pair in E_6

Suppose first that $S = E_6$ so that $G' \times G = PGL_3 \times G_2$. In this case, the natural $L \times G_2$-equivariant projection $\tau : \bar{U}_S \longrightarrow \bar{U}_0$ is given by

$$\tau(x, y) = (\mathrm{Tr}(x), \mathrm{Tr}(y)).$$

The nonzero elements in $\bar{U}_0 \cong k^2$ are in one orbit of L_0; we fix a representative $(0, 1) \in k^2$ and note that its stabilizer in L_0 is the "mirabolic" subgroup P_{L_0} of $L_0 \cong GL_2$. Then the fiber over $(0, 1)$ is given by

$$\{(x, y) \in \mathbb{O}^2 : N(x) = N(y) = \mathrm{Tr}(x) = 0,\ \mathrm{Tr}(y) = 1,\ x \cdot \bar{y} = 0\},$$

and carries a natural action of $P_{L_0} \times G_2$. We note:

Lemma 2. *(i) The group G_2 acts transitively on the fiber $\tau^{-1}(0, 1)$ and the stabilizer of a point (x_0, y_0) is isomorphic to the subgroup $[L, L] \cdot Z(U) \subset J$.*

(ii) If we consider the subset $\{(x_0, y_0 + \lambda x_0) : \lambda \in k\} \subset \tau^{-1}(0, 1)$, then the subgroup of $P_{L_0} \times G_2$ stabilizing this subset is isomorphic to

$$(P_{L_0} \times L \cdot [U, U])^0 = \{(h, g \cdot u) : \det h = \det g\}.$$

The action of the element

$$\begin{pmatrix} a & b \\ 0 & 1 \end{pmatrix} \times g \cdot u \in (P_{L_0} \times L \cdot [U, U])^0$$

is by

$$(x_0, y_0 + \lambda x_0) \mapsto (x_0, y_0 + a^{-1} \cdot (\lambda + b - p(u))x_0)$$

where $p : J \longrightarrow k \cong J/[L, L] \cdot Z(U)$ is the natural projection. Thus, there is a unique generic $L_0 \times G_2$ orbit on Ω_Q given by

$$(L_0 \times G_2) \times_{(P_{L_0} \times L \cdot [U,U])^0} k.$$

6.2 Dual pairs in E_7

Now suppose that $S = E_7$. As above, we first determine the generic L_0-orbits on $\bar{U}_0$. For each generic L_0-orbit in $\bar{U}_0$, let us take a representative χ and let Z_χ denote its stabilizer in L_0. Then the fiber $\tau^{-1}(\chi)$ is preserved by $Z_\chi \times G$. In each case, it follows by Prop. 5(ii) and Prop. 7 that G acts transitively on $\tau^{-1}(\chi)$. Denote the stabilizer in G of $\widetilde{\chi} \in \tau^{-1}(\chi)$ by H_χ. Then under the action of $Z_\chi \times G$, the stabilizer group $\widetilde{H}_\chi$ of $\widetilde{\chi}$ sits in a short exact sequence

$$1 \longrightarrow H_\chi \longrightarrow \tilde{H}_\chi \xrightarrow[p]{} Z_\chi \longrightarrow 1.$$

Thus, the generic $L_0 \times G$-orbits are given by the disjoint union

$$\bigcup_{\text{generic } \chi} (Z_\chi \times G) \times_{\widetilde{H}_\chi} \widetilde{\chi},$$

where the union runs over the generic L_0-orbits on $\bar{U}_0$ and $\widetilde{\chi}$ is an element in $\tau^{-1}(\chi)$ with stabilizer $\widetilde{H}_\chi$. We summarize this discussion in the following table.

G	generic L_0-orbits	$\tau^{-1}(\chi)$	Z_χ	H_χ
F_4	singleton	$\alpha \in J(\mathbb{O})$ rank 1, trace 1	trivial	$Spin_9$
$Spin_8$	(a, b, c) $\in (\mathfrak{k}^\times/\mathfrak{k}^{\times 2})^3/\Delta\mathfrak{k}^\times$	$\alpha \in J(\mathbb{O})$ rank 1, trace 1	center of G' $= \mu_2 \times \mu_2$	G_2
$Spin_7$	$A = \text{diag}(b, c)$ $\in Sym^2\mathfrak{k}^2/GL_2(\mathfrak{k})$	$\alpha \in J(\mathbb{O})$ diagonal$= (1, b, c)$ $\text{Tr}(z) = 0$	O_2	SU_3

7 Minimal representation

In this section, we introduce the (unitary) minimal representation Π of S and describe some models for Π. Note that when $S = F_4$, Π is actually a representation of the double cover of F_4. When S is of type E, then Π is a representation of S.

7.1 Schrödinger model

Because the groups $S = E_6$ and E_7 have a Siegel parabolic subgroup, there is an analog of the Schrödinger model for the minimal representation Π of S. By [6], the representation Π can be realized on the space $L^2(\Omega_Q, \mu_Q)$ of square-integrable functions on Ω_Q with respect to a L_S-equivariant measure μ_Q on Ω_Q. This is analogous to the Schrödinger model of the Weil representation. In particular, we have the following action of Q_S on Π:

$$\begin{cases} (l \cdot f)(\chi) = \delta_{Q_S}(l)^r \cdot f(l^{-1} \cdot \chi) \\ (u \cdot f)(\chi) = \chi(u) \cdot f(\chi), \end{cases}$$

where $r = 1/4$ (resp. $2/9$) if S is of type E_6 (resp. E_7).

7.2 Mixed model

For general $S = S_{\mathcal{R}}$, one has the analog of the mixed model, on which the action of the Heisenberg group P_S is quite transparent. Recall that $N_S/Z_S = \mathfrak{k} \oplus \mathcal{R} \oplus \mathcal{R}' \oplus \mathfrak{k}$ and one has an embedding

$$P_S^1 = [P_S, P_S] \hookrightarrow \mathrm{Sp}(N_S/Z_S) \ltimes N_S.$$

Then by [14], the mixed model of the minimal representation is realized on the Hilbert space

$$\mathrm{Ind}_{P_S^1}^{P_S} L^2(\mathcal{R}' \oplus \mathfrak{k}') \cong L^2(\mathfrak{k}^\times \oplus \mathcal{R} \oplus \mathfrak{k}),$$

where the action of P_S^1 on $L^2(\mathcal{R} \oplus \mathfrak{k})$ is via the Heisenberg–Weil representation (associated to any fixed additive character ψ of $\mathfrak{k}$). The explicit formula can be found in [22, Prop. 43].

In fact, one can describe the full action of S on Π by giving the action of an extra Weyl group element. More precisely, if w_β is the standard Weyl group element in SL_3 associated to the root β (see §5.1), then by [22, Prop. 47], one has

$$(w_\beta \cdot f)(t, x, a) = \psi(\det(x)/a) \cdot f(-a/t, x, -a).$$

Since S is generated by P_S and the element w_β, this completely determines the representation Π.

For example, one may work out the action of an element $u_{-\beta}(b) = w_\beta u_\beta(b) w_\beta^{-1}$ (see §5.1). A short computation gives

$$(u_{-\beta}(b) \cdot f)(t, x, a) = \psi\left(\frac{b\det(x)}{a - t^2}\right) \cdot f(t - \frac{ab}{t}, a - \frac{a^2 b}{t^2}, x).$$

If f is continuous, then the above formula gives

$$(u_{-\beta}(b) \cdot f)(1, x, 0) = \psi(-b\det(x)) \cdot f(1, x, 0). \tag{24}$$

This formula will be useful in the last section.

8 Exceptional theta correspondences: $G \times G'$

Now we may study the restriction of the minimal representation Π to the dual pairs introduced earlier. In this section, we shall treat the family of dual pairs $G \times G'$ given in (23). For simplicity, we shall consider the case when $S = E_6$ and E_7 separately.

8.1 Restriction to $G \times G' \subset E_7$

Suppose first that S is of type E_7, so that Ω_Q is the set of rank-1 elements in $J = J(\mathbb{O})$. Consider the Schrödinger model for Π. On restricting Π to $Q_0 \times G$, we have the following formulae:

$$\begin{cases} (g \cdot f)(\alpha) = f(g^{-1} \cdot \alpha) & \text{for } g \in G; \\ (u(a) \cdot f)(\alpha) = \psi(\mathrm{Tr}(a \cdot \alpha)) \cdot f(\alpha) & \text{for } u(a) \in U_0; \\ (l \cdot f)(\alpha) = |\det(l)|^s \cdot f(l^{-1} \cdot \alpha) & \text{for } l \in L_0, \end{cases}$$

where s is a real number whose precise value will not be important to us here.

From our description of generic $L_0 \times G$-orbits given in §6.2, we deduce as in the derivation of (12) that as a $Q_0 \times G$-module,

$$\Pi \cong \bigoplus_{\chi \text{ generic}} \mathrm{Ind}_{U_0 \times \widetilde{H}_\chi}^{Q_0 \times G} \chi \otimes 1 \cong \bigoplus_{\chi \text{ generic}} \mathrm{Ind}_{Z_\chi \cdot U_0}^{Q_0} L^2(H_\chi \backslash G). \tag{25}$$

Here, G and $Z_\chi \cong \widetilde{H}_\chi / H_\chi$ act on $L^2(H_\chi \backslash G)$ by right and left translation respectively, and U_0 acts by χ.

8.2 Abstract decomposition

On the other hand, there is an abstract direct integral decomposition

$$\Pi = \int_{\widehat{G'}} \pi \otimes \Theta(\pi) \, d\nu_\Theta(\pi).$$

Restricting to Q_0, we may write

$$\pi|_{Q_0} \cong \bigoplus_{\chi} \mathrm{Ind}_{Z_\chi \cdot U_0}^{Q_0} W_\chi(\pi)$$

for some $Z_\chi \cdot U_0$-module $W_\chi(\pi)$ with U_0 acting via χ. Thus,

$$\Pi \cong \bigoplus_{\chi} \int_{\widehat{G'}} \mathrm{Ind}_{Z_\chi \cdot U_0}^{Q_0} W_\chi(\pi) \otimes \Theta(\pi) \, d\nu_\Theta(\pi). \tag{26}$$

8.3 Comparison

Comparing (25) and (26), we deduce that there is an isomorphism of G-modules:

$$L^2(H_\chi \backslash G) \cong \int_{\widehat{G'}} W_\chi(\pi) \otimes \Theta(\pi)\, d\nu_\Theta(\pi). \tag{27}$$

If G' is isogenous to a product of SL_2, the space $W_\chi(\pi) = Wh_\chi(\pi)$ has been determined in Theorem 2(3) and is at most 1-dimensional. If G' is $(Spin_3 \times Spin_5)/\Delta\mu_2$, Theorem 2(1, 2, 4), still gives some partial results on $W_\chi(\pi)$.

8.4 Mixed model

To explain the measure $d\nu_\Theta(\pi)$, we consider the mixed model of Π restricted to $P \times G'$. Since

$$N/Z_S = \mathfrak{k} \oplus \mathcal{R}_0 \oplus \mathcal{R}_0' \oplus \mathfrak{k} \subset N_S/Z_S.$$

Under its adjoint action on $\mathcal{R} \oplus \mathfrak{k}$, G' fixes $\mathcal{R}_0 \oplus \mathfrak{k}$ pointwise, and its action on $\mathcal{R}_0^\perp$ is described in the following table.

G'	$\mathcal{R}_0$	$\mathcal{R}_0^\perp$
PGL_2	$J(\mathfrak{k})$	$adjoint^{\oplus 3}$
$SL_2^3/\Delta\mu_2$	$\mathfrak{k}^3$	$\oplus_{i=1}^3 std_i \otimes std_{i+1}^\vee$
$(Spin_3 \times Spin_5)/\Delta\mu_2$	$\mathfrak{k} \times \mathfrak{k}$	$(Spin \otimes Spin) \oplus (1 \otimes std)$

Thus as a representation of G', we have

$$\Pi \cong L^2(\mathfrak{k}^\times) \otimes L^2(\mathcal{R}_0 \oplus \mathfrak{k}) \otimes L^2(\mathcal{R}_0^\perp)$$

where G' acts only on $L^2(\mathcal{R}_0^\perp)$ and the action is geometric. Thus, Π is weakly equivalent to $L^2(\mathcal{R}_0^\perp)$ as a representation of G'. By our description of the G'-module $\mathcal{R}_0^\perp$, we have:

Lemma 3. *(i) If $G' = PGL_2$ or $SL_2^3/\Delta\mu_2$, the representation $L^2(\mathcal{R}_0^\perp)$ (and hence Π) is weakly equivalent to the regular representation $L^2(G')$.*

(ii) If $G' = (Spin_3 \times Spin_5)/\Delta\mu_2$, the representation $L^2(\mathcal{R}_0^\perp)$ (and hence Π) is weakly contained in the regular representation $L^2(G')$.

Proof. (i) When $G' = PGL_2$, this follows from Corollary 1. When $G' = SL_2^3/\Delta\mu_2$, the representation of G' on $E_A^\perp$ is the restriction of a representation of $\widetilde{G'} = GL_2^3/\Delta\mathfrak{k}^\times$ (by the same formula). Now the action of $\widetilde{G'}$ on $\mathcal{R}_0^\perp$

has finitely many open orbits with representatives $(1, 1, g) \in GL_2^3$ with g regular semisimple, and the stabilizer of such a representative is ΔT with T a maximal torus in PGL_2. Hence, as a representation of $\widetilde{G'}$, $L^2(\mathcal{R}_0^{\perp})$ is weakly equivalent to

$$\bigoplus_T \mathrm{Ind}_{\Delta T}^{\widetilde{G'}} \mathbb{C} \cong \bigoplus_T \mathrm{Ind}_{\Delta PGL_2}^{\widetilde{G'}} L^2(T\backslash PGL_2)$$

as T runs over conjugacy classes of maximal tori in PGL_2. By Corollary 1 and the continuity of induction, we deduce that $L^2(\mathcal{R}_0^{\perp})$ is weakly equivalent to $L^2(\widetilde{G'})$. Thus, on restriction to G', $L^2(E^{\perp})$ is weakly equivalent to $L^2(G')$, as desired.

(ii) We shall only give a sketch in this case. By considering the generic orbits of G' on $\mathcal{R}_0^{\perp}$ as in (i), one shows that $L^2(\mathcal{R}_0^{\perp})$ is weakly equivalent to the representation $L^2(\Delta Spin_3\backslash Spin_3 \times Spin_5)$. One then checks that tempered matrix coefficients on $Spin_3 \times Spin_5$ are absolutely integrable on the subgroup $\Delta Spin_3$. Using the same argument as in [23, §6], one deduces that the spectral measure of $L^2(\Delta Spin_3\backslash Spin_3 \times Spin_5)$ is absolutely continuous with respect to the Plancherel measure of G', whence the result.

Concluding, we have

Theorem 4. *There is an isomorphism of G-modules:*

$$L^2(H_\chi\backslash G) \cong \int_{\widehat{G'}} W_\chi(\pi) \otimes \Theta(\pi)\, d\mu_{G'}(\pi),$$

where $W_\chi(\pi)$ is some multiplicity space and $\mu_{G'}$ is the Plancherel measure. When $G' = PGL_2$ or $SL_2^3/\Delta\mu_2$, $W_\chi(\pi) = Wh_\chi(\pi)$ as given in Theorem 2(3).

In addition, as we discussed in §3.10, the smooth analog of our argument in this section implies that when $G' = PGL_2$ or $SL_2^3/\Delta\mu_2$,

$$W_\chi(\pi) = Wh_\chi(\pi) \cong \mathrm{Hom}_{H_\chi}(\Theta^\infty(\pi^\infty), \mathbb{C}) = \mathrm{Hom}_{H_\chi}(\Theta(\pi)^\infty, \mathbb{C})$$

for $\mu_{G'}$-almost all π.

8.5 Restriction to $PGL_3 \times G_2$

We now treat the dual pair $PGL_3 \times G_2$ in $S = E_6$, which can be done by a similar analysis. In this case, $\Omega_Q \subset \mathbb{O}^2$. If we restrict the action of S to $Q_0 \times G_2$, we deduce by Lemma 2(ii) that as a representation of $Q \times G_2$,

$$\Pi \cong \mathrm{Ind}_{(P_{L_0} \times L\cdot[U,U])^0\cdot U}^{Q_0 \times G_2} L^2(\mathfrak{k}),$$

where the action of $(P_{L_0} \times L \cdot [U,U])^0$ on $L^2(\mathfrak{k})$ is given through the geometric action described in Lemma 2(ii) and the action of U_0 is by a nontrivial character fixed by P_{L_0}.

By using the Fourier transform on $L^2(\mathfrak{k})$, we deduce that as a representation of $(P_{L_0} \times L \cdot [U,U])^0$,

$$L^2(\mathfrak{k}) \cong \mathrm{Ind}_{U_{L_0} \times J}^{(P_{L_0} \times L \cdot [U,U])^0} \psi^{-1} \otimes \psi.$$

Hence, as a representation of $Q_0 \times G_2$

$$\Pi \cong \mathrm{Ind}_{N_0}^{Q_0} \chi \otimes \mathrm{Ind}_J^{G_2} \psi, \tag{28}$$

where $N_0 = U_{L_0} \cdot U_0$ is the unipotent radical of a Borel subgroup of PGL_3 and χ is a generic character of N_0.

On the other hand, we have abstractly

$$\Pi \cong \int_{\widehat{PGL_3}} \pi|_{Q_0} \otimes \Theta(\pi)\, d\nu_{\Theta}(\pi). \tag{29}$$

We note that if π is tempered, then

$$\pi|_{Q_0} \cong \mathrm{Ind}_{N_0}^{Q_0} \chi,$$

in which case we deduce on comparing (28) and (29) that

$$L^2((J,\psi)\backslash G_2) = \mathrm{Ind}_J^{G_2} \psi \cong \int_{\widehat{PGL_3}} \Theta(\pi)\, d\nu_{\Theta}(\pi). \tag{30}$$

For (30) to hold, we thus need to show that ν_Θ is absolutely continuous with respect to the Plancherel measure of PGL_3.

For this, we examine the mixed model of Π which is realized on $L^2(\mathfrak{k}^\times \times J(\mathfrak{k}^2) \times \mathfrak{k})$. Noting that $J(\mathfrak{k}^2) \cong \mathfrak{gl}_3$ as PGL_3-module [19], we deduce that as a representation of PGL_3, Π is weakly equivalent to the representation on $L^2(\mathfrak{sl}_3)$ associated to the adjoint action on $\mathfrak{sl}_3$. As in Corollary 1, we know that $L^2(\mathfrak{sl}_3)$ is weakly equivalent to $\bigoplus_T L^2(T\backslash PGL_3)$, with T running over conjugacy classes of maximal tori in PGL_3.

Using the same argument as in [23, §6], one can show that for each T, the spectral measure for $L^2(T\backslash PGL_3)$ is absolutely continuous with respect to the Plancherel measure of PGL_3, and hence so is the spectral measure of $L^2(\mathfrak{sl}_3)$; this justifies (30) and shows that

$$L^2((J,\psi)\backslash G_2) = \mathrm{Ind}_J^{G_2} \psi \cong \int_{\widehat{PGL_3}} W(\pi) \otimes \Theta(\pi)\, d\mu_{PGL_3}(\pi)$$

for some multiplicity space $W(\pi)$ of dimension ≤ 1.

It is natural to ask:

Question 1. For which adjoint simple algebraic group G is the representation $L^2(\mathfrak{g})$ of G weakly equivalent to the regular representation $L^2(G)$?

Corollary 1 verifies this conjecture for PGL_2. If the conjecture holds for PGL_3, one can then take $W(\pi)$ to be $\mathbb{C}$ for all π. In general, it is not difficult to show (using the argument in [23, §6] for example) that the support of the spectral measure of $L^2(\mathfrak{g})$ is contained in the tempered spectrum of G. We had initially conjectured that $L^2(\mathfrak{g})$ is weakly equivalent to $L^2(G)$. However, Kobayashi has explained to us that for the adjoint group $G = PU(n,1)$ (over $\mathbb{R}$), there is a family of holomorphic discrete series representations which does not occur in $L^2(\mathfrak{g})$. Since it is unclear what the correct statement is in general, we decided to formulate the above question.

9 Exceptional theta correspondence: $SL_3 \times \mathrm{Aut}(\mathcal{R}, \det)$

Finally we come to the family of dual pairs $SL_3 \times \mathrm{Aut}(\mathcal{R}, \det) \subset S = S_{\mathcal{R}}$ given by (22). What is interesting about this situation is that the group S may have no Siegel parabolic subgroup, so that the argument below is not the analog of that in the classical cases of §3. To simplify notation, we shall set $G = \mathrm{Aut}(\mathcal{R}, \det)$. Note that in the case of F_4, S is the double cover of F_4 and the dual pair is $\widetilde{SL}_3 \times G = \widetilde{SL}_3 \times SL_3$.

Let $Q_0 = L_0 \cdot U_0 \subset SL_3$ be the maximal parabolic subgroup stabilizing the subspace $\Bbbk e_1 + \Bbbk e_2$, so that

$$L_0 \cong GL_2 \quad \text{and} \quad U_0 = u_{\beta_0 - \beta}(\Bbbk) \times u_{\beta_0}(\Bbbk).$$

Let χ be a generic character of U_0 trivial on $u_{\beta_0-\beta}(\Bbbk)$. The stabilizer in L_0 of χ is a subgroup of the form $T_0 \ltimes U_{L_0}$ with $T_0 \cong \Bbbk^\times$ contained in the diagonal torus and $U_{L_0} = u_{-\beta}(\Bbbk)$. On restricting the minimal representation Π to $Q_0 \times G$, we may write

$$\Pi \cong \mathrm{Ind}_{P_{L_0} U_0 \times G}^{Q_0 \times G} \Pi_\chi$$

for some representation Π_χ of $P_{L_0} U_0 \times G$ with U_0 acting by χ. Here, we have used the theorem of Howe–Moore which ensures that the trivial character of U_0 does not intervene.

Now we can describe the $P_{L_0} U_0 \times G$-module Π_χ using the mixed model of Π. Recall that this mixed model of Π is realized on $L^2(\Bbbk^\times \times \mathcal{R} \times \Bbbk)$. Moreover, the action of $U_0 = u_{\beta_0-\beta}(\Bbbk) \times u_{\beta_0}(\Bbbk)$ in this model is

$$\begin{cases} (u_{\beta_0}(z) f)(t, x, a) = \psi(tz) \cdot f(t, x, a) \\ (u_{\beta_0-\beta}(y) f)(t, x, a) = \psi(ay) \cdot f(t, x, a). \end{cases}$$

As such, Π_χ is the representation obtained from Π by specializing (continuous) functions $f \in \Pi$ to the function $x \mapsto f(1,x,0)$ of $\mathcal{R}$. Thus

$$\Pi_\chi = L^2(\mathcal{R}),$$

where the action of $T_0 \times G$ is geometric, with T_0 acting by scaling on $\mathcal{R}$. Moreover, it follows by (24) that the action of $u_{-\beta}(b) \in U_{L_0}$ is

$$(u_{-\beta}(b) \cdot f)(x) = \psi(-b \cdot \det(x)) \cdot f(x).$$

Now the set $\{x \in \mathcal{R} : \det(x) \neq 0\}$ is open dense and by Proposition 5(i), it is the union of finitely many generic orbits of $T_0 \times G$ indexed by $\mathfrak{k}^\times/(\mathfrak{k}^\times)^3$. For each $a \in \mathfrak{k}^\times/(\mathfrak{k}^\times)^3$, let H_a be the corresponding stabilizer group whose type is described in Table 4 in §4.2. Then

$$\Pi \cong \bigoplus_a \mathrm{Ind}_{N_0 \times H_a}^{Q_0 \times G} \chi_a \otimes \mathbb{C} \cong \bigoplus_a \mathrm{Ind}_{N_0}^{Q_0} \chi_a \otimes L^2(H_a \backslash G).$$

On the other hand, one has abstractly

$$\Pi \cong \int_{\widehat{SL_3}} \pi|_{Q_0} \otimes \Theta(\pi)\, d\nu_\theta(\pi).$$

Now we note:

Lemma 4. *As a representation of SL_3, Π is weakly equivalent to $L^2(SL_3)$.*

Proof. If S is of type E, the group SL_3 is contained in a conjugate of the Heisenberg parabolic subgroup P_S. Indeed, after an appropriate conjugation, we may assume that

$$SL_3 \subset \mathrm{Aut}(J(\mathfrak{k}^2), \det) = SL_3 \times_{\mu_3} SL_3 \subset \mathrm{Aut}(J(B), \det),$$

where $B = \mathfrak{k}^2$, $M_2(\mathfrak{k})$ or the split octonion algebra $\mathbb{O}$ in the respective case. From the description of the mixed model, one sees that Π is nearly equivalent to the representation of SL_3 on $L^2(J(B)) = L^2(J(\mathfrak{k}^2)) \otimes L^2(J(\mathfrak{k}^2)^\perp)$. Since $J(\mathfrak{k}^2) \cong M_3(\mathfrak{k})$ with SL_3 acting by left multiplication, we see that $J(\mathfrak{k}^2)$ is weakly equivalent to the regular representation of SL_3. This implies that Π is weakly equivalent to the regular representation of SL_3.

The case when $S = F_4$ is a bit more intricate; we omit the details here.

Thus $\nu_\theta = \mu_{SL_3}$ and every π in the support of ν_θ is tempered, so that

$$\pi|_{Q_0} = \bigoplus_{a \in \mathfrak{k}^\times/(\mathfrak{k}^\times)^3} Wh_{\chi_a}(\pi) \otimes \mathrm{Ind}_{N_0}^{Q_0} \chi_a.$$

Comparing, we see that

$$L^2(H_a\backslash G) \cong \int_{\widehat{SL_3}} Wh_{\chi_a}(\pi) \otimes \Theta(\pi) d\mu_{SL_3}(\pi),$$

as desired.

References

1. van den Ban E. and Schlichtkrull H., *The Plancherel theorem for a reductive symmetric space I. Spherical functions, and II. Representation theory.* Invent. Math. 161 (2005), 453–566 and 567–628.
2. Bernstein J., *On the support of Plancherel measure*, J. Geom. Phys., 5(4) (1989), 663–710.
3. Delorme P., *Formule de Plancherel pour les espaces symmétriques réductifs*, Annals of Math. 147 (1998), 417–452.
4. Delorme P., *Formule de Plancherel pour les fonctions de Whittaker sur un groupe réductif p-adique.* Ann. Inst. Fourier (Grenoble) 63 (2013), no. 1, 155–217.
5. Delorme P., *Théorème de Paley-Wiener pour les fonctions de Whittaker sur un groupe réductif p-adique.* J. Inst. Math. Jussieu 11 (2012), no. 3, 501–568.
6. Dvorsky A. and Sahi S., *Explicit Hilbert spaces for certain unipotent representations II*, Invent. Math. 138 (1999), 203–224.
7. Flensted-Jensen M., *Discrete series for semisimple symmetric spaces.* Ann. of Math. 111(1980), 253–311.
8. W. T. Gan and N. Gurevich, *Non-tempered Arthur packets of G_2: liftings from $\widetilde{SL}_2$*, American Journal of Math 128, No. 5 (2006), 1105–1185.
9. Gaitsgory, D. and Nadler, D., *Spherical varieties and Langlands duality.* Mosc. Math. J., 10 (2010), 65–137.
10. Gomez R., *The Bessel–Plancherel theorem and applications.* UCSD PhD thesis.
11. Gomez R. and Wallach N., *Holomorphic continuation of Bessel integrals for general admissible induced representations: the case of compact stabilizer.* Selecta Math. (N.S.) 18 (2012), no 1, 1–26.
12. Howe R., *On some results of Strichartz and Rallis and Schiffman.* J. Funct. Anal. 32 (1979), no. 3, 297–303
13. Knus M.-A., Merkejev A., Rost M. and Tignol J. P., *The book of involutions.* A.M.S. Colloquium Publications Vol. 44 (1998).
14. Kazhdan D. and Savin G., *The smallest representation of simply-laced groups*, in *Israel Math. Conference Proceedings, Piatetski-Shapiro Festschrift*, Vol. 2 (1990), 209–233.
15. Kobayashi T., *Singular unitary representations and discrete series for indefinite Stiefel manifolds $U(p,q;F)/U(p-m,q;F)$*, Mem. Amer. Math. Soc., Vol. 462, Amer. Math. Soc., 1992, 106 pp.
16. Kobayashi T., *Discrete decomposability of the restriction of $A_{\mathfrak{q}}(\lambda)$ with respect to reductive subgroups and its applications*, Invent. Math. 117 (1994), 181–205.
17. Li J. S., *On the discrete series of generalized Stiefel manifolds.* Trans. Amer. Math. Soc. 340 (1993), no. 2, 753–766.
18. Li, J. S., *Singular unitary representations of classical groups.* Invent. Math. 97 (1989), no. 2, 237–255.
19. Magaard K. and Savin G., *Exceptional Θ-correspondences I.* Compositio Math. 107 (1997), no. 1, 89–123.
20. Oshimi T. and Matsuki T., *A description of discrete series for semisimple symmetric spaces.* Adv. Stud. Pure Math. 4 (1984), 331–390.

21. Rallis S. and Schiffmann G., *Weil representation. I. Intertwining distributions and discrete spectrum.* Mem. Amer. Math. Soc. 25 (1980), no. 231, iii+203 pp
22. Rumelhart K., *Minimal representations of exceptional p-adic groups.* Represent. Theory 1 (1997), 133–181.
23. Sakellaridis Y. and Venkatesh A., *Periods and harmonic analysis on spherical varieties*, preprint, arXiv:1203.0039.
24. Savin, G., *Dual pair $G_J \times PGL_2$: G_J is the automorphism group of the Jordan algebra J*, Invent. Math. 118 (1994), 141–160.
25. Tang U.-L., *The Plancherel Formula of $L^2(N_0\backslash G;\psi)$*. Ph. D. Thesis. arXiv:1102.2022.
26. Vogan D., *Irreducibility of discrete series representation for semisimple symmetric spaces.* Adv. Stud Pure Math. 14 (1988), 191–221.
27. Wallach N., *Real Reductive Groups II,* Academic Press Pure and Applied Mathematics, Boston, 132 (1992).
28. Weissman, M., *D_4 modular forms.* American J. of Math. 128 (2006), no. 4, 849–898.
29. Ørsted B. and Zhang G. K., *L^2-versions of the Howe correspondence. I.* Math. Scand. 80 (1997), no. 1, 125–160.
30. Ørsted B. and Zhang G. K., *L^2-versions of the Howe correspondence. II.* J. Math. Pures Appl. (9) 74 (1995), no. 2, 165–183.

Proof of the 2-part compositional shuffle conjecture

Adriano M. Garsia, Gouce Xin, and Mike Zabrocki

In honor of Nolan Wallach

Abstract In a recent paper [9] J. Haglund, J. Morse and M. Zabrocki advanced a refinement of the Shuffle Conjecture of Haglund et. al. [8]. They introduce the notion of "touch composition" of a Dyck path, whose parts yield the positions where the path touches the diagonal. They conjectured that the polynomial $\langle \nabla \mathbf{C}_{p_1}\mathbf{C}_{p_2}\cdots\mathbf{C}_{p_k}\, 1\ ,\ h_{\mu_1}h_{\mu_2}\cdots h_{\mu_l}\rangle$, where $\mathbf{C}_{p_1}\mathbf{C}_{p_2}\cdots\mathbf{C}_{p_k}\, 1$ is essentially a rescaled Hall–Littlewood polynomial and ∇ is the Macdonald eigen-operator introduced in [1], enumerates by $t^{\mathrm{area}}q^{\mathrm{dinv}}$ the parking functions whose Dyck paths hit the diagonal by $(p_1, p_2, \ldots, p_k)$ and whose diagonal word is a shuffle of l increasing words of lengths $\mu_1, \mu_2, \ldots, \mu_l$. In this paper we prove the case $l = 2$ of this conjecture.

Keywords: Symmetric functions • Macdonald polynomials • Parking functions

Mathematics Subject Classification: 05E05

The first author was supported by NSF Grant DMS10-68883, the second author was supported by NSF of China Grant 11171231, and the third author was supported by NSERC.

A.M. Garsia
UC San Diego, La Jolla, CA 92037-0112, USA
e-mail: garsia@math.ucsd.edu

G. Xin
Capital Normal University, Beijing, People's Republic of China
e-mail: guoce.xin@gmail.com

M. Zabrocki
York University, Toronto, ON, Canada M3J 1P3
e-mail: zabrocki@mathstat.yorku.ca

R. Howe et al. (eds.), *Symmetry: Representation Theory and Its Applications*, Progress in Mathematics 257, DOI 10.1007/978-1-4939-1590-3_8

1 Introduction

Parking functions are endowed by a colorful history and jargon (see for instance [7]) that is very helpful in dealing with them combinatorially as well as analytically. Here we will represent them interchangeably as two line arrays or as tableaux. A single example of this correspondence should be sufficient for our purposes. In the figure below we have on the left the two-line array, with the list of cars $V = (v_1, v_2, \ldots, v_n)$ on top and their diagonal numbers $U = (u_1, u_2, \ldots, u_n)$ on the bottom. In the corresponding $n \times n$ tableau of lattice cells we have shaded the *main diagonal* (or 0-diagonal) and drawn the *supporting* Dyck path. The component u_i gives the number of lattice cells EAST of the ith NORTH step and WEST of the main diagonal. The cells adjacent to the NORTH steps of the path are filled with the corresponding cars from bottom to top.

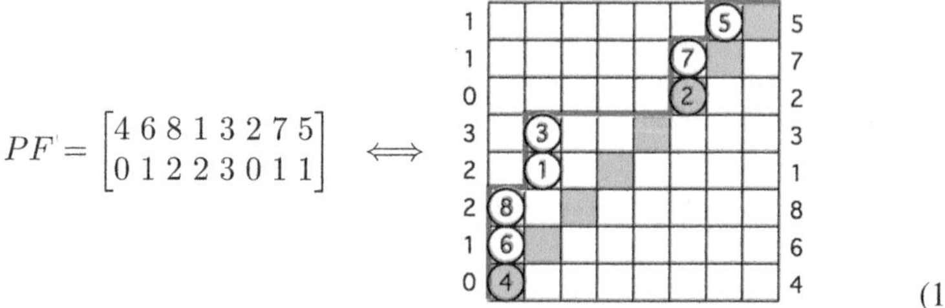

$PF = \begin{bmatrix} 4\,6\,8\,1\,3\,2\,7\,5 \\ 0\,1\,2\,2\,3\,0\,1\,1 \end{bmatrix} \iff$

(1)

The resulting tableau uniquely represents a parking function if and only if the cars increase up the columns.

A necessary and sufficient condition for the vector U to give a Dyck path is that

$$u_1 = 0 \qquad \text{and} \qquad 0 \leq u_i \leq u_{i-1} + 1.$$

Given this, the column increasing property of the corresponding tableau is assured by the requirement that $V = (v_1, v_2, \ldots, v_n)$ is a permutation in S_n satisfying

$$u_i = u_{i-1} + 1 \implies v_i > v_{i-1}.$$

We should mention that the component u_i may also be viewed as the order of the diagonal supporting car v_i. In the example above, car 3 is in the third diagonal, 1 and 8 are in the second diagonal, 5, 7 and 6 are in the first diagonal, and 2 and 4 are in the main diagonal. We have purposely listed the cars by diagonals from right to left starting with the highest diagonal. This gives the *diagonal word* of PF which we will denote $\sigma(PF)$. It is easily seen that $\sigma(PF)$ can also be obtained directly from the 2-line array by successive right to left readings of the components of the vector $V = (v_1, v_2, \ldots, v_n)$ according to decreasing values of $u_1, u_2, \ldots, u_n$. In previous work, each parking function is assigned a *weight*

$$w(PF) = t^{\text{area}(PF)} q^{\text{dinv}(PF)},$$

where

$$\text{area}(PF) = u_1 + u_2 + \cdots + u_n \tag{2}$$

and

$$\text{dinv}(PF) = \sum_{1 \le i < j \le n} \Big(\chi(u_i = u_j \,\&\, v_i < v_j) + \chi(u_i = u_j + 1 \,\&\, v_i > v_j) \Big).$$

It is clear from this imagery that the sum in Equation 2 gives the total number of cells between the supporting Dyck path and the main diagonal. We also see that two cars in the same diagonal with the car on the left smaller than the car on the right will contribute a unit to $\text{dinv}(PF)$, we call this a *primary diagonal inversion*. The same holds true when a car on the left is bigger than a car on the right with the latter in the adjacent lower diagonal, we call this a *secondary diagonal inversion*. For instance, in the example above we see $(6, 7)$ as the only primary diagonal inversion and $(6, 2)$, $(8, 7)$, $(8, 5)$ as the secondary ones. Thus, in the present example we have

$$\text{area}(PF) = 10, \;\; \text{dinv}(PF) = 4, \;\; \sigma(PF) = 31857624,$$

yielding

$$w(PF) = t^{10} q^4.$$

Here and thereafter, the vectors U and V in the two-line representation will be also referred to as $U(PF)$ and $V(PF)$. It will also be convenient to denote by $\mathcal{PF}_n$ the collection of parking functions in the $n \times n$ lattice square.

In [9], J. Haglund, M. Morse and M. Zabrocki introduced a new parking function statistic called *touch composition*. This is the composition $p(PF)$ whose parts give the sizes of the intervals between successive 0's of the vector $U(PF)$. Geometrically the parts of $p(PF)$ yield the places where the supporting Dyck path hits the main diagonal. For instance, for the PF in the example we have $p(PF) = (5, 3)$.

The *Compositional Shuffle Conjecture* [9] states that for any composition $(p_1, p_2, \ldots, p_k) \models n$ and any partition $\mu = (\mu_1, \mu_2, \ldots, \mu_l) \vdash n$ we have the identity

$$\begin{aligned} &\big\langle \nabla \mathbf{C}_{p_1} \mathbf{C}_{p_2} \cdots \mathbf{C}_{p_k} 1 \,,\, h_{\mu_1} h_{\mu_2} \cdots h_{\mu_l} \big\rangle \\ &\quad = \sum_{\substack{PF \in \mathcal{PF}_n \\ p(PF) = (p_1, p_2, \ldots, p_k)}} t^{\text{area}(PF)} q^{\text{dinv}(PF)} \chi\big(\sigma(PF) \in E_1 \sqcup\!\sqcup E_2 \sqcup\!\sqcup \cdots \sqcup\!\sqcup E_l\big), \end{aligned} \tag{3}$$

where ∇ is the Macdonald eigen-operator introduced in [1], $h_{\mu_1} h_{\mu_2} \cdots h_{\mu_l}$ is the complete homogeneous symmetric function basis indexed by μ, E_1, E_2, ..., E_l are successive segments of the word $1234 \cdots n$ of respective lengths $\mu_1, \mu_2, \ldots, \mu_l$

and the symbol $\chi(\sigma(PF) \in E_1 \sqcup\!\sqcup E_2 \sqcup\!\sqcup \cdots \sqcup\!\sqcup E_l)$ is to indicate that the sum is to be carried out over parking functions in $\mathcal{PF}_n$ whose diagonal word is a shuffle of the words $E_1, E_2, \ldots, E_l$. Last but not least the operator $\mathbf{C}_a$ acts on a symmetric polynomial $F[X]$ according to the plethystic formula

$$\mathbf{C}_a F[X] = (-\tfrac{1}{q})^{a-1} F\big[X - \tfrac{1-1/q}{z}\big] \sum_{m \geq 0} z^m h_m[X] \Big|_{z^a} . \tag{4}$$

In this paper we show that the symmetric function methods developed in [5] can be used to prove the $l = 2$ case of Equation 3, that is the identity

$$\begin{aligned} &\big\langle \nabla \mathbf{C}_{p_1} \mathbf{C}_{p_2} \cdots \mathbf{C}_{p_k} 1 \, , \, h_r h_{n-r} \big\rangle \\ &\quad = \sum_{\substack{PF \in \mathcal{PF}_n \\ p(PF)=(p_1,p_2,\ldots,p_k)}} t^{\mathrm{area}(PF)} q^{\mathrm{dinv}(PF)} \chi\big(\sigma(PF) \in 12 \cdots r \sqcup\!\sqcup r+1 \cdots n\big). \end{aligned} \tag{5}$$

Since in [9] it is shown that

$$\sum_{p \models n} \mathbf{C}_{p_1} \mathbf{C}_{p_2} \cdots \mathbf{C}_{p_k} 1 = e_n ,$$

summing Equation 5 over all compositions of n we obtain that

$$\begin{aligned} &\big\langle \nabla e_n \, , \, h_r h_{n-r} \big\rangle \\ &\quad = \sum_{PF \in \mathcal{PF}_n} t^{\mathrm{area}(PF)} q^{\mathrm{dinv}(PF)} \chi\big(\sigma(PF) \in 12 \cdots r \sqcup\!\sqcup r+1 \cdots n\big) \end{aligned} \tag{6}$$

which is the two-part case of the original Shuffle Conjecture. The identity in Equation 6 was, in fact, established, in a 2004 paper [6], by Haglund as the ultimate bi-product of an intricate variety of new identities of Macdonald Polynomial Theory. Our proof of Equation 5 turns out to be much simpler and uses even less machinery than the simplified version of Haglund's original proof given in [4]. Basically, as was done in [5], we only use a small collection of Macdonald polynomial identities established much earlier in [2] and [3] to prove a recursion satisfied by the left-hand side of Equation 5. Then show that the right-hand side satisfies the same recursion, with equality in the base cases.

This recursion, which is the crucial result of this paper, may be stated as the following theorem.

Theorem 1. *For all compositions $p = (p_1, p_2, \ldots, p_k)$ and $0 < r < n$ we have*

$$\begin{aligned} &\big\langle \nabla \mathbf{C}_{p_1} \mathbf{C}_{p_2} \cdots \mathbf{C}_{p_k} 1 \, , \, h_r h_{n-r} \big\rangle = t^{p_1 - 1} \big\langle \nabla \mathbf{B}_{p_1 - 2} \mathbf{C}_{p_2} \cdots \mathbf{C}_{p_k} 1 \, , \, h_{r-1} h_{n-1-r} \big\rangle \\ &\quad + \chi(p_1 = 1) \Big(\big\langle \nabla \mathbf{C}_{p_2} \cdots \mathbf{C}_{p_k} 1 \, , \, h_r h_{n-1-r} \big\rangle + \big\langle \nabla \mathbf{C}_{p_2} \cdots \mathbf{C}_{p_k} 1 \, , \, h_{r-1} h_{n-r} \big\rangle \Big) \end{aligned} \tag{7}$$

with $\mathbf{B}_a = \omega\widetilde{\mathbf{B}}_a\omega$ *and for any symmetric function* $F[X]$

$$\widetilde{\mathbf{B}}_a F[X] = F\big[X - \tfrac{1-q}{z}\big]\sum_{m\geq 0} z^m h_m[X]\Big|_{z^a}. \tag{8}$$

What is remarkably different in this case in contrast with the developments in [5] is that here the symmetric function side guided us on what had to be done in the combinatorial side. In fact we shall see that Equation 5 unravels in a totally unexpected manner some surprising inclusion-exclusions of parking functions.

The reader is advised to have at hand a copy of [5] in reading this work not only for specific references to the identities we use here but also for the notation and definitions of the various symmetric function constructs we deal with in this writing. We already gave in Section 2 of [5] titled a *A Macdonald Polynomial Kit* a detailed list of Macdonald polynomial theory identities that play an essential role in this branch of algebraic combinatorics, and thus we will not repeat it here.

This paper is divided into three sections; in the first section we prove some auxiliary symmetric function identities we use here that are not in [5], in the second section we prove Theorem 1, and in the third section we derive its combinatorial consequences.

Acknowledgment. The authors are indebted to Angela Hicks for helpful guidance in the combinatorial part of this work.

2 Auxiliary symmetric function identities

As we mentioned in the introduction, this section makes heavy use of the notation, definitions and identities listed in Section 2 of [5]. We will present this auxiliary material as a sequence of propositions.

The first obstacle that is encountered in dealing with the shuffle conjecture is to obtain a useable expression for the scalar product of a Macdonald polynomial with a homogeneous basis element; in the two-part case this obstacle can be overcome by means of the following identity proved in [3].

Proposition 1. *For all* $f \in \Lambda^{=r}$ *and* $\mu \vdash n$ *we have*

$$\langle f h_{n-r}\,,\, \widetilde{H}_\mu\rangle \;=\; \nabla^{-1}\big(\omega f\,[\tfrac{X-\epsilon}{M}]\big)\big|_{X\to MB_\mu-1}\,. \tag{9}$$

Given this, we obtain the following proposition.

Proposition 2. *For* $\mu \vdash n$ *and* $0 < r < n$,

$$\langle h_r h_{n-r}\,,\, \widetilde{H}_\mu\rangle \;=\; F_r[MB_\mu - 1] \tag{10}$$

with

$$F_r[X] = \sum_{k=0}^{r} h_{r-k}[\tfrac{1}{M}]\nabla^{-1} e_k[\tfrac{X}{M}]. \tag{11}$$

Proof. From Equation 9 with $f = h_r$ we derive that

$$\langle h_r h_{n-r}\,,\ \widetilde{H}_\mu \rangle = \nabla^{-1}\big(e_r[\tfrac{X-\epsilon}{M}]\big)\big|_{X\to MB_\mu-1}\,.$$

But

$$\nabla^{-1}\big(e_r[\tfrac{X-\epsilon}{M}]\big) = \sum_{k=0}^{r} e_{r-k}[\tfrac{-\epsilon}{M}]\nabla^{-1} e_k[\tfrac{X}{M}] = \sum_{k=0}^{r} h_{r-k}[\tfrac{1}{M}]\nabla^{-1} e_k[\tfrac{X}{M}]$$

and Equation 10 is thus a consequence of Proposition 1. □

Proposition 3. *With n factors* $\mathbf{C}_1$ *we have*

$$\mathbf{C}_1\mathbf{C}_1\cdots\mathbf{C}_1 1 = q^{-\binom{n}{2}}(q,q)_n h_n\big[\tfrac{X}{1-q}\big] = q^{-\binom{n}{2}}\widetilde{H}_n[X;q]\,. \tag{12}$$

In particular it follows that

$$\nabla\mathbf{C}_1\mathbf{C}_1\cdots\mathbf{C}_1 1 = (q,q)_n h_n\big[\tfrac{X}{1-q}\big]\,. \tag{13}$$

Proof. From the definition in Equation 4 it follows that

$$\mathbf{C}_1 1 = e_1[X] = (1-q)e_1[\tfrac{X}{1-q}]$$

which is the case $n = 1$ of Equation 12. So we will proceed by induction and assume that we have

$$\mathbf{C}_1^{n-1} 1 = q^{-\binom{n-1}{2}}(q,q)_{n-1} h_{n-1}\big[\tfrac{X}{1-q}\big].$$

Given this, applying $\mathbf{C}_1$ to both sides and using Equation 4 again we get

$$\begin{aligned}
\frac{q^{\binom{n-1}{2}}}{(q,q)_{n-1}}\mathbf{C}_1^n 1 &= h_{n-1}\Big[\frac{(X-\frac{1-1/q}{z})}{1-q}\Big]\sum_{m\geq 0} z^m h_m[X]\Big|_z \\
&= h_{n-1}\big[\tfrac{X}{1-q}+(X-\tfrac{1-1/q}{z(1-q)})\big]\sum_{m\geq 0} z^m h_m[X]\Big|_z \\
&= h_{n-1}\big[\tfrac{X}{1-q}+\tfrac{1}{qz}\big]\sum_{m\geq 0} z^m h_m[X]\Big|_z \\
&= \sum_{k=0}^{n-1} h_{n-1-k}\big[\tfrac{X}{1-q}\big]\tfrac{1}{q^k} h_{k+1}[X]
\end{aligned}$$

$$= q \sum_{k=0}^{n-1} h_{n-1-k}[\tfrac{X}{1-q}] h_{k+1}[\tfrac{X}{q}]$$

$$= q \sum_{k=0}^{n} h_{n-k}[\tfrac{X}{1-q}] h_k[\tfrac{X}{q}] - q h_n[\tfrac{X}{1-q}]$$

$$= q h_n\big[\tfrac{X}{1-q} + \tfrac{X}{q}\big] - q h_n[\tfrac{X}{1-q}]$$

$$= q h_n\big[\tfrac{X(q+(1-q))}{q(1-q)}\big] - q h_n[\tfrac{X}{1-q}]$$

$$= \tfrac{1}{q^{n-1}} h_n\big[\tfrac{X}{(1-q)}\big] - q h_n[\tfrac{X}{1-q}]$$

$$= \tfrac{1-q^n}{q^{n-1}} h_n[\tfrac{X}{1-q}]\,.$$

This completes the induction and proves the first equality in Equation 12. The second equality in Equation 12 results from a well-known formula for the Macdonald polynomial $\widetilde{H}_\mu$ when $\mu = (n)$. The equality in Equation 13 follows then from the definition of the operator ∇. □

Our next auxiliary result shows how the **C** and **B** operators commute, but to prove it we need some notation. For given expressions $E_1, E_2, \ldots, E_k$ and $P[X]$ a symmetric polynomial we set

$$P^{(r_1,r_2,\ldots,r_k)}[X] = P[X + E_1 u_1 + E_2 u_2 + \cdots + E_k u_k]\Big|_{u_1^{r_1} u_2^{r_2} \cdots u_k^{r_k}}\,.$$

The important property is that if

$$Q^{(r_1)}[X] = P[X + E_1 u_1]\Big|_{u_1^{r_1}},$$

then

$$Q^{(r_1)}[X + E_2 u_2]\Big|_{u_2^{r_2}} = P[X + E_1 u_1 + E_2 u_2]\Big|_{u_1^{r_1} u_2^{r_2}} = P^{(r_1,r_2)}[X].$$

Proposition 4.

$$\big(q\, \mathbf{C}_b \mathbf{B}_a - \mathbf{B}_a \mathbf{C}_b\big) P[X]$$

$$= (q-1)(-1)^{a+b-1} q^{1-b} \times \begin{cases} 0 & \text{if } a+b > 0 \\ P[X] & \text{if } a+b = 0 \\ \sum_{r_1+r_2=-(a+b)} P^{(r_1,r_2)}[X] & \text{if } a+b < 0. \end{cases} \quad (14)$$

Proof. Using Equation 4 we get (with $E_1 = \epsilon(1-q)$)

$$(-q)^{b-1}\mathbf{C}_b P[X] = \sum_{r_1=0}^{d} P^{(r_1)}[X]\tfrac{1}{z^{r_1}} \sum_{m\geq 0} z^m h_m[X]\Big|_{z^b} = \sum_{r_1=0}^{d} P^{(r_1)}[X] h_{b+r_1}[X] \tag{15}$$

and Equation 15 gives (with $E_2 = \epsilon(1-q)$)

$$(-q)^{b-1}\mathbf{B}_a\mathbf{C}_b P[X] = \sum_{r_1=0}^{d} P^{(r_1)}\big[X+\epsilon\tfrac{1-q}{z_2}\big] h_{b+r_1}\big[X+\epsilon\tfrac{1-q}{z_2}\big]\Omega[-\epsilon z_2 X]\Big|_{z_2^a}$$

$$= \sum_{r_1,r_2=0}^{d} P^{(r_1,r_2)}(\tfrac{1}{z_2})^{r_2} \sum_{s=0}^{b+r_1} h_{b+r_1-s}[X] h_s\big[\epsilon\tfrac{1-q}{z_2}\big]\Omega[-\epsilon z_2 X]\Big|_{z_2^a}$$

$$= \sum_{r_1,r_2=0}^{d}\sum_{s=0}^{b+r_1} P^{(r_1,r_2)} h_{b+r_1-s}[X] h_s\big[\epsilon(1-q)\big]\Omega[-\epsilon z_2 X]\Big|_{z_2^{a+r_2+s}}$$

$$= \sum_{r_1,r_2=0}^{d}\sum_{s=0}^{b+r_1} P^{(r_1,r_2)}[X] h_{b+r_1-s}[X](-1)^s h_s\big[(1-q)\big] h_{a+r_2+s}[-\epsilon X]\,.$$

Now note that Equation (2.24) of [5] for $r=0$ and $u=q$ gives

$$h_s\big[1-q\big] = \begin{cases} 1 & \text{if } s=0, \\ 1-q & \text{if } s>0. \end{cases} \tag{16}$$

We can thus write

$$(-q)^{b-1}\mathbf{B}_a\mathbf{C}_b P[X] = \sum_{r_1,r_2=0}^{d} P^{(r_1,r_2)}[X] h_{b+r_1}[X] h_{a+r_2}[-\epsilon X]$$

$$+ (1-q)\sum_{r_1,r_2=0}^{d}\sum_{s=1}^{b+r_1} P^{(r_1,r_2)}[X] h_{b+r_1-s}[X](-1)^s h_{a+r_2+s}[-\epsilon X]$$

and the change of summation index $u = a + r_2 + s$ gives

$$(-q)^{b-1}\mathbf{B}_a\mathbf{C}_b P[X] = \sum_{r_1,r_2=0}^{d} P^{(r_1,r_2)}[X] h_{b+r_1}[X] h_{a+r_2}[-\epsilon X]$$

$$+ (1-q)\sum_{r_1,r_2=0}^{d}\sum_{u=a+r_2+1}^{a+b+r_1+r_2} P^{(r_1,r_2)}[X] h_{a+b+r_1+r_2-u}[X](-1)^{u-a-r_2} h_u[-\epsilon X]. \tag{17}$$

Similarly we get (with $E_1 = \epsilon(1-q)$)

$$\mathbf{B}_a P[X] = \sum_{r_2=0}^{d} P^{(r_2)}[X](\tfrac{1}{z_2})^{r_2} \sum_{u\geq 0} z_2^u h_u[-\epsilon z_2 X]\Big|_{z_2^a} = \sum_{r_2=0}^{d} P^{(r_2)}[X] h_{r_2+a}[-\epsilon X] .$$

Thus (with $E_2 = \epsilon(1-q)$)

$$\begin{aligned}
&(-q)^{b-1}\mathbf{C}_b\mathbf{B}_a P[X] \\
&\quad = \sum_{r_2=0}^{d} P^{(r_2)}\big[X - \tfrac{1-1/q}{z}\big] h_{r_2+a}\big[-\epsilon\big(X - \tfrac{1-1/q}{z_1}\big)\big]\,\Omega[z_1 X]\Big|_{z_1^b} \\
&\quad = \sum_{r_1,r_2=0}^{d} P^{(r_1,r_2)}[X](\tfrac{1}{z_1})^{r_1} \sum_{s=0}^{r_2+a} h_{r_2+a-s}[-\epsilon X]\,(\tfrac{1}{z_1})^s h_s\big[\epsilon(1-1/q)\big]\,\Omega[z_1 X]\Big|_{z_1^b} \\
&\quad = \sum_{r_1,r_2=0}^{d} P^{(r_1,r_2)}[X] \sum_{s=0}^{r_2+a} h_{r_2+a-s}[-\epsilon X]\,(-1)^s h_s\big[1-1/q\big]\,h_{r_1+s+b}[X].
\end{aligned}$$

Note that now Equation 16 gives

$$h_s\big[1-\frac{1}{q}\big] = \begin{cases} 1 & \text{if } s=0, \\ 1-\dfrac{1}{q} & \text{if } s>0. \end{cases}$$

Thus

$$\begin{aligned}
(-q)^{b-1}\mathbf{C}_b\mathbf{B}_a P[X] &= \big(1-(1-\frac{1}{q})\big) \sum_{r_1,r_2=0}^{d} P^{(r_1,r_2)}[X] h_{r_2+a}[-\epsilon X]\,h_{r_1+b}[X] \\
&+ \big(1-\frac{1}{q}\big) \sum_{r_1,r_2=0}^{d} P^{(r_1,r_2)}[X] \sum_{s=0}^{r_2+a} h_{r_2+a-s}[-\epsilon X]\,(-1)^s\,h_{r_1+s+b}[X]
\end{aligned}$$

and the change of summation index $u = r_2 + a - s$ gives

$$\begin{aligned}
(-q)^{b-1}\mathbf{C}_b\mathbf{B}_a P[X] &= \frac{1}{q} \sum_{r_1,r_2=0}^{d} P^{(r_1,r_2)}[X] h_{r_2+a}[-\epsilon X]\,h_{r_1+b}[X] \\
&+ \big(1-\frac{1}{q}\big) \sum_{r_1,r_2=0}^{d} P^{(r_1,r_2)}[X] \sum_{u=0}^{a+r_2} h_u[-\epsilon X]\,(-1)^{r_2+a-u}\,h_{a+b+r_1+r_2-u}[X].
\end{aligned} \tag{18}$$

In summary we get

$$(-q)^{b-1}\ q\,\mathbf{C}_b\mathbf{B}_a P[X] = \sum_{r_1,r_2=0}^{d} P^{(r_1,r_2)}[X]h_{r_2+a}[-\epsilon X]\,h_{r_1+b}[X] \tag{19}$$

$$+\big(q-1\big)(-1)^a \sum_{r_1,r_2=0}^{d} P^{(r_1,r_2)}[X]\sum_{u=0}^{a+r_2} h_u[-X]\,(-1)^{r_2}\,h_{a+b+r_1+r_2-u}[X].$$

On the other hand Equation 17 can also be written as

$$(-q)^{b-1}\mathbf{B}_a\mathbf{C}_b P[X] = \sum_{r_1,r_2=0}^{d} P^{(r_1,r_2)}[X]h_{r_2+a}[-\epsilon X]\,h_{r_1+b}[X] \tag{20}$$

$$+(-1)^a(1-q)\sum_{r_1,r_2=0}^{d}\ \sum_{u=a+r_2+1}^{a+b+r_1+r_2} P^{(r_1,r_2)}[X]h_{a+b+r_1+r_2-u}[X](-1)^{r_2}h_u[-X]$$

and thus subtraction gives

$$(-q)^{b-1}\big(q\,\mathbf{C}_b\mathbf{B}_a - \mathbf{B}_a\mathbf{C}_b\big)P[X] \tag{21}$$

$$= \big(q-1\big)(-1)^a \sum_{r_1,r_2=0}^{d} P^{(r_1,r_2)}[X] \sum_{u=0}^{a+b+r_1+r_2} h_u[-X]\,(-1)^{r_2}\,h_{a+b+r_1+r_2-u}[X]$$

$$= \big(q-1\big)(-1)^a \sum_{r_1,r_2=0}^{d} P^{(r_1,r_2)}[X](-1)^{r_2}\,h_{a+b+r_1+r_2}[X-X].$$

Carrying out the summations and using the definition of $P^{(r_1,r_2)}[X]$ we finally obtain

$$(-q)^{b-1}\big(q\,\mathbf{C}_b\mathbf{B}_a{-}\mathbf{B}_a\mathbf{C}_b\big)P[X]$$

$$= (q-1)(-1)^a \times \begin{cases} 0 & \text{if } a+b>0, \\ P[X] & \text{if } a+b=0, \\ \sum_{r_1+r_2=-(a+b)} P^{(r_1,r_2)}[X] & \text{if } a+b<0, \end{cases}$$

which is easily seen to be Equation 14, completing the proof. □

In particular we have shown that

Theorem 2. *For all $a+b>0$, our Hall–Littlewood operators have the following commutativity property*

$$\mathbf{B}_a\,\mathbf{C}_b = q\,\mathbf{C}_b\,\mathbf{B}_a. \tag{22}$$

An important ingredient in Macdonald polynomial theory is a modified symmetric function scalar product we will refer to as the $*$-*scalar product* which makes the basis $\{\widetilde{H}_\mu[X;q,t]\}_\mu$ an orthogonal set. More precisely, we have the basic identities

$$\langle \widetilde{H}_\lambda, \widetilde{H}_\mu \rangle_* = \begin{cases} 0 & \text{if } \lambda \neq \mu \\ w_\mu(q,t) & \text{if } \lambda = \mu \end{cases} \tag{23}$$

where the $w_\mu(q,t)$ are polynomials in $\mathbf{N}[q,t]$ whose precise definition can be found in Section 2 of [5].

The $*$-scalar product and the Hall scalar product are related by the identity ([5] Equation (2.16)),

$$\langle f \, , \, g \rangle = \langle f \, , \, \omega g^* \rangle_* \tag{24}$$

where for convenience, for any symmetric function $g[X]$ we set

$$g^*[X] = g\left[\tfrac{X}{M}\right] \qquad \text{(with } M = (1-q)(1-t)\text{)}. \tag{25}$$

To compute the action of ∇ on a symmetric function we need to expand that function in terms of the basis $\{\widetilde{H}_\mu[X;q,t]\}_\mu$ and Equation 23 is the tool we need to carry this out. In the remainder of the paper we will make use of the following expansions.

Proposition 5. *For all $n \geq 1$ and $0 < r < n$ we have*

(a) $e_n\left[\frac{X}{M}\right] = \sum_{\mu \vdash n} \frac{\widetilde{H}_\mu[X;q.t]}{w_\mu}$,

(b) $h_n\left[\frac{X}{M}\right] = \sum_{\mu \vdash n} \frac{T_\mu \widetilde{H}_\mu[X;q.t]}{w_\mu}$

(c) $e_r\left[\frac{X}{M}\right] e_{n-r}\left[\frac{X}{M}\right] = \sum_{\mu \vdash n} \frac{\widetilde{H}_\mu[X;q.t]}{w_\mu} F_r[MB_\mu - 1]$

with $F_r[X]$ given by Equation 11.

Proof. Using Equation 23 and Equation 24 we obtain

$$e_n^* = \sum_{\mu \vdash n} \frac{\widetilde{H}_\mu[X;q.t]}{w_\mu} \langle \widetilde{H}_\mu \, , \, h_n \rangle$$

and Proposition 5(a) follows since it is well known, ([5] Equation (2.25)), that

$$\langle \widetilde{H}_\mu \, , \, h_n \rangle = 1. \tag{26}$$

Similarly, we get

$$h_n^* = \sum_{\mu \vdash n} \frac{\widetilde{H}_\mu[X;q.t]}{w_\mu} \langle \widetilde{H}_\mu \, , \, e_n \rangle \tag{27}$$

and Proposition 5(b) follows since it is well known, ([5] Equation (2.25)), that

$$\big\langle \widetilde{H}_\mu \,,\, e_n \big\rangle = T_\mu. \tag{28}$$

The formula in Proposition 5(c) is less immediate. We can again start by writing

$$e_r^* e_{n-r}^* = \sum_{\mu \vdash n} \frac{\widetilde{H}_\mu[X;q.t]}{w_\mu} \big\langle \widetilde{H}_\mu \,,\, h_r h_{n-r} \big\rangle.$$

However, we have no simple evaluation for the scalar product $\big\langle \widetilde{H}_\mu \,,\, h_r h_{n-r} \big\rangle$ other than resorting to the identity in Equation 10 which gives

$$e_r^* e_{n-r}^* = \sum_{\mu \vdash n} \frac{\widetilde{H}_\mu[X;q.t]}{w_\mu} F_r[MB_\mu - 1] \tag{29}$$

with F_r given by Equation 11. This proves Proposition 5(c) and completes our proof. □

Remark 1. As we will shortly see, our proof of Theorem 1 will require working with the polynomial $\nabla e_r^* e_{n-r}^*$. Since ∇ is defined [1], by setting for the Macdonald basis

$$\nabla \widetilde{H}_\mu = T_\mu \widetilde{H}_\mu, \tag{30}$$

the formula in Proposition 5 (c) gives

$$\nabla e_r\big[\tfrac{X}{M}\big] e_{n-r}\big[\tfrac{X}{M}\big] = \sum_{\mu \vdash n} \frac{T_\mu \widetilde{H}_\mu[X;q.t]}{w_\mu} F_r[MB_\mu - 1]. \tag{31}$$

Introducing the operator θ_r by setting for the Macdonald basis

$$\theta_r \widetilde{H}_\mu = F_r[MB_\mu - 1] \widetilde{H}_\mu. \tag{32}$$

The formula in Proposition 5(b) allows us to write Equation 31 in the form

$$\nabla e_r^* e_{n-r}^* = \theta_r h_n^*. \tag{33}$$

We surprised ourselves in discovering that such a simple idea allows us to get around the unavailability of a simple evaluation for the scalar product $\big\langle \widetilde{H}_\mu \,,\, h_r h_{n-r} \big\rangle$. By delivering an expression for $\nabla e_r^* e_{n-r}^*$ that we can work within our calculations, this idea made possible all the results of the present paper.

3 Proof of the symmetric function recursion

We begin by showing the following basic reduction.

Theorem 3. *The identity*

$$\begin{aligned}\langle \nabla \mathbf{C}_{p_1}\mathbf{C}_{p_2}\cdots\mathbf{C}_{p_k}1\,,\ h_r h_{n-r}\rangle &= t^{p_1-1}\langle \nabla \mathbf{B}_{p_1-2}\mathbf{C}_{p_2}\cdots\mathbf{C}_{p_k}1\,,\ h_{r-1}h_{n-1-r}\rangle\\ &+\chi(p_1=1)\big(\langle \nabla \mathbf{C}_{p_2}\cdots\mathbf{C}_{p_k}1\,,\ h_r h_{n-1-r}\rangle+\langle \nabla \mathbf{C}_{p_2}\cdots\mathbf{C}_{p_k}1\,,\ h_{r-1}h_{n-r}\rangle\big)\end{aligned} \tag{34}$$

holds for all compositions $(p_1, p_2, \ldots, p_k) \models n$ and all $0 < r < n$, if and only if the following symmetric function identity holds for all $0 < r < n$ and $a \geq 1$:

$$\mathbf{C}_a^*\theta_r h_n^*[X] = t^{a-1}\mathbf{B}_{a-2}^*\theta_{r-1}h_{n-2}^*[X] + \chi(a=1)\Big(\theta_r h_{n-1}^*[X] + \theta_{r-1}h_{n-1}^*[X]\Big), \tag{35}$$

where the operators $\mathbf{C}_a^*$ and $\mathbf{B}_a^*$ are the $*$-scalar product adjoints of $\mathbf{C}_a$ and $\mathbf{B}_a$, respectively.

Proof. Note first that since the polynomials $\mathbf{C}_{p_2}\cdots\mathbf{C}_{p_k}1$ are essentially only a rescaled version of the Hall–Littlewood polynomials, they span the space $\Lambda^{=(n-p_1)}$. Thus (34) can hold true as asserted if and only if for all $F[X] \in \Lambda^{=(n-p_1)}$ we have

$$\begin{aligned}\langle \nabla \mathbf{C}_{p_1}F[X]\,,\ h_r h_{n-r}\rangle &= t^{p_1-1}\langle \nabla \mathbf{B}_{p_1-2}F[X]\,,\ h_{r-1}h_{n-1-r}\rangle\\ &+\chi(p_1=1)\Big(\langle \nabla F[X]\,,\ h_r h_{n-1-r}\rangle+\langle \nabla F[X]\,,\ h_{r-1}h_{n-r}\rangle\Big).\end{aligned}$$

Now passing to $*$-scalar products we may rewrite this identity in the form

$$\langle \nabla \mathbf{C}_{p_1}F[X]\,,\ e_r^* e_{n-r}^*\rangle_* = t^{p_1-1}\langle \nabla \mathbf{B}_{p_1-2}F[X]\,,\ e_{r-1}^* e_{n-1-r}^*\rangle_* \tag{36}$$

$$+\chi(p_1=1)\Big(\langle \nabla F[X]\,,\ e_r^* e_{n-1-r}^*\rangle_* + \langle \nabla F[X]\,,\ e_{r-1}^* e_{n-r}^*\rangle_*\Big). \tag{37}$$

Next we move all the operators acting on $F[X]$ to the other side of their respective $*$-scalar products and obtain

$$\begin{aligned}\langle F[X]\,, \mathbf{C}_{p_1}^*\nabla e_r^* e_{n-r}^*\rangle_* &= t^{p_1-1}\langle F[X]\,,\ \mathbf{B}_{p_1-2}^*\nabla e_{r-1}^* e_{n-1-r}^*\rangle_*\\ &+\chi(p_1=1)\Big(\langle F[X]\,,\ \nabla e_r^* e_{n-1-r}^*\rangle_* + \langle F[X]\,,\ \nabla e_{r-1}^* e_{n-r}^*\rangle_*\Big).\end{aligned} \tag{38}$$

Of course ∇ does not get a "$*$" since, by Equation 23, all Macdonald polynomial eigen-operators are necessarily self-adjoint with respect to the $*$-scalar product.

But now the arbitrariness of $F[X]$ shows that Equation 38 can be true if and only if we have the following symmetric function equality

$$\mathbf{C}^*_{p_1} \nabla e^*_r e^*_{n-r} = t^{p_1-1} \mathbf{B}^*_{p_1-2} \nabla e^*_{r-1} e^*_{n-1-r} + \chi(p_1=1)\big(\nabla e^*_r e^*_{n-1-r} + \nabla e^*_{r-1} e^*_{n-r}\big).$$

Replacing p_1 by a and using Equation 33 for various values of r and n yields Equation 35 and completes our proof. □

Our next task is now to prove Equation 35. To begin we will need the following expansion.

Proposition 6.

$$\theta_r h^*_n[X] = \sum_{k=0}^{r} h_{r-k}[\tfrac{1}{M}](-1)^k \sum_{\nu \vdash k} \frac{1}{w_\nu} h_n\big[X(\tfrac{1}{M} - B_\nu)\big]. \tag{39}$$

Proof. Note first that Equations 31 and 33 give

$$\theta_r h^*_n[X] = \sum_{\mu \vdash n} \frac{T_\mu \widetilde{H}_\mu[X;q.t]}{w_\mu} F_r[MB_\mu - 1]. \tag{40}$$

Recall from Equation 11 that

$$F_r[X] = \sum_{k=0}^{r} h_{r-k}[\tfrac{1}{M}] \nabla^{-1} e_k[\tfrac{X}{M}], \tag{41}$$

and since Equations 5 and 30 give

$$\nabla^{-1} e_k[\tfrac{X}{M}] = \sum_{\nu \vdash k} \frac{T_\nu^{-1} \widetilde{H}_\nu[X;q,t]}{w_\nu}, \tag{42}$$

we can write

$$F_r[X] = \sum_{k=0}^{r} h_{r-k}[\tfrac{1}{M}] \sum_{\nu \vdash k} \frac{T_\nu^{-1} \widetilde{H}_\nu[X;q,t]}{w_\nu}; \tag{43}$$

Equation 40 becomes

$$\begin{aligned}
\theta_r h^*_n[X] &= \sum_{\mu \vdash n} \frac{T_\mu \widetilde{H}_\mu[X;q,t]}{w_\mu} \sum_{k=0}^{r} h_{r-k}[\tfrac{1}{M}] \sum_{\nu \vdash k} \frac{T_\nu^{-1} \widetilde{H}_\nu[MB_\mu - 1];q,t]}{w_\nu} \\
&= \sum_{k=0}^{r} h_{r-k}[\tfrac{1}{M}] \sum_{\nu \vdash k} \frac{1}{w_\nu} \sum_{\mu \vdash n} \frac{T_\mu \widetilde{H}_\mu[X;q,t]}{w_\mu} \frac{\widetilde{H}_\nu[MB_\mu - 1];q,t]}{T_\nu}.
\end{aligned} \tag{44}$$

We now use the Macdonald reciprocity formula from Equation (2.21) of [5],

$$\frac{\widetilde{H}_\nu[MB_\mu - 1;q,t]}{T_\nu} = (-1)^{n-k} \frac{\widetilde{H}_\mu[MB_\nu - 1];q,t]}{T_\mu}, \tag{45}$$

and Equation 44 becomes

$$\theta_r h_n^*[X] = \sum_{k=0}^{r} h_{r-k}[\tfrac{1}{M}](-1)^{n-k} \sum_{\nu \vdash k} \frac{1}{w_\nu} \sum_{\mu \vdash n} \frac{\widetilde{H}_\mu[X;q,t]\widetilde{H}_\mu[MB_\nu - 1;q,t]}{w_\mu}; \tag{46}$$

a use of the Macdonald–Cauchy identity from Equation (2.17) of [5]

$$\sum_{\mu \vdash n} \frac{\widetilde{H}_\mu[X;q,t]\widetilde{H}_\mu[Y;q,t]}{w_\mu} = e_n\big[\tfrac{XY}{M}\big]$$

gives

$$\theta_r h_n^*[X] = \sum_{k=0}^{r} h_{r-k}[\tfrac{1}{M}](-1)^{n-k} \sum_{\nu \vdash k} \frac{1}{w_\nu} e_n\big[X(B_\nu - \tfrac{1}{M})\big],$$

which is Equation 39 because of the relation (see Equation (2.6) of [5])

$$e_n\big[X(B_\nu - \tfrac{1}{M})\big] = (-1)^n h_n[X(\tfrac{1}{M} - B_\nu)].$$

Our proof is now complete. □

We are now ready to start working on the identity in Equation 35. We will start with the term

$$\mathbf{B}_{a-2}^* \theta_{r-1} h_{n-2}^*[X]$$

which, using Equation 39 with $r \to r-1$ and $n \to n-2$, becomes

$$\mathbf{B}_{a-2}^* \theta_{r-1} h_{n-2}^*[X] = \sum_{k=0}^{r-1} h_{r-1-k}[\tfrac{1}{M}](-1)^k \sum_{\nu \vdash k} \frac{1}{w_\nu} \mathbf{B}_{a-2}^* h_{n-2}\big[X(\tfrac{1}{M} - B_\nu)\big]. \tag{47}$$

Let us now recall that in [5] (Theorem 3.6) it was shown that the action of the operators $\mathbf{B}_a^*$ and $\mathbf{C}_a^*$ on a symmetric polynomial $P[X]$ may be computed by means of the two plethystic formulas:

$$\mathbf{B}_a^* P[X] = P\big[X + \tfrac{M}{z}\big] \sum_{m \geq 0} z^m h_m\big[\tfrac{-X}{1-t}\big]\Big|_{z^{-a}} \tag{48}$$

and

$$\mathbf{C}_a^* P[X] = (\tfrac{-1}{q})^{a-1} P\big[X - \tfrac{\epsilon M}{z}\big] \sum_{m \geq 0} (-\tfrac{z}{q})^m h_m\big[\tfrac{-X}{1-t}\big]\Big|_{z^{-a}}. \tag{49}$$

We can thus use Equation 48 to get

$$\begin{aligned}
&\mathbf{B}^*_{a-2} h_{n-2}\left[X(\tfrac{1}{M} - B_\nu)\right] \\
&\quad = h_{n-2}\left[(X + M/z)(\tfrac{1}{M} - B_\nu)\right] \sum_{m\geq 0} z^m h_m\left[\tfrac{-X}{1-t}\right]\Big|_{z^{-a+2}} \\
&\quad = \sum_{s=0}^{n-2} h_{n-2-s}\left[X(\tfrac{1}{M} - B_\nu)\right] h_s\left[M(\tfrac{1}{M} - B_\nu)\right] \tfrac{1}{z^s} \sum_{m\geq 0} z^m h_m\left[\tfrac{-X}{1-t}\right]\Big|_{z^{-a+2}} \\
&\quad = \sum_{s=a-2}^{n-2} h_{n-2-s}\left[X(\tfrac{1}{M} - B_\nu)\right] h_s\left[1 - MB_\nu\right] h_{s-a+2}[-X/(1-t)]. \qquad (50)
\end{aligned}$$

Using this in Equation 47 gives

$$\begin{aligned}
\mathbf{B}^*_{a-2}\theta_{r-1}h^*_{n-2}[X] &= \sum_{k=0}^{r-1} h_{r-1-k}[\tfrac{1}{M}](-1)^k \sum_{\nu\vdash k} \frac{1}{w_\nu}\times \\
&\quad \sum_{s=a-2}^{n-2} h_{n-2-s}\left[X(\tfrac{1}{M} - B_\nu)\right] h_s\left[1 - MB_\nu\right] h_{s-a+2}[-X/(1-t)] \\
&= \sum_{k=0}^{r-1} h_{r-1-k}[\tfrac{1}{M}](-1)^k \sum_{s=a-2}^{n-2} h_{s-a+2}[-X/(1-t)]\times \\
&\quad \sum_{\nu\vdash k} \frac{1}{w_\nu} h_{n-2-s}\left[X(\tfrac{1}{M} - B_\nu)\right] h_s\left[1 - MB_\nu\right] \qquad (51)
\end{aligned}$$

and a change $s\to s-2$ of summation index finally gives

$$\begin{aligned}
\mathbf{B}^*_{a-2}\theta_{r-1}h^*_{n-2}[X] &= \sum_{k=0}^{r-1} h_{r-1-k}[\tfrac{1}{M}](-1)^k \sum_{s=a}^{n} h_{s-a}[\tfrac{-X}{1-t}]\times \\
&\quad \sum_{\nu\vdash k} \frac{1}{w_\nu} h_{n-s}\left[X(\tfrac{1}{M} - B_\nu)\right] h_{s-2}\left[1 - MB_\nu\right]. \qquad (52)
\end{aligned}$$

Let us now work on the left-hand side of Equation 35, which, using Equation 39 is simply

$$\mathbf{C}^*_a\theta_r h^*_n[X] = \sum_{k=0}^{r} h_{r-k}[\tfrac{1}{M}](-1)^k \sum_{\nu\vdash k} \frac{1}{w_\nu} \mathbf{C}^*_a h_n\left[X(\tfrac{1}{M} - B_\nu)\right]. \qquad (53)$$

Now Equation 49 gives

$$(-q)^{a-1}\mathbf{C}_a^* h_n\left[X(\tfrac{1}{M}-B_\nu)\right] = h_n\left[(X-\epsilon M/z)(\tfrac{1}{M}-B_\nu)\right]\sum_{m\geq 0}(-\tfrac{z}{q})^m h_m\left[\tfrac{-X}{1-t}\right]\Big|_{z^{-a}}$$

$$= \sum_{s=0}^{n} h_{n-s}\left[X(\tfrac{1}{M}-B_\nu)\right] h_s\left[-\epsilon M(\tfrac{1}{M}-B_\nu)\right]\tfrac{1}{z^s}\sum_{m\geq 0}(-\tfrac{z}{q})^m h_m\left[\tfrac{-X}{1-t}\right]\Big|_{z^{-a}}$$

$$= \sum_{s=0}^{n} h_{n-s}\left[X(\tfrac{1}{M}-B_\nu)\right](-1)^s h_s\left[-1+MB_\nu\right](-\tfrac{1}{q})^{s-a} h_{s-a}\left[\tfrac{-X}{1-t}\right].$$

Thus

$$\mathbf{C}_a^* h_n\left[X(\tfrac{1}{M}-B_\nu)\right] = (-q)\sum_{s=0}^{n} h_{n-s}\left[X(\tfrac{1}{M}-B_\nu)\right] h_s\left[-1+MB_\nu\right](-\tfrac{1}{q})^s h_{s-a}\left[\tfrac{-X}{1-t}\right]$$

and Equation 53 becomes

$$\mathbf{C}_a^*\theta_r h_n^*[X] = \sum_{k=0}^{r} h_{r-k}[\tfrac{1}{M}](-1)^k \sum_{\nu\vdash k}\frac{1}{w_\nu}\times$$

$$(-q)\sum_{s=0}^{n} h_{n-s}\left[X(\tfrac{1}{M}-B_\nu)\right] h_s\left[-1+MB_\nu\right](-\tfrac{1}{q})^s h_{s-a}\left[\tfrac{-X}{1-t}\right] \tag{54}$$

$$= \sum_{k=0}^{r} h_{r-k}[\tfrac{1}{M}](-1)^k\,(-q)\sum_{s=a}^{n}(-\tfrac{1}{q})^s\times$$

$$h_{s-a}\left[\tfrac{-X}{1-t}\right]\sum_{\nu\vdash k}\frac{1}{w_\nu}\,h_{n-s}\left[X(\tfrac{1}{M}-B_\nu)\right] h_s\left[-1+MB_\nu\right]. \tag{55}$$

We now will make use of the following two summation formulas (Equations (2.28) and (2.29) of [3], see also [10]):

$$\sum_{\nu\to\mu} c_{\mu\nu}(q,t)\,(T_\mu/T_\nu)^k = \begin{cases} \frac{tq}{M}\,h_{k+1}\left[(-1+MB_\nu)/tq\right] & \text{if } k\geq 1\,, \\ \\ B_\mu(q,t) & \text{if } k=0\,. \end{cases} \tag{56}$$

$$\sum_{\mu\leftarrow\nu} d_{\mu\nu}(q,t)\,(T_\mu/T_\nu)^k = \begin{cases} (-1)^{k-1}\,e_{k-1}\left[-1+MB_\nu\right] & \text{if } k\geq 1, \\ \\ 1 & \text{if } k=0\,. \end{cases} \tag{57}$$

We will start by using Equation 56 in the form

$$h_s\left[-1+MB_\nu\right] = (tq)^{s-1}M\sum_{\tau\to\nu} c_{\nu\tau}(\tfrac{T_\nu}{T_\tau})^{s-1} - \chi(s=1)$$

and obtain

$$
\begin{aligned}
\mathbf{C}_a^*\theta_r h_n^*[X] &= \sum_{k=0}^{r} h_{r-k}[\tfrac{1}{M}](-1)^k(-q)\sum_{s=a}^{n}(-\tfrac{1}{q})^s h_{s-a}\big[\tfrac{-X}{1-t}\big]\times\\
&\sum_{\nu\vdash k}\frac{1}{w_\nu}h_{n-s}\big[X(\tfrac{1}{M}-B_\nu)\big]\Big((tq)^{s-1}M\sum_{\tau\to\nu}c_{\nu\tau}(\tfrac{T_\nu}{T_\tau})^{s-1}-\chi(s=1)\Big)\\
&=\sum_{k=0}^{r} h_{r-k}[\tfrac{1}{M}](-1)^{k-1}\sum_{s=a}^{n}t^{s-1}h_{s-a}[\tfrac{-X}{1-t}]\sum_{\nu\vdash k}\frac{1}{w_\nu}\,h_{n-s}\big[X(\tfrac{1}{M}-B_\nu)\big]\times\\
&\Big(M\sum_{\tau\to\nu}c_{\nu\tau}(\tfrac{T_\nu}{T_\tau})^{s-1}\Big)\\
&\quad+\chi(a=1)\sum_{k=0}^{r} h_{r-k}[\tfrac{1}{M}](-1)^k\sum_{\nu\vdash k}\frac{1}{w_\nu}\,h_{n-1}\big[X(\tfrac{1}{M}-B_\nu)\big]\\
&=\sum_{k=0}^{r} h_{r-k}[\tfrac{1}{M}](-1)^{k-1}\sum_{s=a}^{n}t^{s-1}h_{s-a}[\tfrac{-X}{1-t}]\sum_{\nu\vdash k}\frac{1}{w_\nu}\,h_{n-s}\big[X(\tfrac{1}{M}-B_\nu)\big]\times\\
&\Big(M\sum_{\tau\to\nu}c_{\nu\tau}(\tfrac{T_\nu}{T_\tau})^{s-1}\Big)+\chi(a=1)\theta_r h_{n-1}^*[X] \qquad (58)
\end{aligned}
$$

by applying Equation 39.

Now, changing the order of ν and τ summations and using the relation (from Equation (2.30) of [5])

$$\frac{w_\tau}{w_\nu}Mc_{\nu\tau}=d_{\nu\tau},$$

we may rewrite Equation 58 as

$$
\begin{aligned}
&\mathbf{C}_a^*\theta_r h_n^*[X]-\chi(a=1)\theta_r h_{n-1}^*[X]\\
&\quad=\sum_{k=0}^{r} h_{r-k}[\tfrac{1}{M}](-1)^{k-1}\sum_{s=a}^{n}t^{s-1}h_{s-a}[-\tfrac{X}{1-t}]\sum_{\tau\vdash k-1}\frac{1}{w_\tau}\times\\
&\qquad\sum_{\nu\leftarrow\tau}\frac{w_\tau}{w_\nu}\,h_{n-s}\big[X(\tfrac{1}{M}-B_\nu)\big]Mc_{\nu\tau}(\tfrac{T_\nu}{T_\tau})^{s-1}\\
&\quad=\sum_{k=0}^{r} h_{r-k}[\tfrac{1}{M}](-1)^{k-1}\sum_{s=a}^{n}t^{s-1}h_{s-a}[-\tfrac{X}{1-t}]\sum_{\tau\vdash k-1}\frac{1}{w_\tau}\times\\
&\qquad\times\sum_{\nu\leftarrow\tau}h_{n-s}\big[X(\tfrac{1}{M}-B_\nu)\big]d_{\nu\tau}(\tfrac{T_\nu}{T_\tau})^{s-1}. \qquad (59)
\end{aligned}
$$

Next we split $B_\nu(q,t)$ into the sum $B_\nu(q,t) = B_\tau(q,t) + \frac{T_\nu}{T_\tau}$ and apply Equation 57 to get

$$\sum_{\nu\leftarrow\tau} d_{\nu\tau}(\tfrac{T_\nu}{T_\tau})^{s-1} h_{n-s}\left[X(\tfrac{1}{M} - B_\nu)\right]$$

$$= \sum_{u=0}^{n-s} h_{n-u-s}\left[X(\tfrac{1}{M} - B_\tau)\right] h_u[-X] \sum_{\nu\leftarrow\tau} d_{\nu\tau}\left(\tfrac{T_\nu}{T_\tau}\right)^{u+s-1}$$

$$= \sum_{u=0}^{n-s} h_{n-u-s}\left[X(\tfrac{1}{M} - B_\tau)\right] h_u[-X]\times$$

$$\Big((-1)^{u+s-2} e_{u+s-2}\big[MB_\tau - 1\big] + \chi(u+s=1)\Big)$$

$$= \sum_{v=s}^{n} h_{n-v}\left[X(\tfrac{1}{M} - B_\tau)\right] h_{v-s}[-X]\Big(h_{v-2}\big[1 - MB_\tau\big] + \chi(v=1)\Big).$$

Using this in Equation 59 gives

$$\mathbf{C}_a^*\theta_r h_n^*[X] - \chi(a=1)\theta_r h_{n-1}^*[X]$$

$$= \sum_{k=0}^{r} h_{r-k}[\tfrac{1}{M}](-1)^{k-1} \sum_{s=a}^{n} t^{s-1} h_{s-a}[-\tfrac{X}{1-t}] \sum_{\tau\vdash k-1} \frac{1}{w_\tau}\times$$

$$\times \sum_{v=s}^{n} h_{n-v}\left[X(\tfrac{1}{M} - B_\tau)\right] h_{v-s}[-X]\Big(h_{v-2}\big[1 - MB_\tau\big] + \chi(v=1)\Big).$$

Since there are no partitions of $k-1$ for $k=0$ we make the change of variable $k\to k+1$ and obtain

$$\mathbf{C}_a^*\theta_r h_n^*[X] - \chi(a=1)\theta_r h_{n-1}^*[X]$$

$$= \sum_{k=0}^{r-1} h_{r-1-k}[\tfrac{1}{M}](-1)^{k} \sum_{s=a}^{n} t^{s-1} h_{s-a}[-\tfrac{X}{1-t}] \sum_{\tau\vdash k} \frac{1}{w_\tau}\times$$

$$\sum_{v=s}^{n} h_{n-v}\left[X(\tfrac{1}{M} - B_\tau)\right] h_{v-s}[-X]\times$$

$$\Big(h_{v-2}\big[1 - MB_\tau\big] + \chi(v=1)\Big). \tag{60}$$

Now the term multiplying $\chi(v=1)$ on the right-hand side is

$$\sum_{k=0}^{r-1} h_{r-1-k}[\tfrac{1}{M}](-1)^{k} \sum_{s=a}^{n} t^{s-1} h_{s-a}[-\tfrac{X}{1-t}]\times$$

$$\times \sum_{\tau\vdash k} \frac{1}{w_\tau} \sum_{v=s}^{n} h_{n-v}\left[X(\tfrac{1}{M} - B_\tau)\right] h_{v-s}[-X].$$

Now $v = 1$ forces $s = 1$, which in turn forces $a = 1$. So this term reduces to

$$\sum_{k=0}^{r-1} h_{r-1-k}[\tfrac{1}{M}](-1)^k \sum_{\tau \vdash k} \frac{1}{w_\tau} h_{n-1}\left[X(\tfrac{1}{M} - B_\tau)\right]$$

which we recognize as $\theta_{r-1} h^*_{n-1}[X]$. Thus Equation 60 reduces to

$$\begin{aligned} &\mathbf{C}^*_a \theta_r h^*_n[X] - \chi(a=1)\Big(\theta_r h^*_{n-1}[X] + \theta_{r-1} h^*_{n-1}[X]\Big) \\ &\quad = \sum_{k=0}^{r-1} h_{r-1-k}[\tfrac{1}{M}](-1)^k \sum_{s=a}^{n} t^{s-1} h_{s-a}[-\tfrac{X}{1-t}] \sum_{\tau \vdash k} \frac{1}{w_\tau} \times \\ &\qquad \sum_{v=s}^{n} h_{n-v}\left[X(\tfrac{1}{M} - B_\tau)\right] h_{v-s}[-X] h_{v-2}\big[1 - MB_\tau\big]. \end{aligned} \tag{61}$$

Calling this last factor LF we have

$$\begin{aligned} LF &= t^{a-1} \sum_{s=a}^{n} \sum_{v=s}^{n} h_{v-s}[-X] h_{s-a}[-\tfrac{tX}{1-t}] \times \\ &\qquad \times \sum_{\tau \vdash k} \frac{1}{w_\tau} h_{n-v}\left[X(\tfrac{1}{M} - B_\tau)\right] h_{v-2}\big[1 - MB_\tau\big] \\ &= t^{a-1} \sum_{v=a}^{n} \sum_{s=a}^{v} h_{v-s}[-X] h_{s-a}[-\tfrac{tX}{1-t}] \times \\ &\qquad \times \sum_{\tau \vdash k} \frac{1}{w_\tau} h_{n-v}\left[X(\tfrac{1}{M} - B_\tau)\right] h_{v-2}\big[1 - MB_\tau\big]. \end{aligned}$$

But making the substitution $s - a = u$ we get

$$\sum_{s=a}^{v} h_{v-s}[-X] h_{s-a}[-\tfrac{tX}{1-t}] = \sum_{u=0}^{v-a} h_{v-a-u}[-X] h_u[-\tfrac{tX}{1-t}] = h_{v-a}[-\tfrac{X}{1-t}].$$

This gives

$$LF = t^{a-1} \sum_{v=a}^{n} h_{v-a}[-\tfrac{X}{1-t}] \sum_{\tau \vdash k} \frac{1}{w_\tau} h_{n-v}\left[X(\tfrac{1}{M} - B_\tau)\right] h_{v-2}\big[1 - MB_\tau\big]$$

and Equation 61 becomes

$$\mathbf{C}^*_a \theta_r h^*_n[X] - \chi(a=1)\Big(\theta_r h^*_{n-1}[X] + \theta_{r-1} h^*_{n-1}[X]\Big)$$

$$= t^{a-1} \sum_{k=0}^{r-1} h_{r-1-k}[\tfrac{1}{M}](-1)^k \sum_{v=a}^{n} h_{v-a}[-\tfrac{X}{1-t}] \times$$
$$\sum_{\tau \vdash k} \frac{1}{w_\tau} h_{n-v}\left[X(\tfrac{1}{M} - B_\tau)\right] h_{v-2}\left[1 - MB_\tau\right].$$

A look at Equation 52 reveals that this last expression is none other than $t^{a-1}\mathbf{B}^*_{a-2}\theta_{r-1}h^*_{n-2}[X]$. In other words we have proved the identity

$$\mathbf{C}^*_a \theta_r h^*_n[X] = t^{a-1}\mathbf{B}^*_{a-2}\theta_{r-1}h^*_{n-2}[X]$$
$$+ \chi(a = 1)\Big(\theta_r h^*_{n-1}[X] + \theta_{r-1}h^*_{n-1}[X]\Big)$$

and our proof of Theorem 1 is thus complete.

4 Combinatorial consequences

Let us denote by $\mathcal{PF}_{p_1,p_2,\ldots,p_k}(r)$, for $0 < r < n$, the collection of parking functions with composition $(p_1, p_2, \ldots, p_k) \models n$ and diagonal word a shuffle of $12\cdots r$ with $r + 1 \cdots n$. In symbols

$$\mathcal{PF}_{p_1,p_2,\ldots,p_k}(r) = \left\{ PF \in \mathcal{PF}_n \ : \ p(PF) = (p_1,p_2,\ldots,p_k) \ \& \ \sigma(PF) \in 12\cdots r \sqcup\!\sqcup\, r+1\cdots n \right\}$$

and set

$$\Pi_{(p_1,p_2,\ldots,p_k)}(r; q, t) = \sum_{PF \in \mathcal{PF}_{p_1,p_2,\ldots,p_k}(r)} t^{\mathrm{area}(PF)} q^{\mathrm{dinv}(PF)}.$$

Our basic goal in this section is to prove the identity in Equation 5 which can be written as

$$\Pi_{(p_1,p_2,\ldots,p_k)}(r; q, t) = \big\langle \nabla \mathbf{C}_{p_1}\mathbf{C}_{p_2}\cdots\mathbf{C}_{p_k} 1 \ , \ h_r h_{n-r} \big\rangle. \tag{62}$$

Our plan is to verify that both sides satisfy the same recursion and that they are equal for all the base cases. Now we proved (by Theorem 1) that the right-hand side satisfies

$$\big\langle \nabla \mathbf{C}_{p_1}\mathbf{C}_{p_2}\cdots\mathbf{C}_{p_k} 1 \ , \ h_r h_{n-r} \big\rangle$$
$$= t^{p_1-1}\big\langle \nabla \mathbf{B}_{p_1-2}\mathbf{C}_{p_2}\cdots\mathbf{C}_{p_k} 1 \ , \ h_{r-1}h_{n-1-r} \big\rangle + \chi(p_1 = 1) \times$$
$$\Big(\big\langle \nabla \mathbf{C}_{p_2}\cdots\mathbf{C}_{p_k} 1 \ , \ h_r h_{n-1-r} \big\rangle + \big\langle \nabla \mathbf{C}_{p_2}\cdots\mathbf{C}_{p_k} 1 \ , \ h_{r-1}h_{n-r} \big\rangle \Big). \tag{63}$$

To extract information from this recursion we need to rewrite it in a combinatorially more revealing form. Proposition 4 was included precisely for this purpose. In fact, the Haglund-Morse-Zabrocki conjectures suggest that the operator $\mathbf{B}_{p_1-2}$ in the expression

$$\mathbf{B}_{p_1-2}\mathbf{C}_{p_2}\cdots\mathbf{C}_{p_k}1$$

must be moved to the right passed all operators $\mathbf{C}_{p_i}$ to act on 1. This requires using Equation 14, but only for $b \geq 1$ and $a \geq -1$. This reduces it to the following two cases:

$$\mathbf{B}_a\mathbf{C}_b = q\mathbf{C}_b\mathbf{B}_a \quad (\text{for } a \geq 0 \;\&\; b \geq 1) \tag{64}$$

$$\mathbf{B}_{-1}\mathbf{C}_1 = q\mathbf{C}_1\mathbf{B}_{-1}+(q-1)I \quad (\text{for } a = -1 \;\&\; b = 1) \tag{65}$$

with I the identity operator.

We are thus led to the following version of Equation 63.

Proposition 7. *The right-hand side of Equation 62 satisfies the following recursions:*

(a) *When* $p_1 > 1$,

$$\begin{aligned}\langle \nabla\mathbf{C}_{p_1}\mathbf{C}_{p_2}\cdots\mathbf{C}_{p_k}1\,,\;h_rh_{n-r}\rangle \\ = t^{p_1-1}q^{k-1}\langle \nabla\mathbf{C}_{p_2}\cdots\mathbf{C}_{p_k}\mathbf{B}_{p_1-2}1\,,\;h_{r-1}h_{n-1-r}\rangle\end{aligned} \tag{66}$$

(b) *When* $p_1 = 1$,

$$\begin{aligned}&\langle \nabla\mathbf{C}_1\mathbf{C}_{p_2}\cdots\mathbf{C}_{p_k}1\,,\;h_rh_{n-r}\rangle \\ &= \langle \nabla\mathbf{C}_{p_2}\cdots\mathbf{C}_{p_k}1\,,\;h_rh_{n-1-r}+h_{r-1}h_{n-r}\rangle \\ &\quad + (q-1)\sum_{i=2}^{k}{}^{(p_i=1)}q^{i-2}\langle \nabla\mathbf{C}_{p_2}\cdots\cancel{\mathbf{C}_{p_i}}\cdots\mathbf{C}_{p_k}1\,,\;h_{r-1}h_{n-1-r}\rangle.\end{aligned} \tag{67}$$

Proof. Note that when $p_1 > 1$ then $p_1 - 2 \geq 0$ and since all parts of a composition are ≥ 1 we can use Equation 64 $k-1$ times and immediately obtain Equation 66 from Equation 63. Next note that for $p_1 = 1$ we need to move $\mathbf{B}_{-1}$ passed all $\mathbf{C}_{p_i}$ in the expression

$$\mathbf{B}_{-1}\mathbf{C}_{p_2}\cdots\mathbf{C}_{p_k}1.$$

To see how Equation 67 comes out of this operation we need only work it out in a special case. Let us take $k = 4$. Given this, we have, by repeated uses of 64 and 65,

$$
\begin{aligned}
\mathbf{B}_{-1}\mathbf{C}_{p_2}\mathbf{C}_{p_3}\mathbf{C}_{p_4}1 &= q\mathbf{C}_{p_2}\mathbf{B}_{-1}\mathbf{C}_{p_3}\mathbf{C}_{p_4}1 + \chi(p_2=1)(q-1)\mathbf{C}_{p_3}\mathbf{C}_{p_4}1\\
&= q^2\mathbf{C}_{p_2}\mathbf{C}_{p_3}\mathbf{B}_{-1}\mathbf{C}_{p_4}1 + \chi(p_3=1)q(q-1)\mathbf{C}_{p_2}\mathbf{C}_{p_4}1\\
&\quad + \chi(p_2=1)(q-1)\mathbf{C}_{p_3}\mathbf{C}_{p_4}1\\
&= q^3\mathbf{C}_{p_2}\mathbf{C}_{p_3}\mathbf{C}_{p_4}\mathbf{B}_{-1}1 + \chi(p_4=1)q^2(q-1)\mathbf{C}_{p_2}\mathbf{C}_{p_3}1+\\
&\quad + \chi(p_3=1)q(q-1)\mathbf{C}_{p_2}\mathbf{C}_{p_4}1 + \chi(p_2=1)(q-1)\mathbf{C}_{p_3}\mathbf{C}_{p_4}1.
\end{aligned}
$$

But the first term vanishes, since the operator $\mathbf{B}_{-a}$ decreases degrees by a, and we get that

$$
\begin{aligned}
&\left\langle \nabla\mathbf{B}_{-1}\mathbf{C}_{p_2}\mathbf{C}_{p_3}\mathbf{C}_{p_4}1\,,\ h_r h_{n-r}\right\rangle\\
&\qquad = (q-1)\sum_{i=2}^{4}{}^{(p_i=1)} q^{i-2}\left\langle \nabla\mathbf{C}_{p_2}\cdots\cancel{\mathbf{C}_{p_i}}\cdots\mathbf{C}_{p_4}1\,,\ h_{r-1}h_{n-1-r}\right\rangle.
\end{aligned}
$$

Of course we can complete the proof of Equation 67 by an induction argument, but it wouldn't add anything to what we have just seen. □

Now it was shown in [9] that

$$
\mathbf{B}_a 1 = \sum_{p\models a}\mathbf{C}_{p_1}\mathbf{C}_{p_2}\cdots\mathbf{C}_{p_{l(p)}}1, \tag{68}
$$

where $l(p)$ denotes the length of the composition p.

Given this, by combining Proposition 7 and Equation 68 we obtain the following theorem.

Theorem 1. *The two sides of Equation 62 satisfy the same recursion if for all* $(p_1,p_2,\ldots,p_k)\models n$, *and* $0<r<n$,

(a) when $p_1>1$

$$
\begin{aligned}
&\Pi_{(p_1,p_2,\ldots,p_k)}(r;q,t)\\
&\qquad = t^{p_1-1}q^{k-1}\sum_{z\models p_1-2}\Pi_{(p_2,\ldots,p_k,z_1,z_2,\ldots,z_{l(z)})}(r-1;q,t)
\end{aligned}
\tag{69}
$$

(b) when $p_1=1$

$$
\begin{aligned}
\Pi_{(1,p_2,\ldots,p_k)}(r;q,t) &= \Pi_{(p_2,\ldots,p_k)}(r;q,t)+\Pi_{(p_2,\ldots,p_k)}(r-1;q,t)\\
&\quad + (q-1)\sum_{i=2}^{k}{}^{(p_i=1)}q^{i-2}\Pi_{(p_2,\ldots,\cancel{p_i}\ldots p_k)}(r-1;q,t).
\end{aligned}
\tag{70}
$$

To verify these two identities we need some observations. To begin, for a $PF\in\mathcal{PF}_{p_1,p_2,\ldots,p_k}(r)$, it will be convenient to refer to $1,2,\cdots r$ as the *small cars* and to $r+1,\ldots,n$ as *big cars*. Now the condition that $\sigma(PF)$ is a shuffle of increasing small cars with increasing big cars forces small cars as well as big cars to be increasing from higher to lower diagonals and from right to left along diagonals. Thus there will never be a small car on top of a small car or a big car on top of a big car. This implies that the Dyck paths supporting our parking functions will necessarily have only columns of NORTH steps of length at most 2. For the same reason, primary diagonal inversions will occur only when a small car is to the left of a big car in the same diagonal. Likewise a secondary diagonal inversion occurs only when a big car is to the left of a small car in the adjacent lower diagonal.

Given this, it will be convenient to represent a $PF \in \mathcal{PF}_{p_1,p_2,\ldots,p_k}(r)$ by the *reduced* tableau obtained by replacing all the small cars by a 1 and all big cars by a 2. Clearly, to recover PF from such a tableau we need only replace all the $1's$ by $1, 2, \ldots, r$ and all the $2's$ by $r+1, \ldots, n$ proceeding by diagonals, from the highest to the lowest and within diagonals from right to left.

More precisely, we will work directly with the corresponding two-line array viewed as a sequence of columns which we will call *dominos* and refer to it as $\mathrm{dom}(PF)$.

For instance on the left in the display below, we have a $PF \in \mathcal{PF}_{6,3,1}(5)$. We purposely depicted the big cars $6, 7, 8, 9, 10$ in a larger size than the small cars $1, 2, 3, 4, 5$. On the right we have its reduced tableau with the adjacent column of diagonal numbers. On the bottom we display $\mathrm{dom}(PF)$.

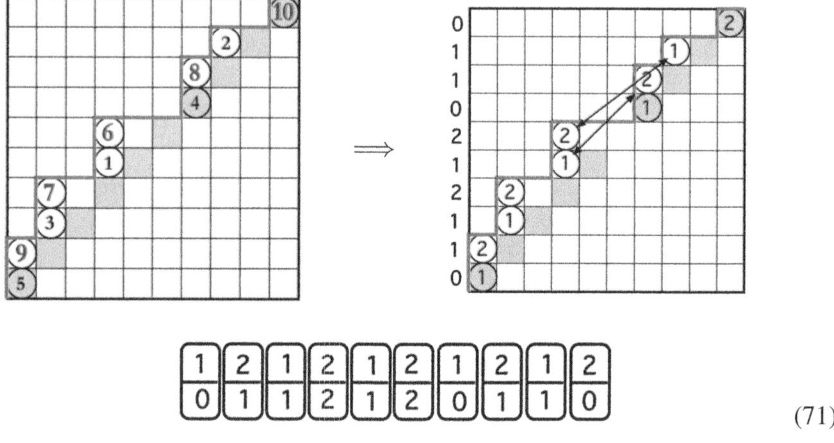

(71)

It may be good to say a few words about the manner in which the parking functions of a family $\mathcal{PF}_{p_1,p_2,\ldots,p_k}(r)$ can be constructed. First, we create all the Dyck paths which hit the diagonal according to the composition $(p_1, p_2, \ldots, p_k)$ and have no more than $\min(r, n-r)$ columns of length two and all remaining columns of length one. Then, for each of these Dyck paths, we fill the lattice cells adjacent to its columns of length two with a 1 below a 2, then place, along the columns of length one, the remaining $r - \min(r, n-r)$ $1's$ and $n - r - \min(r, n-r)$ $2's$ in all possible ways.

Now that we are familiarized with these parking functions we can proceed to establish the identities in Theorem 1. To verify Equation 69 we need only construct a bijection

$$\Phi : \mathcal{PF}_{p_1,p_2,\ldots,p_k}(r) \iff \bigcup_{z \models p_1 - 2} \mathcal{PF}_{(p_2,\ldots,p_k,z_1,z_2,\ldots,z_{l(z)})}(r-1) \tag{72}$$

such that

$$\text{area}(PF) = p_1 - 1 + \text{area}(\Phi(PF)) \text{ and } \text{dinv}(PF) = k - 1 + \text{dinv}(\Phi(PF)). \tag{73}$$

The combinatorial interpretation of these equalities is very suggestive.

- some NORTH steps of the supporting Dyck path must be shifted to the right to cause a loss of area of $p_1 - 1$.
- Note that $p_1 > 1$ forces a $PF \in \mathcal{PF}_{p_1,p_2,\ldots,p_k}(r)$ to start with a column of length 2. If we could remove this column we will cause a loss of one diagonal inversion for each of the remaining cars in the main diagonal, thereby satisfying the required dinv loss of $k - 1$.

Led by these two observations and the experience gained in previous work [5] we construct the map Φ as follows .

Given a $PF \in \mathcal{PF}_{p_1,p_2,\ldots p_k}(r)$ with $p_1 > 1$, we apply to dom(PF) the following 4 step procedure, and then let $\Phi(PF)$ be the parking function corresponding to the resulting domino sequence.

Step 1: Cut dom(PF) in two sections, the first containing its first p_1 dominos and the second containing the remaining $n - p_1$.
Step 2: Remove from the first section the first two dominos.
Step 3: Decrease by 1 the diagonal number of every domino remaining in the first section.
Step 4: Cycle the processed first section to the end of the second section.

For instance in the display below, we have first the result of applying Steps 1,2,3 to the domino sequence in Equation 4 and then below it we give the domino sequence resulting from Step 4 together with the corresponding reduced tableau and the image by Φ of the parking function in 71.

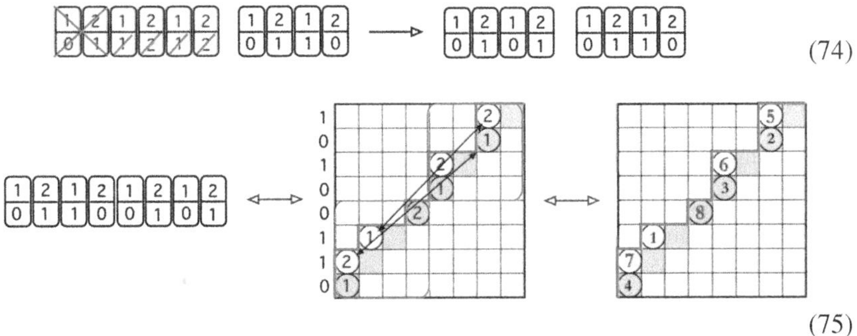

(74)

(75)

To complete the proof of Equation 69 we need to show that Φ is a bijection as stated in Equation 72 and that the requirements in Equation 73 are satisfied.

To begin, $\Phi(PF)$ is in $\mathcal{PF}_{(p_2,\ldots,p_k,z_1,z_2,\ldots,z_{l(z)})}(r-1)$ with

$$z = (z_1, z_2, \ldots, z_{l(z)}) \models p_1 - 2,$$

since decreasing by 1 the diagonal numbers in step 3 will cause the $p_1 - 2$ last cars of $\Phi(PF)$ to have a supporting Dyck path that hits the diagonal according to a composition of $p_1 - 2$. Conversely, given a $PF' \in \mathcal{PF}_{(p_2,\ldots,p_k,z_1,z_2,\ldots,z_{l(z)})}(r-1)$ we can reconstruct the domino sequence of the parking function PF that Φ maps to PF', by applying the following sequence of steps to dom(PF').

Step -1: Cut dom(PF') into two successive sections of respective lengths $p_2 + \cdots + p_k$ and $p_1 - 2$.
Step -2: Add 1 to the area numbers of the dominos in the second section.
Step -3: Cycle back the resulting second section to precede the first section.
Step -4: Prepend the resulting domino sequence by the pair $[{}^1_0][{}^2_1]$.

It is not difficult to see that this construction always yields a legitimate domino sequence of a reduced parking function. For instance, note that since the second section of dom(PF') will necessarily start with one $[{}^1_0]$ or $[{}^2_0]$, then after Step -2 these will become $[{}^1_1]$ or $[{}^2_1]$ and thus we are always able to precede any one of them by the pair $[{}^1_0][{}^2_1]$ and obtain a domino sequence of a parking function with diagonal composition $(p_1, p_2, \ldots p_k)$. This should make it clear that the Φ is a bijection as stated in Equation 72.

It remains to verify the equalities in Equation 73.

It is quite evident that the area equality is guaranteed by Step 3 together with the removal of the domino $[{}^2_1]$ in Step 2. Moreover, the dinv equality holds true for two reasons:

- Before we removed the pair $[{}^1_0][{}^2_1]$, every domino $[{}^1_0]$ to the right contributed a secondary diagonal inversion with the removed $[{}^2_1]$ and every domino $[{}^2_0]$ to the right contributed a primary diagonal inversion with the removed $[{}^1_0]$. Thus the loss of dinv is precisely $k-1$ which is the number of times the Dyck path of PF returns to the main diagonal.
- No dinv gains or losses are produced by the reversal of the sequence orders in Step 4, since the combination of Step 3 and Step 4 causes all the primary diagonal inversions to become secondary and all the secondary to become primary (a pair of examples are illustrated in Equation 71 and 75 by corresponding arrows in the reduced tableaux).

This completes our proof of Equation 69. □

Our next task is to verify Equation 70. We will start with some auxiliary observations. Note first that for any $PF \in \mathcal{PF}_{p_1,p_2,\ldots,p_k}(r)$ we may regard dom(PF) as a sequence of sections of lengths $p_1, p_2, \ldots, p_k$. Each section starts with a *small car* domino $[{}^1_0]$ or a *big car* domino $[{}^2_0]$. We will call them *main diagonal dominos*.

More precisely, for any $p_i = 1$, its corresponding section reduces to a single main diagonal domino. Conversely, each $[{2 \atop 0}]$ occurring in dom(PF) must be the sole element of sections of length 1. This is due to the fact each section of length greater than 1 must start with the pair of dominos $[{1 \atop 0}][{2 \atop 1}]$.

Now let $PF \in \mathcal{PF}_{1,p_2,\dots,p_k}(r)$ and let PF' be the parking function whose domino sequence is dom(PF) with its initial domino removed. Suppose first that dom(PF) starts with a big car domino $[{2 \atop 0}]$. In that case we set $\Phi(PF) = PF'$ and we are done, since $PF' \in \mathcal{PF}_{p_2,\dots,p_k}(r)$. Moreover, there is no area loss, and since the removed $[{2 \atop 0}]$ did not make any diagonal inversions with any of the succeeding dominos, we have

$$t^{\text{area}(PF)} q^{\text{dinv}(PF)} = t^{\text{area}(\Phi(PF))} q^{\text{dinv}(\Phi(PF))} .$$

Suppose next that dom(PF) starts with a $[{1 \atop 0}]$. Note that in this case $PF' \in \mathcal{PF}_{p_2,\dots,p_k}(r-1)$. However, here the removed $[{1 \atop 0}]$ is used to make a diagonal inversion with every main diagonal domino $[{2 \atop 0}]$ of dom(PF). Thus in this case we have

$$t^{\text{area}(PF)} q^{\text{dinv}(PF)} = t^{\text{area}(PF')} q^{\text{dinv}(PF')} q^m, \tag{76}$$

where m gives the number of $[{2 \atop 0}]$ in dom(PF). We certainly cannot set $\Phi(PF) = PF'$ here, since the weight of PF' occurs with coefficient 1 in the second term on the right-hand side of Equation 70. It turns out that the sum on the right-hand side of Equation 70 is precisely what is needed to perform the necessary correction when $m > 0$. To see how this comes about, we start by writing q^m as the sum

$$q^m = 1 + (q-1) + (q-1)q + (q-1)q^2 + \cdots + (q-1)q^{m-1},$$

so that Equation 76 may be rewritten as

$$\begin{aligned} &t^{\text{area}(PF)} q^{\text{dinv}(PF)} \\ &\quad = t^{\text{area}(PF')} q^{\text{dinv}(PF')} + (q-1)\sum_{s=1}^{m} t^{\text{area}(PF')} q^{\text{dinv}(PF')+s-1} . \end{aligned} \tag{77}$$

Now, suppose that the dominos $[{2 \atop 0}]$ occur in dom(PF) in positions

$$1 < i_1 < i_2 < \cdots < i_m \leq n .$$

Note that, by one of our prior observations, we must have $p_{i_s} = 1$ for all $1 \leq s \leq m$. Given this, let $PF^{(i_s)}$ be the parking function whose domino sequence is obtained by removing from dom(PF) the initial domino $[{1 \atop 0}]$ together with the domino $[{2 \atop 0}]$ in position i_s. Now since every main diagonal domino in dom(PF) located between the two removed dominos used to make a primary or secondary diagonal inversion with one or the other of the removed dominos, and the initial domino $\left[\begin{smallmatrix}1\\0\end{smallmatrix}\right]$ made a diagonal inversion with the removed $\left[\begin{smallmatrix}2\\0\end{smallmatrix}\right]$ as well as with all the big car dominos in position $i_{s+1}, \ldots, i_m$, we derive that

$$\text{dinv}(PF') + m = \text{dinv}(PF) = \text{dinv}(PF^{(i_s)}) + i_s - 2 + 1 + m - s.$$

This gives

$$q^{\text{dinv}(PF')+s-1} = q^{\text{dinv}(PF^{(i_s)})+i_s-2}.$$

Using this in Equation 77 rewrites it in the more suggestive form

$$t^{\text{area}(PF)} q^{\text{dinv}(PF)} \\ = t^{\text{area}(PF')} q^{\text{dinv}(PF')} + (q-1) \sum_{s=1}^{m} q^{i_s-2} t^{\text{area}(PF^{(i_s)})} q^{\text{dinv}(PF^{(i_s)})}. \tag{78}$$

Let us now set for any $PF \in \mathcal{PF}_{1,p_2,\ldots,p_k}(r)$

$$\Phi(PF) = \begin{cases} PF' & \text{if dom}(PF) \text{ starts with a } \left[\begin{smallmatrix}2\\0\end{smallmatrix}\right], \\ \left(PF', PF^{(i_1)}, PF^{(i_2)}, \ldots, PF^{(i_m)}\right) & \text{if dom}(PF) \text{ starts with a } \left[\begin{smallmatrix}1\\0\end{smallmatrix}\right], \end{cases}$$

and note that Equation 78 shows that this defines a weight preserving, bijective map of $\mathcal{PF}_{1,p_2,\ldots,p_k}(r)$ onto a disjoint family of subsets covering the union

$$\mathcal{PF}_{p_2,\ldots,p_k}(r) \bigcup \mathcal{PF}_{p_2,\ldots,p_k}(r-1) \bigcup_{p_i=1} \mathcal{PF}_{(p_2,\ldots,\not{p_i},\ldots p_k)}(r-1)$$

Indeed this map is onto since

1. if $PF' \in \mathcal{PF}_{p_2,\ldots,p_k}(r)$, then it is the image by Φ of the PF whose domino sequence is obtained by prepending dom(PF') by a $\left[\begin{smallmatrix}2\\0\end{smallmatrix}\right]$;
2. if $PF' \in \mathcal{PF}_{p_2,\ldots,p_k}(r-1)$, then it is in the image by Φ of the PF whose domino sequence is obtained by prepending dom(PF') by a $\left[\begin{smallmatrix}1\\0\end{smallmatrix}\right]$;
3. if $PF^{(i)} \in \mathcal{PF}_{(p_2,\ldots,\not{p_i},\ldots p_k)}(r-1)$ with $p_i = 1$, then it is in the image by Φ of the PF whose domino sequence is obtained by prepending dom$(PF^{(i)})$ by a $\left[\begin{smallmatrix}1\\0\end{smallmatrix}\right]$ and inserting a $\left[\begin{smallmatrix}2\\0\end{smallmatrix}\right]$ in position i.

Given this, Equation 70 is simply obtained by summing Equation 77 over all $PF \in \mathcal{PF}_{1,p_2,\ldots,p_k}(r)$. This completes our proof that the two sides of Equation 62 satisfy the same recursion.

We are left to verify the equality in the base cases. To this end note that since at each use of the recursion one or more of the following happens:

- r is decreased.
- The composition p is getting finer,
- The number of parts of p decreases.

From the beginning we have required that $0 < r < n$, simply because when $r = 0$ or $r = n$ the family $\mathcal{PF}_{p_1,p_2,\dots,p_k}(r)$ reduces to a triviality. In fact, if there are no small cars, or no big cars, the family is empty unless p reduces to a string of $1's$ and if that happens then there is only one parking function with no area and no dinv. Thus the polynomial $\Pi_{p_1,p_2,\dots,p_k}(r)$ either vanishes or it is equal to 1. Not withstanding this, the recursion forces us to include all the degenerate cases. Omitting some trivial cases in which both the combinatorial side as well as the symmetric function side are easily shown to vanish. The only significant basic cases are when p reduces to a string of $1's$ and $0 < r < n$. In this case the family $\mathcal{PF}_{1^n}(r)$ consists of all the parking functions obtained by placing along the main diagonal and from right to left all the shuffles of $12\cdots r$ with $r+1\cdots n$. In this case there is no area and the dinv statistic reduces to an inversion count. The corresponding polynomial then is none other than the q-binomial coefficient

$$\Pi_{1^n}(r) = \begin{bmatrix} n \\ r \end{bmatrix}_q .$$

We need to show that the symmetric function side yields the same result. That is with n occurrences of $\mathbf{C}_1$ we have

$$\langle \nabla \mathbf{C}_1 \mathbf{C}_1 \cdots \mathbf{C}_1 1 \, , \, h_r h_{n-r} \rangle = \begin{bmatrix} n \\ r \end{bmatrix}_q . \tag{79}$$

Now we have shown (Proposition 3) that in this case we have

$$\nabla \mathbf{C}_1 \mathbf{C}_1 \cdots \mathbf{C}_1 1 = (q,q)_n h_n \left[\tfrac{X}{1-q}\right]. \tag{80}$$

However we obtain from the Cauchy identity that

$$\langle h_n \left[\tfrac{X}{1-q}\right] , \, h_\mu[X] \rangle = h_\mu \left[\tfrac{1}{1-q}\right]$$

which combined with Equation 80 gives

$$\langle \nabla \mathbf{C}_1 \mathbf{C}_1 \cdots \mathbf{C}_1 1 \, , \, h_r h_{n-r} \rangle = \frac{(q,q)_n}{(q,q)_r (q,q)_{n-r}} \tag{81}$$

which is another way of writing Equation 79.

This completes our proof of Equation 5. □

Remark 2. We must note that the very nature of case $p_1 = 1$ in Equation 69 makes it stand apart from any of the recursions encountered in all previous parking function literature. For this reason there is no way we could have discovered what to do with our parking functions in this degenerate case without help from the symmetric function side. What is fascinating is that the intricacy of this case and the parking function magic that takes place is none other but a side product of the commutativity relations afforded by the **C** and **B** operators. In this context it is interesting to see that Equation 70 tells us about q-binomial coefficients. In fact, when p reduces to a string of $1's$, using Equation 81, then Equation 70 states that

$$\frac{(q,q)_n}{(q,q)_r(q,q)_{n-r}} = \frac{(q,q)_{n-1}}{(q,q)_r(q,q)_{n-1-r}} + \frac{(q,q)_{n-1}}{(q,q)_{r-1}(q,q)_{n-r}} + (q-1)\sum_{i=2}^{n} q^{i-2} \frac{(q,q)_{n-2}}{(q,q)_{r-1}(q,q)_{n-1-r}} \,. \tag{82}$$

Since

$$(q-1)\sum_{i=2}^{n} q^{i-2} = (q-1)\sum_{i=0}^{n-2} q^{i} = (q-1)\frac{1-q^{n-1}}{1-q} = q^{n-1} - 1,$$

Equation 82 becomes

$$\frac{(q,q)_n}{(q,q)_r(q,q)_{n-r}} = \frac{(q,q)_{n-1}}{(q,q)_r(q,q)_{n-1-r}} + \frac{(q,q)_{n-1}}{(q,q)_{r-1}(q,q)_{n-r}} + (q^{n-1}-1)\frac{(q,q)_{n-2}}{(q,q)_{r-1}(q,q)_{n-1-r}}$$

or better

$$\frac{(q,q)_n}{(q,q)_r(q,q)_{n-r}} = \frac{(q,q)_{n-1}}{(q,q)_r(q,q)_{n-1-r}} + \frac{(q,q)_{n-1}}{(q,q)_{r-1}(q,q)_{n-r}} \frac{-(1-q^{n-r})(q,q)_{n-1}}{(q,q)_{r-1}(q,q)_{n-r}}$$

which is just another way of writing the classical recursion

$$\frac{(q,q)_n}{(q,q)_r(q,q)_{n-r}} = \frac{(q,q)_{n-1}}{(q,q)_r(q,q)_{n-1-r}} + q^{n-r} \frac{(q,q)_{n-1}}{(q,q)_{r-1}(q,q)_{n-r}} \,.$$

This fact brings us to view these ramifications of the shuffle conjecture as parking function versions of q-binomial identities.

References

1. F. Bergeron and A. M. Garsia, *Science fiction and Macdonald's polynomials*, Algebraic methods and q-special functions (Montréal, QC, 1996), CRM Proc. Lecture Notes, Vol. 22, Amer. Math. Soc., Providence, RI, 1999, pp. 1–52.
2. F. Bergeron, A. M. Garsia, M. Haiman, and G. Tesler, *Identities and positivity conjectures for some remarkable operators in the theory of symmetric functions*, Methods in Appl. Anal. 6 (1999), 363–420.
3. A. Garsia, M. Haiman, and G. Tesler, *Explicit plethystic formulas for the Macdonald q,t-Kostka coefficients*, Séminaire Lotharingien de Combinatoire, B42m (1999), 45 pp.
4. A. M. Garsia, A. Hicks, and A. Stout, *The case $k = 2$ of the shuffle conjecture*, J. Comb. 2 (2011), no. 2, 193–229.
5. A. M. Garsia, G. Xin, and M. Zabrocki, *Hall–Littlewood operators in the theory of parking functions and diagonal harmonics*, Int. Math. Res. Not. 2012, no. 6, 1264–1299.
6. J. Haglund, *A proof of the q,t-Schröder conjecture*, Internat. Math. Res. Notices 11 (2004), 525–560.
7. J. Haglund, *The q,t-Catalan numbers and the space of diagonal harmonics*, AMS University Lecture Series, Vol. 41 (2008) pp. 167.
8. J. Haglund, M. Haiman, N. Loehr, J. B. Remmel, and A. Ulyanov, *A combinatorial formula for the character of the diagonal coinvariants*, Duke J. Math. 126 (2005), 195–232.
9. J. Haglund, J. Morse, and M. Zabrocki, *A compositional shuffle conjecture specifying touch points of the Dyck path* , Canad. J. Math. 64 (2012), no. 4, 822–844.
10. M. Zabrocki, UCSD Advancement to Candidacy Lecture Notes, Posted in http://www.math.ucsd.edu/~garsia/somepapers/.

On symmetric SL-invariant polynomials in four qubits

Gilad Gour and Nolan R. Wallach

To Nolan Wallach, I am indebted for the many hours and patience you invested in explaining representation and invariant theory for a physicist. For an expert mathematician, you are remarkably fluent in the language of a physicist.

— Gilad Gour

Abstract We find the generating set of SL-invariant polynomials in four qubits that are also invariant under permutations of the qubits. The set consists of four polynomials of degrees 2, 6, 8, and 12, for which we find an elegant expression in the space of critical states. These invariants are the degrees if the basic invariants of the invariants for F_4, and in fact, the group plays an important role in this note. In addition, we show that the hyperdeterminant in four qubits is the only SL-invariant polynomial (up to powers of itself) that is non-vanishing precisely on the set of generic states.

Keywords: Quantum entanglement • SL-invariant polynomials • Four qubits • Permutation invariant

Mathematics Subject Classification: 22, 81

Research of the first author has been supported by NSERC. Research of the second author has been partially supported by NSF grant DMS-0963036.

G. Gour
Department of Mathematics and Statistics, University of Calgary, Calgary, AB, Canada T2N 1N4
e-mail: gour@ucalgary.ca

N.R. Wallach
Department of Mathematics, University of California at San Diego, La Jolla, CA 92093-0112, USA
e-mail: nwallach@ucsd.edu

R. Howe et al. (eds.), *Symmetry: Representation Theory and Its Applications*, Progress in Mathematics 257, DOI 10.1007/978-1-4939-1590-3_9

1 Introduction

With the emergence of quantum information science in recent years, much effort has been given to the study of entanglement [7]: in particular, to its characterization, manipulation and quantification [15]. It was realized that highly entangled states are the most desirable resources for many quantum information processing tasks. While two-party entanglement has been very well studied, entanglement in multi-qubits systems is far less understood. Perhaps one of the reasons is that n qubits (with $n > 3$) can be entangled in an uncountable number of ways [5, 6, 18] with respect to stochastic local operations assisted by classical communication (SLOCC). It is therefore not very clear what role entanglement measures can play in multi-qubits or multi-qudits systems unless they are defined operationally. One exception from this conclusion are entanglement measures that are defined in terms of the absolute value of SL-invariant polynomials [3, 6, 10, 12, 14, 16, 18, 19].

Two important examples are the concurrence [20] and the square root of the 3-tangle (SRT) [3]. The concurrence and the SRT, respectively, are the only $\mathrm{SL}(2,\mathbb{C}) \otimes \mathrm{SL}(2,\mathbb{C})$ and $\mathrm{SL}(2,\mathbb{C}) \otimes \mathrm{SL}(2,\mathbb{C}) \otimes \mathrm{SL}(2,\mathbb{C})$ invariant measures of entanglement that are homogenous of degree 1. The reason for that is that in two or three qubit-systems there exists a unique SL-invariant polynomial. However, for 4-qubits or more, the picture is different since there are many homogenous SL-invariant polynomials such as the 4-tangle [16] or the hyperdeterminant [12].

In this paper, we find the generating set of all SL-invariant polynomials with the property that they are also invariant under any permutation of the four qubits. Such polynomials yield a measure of entanglement that capture genuine 4 qubits entanglement. In addition, we show that the 4-qubit hyperdeterminant [12] is the only homogeneous SL-invariant polynomial (of degree 24) that is non-vanishing precisely on generic states.

This paper is written with a variety of audiences in mind. First and foremost are the researchers who study quantum entanglement. We have therefore endeavored to keep the mathematical prerequisites to a minimum and have opted for proofs that emphasize explicit formulas for the indicated SL-invariant polynomials. We are aware that there are shorter proofs of the main results using the important work of Vinberg [17]. However, to us, the most important aspect of the paper is that the Weyl group of F_4 is built into the study of entanglement for 4 qubits. Indeed, the well-known result of Shepherd–Todd on the invariants for the Weyl group of F_4 gives an almost immediate proof of Theorem 2.1. The referee has indicated a short proof of Theorem 3.3 using more algebraic geometry. Although our proof is longer, we have opted to keep it since it is more elementary. We should also point out that in the jargon of Lie theory the hyperdeterminant is just the discriminant for the symmetric space corresponding to $\mathrm{SO}(4,4)$.

2 Symmetric invariants

Let $\mathcal{H}_n = \otimes^n \mathbb{C}^2$ denote the space of n-qubits, and let $G = \mathrm{SL}(2,\mathbb{C})^{\otimes n}$ act on $\mathcal{H}_n$ by the tensor product action. An SL-invariant polynomial, $f(\psi)$, is a polynomial in the components of the vector $\psi \in \mathcal{H}_n$, which is invariant under the action of the group G. That is, $f(g\psi) = f(\psi)$ for all $g \in G$. In the case of two qubits there exists only one unique SL-invariant polynomial. It is homogeneous of degree 2 and is given by the bilinear form (ψ, ψ):

$$f_2(\psi) \equiv (\psi, \psi) \equiv \langle \psi^* | \sigma_y \otimes \sigma_y | \psi \rangle \ , \ \ \psi \in \mathbb{C}^2 \otimes \mathbb{C}^2 \, ,$$

where σ_y is the second 2×2 Pauli matrix with i and $-i$ on the off-diagonal terms. Its absolute value is the celebrated concurrence [20].

Also in three qubits there exists a unique SL-invariant polynomial. It is homogeneous of degree 4 and is given by

$$f_4(\psi) = \det \begin{bmatrix} (\varphi_0, \varphi_0) & (\varphi_0, \varphi_1) \\ (\varphi_1, \varphi_0) & (\varphi_1, \varphi_1) \end{bmatrix} \, ,$$

where the two qubits states φ_i for $i = 0, 1$ are defined by the decomposition $|\psi\rangle = |0\rangle|\varphi_0\rangle + |1\rangle|\varphi_1\rangle$, and the bilinear form (φ_i, φ_j) is defined above for two qubits. The absolute value of f_4 is the celebrated 3-tangle [3].

In four qubits, however, there are many SL-invariant polynomials and it is possible to show that they are generated by four SL-invariant polynomials (see e.g., [6] for more details and references). Here we are interested in SL-invariant polynomials that are also invariant under the permutation of the qubits.

Consider the permutation group S_n acting by the interchange of the qubits. Let $\widetilde{G}$ be the group $S_n \ltimes G$. That is, the set $S_n \times G$ with multiplication

$$(s, g_1 \otimes \cdots \otimes g_n)(t, h_1 \otimes \cdots \otimes h_n) = (st, g_{t^{-1}1} h_1 \otimes \cdots \otimes g_{t^{-1}n} h_n).$$

Then $\widetilde{G}$ acts on $\mathcal{H}_n$ by these two actions. We are interested in the polynomial invariants of this group action.

One can easily check that f_2 and f_4 above are also $\widetilde{G}$-invariant. However, this automatic $\widetilde{G}$-invariance of G-invariants breaks down for $n = 4$. As is well known [6], the polynomials on $\mathcal{H}_\triangle$ that are invariant under G are generated by four polynomials of respective degrees 2, 4, 4, 6. For $\widetilde{G}$ we have the following theorem:

Theorem 2.1. *The $\widetilde{G}$-invariant polynomials on $\mathcal{H}_\triangle$ are generated by four algebraically independent homogeneous polynomials h_1, h_2, h_3 and h_4 of respective degrees $2, 6, 8$ and 12. Furthermore, the polynomials can be taken to be $\mathcal{F}_1(z), \mathcal{F}_3(z), \mathcal{F}_4(z), \mathcal{F}_6(z)$ as given explicitly in Eq. (1) of the proof.*

Proof. To prove this result we will use some results from [6]. Let

$$u_0 = \frac{1}{2}(|0000\rangle + |0011\rangle + |1100\rangle + |1111\rangle),$$

$$u_1 = \frac{1}{2}(|0000\rangle - |0011\rangle - |1100\rangle + |1111\rangle),$$

$$u_2 = \frac{1}{2}(|0101\rangle + |0110\rangle + |1001\rangle + |1010\rangle),$$

$$u_3 = \frac{1}{2}(|0101\rangle - |0110\rangle - |1001\rangle + |1010\rangle).$$

Let A be the vector subspace of $\mathcal{H}_\triangle$ generated by the u_j. Then GA contains an open subset of $\mathcal{H}_\triangle$ and is dense. This implies that any G-invariant polynomial on $\mathcal{H}_\triangle$ is determined by its restriction to A. Writing a general state in A as $z = \sum z_i u_i$, we can choose the invariant polynomials such that their restrictions to A are given by

$$\mathcal{E}_0(z) = z_0 z_1 z_2 z_3, \quad \mathcal{E}_j(z) = z_0^{2j} + z_1^{2j} + z_2^{2j} + z_3^{2j}, \quad j = 1, 2, 3.$$

In [6] we give explicit formulas for their extensions to $\mathcal{H}_\triangle$.

Also, let W be the group of transformations of A given by

$$\{g \in G \mid gA = A\}_{|A}.$$

Then W is the finite group of linear transformations of the form

$$u_i \longmapsto \varepsilon_i u_{s^{-1}i}$$

with $\varepsilon_i = \pm 1$, $s \in S_4$ and $\varepsilon_0\varepsilon_1\varepsilon_2\varepsilon_3 = 1$. One can show [6, 19] that every W-invariant polynomial can be written as a polynomial in $\mathcal{E}_0, \mathcal{E}_1, \mathcal{E}_2, \mathcal{E}_3$. We now look at the restriction of the S_4 that permutes the qubits to A. Set $\sigma_i = (i, i+1), i = 1, 2, 3$ where $(2, 3)$ corresponds to fixing the first and last qubit and interchanging the second and third. Then they have matrices relative to the basis u_j :

$$\sigma_{1|A} = \begin{bmatrix} 1 & 0 & 0 & 0 \\ 0 & 1 & 0 & 0 \\ 0 & 0 & 1 & 0 \\ 0 & 0 & 0 & -1 \end{bmatrix}, \quad \sigma_{2|A} = \frac{1}{2}\begin{bmatrix} 1 & 1 & 1 & 1 \\ 1 & 1 & -1 & -1 \\ 1 & -1 & -1 & 1 \\ 1 & -1 & 1 & -1 \end{bmatrix}, \quad \sigma_{3|A} = \begin{bmatrix} 1 & 0 & 0 & 0 \\ 0 & 1 & 0 & 0 \\ 0 & 0 & 1 & 0 \\ 0 & 0 & 0 & -1 \end{bmatrix}.$$

Since S_4 is generated by $(1, 2), (2, 3), (3, 4)$, it is enough to find those W-invariants that are also $\sigma_{i|A}$ invariant for $i = 1, 2$. We note that the only one of the $\mathcal{E}_j$ that is not invariant under $\sigma_{1|A}$ is $\mathcal{E}_0$ and $\mathcal{E}_0(\sigma_1 z) = -\mathcal{E}_0(z)$ for $z \in A$. Thus if $F(x_0, x_1, x_2, x_3)$ is a polynomial in the indeterminates x_j, then $F(\mathcal{E}_0, \mathcal{E}_1, \mathcal{E}_2, \mathcal{E}_3)$ is invariant under $\nu = \sigma_{1|A}$ if and only if x_0 appears to even powers. It is an easy exercise to show that if $\mathcal{E}_4 = z_0^8 + z_1^8 + z_2^8 + z_3^8$, then any polynomial in $\mathcal{E}_0^2, \mathcal{E}_1, \mathcal{E}_2, \mathcal{E}_3$ is a polynomial in $\mathcal{E}_1, \mathcal{E}_2, \mathcal{E}_3, \mathcal{E}_4$ and conversely (see the argument in the very beginning of the next section). Thus we need only find the polynomials in $\mathcal{E}_1, \mathcal{E}_2, \mathcal{E}_3, \mathcal{E}_4$ that are invariant under $\tau = \sigma_{2|A}$. A direct calculation shows that $\mathcal{E}_1$ is invariant under τ. Also,

$$\mathcal{E}_1^2 \circ \tau = \mathcal{E}_1^2\,, \quad \mathcal{E}_2 \circ \tau = \frac{3}{4}\mathcal{E}_1^2 - \frac{1}{2}\mathcal{E}_2 + 6\mathcal{E}_0\,, \quad \mathcal{E}_0 \circ \tau = -\frac{1}{16}\mathcal{E}_1^2 + \frac{1}{8}\mathcal{E}_2 + \frac{1}{2}\mathcal{E}_0.$$

Since $\mathcal{E}_0, \mathcal{E}_1^2, \mathcal{E}_2$ forms a basis of the W-invariant polynomials of degree 4 this calculation shows that the space of polynomials of degree 4 invariant under τ, ν and W (hence under $\widetilde{W}$) consists of the multiples of $\mathcal{E}_1^2$. The space of polynomials invariant under W and ν and homogeneous of degree 6 is spanned by $\mathcal{E}_1^3, \mathcal{E}_1\mathcal{E}_2$, and $\mathcal{E}_3$. From this it is clear that the space of homogeneous degree 6 polynomials that are invariant under $\widetilde{W}$ is two dimensional. Since $\mathcal{E}_1^3$ is clearly $\widetilde{W}$-invariant there is one new invariant of degree 6. Continuing in this way we find that to degree 12 there are invariants h_1, h_2, h_3, h_4 of degrees 2, 6, 8 and 12, respectively, such that: (1) the invariant of degree 8, h_3, is not of the form $ah_1^4 + bh_1h_2$, (2) there is no new invariant of degree 10, and (3) the invariant of degree 12, h_4, cannot be written in the form $ah_1^6 + bh_1^3h_2 + ch_1^2h_3$. To describe these invariants we write out a new set of invariants. We put

$$\mathcal{F}_k(z) = \frac{1}{6}\sum_{i<j}\left(z_i - z_j\right)^{2k} + \frac{1}{6}\sum_{i<j}\left(z_i + z_j\right)^{2k}. \tag{1}$$

We note that $\mathcal{F}_1 = \mathcal{E}_1$, $\mathcal{F}_2 = \mathcal{E}_1^2$. A direct check shows that these polynomials are invariant under $\widetilde{W}$. Since $\mathcal{F}_3(z) \neq c\mathcal{E}_1^3$ we can use it as the "missing polynomial". If one calculates the Jacobian determinant of $\mathcal{F}_1(z), \mathcal{F}_3(z), \mathcal{F}_4(z), \mathcal{F}_6(z)$, then it is not 0. This implies that none of these polynomials can be expressed as a polynomial in the others. Thus they can be taken to be h_1, h_2, h_3, h_4.

Let $A_{\mathbb{R}}$ denote the vector space over $\mathbb{R}$ spanned by the u_j. If $\lambda \in A_{\mathbb{R}}$ is non-zero, then we set for $a \in A$, $s_\lambda a = a - \frac{2\langle\lambda|a\rangle}{\langle\lambda|\lambda\rangle}\lambda$. Then such a transformation is called a reflection. It is the reflection about the hyperplane perpendicular to λ. The obvious calculation shows that $\nu = s_{u_3}$and if $\alpha = \frac{1}{2}(u_0 - u_1 - u_2 - u_3)$, then $\tau = s_\alpha$. We note that W is generated by the reflections corresponding to $u_0 - u_1$, $u_1 - u_2, u_2 - u_3$ and $u_2 + u_3$. This implies that the group $\widetilde{W}$ is generated by reflections. One also checks that it is finite (actually of order 576). The general theory (cf. J. E. Humphreys [8, Thm 3.5, p. 54]) implies that the algebra of invariants is generated by algebraically independent homogeneous polynomials. Using this it is easy to see that $\mathcal{F}_1(z), \mathcal{F}_3(z), \mathcal{F}_4(z), \mathcal{F}_6(z)$ generate the algebra of invariants.

Remark 2.2. Alternatively, we note that $\widetilde{W}$ is isomorphic with the Weyl group of the exceptional group F_4 (see Bourbaki, Chapitres 4, 5, et 6, Planche VIII pp. 272, 273). The exponents (on p. 273) are 1, 5, 7, 11. This implies that the algebra of invariants is generated by algebraically independent homogeneous polynomials of degrees one more than the exponents, so 2, 6, 8, 12. We also note that the basic invariants for F_4 were given as $\mathcal{F}_1(z), \mathcal{F}_3(z), \mathcal{F}_4(z), \mathcal{F}_6(z)$ for the first time by M. L. Mehta [11].

For $n \geq 4$ qubits the analogue of the space A would have to be of dimension $2^n - 3n$. Thus even if there were a good candidate one would be studying, say, for 5 qubits, a space of dimension 17 and an immense finite group that cannot be generated by reflections.

3 A special invariant (hyperdeterminant) for 4 qubits

In this section we show that the hyperdeterminant for qubits is the only polynomial that quantifies genuine 4-way generic entanglement. We start by observing that Newton's formulas (relating power sums to elementary symmetric functions) imply that if $a_1, \ldots, a_n$ are elements of an algebra over $\mathbb{Q}$ (the rational numbers), then

$$a_1 a_2 \cdots a_n = f_n(p_1(a_1, \ldots, a_n), \ldots, p_n(a_1, \ldots, a_n))$$

with f_n a polynomial with rational coefficients in n indeterminates and $p_i(x_1, \ldots, x_n) = \sum x_j^i$. This says that in the notation of the previous theorem

$$\gamma(z) = \prod_{i<j} (z_i - z_j)^2 (z_i + z_j)^2$$

is $\widetilde{W}$-invariant. Indeed, take $a_1, \ldots, a_{n(n-1)}$ to be

$$\{(z_i - z_j)^2 \mid i < j\} \cup \{(z_i + z_j)^2 \mid i < j\}$$

in some order. We will also use the notation γ for the corresponding polynomial of degree 24 on $\mathcal{H}_\triangle$.

We define the generic set, Ω, in $\mathcal{H}_\triangle$ to be the set of elements, v, such that Gv is closed and $\dim Gv$ is maximal (that is, 12). Then every such element can be conjugated to an element of A by an element of G. It is easily checked that

$$\Omega \cap A = \{\textstyle\sum z_i u_i \mid z_i \neq \pm z_j \text{ if } i \neq j\}.$$

This implies that $\Omega = \{\phi \in \mathcal{H}_\triangle \mid \gamma(\phi) \neq 0\}$.

Proposition 3.1. *If f is a polynomial on $\mathcal{H}_\triangle$ that is invariant under the action of G and is such that $f(\mathcal{H}_\triangle - \Omega) = 0$, then f is divisible by γ.*

Proof. Since $f(z) = 0$ if $z_i = \pm z_j$ for $i \neq j$ we see that f is divisible by $z_i - z_j$ and $z_i + z_j$ for $i < j$. Thus if

$$\Delta(z) = \prod_{i<j} (z_i - z_j)(z_i + z_j),$$

then $f = \Delta g$ with g a polynomial on A. One checks that $\Delta(sz) = \det(s)\Delta(z)$ for $s \in W$ (see the notation in the previous section). Since $f(sz) = f(z)$ for $s \in W$ we see that $g(sz) = \det(s)g(z)$ for $s \in W$. But this implies that $g(z) = 0$ if $z_i = \pm z_j$ for $i \neq j$. So g is also divisible by Δ. We conclude that f is divisible by Δ^2. This is the content of the theorem.

Lemma 3.2. *γ is an irreducible polynomial.*

Proof. Let $\gamma = \gamma_1\gamma_2\cdots\gamma_m$ be a factorization into irreducible (non-constant) polynomials. If $g \in G$, then since the factorization is unique up to order and scalar multiple there is for each $g \in G$, a permutation $\sigma(g) \in S_m$ and $c_i(g) \in \mathbb{C} - \{0\}$, $i = 1,\ldots,m$ such that $\gamma_j \circ g^{-1} = c_j(g)\gamma_{\sigma(g)j}$ for $j = 1,\ldots,m$. The map $g \longmapsto \sigma(g)$ is a group homomorphism. The kernel of σ is a closed subgroup of G. Thus $G/\ker\sigma$ is a finite group that is a continuous image of G. So it must be the group with one element since G is connected. This implies that each γ_j satisfies $\gamma_j \circ g^{-1} = c_j(g)\gamma_j$ for all $g \in G$. We therefore see that $c_j : G \to \mathbb{C}-\{0\}$ is a group homomorphism for each j. But the commutator group of G is G. Thus $c_j(g) = 1$ for all g. This implies that each of the factors γ_j is invariant under G. Now each $\gamma_{j|A}$ divides $\gamma_{|A}$ thus it must be a product

$$\prod_{i<j}(z_i - z_j)^{a_{ij}}(z_i + z_j)^{b_{ij}}$$

by unique factorization. We note that if $i < j$, then $\{(z_i + z_j) \circ s \mid s \in W\} = \{(z_i - z_j) \circ s \mid s \in W\} = \{\varepsilon(z_i + z_j) \mid i < j, \varepsilon \in \{\pm 1\}\} \cup \{z_i - z_j \mid i \neq j\}$. This implies that since $\gamma_{j|A}$ is non-constant and W-invariant that each $\gamma_{j|A}$ must be divisible by $\Delta_{|A}$. Now arguing as in the previous proposition, the invariance implies that γ_j is divisible by $\Delta^2 = \gamma$. This implies that $m = 1$.

Theorem 3.3. *If f is a polynomial on $\mathcal{H}_\triangle$ such that $f(\phi) \neq 0$ for $\phi \in \Omega$, then there exists $c \in \mathbb{C}$, $c \neq 0$ and r such that $f = c\gamma^r$.*

Proof. We may assume that f is non-constant. Let h be an irreducible factor of f. Then $h(\phi) \neq 0$ if $\phi \in \Omega$. This implies that the irreducible variety

$$Y = \{x \in \mathcal{H}_\triangle \mid h(x) = 0\} \subset \mathcal{H}_\triangle - \Omega = \{x \in \mathcal{H}_\triangle \mid \gamma(x) = 0\}.$$

Since both varieties are of dimension 15 over $\mathbb{C}$ they must be equal. This implies that h must be a multiple of γ. Since f factors into irreducible non-constant factors the theorem follows.

4 Discussion

In this paper we have shown that the set of all 4-qubit SL-invariant polynomials that are also invariant under permutations of the qubits is generated by four polynomials of degrees 2, 6, 8, 12. Using a completely different approach, in [14] it was also shown that these polynomials exist, but they were not given elegantly as in Eq. (1). In addition, we have shown here that the hyperdeterminant [12] is the *only* SL-invariant polynomial (up to its powers) that is not vanishing precisely on the set of generic states.

Since the hyperdeterminant (in our notations $\gamma(z)$) quantifies generic entanglement, a state with the most amount of generic entanglement can be defined as a state, z, that maximizes $|\gamma(z)|$. We are willing to conjecture that the state

$$|L\rangle = \frac{1}{\sqrt{3}}(u_0 + \omega u_1 + \bar{\omega} u_2) \ , \quad \omega \equiv e^{i\pi/3} \ ,$$

is the unique state (up to a local unitary transformation) that maximizes $|\gamma(z)|$. It was shown in [6] that the state $|L\rangle$ maximizes uniquely many measures of 4 qubits entanglement. Moreover, one can easily check that the state $|L\rangle$ is the only state for which $\mathcal{E}_0 = \mathcal{E}_1 = \mathcal{E}_2 = 0$ while $\mathcal{E}_3(|L\rangle) = 1/9$. It is known that a state with such a property is unique [9]. Similarly, we found out the unique state, $|F\rangle$ for which $\mathcal{F}_1(|F\rangle) = \mathcal{F}_3(|F\rangle) = \mathcal{F}_4(|F\rangle) = 0$ but $\mathcal{F}_6(|F\rangle) \neq 0$. The (non-normalized) unique state (up to a local unitary transformation) is

$$|F\rangle = (3-\sqrt{3})u_0 + (1+i)\sqrt{3}u_1 + (1-i)\sqrt{3}u_2 - i(3-\sqrt{3})u_3 \ .$$

Acknowledgments: GG research is supported by NSERC, NW was partially supported by an NSF Summer Grant

References

1. N. Bourbaki, *Éléments de mathématique. Fasc. XXXIV. Groupes et algèbres de Lie. Chapitre IV: Groupes de Coxeter et systèmes de Tits. Chapitre V: Groupes engendrés par des réflexions. Chapitre VI: systèmes de racines* (French) Actualités Scientifiques et Industrielles, No. 1337, Hermann, Paris, 1968, 288 pp.
2. S. S. Bullock and G. K. Brennen, *Canonical decompositions of n-qubit quantum computations and concurrence*, J. Math. Phys. **45** (2004), 2447–2467.
3. V. Coffman, J. Kundu, and W. K. Wootters, *Distributed entanglement*, Phys. Rev. A **61** (2000), 052306, 5 pp.
4. D. Ž. Doković and A. Osterloh, *On polynomial invariants of several qubits*, J. Math. Phys. **50** (2009), 033509, 23 pp.
5. W. Dür, G. Vidal, and J. I. Cirac, *Three qubits can be entangled in two inequivalent ways*, Phys. Rev. A (3) **62** (2000), 062314, 12 pp.
6. G. Gour and N. R. Wallach, *All maximally entangled four-qubit states*, J. Math. Phys. **51** (2010), 112201, 24 pp.
7. R. Horodecki, P. Horodecki, M. Horodecki, and K. Horodecki, *Quantum entanglement*, Rev. Modern Phys. **81** (2009), 865–942.
8. J. E. Humphreys, *Reflection groups and Coxeter groups*. Cambridge Studies in Advanced Mathematics, 29. Cambridge University Press, Cambridge, 1990. xii+204 pp.
9. B. Kostant, *Lie group representations on polynomial rings*, Amer. J. Math. **85** (1963) 327–404.

10. J.-G. Luque and J.-Y. Thibon, *Polynomial invariants of four qubits*, Phys. Rev. A **67** (2003), 042303, 5 pp.
11. M. L. Mehta, *Basic sets of invariant polynomials for finite reflection groups*, Comm. Algebra **16** (1988), 1083–1098.
12. A. Miyake, *Classification of multipartite entangled states by multidimensional determinants*, Phys. Rev. A **67** (2003), 012108, 10 pp.
13. A. Miyake and M. Wadati, *Multipartite entanglement and hyperdeterminants*, ERATO Workshop on Quantum Information Science (Tokyo, 2002), Quantum Inf. Comput. **2** (2002), suppl., 540–555.
14. A. Osterloh and J. Siewert, *Constructing N-qubit entanglement monotones from antilinear operators*, Phys. Rev. A. **72** (2005), 012337, 4 pp.
15. M. B. Plenio and S. Virmani, *An introduction to entanglement measures*, Quantum Inf. Comput. **7** (2007), 1–51.
16. A. Uhlmann, *Fidelity and concurrence of conjugated states*, Phys. Rev. A **62** (2000), 032307, 9 pp.
17. È. B. Vinberg, *The Weyl group of a graded Lie algebra*, (Russian) Izv. Akad. Nauk SSSR Ser. Mat. **40** (1976), 488–526, 709.
18. F. Verstraete, J. Dehaene, and B. De Moor, *Normal forms and entanglement measures for multipartite quantum states*, Phys. Rev. A **68** (2003), 012103, 7 pp.
19. N. R. Wallach, *Quantum computing and entanglement for mathematicians*, Notes from Venice C.I.M.E. June 2004, http://www.math.ucsd.edu/~nwallach/venice.pdf, 29 pp.
20. W. K. Wootters, *Entanglement of formation of an arbitrary state of two qubits*, Phys. Rev. Lett. **80** (1998), 2245–2248.

Finite maximal tori

Gang Han and David A. Vogan Jr.

Dedicated with admiration and affection to Nolan Wallach, who has helped us all to see Lie more clearly.

Abstract We define a "finite maximal torus" of a compact Lie group G to be a maximal finite abelian subgroup A of G. We introduce structure for finite maximal tori parallel to the classical structure for maximal tori, like roots and the Weyl group; and we recall a large number of (previously known) examples.

Keywords: Compact group • Maximal finite abelian subgroup

Mathematics Subject Classification: 17B22, 20G15, 22C05

1 Introduction

Suppose G is a compact Lie group, with identity component G_0. There is a beautiful and complete structure theory for G_0, based on the notion of maximal tori and root systems introduced by Élie Cartan and Hermann Weyl. The purpose of this paper is

---

The first author was supported by NSFC Grant No. 10801116, and the second author in part by NSF grant DMS-0967272.

G. Han
Department of Mathematics, Zhejiang University, Hangzhou 310027, China
e-mail: mathhgg@zju.edu.cn

D.A. Vogan Jr.
Department of Mathematics, Massachusetts Institute of Technology,
Cambridge, MA 02139, USA
e-mail: dav@math.mit.edu

R. Howe et al. (eds.), *Symmetry: Representation Theory and Its Applications*,
Progress in Mathematics 257, DOI 10.1007/978-1-4939-1590-3_10

to introduce a parallel structure theory using "finite maximal tori." A *maximal torus* is by definition a maximal *connected* abelian subgroup of G_0. We define a *finite maximal torus* to be a maximal *finite* abelian subgroup of G.

It would be etymologically more reasonable to use the term *finite maximally diagonalizable subgroup*, but this name seems not to roll easily off the tongue. A more restrictive notion is that of *Jordan subgroup*, introduced by Alekseevskiĭ in [1]; see also [6], Definition 3.18. Also very closely related is the notion of *fine grading* of a Lie algebra introduced in [14], and studied extensively by Patera and others.

The classical theory of root systems and maximal tori displays very clearly many interesting structural properties of G_0. The central point is that root systems are essentially finite combinatorial objects. Subroot systems can easily be exhibited by hand, and they correspond automatically to (compact connected) subgroups of G_0; so many subgroups can be described in a combinatorial fashion. A typical example visible in this fashion is the subgroup $U(n) \times U(m)$ of $U(n+m)$. A more exotic example is the subgroup $(E_6 \times SU(3))/\mu_3$ of E_8 (with μ_3 the cyclic group of third roots of 1).

Unfortunately, there is so far no general "converse" to this correspondence: it is not known how to relate the root system of an arbitrary (compact connected) subgroup of G_0 to the root system of G_0. (Many powerful partial results in this direction were found by Dynkin in [5].) Consequently there are interesting subgroups that are more or less invisible to the theory of root systems.

In this paper we will describe an analogue of root systems for finite maximal tori. Again these will be finite combinatorial objects, so it will be easy to describe subroot systems by hand, which must correspond to subgroups of G. The subgroups arising in this fashion are somewhat different from those revealed by classical root systems. A typical example is the subgroup $PU(n)$ of $PU(nm)$, arising from the action of $U(n)$ on $\mathbb{C}^n \otimes \mathbb{C}^m$. A more exotic example is the subgroup $F_4 \times G_2$ of E_8 (see Example 4.5).

In Section 2 we recall Grothendieck's formulation of the Cartan–Weyl theory in terms of "root data." His axiomatic characterization of root data is a model for what we seek to do with finite maximal tori.

One of the fundamental classical theorems about maximal tori is that if T is a maximal torus in G_0, and $\widetilde{G}_0$ is a finite covering of G_0, then the preimage $\widetilde{T}$ of T in $\widetilde{G}_0$ is a maximal torus in $\widetilde{G}_0$. The corresponding statement about finite maximal tori is *false* (see Example 4.1); the preimage often fails to be abelian. In order to keep this paper short, we have avoided any serious discussion of coverings.

In Section 3 we define root data and Weyl groups for finite maximal tori. We will establish analogues of Grothendieck's axioms for these finite root data, but we do not know how to prove an existence theorem like Grothendieck's (saying that every finite root datum arises from a compact group).

In Section 4 we offer a collection of examples of finite maximal tori. The examples (none of which is original) are the main point of this paper, and are what interested us in the subject. Reading this section first is an excellent way to approach the paper.

Of course it is possible and interesting to work with maximal abelian subgroups which may be neither finite nor connected. We have done nothing about this.

Grothendieck's theory of root data was introduced not for compact Lie groups but for reductive groups over algebraically closed fields. The theory of finite maximal tori can be put into that setting as well, and this seems like an excellent exercise. It is not clear to us (for example) whether one should allow p-torsion in a "finite maximal torus" for a group in characteristic p; excluding it would allow the theory to develop in a straightforward parallel to what we have written about compact groups, but allowing p-torsion could lead to very interesting examples of finite maximal tori.

Much of the most interesting structure and representation theory for a connected reductive algebraic group G_0 (over an algebraically closed field) can be expressed in terms of (classical) root data. For example, the irreducible representations of G_0 are indexed (following Cartan and Weyl) by orbits of the Weyl group on the character lattice; and Lusztig has defined a surjective map from conjugacy classes in the Weyl group to unipotent classes in G. It would be fascinating to rewrite such results in terms of finite root data; but we have done nothing in this direction.

2 Root data

In this section we introduce Grothendieck's root data for compact connected Lie groups. As in the introduction, we begin with

$$G = \text{compact connected Lie group.} \tag{2.1a}$$

The real Lie algebra of G and its complexification are written

$$\mathfrak{g}_0 = \mathrm{Lie}(G), \qquad \mathfrak{g} = \mathfrak{g}_0 \otimes_{\mathbb{R}} \mathbb{C}. \tag{2.1b}$$

The conjugation action of G on itself is written Ad:

$$\mathrm{Ad}\colon G \to \mathrm{Aut}(G), \qquad \mathrm{Ad}(g)(x) = gxg^{-1} \quad (g, x \in G). \tag{2.1c}$$

The differential (in the target variable) of this action is an action of G on $\mathfrak{g}_0$ by Lie algebra automorphisms

$$\mathrm{Ad}\colon G \to \mathrm{Aut}(\mathfrak{g}_0). \tag{2.1d}$$

This differential of this action of G is a Lie algebra homomorphism

$$\mathrm{ad}\colon \mathfrak{g}_0 \to \mathrm{Der}(\mathfrak{g}_0), \qquad \mathrm{ad}(X)(Y) = [X, Y] \quad (X, Y \in \mathfrak{g}_0). \tag{2.1e}$$

Analogous notation will be used for arbitrary real Lie groups. The kernel of the adjoint action of G on G or on $\mathfrak{g}_0$ is the center $Z(G)$:

$$\begin{aligned} Z(G) &= \{g \in G \mid gxg^{-1} = x, \quad \text{all } x \in G\} \\ &= \{g \in G \mid \mathrm{Ad}(g)(Y) = Y, \quad \text{all } Y \in \mathfrak{g}_0\}. \end{aligned} \tag{2.1f}$$

So far all of this applies to arbitrary connected Lie groups G. We will also have occasion to use the existence of a nondegenerate symmetric bilinear form B on $\mathfrak{g}_0$, with the invariance properties

$$B(\mathrm{Ad}(g)X, \mathrm{Ad}(g)Y) = B(X, Y) \qquad (X, Y \in \mathfrak{g}_0, g \in G). \tag{2.1g}$$

We may arrange for this form to be negative definite: if for example G is a group of unitary matrices, so that the Lie algebra consists of skew-Hermitian matrices, then

$$B(X, Y) = \mathrm{tr}(XY)$$

will serve. (Since X has purely imaginary eigenvalues, the trace of X^2 is negative.) We will also write B for the corresponding (nondegenerate) complex-linear symmetric bilinear form on $\mathfrak{g}$.

Definition 2.2. A *maximal torus* of a compact connected Lie group G is a maximal connected abelian subgroup T of G.

We now fix a maximal torus $T \subset G$. Because of [11], Corollary 4.52, T is actually a maximal abelian subgroup of G, and therefore equal to its own centralizer in G:

$$T = Z_G(T) = G^T = \{g \in G \mid \mathrm{Ad}(t)(g) = g, \text{ (all } t \in T)\}. \tag{2.3a}$$

Because T is a compact connected abelian Lie group, it is isomorphic to a product of copies of the unit circle

$$S^1 = \{e^{2\pi i\theta} | \theta \in \mathbb{R}\}, \qquad \mathrm{Lie}(S^1) = \mathbb{R}; \tag{2.3b}$$

the identification of the Lie algebra is made using the coordinate θ. The *character lattice of* T is

$$\begin{aligned} X^*(T) &= \mathrm{Hom}(T, S^1) \\ &= \{\lambda : T \to S^1 \text{ continuous}, \lambda(st) = \lambda(s)\lambda(t) \quad (s, t \in T)\}. \end{aligned} \tag{2.3c}$$

The character lattice is a (finitely generated free) abelian group, written additively, under multiplication of characters:

$$(\lambda + \mu)(t) = \lambda(t)\mu(t) \qquad (\lambda, \mu \in X^*(T)).$$

The functor X^* is a contravariant equivalence of categories from compact connected abelian Lie groups to finitely generated torsion-free abelian groups. The inverse functor is given by Hom into S^1:

$$T \simeq \mathrm{Hom}(X^*(T), S^1), \qquad t \mapsto (\lambda \mapsto \lambda(t)). \tag{2.3d}$$

The *cocharacter lattice of* T is

$$\begin{aligned} X_*(T) &= \mathrm{Hom}(S^1, T) \\ &= \{\xi \colon S^1 \to T \text{ continuous}, \xi(zw) = \xi(z)\xi(w) \quad (z, w \in S^1)\}. \end{aligned} \tag{2.3e}$$

There are natural identifications

$$X^*(S^1) = X_*(S^1) = \mathrm{Hom}(S^1, S^1) \simeq \mathbb{Z}, \qquad \lambda_n(z) = z^n \quad (z \in S^1).$$

The composition of a character with a cocharacter is a homomorphism from S^1 to S^1, which is therefore some nth power map. In this way we get a biadditive pairing

$$\langle \cdot, \cdot \rangle \colon X^*(T) \times X_*(T) \to \mathbb{Z}, \tag{2.3f}$$

defined by

$$\langle \lambda, \xi \rangle = n \Leftrightarrow \lambda(\xi(z)) = z^n \quad (\lambda \in X^*(T), \xi \in X_*(T), z \in S^1). \tag{2.3g}$$

This pairing identifies each of the lattices as the dual of the other:

$$X_* \simeq \mathrm{Hom}_{\mathbb{Z}}(X^*, \mathbb{Z}), \qquad X^* \simeq \mathrm{Hom}_{\mathbb{Z}}(X_*, \mathbb{Z}). \tag{2.3h}$$

The functor X_* is a covariant equivalence of categories from compact connected abelian Lie groups to finitely generated torsion-free abelian groups. The inverse functor is given by tensoring with S^1:

$$X_*(T) \otimes_{\mathbb{Z}} S^1 \simeq T, \qquad \xi \otimes z \mapsto \xi(z). \tag{2.3i}$$

The action by Ad of T on the complexified Lie algebra $\mathfrak{g}$ of G, like any complex representation of a compact group, decomposes into a direct sum of copies of irreducible representations; in this case, of characters of T:

$$\mathfrak{g} = \sum_{\lambda \in X^*(T)} \mathfrak{g}_\lambda, \qquad \mathfrak{g}_\lambda = \{Y \in \mathfrak{g} \mid \mathrm{Ad}(t)Y = \lambda(t)Y \ (t \in T)\}. \tag{2.3j}$$

In particular, the zero weight space is

$$\begin{aligned} \mathfrak{g}_{(0)} &= \{Y \in \mathfrak{g} \mid \mathrm{Ad}(t)Y = 0 \ (t \in T)\} \\ &= \mathrm{Lie}(G^T) \otimes_{\mathbb{R}} \mathbb{C} = \mathrm{Lie}(T) \otimes_{\mathbb{R}} \mathbb{C} = \mathfrak{t}, \end{aligned} \tag{2.3k}$$

the complexified Lie algebra of T; the last two equalities follow from (2.3a). We define the *roots of T in G* to be the nontrivial characters of T appearing in the decomposition of the adjoint representation:

$$R(G,T) = \{\alpha \in X^*(T) - \{0\} \mid \mathfrak{g}_\alpha \neq 0\}. \tag{2.3l}$$

Because of the description of the zero weight space in (2.3k), we have

$$\mathfrak{g} = \mathfrak{t} \oplus \sum_{\alpha \in R(G,T)} \mathfrak{g}_\alpha, \tag{2.3m}$$

We record two elementary facts relating the root decomposition to the Lie bracket and the invariant bilinear form B:

$$[\mathfrak{g}_\alpha, \mathfrak{g}_\beta] \subset \mathfrak{g}_{\alpha+\beta}, \tag{2.3n}$$

$$B(\mathfrak{g}_\alpha, \mathfrak{g}_\beta) = 0, \qquad \alpha + \beta \neq 0. \tag{2.3o}$$

Example 2.4. Suppose $G = SU(2)$, the group of 2×2 unitary matrices of determinant 1. We can choose as a maximal torus

$$SD(2) = \left\{ \begin{pmatrix} e^{2\pi i\theta} & 0 \\ 0 & e^{-2\pi i\theta} \end{pmatrix} \mid \theta \in \mathbb{R} \right\} \simeq S^1.$$

We have given this torus a name in order to be able to formulate the definition of coroot easily. The "D" is meant to stand for "diagonal," and the "S" for "special" (meaning determinant one, as in the "special unitary group"). The last identification gives canonical identifications

$$X^*(SD(2)) \simeq \mathbb{Z}, \qquad X_*(SD(2)) \simeq \mathbb{Z}.$$

The Lie algebra of G is

$$\mathfrak{su}(2) = \{2 \times 2 \text{ complex matrices } X \mid {}^t\overline{X} = -X, \operatorname{tr}(X) = 0\}.$$

The obvious map identifies

$$\mathfrak{su}(2)_{\mathbb{C}} \simeq \{2 \times 2 \text{ complex matrices } Z \mid \operatorname{tr}(Z) = 0\} = \mathfrak{sl}(2,\mathbb{C}).$$

The adjoint action of $SD(2)$ on $\mathfrak{g}$ is

$$\operatorname{Ad}\begin{pmatrix} e^{2\pi i\theta} & 0 \\ 0 & e^{-2\pi i\theta} \end{pmatrix}\begin{pmatrix} a & b \\ c & -a \end{pmatrix} = \begin{pmatrix} a & e^{4\pi i\theta} b \\ e^{-4\pi i\theta} c & -a \end{pmatrix}.$$

This formula shows at once that the roots are

$$R(SU(2), SD(2)) = \{\pm 2\} \subset X^*(SD(2)) \simeq \mathbb{Z},$$

with root spaces

$$\mathfrak{sl}(2)_2 = \left\{ \begin{pmatrix} 0 & t \\ 0 & 0 \end{pmatrix} \mid t \in \mathbb{C} \right\}, \qquad \mathfrak{sl}(2)_{-2} = \left\{ \begin{pmatrix} 0 & 0 \\ s & 0 \end{pmatrix} \mid s \in \mathbb{C} \right\}.$$

When we define coroots in a moment, it will be clear that

$$R^\vee(SU(2), SD(2)) = \{\pm 1\} \subset X_*(SD(2)) \simeq \mathbb{Z}.$$

Now we are ready to define coroots in general. For every root α, we define

$$\mathfrak{g}^{[\alpha]} = \text{Lie subalgebra generated by root spaces } \mathfrak{g}_{\pm\alpha}. \tag{2.5a}$$

It is easy to see that $\mathfrak{g}^{[\alpha]}$ is the complexification of a real Lie subalgebra $\mathfrak{g}_0^{[\alpha]}$, which in turn is the Lie algebra of a compact connected subgroup

$$G^{[\alpha]} \subset G. \tag{2.5b}$$

This subgroup meets the maximal torus T in a one-dimensional torus $T^{[\alpha]}$, which is maximal in $G^{[\alpha]}$. There is a continuous surjective group homomorphism

$$\phi_\alpha \colon SU(2) \to G^{[\alpha]} \subset G, \tag{2.5c}$$

which we may choose to have the additional properties

$$\phi_\alpha(SD(2)) = T_\alpha \subset T \quad (d\phi_\alpha)_{\mathbb{C}}(\mathfrak{sl}(2)_2) = \mathfrak{g}_\alpha. \tag{2.5d}$$

The homomorphism ϕ_α is then unique up to conjugation by T in G (or by $SD(2)$ in $SU(2)$). In particular, the restriction to $SD(2) \simeq S^1$, which we call $\alpha^\vee$, is a uniquely defined cocharacter of T:

$$\alpha^\vee \colon S^1 \to T, \qquad \alpha^\vee(e^{2\pi i\theta}) = \phi_\alpha \begin{pmatrix} e^{2\pi i\theta} & 0 \\ 0 & e^{-2\pi i\theta} \end{pmatrix}. \tag{2.5e}$$

The element $\alpha^\vee \in X_*(T)$ is called the *coroot corresponding to the root* α. We write

$$R^\vee(G, T) = \{\alpha^\vee \mid \alpha \in R(G, T)\} \subset X_*(T) - \{0\}. \tag{2.5f}$$

Essentially because the positive root in $SU(2)$ is $+2$, we see that

$$\langle \alpha, \alpha^\vee \rangle = 2 \qquad (\alpha \in R(G, T)). \tag{2.5g}$$

We turn now to a description of the Weyl group (of a maximal torus in a compact group). Here is the classical definition.

Definition 2.6. Suppose G is a compact connected Lie group, and T is a maximal torus in G. The *Weyl group of T in G* is

$$W(G,T) = N_G(T)/T,$$

the quotient of the normalizer of T by its centralizer. From this definition, it is clear that $W(G,T)$ acts faithfully on T by automorphisms (conjugation):

$$W(G,T) \hookrightarrow \mathrm{Aut}(T).$$

By acting on the range of homomorphisms, $W(G,T)$ may also be regarded as acting on cocharacters:

$$W(G,T) \hookrightarrow \mathrm{Aut}(X_*(T)), \qquad (w \cdot \xi)(z) = w \cdot (\xi(z))$$
$$(w \in W, \quad \xi \in X_*(T) = \mathrm{Hom}(S^1, T), z \in S^1).$$

Similarly,

$$W(G,T) \hookrightarrow \mathrm{Aut}(X^*(T)), \qquad (w \cdot \lambda)(t) = \lambda(w^{-1} \cdot t))$$
$$(w \in W, \quad \lambda \in X^*(T) = \mathrm{Hom}(T, S^1), t \in T).$$

The actions on the dual lattices $X_*(T)$ and $X^*(T)$ are inverse transposes of each other. Equivalently, for the pairing of (2.3g),

$$\langle w \cdot \lambda, \xi \rangle = \langle \lambda, w^{-1} \cdot \xi \rangle \qquad (\lambda \in X^*(T), \xi \in X_*(T)).$$

We recall now how to construct the Weyl group from the roots and the coroots; this is the construction that we will seek to extend to finite maximal tori. We begin with an arbitrary root $\alpha \in R(G,T)$, and ϕ_α as in (2.5c). The element

$$\sigma_\alpha = \phi_\alpha \begin{pmatrix} 0 & 1 \\ -1 & 0 \end{pmatrix} \in N_G(T) \tag{2.7a}$$

is well defined (that is, independent of the choice of ϕ_α) up to conjugation by $T \cap G^{[\alpha]}$. Consequently the coset

$$s_\alpha = \sigma_\alpha T \in N_G(T)/T = W(G,T) \tag{2.7b}$$

is well defined; it is called the *reflection in the root α*. Because it is constructed from the subgroup G_α, we see that

$$\sigma_\alpha \text{ commutes with } \ker(\alpha) \subset T, \tag{2.7c}$$

and therefore that

$$s_\alpha \text{ acts trivially on } \ker(\alpha) \subset T. \tag{2.7d}$$

A calculation in $SU(2)$ shows that

$$\begin{pmatrix} 0 & 1 \\ -1 & 0 \end{pmatrix} \text{ acts by inversion on } SD(2), \tag{2.7e}$$

and therefore that

$$s_\alpha \text{ acts by inversion on } \operatorname{im}(\alpha^\vee) \subset T. \tag{2.7f}$$

The two properties (2.7d) and (2.7f) are equivalent to

$$s_\alpha(t) = t \cdot \alpha^\vee(\alpha(t))^{-1} \qquad (t \in T). \tag{2.7g}$$

From this formula we easily deduce that

$$s_\alpha(\lambda) = \lambda - \langle \lambda, \alpha^\vee \rangle \alpha \qquad (\lambda \in X^*(T)) \tag{2.7h}$$

and similarly

$$s_\alpha(\xi) = \xi - \langle \alpha, \xi \rangle \alpha^\vee \qquad (\xi \in X_*(T)). \tag{2.7i}$$

Here is the basic theorem about the Weyl group.

Theorem 2.8. *Suppose G is a compact connected Lie group, and $T \subset G$ is a maximal torus. Then the Weyl group of T in G (Definition 2.6) is generated by the reflections described by any of the equivalent conditions* (2.7g), (2.7h), *or* (2.7i):

$$W(G,T) = \langle s_\alpha \mid \alpha \in R(G,T) \rangle .$$

The automorphisms s_α of $X^*(T)$ must permute the roots $R(G,T)$, and the automorphisms s_α of $X_*(T)$ must permute the coroots $R^\vee(G,T)$.

Grothendieck's understanding of the classification of compact Lie groups by Cartan and Killing is that the combinatorial structure of roots and Weyl group determines G completely. Here is a statement.

Definition 2.9 (Root datum; see [16], 7.4). An *abstract (reduced) root datum* is a quadruple $\Psi = (X^*, R, X_*, R^\vee)$, subject to the requirements

(a) X^* and X_* are lattices (finitely generated torsion-free abelian groups), dual to each other (cf. (2.3h)) by a specified pairing

$$\langle , \rangle \colon X^* \times X_* \to \mathbb{Z};$$

(b) $R \subset X^*$ and $R^\vee \subset X_*$ are finite subsets, with a specified bijection $\alpha \mapsto \alpha^\vee$ of R onto $R^\vee$.

These data define lattice endomorphisms (for every root $\alpha \in R$)

$$s_\alpha \colon X^* \to X^*, \qquad s_\alpha(\lambda) = \lambda - \langle \lambda, \alpha^\vee \rangle \alpha,$$
$$s_\alpha \colon X_* \to X_*, \qquad s_\alpha(\xi) = \xi - \langle \alpha, \xi \rangle \alpha^\vee,$$

called *root reflections*. It is easy to check that each of these endomorphisms is the transpose of the other with respect to the pairing $\langle , \rangle$. We impose the axioms

(RD 0) if $\alpha \in R$, then $2\alpha \notin R$;
(RD 1) $\langle \alpha, \alpha^\vee \rangle = 2 \qquad (\alpha \in R)$; and
(RD 2) $s_\alpha(R) = R, \quad s_\alpha(R^\vee) = R^\vee \qquad (\alpha \in R)$.

Axiom (RD 0) is what makes the root datum *reduced.* Axiom (RD 1) implies that $s_\alpha^2 = 1$, so s_α is invertible. The *Weyl group of the root datum* is the group generated by the root reflections:

$$W(\Psi) = \langle s_\alpha \quad (\alpha \in R) \rangle \subset \operatorname{Aut}(X^*).$$

The definition of root datum is symmetric in the two lattices: the *dual root datum* is

$$\Psi^\vee = (X_*, R^\vee, X^*, R).$$

The inverse transpose isomorphism identifies the Weyl group with

$$W(\Psi^\vee) = \langle s_\alpha \quad (\alpha^\vee \in R^\vee) \rangle \simeq W(\Psi).$$

Theorem 2.8 (and the material leading to its formulation) show that if T is a maximal torus in a compact connected Lie group G, then

$$\Psi(G,T) = (X^*(T), R(G,T), X_*(T), R^\vee(G,T)) \tag{2.10}$$

is an abstract reduced root datum. The amazing fact—originating in the work of Cartan and Killing, but most beautifully and perfectly formulated by Grothendieck—is that the root datum determines the group, and that every root datum arises in this way. Here is a statement.

Theorem 2.11 ([4], exp. XXV; also [16], Theorems 9.6.2 and 10.1.1). *Suppose $\Psi = (X^*, R, X_*, R^\vee)$ is an abstract reduced root datum. Then there is a maximal torus in a compact connected Lie group $T \subset G$ so that*

$$\Psi(G,T) \simeq \Psi$$

(notation (2.10)). The pair (G, T) is determined by these requirements up to an inner automorphism from T. We have

$$W(G, T) = N_G(T)/T \simeq W(\Psi).$$

Sketch of proof. What is proved in [4] is that to Ψ there corresponds a complex connected reductive algebraic group $\mathbf{G}(\Psi)$. There is a correspondence between complex connected reductive algebraic groups and compact connected Lie groups obtained by passage to a compact real form (see for example [13], Theorem 5.12 (page 247)). Combining these two facts proves the theorem.

□

3 Finite maximal tori

Throughout this section we write

$$\begin{aligned} G &= \text{(possibly disconnected) compact Lie group} \\ G_0 &= \text{identity component of } G. \end{aligned} \tag{3.1}$$

We use the notation of (2.1), especially for the identity component G_0.

Definition 3.2. A *finite maximal torus* for G is a finite maximal abelian subgroup

$$A \subset G. \tag{3.3}$$

The definition means that the centralizer in G of A is equal to A:

$$Z_G(A) = A. \tag{3.3a}$$

The differentiated version of this equation is

$$Z_{\mathfrak{g}}(A) = \{X \in \mathfrak{g} \mid \mathrm{Ad}(a)X = X, \text{all } a \in A\} = \mathrm{Lie}(A) = \{0\}; \tag{3.3b}$$

the last equality is because A is finite. We define the *large Weyl group of A in G_0* to be

$$W_{\text{large}}(G_0, A) = N_{G_0}(A)/Z_{G_0}(A) = N_{G_0}(A)/(A \cap G_0). \tag{3.3c}$$

Clearly

$$W_{\text{large}}(G_0, A) \subset \mathrm{Aut}(A), \tag{3.3d}$$

a finite group.

The term "large" should be thought of as temporary. We will introduce in Definition 3.8 a subgroup $W_{\mathrm{small}}(G, A)$, given by generators analogous to the root reflections in a classical Weyl group. We believe that the two groups are equal; but until that is proved, we need terminology to talk about them separately.

In contrast to classical maximal tori, finite maximal tori need not exist. For example, if $G = U(n)$, then any abelian subgroup must (after change of basis) consist entirely of diagonal matrices; so the only maximal abelian subgroups are the connected maximal tori, none of which is finite.

For the rest of this section we fix a (possibly disconnected) compact Lie group G, and a finite maximal torus

$$A \subset G. \tag{3.4a}$$

Our goal in this section is to introduce roots, coroots, and root transvections, all by analogy with the classical case described in Section 2.

The *character group of* A is

$$\begin{aligned} X^*(A) &= \mathrm{Hom}(A, S^1) \\ &= \{\lambda\colon A \to S^1\ \lambda(ab) = \lambda(a)\lambda(b) \quad (a, b \in A)\}. \end{aligned} \tag{3.4b}$$

The character group is a finite abelian group, written additively, under multiplication of characters:

$$(\lambda + \mu)(a) = \lambda(a)\mu(a) \qquad (\lambda, \mu \in X^*(A)).$$

In particular, we write 0 for the trivial character of A. We can recover A from $X^*(A)$ by a natural isomorphism

$$A \simeq \mathrm{Hom}(X^*(A), S^1), \qquad a \mapsto [\lambda \mapsto \lambda(a)]. \tag{3.4c}$$

As a consequence, the functor X^* is a contravariant exact functor from the category of finite abelian groups to itself. The group $X^*(A)$ is always isomorphic to A, but not canonically.

For any positive integer n, define

$$\mu_n = \{z \in \mathbb{C} \mid z^n = 1\}, \tag{3.4d}$$

the group of *nth roots of unity in* $\mathbb{C}$. We identify

$$X^*(\mathbb{Z}/n\mathbb{Z}) \simeq \mu_n, \qquad \lambda_\omega(m) = \omega^m \quad (\omega \in \mu_n, m \in \mathbb{Z}/n\mathbb{Z}). \tag{3.4e}$$

Similarly we identify

$$X^*(\mu_n) \simeq \mathbb{Z}/n\mathbb{Z}, \qquad \lambda_m(\omega) = \omega^m \quad (m \in \mathbb{Z}/n\mathbb{Z}, \omega \in \mu_n). \tag{3.4f}$$

Of course we can write

$$\mu_n = \{e^{2\pi i\theta} \mid \theta \in \mathbb{Z}/n\mathbb{Z}\} \simeq \mathbb{Z}/n\mathbb{Z},$$

and so identify a particular generator of the cyclic group μ_n; but (partly with the idea of working with reductive groups over other fields, and partly to see what is most natural) we prefer to avoid using this identification.

A character $\lambda \in X^*(A)$ is said to be *of order dividing n* if $n\lambda = 0$; equivalently, if

$$\lambda\colon A \to \mu_n. \tag{3.4g}$$

We will say "character of order n" to mean a character of order dividing n. We write

$$\begin{aligned} X^*(A)(n) &= \{\lambda \in X^*(A) \mid n\lambda = 0\} \\ &= \operatorname{Hom}(A, \mu_n), \end{aligned} \tag{3.4h}$$

for the *group of characters of order n*. Therefore

$$X^*(A) = \bigcup_{n\geq 1} X^*(A)(n). \tag{3.4i}$$

We say that λ has *order exactly n* if n is the smallest positive integer such that $n\lambda = 0$; equivalently, if

$$\lambda\colon A \twoheadrightarrow \mu_n \tag{3.4j}$$

is surjective.

The action by Ad of A on the complexified Lie algebra $\mathfrak{g}$ of G decomposes into a direct sum of characters:

$$\mathfrak{g} = \sum_{\lambda\in X^*(A)} \mathfrak{g}_\lambda, \qquad \mathfrak{g}_\lambda = \{Y \in \mathfrak{g} \mid \operatorname{Ad}(a)Y = \lambda(a)Y \ (a \in A)\}. \tag{3.4k}$$

According to (3.3b), the trivial character does not appear in this decomposition; that is, $\mathfrak{g}_0 = 0$. We define the *roots of A in G* to be the characters of A that do appear:

$$R(G, A) = \{\alpha \in X^*(A)\} \mid \mathfrak{g}_\alpha \neq 0\} \subset X^*(A) - \{0\}. \tag{3.4l}$$

The analogue of the root decomposition (2.3m) has no term like the Lie algebra of the maximal torus:

$$\mathfrak{g} = \sum_{\alpha\in R(G,A)} \mathfrak{g}_\alpha. \tag{3.4m}$$

Just as for classical roots, we see immediately that

$$[\mathfrak{g}_\alpha, \mathfrak{g}_\beta] \subset \mathfrak{g}_{\alpha+\beta}, \tag{3.4n}$$

and

$$B(\mathfrak{g}_\alpha, \mathfrak{g}_\beta) = 0, \qquad \alpha + \beta \neq 0. \tag{3.4o}$$

Fix a positive integer n. A *cocharacter of order dividing n* is a homomorphism

$$\xi \colon \mu_n \to A. \tag{3.5a}$$

We will say "cocharacter of order n" to mean a cocharacter of order dividing n. The cocharacter has order exactly n if and only if ξ is injective.

If we fix a primitive nth root ω, then a cocharacter ξ of order n is the same thing as an element $x \in A$ of order n by the correspondence

$$x = \xi(\omega). \tag{3.5b}$$

We write

$$X_*(A)(n) = \mathrm{Hom}(\mu_n, A) \tag{3.5c}$$

for the group of cocharacters of order n. The natural surjection

$$\mu_{mn} \twoheadrightarrow \mu_n, \qquad \omega \mapsto \omega^m$$

gives rise to a natural inclusion

$$\mathrm{Hom}(\mu_n, A) \hookrightarrow \mathrm{Hom}(\mu_{mn}, A), \qquad X_*(A)(n) \hookrightarrow X_*(A)(mn). \tag{3.5d}$$

Using these inclusions, we can define the *cocharacter group of A*

$$\bigcup_n X_*(A)(n). \tag{3.5e}$$

The functor X_* is a covariant equivalence of categories from the category of finite abelian groups to itself; but the choice of a functorial isomorphism $A \simeq X_*(A)$ requires compatible choices of primitive nth roots ω_n for every n. (The compatibility requirement is $\omega_n = \omega_{mn}^m$.) Partly to maintain the analogy with cocharacters of connected tori, and partly for naturality, we prefer not to make such choices, and to keep $X_*(A)$ as a group distinct from A.

Suppose $\lambda \in X^*(A)$ is any character and $\xi \in X_*(A)(n)$ is an order n cocharacter. The composition $\lambda \circ \xi$ is a homomorphism $\mu_n \to S^1$. Such a homomorphism must take values in μ_n, and is necessarily raising to the mth power for a unique $m \in \mathbb{Z}/n\mathbb{Z} \simeq (1/n)\mathbb{Z}/\mathbb{Z}$. In this way we get a natural pairing

$$X^*(A) \times X_*(A)(n) \to (1/n)\mathbb{Z}/\mathbb{Z},$$
$$\lambda(\xi(\omega)) = \omega^{n\langle\lambda,\xi\rangle} \qquad (\lambda \in X^*(A)(n),\ \xi \in X_*(A)(n),\ \omega \in \mu_n). \tag{3.5f}$$

Taking the union over n defines a biadditive pairing

$$X^*(A) \times X_*(A) \to \mathbb{Q}/\mathbb{Z} \tag{3.5g}$$

which identifies

$$X_*(A) \simeq \mathrm{Hom}(X^*(A), \mathbb{Q}/\mathbb{Z}), \qquad X^*(A) \simeq \mathrm{Hom}(X_*(A), \mathbb{Q}/\mathbb{Z}). \tag{3.5h}$$

Before defining coroots in general, we need an analog of Example 2.4.

Example 3.6. Suppose $A \subset H$ is a finite maximal torus, in a compact Lie group H of strictly positive dimension N. Assume that the roots of A in H lie on a single line; that is, that there is a character α so that

$$R(H, A) \subset \{m\alpha \mid m \in \mathbb{Z}\} \subset X^*(A). \tag{3.6a}$$

(We do not assume that α itself is a root.) Since H is assumed to have positive dimension, there must be *some* (necessarily nonzero) roots; so α must have some order exactly $n > 0$:

$$\alpha\colon A \twoheadrightarrow \mu_n. \tag{3.6b}$$

Fix now a primitive nth root of unity ω, and an element $y \in A$ so that

$$\mu(y) = \omega. \tag{3.6c}$$

Then

$$\mathfrak{h}_{m\alpha} = \{X \in \mathfrak{h} \mid \mathrm{Ad}(y)X = \omega^m X\}. \tag{3.6d}$$

From this description (or indeed from (3.4m) and (3.4n)) it is clear that $\mathfrak{h}[m] = \mathfrak{h}_{m\alpha}$ is a $\mathbb{Z}/n\mathbb{Z}$-grading of the complex reductive Lie algebra $\mathfrak{h}$, and that $\mathfrak{h}[0] = 0$. According to the Kač classification of automorphisms of finite order (see for example [10], pp. 490–515; what we need is Lemma 10.5.3 on page 492)

$$\mathfrak{h} \text{ is necessarily abelian,} \tag{3.6e}$$

so the identity component H_0 is a compact torus.

We chose y so that $\alpha(y)$ generates the image of α. From this it follows immediately that A is generated by y and the kernel of α:

$$A = \langle \ker(\alpha), y \rangle. \tag{3.6f}$$

Because of the definition of roots and (3.6a), $\mathrm{Ad}(\ker(\alpha))$ must act trivially on $\mathfrak{h}$, and therefore on H_0. It follows that

$$Z_{H_0}(A) = H_0^y,$$

the fixed points of the automorphism $\mathrm{Ad}(y)$ on the torus H_0. Since A is assumed to be maximal abelian, we deduce that

$$A \cap H_0 = H_0^y. \tag{3.6g}$$

We are therefore going to analyze this fixed point group.

Because

$$\mathrm{Aut}(H_0) = \mathrm{Aut}(X_*(H_0)), \tag{3.6h}$$

the automorphism $\mathrm{Ad}(y)$ is represented by a lattice automorphism

$$y_* \in \mathrm{Aut}(X_*(H_0)), \qquad (y_*)^n = 1 \tag{3.6i}$$

and therefore (after choice of a lattice basis) by an invertible $N \times N$ integer matrix

$$Y_* \in GL(N, \mathbb{Z}), \qquad (Y_*)^n = I. \tag{3.6j}$$

Because 0 is not a root, the matrix Y_* does not have one as an eigenvalue. Every eigenvalue must be a primitive dth root of 1 for some d dividing n (and not equal to 1). Define

$$\begin{aligned} m_y(d) &= \text{multiplicity of primitive } d\text{th roots as eigenvalues of } Y_* \\ &= \dim(\mathfrak{h}_{m\alpha}), \qquad \text{all } m \text{ such that } \gcd(m,n) = n/d. \end{aligned} \tag{3.6k}$$

Then the characteristic polynomial of the matrix Y_* is

$$\det(xI - Y_*) = \prod_{d|n,\ d>1} \Phi_d(x)^{m_y(d)}. \tag{3.6l}$$

The lattice basis chosen to get the matrix Y_* identifies

$$H_0 \simeq \mathbb{C}^N/\mathbb{Z}^N. \tag{3.6m}$$

In this picture, the fixed points of $\mathrm{Ad}(y)$ correspond to

$$\begin{aligned} \{v \in \mathbb{C}^N \mid Y_*(v) - v \in \mathbb{Z}^N\}/\mathbb{Z}^N &= (Y_* - I)^{-1}\mathbb{Z}^N/\mathbb{Z}^N \\ &\simeq \mathbb{Z}^N/(Y_* - I)\mathbb{Z}^N. \end{aligned} \tag{3.6n}$$

If D is any nonsingular $N \times N$ integer matrix, then

$$|\mathbb{Z}^N/D\mathbb{Z}^N| = |\det D|,$$

(as follows for example from looking at Riemann sums and the linear change of variable formula for integrating over a cube in $\mathbb{R}^N$). Consequently the number of fixed points of $\mathrm{Ad}(y)$ is

$$\begin{aligned} |H_0^y| &= |\det(I - Y_*)| \\ &= \prod_{d|n,\ d>1} \Phi_d(1)^{m_y(d)}. \end{aligned} \tag{3.6o}$$

Here Φ_d is the dth cyclotomic polynomial

$$\Phi_d(x) = \prod_{\omega \in \mu_d \text{ primitive}} (x - \omega). \tag{3.6p}$$

Evaluating cyclotomic polynomials at 1 is standard and easy:

$$\Phi_d(1) = \begin{cases} 0 & d = 1 \\ p & d = p^m,\ (p \text{ prime},\ m \geq 1) \\ 1 & d \text{ divisible by at least two primes.} \end{cases} \tag{3.6q}$$

Inserting these values (3.6o) gives

$$|H_0^y| = \prod_{\substack{p^m|n \\ p \text{ prime},\ m \geq 1}} p^{\dim \mathfrak{h}_{(n/p^m)\alpha}}. \tag{3.6r}$$

It is easy to see that every element of H_0^y has order dividing n.

Definition 3.7. Suppose A is a finite maximal torus in the compact Lie group G, and $\alpha \in R(G, A)$ is a root of order exactly n:

$$1 \longrightarrow \ker\alpha \longrightarrow A \xrightarrow{\alpha} \mu_n \longrightarrow 1. \tag{3.7a}$$

The characters of A that are trivial on $\ker\alpha$ are precisely the multiples of α. (The reason is that the characters of μ_n are precisely the integer multiples of the "tautological" character sending each nth root of 1 to itself; and this in turn is a consequence of the fact that the group μ_n is cyclic.) If we define

$$G^{[\alpha]} = Z_G(\ker\alpha), \tag{3.7b}$$

(a compact subgroup of G), then its complexified Lie algebra is

$$\mathfrak{g}^{[\alpha]} = \sum_{m\in\mathbb{Z}} \mathfrak{g}_{m\alpha}. \tag{3.7c}$$

That is, the pair $(G^{[\alpha]}, A)$ is of the sort considered in Example 3.6. We now use the notation of that example, choosing in particular a primitive nth root $\omega \in \mu_n$, and an element $y \in A$ so that

$$\alpha(y) = \omega. \tag{3.7d}$$

As we saw in the example, $(G^{[\alpha]})_0$ is a (connected) torus, on which y acts as an automorphism of order n, and

$$A \cap (G^{[\alpha]})_0 = ((G^{[\alpha]})_0)^y. \tag{3.7e}$$

The outer parentheses are included for clarity: first take the identity component, then compute the fixed points of $\mathrm{Ad}(y)$. Reversing this order would give $(G^{[\alpha]})^y = A$, which has a trivial identity component. But we will omit them henceforth. We define the group of *coroots for* α to be

$$R^\vee(\alpha) = \{\xi\colon \mu_n \to (G^{[\alpha]})_0^y \subset \ker\alpha \subset A\} \subset X_*(\ker\alpha)(n) \subset X_*(n), \tag{3.7f}$$

the cocharacters taking values in the group of fixed points of $\mathrm{Ad}(y)$ on $(G^{[\alpha]})_0$. Choosing a primitive nth root of 1 identifies $R^\vee(\alpha)$ with $(G^{[\alpha]})_0^y$. Its cardinality may therefore be computed in terms of root multiplicities using (3.6r):

$$\begin{aligned} |R^\vee(\alpha)| &= |(G^{[\alpha]})_0^y| \\ &= \prod_{\substack{p^m | n \\ p \text{ prime},\ m\geq 1}} p^{\dim \mathfrak{g}_{(n/p^m)\alpha}}. \end{aligned} \tag{3.7g}$$

There are nontrivial coroots for α if and only if there is a nontrivial prime power p^m dividing n so that $(n/p^m)\alpha$ is a root.

Definition 3.8. Suppose A is a finite maximal torus in the compact Lie group G, and $\alpha \in R(G, A)$ is a root of order n, and

$$\xi\colon \mu_n \to (G^{[\alpha]})_0^y \subset \ker\alpha \subset A \tag{3.8a}$$

is a coroot for α (Definition 3.7). A *transvection generator for* (α, ξ) is an element

$$\sigma(\alpha,\xi) \in (G^{[\alpha]})_0 \tag{3.8b}$$

with the property that

$$\mathrm{Ad}(y^{-1})(\sigma(\alpha,\xi)) = \sigma(\alpha,\xi)\xi(\omega). \tag{3.8c}$$

We claim that there is a transvection generator for each coroot. To see this, write the abelian group $(G^{[\alpha]})_0$ additively. Then the equation we want to solve looks like

$$[\mathrm{Ad}(y^{-1}) - I]\sigma = \xi(\omega).$$

Because the determinant of the Lie algebra action is

$$|\det(\mathrm{Ad}(y^{-1}) - I)| = |\det(I - \mathrm{Ad}(y))| = |\det(I - Y_*)|$$

which is equal to the number of coroots (see (3.6o)), we see that (3.8c) has a solution $\sigma(\alpha, \xi)$, and that in fact $\sigma(\alpha, \xi)$ is unique up to a factor from $(G^{[\alpha]})_0^y$.

The defining equation for a transvection generator may be rewritten as

$$\sigma(\alpha, \xi)\, y\, \sigma(\alpha, \xi)^{-1} = y\, \xi(\alpha(y)). \tag{3.8d}$$

Because $\sigma(\alpha, \xi)$ is built from exponentials of root vectors for roots that are multiples of α, $\sigma(\alpha, \xi)$ must commute with $\ker \alpha$:

$$\sigma(\alpha, \xi)\, a_0\, \sigma(\alpha, \xi)^{-1} = a_0 \qquad (a_0 \in \ker \alpha \subset A). \tag{3.8e}$$

Combining the last two formulas, and the fact that A is generated by $\ker \alpha$ and y, we find that

$$\sigma(\alpha, \xi)\, a\, \sigma(\alpha, \xi)^{-1} = a\, \xi(\alpha(a)) \qquad (a \in A). \tag{3.8f}$$

In particular,

$$\sigma(\alpha, \xi) \in N_{G_0}(A), \tag{3.8g}$$

and the *root transvection* is the coset

$$s(\alpha, \xi) = \sigma(\alpha, \xi)(A \cap G_0) \in N_{G_0}(A)/(A \cap G_0) = W_{\mathrm{large}}(G, A). \tag{3.8h}$$

We define the *small Weyl group of A in G* to be the subgroup

$$W_{\mathrm{small}}(G, A) = \langle s(\alpha, \xi) \quad (\alpha \in R(G, A), \xi \in R^\vee(\alpha) \rangle, \tag{3.8i}$$

generated by root transvections.

Conjecture 3.9. If A is a finite maximal torus in a compact Lie group G (Definition 3.2), the normalizer of A in G_0 is generated by $A \cap G_0$ and the transvection generators $\sigma(\alpha, \xi)$ described in Definition 3.8. Equivalently,

$$W_{\mathrm{small}}(G, A) = W_{\mathrm{large}}(G, A)$$

(Definitions 3.2 and 3.8).

In case G is the projective unitary group $PU(n)$, then it is shown in [9] that A must be one of the subgroups described in (4.3) below. In these cases the conjecture is established in [8].

We want to record explicitly one of the conclusions of Example 3.6.

Proposition 3.10. *Suppose A is a finite maximal torus in a compact Lie group G (Definition 3.2), and that α and β are roots of A in G.*

1. *If α and β are both multiples of the same root γ, then $[\mathfrak{g}_\alpha, \mathfrak{g}_\beta] = 0$.*
2. *If α and β have relatively prime orders, then $[\mathfrak{g}_\alpha, \mathfrak{g}_\beta] = 0$.*

Proof. Part (1) is (3.6e) (together with the argument used in Definition 3.7 to get into the setting of Example 3.6). If α and β have orders m and n, then the hypothesis of (2) produces integers x and y so that $mx + ny = 1$. Consequently

$$\beta = (mx + ny)\beta = mx\beta = mx(\alpha + \beta),$$

and similarly

$$\alpha = ny(\alpha + \beta).$$

So (2) follows from (1) (with $\gamma = \alpha + \beta$).

We conclude this section with a (tentative and preliminary) analogue of Grothendieck's notion of root datum.

Definition 3.11 (Finite root datum). An *abstract finite root datum* is a quadruple $\Psi = (X^*, R, X_*, R^\vee)$, subject to the requirements

(a) X^* and X_* are finite abelian groups, dual to each other (cf. (3.5h)) by a specified pairing

$$\langle , \rangle : X^* \times X_* \to \mathbb{Q}/\mathbb{Z};$$

(b) $R \subset X^* - \{0\}$

(c) $R^\vee$ is a map from R to subgroups of X_*; we call $R^\vee(\alpha)$ the group of *coroots for α*.

We impose first the axioms

(FRD 0) If α has order n, and k is relatively prime to n, then $k\alpha$ is also a root and $R^\vee(\alpha) = R^\vee(k\alpha) \subset X_*(n)$; and

(FRD 1) If $\xi \in R^\vee(\alpha)$, then $\langle \alpha, \xi \rangle = 0$.

(The condition about k is a rationality hypothesis, corresponding to some automorphism being defined over $\mathbb{Q}$. The rest of (FRD 0) says that the order of a coroot must divide the order of the corresponding root. Axiom (FRD 1) says that a coroot must take values in the kernel of the corresponding root.)

For each root α of order n and coroot $\xi \in R^\vee(\alpha)$ we get homomorphisms of abelian groups

$$s(\alpha,\xi)\colon X^* \to X^*, \qquad s(\alpha,\xi)(\lambda) = \lambda - n\langle\lambda,\xi\rangle\alpha,$$
$$s(\alpha,\xi)\colon X_* \to X_*, \qquad s(\alpha,\xi)(\tau) = \tau - n\langle\alpha,\tau\rangle\xi,$$

called *root transvections*. The coefficients of α and of ξ in these formulas are integers because of axiom (FRD)(0); so the formulas make sense. It is easy to check that each of these endomorphisms is the transpose of the other with respect to the pairing $\langle,\rangle$. The axiom (FRD)(1) means that $s(\alpha,\xi)$ is the identity on multiples of α, and clearly $s(\alpha,\xi)$ induces the identity on the quotient $X^*/\langle\alpha\rangle$. Therefore $s(\alpha,\xi)$ is a transvection, and

$$s(\alpha,\cdot)\colon R^\vee(\alpha) \hookrightarrow \mathrm{Aut}(X^*)$$

is a group homomorphism.

We impose in addition the axioms

(FRD 2) If $\alpha \in R$ and $\xi \in R^\vee(\alpha)$, then $s(\alpha,\xi)(R) = R$.
(FRD 3) If $\alpha \in R$ and $\xi \in R^\vee(\alpha)$, then $s(\alpha,\xi)(R^\vee(\beta)) = R^\vee(s(\alpha,\xi)(\beta))$.

The *Weyl group of the root datum* is the group generated by the root transvections:

$$W(\Psi) = \langle s(\alpha,\xi) \ \ (\alpha \in R,\ \xi \in R^\vee(\alpha))\rangle \subset \mathrm{Aut}(X^*).$$

The inverse transpose isomorphism identifies the Weyl group with a group of automorphisms of X_*.

We have shown in this section that the root datum

$$\Psi(G,A) = (X^*(A), R(G,A), X_*(A), R^\vee) \tag{3.12}$$

of a finite maximal A torus in a compact Lie group G is an abstract finite root datum. The point of making these observations is the hope of finding and proving a result analogous to Theorem 2.11: that an abstract finite root datum determines a pair (G,A) uniquely.

We do not yet understand precisely how to formulate a reasonable conjecture along these lines. First, in order to avoid silly counterexamples from finite groups, we should assume

$$G = G_0A; \tag{3.13}$$

that is, that A meets every component of G.

To see a more serious failure of the finite root datum to determine G, consider the finite root datum

$$(\mathbb{Z}/6\mathbb{Z}, \{1,5\}, (1/6)\mathbb{Z}/\mathbb{Z}, R^\vee), \tag{3.14a}$$

in which $R^\vee(1) = R^\vee(5) = \{0\}$. Write

$$\mathbb{A} = \mathbb{Z}[x]/\langle\Phi_6(x)\rangle = \mathbb{Z}[x]/\langle x^2 - x + 1\rangle, \tag{3.14b}$$

the ring of integers of the cyclotomic field $\mathbb{Q}[\omega_6]$, with ω_6 a primitive sixth root of unity. The choice of ω_6 defines an inclusion $\mu_6 \hookrightarrow \mathbb{A}$ sending ω_6 to the image of x. Therefore the rank two free abelian group $\mathbb{A}$ acquires an action of $A = \mu_6$. If we write T^1 for the two-dimensional torus with

$$X_*(T^1) = \mathbb{A}, \tag{3.14c}$$

then the equivalence of categories (2.3i) provides an action of μ_6 on T^1. Explicitly, the action of ω_6 on T^1 is

$$\omega_6 \cdot (z, w) = (w^{-1}, zw) \qquad (z, w \in S^1). \tag{3.14d}$$

The roots for this action are 1 and 5. According to (3.6r) (or by inspection of (3.14d)) the action of A on T^1 has no fixed points. It follows that A is a maximal abelian subgroup of

$$G^1 = T^1 \rtimes A, \tag{3.14e}$$

and that the corresponding finite root datum is exactly the one described by (3.14a).

So far so good. But we could equally well use the $2m$-dimensional torus

$$T^m = T^1 \times \cdots \times T^1$$

with the diagonal action of μ_6, and define

$$G^m = T^m \rtimes A. \tag{3.14f}$$

Again $A = \mu_6$ is a maximal abelian subgroup, and the root datum is exactly (3.14a). So in this case there are many different G, of different dimensions, with the same finite root datum.

The most obvious way to address this particular family of counterexamples is to include root multiplicities as part of the finite root datum, and to require that they compute the cardinalities of the coroot groups $R^\vee(\alpha)$ by (3.7g). (If A is an elementary abelian p-group, then the coroot groups determine the root multiplicities: if α has multiplicity m, then $|R^\vee(\alpha)| = p^m$. That is why we needed A of order 6 to have make an easy example where many multiplicities are possible.) But the root multiplicities alone do not determine G; one can make counterexamples with G_0 a torus and A cyclic using cyclotomic fields of class number greater than one. Perhaps the finite root datum should be enlarged to include the tori $G_0^{[\alpha]}$ (or rather the corresponding lattices), equipped with the action of μ_n constructed in Definition 3.7.

4 Examples

Example 4.1. The simplest example of a finite maximal torus is in the three-dimensional compact group

$$G = SO(3)$$

of three by three real orthogonal matrices. We can choose

$$A = \left\{ \begin{pmatrix} \epsilon_1 & 0 & 0 \\ 0 & \epsilon_2 & 0 \\ 0 & 0 & \epsilon_3 \end{pmatrix} \mid \epsilon_i = \pm 1, \prod_i \epsilon_i = 1 \right\}.$$

This is the "Klein four-group," the four-element group in which each non-identity element has order 2. We can identify characters with subsets of $S \subset \{1, 2, 3\}$, modulo the equivalence relation that each subset is equivalent to its complement: $S \sim S^c$. The formula is

$$\lambda_S(\epsilon_1, \epsilon_2, \epsilon_3) = \prod_{i \in S} \epsilon_i.$$

Thus the trivial character of A is $\lambda_\emptyset = \lambda_{\{1,2,3\}}$, and the three nontrivial characters correspond to the three two-element subsets $\{i, j\}$ (or equivalently to their three one-element complements):

$$\lambda_{\{i,j\}}(\epsilon_1, \epsilon_2, \epsilon_3) = \epsilon_i \epsilon_j.$$

The Lie algebra $\mathfrak{g} = \mathfrak{so}(3)$ consists of 3×3 skew-symmetric matrices. The root spaces of A are one-dimensional:

$$\mathfrak{g}_{\lambda_{\{i,j\}}} = \mathbb{C}(e_{ij} - e_{ji}) \qquad (1 \leq i \neq j \leq 3),$$

the most natural and obvious lines of skew-symmetric matrices. Therefore

$$R(G, A) = \{\lambda_{\{i,j\}} \in X^*(A) \mid (1 \leq i \neq j \leq 3)\},$$

the set of all three nonzero characters of A.

Each root space is the Lie algebra of one of the three obvious $SO(2)$ subgroups of $SO(3)$, and these are the tori $G_0^{[\alpha]}$ used in Definition 3.7. The automorphism y of each torus is inversion, so the coroots are the two elements of order (1 or) 2 in each torus. If ξ is the nontrivial coroot attached to the root (i, j), then the transvection $s(\alpha, \xi)$ acts on $\{1, 2, 3\}$ by transposition of i and j. The (small) Weyl group is therefore

$$W_{\text{small}}(G, A) = S_3.$$

Since this is the full automorphism group of A, it is also equal to the large Weyl group.

It is a simple and instructive matter to make a similar definition for $O(n)$, taking for A the group of 2^n diagonal matrices. The whole calculation is exactly parallel to that for the root system of $U(n)$, with the role of the complex units S^1 played by the real units $\{\pm 1\}$, or, on the level of X^*, with $\mathbb{Z}$ replaced by $\mathbb{Z}/2\mathbb{Z}$.

Example 4.2. We begin with the unitary group

$$\widetilde{G} = U(n) = n \times n \text{ unitary matrices.} \tag{4.2a}$$

The center of $\widetilde{G}$ consists of the scalar matrices

$$Z(n) = \{zI \mid z \in S^1\} \simeq S^1 \tag{4.2b}$$

(notation (2.3b)). We are going to construct a finite maximal torus A inside the projective unitary group

$$G = PU(n) = U(n)/Z(n). \tag{4.2c}$$

It is convenient to construct a preimage $\widetilde{A} \subset \widetilde{G} = U(n)$.

The Lie algebra of $U(n)$ consists of skew-Hermitian $n \times n$ complex matrices:

$$\widetilde{\mathfrak{g}}_0 = \mathfrak{u}(n) = \{X \in M_n(\mathbb{C} \mid {}^t\overline{X} = -X\}. \tag{4.2d}$$

An obvious map identifies its complexification with all $n \times n$ matrices:

$$\widetilde{\mathfrak{g}} = M_n(\mathbb{C}) = \mathfrak{gl}(n, \mathbb{C}). \tag{4.2e}$$

The adjoint action is given by conjugation of matrices. Dividing by the center gives

$$\mathfrak{g}_0 = \mathfrak{pu}(n) = \mathfrak{u}(n)/i\mathbb{R}I, \tag{4.2f}$$

$$\mathfrak{g} = \mathfrak{pgl}(n, \mathbb{C}) = M_n(\mathbb{C})/\mathbb{C}I. \tag{4.2g}$$

It will be convenient to think of $U(n)$ as acting on the vector space

$$\mathbb{C}^n = \text{functions on } \mathbb{Z}/n\mathbb{Z},$$

functions on the cyclic group of order n. We will call the standard basis

$$e_0, e_1, \ldots, e_{n-1}$$

with e_i the delta function at the group element $i + n\mathbb{Z}$. It is therefore often convenient to regard the indices as belonging to $\mathbb{Z}/n\mathbb{Z}$.

We are going to define two cyclic subgroups

$$\tau : \mathbb{Z}/n\mathbb{Z} \to U(n), \qquad \sigma : \mu_n \to U(n) \tag{4.2h}$$

of $U(n)$. We will also be interested in

$$\zeta\colon \mu_n \to Z(n), \qquad \zeta(\omega) = \omega I. \tag{4.2i}$$

The map τ comes from the action of $\mathbb{Z}/n\mathbb{Z}$ on itself by translation; the generator $1 = 1 + n\mathbb{Z}$ acts by

$$\tau(1) = \begin{pmatrix} 0 & 1 & 0 & \cdots & 0 \\ 0 & 0 & 1 & \cdots & 0 \\ & & & \vdots & \\ 0 & 0 & 0 & \cdots & 1 \\ 1 & 0 & 0 & \cdots & 0 \end{pmatrix}. \tag{4.2j}$$

Often it is convenient to compute with the action on basis vectors:

$$\tau(m)e_i = e_{i-m},$$

as usual with the subscripts interpreted modulo n. The map σ is from the character group of $\mathbb{Z}/n\mathbb{Z}$. The element $\omega \in \mu_n$ is realized as multiplication by the character $m \mapsto \omega^m$:

$$\sigma(\omega) = \begin{pmatrix} 1 & 0 & \cdots & 0 \\ 0 & \omega & \cdots & 0 \\ & & \vdots & \\ 0 & 0 & \cdots & \omega^{n-1} \end{pmatrix}. \tag{4.2k}$$

This time the formula on basis vectors is

$$\sigma(\omega)e_i = \omega^i e_i.$$

Each of σ and τ has order n, and their commutator is

$$\sigma(\omega)\tau(m)\sigma(\omega^{-1})\tau(-m) = \omega^m I = \zeta(\omega^m) \in Z(G).$$

The three cyclic groups σ, τ, and ζ generate a group

$$\widetilde{A} = \langle \tau(\mathbb{Z}/n\mathbb{Z}), \sigma(\mu_n), \zeta(\mu_n)\rangle \tag{4.2l}$$

of order n^3, with defining relations

$$\sigma(\omega)\tau(m)\sigma(\omega^{-1})\tau(-m) = \zeta(\omega^m), \qquad \sigma\zeta = \zeta\sigma, \qquad \tau\zeta = \zeta\tau; \tag{4.2m}$$

this group is a *finite Heisenberg group* of order n^3. (One early appearance of such groups is in [12], pp. 294–297. There is an elementary account of their representation theory in [17], Chapter 19.)

The "finite maximal torus" we consider is

$$\begin{aligned} A &= \text{image of } \widetilde{A} \text{ in } PU(n) \\ &= \widetilde{A}/\zeta(\mu_n) \simeq (\mathbb{Z}/n\mathbb{Z}) \times \mu_n. \end{aligned} \tag{4.2n}$$

We will explain in a moment why A is *maximal* abelian. The adjoint action of A on the Lie algebra is easily calculated to be

$$\mathrm{Ad}(\tau(m))(e_{rs}) = e_{r-m,s-m}, \tag{4.2o}$$

with the subscripts interpreted modulo n. Similarly

$$\mathrm{Ad}(\sigma(\omega))(e_{rs}) = \omega^{r-s} e_{rs}. \tag{4.2p}$$

The character group of A is

$$X^*(A) \simeq \mu_n \times \mathbb{Z}/n\mathbb{Z} \qquad \lambda_{\phi,j}(\tau(m)\sigma(\omega)) = \phi^m \omega^j.$$

We now describe the roots of A in the Lie algebra $\mathfrak{g} = M_n(\mathbb{C})/\mathbb{C}I$. Fix

$$j \in \mathbb{Z}/n\mathbb{Z}, \qquad \phi \in \mu_n,$$

and define

$$X_{\phi,j} = \sum_{r-s=j} \phi^r e_{rs} = \sigma(\phi)\tau(-j). \tag{4.2q}$$

(That is, the root vectors as matrices can be taken equal to the group elements as matrices.) It follows from (4.2o) and (4.2p) that

$$\mathrm{Ad}(\tau(m))(X_{\phi,j}) = \phi^m X_{\phi,j}, \qquad \mathrm{Ad}(\sigma(\omega))(X_{\phi,j}) = \omega^j X_{\phi,j}. \tag{4.2r}$$

That is, $X_{\phi,j}$ is a weight vector for the character $(\phi, j) \in X^*(A)$. The weight vector $X_{1,0}$ is the identity matrix, by which we are dividing to get $\mathfrak{g}$; so $(1,0)$ is not a root. Therefore

$$R(G, A) = \{(\phi, j) \neq (1,0) \in X^*(A)\} \simeq [\mu_n \times \mathbb{Z}/n\mathbb{Z}] - (1,0), \tag{4.2s}$$

the set of $n^2 - 1$ nontrivial characters of $\mathbb{Z}/n\mathbb{Z} \times \mu_n$.

There remains the question of why A is *maximal* abelian in $PU(n)$. Suppose $g \in PU(n)^A$. Choose a preimage $\widetilde{g} \in U(n) \subset M_n(\mathbb{C})$. Then the fact that g commutes with the images of τ and σ in $PU(n)$ means that

$$\tau(m)\widetilde{g}\tau(-m) = b(m)\widetilde{g}, \qquad \sigma(\omega)\widetilde{g}\sigma(\omega^{-1}) = c(\omega)\widetilde{g}.$$

If we write $\widetilde{g}$ in the matrix basis $X_{\phi,j}$ of (4.2q), the conclusion is that $b(m) = \phi$ is an nth root of unity, that $c(\omega) = \omega^j$, and that

$$\widetilde{g} = zX_{\phi,j} = z\sigma(\phi)\tau(-j). \tag{4.2t}$$

Therefore $g = \sigma(\phi)\tau(-j) \in A$, as we wished to show.

We want to understand, or at least to count, the coroots corresponding to each root α of order d. Since every (nontrivial) character of A has multiplicity one as a root, we conclude from (3.7g) that there are precisely d coroots ξ attached to α. In particular, if p^m is the largest power of some prime dividing n, and α has order p^m, then (since $A \simeq (\mathbb{Z}/n\mathbb{Z})^2$)

$$\ker\alpha \simeq (\mathbb{Z}/n\mathbb{Z}) \times \mathbb{Z}/(n/p^m)\mathbb{Z}.$$

So in this case there are exactly p^m homomorphisms from μ_{p^m} into $\ker\alpha$, and all of them must be coroots. We conclude that the root transvections include all transvections of A attached to characters of order exactly p^m. One can show that these transvections generate $SL(2, \mathbb{Z}/n\mathbb{Z})$, so

$$W_{\text{small}}(G, A) \simeq SL(2, \mathbb{Z}/n\mathbb{Z}). \tag{4.2u}$$

We conclude this example by calculating the structure constants of $\mathfrak{g}$ in the root basis. Using the relation $X_{\phi,j} = \sigma(\phi)\tau(-j)$ from (4.2q), and the commutation relation (4.2m), we find that

$$\begin{aligned} X_{\phi,j} X_{\psi,k} &= \sigma(\phi)\tau(-j)\sigma(\psi)\tau(-k) \\ &= \psi^j \sigma(\phi)\sigma(\psi)\tau(-j)\tau(-k) \\ &= \psi^j \sigma(\phi\psi)\tau(-j-k) = \psi^j X_{\phi\psi,j+k}. \end{aligned}$$

Similarly

$$X_{\psi,k} X_{\phi,j} = \phi^k X_{\phi\psi,j+k}.$$

Therefore

$$[X_{\phi,j}, X_{\psi,k}] = (\psi^j - \phi^k) X_{\phi\psi,j+k}. \tag{4.2v}$$

A fundamental fact about classical roots in reductive Lie algebras (critical to Chevalley's construction of reductive groups over arbitrary fields; see [16], Chapter 10) is that the structure constants may be chosen to be integers. Here we see that the structure constants are integers in the cyclotomic field $\mathbb{Q}[\mu_n]$.

The preceding example can be generalized by replacing the cyclic group $\mathbb{Z}/n\mathbb{Z}$ with any abelian group F of order n, and A by $F \times X^*(F)$ ([18], page 148). If D is the largest order of an element of F, then the symplectic form takes values in μ_D:

$$\Sigma((f_1, \lambda_1), (f_2, \lambda_2)) = \lambda_1(f_2)[\lambda_2(f_1)]^{-1}. \tag{4.3a}$$

Characters of A may be indexed by elements of A using the symplectic form:

$$\alpha_x(a) = \Sigma(a, x).$$

The transvection generators are precisely the symplectic transvections

$$a \mapsto a + \xi(\langle a, x \rangle) \tag{4.3b}$$

on A. Here x is any element of A of order d, and

$$\xi \colon \mu_d \to \langle x \rangle$$

is any homomorphism. They generate the full symplectic group

$$W_{\text{small}}(G, A) = Sp(A, \Sigma); \tag{4.3c}$$

a proof may be found in [8], Theorem 3.16. (When F is a product of elementary abelian p groups for various primes p, the assertion that symplectic transvections generate the full symplectic group comes down to the (finite) field case, and there it is well known.)

Here is a very different example.

Example 4.4. We begin with the compact connected Lie group G of type E_8; this is a simple group of dimension 248, with trivial center. We are going to describe a finite maximal torus

$$A = \mathbb{Z}/5\mathbb{Z} \times \mu_5 \times \mu_5. \tag{4.4a}$$

The roots will be the 124 nontrivial characters of A, each occurring with multiplicity 2. The group A is described in detail in [7], Lemma 10.3. We present here another description, taken from [2], p. 231.

An element of the maximal torus T of G may be specified by specifying its eigenvalue γ_i (a complex number of absolute value 1) on each of the eight simple roots α_i (the white vertices in Figure 1. Then the eigenvalue γ_0 on the lowest root α_0 (the black vertex) is specified by the requirement

$$\gamma_0^{-1} = \prod_{i=1}^{8} \gamma_i^{n_i}, \tag{4.4b}$$

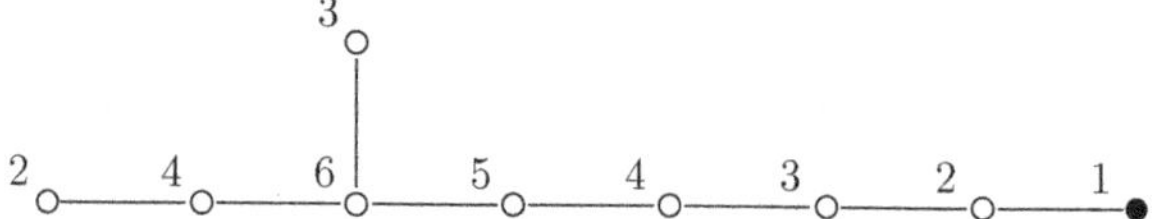

Figure 1 Extended Dynkin diagram for E_8

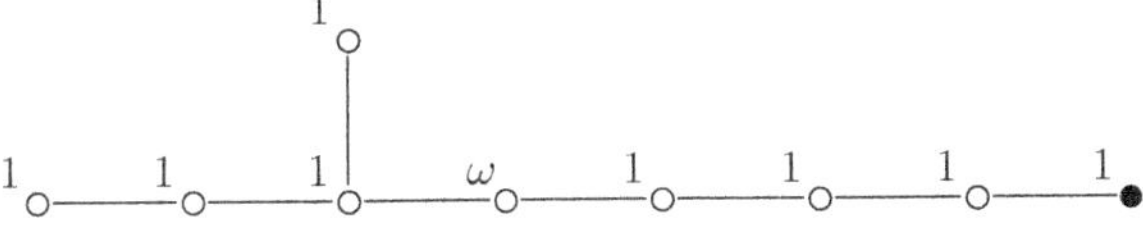

Figure 2 Toral subgroup $\mu_5 \subset E_8$

with n_i the coefficient of α_i in the highest root (the vertex labels in the figure). Equivalently, we require

$$\prod_{i=0}^{8} \gamma_i^{n_i} = 1. \tag{4.4c}$$

There is a map

$$\rho\colon \mu_5 \to T \tag{4.4d}$$

in which the element $\rho(\omega)$ corresponds to the diagram of Figure 2. The eight roots labeled 1 in this diagram are simple roots for a subsystem of type $A_4 \times A_4$. As is explained in [3], page 219, this subsystem corresponds to a subgroup

$$H = (SU(5) \times SU(5))/(\mu_5)_\Delta, \tag{4.4e}$$

the quotient by the diagonal copy of μ_5 in the center.

We have

$$G^{\rho(\mu_5)} = H, \qquad \rho(\mu_5) = Z(H). \tag{4.4f}$$

Because of this, the rest of the calculations we want to do can be performed inside H. It is convenient to label the two $SU(5)$ factors as L and R (for "left" and "right").

We now recall the maps σ, τ, and ζ of Example 4.2. Because 5 is odd, they are actually maps into $SU(5)$ (rather than just $U(5)$). We use subscripts L and R to denote the maps into the two factors of H, so that for example

$$\sigma_L \times \sigma_R\colon \mu_5 \times \mu_5 \to H.$$

Taking the diagonal copies of these maps gives

$$\sigma_\Delta\colon \mu_5 \to SU(5) \times SU(5), \quad \tau_\Delta\colon \mathbb{Z}/5\mathbb{Z} \to SU(5) \times SU(5). \tag{4.4g}$$

The diagonal map ζ_Δ is trivial. According to (4.2m), we have

$$\sigma_\Delta(\omega)\tau_\Delta(m)\sigma_\Delta(\omega^{-1})\tau_\Delta(-m) = \zeta_\Delta(\omega^m) \tag{4.4h}$$

in $SU(5) \times SU(5)$; so in the quotient group H, we get

$$\tau_\Delta \times \sigma_\Delta \times \rho\colon \mathbb{Z}/5\mathbb{Z} \times \mu_5 \times \mu_5 \to H \subset G; \tag{4.4i}$$

the image is our abelian group A of order 125. Because of (4.4f), we have

$$G^A = H^{\tau_\Delta, \sigma_\Delta}, \tag{4.4j}$$

and an easy calculation in $SU(5) \times SU(5)$ (parallel to the one leading to (4.2t)) shows that this is exactly A. So A is indeed a finite maximal torus.

We turn next to calculation of the roots. The character group of A is

$$X^*(A) = \mu_5 \times \mathbb{Z}/5\mathbb{Z} \times \mathbb{Z}/5\mathbb{Z}.$$

The roots $(\phi, j, 0)$ are those in the centralizer H of $\rho(\mu_5)$; so they are the roots of A in

$$\mathfrak{h} = \mathfrak{sl}(5, \mathbb{C})_L \times \mathfrak{sl}(5, \mathbb{C})_R.$$

These were essentially calculated in Example 4.2. We have

$$\mathfrak{g}_{\phi,j,0} = \langle X^L_{\phi,j}, X^R_{\phi,j} \rangle \qquad ((\phi, j) \neq (1, 0)).$$

Here the root vectors are the ones defined in (4.2q). In particular, each of these 24 roots has multiplicity two.

To study the root vectors for the 100 roots (ϕ, j, k) with $k \neq 0$ modulo 5, one can analyze the representation of H on $\mathfrak{g}/\mathfrak{h}$, which has dimension 200. We will not do this here; the conclusion is that every root space has dimension two.

Since all 124 characters of A have multiplicity two, it follows from (3.7g) that each root α has exactly 25 coroots; these are all the homomorphisms

$$\xi\colon \mu_5 \to \ker \alpha \subset A.$$

The corresponding transvections

$$s(\alpha, \xi)(\lambda) = \lambda - \langle \lambda, \xi \rangle \alpha \tag{4.4k}$$

are all the transvections moving λ by a multiple of α; the multiple is given by the linear functional ξ which is required only to vanish on α. The (small) Weyl group generated by all of these transvections is therefore

$$W_{\text{small}}(G, A) = SL(A) \simeq SL(3, \mathbb{F}_5), \tag{4.4l}$$

the special linear group over the field with five elements. Its cardinality is

$$|W_{\text{small}}(G, A)| = (5^2 + 5 + 1)(5 + 1)(1)(5^3)(5 - 1)^2 = 372000.$$

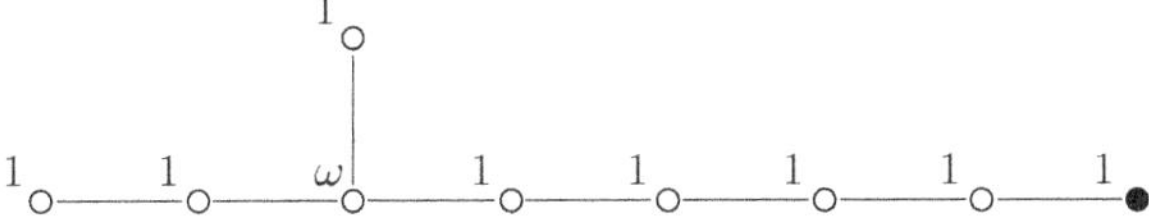

Figure 3 Toral subgroup $\mu_6 \subset E_8$

Precisely parallel discussions can be given for $G = F_4$, $A = \mathbb{Z}_3 \times \mu_3 \times \mu_3$, and for $G = G_2$, $A = \mathbb{Z}_2 \times \mu_2 \times \mu_2$. We omit the details. The next example, however, is sufficiently different to warrant independent discussion.

Example 4.5. We begin as in Example 4.4 with G a compact connected group of type E_8. We are going to describe a finite maximal torus

$$A = \mathbb{Z}/6\mathbb{Z} \times \mu_6 \times \mu_6. \tag{4.5a}$$

The roots will be the 215 nontrivial characters of A. The 7 characters of order 2 will have multiplicity two; the 26 characters of order 3 will have multiplicity two; and the 182 characters of order 6 will have multiplicity one. To begin, we define

$$\rho\colon \mu_6 \to T \tag{4.5b}$$

so that the element $\rho(\omega)$ corresponds to the diagram of Figure 3. This time the eight roots labeled 1 are those for a subsystem of type $A_5 \times A_2 \times A_1$. As we learn in [3], page 220, the corresponding subgroup of G is

$$H = (SU(6) \times SU(3) \times SU(2))/\zeta_\Delta(\mu_6); \tag{4.5c}$$

here

$$\zeta_\Delta\colon \mu_6 \to \mu_6 \times \mu_3 \times \mu_2 = Z(SU(6) \times SU(3) \times SU(2)), \qquad \zeta_\Delta(\omega) = (\omega, \omega^2, \omega^3).$$

Because the centralizer of a single element of a compact simply connected Lie group is connected, we conclude that

$$G^{\rho(\mu_6)} = H, \qquad \rho(\mu_6) = Z(H). \tag{4.5d}$$

Again we want to make use of the maps defined in Example 4.2. The first difficulty is that because 6 and 2 are even, the maps $\sigma_{SU(2)}$, $\tau_{SU(2)}$, $\sigma_{SU(6)}$, and $\tau_{SU(6)}$ take some of their values in matrices of determinant -1. In order to correct this, we fix a primitive twelfth root γ of 1, and define

$$\begin{aligned} \widetilde{\tau}_{SU(6)}&\colon \mathbb{Z}/12\mathbb{Z} \to SU(6), & \widetilde{\tau}_{SU(6)}(m) &= \gamma^m \cdot \tau_{U(6)}(2m), \\ \widetilde{\tau}_{SU(3)}&\colon \mathbb{Z}/12\mathbb{Z} \to SU(3), & \widetilde{\tau}_{SU(3)}(m) &= \tau_{U(3)}(4m), \\ \widetilde{\tau}_{SU(2)}&\colon \mathbb{Z}/12\mathbb{Z} \to SU(2), & \widetilde{\tau}_{SU(2)}(m) &= \gamma^{3m} \cdot \tau_{U(2)}(6m). \end{aligned} \tag{4.5e}$$

It is easy to check that these three maps are well defined. If we form the diagonal

$$\widetilde{\tau}_\Delta\colon \mathbb{Z}/12\mathbb{Z} \to SU(6) \times SU(3) \times SU(2), \tag{4.5f}$$

then

$$\widetilde{\tau}_\Delta(6) = (\gamma^6, 1, \gamma^{18}) = (-1, 1, -1) = \zeta_\Delta(-1).$$

The image in H of this element is trivial; so $\widetilde{\tau}_\Delta$ descends to

$$\tau_\Delta\colon \mathbb{Z}/6\mathbb{Z} \to H. \tag{4.5g}$$

In exactly the same way we can define

$$\widetilde{\sigma}_\Delta\colon \mu_{12} \to SU(6) \times SU(3) \times SU(2), \tag{4.5h}$$

descending to

$$\sigma_\Delta\colon \mu_6 \to H. \tag{4.5i}$$

Just as in Example 4.4, we find a group homomorphism

$$\tau_\Delta \times \sigma_\Delta \times \rho\colon \mathbb{Z}/6\mathbb{Z} \times \mu_6 \times \mu_6 \to H \subset G; \tag{4.5j}$$

the image is our abelian group A of order 216. Because of (4.5d), we have

$$G^A = H^{\tau_\Delta, \sigma_\Delta}, \tag{4.5k}$$

and a calculation in $SU(6) \times SU(3) \times SU(2)$ (parallel to the one leading to (4.2t)) shows that this is exactly A. So A is indeed a finite maximal torus.

We turn next to the roots. The character group of A is

$$X^*(A) = \mu_6 \times \mathbb{Z}/6\mathbb{Z} \times \mathbb{Z}/6\mathbb{Z}. \tag{4.5l}$$

The roots $(\phi, j, 0)$ are those in the centralizer H of $\rho(\mu_6)$; so they are the roots of A in

$$\mathfrak{h} = \mathfrak{sl}(6, \mathbb{C}) \times \mathfrak{sl}(3, \mathbb{C}) \times \mathfrak{sl}(2, \mathbb{C}).$$

These were essentially calculated in Example 4.2. We have

$$\mathfrak{g}_{\phi,j,0} = \left\langle X_{\phi,j}^{\mathfrak{sl}(6)}, X_{\phi,j}^{\mathfrak{sl}(3)}, X_{\phi,j}^{\mathfrak{sl}(2)} \right\rangle \qquad (\phi, j) \neq (1, 0). \tag{4.5m}$$

The meaning of the first of these root vectors (defined in (4.2q) for $n = 6$) is clear. The second root vector makes sense if both ϕ and j have order 3, that is, if ϕ is the square of a sixth root of 1, and j is twice an integer modulo 6. Similarly, the third root vector makes sense if ϕ and j both have order 2. The second and third root vectors cannot both make sense, for in that case (ϕ, j) would be trivial.

We have therefore shown that, among the 35 roots α vanishing on $\rho(\mu_6)$,

$$\dim \mathfrak{g}_\alpha = \begin{cases} 1 & \text{if } \alpha \text{ has order } 6 \\ 2 & \text{if } \alpha \text{ has order } 3 \\ 2 & \text{if } \alpha \text{ has order } 2. \end{cases} \tag{4.5n}$$

By analyzing the action of A in the 202-dimensional representation of H on $\mathfrak{g}/\mathfrak{h}$, one can see that the same statements hold for all 215 roots.

We now calculate the coroots. If α is a root of order 6, then (3.6r) says that the number of coroots is

$$2^{\dim \mathfrak{g}_{(6/2)\alpha}} \cdot 3^{\dim \mathfrak{g}_{(6/3)\alpha}} = 2^2 \cdot 3^2 = 36;$$

so the coroots are all the 36 homomorphisms

$$\xi \colon \mu_6 \to \ker \alpha \simeq (\mathbb{Z}/6\mathbb{Z})^2. \tag{4.5o}$$

The corresponding root transvections are all the transvections associated to the character α.

If β is a root of order 3, then (3.6r) says that the number of coroots is

$$3^{\dim \mathfrak{g}_{(3/3)\beta}} = 3^2 = 9;$$

so the coroots are all the 9 homomorphisms

$$\xi \colon \mu_3 \to \ker \beta \simeq (\mathbb{Z}/3\mathbb{Z})^2 \times (\mathbb{Z}/2\mathbb{Z})^3. \tag{4.5p}$$

The corresponding root transvections are all the transvections associated to the character β.

Similarly, if γ is a root of order 2, there are 4 coroots, and the root transvections are all of the 4 transvections associated to γ.

We see therefore that the (small) Weyl group of A in G contains *all* the transvection automorphisms of $A \simeq (\mathbb{Z}/6\mathbb{Z})^3$; so

$$W_{\text{small}}(G, A) \simeq SL(3, \mathbb{Z}/6\mathbb{Z}) \simeq SL(3, \mathbb{Z}/2\mathbb{Z}) \times SL(3, \mathbb{Z}/3\mathbb{Z}), \tag{4.5q}$$

a group of order $168 \cdot 13392 = 2249856$.

The 27 characters of A of order 3 are exactly the characters trivial on the 8-element subgroup $A[2]$ of elements of order 2 in A; so $G[3] = G^{A[2]}$ has Lie algebra

$$\mathfrak{g}[3] = \sum_{3\beta = 0} \mathfrak{g}_\beta. \tag{4.5r}$$

The 26 roots here all have multiplicity 2, so $G[3]$ has dimension 52. It turns out that

$$G[3] \simeq F_4 \times A[2]. \tag{4.5s}$$

Similarly, if we define $G[2] = G^{A[3]}$, then

$$G[2] \simeq G_2 \times A[3]. \tag{4.5t}$$

Now Proposition 3.10 guarantees that $G[2]$ and $G[3]$ commute with each other, so we get a subgroup

$$G[2] \times G[3] \subset G, \qquad G_2 \times F_4 \subset E_8. \tag{4.5u}$$

Perhaps most strikingly

$$W(G, A) = W(G[2], A) \times W(G[3], A); \tag{4.5v}$$

this is just the product decomposition noted in (4.5q).

The construction also shows (since $A[2] \subset G_2$ and $A[3] \subset F_4$) that each of the subgroups G_2 and F_4 is the centralizer of the other in E_8. The existence of these subgroups has been known for a long time (going back at least to [5], Table 39 on page 233; see also [15], pages 62–65); but it is not easy to deduce from the classical theory of root systems alone.

References

1. A. V. Alekseevskiĭ, *Jordan finite commutative subgroups of simple complex Lie groups*, Funkcional. Anal. i Priložen. **8** (1974), no. 4, 1–4 (Russian); English transl., Functional Anal. Appl. **8** (1974), 277–279.
2. A. Borel, *Sous-groupes commutatifs et torsion des groupes de Lie compacts connexes*, Tohoku Math. J.(2) **13** (1961), 216–240.
3. A. Borel and J. De Siebenthal, *Les sous-groupes fermés de rang maximum des groupes de Lie clos*, Comment. Math. Helv. **23** (1949), 200–221 (French).
4. M. Demazure and A. Grothendieck *et al., Schémas en Groupes. III: Structure des groupes réductifs*, Séminaire de Géométrie Algébrique du Bois Marie 1962/64 (SGA 3). Dirigé par M. Demazure et A. Grothendieck. Lecture Notes in Mathematics, Vol. 153, Springer-Verlag, Berlin, Heidelberg, New York, 1970.
5. E.B. Dynkin, *Semisimple subalgebras of semisimple Lie algebras*, Mat. Sbornik N.S. **30(72)** (1952), 349–462 (Russian); English transl., Amer. Math. Soc. Transl., Ser. 2 **6** (1957), 111–245.
6. V. V. Gorbatsevich, A. L. Onishchik, and È. B. Vinberg, *Structure of Lie groups and Lie algebras*, Current problems in mathematics. Fundamental directions, Vol. 41, Akad. Nauk SSSR Vsesoyuz. Inst. Nauchn. i Tekhn. Inform., Moscow (Russian); English transl. in translated by V. Minachin, Encyclopedia of Mathematical Sciences, Vol. 41, Springer-Verlag, Berlin, New York, 1994.
7. Robert L. Griess Jr., *Elementary abelian p-subgroups of algebraic groups*, Geom. Dedicata **39** (1991), no. 3, 253–305.

8. Gang Han, *The Weyl group of the fine grading of* $\mathrm{sl}(n, \mathbb{C})$ *associated with tensor product of generalized Pauli matrices*, J. Math. Phys. **52** (2011), no. 4, 042109, 18.
9. Miloslav Havlícek, Jiří Patera, and Edita Pelantova, *On Lie gradings. II*, Linear Algebra Appl. **277** (1998), no. 1–3, 97–125.
10. S. Helgason, *Differential Geometry, Lie Groups, and Symmetric Spaces*, Academic Press, New York, San Francisco, London, 1978.
11. Anthony W. Knapp, *Lie Groups Beyond an Introduction*, Second Edition, Progress in Mathematics, Vol. 140, Birkhäuser, Boston, 2002.
12. D. Mumford, *On the equations defining abelian varieties. I*, Invent. Math. **1** (1966), 287–354.
13. A. L. Onishchik and È. B. Vinberg, *Lie groups and algebraic groups*, Springer Series in Soviet Mathematics, Springer-Verlag, Berlin, 1990. Translated from the Russian and with a preface by D. A. Leites.
14. J. Patera and H. Zassenhaus, *On Lie gradings. I*, Linear Algebra Appl. **112** (1989), 87–159.
15. H. Rubenthaler, *Les paires duales dans les algèbres de Lie réductives*, Astérisque **219** (1994) (French, with English and French summaries).
16. T. A. Springer, *Linear Algebraic Groups*, 2nd ed., Progress in Mathematics, Vol. 9, Birkhäuser Boston Inc., Boston, MA, 1998.
17. A. Terras, *Fourier analysis on finite groups and applications*, London Mathematical Society Student Texts, Vol. 43, Cambridge University Press, Cambridge, 1999.
18. A. Weil, *Sur certains groupes d'opérateurs unitaires*, Acta Math. **111** (1964), 143–211.

Sums of squares of Littlewood–Richardson coefficients and GL_n-harmonic polynomials

Pamela E. Harris and Jeb F. Willenbring

To Nolan Wallach, who has influenced generations of mathematicians

Abstract We consider the example from invariant theory concerning the conjugation action of the general linear group on several copies of the $n \times n$ matrices, and examine a symmetric function which stably describes the Hilbert series for the invariant ring with respect to the multigradation by degree. The terms of this Hilbert series may be described as a sum of squares of Littlewood–Richardson coefficients. A "principal specialization" of the gradation is then related to the Hilbert series of the K-invariant subspace in the GL_n-harmonic polynomials (in the sense of Kostant), where K denotes a block diagonal embedding of a product of general linear groups. We also consider other specializations of this Hilbert series.

Keywords: Littlewood–Richardson coefficients • Harmonic functions

Mathematics Subject Classification: 05E05, 17B10, 22E46

1 Introduction

We let F_n^λ denote the finite-dimensional irreducible representation of the general linear group, GL_n, (over $\mathbb{C}$) whose highest weight is indexed by the integer partition $\lambda = (\lambda_1 \geq \lambda_2 \geq \cdots \geq \lambda_n)$ (in standard coordinates). Given a finite sequence of

---

This research was supported by the National Security Agency grant # H98230-09-0054.

P.E. Harris
MSCS Department, Marquette University, P.O. Box 1881, Milwaukee, WI 53201, USA
e-mail: pamela.harris@marquette.edu

J.F. Willenbring
Department of Mathematical Sciences, University of Wisconsin-Milwaukee,
P.O. Box 0413, Milwaukee, WI 53211, USA
e-mail: jw@uwm.edu

R. Howe et al. (eds.), *Symmetry: Representation Theory and Its Applications*,
Progress in Mathematics 257, DOI 10.1007/978-1-4939-1590-3_11

integer partitions $\boldsymbol{\mu} = (\mu^{(1)}, \ldots, \mu^{(m)})$, we will let $c_{\boldsymbol{\mu}}^{\lambda}$ be the multiplicity of F_n^{λ} in the m-fold tensor product $F_n^{\mu^{(1)}} \otimes \cdots \otimes F_n^{\mu^{(m)}}$ under the diagonal action of GL_n. That is, $c_{\boldsymbol{\mu}}^{\lambda}$ denotes a (generalized) Littlewood–Richardson coefficient.

We remark that the usual exposition of Littlewood–Richardson coefficients (see [4, 5, 9–11, 13]) concerns the case where $m = 2$. However, by iterating the Littlewood–Richardson rule (or its equivalents) one obtains several effective combinatorial interpretations of our $c_{\boldsymbol{\mu}}^{\lambda}$.

The subject of this exposition concerns some interpretations of the positive integer $\sum \left(c_{\boldsymbol{\mu}}^{\lambda}\right)^2$ where the sum is over certain finite subsets of nonnegative integer partitions. We believe that such sums have under-appreciated combinatorial significance. For example, one immediately observes the very simple specialization to the case where $\mu^{(j)} = (1)$ for all $j = 1, \ldots, m$, in which case the sum of squares is $m!$, which may be viewed as a consequence of Schur–Weyl duality. More generally, if $\nu = (\nu_1 \geq \cdots \geq \nu_m \geq 0)$ is a partition and $\mu^{(j)} = (\nu_j)$, then $c_{\boldsymbol{\mu}}^{\lambda}$ is equal to the Kostka number $K_{\lambda\nu}$ (i.e., the multiplicity of the weight indexed by ν in F_n^{λ}).

Our motivation for considering these numbers comes from invariant theory. On the one hand, we consider the conjugation action of the general linear group on several copies of the $n \times n$ matrices. On the other hand, we consider the K-conjugation action on one copy of the $n \times n$ matrices, where K denotes a block diagonal embedding of a product of general linear groups. These problems are related, and have been studied extensively. We make no attempt to survey the literature, but recommend [3].

Central to this work is the notion of a *Hilbert series*. Let V be a graded vector space. That is, $V = \bigoplus_{d=0}^{\infty} V_d$ where V_d is a finite-dimensional subspace. The Hilbert series, $\sum_{d=0}^{\infty} (\dim V_d) q^d$, formally records the dimensions of the graded components. Here q is an indeterminate. We also consider multivariate generalizations corresponding to situations where V is graded by a cone in a lattice.

In our setting, V will be a space of invariant polynomial functions on m copies of the $n \times n$ matrices. For fixed values of the parameters, the Hilbert series is the Taylor expansion of a rational function around zero. When these parameters are small, one can expect to write down the numerator and denominator explicitly. These polynomials encode structural information about the invariants. However, as the size of the matrix becomes large, these rational functions are difficult to compute explicitly. This motivates reorganizing the data by studying the coefficients of the Hilbert series of a fixed degree as the size of the matrix goes to infinity. The limit exists. The formal series recording this information will be referred to as the *stable Hilbert series*.

Certain sums of squares of Littlewood–Richardson coefficients describe the coefficients of the (stable) Hilbert series for the invariant algebra in each case. These Hilbert series, stably, may be expressed as a product. Furthermore, a "principal specialization" of this product is then related to the Hilbert series of the K-invariant subspace in the GL_n-harmonic polynomials. [Harmonic in the sense of Kostant (see [12]), which generalizes the usual notion of a harmonic polynomial.]

Unless otherwise stated, we will only need notation for representations with polynomial matrix coefficients, which are indexed by partitions with *nonnegative* integer components. The sum of the parts of a partition ν will be called the *size* (denoted $|\nu|$), while the number of parts will be called the *length* denoted $\ell(\nu)$. As usual, we will also write $\lambda \vdash d$ to mean $|\lambda| = d$. Furthermore, we also adapt the (non-standard) notation that if $\boldsymbol{\mu} = (\mu^{(1)}, \ldots, \mu^{(m)})$ is a finite sequence of partitions, we set $|\boldsymbol{\mu}| = \sum_{j=1}^{m} |\mu^{(j)}|$, and write $\boldsymbol{\mu} \vdash d$ to mean $|\boldsymbol{\mu}| = d$. From a combinatorial point of view, the results involve a specialization of the following

Theorem 1.1 (Main Formula). *Let $t_1, t_2, t_3, \ldots$ denote a countably infinite set of indeterminates. We have*

$$\prod_{k=1}^{\infty} \frac{1}{1 - \left(t_1^k + t_2^k + t_3^k + \cdots\right)} = \sum_{\lambda} \sum_{\boldsymbol{\mu}} \left(c_{\boldsymbol{\mu}}^{\lambda}\right)^2 \boldsymbol{t}^{\boldsymbol{\mu}},$$

where the outer sum is over all partitions λ and the inner sum is over all finite sequences of partitions $\boldsymbol{\mu} = (\mu^{(1)}, \mu^{(2)}, \mu^{(3)}, \ldots)$ with $\boldsymbol{t}^{\boldsymbol{\mu}} = t_1^{|\mu^{(1)}|} t_2^{|\mu^{(2)}|} t_3^{|\mu^{(3)}|} \cdots$.

Proof. See Section 7. □

As an application of the main formula, we turn to the space, $\mathcal{H}(\mathfrak{gl}_n)$, of GL_n-harmonic polynomial functions on the adjoint representation (with its natural gradation) by polynomial degree, $\mathcal{H}(\mathfrak{gl}_n) = \bigoplus_{d=0}^{\infty} \mathcal{H}^d(\mathfrak{gl}_n)$. The group GL_n acts on $\mathcal{H}(\mathfrak{gl}_n)$. Note that the constant functions are the only GL_n-invariant harmonic functions. However, if K is a reductive subgroup of GL_n, the space of K-invariant functions is much larger. Consider the example when the group K is the block diagonally embedded copy of $\mathrm{GL}_{n_1} \times \cdots \times \mathrm{GL}_{n_m}$ in GL_n, with $n_1 + \cdots + n_m = n$. We will denote this group by $K(\boldsymbol{n})$ where $\boldsymbol{n} = (n_1, \ldots, n_m)$. The purpose of this paper is to relate the dimension of the $K(\boldsymbol{n})$-invariant polynomials in $\mathcal{H}^d(\mathfrak{gl}_n)$ to a sum of squares of Littlewood–Richardson coefficients. See Theorem 5.1 for the precise statement.

We consider this question because the related algebraic combinatorics are particularly elegant, and hence have expository value in connecting harmonic analysis with algebraic combinatorics. However, this example is the tip of an iceberg. Indeed, one can replace GL_n with any algebraic group G (with $\mathfrak{g}$ =Lie(G)) and $K(\boldsymbol{n})$ with any subgroup of G. This area of investigation is wide open and well motivated as an examination of the special symmetries of harmonic polynomials.

In Section 2, we describe a general "answer" to this question when K is a symmetric subgroup of a reductive group G. This answer is not as explicit as would be desired, but applies to any symmetric pair (G, K). The remainder of the paper is related to the $G = \mathrm{GL}_n$ example with $K = K(\boldsymbol{n})$. Note that when $m = 2$ (i.e., $n = (n_1, n_2)$) the pair (G, K) is symmetric, but for $m > 2$ is not. We remark that in the $m = 2$ case, the results presented here were first described in [18]. Our present discussion amounts to a generalization to $m > 2$.

After setting up appropriate notation in Section 3 we provide an interpretation for a description of the Hilbert series of the $K(\boldsymbol{n})$-invariants in the GL_n-harmonic polynomials on $\mathfrak{gl}_n$ in Sections 4 and 5. Chief among these involves sums of squares of

Littlewood–Richardson coefficients. We recall other combinatorial interpretations in Section 6. These interpretations involve counting the conjugacy classes in the general linear group over a finite field.

Acknowledgements. The first author wishes to thank Marquette University for support during the preparation of this article. The second author wishes to thank the National Security Agency for support. We also would like to thank both Lindsey Mathewson and the referee for pointing out several references that improved the exposition.

2 The case of a symmetric pair

Let G denote a connected reductive linear algebraic group over the complex numbers and let $\mathfrak{g}$ be its complex Lie algebra. We have $\mathfrak{g} = \mathfrak{z}(\mathfrak{g}) \oplus \mathfrak{g}_{ss}$, where $\mathfrak{z}(\mathfrak{g})$ denotes the center of $\mathfrak{g}$, while $\mathfrak{g}_{ss} = [\mathfrak{g}, \mathfrak{g}]$ denotes the semisimple part of $\mathfrak{g}$. A celebrated result of Kostant (see [12]) is that the polynomial functions on $\mathfrak{g}$, denoted $\mathbb{C}[\mathfrak{g}]$, are a free module over the invariant subalgebra, $\mathbb{C}[\mathfrak{g}]^G$, under the adjoint action. Choose a Cartan subalgebra $\mathfrak{h}$ of $\mathfrak{g}$, and let Φ and W denote the corresponding set of roots and Weyl group, respectively. Choose a set of positive roots Φ^+, and let $\Phi^- = -\Phi^+$ denote the negative roots. Set $\rho = \frac{1}{2}\sum_{\alpha\in\Phi^+}\alpha$. For $w \in W$, let $l(w)$ denote the number of positive roots sent to negative roots by w. Fix an indeterminate t. There exist positive integers $e_1 \leq e_2 \leq \cdots \leq e_r$ such that $\sum_{w\in W} t^{l(w)} = \prod_{j=1}^{r} \frac{1-t^{e_j}}{1-t}$ where r is the rank of $\mathfrak{g}_{ss}$. A consequence of the Chevalley restriction theorem ([2]) is that $\mathbb{C}[\mathfrak{g}]^G$ is freely generated, as a commutative ring, by $\dim \mathfrak{z}(\mathfrak{g})$ polynomials of degree 1, and r polynomials of degree $e_1, \ldots, e_r$. These polynomials are the *basic invariants*, while $e_1, \ldots, e_r$ are called the *exponents* of G.

2.1 Harmonic polynomials

We define the *harmonic polynomials* on $\mathfrak{g}$ by

$$\mathcal{H}_{\mathfrak{g}} = \left\{ f \in \mathbb{C}[\mathfrak{g}] \mid \Delta(f) = 0 \text{ for all } \Delta \in \mathbb{D}[\mathfrak{g}]^G \right\},$$

where $\mathbb{D}[\mathfrak{g}]^G$ is the space of constant coefficient G-invariant differential operators on $\mathfrak{g}$. In [12], it is shown that as a G-representation $\mathcal{H}_{\mathfrak{g}}$ is equivalent to the G-representation algebraically induced from the trivial representation of a maximal algebraic torus T in G. Thus, by Frobenius reciprocity, the irreducible rational representations of G occur with multiplicity equal to the dimension of their zero weight space. Moreover, as a representation of G, the harmonic polynomials are equivalent to the regular functions on the nilpotent cone in $\mathfrak{g}$.

The harmonic polynomials inherit a gradation by degree from $\mathbb{C}[\mathfrak{g}]$. Set $\mathcal{H}_{\mathfrak{g}}^d = \mathcal{H}_{\mathfrak{g}} \cap \mathbb{C}[\mathfrak{g}]_d$. Thus, $\mathcal{H}_{\mathfrak{g}} = \bigoplus_{d=0}^{\infty} \mathcal{H}_{\mathfrak{g}}^d$. We next consider the distribution of the multiplicity of an irreducible G-representation among the graded components of $\mathcal{H}_{\mathfrak{g}}$. A solution to this problem was originally due to Hesselink [8], which we recall next.

Let $\wp_t : \mathfrak{h}^* \to \mathbb{N}$ denote Lusztig's q-analog of Kostant's partition function and as always $\mathbb{N} = \{0, 1, 2, 3, \ldots\}$ is the set of nonnegative integers. That is $\wp_t$ is defined by the equation

$$\prod_{\alpha \in \Phi+} \frac{1}{1 - te^{\alpha}} = \sum_{\xi \in Q(\mathfrak{g},\mathfrak{h})} \wp_t(\xi) e^{\xi}$$

where $Q(\mathfrak{g}, \mathfrak{h}) \subseteq \mathfrak{h}^*$ denotes the lattice defined by the integer span of the roots. As usual, e^{ξ} denotes the corresponding character of T, with $\mathrm{Lie}(T) = \mathfrak{h}$. As usual, we set $\wp(\xi) = 0$ for $\xi \notin Q(\mathfrak{g}, \mathfrak{h})$.

Let $P(\mathfrak{g})$ denote the integral weights corresponding to the pair $(\mathfrak{g}, \mathfrak{h})$. The dominant integral weights corresponding to Φ^+ will be denoted by $P_+(\mathfrak{g})$. Let $L(\lambda)$ denote the (finite-dimensional) irreducible highest weight representation with highest weight $\lambda \in P_+(\mathfrak{g})$. The multiplicity of $L(\lambda)$ in the degree d harmonic polynomials $\mathcal{H}^d(\mathfrak{g})$ will be denoted by $m_d(\lambda)$. Set $m_\lambda(t) = \sum_{d=0}^{\infty} m_d(\lambda) t^d$. Hesselink's theorem asserts that

$$m_\lambda(t) = \sum_{w \in W} (-1)^{l(w)} \wp_t(w(\lambda + \rho) - \rho).$$

See [17] for a generalization of this result.

We remark that the above formula is very difficult to implement in practice. This is in part due to the fact that the order of W grows exponentially with the Lie algebra rank. Thankfully, only a small number of terms actually contribute to the overall multiplicity. See [7] for a very interesting special case where the number of contributed terms is shown to be a Fibonacci number.

2.2 *The K-spherical representations of G*

Let (G, K) be a symmetric pair. That is, G is a connected reductive linear algebraic group over $\mathbb{C}$ and $K = \{g \in G \mid \theta(g) = G\}$, where θ is a regular automorphism of G of order two. Since K will necessarily be reductive the quotient, G/K is an affine variety and the $\mathbb{C}$-algebra of regular function $\mathbb{C}[G/K]$ is multiplicity-free as a representation of G. This fact follows from the (complexified) Iwasawa decomposition of G. Put another way, there exists $S \subseteq P_+(\mathfrak{g})$ such that for all $\lambda \in P_+(\mathfrak{g})$ we have

$$\dim L(\lambda)^K = \begin{cases} 1, \lambda \in S, \\ 0, \lambda \notin S. \end{cases}$$

Note that the subset S may be read off of the data encoded in the Satake diagram associated to the pair (G, K). The above fact implies that the Hilbert series $H_t(G, K) = \sum_{d=0}^{\infty} h_d t^d$ with $h_d = \dim \mathcal{H}^d(\mathfrak{g})^K$ has the following formal expression:

$$H_t(G, K) = \sum_{w \in W} (-1)^{l(w)} \left(\sum_{\lambda \in S} \wp_t(w(\lambda + \rho) - \rho) \right).$$

This formula seems rather encouraging. Unfortunately, the inner sum is very difficult to put into a closed form for general $w \in W$. This is, in part, a reflection of the fact that the values of $\wp_t$ cannot be determined from a "closed form" expression. However, note that $w(\lambda+\rho)-\rho$ often falls outside of the support of $\wp_t$, and therefore it may be possible to obtain explicit results along these lines. Moreover, the point of this exposition is to advertise that combinatorially elegant expressions may exist. At least this is the case for the pair $(\mathrm{GL}_{n_1+n_2}, \mathrm{GL}_{n_1} \times \mathrm{GL}_{n_2})$, as we shall see.

3 Preliminaries

We let $\mathfrak{gl}_n$ denote the complex Lie algebra of $n \times n$ matrices with the usual bracket, $[X, Y] = XY - YX$, for $X, Y \in \mathfrak{gl}_n$. Let E_{ij} denote the $n \times n$ matrix with a 1 in row i and column j, and 0 everywhere else. The Cartan subalgebra will be chosen to be the span of $\{E_{ii} \mid 1 \le i \le n\}$. The dual basis in $\mathfrak{h}^*$ to $(E_{11}, E_{22}, \ldots, E_{nn})$ will be denoted $(\epsilon_1, \ldots, \epsilon_n)$. Choose the simple roots as usual, $\Pi = \{\epsilon_i - \epsilon_{i+1} \mid 1 \le i < n\}$. Let Φ (resp. Φ^+) denote the roots (resp. positive roots). We will identify $\mathfrak{h}$ with $\mathfrak{h}^*$ using the trace form $(H_1, H_2) = \mathrm{Tr}(H_1 H_2)$ (for $H_1, H_2 \in \mathfrak{h}$). The fundamental weights are $\omega_i = \sum_{j=1}^{i} \epsilon_j \in \mathfrak{h}^*$ for $1 \le i \le n-1$. We also set $\omega_n = \sum_{j=1}^{n} \epsilon_j \in \mathfrak{h}^*$. Let $P(\mathrm{GL}_n) = \sum_{j=1}^{n} \mathbb{Z}\omega_j$, and $P_+(\mathrm{GL}_n) = \mathbb{Z}\omega_n + \sum_{j=1}^{n-1} \mathbb{N}\omega_j$.

From this point on, we will write $(a_1, \ldots, a_n)$ for $\sum a_i \epsilon_i$. Thus, we have $\lambda = (\lambda_1, \ldots, \lambda_n) \in P_+(\mathrm{GL}_n)$ iff each λ_i in an integer and $\lambda_1 \ge \cdots \ge \lambda_n$. The finite-dimensional irreducible representation of GL_n with highest weight λ will be denoted $(\pi_\lambda, \mathrm{F}_n^\lambda)$ where

$$\pi_\lambda : \mathrm{GL}_n \to \mathrm{GL}(\mathrm{F}_n^\lambda).$$

To simply notation will write F_n^λ for $(\pi_\lambda, \mathrm{F}_n^\lambda)$.

Throughout this article, the representations of GL_n which we will consider have polynomial matrix coefficients. Thus the components of the highest weight λ will be nonnegative integers. Therefore, if λ is a (nonnegative integer) partition with at most ℓ parts ($\ell \le n$), then the n-tuple, $(\lambda_1, \ldots, \lambda_\ell, 0, \ldots, 0)$, corresponds to the highest weight of a finite-dimensional irreducible representation of GL_n (with polynomial matrix coefficients).

Under the diagonal action, GL_n acts on the d-fold tensor product $\otimes^d \mathbb{C}^n$. Schur–Weyl duality (see [5] Chapter 9) asserts that the full commutant to the GL_n-action is generated by the symmetric group action defined by permutation of factors. Consequently, one has a multiplicity-free decomposition with respect to the joint action of $\mathrm{GL}_n \times S_d$. Moreover, if $n \geq d$, then every irreducible representation of S_d occurs. The irreducible representation of S_d paired with F_n^λ will be denoted by V_d^λ. The full decomposition into the irreducible $GL_n \times S_d$-representation is

$$\bigotimes^d \mathbb{C}^n \cong \bigoplus_\lambda \mathrm{F}_n^\lambda \otimes V_d^\lambda,$$

where the sum is over all nonnegative integer partitions λ of size d and length at most n. Note that when $n = d$, then the condition on $\ell(\lambda)$ is automatic. Thus, all irreducible representations of V_d^λ occur. In this manner, the highest weights of GL_n-representations provide a parametrization of the S_d-representations.

3.1 Littlewood–Richardson coefficients

Let $\boldsymbol{d} = (d_1, \ldots, d_m)$ denote a tuple of positive integers with $d = d_1 + \cdots + d_m$. Let $S_{\boldsymbol{d}}$ denote the subgroup of S_d consisting of permutations that stabilize the sets permuting the first m_1 indices, then the second m_2 indices, etc. Clearly, we have

$$S_{\boldsymbol{d}} \cong S_{d_1} \times \cdots \times S_{d_m}.$$

The irreducible representations of $S_{\boldsymbol{d}}$ are of the form

$$V(\boldsymbol{\mu}) = V_{d_1}^{\mu^{(1)}} \otimes \cdots \otimes V_{d_m}^{\mu^{(m)}},$$

where $\mu^{(j)}$ is a partition of size d_j. It is well known that if an irreducible representation V_d^λ of S_d is restricted to $S_{\boldsymbol{d}}$, then the multiplicity of $V(\boldsymbol{\mu})$ in V_d^λ is given by the Littlewood–Richardson coefficient $c_{\boldsymbol{\mu}}^\lambda$. This fact is a consequence of Schur–Weyl duality.

4 Invariant polynomials on matrices

A permutation of $\{1, 2, \ldots, m\}$ may be written as a product of disjoint cycles. This result is fundamental to combinatorial properties of the symmetric group S_m. Keeping this elementary fact in mind, let $\boldsymbol{X} = (X_1, X_2, \ldots, X_m)$ be a list of complex $n \times n$ matrices. Let Tr denote the trace of a matrix and define

$$\mathrm{Tr}_\sigma(X) = \mathrm{Tr}(X_{\sigma_1^{(1)}} X_{\sigma_2^{(1)}} \cdots X_{\sigma_{m_1}^{(1)}})\ \mathrm{Tr}(X_{\sigma_1^{(2)}} X_{\sigma_2^{(2)}} \cdots X_{\sigma_{m_2}^{(2)}}) \cdots$$
$$\cdots \mathrm{Tr}(X_{\sigma_1^{(k)}} X_{\sigma_2^{(k)}} \cdots X_{\sigma_{m_k}^{(k)}})$$

where $\sigma = (\sigma_1^{(1)}\sigma_2^{(1)} \cdots \sigma_{m_1}^{(1)})(\sigma_1^{(2)}\sigma_2^{(2)} \cdots \sigma_{m_2}^{(2)}) \cdots (\sigma_1^{(k)}\sigma_2^{(k)} \cdots \sigma_{m_k}^{(k)})$ is a permutation of $m = \sum m_i$ written as a product of k disjoint cycles. Observe that the cycles of σ may be permuted and rotated without changing the permutation. In turn, the function Tr_σ displays these same symmetries.

Chief among the significance of Tr_σ is the fact that they are invariant under the conjugation action $g \cdot (X_1, \ldots, X_m) = (gX_1g^{-1}, \ldots, gX_mg^{-1})$, where $g \in \mathrm{GL}_n$. By setting some of the components of $(X_1, \ldots, X_m)$ equal, one defines a polynomial of equal degree but on fewer than m copies of M_n. Intuitively, this fact may be described by allowing equalities in the components of the cycles of σ. That is (formally), we consider σ up to conjugation by Levi subgroup of S_m.

In [14,15], C. Procesi described these generators for the algebra of GL_n-invariant polynomials, denoted $\mathbb{C}[V]^{\mathrm{GL}_n}$, on $V = M_n^m$, and provided a proof that these polynomials span the invariants. [Here we let M_n denote the complex vector spaces of $n \times n$ matrices (with entries from $\mathbb{C}$).] Hilbert tells us that the ring of invariants must be finitely generated. Thus, there must necessarily be algebraic relations among this (infinite) set of generators. In light of Procesi's work, these generators and relations are understood. However, recovering the Hilbert series from these data is not automatic.

In order to precisely quantify the failure of the Tr_σ being independent, we introduce the formal power series $A_n(\boldsymbol{t}) = A_n(t_1, t_2, \ldots, t_m) = \sum a_n(\boldsymbol{d})\boldsymbol{t}^{\boldsymbol{d}}$, where we will use the notation $\boldsymbol{d} = (d_1, \ldots, d_m)$, $\boldsymbol{t}^{\boldsymbol{d}} = t_1^{d_1} \cdots t_m^{d_m}$ and the coefficient are defined as $a_n(\boldsymbol{d}) = \dim \mathbb{C}[M_n^m]_{\boldsymbol{d}}^{\mathrm{GL}_n}$, where $\mathbb{C}[M_n^m]_{\boldsymbol{d}} = \mathbb{C}[M_n \oplus \cdots \oplus M_n]_{(d_1, \cdots, d_m)}$ denotes the homogeneous polynomials of degree d_i on the ith copy of M_n. The multivariate series, $A_n(\boldsymbol{t})$, is called the *Hilbert Series* of the invariant ring. Except in some simple cases, a closed form for $A_n(\boldsymbol{t})$ is not known. Part of this exposition is to point out the rather simple fact that $a_n(\boldsymbol{d})$ may be expressed in terms of the squares of Littlewood–Richardson coefficients. Furthermore, we prove

Proposition 4.1. *For any natural numbers m and $\boldsymbol{d} = (d_1, \ldots, d_m)$ the limit $\lim_{n\to\infty} a_n(\boldsymbol{d})$ exists. If we call the limiting value $a(\boldsymbol{d})$ and set $\widetilde{A}(\boldsymbol{t}) = \sum a(\boldsymbol{d})\boldsymbol{t}^{\boldsymbol{d}}$, then*

$$\widetilde{A}(\boldsymbol{t}) = \prod_{k=1}^{\infty} \frac{1}{1 - (t_1^k + t_2^k + \cdots + t_m^k)}.$$

Proof. The polynomial functions on M_n are multiplicity-free under the action of $\mathrm{GL}_n \times \mathrm{GL}_n$ given by $(g_1, g_2)f(X) = f(g_1^{-1}Xg_2)$ for $g_1, g_2 \in \mathrm{GL}_n$, $X \in M_n$ and $f \in \mathbb{C}[M_n]$. The decomposition is a "Peter–Weyl" type, $\mathbb{C}[M_n] \cong \bigoplus (\mathrm{F}_n^\lambda)^* \otimes \mathrm{F}_n^\lambda$, where the sum is over all nonnegative integer partitions λ with $\ell(\lambda) \le n$. We have $\mathbb{C}[\oplus_{i=1}^m M_n] \cong \otimes_{i=1}^m \mathbb{C}[M_n]$. Thus

$$\mathbb{C}[M_n^m] \cong \left(\bigoplus_{\mu^{(1)}} \left(\mathrm{F}_n^{\mu^{(1)}}\right)^* \otimes \mathrm{F}_n^{\mu^{(1)}}\right) \otimes \cdots \otimes \left(\bigoplus_{\mu^{(m)}} \left(\mathrm{F}_n^{\mu^{(m)}}\right)^* \otimes \mathrm{F}_n^{\mu^{(m)}}\right)$$

$$\cong \bigoplus_{\alpha,\beta} \left(\sum_{\boldsymbol{\mu}=(\mu^{(1)},\cdots,\mu^{(m)})} c_{\boldsymbol{\mu}}^{\alpha} c_{\boldsymbol{\mu}}^{\beta}\right) (\mathrm{F}_n^{\alpha})^* \otimes \mathrm{F}_n^{\beta},$$

with respect to the $\mathrm{GL}_n \times \mathrm{GL}_n$ action on the diagonal of M_n^m. Note that in multidegree $(d_1,\ldots,d_m)$ the sum is over all $\boldsymbol{\mu}$ with $|\boldsymbol{\mu}^{(j)}| = d_j$. If we restrict to the subgroup $\{(g,g) \mid g \in \mathrm{GL}_n\}$ of $\mathrm{GL}_n \times \mathrm{GL}_n$, we obtain an invariant exactly when $\alpha = \beta$. The dimension of the GL_n-invariants in the degree d homogeneous polynomials on M_n^m is therefore $\sum \left(c_{\boldsymbol{\mu}}^{\lambda}\right)^2$ where the sum is over all $\boldsymbol{\mu} = (\mu^{(1)},\ldots,\mu^{(m)})$ and $\lambda \vdash m$ with length at most n. The degree d component decomposes into multi-degree components $(d_1,\ldots,d_m)$ with $d = \sum d_j$.

If $d \geq n$, then the condition that $\ell(\lambda) \leq n$ is automatic, and this fact implies that if $c_{\boldsymbol{\mu}}^{\lambda} \neq 0$, then $\ell(\mu^{(j)}) \leq n$ for all j. Thus, the dimension of the degree d invariants in $\mathbb{C}[M_n^m]$ is $\sum \left(c_{\boldsymbol{\mu}}^{\lambda}\right)^2$ where the sum is over *all* partitions of size d. If we specialize the main formula so that $t_j = 0$ for $j > m$, then the sums of squares of Littlewood–Richardson coefficients agree with $\widetilde{A}(\boldsymbol{t})$. □

For our purposes, we will specialize the multigradation on the invariants in $\mathbb{C}[M_n^m]$ to one that is more coarse. From this process, we can relate the stabilized Hilbert series of the invariants in $\mathcal{H}[M_n]$. This specialization will be the subject of the next section.

We now turn to another problem that we shall see is surprisingly related. Consider the $n \times n$ complex matrix

$$X = \begin{bmatrix} X(1,1) & X(1,2) & \cdots & X(1,m) \\ X(2,1) & X(2,2) & \cdots & X(2,m) \\ \vdots & \vdots & \ddots & \vdots \\ X(m,1) & X(m,2) & \cdots & X(m,m) \end{bmatrix}$$

where $X(i,j)$ is an $n_i \times n_j$ complex matrix with $n = \sum n_j$. Define

$$\mathrm{Tr}^{\sigma}(X) = \prod_{j=1}^{k} \mathrm{Tr}\left(X(\sigma_1^{(j)},\sigma_2^{(j)})X(\sigma_2^{(j)},\sigma_3^{(j)})X(\sigma_3^{(j)},\sigma_4^{(j)})\cdots X(\sigma_{m_j}^{(j)},\sigma_1^{(j)})\right).$$

Let $K(\boldsymbol{n})$ denote the block diagonal subgroup of GL_n of the form

$$K(\boldsymbol{n}) = \begin{bmatrix} \mathrm{GL}_{n_1} & & \\ & \ddots & \\ & & \mathrm{GL}_{n_m} \end{bmatrix}.$$

The group $K(\boldsymbol{n})$ acts on M_n by restricting the adjoint action of GL_n. The $K(\boldsymbol{n})$-invariant subring of $\mathbb{C}[M_n]$ (denoted $\mathbb{C}[M_n]^{K(\boldsymbol{n})}$) is spanned by $\mathrm{Tr}^\sigma(X)$. For small values of the parameter space, these expressions are far from linearly independent (as in the last example). Formally, one cannot help but notice the symbolic map $\mathrm{Tr}^\sigma \mapsto \mathrm{Tr}_\sigma$. We will try next to make a precise statement along these lines.

Define $\mathbb{C}[M_n]_d$ to be the homogeneous degree d polynomials on M_n, and let $\mathbb{C}[M_n]_d^{K(\boldsymbol{n})}$ denote the $K(\boldsymbol{n})$-invariant subspace. Set $a^{(\boldsymbol{n})}(d) = \dim \mathbb{C}[M_n]_d^{K(\boldsymbol{n})}$, and $A^{(\boldsymbol{n})}(t) = \sum_{d=0}^\infty a^{(\boldsymbol{n})}(d)\, t^d$. Analogous to Proposition 4.1, we have

Proposition 4.2. *For any nonnegative integer d, the limit*

$$\lim_{n_1\to\infty} \lim_{n_2\to\infty} \cdots \lim_{n_m\to\infty} a^{(n_1,\cdots,n_m)}(d)$$

exists. Denote the limiting value $a(d)$ and set $A(t) = \sum_{d=0}^\infty a(d) t^d$. We have

$$A(t) = \prod_{k=1}^{\infty} \frac{1}{1 - m\, t^k}.$$

Proof. We begin with the $\mathrm{GL}_n \times \mathrm{GL}_n$-decomposition $\mathbb{C}[M_n] = \bigoplus \left(\mathrm{F}_n^\lambda\right)^* \otimes \mathrm{F}_n^\lambda$ with respect to the action in the proof of Proposition 4.1. The irreducible GL_n-representation F_n^λ is reducible upon restriction to $K(\boldsymbol{n})$. The decomposition is given in terms of Littlewood–Richardson coefficients

$$\mathrm{F}_n^\lambda \cong \bigoplus_{\boldsymbol{\mu}=(\mu^{(1)},\cdots,\mu^{(m)})} c_{\boldsymbol{\mu}}^\lambda\, \mathrm{F}_n^{\mu^{(1)}} \otimes \cdots \otimes \mathrm{F}_n^{\mu^{(m)}}.$$

Therefore, as a $K(\boldsymbol{n}) \times K(\boldsymbol{n})$-representation we have

$$\mathbb{C}[M_n]_d = \sum_{\boldsymbol{\mu},\boldsymbol{\nu}} \left(\sum_\lambda c_{\boldsymbol{\mu}}^\lambda c_{\boldsymbol{\nu}}^\lambda \right) \left(\otimes_{j=1}^m \mathrm{F}_{n_j}^{\mu^{(j)}} \right)^* \otimes \left(\otimes_{j=1}^m \mathrm{F}_{n_j}^{\nu^{(j)}} \right),$$

where the sum is over all λ with $|\lambda| = d$, $\ell(\lambda) \le n$ and $\ell(\mu^{(j)}), \ell(\nu^{(j)}) \le n_j$. If we restrict to the diagonally embedded $K(\boldsymbol{n})$-subgroup, we obtain an invariant exactly when $\boldsymbol{\mu} = \boldsymbol{\nu}$. Thus, $a^{(\boldsymbol{n})}(d) = \sum \left(c_{\boldsymbol{\mu}}^\lambda\right)^2$, with the appropriate restrictions on λ and $\boldsymbol{\mu}$.

If all $n_j \ge d$, then the condition on the lengths of partitions disappears, and we may sum over all $\lambda \vdash d$. The result follows by specializing the main formula by setting $t_j = t$ for $1 \le j \le m$ and $t_j = 0$ for $j > m$. □

Although we will not need it for our present purposes, it is worth pointing out that the algebra $\mathbb{C}[M_n]$ has a natural $\mathbb{N}^m$ gradation defined by the action of the center of $K(\boldsymbol{n})$. This multigradation refines the gradation by degree. The limiting multigraded

Hilbert series is the same as that of Proposition 4.1. Upon specialization to $t_1 = t_2 = \cdots = t_m = t$ we obtain the usual gradation by degree (in both situations). The advantage of considering the more refined gradation is that one can consider the direct limit as m and n go to infinity. This will be relevant in the next section.

5 Harmonic polynomials on matrices

A specific goal of this article is to understand the dimension of the space of degree d homogeneous $K(\boldsymbol{n})$-invariant harmonic polynomials on M_n. With this fact in mind, we observe the following specialization of the product in the main formula. Let $t_j = t^j$. Then we obtain

$$\prod_{k=1}^{\infty} \frac{1}{1-(t^k+t^{2k}+t^{3k}+\cdots)} = \prod_{k=1}^{\infty} \frac{1}{1-\frac{t^k}{1-t^k}} = \prod_{k=1}^{\infty} \frac{1-t^k}{1-2t^k}.$$

For a sequence $\boldsymbol{\mu} = (\mu^{(1)}, \mu^{(2)}, \ldots)$ set $\mathrm{gr}(\boldsymbol{\mu}) = \sum_{j=1}^{\infty} j\,|\mu^{(j)}|$. The equation in the main formula becomes

$$\prod_{k=1}^{\infty} \frac{1-t^k}{1-2t^k} = \sum_{\lambda} \sum_{\boldsymbol{\mu}} \left(c_{\boldsymbol{\mu}}^{\lambda}\right)^2 t^{\mathrm{gr}(\boldsymbol{\mu})}.$$

The notation gr is used for the word *grade*. We explain this choice next. Let $\mathbb{C}[M_n; d]$ denote the polynomials functions on the $n \times n$ complex matrices together with the gradation defined by d times the usual degree. That is, $\mathbb{C}[M_n; d_1]_{(d_2)}$ denotes the degree d_2 homogeneous polynomials, but regarded as the $d_1 d_2$ graded component in $\mathbb{C}[M_n; d_1]$. We consider the $\mathbb{N}$-graded complex algebra $\mathcal{A}_n$ defined as

$$\begin{aligned} \mathcal{A}_n &= \mathbb{C}[M_n;1] \otimes \mathbb{C}[M_n;2] \otimes \mathbb{C}[M_n;3] \otimes \cdots \\ &= \sum_{\delta=0}^{\infty} \mathcal{A}_n[\delta], \end{aligned}$$

where $\mathcal{A}_n[\delta]$ is the graded $\delta \in \mathbb{N}$ component (with the usual grade defined on a tensor product of algebras). The group GL_n acts on each $\mathbb{C}[M_n; d]$ by the adjoint action, and respects the grade. Under the diagonal action on the tensors, the GL_n-invariants, $\mathcal{A}_n^{\mathrm{GL}_n}$, have $A_n(\boldsymbol{t})$ as the Hilbert series when $\boldsymbol{t}$ is specialized to $(t_1, t_2, t_3, \ldots) = (t, t^2, t^3, \ldots)$. This GL_n-action respects the δ component in the gradation. That is, that the Hilbert series is

$$A_n(t, t^2, t^3, \ldots) = \sum_{\delta=0}^{\infty} \dim \left(\mathcal{A}_n[\delta]\right)^{\mathrm{GL}_n} t^{\delta}.$$

For $d = 0, \ldots, n$ the coefficient of t^d in $A_n(t, t^2, \ldots)$ is the same as the coefficient of t^d in $\widetilde{A}(t, t^2, \ldots)$. Summarizing, we can say that stably, as $n \to \infty$, the Hilbert series of $\mathcal{A}_n^{\mathrm{GL}_n}$ is $\prod_{k=1}^{\infty} \frac{1-t^k}{1-2t^k}$.

We turn now to the GL_n-harmonic polynomials on M_n together with its usual gradation by degree. Kostant's theorem [12] tells us that

$$\mathbb{C}[M_n] \cong \mathbb{C}[M_n]^{\mathrm{GL}_n} \otimes \mathcal{H}(M_n).$$

As before let $K = K(n_1, n_2)$ denote the copy of $\mathrm{GL}_{n_1} \times \mathrm{GL}_{n_2}$ (symmetrically) embedded in $\mathrm{GL}_{n_1+n_2}$. Passing to the $K(n_1, n_2)$-invariant subspaces we obtain

$$\mathbb{C}[M_n]^K = \mathbb{C}[M_n]^{\mathrm{GL}_n} \otimes \mathcal{H}(M_n)^K.$$

The Hilbert series of $\mathbb{C}[M_n]^{\mathrm{GL}_n}$ is well known to be $\prod_{k=1}^{n} \frac{1}{1-t^k}$, while the Hilbert series $\mathbb{C}[M_n]^K$ is $A_n(t,t)$. These facts imply that the dimension of the degree d homogenous K-invariant harmonic polynomials is the coefficient of t^d in $F_n(t) = A_n(t,t) \prod_{j=1}^{n}(1 - t^j)$. We have that the coefficient of t^d in $F_n(t)$ for $d = 0, \ldots, \min(n_1, n_2)$ agrees with the coefficient of t^d in

$$\widetilde{A}(t, t^2, t^3, \cdots) \prod_{k=1}^{\infty}(1 - t^k) = \prod_{k=1}^{\infty} \frac{1 - t^k}{1 - 2t^k}.$$

Again summarizing, we say that stably, as $n_1, n_2 \to \infty$, the Hilbert series of $\mathcal{H}(M_{n_1+n_2})^{K(n_1,n_2)}$ is the same as $\mathcal{A}_n^{\mathrm{GL}_n}$ as $n, n_1, n_2 \to \infty$. That is, for *fixed* δ we have

$$\lim_{n_1,n_2 \to \infty} \dim \mathcal{H}(M_{n_1+n_2})_\delta^{K(n_1,n_2)} = \lim_{n \to \infty} \dim \mathcal{A}_n[\delta]^{\mathrm{GL}_n}.$$

Observe that this procedure generalizes. Let $m \geq 2$. Analogous to before, let $\mathbb{C}[M_n^m; d]$ denote the polynomial function on $M_n^m = M_n \oplus \cdots \oplus M_n$ (m-copies) together with the $\mathbb{N}$-gradation defined such that $\mathbb{C}[M_n^m; d_1]_{d_2}$ consists of the degree d_2 homogeneous polynomials but regarded as being the $d_1 d_2$-th graded component. Note that the δ-th component in the grade is (0) if δ is not a multiple of d_1.

Let $\mathcal{A}_n^m$ be the $\mathbb{N}$-graded algebra defined as

$$\mathcal{A}_n^m = \mathbb{C}[M_n^{m-1}; 1] \otimes \mathbb{C}[M_n^{m-1}; 2] \otimes \mathbb{C}[M_n^{m-1}; 3] \otimes \cdots.$$

Since $\mathcal{A}_n^m$ is a tensor product of $\mathbb{N}$-graded algebras, it has the structure of an $\mathbb{N}$-graded algebra. As before let $\mathcal{A}_n^m[\delta]$ be the δ-th graded component. The group GL_n acts on each $\mathbb{C}[M_n^m; d]$ by the adjoint action, and respects the grade.

Next, set $t_1 = t_2 = \cdots = t_{m-1} = t$, then $t_m = t_{m+1} = \cdots = t_{2m-2} = t^2$, then $t_{2m-1} = \cdots = t_{3m-3} = t^3$, and so on. The result of this procedure is

Theorem 5.1. *For all* $\boldsymbol{n} = (n_1, \ldots, n_m) \in \mathbb{Z}_+^m$ *and* $d \in \mathbb{N}$, *let*

$$h_d(\boldsymbol{n}) = \dim \left(\mathcal{H}^d(M_n)\right)^{K(\boldsymbol{n})}.$$

Then for fixed d, *the limit*

$$\lim_{n_1\to\infty} \cdots \lim_{n_m\to\infty} h_d(n_1, \ldots, n_m)$$

exists. Let the limiting value be h_d. *Then*

$$h_d = \lim_{n\to\infty} \dim (\mathcal{A}_n^m[d])^{\mathrm{GL}_n}.$$

Proof. After the specialization we obtain

$$\prod_{k=1}^{\infty} \frac{1}{1-(\underbrace{t^k+\cdots+t^k}_{m-1 \text{ copies}}+\underbrace{t^{2k}+\cdots+t^{2k}}_{m-1 \text{ copies}}+\cdots)} = \prod_{k=1}^{\infty} \frac{1}{1-(m-1)(t^k+t^{2k}+\cdots)}$$

$$= \prod_{k=1}^{\infty} \frac{1}{1-(m-1)\frac{t^k}{1-t^k}}$$

$$= \prod_{k=1}^{\infty} \frac{1}{\frac{1-t^k-(m-1)t^k}{1-t^k}}$$

$$= \prod_{k=1}^{\infty} \frac{1-t^k}{1-m\,t^k}.$$

The significance of this calculation is that it allows for another interpretation of sums of Littlewood–Richardson coefficients. The rest of the proof is identical to the $m = 2$ case in the preceding discussion. □

5.1 A bigraded algebra and a specialization of the main formula

As before, the group GL_n acts on M_n by conjugation, and then in turn acts diagonally on $M_n \oplus M_n$. That is, given $(X, Y) \in M_n \oplus M_n$, and $g \in \mathrm{GL}_n$, we have $g \cdot (X, Y) = (gXg^{-1}, gYg^{-1})$. We then obtain an action on $\mathbb{C}[M_n \oplus M_n]$.

We have already observed that the algebra $\mathbb{C}[M_n \oplus M_n]$ is bigraded. That is, let $\mathbb{C}[M_n \oplus M_n](i, j)$ denote the homogenous polynomial functions on $M_n \oplus M_n$ of degree i in the first copy of M_n and degree j in the second copy of M_n. Let a and b be positive integers. We set

$$\mathbb{C}[M_n \oplus M_n; (a,b)](ai, bj) = \mathbb{C}[M_n \oplus M_n](i, j)$$

with the other components zero. Put another way, $\mathbb{C}[M_n \oplus M_n; (a,b)]$ is the algebra of polynomial function on $M_n \oplus M_n$ together with the bi-gradation defined by (a,b) times the usual degree. As before we consider the infinite tensor product

$$\mathcal{B}_n = \bigotimes_{a=1}^{\infty} \bigotimes_{b=1}^{\infty} \mathbb{C}[M_n \oplus M_n; (a,b)].$$

Next, note the following, obvious, identity:

$$\prod_{k=1}^{\infty} \frac{1}{1 - \frac{x^k y^k}{(1-x^k)(1-y^k)}} = \prod_{k=1}^{\infty} \frac{(1-x^k)(1-y^k)}{1-(x^k+y^k)}.$$

We observe that this is a specialization of the product in the main formula. Specifically, let $q^i t^j = z_s$ where z_s is given as the (i,j) entry in the following table.

$q^i t^j$	q	q^2	q^3	q^5	q^6	$\cdots$
t	z_1	z_2	z_4	z_7	z_{11}	$\cdots$
t^2	z_3	z_5	z_8	z_{12}	z_{17}	$\cdots$
t^3	z_6	z_9	z_{13}	z_{18}	z_{24}	$\cdots$
t^4	z_{10}	z_{14}	z_{19}	z_{25}	z_{32}	$\cdots$
t^5	z_{15}	z_{20}	z_{26}	z_{33}	z_{41}	$\cdots$
$\vdots$	$\vdots$	$\vdots$	$\vdots$	$\vdots$	$\vdots$	$\ddots$

Then,

$$\begin{aligned}\prod_{k=1}^{\infty} \frac{1}{1-(z_1^k + z_2^k + z_3^k + \cdots)} &= \prod_{k=1}^{\infty} \frac{1}{1 - \sum_{i,j\geq 1} (q^i t^j)^k} \\ &= \prod_{k=1}^{\infty} \frac{1}{1 - \frac{q^k t^k}{(1-q^k)(1-t^k)}} \\ &= \prod_{k=1}^{\infty} \frac{(1-q^k)(1-t^k)}{1-(q^k+t^k)}.\end{aligned}$$

In this way, the GL_n-invariants in the harmonic polynomials on $M_n \oplus M_n$ may be related to the multigraded algebra structure $\mathcal{B}$, as in the singly graded case.

This identity becomes significantly more complicated when generalized to harmonic polynomials on more than two copies of the matrices. This fact will be the subject of future work.

6 Some combinatorics related to finite fields

In this section, we collect remarks of a combinatorial nature that provide a more concrete understanding of the sum of squares that we consider in this paper.

From elementary combinatorics one knows that the infinite product

$$\prod_{k=1}^{\infty} \frac{1}{1-q\,t^k} = \sum_{n=0}^{\infty}\sum_{\ell=0}^{\infty} p_{n,\ell}\, q^{\ell} t^{n},$$

where $p_{n,\ell}$ is the number of partitions of n with exactly ℓ parts. When q is specialized to a positive integer the coefficients of this series in t has many interpretations.

6.1 *The symmetric group*

Fix a positive integer d, and a tuple of positive integers $\boldsymbol{d} = (d_1, d_2, \ldots, d_m)$ with $d_1 + \cdots + d_m = d$. As before, let $S_{\boldsymbol{d}}$ denote the subgroup of S_d isomorphic to $S_{d_1} \times \cdots \times S_{d_m}$ embedded by letting the ith factor permute the set J_i where $\{J_1, \ldots, J_m\}$ is the partition of $\{1, \ldots, d\}$ into the m contiguous intervals with $|J_i| = d_i$. That is, $J_1 = \{1, 2, \ldots, d_1\}$, $J_2 = \{d_1 + 1, \ldots, d_1 + d_2\}$, etc.

The group $S_{\boldsymbol{d}}$ acts on S_d by conjugation. Let the set of orbits be denoted by $\mathcal{O}(\boldsymbol{d})$. We have

Proposition 6.1. *For any $\boldsymbol{d} = (d_1, \ldots, d_m)$,*

$$|\mathcal{O}(\boldsymbol{d})| = \sum_{\lambda \vdash m} \sum_{\boldsymbol{\mu}=(\mu^{(1)},\cdots,\mu^{(m)})} \left(c^{\lambda}_{\boldsymbol{\mu}}\right)^2$$

where the inner sum is over all tuples of partitions with $\mu^{(j)} \vdash d_j$.

Proof. We begin with the Peter–Weyl type decomposition of $\mathbb{C}[S_d]$

$$\mathbb{C}[S_d] \cong \bigoplus_{\lambda \vdash d} \left(V^{\lambda}_d\right)^* \otimes V^{\lambda}_d .$$

We then recall that Littlewood–Richardson coefficients describe the branching rule from S_d to $S_{\boldsymbol{d}}$:

$$V^{\lambda} = \bigoplus_{\boldsymbol{\mu}} c^{\lambda}_{\boldsymbol{\mu}} V^{\mu^{(1)}} \otimes \cdots \otimes V^{\mu^{(m)}}.$$

Combining the above decompositions, we obtain the result from Schur's Lemma. □

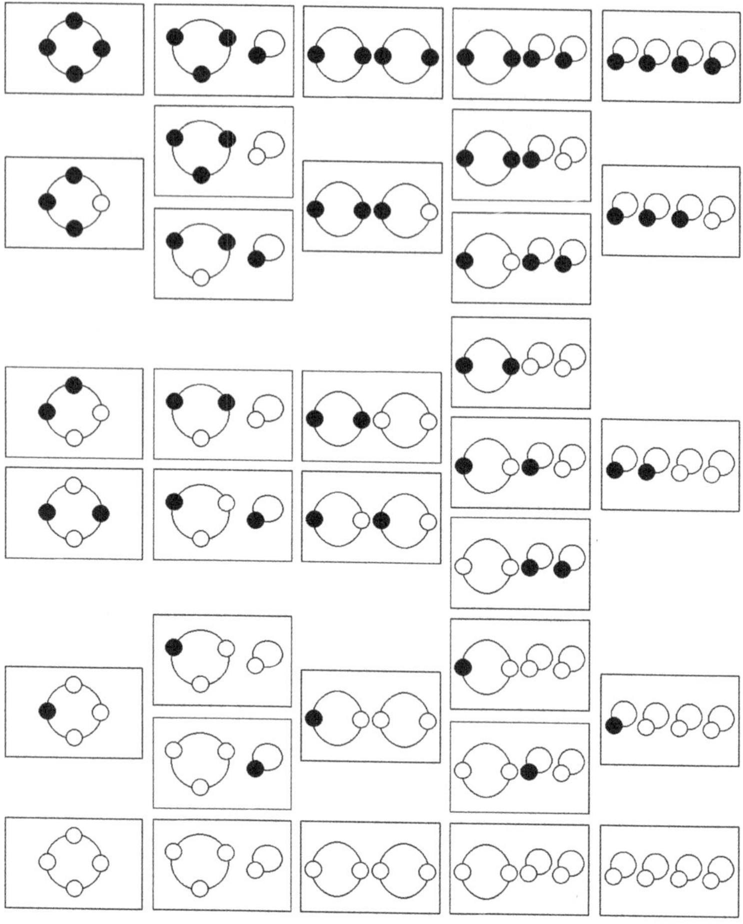

Figure 1 2-colored necklaces

It is an elementary fact that the S_d-conjugacy class of a permutation σ in S_d is determined by the lengths of the disjoint cycles of σ. A slightly more general statement is that the S_d-conjugacy class of $\sigma \in S_d$ is determined by a union of cycles in which each element of a cycle is "colored" by colors corresponding to $J_1, \ldots, J_m$. It is not difficult to write down a proof of this fact, but we omit it here for space considerations.

If one fixes d, then the sum over all d-compositions, $\sum_{\boldsymbol{d}} |\mathcal{O}(\boldsymbol{d})|$, may be interpreted as the number of d-bead unions of necklaces with each bead colored by m colors. For example if $m = 2$, and $d = 4$, the resulting set may be depicted as noted in Figure 1.

A single k-bead necklace has $\mathbb{Z}/k\mathbb{Z}$-symmetry. If the beads of this necklace are colored with m colors, then the resulting colored necklace may have smaller group of symmetries. Using Polya enumeration (i.e., "Burnside's Lemma"), one can count such necklaces by the formula

$$N_k(m) = \frac{1}{k}\sum_{r|k}\phi(r)m^{\frac{k}{r}},$$

where ϕ denotes the Euler totient function. That is, $\phi(r)$ is the number of positive integers relatively prime to r. The theory forming the underpinnings of the above formula may be put in a larger context of *cycle index polynomials*. We refer the reader to Doron Zeilberger's survey article (IV.18 of [6]).

The generating function for disjoint unions of such necklaces can be given by the product

$$\eta_m(t) = \prod_{k=1}^{\infty}\left(\frac{1}{1-t^k}\right)^{N_k(m)}$$

That is, the number of d-bead necklaces, counted up to cyclic symmetry, is equal to the coefficient of t^d in $\eta(t)$. The main formula specializes to

$$\eta_m(t) = \sum_{\lambda,\boldsymbol{\mu}\vdash d}\left(c_{\boldsymbol{\mu}}^{\lambda}\right)^2 t^{|\lambda|}, \tag{1}$$

where the sum is over all partitions λ and over all m-tuples of partitions $\boldsymbol{\mu} = (\mu^{(1)}, \ldots, \mu^{(m)})$ such that $\sum_{i=1}^{m}|\mu^{(i)}| = d$.

Equation 1 begs for an (explicit) bijective proof which is no doubt obtained by merging both the Littlewood–Richardson rule and the Robinson–Schensted–Knuth bijection. It is likely that more than one "natural" bijection exists.

6.2 The general linear group over a finite field

Let p denote a prime number, and $v \in \mathbb{Z}^+$. Set $q = p^v$. Let $\mathrm{GL}_m(q)$ denote the general linear group of invertible $m \times m$ matrices over the field with q elements. The set, $\mathcal{C}_m(q)$, of conjugacy classes of $\mathrm{GL}_m(q)$ has a cardinality of note in that the infinite products has the following expansion:

$$\prod_{k=1}^{\infty}\frac{1-t^k}{1-qt^k} = \sum_{m=0}|\mathcal{C}_m(q)|t^m,$$

see [1, 16], and the references within.

Note that from Equation 1 we obtain a new formula for the number of conjugacy classes of $\mathrm{GL}_m(q)$, namely,

$$\left(\prod_{k=1}^{\infty}(1-t^k)\right)\eta_q(t) = \prod_{k=1}^{\infty}\left(\frac{1}{1-t^k}\right)^{N_k(q)-1}.$$

7 Proof of the main formula

Before proceeding, we require some more notation. Given a partition $\lambda = (\lambda_1 \geq \lambda_2 \geq \cdots)$ with v_1 ones, and v_2 twos, etc., we will call the sequence $(v_1, v_2, \ldots)$ the *type vector* of λ. Note that $\ell(\lambda) = \sum v_i$, while the size of λ is $m = |\lambda| = \sum \lambda_i = \sum i v_i$. As is standard, we set

$$z_\lambda = (v_1! 1^{v_1})(v_2! 2^{v_2})(v_3! 3^{v_3}) \cdots .$$

It is elementary that the cardinality of a conjugacy class of a permutation with cycle type λ is $\frac{m!}{z_\lambda}$. (Equivalently, the centralizer subgroup has order z_λ.)

We now prove the main formula. The product on the left side (LS) may be expanded using the sum of a geometric series, the multinomial theorem, and then the sum-product formula, as follows

$$\begin{aligned}
LS &= \prod_{k=1}^{\infty} \frac{1}{1 - (t_1^k + t_2^k + t_3^k + \cdots)} \\
&= \prod_{k=1}^{\infty} \sum_{u=0}^{\infty} (t_1^k + t_2^k + \cdots)^u \\
&= \prod_{k=1}^{\infty} \sum_{u=0}^{\infty} \sum_{u_1 + u_2 + \cdots = u} \frac{u!}{u_1! u_2! \cdots} t_1^{k u_1} t_2^{k u_2} \cdots \\
&= \prod_{k=1}^{\infty} \sum_{u_1, u_2, \cdots} \frac{(u_1 + u_2 + \cdots)!}{u_1! u_2! \cdots} t_1^{k u_1} t_2^{k u_2} \cdots \\
&= \sum_{u_j^{(i)}, \cdots} \prod_{i=1}^{\infty} \frac{(u_1^{(i)} + u_2^{(i)} + \cdots)!}{u_1^{(i)}! u_2^{(i)}! \cdots} t_1^{i u_1^{(i)}} t_2^{i u_2^{(i)}} \cdots
\end{aligned}$$

(with all sequences having finite support.) We will introduce another sequence, $\mathbf{a} = (a_1, a_2, \ldots)$ and extract the coefficient of $\boldsymbol{t}^{\mathbf{a}} = \prod t_i^{a_i}$ in the above formal expression to obtain

$$LS = \sum_{\mathbf{a}} \left(\sum_{u_j^{(i)} : \forall j, \sum_i i u_j^{(i)} = a_j} \prod_{i=1}^{\infty} \frac{(u_1^{(i)} + u_2^{(i)} + \cdots)!}{u_1^{(i)}! u_2^{(i)}! \cdots} \right) \boldsymbol{t}^{\mathbf{a}}$$

The above is not such a complicated expression, although these formal manipulations may, at first, seem daunting. Observe that the sequence $u_j^{(1)}, u_j^{(2)}, \ldots$ with $\sum_i i u_j^{(i)} = a_j$ encodes a partition of a_j with the number i occurring exactly $u_j^{(i)}$

times, while $\sum_{i\geq 1} u_j^{(i)}$ is the length of the partition. We will call this partition $\mu^{(j)}$. That is, $\mu^{(j)}$ has type vector $(u_j^{(1)}, u_j^{(2)}, \ldots)$. The coefficient of $\boldsymbol{t}^{\mathbf{a}}$ may be rewritten as a sum over all double sequences $u_j^{(i)}$ such that for all j, $\sum_i i u_j^{(i)} = a_j$ of

$$\prod_{i=1}^{\infty} \frac{(u_1^{(i)} + u_2^{(i)} + \cdots)!}{u_1^{(i)}!u_2^{(i)}!\cdots} = \prod_{i=1}^{\infty} \frac{(u_1^{(i)} + u_2^{(i)} + \cdots)! i^{\sum u_j^{(i)}}}{(u_1^{(i)}! i^{u_1^{(i)}})(u_2^{(i)}! i^{u_2^{(i)}})\cdots} . \tag{2}$$

Given a (finitely supported) sequence of partitions $\boldsymbol{\mu} = (\mu^{(1)}, \mu^{(2)}, \ldots)$, we denote the partition obtained from the (sorted) concatenation of all $\mu^{(j)}$ by $\cup\mu^{(j)}$. Thus, if $\mu^{(j)}$ has type vector $(u_1^{(i)}, u_2^{(i)}, \ldots)$, then the number of i's in $\cup\mu^{(j)}$ is $u_1^{(i)} + u_2^{(i)} + \cdots$. It therefore follows that the numerator of the right-hand side of Equation 2 is z_λ when $\lambda = \cup\mu^{(j)}$. The denominator can easily be seen to be $z_{\mu^{(j)}}$. From this observation we obtain

$$LS = \sum_{\boldsymbol{\mu}} \frac{z_{\cup_{j=1}^{\infty}\mu^{(j)}}}{\prod_{j=1}^{\infty} z_{\mu^{(j)}}} z_1^{|\mu^{(1)}|} z_2^{|\mu^{(2)}|} z_3^{|\mu^{(3)}|} \cdots .$$

7.1 *An application of the Hall scalar product*

Let Λ_n denote the S_n-invariant polynomials (over $\mathbb{C}$ as usual) in the indeterminates $x_1, \ldots, x_n$, let $\Lambda[\boldsymbol{x}] = \lim_{\leftarrow} \Lambda_n$ denote the inverse limit. Thus, $\Lambda[\boldsymbol{x}]$ is the algebra of symmetric functions. For a nonnegative integer partition λ, we let $s_\lambda(\boldsymbol{x})$ denote the Schur function. That is, for each n, $s_\lambda(\boldsymbol{x})$ projects to the polynomial $s_\lambda(x_1, \ldots, x_n)$ which, as a function on the diagonal matrices, coincides with the character of the GL_n-irrep F_n^λ. The set $\{s_\lambda(\boldsymbol{x}) : \lambda \vdash d\}$ is a $\mathbb{C}$-vector space basis of the homogeneous degree d symmetric functions. We will define a nondegenerate symmetric bilinear form $\langle\cdot,\cdot\rangle$ by declaring $\langle s_\alpha(\boldsymbol{x}), s_\beta(\boldsymbol{x})\rangle = \delta_{\alpha\beta}$ for all nonnegative integer partitions α, β. This form is the *Hall scalar product*.

Given an integer m, define $p_m(\boldsymbol{x}) = x_1^m + x_2^m + \cdots$ to denote the power sum symmetric function. Given $\nu \vdash N$, set $p_\nu(\boldsymbol{x}) = \prod p_{\nu_j}(\boldsymbol{x})$. We remark that the left side of the main formula is easily seen to be $\sum_\nu p_\nu(\boldsymbol{t})$.

It is a consequence of Schur–Weyl duality that the coefficients of the Schur function expansion

$$p_\nu(\boldsymbol{x}) = \sum_\lambda \chi^\lambda(\nu) s_\lambda(\boldsymbol{x})$$

are the characters of the S_N-irrep indexed by λ evaluated at any permutation with cycle type ν. It is a standard exercise to see, from the orthogonality of the character table, that $p_\nu(\boldsymbol{x})$ are an orthogonal basis of $\Lambda[\boldsymbol{x}]$, and $\langle p_\nu(\boldsymbol{x}), p_\nu(\boldsymbol{x})\rangle = z_\nu$.

We next consider another set of indeterminates, $\boldsymbol{y} = y_1, y_2, \ldots$. Set $\Lambda[\boldsymbol{x}, \boldsymbol{y}] = \Lambda[\boldsymbol{x}] \otimes \Lambda[\boldsymbol{y}]$. The Hall scalar product extends to $\Lambda[\boldsymbol{x}, \boldsymbol{y}]$ in the standard way as

$$\langle f(\boldsymbol{x}) \otimes g(\boldsymbol{y}), f'(\boldsymbol{x})g'(\boldsymbol{y})\rangle = \langle f(\boldsymbol{x}), f'(\boldsymbol{x})\rangle\langle g(\boldsymbol{y}), g'(\boldsymbol{y})\rangle,$$

where $f(\boldsymbol{x}), f'(\boldsymbol{x}) \in \Lambda[\boldsymbol{x}]$ and $g(\boldsymbol{y}), g'(\boldsymbol{y}) \in \Lambda[\boldsymbol{y}]$. (We then extend by linearity to all of $\Lambda[\boldsymbol{x}, \boldsymbol{y}]$.)

The character-theoretic consequence of $(\mathrm{GL}_n, \mathrm{GL}_k)$-Howe duality is the Cauchy identity:

$$\prod_{i,j=1}^{\infty} \frac{1}{1 - x_i y_j} = \sum_{\lambda} s_\lambda(\boldsymbol{x}) s_\lambda(\boldsymbol{y}) \tag{3}$$

(in the infinite sets of variables). In fact, for any pair of dual bases, a_λ, b_λ, with respect to the Hall scalar product, one has $\prod_{i,j=1}^{\infty} \frac{1}{1-x_i y_j} = \sum_\lambda a_\lambda(\boldsymbol{x}) b_\lambda(\boldsymbol{y})$. From this fact one obtains

$$\prod_{i,j=1}^{\infty} \frac{1}{1 - x_i y_j} = \sum_{\lambda} p_\lambda(\boldsymbol{x}) p_\lambda(\boldsymbol{y}) / z_\lambda. \tag{4}$$

From our point of view, we will expand the following scalar product

$$\left\langle \prod_{k=1}^{\infty} \prod_{i,j=1}^{\infty} \frac{1}{1 - x_i y_j t_k}, \prod_{i,j=1}^{\infty} \frac{1}{1 - x_i y_j} \right\rangle \tag{5}$$

in two different ways, corresponding to Equations 3 and 4.

First, by homogeneity of the Schur function and Cauchy's identity

$$\prod_{k=1}^{\infty} \sum_{\mu} s_\mu(\boldsymbol{x}) s_\mu(\boldsymbol{y}) t_k^{|\mu|} = \sum_{\mu=(\mu^{(1)},\mu^{(2)},\ldots)} \prod_j s_{\mu^{(j)}}(\boldsymbol{x}) \prod_j s_{\mu^{(j)}}(\boldsymbol{y}) t_1^{|\mu^{(1)}|} t_2^{|\mu^{(2)}|} \cdots .$$

Since the multiplication of characters is the character of the tensor product of the corresponding representations, we have $s_\alpha s_\beta = \sum c_{\alpha\beta}^{\gamma} s_\gamma$ in the $\boldsymbol{x}$ (resp. $\boldsymbol{y}$) variables. Expanding the above product gives

$$\left\langle \prod_{k=1}^{\infty} \prod_{i,j=1}^{\infty} \frac{1}{1 - x_i y_j t_k}, \prod_{i,j=1}^{\infty} \frac{1}{1 - x_i y_j} \right\rangle = \sum_{\mu} \left(c_{\mu}^{\lambda}\right)^2 t_1^{|\mu^{(1)}|} t_2^{|\mu^{(2)}|} \cdots .$$

Secondly, the scalar product in (5) may be expressed as

$$\left\langle \prod_{k=1}^{\infty} \sum_{\mu^{(k)}} p_{\mu^{(k)}}(\boldsymbol{x}) p_{\mu^{(k)}}(\boldsymbol{y}) / z_{\mu^{(k)}} \, t^{|\mu^{(k)}|}, \sum_{\lambda} p_\lambda(\boldsymbol{x}) p_\lambda(\boldsymbol{y}) / z_\lambda \right\rangle.$$

We observe that $\prod_{k=1}^{\infty} p_{\mu^{(k)}} = p_\lambda$ where $\lambda = \cup_{k=1}^{\infty} \mu^{(k)}$.

7.2 A remark from "Macdonald's book"

The results presented in this paper emphasize describing the cardinality of an orbit space by a sum of Littlewood–Richardson coefficients. It is important to note, however, that the main formula can be written simply as

$$\sum_{\nu} p_{\nu}(\boldsymbol{x}) = \sum_{\mu} \left(\sum_{\lambda} \left(c_{\mu}^{\lambda} \right)^2 \right) x_1^{\mu(1)} x_2^{\mu(2)} x_1^{\mu(3)} \cdots$$

So from the point of view of [13], one realizes that the main formula is simply a way of expanding $\sum p_{\nu}(\boldsymbol{x})$. With this remark in mind, we recall the "standard" viewpoint.

For a nonnegative integer partition $\delta = (\delta_1 \geq \delta_2 \geq \cdots)$ let $x^{\delta} = x_1^{\delta_1} x_2^{\delta_2} x_3^{\delta_3} \cdots$. The monomial symmetric function, $m_{\delta}(\boldsymbol{x})$, is the sum over the orbit obtained by all permutations of the variables. The monomial symmetric functions are a basis for the algebra Λ.

The question becomes obtaining an expansion of $\sum_{\gamma} p_{\gamma}(\boldsymbol{x})$ into monomial symmetric functions. This question is answered immediately by observing the expansion of p_{γ} into monomial symmetric functions, which can be found in [13] on page 102 of Chapter I, Section 6. For partitions γ and δ define $L(\gamma, \delta)$ by the expansion

$$p_{\gamma}(\boldsymbol{x}) = \sum_{\delta} L_{\gamma\delta} m_{\delta}(\boldsymbol{x}).$$

We next provide a combinatorial description of $L_{\gamma\delta}$. Let γ denote a partition of length ℓ. Given an integer valued function, f, defined on $\{1, 2, 3, \ldots, \ell\}$, set

$$f(\gamma)_i = \sum_{j:f(j)=i} \nu_j$$

for each $i \geq 1$.

The sequence $(f(\gamma)_1, f(\gamma)_2, f(\gamma)_3, \ldots)$ does not have to be weakly decreasing. For example, if $\gamma = (1, 1, 1)$ and $f(1) = 1$, $f(2) = 4$ and $f(3) = 4$ then $f(\gamma)_1 = 1$, $f(\gamma)_4 = 2$ and $f(\gamma)_k = 0$ for all $k \neq 1, 4$. However, often this sequences does define a partition. We have

Proposition 7.1. *$L_{\gamma,\delta}$ is equal to the number of functions f such that $f(\gamma) = \delta$.*

Proof. See [13] Proposition I (6.9) □

From Proposition 7.1 and the main formula, we obtain

Corollary 1. *Given a partition δ, the cardinality of*

$$\{f \mid \text{for some partition } \gamma,\ f(\gamma) = \delta\}$$

is equal to

$$\sum_{\lambda}\sum_{\mu}\left(c_{\mu}^{\lambda}\right)^{2}$$

where the sum is over all $\boldsymbol{\mu}$ *such that* $|\mu^{(j)}| = \delta_j$ *for all* j *and* $|\lambda| = |\boldsymbol{\mu}|$.

References

1. David Benson, Walter Feit, and Roger Howe, Finite linear groups, the Commodore 64, Euler and Sylvester. *Amer. Math. Monthly*, 93(9):717–719, 1986.
2. C. Chevalley, Sur certains groupes simples. *Tôhoku Math. J. (2)*, 7:14–66, 1955.
3. Vesselin Drensky, Computing with matrix invariants. *Math. Balkanica (N.S.)*, 21(1–2):141–172, 2007.
4. William Fulton, *Young tableaux*, Vol. 35, *London Mathematical Society Student Texts*. Cambridge University Press, Cambridge, 1997. With applications to representation theory and geometry.
5. Roe Goodman and Nolan R. Wallach, *Symmetry, Representations, and Invariants*, Vol. 255, *Graduate Texts in Mathematics*. Springer, New York, 2009.
6. Timothy Gowers, June Barrow-Green, and Imre Leader (editors), *The Princeton Companion to Mathematics*. Princeton University Press, Princeton, NJ, 2008.
7. Pamela E. Harris, On the adjoint representation of $\mathfrak{sl}_n$ and the Fibonacci numbers. *C. R. Math. Acad. Sci. Paris*, 349(17–18):935–937, 2011.
8. W. H. Hesselink, Characters of the nullcone. *Math. Ann.*, 252(3):179–182, 1980.
9. Roger Howe and Soo Teck Lee, Bases for some reciprocity algebras. I. *Trans. Amer. Math. Soc.*, 359(9):4359–4387, 2007.
10. Roger Howe and Soo Teck Lee, Why should the Littlewood-Richardson rule be true? *Bull. Amer. Math. Soc. (N.S.)*, 49(2):187–236, 2012.
11. Roger Howe, Eng-Chye Tan, and Jeb F. Willenbring, Stable branching rules for classical symmetric pairs. *Trans. Amer. Math. Soc.*, 357(4):1601–1626, 2005.
12. Bertram Kostant, Lie group representations on polynomial rings. *Amer. J. Math.*, 85:327–404, 1963.
13. I. G. Macdonald, *Symmetric Functions and Hall Polynomials*. Oxford Mathematical Monographs. The Clarendon Press Oxford University Press, New York, second edition, 1995. With contributions by A. Zelevinsky, Oxford Science Publications.
14. C. Procesi, The invariant theory of $n \times n$ matrices. *Advances in Math.*, 19(3):306–381, 1976.
15. Claudio Procesi, The invariants of $n \times n$ matrices. *Bull. Amer. Math. Soc.*, 82(6):891–892, 1976.
16. Richard Stong, Some asymptotic results on finite vector spaces. *Adv. in Appl. Math.*, 9(2):167–199, 1988.
17. N. R. Wallach and J. Willenbring, On some q-analogs of a theorem of Kostant-Rallis. *Canad. J. Math.*, 52(2):438–448, 2000.
18. Jeb F. Willenbring, Stable Hilbert series of $\mathcal{S}(\mathfrak{g})^K$ for classical groups. *J. Algebra*, 314(2): 844–871, 2007.

Polynomial functors and categorifications of Fock space

Jiuzu Hong, Antoine Touzé, and Oded Yacobi

Dedicated, with gratitude and admiration, to Nolan Wallach on the occasion of his 70th birthday

Abstract Fix an infinite field k of characteristic p, and let $\mathfrak{g}$ be the Kac–Moody algebra $\mathfrak{sl}_\infty$ if $p = 0$ and $\widehat{\mathfrak{sl}}_p$ otherwise. Let $\mathcal{P}$ denote the category of strict polynomial functors defined over k. We describe a categorical $\mathfrak{g}$-action on $\mathcal{P}$ (in the sense of Chuang and Rouquier) categorifying the Fock space representation of $\mathfrak{g}$.

Keywords: Categorification • Fock space

Mathematics Subject Classification: 18D05, 17B67

1 Introduction

Fix an infinite field k of characteristic p. In this work we elaborate on a study, begun in [HY], of the relationship between the symmetric groups $\mathfrak{S}_d$, the general linear groups $GL_n(k)$, and the Kac–Moody algebra $\mathfrak{g}$, where

$$\mathfrak{g} = \begin{cases} \mathfrak{sl}_\infty(\mathbb{C}) & \text{if } p = 0 \\ \widehat{\mathfrak{sl}}_p(\mathbb{C}) & \text{if } p \neq 0. \end{cases}$$

J. Hong
Department of Mathematics, Yale University, New Haven, CT 06520-8283, USA
e-mail: jiuzu.hong@yale.edu

A. Touzé
LAGA Institut Galilée, Université Paris 13, 99 Av. J-B Clément, 93430 Villetaneuse, France
e-mail: touze@math.univ-paris13.fr

O. Yacobi
School of Mathematics and Statistics, University of Sydney, NSW 2006, Australia
e-mail: oded.yacobi@sydney.edu.au

R. Howe et al. (eds.), *Symmetry: Representation Theory and Its Applications*, Progress in Mathematics 257, DOI 10.1007/978-1-4939-1590-3_12

In [HY] Hong and Yacobi defined a category $\mathcal{M}$ constructed as an inverse limit of polynomial representations of the general linear groups. The main result of [HY] is that $\mathfrak{g}$ acts on $\mathcal{M}$ (in the sense of Chuang and Rouquier), and categorifies the Fock space representation of $\mathfrak{g}$.

The result in [HY] is motivated by a well-known relationship between the basic representation of $\mathfrak{g}$ and the symmetric groups. Let $\mathcal{R}_d$ denote the category of representations of $\mathfrak{S}_d$ over k, and let $\mathcal{R}$ denote the direct sum of categories $\mathcal{R}_d$. By work of Lascoux, Leclerc, and Thibon [LLT], it is known that $\mathcal{R}$ is a categorification of the basic representation of $\mathfrak{g}$ (in a weaker sense than the Chuang–Rouquier theory). This means that there are exact endo-functors $E_i, F_i : \mathcal{R} \to \mathcal{R}$ ($i \in \mathbb{Z}/p\mathbb{Z}$) whose induced operators on the Grothedieck group give rise to a representation of $\mathfrak{g}$ isomorphic to its basic representation.

Since $\mathcal{R}$ consists of all representations of all symmetric groups, and the representations of symmetric groups and general linear groups are related via Schur–Weyl duality, it is natural to seek a category which canonically considers all polynomial representations of all general linear groups. This is precisely the limit category of polynomial representations alluded to above.

The limit category $\mathcal{M}$ is naturally equivalent (Lemma B.2, [HY]) to the category $\mathcal{P}$ of "strict polynomial functors of finite degree" introduced by Friedlander and Suslin in [FS] (in characteristic zero the category $\mathcal{P}$ appears in [Mac]). The objects of $\mathcal{P}$ are endo-functors on $\mathcal{V}$ (the category of finite-dimensional vector spaces over k) satisfying natural polynomial conditions, and the morphisms are natural transformations of functors.

Friedlander and Suslin's original motivation was to study the finite generation of affine group schemes. This is related to the study of extensions of representations of general linear groups over fields of positive characteristic (cf. Section A.27, [J]). Since their landmark work, the theory of polynomial functors has developed in many directions. In algebraic topology, the category $\mathcal{P}$ is connected to the category of unstable modules over the Steenrod algebra, to the cohomology of the finite linear groups [FFSS, Ku], and also to derived functors in the sense of Dold and Puppe [T2]. Polynomial functors are also applied to the cohomology of group schemes. For example, the category $\mathcal{P}$ is used in the study of support varieties for finite group schemes [SFB], to compute the cohomology of classical groups [T1], and in the proof of cohomological finite generation for reductive groups [TvdK].

The goal of this paper is to develop an explicit connection relating the category of strict polynomial functors to the affine Kac–Moody algebra $\mathfrak{g}$. We describe a categorical action of $\mathfrak{g}$ on $\mathcal{P}$ (in the sense of [CR, R, KL, KL2, KL3]), which is completely independent of the results or arguments in [HY]. The main advantage of this approach is that the category $\mathcal{P}$ affords a more canonical setting for categorical $\mathfrak{g}$-actions. Indeed, many of the results obtained in [HY] have a simple and natural formulation in this setting. Further, we hope that the ideas presented here will provide new insight to the category of polynomial functors. As an example of this, in the last section of the paper we describe how the categorification theory implies that certain blocks of the category $\mathcal{P}$ are derived equivalent. These kinds of applications are typical in this framework; the main result in [CR] was to establish derived equivalences between blocks of representations of the symmetric groups. Categorical $\mathfrak{g}$-actions have since been used in Lie theory to establish equivalences of abelian categories (see e.g., Theorems 1.1 in [BS1, BS2]).

The category of strict polynomial functors is actually defined over arbitrary fields, but the general definition given in [FS] is more involved than the one we use (the problem comes from the fact that different polynomials might induce the same function over finite fields). All our results remain valid in this general context, but we have opted to work over an infinite field to simplify the exposition. In addition, we assume in our main theorem that $p \neq 2$. The theorem is valid also for $p = 2$, but including this case would complicate our exposition.

In the sequel to this work we continue the study of $\mathcal{P}$ from the point of view of higher representation theory [HY2]. We show that Khovanov's category $\mathcal{H}$ naturally acts on $\mathcal{P}$, and this gives a categorification of the Fock space representation of the Heisenberg algebra when $\text{char}(k) = 0$. When $\text{char}(k) > 0$ the commuting actions of $\mathfrak{g}'$ (the derived algebra of $\mathfrak{g}$) and the Heisenberg algebra are also categorified. Moreover, we formulate Schur–Weyl duality as a functor from $\mathcal{P}$ to the category of linear species. The category of linear species is known to carry actions of $\mathfrak{g}$ and the Heisenberg algebra. We prove that Schur–Weyl duality is a tensor functor which is a morphism of both the categorical $\mathfrak{g}$-action and the categorical Heisenberg action.

Finally, we mention the work by Ariki [A] on qraded q-Schur algebras, and the recent work by Stroppel–Webster on quiver Schur algebras [SW]. These works suggest the existence of a graded version of the polynomial functor, which would gives rise to a natural categorification of the Fock space of the quantum affine algebra $U_q(\widehat{\mathfrak{sl}}_n)$. It would be interesting to pursue this generalization of our present work. We also mention ongoing work of the second author with L. Rigal, where they define a notion of quantum strict polynomial functors, which should also fit well within the categorification scheme.

Acknowledgments. We thank the referee for many helpful comments which greatly improved the exposition of the paper.

2 Type A Kac–Moody algebras

Let $\mathfrak{g}$ denote the following Kac–Moody algebra (over $\mathbb{C}$):

$$\mathfrak{g} = \begin{cases} \mathfrak{sl}_\infty \text{ if } p = 0 \\ \widehat{\mathfrak{sl}}_p \text{ if } p > 0 \end{cases}$$

By definition, the Kac–Moody algebra $\mathfrak{sl}_\infty$ is associated to the Dynkin diagram:

$$\cdots \text{———} \bullet \text{———} \bullet \text{———} \bullet \text{———} \bullet \text{———} \cdots$$

while the Kac–Moody algebra $\widehat{\mathfrak{sl}}_p$ is associated to the diagram with p nodes:

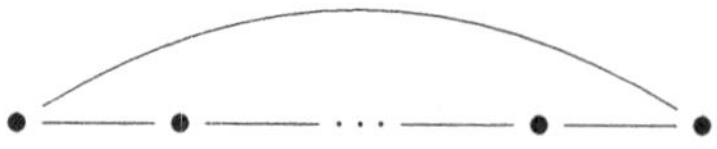

The Lie algebra $\mathfrak{g}$ has standard Chevalley generators $\{e_i, f_i\}_{i \in \mathbb{Z}/p\mathbb{Z}}$. Here and throughout, we identify $\mathbb{Z}/p\mathbb{Z}$ with the prime subfield of k. For the precise relations defining $\mathfrak{g}$, see e.g., [Kac].

We let Q denote the root lattice and P the weight lattice of $\mathfrak{g}$. Let $\{\alpha_i : i \in \mathbb{Z}/p\mathbb{Z}\}$ denote the set of simple roots, and $\{h_i : i \in \mathbb{Z}/p\mathbb{Z}\}$ the simple coroots. The cone of dominant weights is denoted P_+ and denote the fundamental weights $\{\Lambda_i : i \in \mathbb{Z}/p\mathbb{Z}\}$, i.e., $\langle h_i, \Lambda_j \rangle = \delta_{ij}$. When $p > 0$ the Cartan subalgebra of $\mathfrak{g}$ is spanned by the h_i along with an element d. In this case we also let $\delta = \sum_i \alpha_i$; then $\Lambda_0, \ldots, \Lambda_{p-1}, \delta$ form a $\mathbb{Z}$-basis for P. When $p = 0$ the fundamental weights are a $\mathbb{Z}$ basis for the weight lattice.

Let $\mathfrak{S}_n$ denote the symmetric group on n letters. $\mathfrak{S}_n$ acts on the polynomial algebra $\mathbb{Z}[x_1, \ldots, x_n]$ by permuting variables, and we denote by $B_n = \mathbb{Z}[x_1, \ldots, x_n]^{\mathfrak{S}_n}$ the polynomials invariant under this action. There is a natural projection $B_n \twoheadrightarrow B_{n-1}$ given by setting the last variable to zero. Consequently, the rings B_n form a inverse system; let $B_{\mathbb{Z}}$ denote the subspace of finite degree elements in the inverse limit $\varprojlim B_n$. This is the *algebra of symmetric functions* in infinitely many variables $\{x_1, x_2, \ldots\}$. Let $B = B_{\mathbb{Z}} \otimes_{\mathbb{Z}} \mathbb{C}$ denote the *(bosonic) Fock space*.

The algebra $B_{\mathbb{Z}}$ has many well-known bases. Perhaps the nicest is the basis of *Schur functions* (see e.g., [Mac]). Let $\wp$ denote the set of all partitions, and for $\lambda \in \wp$ let $s_\lambda \in B_{\mathbb{Z}}$ denote the corresponding Schur function. Let us review some combinatorial notions related to Young diagrams. Firstly, we identify partitions with their Young diagram (using English notation). For example, the partition $(4, 4, 2, 1)$ corresponds to the diagram

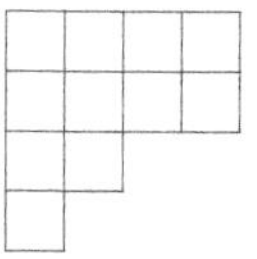

The *content* of a box in position (k, l) is the integer $l - k \in \mathbb{Z}/p\mathbb{Z}$. Given $\mu, \lambda \in \wp$, we write $\mu \longrightarrow \lambda$ if λ can be obtained from μ by adding some box. If the arrow is labelled i, then λ is obtained from μ by adding a box of content i (an i-box, for short). For instance, if $m = 3$, $\mu = (2)$ and $\lambda = (2, 1)$ then $\mu \xrightarrow{2} \lambda$. An i-box of λ is *addable* (resp. *removable*) if it can be added to (resp. removed from) λ to obtain another partition.

Of central importance to us is the *Fock space* representation of $\mathfrak{g}$ on B (or $B_{\mathbb{Z}}$). The Schur functions form a $\mathbb{Z}$-basis of the algebra of symmetric functions:

$$B_{\mathbb{Z}} = \bigoplus_{\lambda \in \wp} \mathbb{Z} s_\lambda.$$

The action of $\mathfrak{g}$ on B is given on this basis by the following formulas: $e_i.s_\lambda = \sum s_\mu$, where the sum is over all μ such that $\mu \xrightarrow{i} \lambda$, and $f_i.s_\lambda = \sum s_\mu$, where the sum is over all μ such that $\lambda \xrightarrow{i} \mu$. Moreover, d acts on s_λ by $m_0(\lambda)$, where $m_0(\lambda)$ is the number of boxes of content zero in λ. These equations define an integral representation of $\mathfrak{g}$ (see e.g., [LLT]).

Note that $s_\emptyset$ is a highest weight vector of highest weight Λ_0. We note also that the standard basis of B is a weight basis. Let $m_i(\lambda)$ denote the number of i-boxes of λ. Then s_λ is of weight $wt(\lambda)$, where

$$wt(\lambda) = \Lambda_0 - \sum_i m_i(\lambda)\alpha_i. \tag{1}$$

For a k-linear abelian category $\mathcal{C}$, let $K_0(\mathcal{C})$ denote the Grothendieck group of $\mathcal{C}$, and let $K(\mathcal{C})$ denote the complexification of $K_0(\mathcal{C})$. If $A \in \mathcal{C}$ we let $[A]$ denote its image in $K_0(\mathcal{C})$. Similarly, for an exact functor $F : \mathcal{C} \to \mathcal{C}'$ we let $[F] : K_0(\mathcal{C}) \to K_0(\mathcal{C}')$ denote the induced operator on the Grothendieck groups. Slightly abusing notation, the complexification of $[F]$ is also denoted by $[F]$.

We will also need the following combinatorial definition: for a partition λ of d, the permutation $\sigma_\lambda \in \mathfrak{S}_d$ is defined as follows. Let t_λ be the Young tableaux with standard filling: $1, \dots, \lambda_1$ in the first row, $\lambda_1 + 1, \dots, \lambda_2$ in the second row, and so forth. Then σ_λ, in one-line notation, is the row-reading of the conjugate tableaux t_λ°. For example, if $\lambda = (3, 1)$, then, $\sigma_\lambda = 1423$, the permutation mapping $1 \mapsto 1, 2 \mapsto 4, 3 \mapsto 2$, and $4 \mapsto 3$.

3 Categorical $\mathfrak{g}$-actions

Higher representation theory concerns the action of $\mathfrak{g}$ on categories rather than on vector spaces. The pioneering work on higher representation theory concerned constructing actions on Grothendieck groups of representation theoretic categories of algebraic or geometric origin; this is known as "weak" categorification. We are concerned with "strong" categorical $\mathfrak{g}$-actions, in a sense to be made precise below. The foundational papers which define this notion are [CR, KL, KL2, KL3, R]. There are great overviews of the theory appearing in [L, Ma].

At the very least, an action of $\mathfrak{g}$ on a k-linear additive category $\mathcal{C}$ consists of the data of exact endo-functors E_i and F_i on $\mathcal{C}$ (for $i \in \mathbb{Z}/p\mathbb{Z}$), such that $\mathfrak{g}$ acts on $K(\mathcal{C})$ via the assignment $e_i \mapsto [E_i]$ and $f_i \mapsto [F_i]$. For instance, if i and j are not connected in the Dynkin diagram of $\mathfrak{g}$ (i.e., $[e_i, f_j] = 0$), then we require that $[[E_i], [F_j]] = 0$ in $\mathrm{End}(K(\mathcal{C}))$. This is known as a "weak categorification".

This notion is qualified as "weak" because the relations defining $\mathfrak{g}$, such as $[e_i, f_j] = 0$, are not lifted to the level of categories. A stronger notion of categorification would require isomorphisms of functors lifting the relations of $\mathfrak{g}$, e.g., functorial isomorphisms $E_i \circ F_j \simeq F_j \circ E_i$. Moreover, these isomorphisms need to be compatible in a suitable sense. Making these ideas precise leads to an enriched theory, which introduces new symmetries coming from an affine Hecke algebra.

To give the definition of categorical $\mathfrak{g}$-action we use here, due to Chuang and Rouquier (a related formulation appears in the works of Khovanov and Lauda [KL]), we first introduce the relevant Hecke algebra.

Definition 1. Let DH_n be the *degenerate affine Hecke algebra* of GL_n. As an abelian group

$$DH_n = \mathbb{Z}[y_1, \dots, y_n] \otimes \mathbb{Z}\mathfrak{S}_n.$$

The algebra structure is defined as follows: $\mathbb{Z}[y_1, \dots, y_n]$ and $\mathbb{Z}\mathfrak{S}_n$ are subalgebras, and the following relations hold between the generators of these subalgebras:

$$\tau_i y_j = y_j \tau_i \text{ if } |i - j| \geq 1$$

and

$$\tau_i y_{i+1} - y_i \tau_i = 1 \tag{2}$$

(here $\tau_1, \dots, \tau_{n-1}$ are the simple generators of $\mathbb{Z}\mathfrak{S}_n$).

Remark 1. One can replace Relation (2) by

$$\tau_i y_i - y_{i+1} \tau_i = 1. \tag{3}$$

These two presentations are equivalent; the isomorphism is given by

$$\tau_i \mapsto \tau_{n-i}, y_i \mapsto y_{n+1-i}.$$

Definition 2. [Definition 5.29 in [R]] Let $\mathcal{C}$ be an abelian k-linear category. A *categorical* $\mathfrak{g}$*-action* on $\mathcal{C}$ is the data of:

1. an adjoint pair (E, F) of exact functors $\mathcal{C} \to \mathcal{C}$,
2. morphisms of functors $X \in \mathrm{End}(E)$ and $T \in \mathrm{End}(E^2)$, and
3. a decomposition $\mathcal{C} = \bigoplus_{\omega \in P} \mathcal{C}_\omega$.

Let $X^\circ \in \mathrm{End}(F)$ be the endomorphism of F induced by adjunction. Then given $a \in k$ let E_a (resp. F_a) be the generalized a-eigensubfunctor of X (resp. X°) acting on E (resp. F). We assume that

4. $E = \bigoplus_{i\in\mathbb{Z}/p\mathbb{Z}} E_i$,
5. the action of $\{[E_i],[F_i]\}_{i\in\mathbb{Z}/p\mathbb{Z}}$ on $K(\mathcal{C})$ gives rise to an integrable representation of $\mathfrak{g}$,
6. for all i, $E_i(\mathcal{C}_\omega) \subset \mathcal{C}_{\omega+\alpha_i}$ and $F_i(\mathcal{C}_\omega) \subset \mathcal{C}_{\omega-\alpha_i}$,
7. the functor F is isomorphic to the left adjoint of E, and
8. the degenerate affine Hecke algebra DH_n acts on $\mathrm{End}(E^n)$ via

$$y_i \mapsto E^{n-i}XE^{i-1} \text{ for } 1 \leq i \leq n, \tag{4}$$

and

$$\tau_i \mapsto E^{n-i-1}TE^{i-1} \text{ for } 1 \leq i \leq n-1. \tag{5}$$

Remark 2. The definition (cf. Definition 5.29 in [R]) uses Relation (2). For our purposes we use Relation (3). On the representations of the symmetric groups (the main example considered in [CR, Section 3.1.2]) another variant of Relation (3) is used.

Remark 3. To clarify notation, the natural endomorphism y_i of E^n assigns to $M \in \mathcal{C}$ an endomorphism of $E^n(M)$ as follows: first evaluate the natural transformation at the object $E^{i-1}(M)$ yielding a morphism $X_{E^{i-1}(M)} : E^i(M) \to E^i(M)$. Applying the functor E^{n-i} to this morphism we obtain the endomorphism $(y_i)_M : E^n(M) \to E^n(M)$. See [BS1, BS2] for a more details on this construction.

The functorial isomorphisms lifting the defining relations of $\mathfrak{g}$ are constructed from the data of categorical $\mathfrak{g}$-action. More precisely, the adjunctions between E and F and the functorial morphisms X and T are introduced precisely for this purpose. The action of DH_n on $\mathrm{End}(E^n)$ in part (8) of Definition 2 is needed in order to express the compatibility between the functorial isomorphisms. See [R] for details.

4 Polynomial functors

4.1 *The category $\mathcal{P}$*

Our main goal in this paper is to define a categorical $\mathfrak{g}$-action on the category $\mathcal{P}$ of *strict polynomial functors of finite degree*, and show that this categorifies the Fock space representation of $\mathfrak{g}$. In this section we define the category $\mathcal{P}$ and recall some of its basic features.

Let $\mathcal{V}$ denote the category of finite-dimensional vector spaces over k. For $V, W \in \mathcal{V}$, *polynomial maps* from V to W are by definition elements of $S(V^*) \otimes W$, where $S(V^*)$ denotes the symmetric algebra of the linear dual of V. Elements of $S^d(V^*) \otimes W$ are said to be *homogeneous* of degree d.

Definition 3. The objects of the category $\mathcal{P}$ are functors $M : \mathcal{V} \to \mathcal{V}$ that satisfy the following properties:

1. for any $V, W \in \mathcal{V}$, the map of vector spaces
$$\mathrm{Hom}_k(V, W) \to \mathrm{Hom}_k(M(V), M(W))$$
is polynomial, and
2. the degree of the map
$$\mathrm{End}_k(V) \to \mathrm{End}_k(M(V))$$
is bounded with respect to $V \in \mathcal{V}$.

The morphisms in $\mathcal{P}$ are natural transformations of functors. For $M \in \mathcal{P}$ we denote by $1_M \in \mathrm{Hom}_{\mathcal{P}}(M, M)$ the identity natural transformation.

Let $I \in \mathcal{P}$ be the identity functor from $\mathcal{V}$ to $\mathcal{V}$ and let $k \in \mathcal{P}$ denote the constant functor with value k. Tensor products in $\mathcal{V}$ define a symmetric monoidal structure $\otimes$ on $\mathcal{P}$, with unit k. The category $\mathcal{P}$ is abelian.

Let $M \in \mathcal{P}$ and $V \in \mathcal{V}$. By functoriality $M(V)$ carries a polynomial action of the linear algebraic group $GL(V)$. We denote this representation by $\pi_{M,V}$, or by π when the context is clear:
$$\pi_{M,V} : GL(V) \to GL(M(V)).$$

Similarly, a morphism $\phi : M \to N$ induces a $GL(V)$-equivariant map $\phi_V : M(V) \to N(V)$. Thus evaluation on V yields a functor from $\mathcal{P}$ to $Pol(GL(V))$, the category of polynomial representations of $GL(V)$.

Remark 4. Given a morphism $\phi : M \to N$ of polynomial functors M, N, one can talk about $\mathrm{im}(\phi) \in \mathcal{P}$. Explicitly, this functor is given on $V \in \mathcal{V}$ by $\mathrm{im}(\phi)(V) = \mathrm{im}(\phi_V)$, and on linear maps $f : V \to W$ by $\mathrm{im}(\phi)(f) = N(f)|_{\mathrm{im}(\phi_V)}$. This is well-defined since ϕ is a natural transformation.

4.2 Degrees and weight spaces

The *degree* of a functor $M \in \mathcal{P}$ is the upper bound of the degrees of the polynomials $\mathrm{End}_k(V) \to \mathrm{End}_k(M(V))$ for $V \in \mathcal{V}$. For example, the functors of degree zero are precisely the functors $\mathcal{V} \to \mathcal{V}$ which are isomorphic to constant functors. A functor $M \in \mathcal{P}$ is *homogeneous of degree* d if all the polynomials $\mathrm{End}_k(V) \to \mathrm{End}_k(M(V))$ are homogeneous polynomials of degree d.

For $M \in \mathcal{P}$, $GL(k)$ acts on $M(V)$ by the formula
$$\lambda \cdot m = \pi_{M,V}(\lambda 1_V)(m), \text{ for } \lambda \in GL(k) \text{ and } m \in M(V).$$

This action is a polynomial action of $GL(k)$, so $M(V)$ splits as a direct sum of weight spaces

$$M(V) = \bigoplus_{d \geq 0} M(V)_d,$$

where

$$M(V)_d = \{m \in M(V) : \lambda \cdot m = \lambda^d m\}.$$

Moreover, if $f : V \to W$ is a linear map, it commutes with homotheties, so $M(f)$ is $GL(k)$-equivariant. Hence $M(f)$ preserves weight spaces, and we denote by $M(f)_d$ its restriction to the d-th weight spaces.

So we can define a strict polynomial functor M_d by letting

$$M_d(V) = M(V)_d, M_d(f) = M(f)_d.$$

A routine check shows that M_d is homogeneous of degree d. Thus, any functor M decomposes as a finite direct sum of homogeneous functors M_d of degree d. Similarly, a morphism $\phi : M \to N$ between strict polynomial functors preserves weight spaces. So it decomposes as a direct sum of morphisms of homogeneous functors $\phi_d : M_d \to N_d$. This can be formulated by saying that the category $\mathcal{P}$ is the direct sum of its subcategories $\mathcal{P}_d$ of homogeneous functors of degree d:

$$\mathcal{P} = \bigoplus_{d \geq 0} \mathcal{P}_d. \tag{6}$$

If $M \in \mathcal{P}$, we define its *Kuhn dual* $M^\sharp \in \mathcal{P}$ by $M^\sharp(V) = M(V^*)^*$, where '$*$' refers to k-linear duality in the category of vector spaces. Since $(M^\sharp)^\sharp \simeq M$, duality yields an equivalence of categories [FS, Prop 2.6]:

$$\sharp : \mathcal{P} \xrightarrow{\simeq} \mathcal{P}^{\mathrm{op}}.$$

A routine check shows that $\sharp$ respects degrees, i.e., $M^\sharp$ is homogeneous of degree d if and only if M also is. Indeed, if $\lambda \in GL(k)$, then for $\ell \in M^\sharp(V)$ and $m \in M(V^*)$, we have that $(\lambda \cdot \ell)(m) = \ell(M(\lambda^*)(m)) = \ell(M(\lambda)(m))$.

The following theorem, due to Friedlander and Suslin [FS], shows the categories $\mathcal{P}_d$ are a model for the stable categories of homogeneous polynomial $GL_n(k)$-modules of degree d. Let $\mathrm{Pol}_d(GL(V))$ denote the category of polynomial representations of $GL(V)$ of degree d.

Theorem 1. *Let $V \in \mathcal{V}$ be a k-vector space of dimension $n \geq d$. The functor induced by evaluation on V:*

$$\mathcal{P}_d \to \mathrm{Pol}_d(GL(V)),$$

is an equivalence of categories.

As a consequence of Theorem 1, we obtain that strict polynomial functors are noetherian objects in the following sense:

Corollary 1. *Let $M \in \mathcal{P}$. Assume there is an increasing sequence of subfunctors of M:*

$$M^0 \subset M^1 \subset \cdots \subset M^i \subset \cdots .$$

Then there exists an integer N such that for all $n \geq N$, $M^n = M^N$.

Let $\otimes^d$ denote the d-th tensor product functor, which sends $V \in \mathcal{V}$ to $V^{\otimes d} \in \mathcal{V}$. Then $\otimes^d \in \mathcal{P}_d$. Let λ be a tuple of nonnegative integers summing to d, and let $\mathfrak{S}_\lambda \subset \mathfrak{S}_d$ denote the associated Young subgroup. We denote by Γ^λ the subfunctor of $\otimes^d$ defined by $\Gamma^\lambda(V) = (V^{\otimes d})^{\mathfrak{S}_\lambda}$.

Proposition 1 (Theorem 2.10, [FS]). *The functor Γ^λ, $\lambda \in \wp$, a partition of d, is a (projective) generator of $\mathcal{P}_d$.*

In other words, the objects $M \in \mathcal{P}_d$ are exactly the functors $M : \mathcal{V} \to \mathcal{V}$ which can be obtained as subquotients of a direct sum of a finite number of copies of the d-th tensor product functor $\otimes^d$.

4.3 Recollections of Schur and Weyl functors

In this section we introduce Schur functors and Weyl functors. These strict polynomial functors are the functorial version of the Schur modules and the Weyl modules, and they were first defined in [ABW].

Let $\wp_d$ denote the partitions of d. For $\lambda \in \wp_d$ let λ° denote the conjugate partition. We define a morphism of polynomial functors d_λ as the composite:

$$d_\lambda \,:\, \Lambda^{\lambda_1^\circ} \otimes \cdots \otimes \Lambda^{\lambda_n^\circ} \hookrightarrow \otimes^d \xrightarrow{\sigma_\lambda} \otimes^d \twoheadrightarrow S^{\lambda_1} \otimes \cdots \otimes S^{\lambda_m}.$$

Here the first map is the canonical inclusion and the last one is the canonical epimorphism. The middle map is the isomorphism of $\otimes^d$ which maps $v_1 \otimes \cdots v_d$ onto $v_{\sigma_\lambda(1)} \otimes \cdots \otimes v_{\sigma_\lambda(d)}$, where $\sigma_\lambda \in \mathfrak{S}_d$ is the permutation defined in the last paragraph of Section 2.

Definition 4. Let $\lambda \in \wp_d$.

1. The *Schur functor* $S_\lambda \in \mathcal{P}_d$ is the image of d_λ (cf. Remark 4).
2. The *Weyl functor* W_λ is defined by duality $W_\lambda := S_\lambda^\sharp$.
3. Let L_λ be the socle of the functor S_λ.

Remark 5. In [ABW, def. II.1.3], Schur functors are defined in the more general setting of "skew partitions" λ/α, (i.e., pairs of partitions (λ, α) with $\alpha \subset \lambda$), and over arbitrary commutative rings. They denote Schur functors by L_{λ°, but we prefer to reserve this notation for simple objects in $\mathcal{P}_d$.

The following statement makes the link between Schur functors and induced modules (also called costandard modules, or Schur modules) and between Weyl functors and Weyl modules (also called standard modules or Verma modules).

Proposition 2. *Let* $\lambda \in \wp_d$.

(i) *There is an isomorphism of* $GL(k^n)$*-modules* $S_\lambda(k^n) \simeq H^0(\lambda)$*, where* $H^0(\lambda) = ind_{B^-}^{GL(k^n)}(k^\lambda)$ *is the induced module from [J, II.2].*
(ii) *There is an isomorphism of* $GL(k^n)$*-modules* $W_\lambda(k^n) \simeq V(\lambda)$*, where*

$$V(\lambda) = H^0(-w_0\lambda)^*$$

is the Weyl module from [J, II.2].

Proof. We observe that (ii) follows from (i). Indeed, we know that $V(\lambda)$ is the transpose dual of $H^0(\lambda)$, and evaluation on k^n changes the duality $^\sharp$ in $\mathcal{P}$ into the transpose duality. To prove (ii), we refer to [Mar]. The Schur module $M(\lambda)$ defined in [Mar, Def 3.2.1] is isomorphic to $H^0(\lambda)$ (this is a theorem of James, cf. [Mar, Thm 3.2.6]). Now, using the embedding of $M(\lambda)$ into $S^{\lambda_1}(k^n)\otimes\cdots\otimes S^{\lambda_m}(k^n)$ of [Mar, Example (1) p.73], and [ABW, Thm II.2.16], we get an isomorphism $S_\lambda(k^n) \simeq M(\lambda)$. □

The following portemanteau theorem collects some of the most important properties of the functors S_λ, W_λ, L_λ, $\lambda \in \wp_d$.

Theorem 2. (i) *The functors* L_λ, $\lambda \in \wp_d$ *form a complete set of representatives for the isomorphism classes of irreducible functors of* $\mathcal{P}_d$.
(ii) *Irreducible functors are self-dual: for all* $\lambda \in \wp_d$, $L_\lambda^\sharp \simeq L_\lambda$.
(iii) *For all* $\lambda \in \wp_d$*, the* L_μ *which appear as composition factors in* S_λ *satisfy* $\mu \leq \lambda$*, where* $\leq$ *denotes the lexicographic order. Moreover, the multiplicity of* L_λ *in* S_λ *is one.*
(iv) *For all* $\lambda, \mu \in \wp_d$,

$$\mathrm{Ext}^i_{\mathcal{P}}(W_\mu, S_\lambda) = \begin{cases} k \text{ if } \lambda = \mu \text{ and } i = 0, \\ 0 \text{ otherwise.} \end{cases}$$

Proof. All these statements have functorial proofs, but for sake of brevity we shall use Proposition 2, together with the fact that evaluation on V for $\dim V \geq d$ is an equivalence of categories. Thus, (i) follows from [Mar, Thm. 3.4.2], (ii) follows from [Mar, Thm. 3.4.9], (iii) follows from [Mar, Thm. 3.4.1(iii)]. Finally, (iv) follows from [J, Prop. 4.13] and [FS, Cor. 3.13]. □

Note that for any $d \geq 0$ the categories $\mathcal{P}_d$ are of finite global dimension (cf. e.g., Theorem 3.3.8, [Mar]). Therefore projective objects descend to a basis of the Grothendieck group. Simple objects of course also descend to a basis.

Corollary 2. *The equivalence classes of the Weyl functors* $[W_\lambda]$ *for* $\lambda \in \wp$ *form a basis of* $K(\mathcal{P})$.

Proof. Order $\wp$ by the lexicographic order, denoted $\leq$. By parts (ii) and (iii) of Theorem 2, the multiplicity of L_λ in W_λ is one, and all other simple objects appearing as composition factors in W_λ are isomorphic to L_μ, where $\mu \leq \lambda$. Form the matrix of the map given by $[L_\lambda] \mapsto [W_\lambda]$ in the basis $[L_\lambda]_{\lambda\in\wp}$ (ordered by $\leq$). This is a lower triangular matrix, with 1's on the diagonal. Hence it is invertible and we obtain the result. □

Corollary 3. *The map $K(\mathcal{P}) \to K(\mathcal{P})$ given by $[M] \mapsto [M^\sharp]$ is the identity. In particular, for all $\lambda \in \wp$, $[W_\lambda] = [S_\lambda]$.*

Proof. By Theorem 2(ii) simple functors are self-dual, hence the result. □

4.4 Polynomial bifunctors

We shall also need the category $\mathcal{P}^{[2]}$ of *strict polynomial bi-functors*. The objects of $\mathcal{P}^{[2]}$ are functors $B : \mathcal{V}\times\mathcal{V} \to \mathcal{V}$ such that for every $V \in \mathcal{V}$, the functors $B(\cdot, V)$ and $B(V, \cdot)$ are in $\mathcal{P}$ and their degrees are bounded with respect to V. Morphisms in $\mathcal{P}^{[2]}$ are natural transformations of functors. The following example will be of particular interest to us.

Exmaple 1. Let $M \in \mathcal{P}$. We denote by $M^{[2]}$ the bifunctor:

$$\begin{aligned} M^{[2]} : \mathcal{V}\times\mathcal{V} &\to \mathcal{V} \\ (V,W) &\mapsto M(V\oplus W) \\ (f,g) &\mapsto M(f\oplus g). \end{aligned}$$

Mapping M to $M^{[2]}$ yields a functor: $\mathcal{P} \to \mathcal{P}^{[2]}$.

If $B \in \mathcal{P}^{[2]}$ and (V, W) is a pair of vector spaces, then functoriality endows $B(V, W)$ with a polynomial $GL(V) \times GL(W)$-action, which we denote by $\pi_{B,V,W}$ (or simply by π if the context is clear):

$$\pi_{B,V,W} : GL(V) \times GL(W) \to GL(B(V, W)).$$

Evaluation on a pair (V, W) of vector spaces yields a functor from $\mathcal{P}^{[2]}$ to $\mathrm{Rep}_{\mathrm{pol}}(GL(V) \times GL(W))$.

A bifunctor B is *homogeneous of bidegree* (d, e) if for all $V \in \mathcal{V}$, $B(V, \cdot)$ (resp. $B(\cdot, V)$) is a homogeneous strict polynomial functor of degree d, (resp. of degree e). The decomposition of strict polynomial functors into a finite direct sums of homogeneous functors generalizes to bifunctors. Indeed, if $B \in \mathcal{P}^{[2]}$, the vector space $B(V, W)$ is endowed with a polynomial action of $GL(k) \times GL(k)$ defined by

$$(\lambda, \mu) \cdot m = \pi_{B,V,W}(\lambda 1_V, \mu 1_W)(m),$$

and pairs of linear maps (f, g) induce $GL(k) \times GL(k)$-equivariant morphisms $B(f, g)$. So for $i, j \geq 0$ we can use the (i, j) weight spaces with respect to the action of $GL(k) \times GL(k)$ to define bifunctors $B_{i,j}$, namely

$$B_{i,j}(V, W) = \{m \in B(V, W) : (\lambda, \mu) \cdot m = \lambda^i \mu^j m\}$$

and $B_{i,j}(f, g)$ is the restriction of $B(f, g)$ to the (i, j)-weight spaces. Functors $B_{i,j}$ are homogenous of bidegree (i, j) and $\mathcal{P}^{[2]}$ splits as the direct sum of its full subcategories $\mathcal{P}^{[2]}_{i,j}$ of homogeneous bifunctors of bidegree (i, j). If $B \in \mathcal{P}^{[2]}$, we denote by $B_{*,j}$ the direct sum

$$B_{*,j} = \bigoplus_{i \geq 0} B_{i,j}. \tag{7}$$

Note that we have also a duality for bifunctors

$$\sharp : \mathcal{P}^{[2]} \xrightarrow{\simeq} \mathcal{P}^{[2]\,\mathrm{op}},$$

which sends B to $B^\sharp$, with $B^\sharp(V, W) = B(V^*, W^*)^*$, and which respects the bidegrees (the same argument as in the previous section for usual polynomial functors works also in the bi-functor case).

The generalization of these ideas to the category of strict polynomial *tri-functors* of finite degree $\mathcal{P}^{[3]}$, which contains the tri-functors $M^{[3]} : (U, V, W) \mapsto M(U \oplus V \oplus W)$, and so on, is straightforward.

We conclude this section by introducing a construction of new functors in $\mathcal{P}$ from old ones that will be used in the next section. Let $M \in \mathcal{P}$ and consider the functor $M^{[2]}(\cdot, k) \in \mathcal{P}$. By (7) we have a decomposition

$$M^{[2]}(\cdot, k) = \bigoplus_{i \geq 0} M^{[2]}_{*,i}(\cdot, k).$$

In other words, $M^{[2]}_{*,i}(V, k)$ is the subspace of weight i of $M(V \oplus k)$ acted on by $GL(k)$ via the composition

$$GL(k) = 1_V \times GL(k) \hookrightarrow GL(V \oplus k) \xrightarrow{\pi_{M,V \oplus k}} GL(M(V \oplus k)).$$

Since evaluation on $V \oplus k$ as well as taking weight spaces are exact, the assignment $M \mapsto M^{[2]}_{*,i}(\cdot, k)$ defines an exact endo-functor on $\mathcal{P}$. Hence it descends to an operator on Grothendieck groups.

5 Categorification data

Having defined the notion of categorical $\mathfrak{g}$-action and the category $\mathcal{P}$, we are now ready to begin the task of defining a categorical $\mathfrak{g}$-action on $\mathcal{P}$. The present section is devoted to introducing the necessary data to construct the categorification (cf. items (1)–(3) of Definition 2. The following section will be devoted to showing that this data satisfies the required properties (cf. items (4)–(8) of Definition 2).

5.1 *The functors* E *and* F

Define $\mathbf{E}, \mathbf{F} : \mathcal{P} \to \mathcal{P}$ by

$$\mathbf{E}(M) = M^{[2]}_{*,1}(\cdot, k)$$
$$\mathbf{F}(M) = M \otimes I$$

for $M \in \mathcal{P}$. These are exact functors ($\mathbf{F}$ is clearly exact; for the exactness of $\mathbf{E}$ see the last paragraph of Section 4.4). We prove that $\mathbf{E}$ and $\mathbf{F}$ are bi-adjoint.

Proposition 3. *The pair* $(\mathbf{F}, \mathbf{E})$ *is an adjoint pair, i.e., we have an isomorphism, natural with respect to* $M, N \in \mathcal{P}$:

$$\beta : \mathrm{Hom}_{\mathcal{P}}(\mathbf{F}(M), N) \simeq \mathrm{Hom}_{\mathcal{P}}(M, \mathbf{E}(N)).$$

Proof. We shall use the category $\mathcal{P}^{[2]}$ of strict polynomial bifunctors. There are functors:

$$\boxtimes : \mathcal{P} \times \mathcal{P} \to \mathcal{P}^{[2]} \qquad \otimes : \mathcal{P} \times \mathcal{P} \to \mathcal{P}$$
$$\Delta : \mathcal{P}^{[2]} \to \mathcal{P} \qquad {}^{[2]} : \mathcal{P} \to \mathcal{P}^{[2]}$$

respectively given by

$$M \boxtimes N(V, W) = M(V) \otimes N(W) \qquad M \otimes N(V) = M(V) \otimes N(V)$$
$$\Delta B(V) = B(V, V) \qquad M^{[2]}(V, W) = M(V \oplus W).$$

We observe that $\Delta(M \boxtimes N) = M \otimes N$. Moreover, we know (cf. [FFSS, Proof of Thm 1.7] or [T1, Lm 5.8]) that Δ and ${}^{[2]}$ are bi-adjoint.

Now we are ready to establish the existence of the adjunction isomorphism. We have the following natural isomorphisms:

$$\begin{aligned}\mathrm{Hom}_{\mathcal{P}}(\mathbf{F}(M), N) &= \mathrm{Hom}_{\mathcal{P}}(M \otimes I, N)\\ &\simeq \mathrm{Hom}_{\mathcal{P}^{[2]}}(M \boxtimes I, N^{[2]})\\ &\simeq \mathrm{Hom}_{\mathcal{P}}(M(\cdot), \mathrm{Hom}_{\mathcal{P}}(I(*), N(\cdot \oplus *))).\end{aligned}$$

Here $\mathrm{Hom}_{\mathcal{P}}(I(*), N(\cdot \oplus *))$ denotes the polynomial functor which assigns to $V \in \mathcal{V}$ the vector space $\mathrm{Hom}_{\mathcal{P}}(I, N(V \oplus *))$. By Yoneda's Lemma [FS, Thm 2.10], for any $F \in \mathcal{P}$, $\mathrm{Hom}_{\mathcal{P}}(I, F) \simeq F(k)$ if F is of degree one, and zero otherwise. In particular, $\mathrm{Hom}_{\mathcal{P}}(I, N(V \oplus *)) \simeq N(V \oplus k)_1 = \mathbf{E}(N)(V)$. Hence, $\mathrm{Hom}_{\mathcal{P}}(I(*), N(\cdot \oplus *)) \simeq \mathbf{E}(N)$ and we conclude that there is a natural isomorphism:

$$\mathrm{Hom}_{\mathcal{P}}(\mathbf{F}(M), N) \simeq \mathrm{Hom}_{\mathcal{P}}(M, \mathbf{E}(N)).$$ □

We are now going to derive the adjunction $(\mathbf{E}, \mathbf{F})$ from proposition 3 and a duality argument. The following lemma is an easy check.

Lemma 1. *For all $M \in \mathcal{P}$, we have isomorphisms, natural with respect to M:*

$$\mathbf{F}(M)^\sharp \simeq \mathbf{F}(M^\sharp)\ ,\ \mathbf{E}(M)^\sharp \simeq \mathbf{E}(M^\sharp).$$

Proof. We have an isomorphism:

$$\mathbf{F}(M)^\sharp = (M \otimes I)^\sharp \simeq M^\sharp \otimes I^\sharp = \mathbf{F}(M^\sharp)\ ,$$

and a chain of isomorphisms:

$$\begin{aligned}\mathbf{E}(M)^\sharp = \left(M^{[2]}_{*,1}(\cdot,k)\right)^\sharp &\simeq (M^{[2]}_{*,1})^\sharp(\cdot,k)\\ &\simeq (M^{[2]\sharp})_{*,1}(\cdot,k)\\ &\simeq (M^\sharp)^{[2]}_{*,1}(\cdot,k) = \mathbf{E}(M^\sharp).\end{aligned}$$

In the chain of isomorphisms, the first isomorphism follows from the isomorphism of vector spaces $k^\vee \simeq k$, the second follows from the fact that duality preserves bidegrees, and the last from the fact that duality of vector spaces commutes with direct sums.

Proposition 4. *The pair $(\mathbf{E},\mathbf{F})$ is an adjoint pair, i.e., we have an isomorphism, natural with respect to $M, N \in \mathcal{P}$:*

$$\alpha : \mathrm{Hom}_{\mathcal{P}}(\mathbf{E}(M), N) \simeq \mathrm{Hom}_{\mathcal{P}}(M, \mathbf{F}(N)).$$

Proof. The adjunction isomorphism of proposition 4 is defined as the composite of the natural isomorphisms:

$$\begin{aligned}&\mathrm{Hom}_{\mathcal{P}}(\mathbf{E}(M), N) \simeq \mathrm{Hom}_{\mathcal{P}}(N^\sharp, \mathbf{E}(M)^\sharp) \simeq \mathrm{Hom}_{\mathcal{P}}(N^\sharp, \mathbf{E}(M^\sharp))\\ &\simeq \mathrm{Hom}_{\mathcal{P}}(\mathbf{F}(N^\sharp), M^\sharp) \simeq \mathrm{Hom}_{\mathcal{P}}(\mathbf{F}(N)^\sharp, M^\sharp) \simeq \mathrm{Hom}_{\mathcal{P}}(M, \mathbf{F}(N)).\end{aligned}$$

□

Remark 6. The unit and counits of the adjunctions appearing in Propositions 3,4 are implicit from the canonical isomorphisms. For an explicit description see [HY2].

5.2 *The operators X and T*

We first introduce the natural transformation $X : \mathbf{E} \to \mathbf{E}$. We assume that $p \neq 2$. For any $V \in \mathcal{V}$, let $U(\mathfrak{gl}(V \oplus k))$ denote the enveloping algebra of $\mathfrak{gl}(V \oplus k)$, and let $X_V \in U(\mathfrak{gl}(V \oplus k))$ be defined as follows. Fix a basis $V = \bigoplus_{i=1}^n ke_i$; this choice induces a basis of $V \oplus k$. Let $x_{i,j} \in \mathfrak{gl}(V \oplus k)$ be the operator mapping e_j to e_i and e_ℓ to zero for all $\ell \neq j$. Then define

$$X_V = \sum_{i=1}^n x_{n+1,i}x_{i,n+1} - n.$$

The element X_V does not depend on the choice of basis. (For a proof of this see Lemma 3.27 in [HY]. Note also the similarity to constructions which appear in [BS1, BS2]. We also remark that this is where the hypothesis that $p \neq 2$ is used.)

The group $GL(V) \times GL(k) \subset GL(V \oplus k)$ acts on the Lie algebra $\mathfrak{gl}(V \oplus k)$ by the adjoint action, hence on the algebra $U(\mathfrak{gl}(V \oplus k)$. By Lemma 4.22 in [HY] we have:

Lemma 2. *Let $V \in \mathcal{V}$. Then X_V commutes with $GL(V) \times GL(k)$, i.e.,*

$$X_V \in U(\mathfrak{gl}(V \oplus k))^{GL(V)\times GL(k)}.$$

The universal enveloping algebra $U(\mathfrak{gl}(V \oplus k))$ acts on $M(V \oplus k)$ via differentiation:

$$d\pi_{M,V} : U(\mathfrak{gl}(V \oplus k)) \to \operatorname{End}(M(V \oplus k)).$$

Exmaple 2. If $M = I$ is the identity functor of $\mathcal{V}$, and $f \in \mathfrak{gl}(V \oplus k)$, then $d\pi_{I,V\oplus k}(f) = f$. More generally, if $d \geq 2$ and $M = \otimes^d$ is the d-th tensor product, then $d\pi_{\otimes^d, V\oplus k}$ sends $f \in \mathfrak{gl}(V \oplus k)$ onto the element

$$\sum_{i=1}^{d} (1_{V\oplus k})^{\otimes i-1} \otimes f \otimes (1_{V\oplus k})^{\otimes d-i} \quad \in \operatorname{End}((V \oplus k)^{\otimes d}).$$

The element X_V acts on the vector space $M(V \oplus k)$ via $d\pi_{M,V}$, and we denote by $X_{M,V}$ the induced k-linear map:

$$X_{M,V} : M(V \oplus k) \to M(V \oplus k).$$

By Lemma 2, $X_{M,V}$ is $GL(V)\times GL(k)$-equivariant. Thus it restricts to the subspaces $\mathbf{E}(M)(V)$ of weight 1 under the action of $\{1_V\} \times GL(k)$. We denote the resulting map also by $X_{M,V}$:

$$X_{M,V} : \mathbf{E}(M)(V) \to \mathbf{E}(M)(V).$$

Proposition 5. *The linear maps $X_{M,V} : \mathbf{E}(M)(V) \to \mathbf{E}(M)(V)$ are natural with respect to M and V. Hence they define a morphism of functors*

$$X : \mathbf{E} \to \mathbf{E}.$$

Proof. The action of $U(\mathfrak{gl}(V \oplus k))$ on $M(V \oplus k)$ is natural with respect to M. Hence the k-linear maps $X_{M,V}$ are natural with respect to M.

So it remains to check the naturality with respect to $V \in \mathcal{V}$. For this, it suffices to check that for all $M \in \mathcal{P}$, and for all $f \in \operatorname{Hom}(V, W)$, diagram (D) below is commutative.

$$\begin{array}{ccc} M(V \oplus k) & \xrightarrow{M(f \oplus 1_k)} & M(W \oplus k) \\ \Big\downarrow x_{V,M} & & \Big\downarrow x_{W,M} \\ M(V \oplus k) & \xrightarrow{M(f \oplus 1_k)} & M(W \oplus k). \end{array} \qquad (D)$$

We observe that if diagram (D) commutes for a given strict polynomial functor M, then by naturality with respect to M, it also commutes for direct sums $M^{\oplus n}$, for $n \geq 1$, for the subfunctors $N \subset M$ and the quotients $M \twoheadrightarrow N$. But as we already explained in Remark 1, every functor $M \in \mathcal{P}$ is a subquotient of a finite direct sum of copies of the tensor product functors $\otimes^d$, for $d \geq 0$. Thus, to prove naturality with respect to V, it suffices to check that diagram (D) commutes for $M = \otimes^d$ for all $d \geq 0$.

In the case of the tensor products $\otimes^d$ the action of $U(\mathfrak{gl}(V \oplus k)$ is explicitly given in Example 2. Using this expression, a straightforward computation shows that diagram (D) is commutative in this case. This finishes the proof. □

We next introduce a natural transformation $T : \mathbf{E}^2 \to \mathbf{E}^2$. Let $M \in \mathcal{P}$ and $V \in \mathcal{V}$. By definition,

$$\mathbf{E}^2(M) = M^{[3]}_{*,1,1}(\cdot, k, k).$$

Consider the map $1_V \oplus \sigma : V \oplus k \oplus k \to V \oplus k \oplus k$ given by: $(v, a, b) \mapsto (v, b, a)$. Applying $M^{[3]}$ to this map we obtain a morphism:

$$T_{M,V} : M^{[3]}_{*,1,1}(V, k, k) \to M^{[3]}_{*,1,1}(V, k, k).$$

Lemma 3. *The linear maps $T_{M,V} : \mathbf{E}^2(M)(V) \to \mathbf{E}^2(M)(V)$ are natural with respect to M and V. Hence they define a morphism of functors*

$$T : \mathbf{E}^2 \to \mathbf{E}^2.$$

Proof. Clearly the maps $T_{M,V}$ are natural with respect to M. Let $f : V \to W$ be a linear operator of vector spaces. We need to show that the following diagram commutes:

$$\begin{array}{ccc} \mathbf{E}^2(M)(V) & \xrightarrow{\mathbf{E}^2(M)(f)} & \mathbf{E}^2(M)(W) \\ \Big\downarrow T_{M,V} & & \Big\downarrow T_{M,W} \\ \mathbf{E}^2(M)(V) & \xrightarrow[\mathbf{E}^2(M)(f)]{} & \mathbf{E}^2(M)(W) \end{array}.$$

On the one hand, $\mathbf{E}^2(M)(f)$ is the restriction of $M^{[3]}(f \oplus 1_k \oplus 1_k)$ to the tri-degrees $(*, 1, 1)$. On the other hand, $T_{M,V}$ is the restriction of $M^{[3]}(1_V \oplus \sigma)$ to the tri-degrees $(*, 1, 1)$. Since $f \oplus 1_k \oplus 1_k$ clearly commutes with $1_V \oplus \sigma$, the above diagram commutes. □

5.3 The weight decomposition of $\mathcal{P}$

As part of the data of categorical $\mathfrak{g}$-action, we need to introduce a decomposition of $\mathcal{P}$ indexed by the weight lattice P of $\mathfrak{g}$. In this section we define such a decomposition via the blocks of $\mathcal{P}$.

We begin by recalling some combinatorial notions. For a nonnegative integer d, let $\wp_d$ denote the set of partitions of d. A partition λ is a *p-core* if there exist no $\mu \subset \lambda$ such that the skew-partition λ/μ is a rim p-hook. By definition, if $p = 0$, then all partitions are p-cores. Given a partition λ, we denote by $\widetilde{\lambda}$ the p-core obtained by successively removing all rim p-hooks. For instance, the 3-core of $(6, 5, 2)$ is $3, 1)$. The *p-weight* of λ is by definition the number $(|\lambda| - |\widetilde{\lambda}|)/p$. The notation $|\lambda|$ denotes the size of the partition λ. Define an equivalence relation $\sim$ on $\wp_d$ by decreeing $\lambda \sim \mu$ if $\widetilde{\lambda} = \widetilde{\mu}$.

Let $\lambda, \mu \in \wp_d$. As a consequence of (11.6) in [Kl] we have

$$\widetilde{\lambda} = \widetilde{\mu} \Longleftrightarrow wt(\lambda) = wt(\mu). \tag{8}$$

(See (1) for the definition of $wt(\lambda)$.) Therefore we index the set of equivalence classes $\wp_d/\sim$ by weights in P, i.e., a weight $\omega \in P$ corresponds to a subset (possibly empty) of $\wp_d$. For a more explicit description of the bijection which associates to a weight of Fock space a p-core partition; see Section 2 of [LM].

Let $\operatorname{Irr}\mathcal{P}_d$ denote the set of simple objects in $\mathcal{P}_d$ up to isomorphism. This set is naturally identified with $\wp_d$. We say two simple objects in $\mathcal{P}_d$ are *adjacent* if they occur as composition factors of some indecomposable object in $\mathcal{P}_d$. Consider the equivalence relation $\approx$ on $\operatorname{Irr}\mathcal{P}_d$ generated by adjacency. Via the identification of $\operatorname{Irr}\mathcal{P}_d$ with $\wp_d$ we obtain an equivalence relation $\approx$ on $\wp_d$.

Theorem 3 (Theorem 2.12, [D]). *The equivalence relations $\sim$ and $\approx$ on $\wp_d$ are the same.*

Given an equivalence class $\Theta \in \operatorname{Irr}\mathcal{P}_d/\approx$, the corresponding *block* $\mathcal{P}_\Theta \subset \mathcal{P}_d$ is the subcategory of objects whose composition factors belong to Θ. The *block decomposition* of $\mathcal{P}$ is given by $\mathcal{P} = \bigoplus \mathcal{P}_\Theta$, where Θ ranges over all classes in $\operatorname{Irr}\mathcal{P}_d/\approx$ and $d \geq 0$.

By the above theorem and Equation (8), we can label the blocks of $\mathcal{P}_d$ by weights $\omega \in P$. Moreover, by Equation (1), $wt(\lambda)$ determines the size of λ. Therefore the block decomposition of $\mathcal{P}$ can be expressed as

$$\mathcal{P} = \bigoplus_{\omega \in P} \mathcal{P}_\omega.$$

The *p-weight* of a block $\mathcal{P}_\omega$ is the p-weight of λ, where $\omega = wt(\lambda)$. This is well-defined since if $wt(\lambda) = wt(\mu)$ then $|\lambda| = |\mu|$ and $\widetilde{\lambda} = \widetilde{\mu}$, and hence the p-weights of λ and μ agree.

6 Categorification of Fock space

In the previous section we defined all the data necessary to formulate the action $\mathfrak{g}$ on $\mathcal{P}$. In this section we prove the main theorem:

Theorem 4. *Suppose $p \neq 2$. The category $\mathcal{P}$ along with the data of adjoint functors $\mathbf{E}$ and $\mathbf{F}$, operators $X \in \mathrm{End}(\mathbf{E})$ and $T \in \mathrm{End}(\mathbf{E}^2)$, and the weight decomposition $\mathcal{P} = \bigoplus_{\omega\in P} \mathcal{P}_\omega$ defines a categorical $\mathfrak{g}$-action (in the sense of Definition 2) which categorifies the Fock space representation of $\mathfrak{g}$.*

Remark 7. The theorem is still true for $p = 2$. We only include this hypothesis for ease of exposition (one can prove the $p = 2$ case using hyperalgebras instead of enveloping algebras).

To prove this theorem we must show that the data satisfies properties (4)–(6), (8) of Definition 2, and that the resulting representation of $\mathfrak{g}$ on $K(\mathcal{P})$ is isomorphic to the Fock space representation (property (7) already appears as Proposition 3).

6.1 *The functors* $\mathbf{E}_i$

In this section we prove property (4) of Definition 2. For all $a \in k$, and $M \in \mathcal{P}$ we can form a nested collection of subspaces of $\mathbf{E}(M)$, natural with respect to M:

$$0 \subset \mathbf{E}_{a,1}(M) \subset \mathbf{E}_{a,2}(M) \subset \cdots \subset \mathbf{E}_{a,n}(M) \subset \cdots \subset \mathbf{E}(M),$$

where $\mathbf{E}_{a,n}(M)$ is the kernel of $(X_M - a)^n : \mathbf{E}(M) \to \mathbf{E}(M)$. We define

$$\mathbf{E}_a(M) = \bigcup_{n\geq 0} \mathbf{E}_{a,n}(M).$$

Since the inclusions $\mathbf{E}_{a,n}(M) \subset \mathbf{E}_{a,n+1}(M)$ are natural with respect to M, the assignment $M \mapsto \mathbf{E}_a(M)$ defines a sub-endofunctor of $\mathbf{E}$.

Lemma 4. *The endofunctor $\mathbf{E} : \mathcal{P} \to \mathcal{P}$ splits as a direct sum of its subfunctors $\mathbf{E}_a$:*

$$\mathbf{E} = \bigoplus_{a\in k} \mathbf{E}_a.$$

Moreover, for all $M \in \mathcal{P}$ there exists an integer N such that for all $n \geq N$, $\mathbf{E}_a(M) = \mathbf{E}_{a,n}(M)$.

Proof. The decomposition as a direct summand of generalized eigenspaces is standard linear algebra. The finiteness of the filtration $(\mathbf{E}_{a,n}(M))_{n\geq 0}$ follows from Corollary 1. □

Proposition 6. *Let $\lambda \in \wp$ be a partition of d and set $W = W_\lambda$.*

(i) *The polynomial functor $\mathbf{E}(W)$ carries a Weyl filtration:*

$$0 = \mathbf{E}(W)^0 \subset \mathbf{E}(W)^1 \subset \cdots \subset \mathbf{E}(W)^N = \mathbf{E}(W).$$

The composition factors which occur in this filtration are isomorphic to W_μ for all μ such that $\mu \longrightarrow \lambda$ and each such factor occurs exactly once.

(ii) *The operator $X_W : \mathbf{E}(W) \to \mathbf{E}(W)$ preserves the filtration of $\mathbf{E}(W)$, and hence it acts on the associated graded object.*

(iii) *Given $0 \leq i \leq N-1$, set $j \in \mathbb{Z}/p\mathbb{Z}$ and $\mu \in \wp$ such that $\mathbf{E}(W)^{i+1}/\mathbf{E}(W)^i \simeq W_\mu$, and $\mu \xrightarrow{j} \lambda$. Then X_W acts on $\mathbf{E}(W)^{i+1}/\mathbf{E}(W)^i$ by multiplication by j.*

In particular $\mathbf{E}_a = 0$ for $a \notin \mathbb{Z}/p\mathbb{Z}$, and hence

$$\mathbf{E} = \bigoplus_{i\in\mathbb{Z}/p\mathbb{Z}} \mathbf{E}_i .$$

Proof. Theorem II.4.11 of [ABW] yields a filtration of the bifunctor $S_\lambda^{[2]}$ with associated graded object $\bigoplus_{\alpha\subset\lambda} S_\alpha \boxtimes S_{\lambda/\alpha}$. Here, $S_{\lambda/\alpha} \in \mathcal{P}_{|\lambda|-|\alpha|}$ refers to the Schur functor associated to the skew partition λ/α and $S_\alpha \boxtimes S_{\lambda/\alpha}$ is the homogeneous bifunctor of bidegree $(|\alpha|, |\lambda| - |\alpha|)$, defined by $(V, U) \mapsto S_\alpha(V) \otimes S_{\lambda/\alpha}(U)$. Thus $(S_\lambda^{[2]})_{*,1}$ has a filtration whose graded object is the sum of the $S_\alpha \boxtimes S_{\lambda/\alpha}$ with $|\lambda| = |\alpha| + 1$. In this case, $S_{\lambda/\alpha}$ is the identity functor of $\mathcal{V}$ by definition. Thus taking $U = k$, we get a filtration of $\mathbf{E}(S_\lambda)$ whose graded object is $\bigoplus S_\alpha$, for all $\alpha \to \lambda$. The first part of the proposition follows by duality $^\sharp$. (For an alternative proof based on [Mar] and [GW, Thm. 8.1.1], see [HY, Lemma A.3].)

For any $V \in \mathcal{V}$, by Lemma 4.22 in [HY] the map $X_{W,V}$ preserves the filtration of $GL(V)$-modules:

$$0 = \mathbf{E}(W)^0(V) \subset \mathbf{E}(W)^1(V) \subset \cdots \subset \mathbf{E}(W)^N(V) = \mathbf{E}(W)(V).$$

Indeed, since Weyl modules are highest weight modules, $C_{V\oplus k}$ acts on

$$W(V \oplus k)$$

by scalar, and C_V acts on the factors of the filtration by scalar as well. Therefore X_W preserves the filtration of $\mathbf{E}(W)$, proving the second part of the proposition.

Finally, let μ and j be chosen as in the third part of the proposition. By Lemma 5.7(1) in [HY], for any $V \in \mathcal{V}$, $X_{W,V}$ acts by j on $\mathbf{E}(W)^{i+1}(V)/\mathbf{E}(W)^i(V)$. Therefore X_W acts on $\mathbf{E}(W)^{i+1}/\mathbf{E}(W)^i$ also by j. □

By the adjunction of $\mathbf{E}$ and $\mathbf{F}$ and the Yoneda Lemma, the operator $X \in \mathrm{End}(\mathbf{E})$ induces an operator $X^\circ \in \mathrm{End}(\mathbf{F})$. The generalized eigenspaces of this operator produce subfunctors $\mathbf{F}_a$ of $\mathbf{F}$, which, by general nonsense, are adjoint to $\mathbf{E}_a$. Therefore we have decompositions

$$\mathbf{E} = \bigoplus_{i \in \mathbb{Z}/p\mathbb{Z}} \mathbf{E}_i\,, \mathbf{F} = \bigoplus_{i \in \mathbb{Z}/p\mathbb{Z}} \mathbf{F}_i\,.$$

6.2 The action of $\mathfrak{g}$ on $K(\mathcal{P})$

In this section we prove property (5) of Definition 2. The functors $\mathbf{E}_i, \mathbf{F}_i$, being exact functors, induce linear operators

$$[\mathbf{E}_i], [\mathbf{F}_i] : K(\mathcal{P}) \to K(\mathcal{P})$$

for all $i \in \mathbb{Z}/p\mathbb{Z}$. Define a map $\varkappa : \mathfrak{g} \to \mathrm{End}(K(\mathcal{P}))$ by $e_i \mapsto [\mathbf{E}_i]$ and $f_i \mapsto [\mathbf{F}_i]$. Let $\Psi : K(\mathcal{P}) \to B$ be given by $\Psi([W_\lambda]) = v_\lambda$.

Proposition 7. *The map $\varkappa$ is a representation of $\mathfrak{g}$ and Ψ is an isomorphism of $\mathfrak{g}$-modules.*

Proof. By Corollary 2 Ψ is a linear isomorphism. By Proposition 6,

$$[\mathbf{E}_i]([W_\lambda]) = \sum_{\mu \xrightarrow{i} \lambda} [W_\mu].$$

Therefore Ψ intertwines e_i and $[\mathbf{E}_i]$, i.e., $\Psi \circ [\mathbf{E}_i] = e_i \circ \Psi$. Consider the bilinear form on $K(\mathcal{P})$ given by

$$\langle M, N \rangle = \sum_{i \geq 0} (-1)^i \dim \mathrm{Ext}^i(M, N).$$

By adjunction $[\mathbf{E}_i]$ and $[\mathbf{F}_i]$ are adjoint operators with respect to $\langle \cdot, \cdot \rangle$, and by Theorem 2(iv), $\langle W_\lambda, S_\mu \rangle = \delta_{\lambda\mu}$. Therefore

$$[\mathbf{F}_i]([S_\lambda]) = \sum_{\lambda \xrightarrow{i} \mu} [S_\mu].$$

Hence by Corollary 3, Ψ also intertwines the operators f_i and $[\mathbf{F}_i]$. Both claims of the proposition immediately follow. □

6.3 Chevalley functors and weight decomposition of $\mathcal{P}$

In this section we prove property (6) of Definition 2.

Proposition 8. *Let $\omega \in P$. For every $i \in \mathbb{Z}/p\mathbb{Z}$, the functors $\mathbf{E}_i, \mathbf{F}_i : \mathcal{P} \to \mathcal{P}$ restrict to $\mathbf{E}_i : \mathcal{P}_\omega \to \mathcal{P}_{\omega+\alpha_i}$ and $\mathbf{F}_i : \mathcal{P}_\omega \to \mathcal{P}_{\omega-\alpha_i}$.*

Proof. We prove that $\mathbf{E}_i(\mathcal{P}_\omega) \subset \mathcal{P}_{\omega+\alpha_i}$ (the proof for $\mathbf{F}_i$ being entirely analogous). Since $\mathbf{E}_i$ is exact it suffices to prove that if $L_\lambda \in \mathcal{P}_\omega$, then $\mathbf{E}_i(L_\lambda) \in \mathcal{P}_{\omega+\alpha_i}$. Then, by the same idea as used in the proof of Lemma 2, it suffices to show that if $W_\lambda \in \mathcal{P}_\omega$, then $\mathbf{E}_i(W_\lambda) \in \mathcal{P}_{\omega+\alpha_i}$. By Proposition 6, $\mathbf{E}_i(W_\lambda)$ has a Weyl filtration with factors all of the form W_μ, where $\mu \xrightarrow{i} \lambda$. But then $\mu \in \omega + \alpha_i$, so $W_\mu \in \mathcal{P}_{\omega+\alpha_i}$. Therefore $\mathbf{E}_i(W_\lambda) \in \mathcal{P}_{\omega+\alpha_i}$. □

6.4 The degenerate affine Hecke algebra action on $\mathbf{E}^n$

In this section we prove property (8) of Definition 2.

Proposition 9. *The assignments*

$$y_i \mapsto \mathbf{E}^{n-i} X \mathbf{E}^{i-1} \text{ for } 1 \le i \le n,$$

$$\tau_i \mapsto \mathbf{E}^{n-i-1} T \mathbf{E}^{i-1} \text{ for } 1 \le i \le n-1$$

define an action of DH_n on End($\mathbf{E}^n$).

Proof. By definition, $\mathbf{E}^n(M)(V)$ is the subspace of $M(V \oplus k^n)$ formed by the vectors of weight $\varpi_n = (1, 1, \dots, 1)$ for the action of $GL(k)^{\times n}$. Here $GL(k)^{\times n}$ acts via the composition:

$$GL(k)^{\times n} = 1_V \times GL(k)^{\times n} \subset GL(V \oplus k^n) \xrightarrow{\pi_{M,V \oplus k^n}} GL(M(V \oplus k^n)).$$

The map $(\tau_{n-i})_{M,V}$ is equal to the restriction of $M(t_i)$ to $\mathbf{E}^n(M)(V)$, where $t_i : V \oplus k^n \to V \oplus k^n$ maps $(v, x_1, \dots, x_n)$ to $(v, x_1, \dots, x_{i+1}, x_i, \dots, x_n)$. To check that the τ_i define an action of $\mathbb{Z}\mathfrak{S}_n$ on $\mathbf{E}^n$, we need to check that the $(\tau_i)_{M,V}$ define an action of the symmetric group on $\mathbf{E}^n(M)(V)$. By Remark 1 it suffices to check this for $M = \otimes^d$, and this is a straightforward computation. Moreover, it is also straightforward from the definition that the y_i commute with each other. Thus they define an action of the polynomial algebra $\mathbb{Z}[y_1, \dots, y_n]$ on $\mathbf{E}^n$. Similarly, τ_i and y_j commute with each other if $|i - j| \ge 1$.

So, to obtain the action of the Hecke algebra on End($\mathbf{E}^n$), it remains to show that $\tau_i y_i - y_{i+1}\tau_i = 1$ (see Remark 2). This will be proved by showing the following identity in End($\mathbf{E}^2$):

$$T \circ \mathbf{E}X - X\mathbf{E} \circ T = 1. \tag{9}$$

To check (9), it suffices to check that for all $M \in \mathcal{P}$ and all $V \in \mathcal{V}$,

$$T_{M,V} \circ \mathbf{E}(X_M)_V - X_{\mathbf{E}(M)} \circ T_{M,V} = 1_{\mathbf{E}^2(M)(V)} \tag{10}$$

If (10) holds for $M \in \mathcal{P}$, then by naturality with respect to M, it also holds for direct sums $M^{\oplus n}$, for subfunctors $N \subset M$, and quotients $M \twoheadrightarrow N$. By Remark 1, every functor $M \in \mathcal{P}$ is a subquotient of a finite direct sum of copies of the tensor product functors $\otimes^d$, for $d \geq 0$. Thus it suffices to check that Equation (10) holds for $M = \otimes^d$ for all $d \geq 0$.

Let $M = \otimes^d$ and let $V \in \mathcal{V}$. Choose a basis $(e_1, \ldots, e_n)$ of V. We naturally extend this to a basis $(e_1, \ldots, e_{n+2})$ of $V \oplus k \oplus k$. By definition, $\mathbf{E}^2(\otimes^d)(V)$ is the subspace of $(V \oplus k \oplus k)^{\otimes d}$ spanned by the vectors of the form $e_{i_1} \otimes \cdots \otimes e_{i_d}$, where exactly one of the e_{i_k} equals e_{n+1} and exactly one of the e_{i_k} equals e_{n+2}. Let us fix a vector $\xi = e_{i_1} \otimes \cdots \otimes e_{i_d}$ with e_{n+1} in a-th position and e_{n+2} in b-th position. We will show that Equation (10) holds for ξ.

First, note that $T_{M,V}(\xi) = e_{i_{(ab)(1)}} \otimes \cdots \otimes e_{i_{(ab)(d)}}$, where (ab) denotes the transposition of $\mathfrak{S}_d$ which exchanges a and b. Then

$$\begin{aligned}(X\mathbf{E})_{M,V} \circ T_{M,V}(\xi) &= \left(\sum_{j=1}^{n} x_{n+1,j} x_{j,n+1} - n\right).(e_{i_{(ab)(1)}} \otimes \cdots \otimes e_{i_{(ab)(d)}}) \\ &= \sum_{\ell \neq a,b} e_{i_{(\ell ba)(1)}} \otimes \cdots \otimes e_{i_{(\ell ba)(d)}}.\end{aligned}$$

Now we compute the other term on the left hand side of (10). Then

$$\begin{aligned}T_{M,V} \circ (\mathbf{E}X)_{M,V}(\xi) &= T_{M,V} \circ \left(\sum_{j=1}^{n+1} x_{n+2,j} x_{j,n+2} - (n+1)\right)(\xi) \\ &= \sum_{\ell \neq a,b} e_{i_{(\ell ba)(1)}} \otimes \cdots \otimes e_{i_{(\ell ba)(d)}} + \xi.\end{aligned}$$

Therefore (10) holds.

This completes the proof of Theorem 4. □

7 Remarks

We conclude the paper by mentioning briefly some consequences of the categorical $\mathfrak{g}$-action on $\mathcal{P}$.

7.1 Derived equivalences

For this discussion we focus on the case $p = \text{char}(k) > 0$. The main motivation for Chuang and Rouquier's original work on categorification was to prove Broué's abelian defect conjecture for the symmetric groups, which can be reduced to showing that any two blocks of symmetric groups of the same p-weight are derived equivalent [CR]. Their technique applies to the setting of $\mathfrak{sl}_2$-categorifications. Since for every simple root α of $\mathfrak{g}$ there is a corresponding root subalgebra of $\mathfrak{g}$ isomorphic to $\mathfrak{sl}_2$, we have in fact defined a family of $\mathfrak{sl}_2$-categorifications on $\mathcal{P}$. To each of these categorifications we can apply the Chuang–Rouqueir machinery.

Let $W^{\text{aff}} = \mathfrak{S}_p \ltimes Q$ denote the affine Weyl group associated to $\mathfrak{g}$, acting on P in the usual way. By [Kac, Section 12], any weight ω appearing in the weight decomposition of Fock space is of the form $\sigma(\omega_0) - \ell\delta$, where $\sigma \in W^{\text{aff}}$ and $\ell \geq 0$. By Proposition 11.1.5 in [Kl], ℓ is exactly the p-weight of the corresponding block. Therefore the weights of any two blocks are conjugate by some element of affine Weyl group if and only if they have the same p-weight. By Theorem 6.4 in [CR] we obtain

Theorem 5. *If two blocks of $\mathcal{P}$ have the same p-weight, then they are derived equivalent.*

7.2 Misra–Miwa crystal

We can also apply the theory of categorical $\mathfrak{g}$-action to crystal basis theory. The crystal structure is a combinatorial structure associated to integrable representations of Kac–Moody algebras, introduced originally by Kashiwara via the theory of quantum groups. From Kashiwara's theory one can construct a canonical basis for the corresponding representations, which agrees with Lusztig's canonical basis of geometric origins.

Loosely speaking, the crystal structure of an integrable representation of some Kac–Moody algebra consists of a set $\mathbb{B}$ in bijection with a basis of the representation, along with Kashiwara operators $\widetilde{e}_i, \widetilde{f}_i$ on $\mathbb{B}$ indexed by the simple roots of the Kac–Moody algebra, along with further data. For a precise definition see [Kas].

From the categorical $\mathfrak{g}$-action on $\mathcal{P}$ we can recover the crystal structure of Fock space as follows. For the set $\mathbb{B}$ we take $\operatorname{Irr}\mathcal{P} \subset K(\mathcal{P})$, the set of equivalence classes of simple objects. We construct Kashiwara operators on $\operatorname{Irr}\mathcal{P}$ by composing the

Chevalley functors with the socle functor:

$$\widetilde{e}_i, \widetilde{f}_i = [[\text{socle} \circ E_i], [\text{socle} \circ F_i] : \operatorname{Irr} \mathcal{P} \to \operatorname{Irr} \mathcal{P}.$$

The other data defining a crystal structure can also be naturally obtained. In Section 5.3 of [HY] it is shown that this data agrees with the crystal of B originally discovered by Misra and Miwa [MM]. In particular, we can construct the *crystal graph* of Fock space by taking the $\mathbb{Z}/p\mathbb{Z}$-colored directed graph whose vertices are $\operatorname{Irr} \mathcal{P}$ and edges are $\mu \xrightarrow{i} \lambda$ if $\widetilde{f}_i(\mu) = \lambda$. This graph is equal to the Misra–Miwa crystal of Fock space.

References

[A] S. Ariki, *Graded q-Schur algebras*, arXiv:0903.3453.

[ABW] K. Akin, D.A. Buchsbaum, J. Weyman, *Schur functors and Schur complexes*, Adv. Math. (1982), 207–278.

[BS1] J. Brundan, C. Stroppel, *Highest weight categories arising from Khovanov's diagram algebra IV: the general linear supergroup*, J. Eur. Math. Soc. (JEMS) 14 (2012), no. 2, 373–419.

[BS2] J. Brundan, C. Stroppel, *Highest weight categories arising from Khovanov's diagram algebra III: category $\mathcal{O}$*, Represent. Theory 15 (2011), 170–243.

[CR] J. Chuang, R. Rouquier, *Derived equivalences for symmetric groups and sl_2- categorification*, Ann. of Math. (2) 167 (2008), no. 1, 245–298.

[D] S. Donkin, *On Schur algebras and related algebras. II*, J. Algebra 111 (1987), no. 2, 354–364.

[FFSS] V. Franjou, E. Friedlander, A. Scorichenko, A. Suslin, *General linear and functor cohomology over finite fields*, Ann. of Math. (2) 150 (1999), no. 2, 663–728.

[FS] E. Friedlander, A. Suslin, *Cohomogy of finite group schemes over a field*, Invent. Math. 127 (1997), 209–270.

[GW] R. Goodman, N. R. Wallach, Symmetry, Representations, and Invariants, Graduate Texts in Mathematics, 255, Springer, Dordrecht, 2009.

[HY] J. Hong, O. Yacobi, *Polynomial representations of general linear groups and categorifications of Fock space*, Algebras and Representation Theory (2012), 1–39.

[HY2] J. Hong, J., O. Yacobi, *Polynomial functors and categorifications of Fock space II*, Adv. Math. 237 (2013), 360–403.

[J] J. C. Jantzen, Representations of Algebraic Groups. Second edition. Mathematical Surveys and Monographs, 107. American Mathematical Society, Providence, RI, 2003.

[Kac] V. G. Kac, Infinite-dimensional Lie algebras. Third edition. Cambridge University Press, Cambridge, 1990.

[Kas] M. Kashiwara, *Crystallizing the q-analogue of universal enveloping algebras*, Comm. Math. Phys. 133 (1990), no. 2, 249–260.

[KL] M. Khovanov, A. Lauda, *A diagrammatic approach to categorification of quantum groups I*, Represent. Theory 13 (2009), 309–347.

[KL2] M. Khovanov, A. Lauda, *A diagrammatic approach to categorification of quantum groups II*, Trans. Amer. Math. Soc. 363 (2011), 2685–2700.

[KL3] M. Khovanov, A. Lauda, *A diagrammatic approach to categorification of quantum groups III*, Quantum Topology, Vol. 1, Issue 1, 2010, 1–92.

[Kl] A. Kleshchev, Linear and Projective Representations of Symmetric Groups. Cambridge Tracts in Mathematics, 163. Cambridge University Press, Cambridge, 2005.

[Ku] N. Kuhn, *Rational cohomology and cohomological stability in generic representation theory*, Amer. J. Math. 120 (1998), 1317–1341.

[L] A. Lauda, *An introduction to diagrammatic algebra and categorified quantum* $\mathfrak{sl}(2)$, Bull. Inst. of Math. Acad. Sin., Vol 7 (2012), No. 2 165–270.

[LLT] A. Lascoux, B. Leclerc, J. Thibon, *Hecke algebras at roots of unity and crystal bases of quantum affine algebras*, Comm. Math. Phys. 181 (1996), no. 1, 205–263.

[LM] B. Leclerc, H. Miyachi, *Constructible characters and canonical bases*, J. Algebra 277 (2004), no. 1, 298–317.

[Ma] V. Mazorchuk, Lectures on Algebraic Categorification. QGM Master Class Series, EMS, Zurich, 2012, 199 pp.

[Mac] I. G. Macdonald, Symmetric Functions and Hall Polynomials. Second edition. With contributions by A. Zelevinsky. Oxford Mathematical Monographs. Oxford Science Publications. The Clarendon Press, Oxford University Press, New York, 1995.

[Mar] S. Martin, Schur Algebras and Representation Theory, Cambridge Tracts in Mathematics, 112, Cambridge, 1993.

[MM] K. C. Misra, T. Miwa, *Crystal base of the basic representation of* $U_q(\widehat{\mathfrak{sl}_n})$, Commun. Math. Phys 134 (1990), 79–88.

[R] R. Rouquier, *2-Kac–Moody algebras*, preprint, arXiv:0812.5023.

[SFB] A. Suslin, E. Friedlander, C. Bendel, *Infinitesimal* 1*-parameter subgroups and cohomology*, J. Amer. Math. Soc. 10 (1997), no. 3, 693–728.

[SW] C. Stroppel, B. Webster, *Quiver Schur algebras and* q*-Fock space*, arXiv: 1110.1115v1.

[T1] A. Touzé, *Cohomology of classical algebraic groups from the functorial viewpoint*, Adv. Math. 225 (2010), no. 1, 33–68.

[T2] A. Touzé, *Ringel duality and derivatives of non-additive functors*, J. Pure Appl. Algebra 217 (2013), no. 9, 1642–1673.

[TvdK] A. Touzé, W. van der Kallen, *Bifunctor cohomology and cohomological finite generation for reductive groups*, Duke Math. J. 151 (2010), no. 2, 251–278.

Pieri algebras and Hibi algebras in representation theory

Roger Howe

For Nolan Wallach, in friendship and respect

Abstract A class of algebras that unify a variety of calculations in the representation theory of classical groups is discussed. Because of their relation to the classical Pieri Rule, these algebras are called *double Pieri algebras*. A generalization of the standard monomial theory of Hodge is developed for double Pieri algebras, that uses pairs of semistandard tableaux, rather than a single one. SAGBI theory and toric deformation are key tools. The deformed double Pieri algebras are described using a doubled version of Gelfand–Tsetlin patterns. The approach allows the discussion to avoid dealing with relations between generators.

Keywords: Pieri Rule • Classical groups • SAGBI theory • Standard monomials • Toric deformation

Mathematics Subject Classification 2010: 13A50, 15A72, 20G05, 22E46, 22E47

1 Introduction

There is a large literature devoted to describing the finite-dimensional irreducible representations of reductive complex algebraic groups, *mutatis mutandis*, the irreducible representations of compact Lie groups. See, for example [17, 21, 27, 30, 42, 43, 44, 56] and their references. In this effort, a substantial collection of combinatorial tools, including Young diagrams and tableaux, the Robinson–Schensted and other correspondences, Gelfand–Tsetlin patterns, canonical and crystal bases, among others, were developed over the course of the 20th century.

R. Howe
Yale University, 442 Dunham Lab, 10 Hillhouse Avenue, New Haven, CT 06511, USA
e-mail: roger.howe@yale.edu

R. Howe et al. (eds.), *Symmetry: Representation Theory and Its Applications*, Progress in Mathematics 257, DOI 10.1007/978-1-4939-1590-3_13

More recently, techniques have been developed that connect some of these combinatorial objects more closely to representations, and some of them have come to be seen as proxies for more fundamental structures. In particular, the idea of describing rings that arise naturally in representation theory in terms of *flat deformation to semigroup rings* (aka, *toric deformation*), has brought new insight into the area, and has provided a context in which the combinatorial objects emerge in a reasonably straightforward way from the structure of the rings in question. For a fair number of the most basic cases, the relevant semigroups are of a special, very pleasant nature, and are naturally defined in terms of partially ordered sets. This is the structure that has given rise to Gelfand-Tsetlin and similar kinds of patterns. Very useful in this approach is the SAGBI theory, which describes a ring of polynomial functions by means of an associated semigroup of highest terms, with respect to a designated term order, of elements of the ring.

The goal of this article is to review some of this development, and in particular, to highlight the role of a certain family of algebras, here termed *double Pieri algebras*, that unify a variety of calculations in the representation theory of the classical groups.

We will assume that the reader is familiar with basic general concepts and constructions in representation theory, including what a group representation is, subrepresentations and quotient representations, irreducible representations, direct sums, complete reducibility, dual or contragredient representations, intertwining operator/G-morphism of representations, equivalence of representations, tensor product of representations, restriction of a representation to a subgroup, multiplicative characters of commutative groups, and the fact that these themselves form a group under pointwise multiplication. Some key ideas and facts, such as matrix coefficients, complete reducibility of representations of reductive groups, and the theory of the highest weight, will be reviewed, because of their salience for setting the context for the central results.

2 Theorem of the highest weight and the flag algebra

Let G be a reductive complex algebraic group [2, 17]. For most of this paper, G will be the group $GL_n(\mathbf{C}) = GL_n$ of invertible complex $n \times n$ matrices, or closely related to it. We want to discuss representations of G. To do this, one has to say something about what sort of representations are allowed.

For a compact group, one can just agree to consider finite-dimensional continuous representations. Since $G = GL_n(\mathbf{C})$ is not compact, one needs to pay a little more attention to the issue.

One solution could be to regard $GL_n(\mathbf{C})$ as the complexification of the unitary group U_n, which is compact. One could then agree to study representations of $GL_n(\mathbf{C})$ that are holomorphic extensions of continuous representations of U_n.

A second solution is to regard GL_n as an algebraic group. It supports a ring of "rregular" functions,[1] generated by the matrix entries $g_{ij}\ :\ 1 \leq i, j \leq n$ of the typical element g of GL_n, together with the reciprocal of the determinant function. This ring is the union of finite-dimensional subspaces invariant by multiplication by elements of GL_n on the right or on the left. We can restrict attention to representations that are "rregular" in the sense that their matrix coefficients (to be defined below) are in this ring.

In fact, these two solutions coincide, and the resulting family of representations are the same. These are the representations we will study.

A fundamental fact is that such a representation is *completely reducible* [2, 17]: it can always be decomposed as a direct sum of irreducible subrepresentations. We will take this for granted in what follows. The collection of irreducible representations of G is denoted $\widehat{G}$.

Up to conjugation, there is a unique maximal connected solvable subgroup $B_G = B$ of G, called the *Borel subgroup* of G [2, 17]. For $G = GL_n$, we may take $B_{GL_n} = B_n$ to be the group of invertible upper triangular $n \times n$ matrices.

In turn, B contains a unique normal unipotent subgroup, $U = U_G$, which in the case of GL_n is $U_{GL_n} = U_n$, the group of upper triangular matrices that are unipotent, that is, have all diagonal entries equal to 1. The group B also contains a *maximal torus* $A = A_G$, an abelian complement to U in B, consisting of diagonalizable operators. The torus A is unique up to conjugation in B. For GL_n, we may take $A = A_{GL_n} = A_n$ as the group of invertible diagonal $n \times n$ matrices.

The irreducible representations of the torus A are all one-dimensional, so they amount to (rregular) homomorphisms

$$\chi : A \to \mathbf{C}^\times \simeq GL_1.$$

In other words, they can be thought of as (rregular, or holomorphic) complex-valued functions χ on A that satisfy $\chi(aa') = \chi(a)\chi(a')$. They are often referred to as *characters*.

The identity map of $GL_1 \simeq \mathbf{C}^\times$ is a character, and all (rregular) characters of GL_1 are just the integer powers of the identity character. Thus, the character group of GL_1 is just $\mathbf{Z}$.

Since $A_n \simeq (GL_1)^n$, the group $\hat{A}_n$ of characters of A_n is isomorphic to $\mathbf{Z}^n$. Explicitly, for an n-tuple $\vec{\beta}$ of integers β_j, the associated character $\chi_{\vec{\beta}}$ in $\widehat{A}_n$ is

$$\chi_{\vec{\beta}}([\mathbf{a}]) = \prod_{i=1}^{n} a_i^{\beta_i}. \tag{1}$$

[1] We use the neologism "irregular," rather than the more usual but overused "regular," to refer to functions in a specified well behaved algebra of nice functions on a given set.

Here

$$[\mathbf{a}] = \begin{bmatrix} a_1 & 0 & 0 & 0 & \dots & 0 \\ 0 & a_2 & 0 & 0 & \dots & 0 \\ 0 & 0 & a_3 & 0 & \dots & 0 \\ & \cdot & \cdot & \cdot & \cdot & \cdot \\ & \cdot & \cdot & \cdot & \cdot & \cdot \\ & \cdot & \cdot & \cdot & \cdot & \cdot \\ 0 & 0 & 0 & 0 & \dots & a_n \end{bmatrix} : a_j \neq 0 \tag{2}$$

Let $R(G)$ be the rregular functions on G. We can define the left and right regular actions on $R(G)$ by the recipes

$$L_h(f)(g) = f(h^{-1}g), \qquad R_h(f)(g) = f(gh), \qquad g, h \in G;\ f \in R(G). \tag{3}$$

Let ρ be a representation of G on a (finite-dimensional, complex) vector space V. Let $\rho*$ be the contragredient representation on the vector space dual V^* to V. Recall that the formula connecting ρ and ρ^* is

$$\rho^*(\lambda)(\vec{v}) = \lambda(\rho(g)^{-1}\vec{v}), \qquad g \in G, \vec{v} \in V,\ \lambda \in V^*. \tag{4}$$

Define a function on G, the *matrix coefficient* $\phi_{\lambda,\vec{v}}$ of V with respect to $\vec{v}$ in V and λ in V^* by

$$\phi_{\lambda,\vec{v}}(g) = \lambda(\rho(g)^{-1}\vec{v}) = \rho^*(g)(\lambda)(\vec{v}). \tag{5}$$

It is straightforward to check that

$$L_h(\phi_{\lambda,\vec{v}}) = \phi_{\lambda,\rho(h)(\vec{v})}; \qquad \text{and} \qquad R_h(\phi_{\lambda,\vec{v}}) = \phi_{\rho^*(h)(\lambda),\vec{v}}. \tag{6}$$

In other words, for a fixed λ, the map

$$\Phi_\lambda : V \to R(G), \qquad \Phi_\lambda(\vec{v}) = \phi_{\lambda,\vec{v}}, \tag{7}$$

is an intertwining operator between V and the left regular representation of G. A nice, standard, easily verified fact [23] is

Proposition 2.1. *The map $\lambda \to \Phi_\lambda$ is an isomorphism from V^* to the vector space $Hom_G(V, R(G))$ of G-intertwining maps from V to the left regular representation.*

The discussion in the previous paragraph was valid for essentially any group G. Return to letting G be a complex reductive group. Let ρ be a (rregular) representation of G on V. Consider the space V^{U_G} of vectors invariant under the unipotent subgroup $U_G = U$. Since $A_G = A$ normalizes U_G, A will leave V^U invariant, as is easily checked. Since A is commutative and reductive, any representation of it breaks up into eigenspaces. Thus, we may write

$$V^{U_G} \simeq \oplus_{\chi \in \hat{A}} (V^U)^{(A,\chi)}, \tag{8}$$

where $(V^U)^{(A,\chi)}$ indicates the χ eigenspace for A, for χ in $\hat{A}$.

Theorem 2.2 (of the Highest Weight [17]).

(a) *Let ρ be any (rregular) representation of (the reductive connected algebraic group) G on a (finite-dimensional, complex) vector space V. Then V^U is nonzero.*

(b) *If ρ is irreducible, then V^U is one-dimensional; and*

(c) *the character χ_ρ of A such that $V^U = (V^U)^{(A,\chi_\rho)}$ determines ρ up to equivalence.*

A vector in V^U is called a *highest weight vector* for V, and when V is irreducible, the character χ_ρ is called the *highest weight* of ρ. Similarly, we write ρ_χ for the irreducible representation with highest weight χ.

If ρ is any finite-dimensional but not necessarily irreducible representation of G on the space V, let $V = \oplus_j V_j$ be a decomposition of V into irreducible summands. Then

$$V^U = \left(\oplus_j V_j\right)^U = \oplus_j V_j^U = \oplus_j (V_j^U)^{(A,\chi_j)} \tag{9}$$

is the corresponding decomposition of the space of highest weight vectors into eigenspaces for A. We see that there is one χ eigenvector for each summand that is isomorphic to the irreducible representation V_χ with highest weight χ. In other words, if we know the decomposition of V^U into A-eigenspaces, we know the decomposition of V into irreducible representations. In fact, we know not just the multiplicities of the irreducible representations, but we know the highest weight vectors in the sum of the irreducible constituents of a given isomorphism type. This principle underlies much of the discussion below.

Not all characters of A can be highest weights. However, the collection of highest weights of irreducible representations do form a semigroup in $\widehat{A}$. This can be seen by taking tensor products. If $\vec{v}$ is a highest weight vector for the representation ρ with weight $\chi_\rho = \chi$, and $\vec{v}'$ is a highest weight vector for the representation ρ', with weight $\chi_{\rho'} = \chi'$, then the vector $\vec{v} \otimes \vec{v}'$ in $V \otimes V'$ will be a highest weight vector with weight $\chi\chi'$.

Characters of A that are highest weights of representations are called *dominant* characters. The semigroup of dominant characters is denoted $\widehat{A}^+$. To explicitly compute the semigroup $\widehat{A}^+$ for $G = GL_n$ is not difficult, but would involve too long a digression. Here we simply record that it is ([17])

$$\widehat{A}_n^+ = \{\chi_{\vec{\beta}} : \beta_i \geq \beta_{i+1},\ 1 \leq i \leq n-1\}. \tag{10}$$

Take an irreducible representation ρ of G on the space V. Let λ_o be the highest weight vector in V^*. Consider the matrix coefficient mapping Φ_{λ_o}. Then the second formula in (6) shows that the functions in $\Phi_{\lambda_o}(V)$ will be invariant under right translation by $U_G = U$. Thus, they will factor to the coset space G/U. We denote by $R(G/U)$ the algebra of functions in $R(G)$ that factor to G/U. Thus we can say that $\Phi_{\lambda_o}(V) \subset R(G/U)$.

Moreover, again using formula (6), we can compute that

$$R_{\mathbf{a}}(\Phi_{\lambda_o}(\vec{v})) = \chi_{\rho^*}(\mathbf{a})\Phi_{\lambda_o}(v)$$

for $\mathbf{a}$ in $A_G = A$. In other words, the functions of $\Phi_{\lambda_o}(V)$ are eigenvectors, with eigencharacter χ_{ρ^*} under right translation by the torus A.

From Proposition 2.1 and the uniqueness of the highest weight vector in an irreducible representation, we see that there is only one embedding (up to scalar multiples) of ρ into $R(G/U)$, namely the one constructed using Φ_{λ_o}. We may conclude therefore that the image of this embedding is the eigenspace for right translations by A, for the character χ_{ρ^*}. We can summarize this discussion in the following statement.

Theorem 2.3.

(a) *The ring $R(G/U)$ of rregular functions on the coset space G/U decomposes into a direct sum of one copy of each rregular irreducible representation of G:*

$$R(G/U) \simeq \sum_{\rho \in \hat{G}} V_\rho = \sum_{\chi \in \hat{A}+} V_\chi. \tag{11}$$

(b) *Moreover, each space V_ρ is an eigenspace for right translations by A, with eigencharacter χ_{ρ^*}. This entails that*

$$V_{\rho_\chi} \cdot V_{\rho_{\chi'}} = V_{\rho_{\chi\chi'}} \qquad \text{in } R(G/U). \tag{12}$$

(c) *Thus, $R(G/U)$ has the structure of an $\widehat{A}$-graded algebra, with the irreducible representation spaces V_ρ being the homogeneous components.*

We call $R(G/U)$ the *flag algebra* for G. It and algebras constructed from it will be the central objects in our discussion.

Remark 2.4. Note that the space V_χ supporting the irreducible representation with highest weight χ is not the χ-eigenspace under right translations by A, but the χ^*-eigenspace, where χ^* denotes the highest weight of the contragredient representation.

3 Branching algebras

Let $H \subset G$ be a reductive subgroup. An important problem in representation theory is to describe, for an irreducible representation V_ρ of G, the restriction $(V_\rho)_{|H}$ to give the decomposition of V_ρ into irreducible representations for H. This is the finite-dimensional version of harmonic analysis; it is often called the *branching problem*. Thanks to the theorem of the highest weight, this problem can be encoded in the structure of an algebra, a subalgebra of the flag algebra $R(G/U_G)$.

Specifically, let $B_H = A_H U_H$ be a Borel subgroup of H. Consider the algebra $R(G/U_G)^{U_H}$ of rregular U_H-invariant functions on G/U_G. We have the decomposition

$$\begin{aligned} R(G/U_G)^{U_H} &\simeq \left(\sum_{\chi \in \hat{A}_G^+} V_\chi\right)^{U_H} \simeq \sum_{\chi \in \hat{A}_G^+} (V_\chi)^{U_H} \\ &\simeq \sum_{\chi \in \hat{A}_G^+} \left(\sum_{\psi \in \hat{A}_H^+} (V_\chi^{U_H})^{(A_H,\psi)}\right) \\ &\simeq \sum_{\chi \in \hat{A}_G^+} \sum_{\psi \in \hat{A}_H^+} (V_\chi^{U_H})^{(A_H,\psi)}. \end{aligned} \tag{13}$$

Here $(V_\chi^{U_H})^{(A_H,\psi)}$ indicates the ψ-eigenspace for A_H acting on the space $V_\chi^{U_H}$ of U_H-invariant vectors in V_χ. According to the theory of the highest weight, the dimension of $(V_\chi^{U_H})^{(A_H,\psi)}$ gives the multiplicity of the irreducible representation W_ψ in the representation $(V_\chi)_{|H}$. On the other hand, we know that $(V_\chi^{U_H})^{(A_H,\psi)}$ is the χ^*-eigenspace for A_G acting on G/U_G on the right, intersected with the ψ-eigenspace for A_H acting on G/U_G on the left. Thus, $R(G/U_G)^{U_H}$ is an $\hat{A}_G \times \hat{A}_H$-graded algebra, whose $\chi \times \psi$ homogeneous component has dimension equal to the multiplicity of W_ψ in V_χ. Thus, this algebra encodes the branching rule from G to H. We therefore call it the (G, H) *branching algebra.*

The branching algebra $R(G/U_G)^{U_H}$ presents a solution to the branching problem from G to H. How practical this solution is depends on how explicitly we can describe the algebra. In the following sections we will discuss some methods for describing algebras, and then will study some examples in which explicit descriptions can be given.

Remarks. (a) Note that, in fact, the algebra gives more than the multiplicities of $(V_\chi)_{|H}$ – it gives explicit highest weight vectors for the H-constituents of V_χ.

(b) The value of finding highest weight vectors is emphasized in [57].

4 Tensor product algebras

Given a group G and two representations ρ and ρ' of G, one can form the *tensor product* $\rho \otimes \rho'$ of the two representations. This construction comes up in the consideration of interacting quantum mechanical systems with symmetry, for example, in the description of multi-q-bit systems in quantum information theory [55]. Typically, even if ρ and ρ' are irreducible, the product $\rho \otimes \rho'$ will not be, and it is a natural problem to find its irreducible components. The point of this section is to note that this problem can be considered as a branching problem.

If G and G' are reductive groups, and ρ is a representation of G on V, and σ is a representation of G' on W, then we can define the *tensor product representation* $\rho \otimes \sigma$ of $G \times G'$ on the space $V \otimes W$ by the recipe

$$(\rho \otimes \sigma)(g, g')(\vec{v} \otimes \vec{w}) = (\rho(g)\vec{v}) \otimes (\sigma(g')\vec{w}) \tag{14}$$

for $g \in G$, $g' \in G'$, $\vec{v} \in V$ and $\vec{w} \in W$. If ρ and σ are irreducible representations of G and G' respectively, then $\rho \otimes \sigma$ is an irreducible representation of $G \times G'$. Moreover the mapping

$$\hat{G} \times \hat{G}' \to \widehat{(G \times G')} \tag{15}$$

given by

$$(\rho, \sigma) \to \rho \otimes \sigma, \tag{16}$$

is a bijection.

In case $G' = G$, then the group G may be embedded diagonally in $G \times G$:

$$g \to (g, g).$$

We call the image of this map ΔG, the *diagonal subgroup* of $G \times G$. The tensor product $\rho \otimes \rho'$ of two irreducible representations to make an irreducible representation of $G \times G$ is sometimes called the *outer tensor product*, and the restriction of $\rho \otimes \rho'$ to ΔG is then referred to as the *inner tensor product*. Decomposing inner tensor products is thus a particular case of decomposing the restriction of representations to a subgroup; i.e., it is a branching problem. Let $\Delta U_G = U_{\Delta G}$ be the maximal unipotent subgroup of the Borel of ΔG. The branching algebra associated with the tensor product problem is

$$\begin{aligned} R((G \times G)/U_{G \times G})^{\Delta U_G} &= R((G \times G)/(U_G \times U_G))^{\Delta U_G} \\ &\simeq (R(G/U_G) \otimes R(R/U_G))^{\Delta U_G}. \end{aligned} \tag{17}$$

We will refer to this algebra as the *tensor product algebra* for G.

5 (GL_n, GL_m)-duality

We now begin closer consideration of our main examples, involving GL_n and closely related subgroups. To prepare, we recall the notion of *polynomial representation* of GL_n, and the associated "diagram notation" (see e.g., [17,20]).

Polynomial representations of GL_n: As mentioned in §3, the highest weights of irreducible representations of GL_n have the form $\chi_{\vec{\beta}}$, where the components β_i of the n-tuple $\vec{\beta}$ satisfy $\beta_i \geq \beta_{i+1}$. A representation whose highest weight satisfies $\beta_j \geq 0$ for all j (equivalently, $\beta_n \geq 0$) is called a *polynomial* representation. It can be shown that the matrix coefficients of a polynomial representation are polynomials in the coordinate functions x_{ij} on the $n \times n$ matrices.

Among the representations of GL_n, we have the distinguished one dimensional *determinant representation*, det, given by $g \to \det g$. The highest weight of det is just restriction of det to A_n. As a character of A_n, we have

$$\det = \chi_{\vec{1}_n}, \tag{18}$$

where $\vec{1}_n$ is the n-tuple all of whose coordinates are equal to 1.

If ρ is any irreducible representation of GL_n, then the tensor product $\det \otimes \rho$ will again be irreducible. If the highest weight of ρ is $\chi_{\vec{\beta}}$, then the highest weight of $\det \otimes \rho$ will just be $\chi_{\vec{1}_n}\chi_{\vec{\beta}} = \chi_{\vec{\beta}+\vec{1}_n}$. Clearly, by tensoring with a suitable power of det (for example, we could take $\det^{-\beta_n}$), we can arrive at a representation with a highest weight corresponding to an n-tuple $\vec{\lambda}$, with entries

$$\lambda_1 \geq \lambda_2 \geq \lambda_3 \geq \ldots \geq \lambda_d > 0 = \lambda_{d+1} = 0 = \ldots = 0.$$

In particular, any irreducible representation of GL_n can be written as the tensor product of a power of det with a polynomial representation.

The polynomial representations of GL_n span a subring of the flag algebra $R(GL_n/U_n)$. We will denote this subalgebra by $R^+(GL_n/U_n)$.

Diagram notation: There is a special notation for polynomial representations that is useful in many contexts. Given any sequence

$$D = \{\lambda_1 \geq \lambda_2 \geq \lambda_3 \geq \ldots \geq \lambda_d > 0\}, \tag{19}$$

we can extend it by zeroes to make an n-tuple for any $n \geq d$, which will then define a dominant weight for GL_n.

The sequence D is called a *partition* of the number $\sum_i \lambda_i$. The λ_i are the *parts* of D, and d is the number of parts, and is also referred to as the *depth* of the partition. The term "depth" comes from the practice of associating to D a (Young or Ferrers) *diagram*, which for us will mean an array of left justified rows of squares or "boxes", with each row stacked one above the next, with λ_i boxes in the i-th row. Later on, we will see that it is useful to label these boxes in various ways to produce what are called *tableaux*.

Given a diagram D of depth d, and any integer $n \geq d$, we denote the representation of GL_n corresponding to D by ρ_n^D. This "diagram notation" for the polynomial representations of GL_n will be what we use from now on.

With diagram notation in hand, we can state a central fact of finite-dimensional representation theory for the classical groups.

Let $GL_n \times GL_m$ act on the $n \times m$ matrices M_{nm} by the recipe

$$(g, g')(T) = (g^t)^{-1} T (g')^{-1}. \tag{20}$$

We get a representation π of $GL_n \times GL_m$ on the ring $P(M_{nm})$ of polynomial functions on M_{nm} in the usual way:

$$\pi(g, g')(p)(T) = p((g, g')^{-1}(T)) = p(g^t T g'). \tag{21}$$

Theorem 5.1 (**(GL_n, GL_m) duality**). *As a $GL_n \times GL_m$ module, we have the decomposition*

$$P(M_{nm}) \simeq \sum_D \rho_n^D \otimes \rho_m^D, \tag{22}$$

where the sum is over all Young diagrams with at most $\min(n, m)$ *rows.*

The key fact here is that the diagram D in all tensor products is the same for GL_n and for GL_m. This result is a simple application of the idea of multiplicity-free action, which provides an effective method for exploiting highest weight theory. See for example [21] or [17] for details.

If we look at the GL_m highest weight vectors in $P(M_{nm})$, then (GL_n, GL_m) duality tells us that

$$P(M_{nm})^{U_m} \simeq \left(\sum_D \rho_n^D \otimes \rho_m^D\right)^{U_m} \simeq \sum_D \left(\rho_n^D \otimes \rho_m^D\right)^{U_m} \simeq \sum_D \rho_n^D \otimes (\rho_m^D)^{U_m}. \tag{23}$$

Highest weight theory says that the dimension of $(\rho_m^D)^{U_m}$ is just 1. Therefore, the sum (23) is a multiplicity free sum of irreducible representations of GL_n. Moreover, the line $(\rho_m^D)^{U_m}$ is an eigenline for the torus A_m in GL_m, with character ψ_D. Also, the space $P(M_{nm})^{U_m}$ is an algebra, on which A_m acts by automorphisms. Thus $P(M_{nm})^{U_m}$ is an $\widehat{A}_m^+$-graded algebra, so that except for the fact that the diagrams D are limited to have at most m rows, $P(M_{nm})^{U_m}$ bears a strong resemblance to $R^+(GL_n/U_n)$. In fact, it is an easy matter [21] to construct an embedding

$$\alpha_{n,m} : P(M_{nm})^{U_m} \to R^+(GL_n/U_n), \tag{24}$$

with image consisting of all V_D where D has at most m rows. We call $\alpha_{n,m}(P(M_{nm})^{U_m})$ a *band-limited subalgebra* of $R^+(GL_n/U_n)$. In what follows, we usually will suppress the embedding $\alpha_{n,m}$.

6 Duality between tensor products and block diagonal restriction

Consider $GL_n \times GL_{m+\ell}$ acting on $M_{n(m+\ell)}$. We have the $GL_n \times GL_{m+\ell}$ module decomposition as in §5:

$$P(M_{n(m+\ell)}) \simeq \sum_D \rho_n^D \otimes \rho_{m+\ell}^D. \tag{25}$$

On the other hand, we can write $M_{n(m+\ell)} \simeq M_{nm} \oplus M_{n\ell}$, which implies that

$$P(M_{n(m+\ell)}) \simeq P(M_{nm}) \otimes P(M_{n\ell}). \tag{26}$$

Applying §5 to each factor, we obtain a decomposition of $P(M_{n(m+\ell)})$ as a module for $(GL_n \times GL_m) \times (GL_n \times GL_\ell) \simeq (GL_n \times GL_n) \times (GL_m \times GL_\ell)$:

$$\begin{aligned} P(M_{n(m+\ell)}) &\simeq \left(\sum_E \rho_n^E \otimes \rho_m^E\right) \otimes \left(\sum_F \rho_n^F \otimes \rho_\ell^F\right) \\ &\simeq \sum_{E,F} \left(\rho_n^E \otimes \rho_n^F\right) \otimes \left(\rho_m^E \otimes \rho_\ell^F\right). \end{aligned} \tag{27}$$

The relationship between the groups involved in these decompositions is described by this diagram.

$$\begin{array}{ccc} GL_n & \times & GL_{m+\ell} \\ \cap & & \cup \\ (GL_n \times GL_n) & \times & (GL_m \times GL_\ell) \end{array} \tag{28}$$

The copy of GL_n in the upper row is embedded diagonally in the factor $GL_n \times GL_n$ in the lower row. Denote this upper copy by ΔGL_n. The product $GL_m \times GL_\ell$ is embedded block diagonally in $GL_{m+\ell}$.

Write $U_{\Delta GL_n} = \Delta U_n$. Consider the algebra $P(M_{n(m+\ell)})^{\Delta U_n \times U_m \times U_\ell}$ of joint highest weight vectors for ΔGL_n, GL_m and GL_ℓ. We can interpret this algebra in two different ways, suggested by the top row and the bottom row of display (28).

Looking at this from the point of view of the top row, we find that

$$P(M_{n(m+\ell)})^{\Delta U_n \times (U_m \times U_\ell)} \simeq \left(P(M_{n(m+\ell)})^{\Delta U_n}\right)^{U_m \times U_\ell}). \tag{29}$$

We know from §5 that the algebra $P(M_{n(m+\ell)})^{\Delta U_n}$ is isomorphic to (a band-limited subalgebra of) the flag algebra $R^+(GL_{m+\ell}/U_{m+\ell})$. The algebra $\left(P(M_{n(m+\ell)})^{\Delta U_n}\right)^{U_m \times U_\ell}$ then appears as the subalgebra of highest weight vectors for the block diagonal subgroup $GL_m \times GL_\ell \subset GL_{m+\ell}$. In other words, we can think of the algebra $P(M_{n(m+\ell)})^{\Delta U_n \times (U_m \times U_\ell)}$ as (a band-limited subalgebra of) the $(GL_{m+\ell}, GL_m \times GL_\ell)$ branching algebra. The corresponding decomposition into submodules is as follows:

$$\left(\sum_D \rho_n^D \otimes \rho_{m+\ell}^D\right)^{\Delta U_n \times (U_m \times U_\ell)} \simeq \sum_D (\rho_n^D)^{\Delta U_n} \otimes (\rho_{m+\ell}^D)^{U_m \times U_\ell}. \tag{30}$$

On the other hand, looking at $P(M_{n(m+\ell)})^{\Delta U_n \times (U_m \times U_\ell)}$ from the point of view of the bottom row, we have

$$\begin{aligned} P(M_{n(m+\ell)})^{\Delta U_n \times (U_m \times U_\ell)} &\simeq \left((P(M_{nm}) \otimes P(M_{n\ell}))^{U_m \times U_\ell}\right)^{\Delta U_n} \\ &\simeq \left(P(M_{nm})^{U_m} \otimes P(M_{n\ell})^{U_\ell}\right)^{\Delta U_n}. \end{aligned} \tag{31}$$

Again from §5, we know that the algebra $P(M_{nm})^{U_m} \otimes P(M_{n\ell})^{U_\ell}$ is isomorphic to (a band-limited subalgebra of) the flag algebra $R^+(GL_n \times GL_n / U_n \times U_n)$. Then the algebra $\left(P(M_{nm})^{U_m} \otimes P(M_{n\ell})^{U_\ell}\right)^{\Delta U_n}$ consists of the subalgebra of ΔGL_n highest weight vectors. In other words, we can think of the algebra $P(M_{n(m+\ell)})^{\Delta U_n \times (U_m \times U_\ell)}$ as (a band-limited subalgebra of) the GL_n tensor product algebra. The corresponding decomposition into submodules is as follows:

$$\begin{aligned} &\left(\sum_{E,F} (\rho_n^E \otimes \rho_n^F) \otimes (\rho_m^E \otimes \rho_\ell^F)\right)^{\Delta U_n \times (U_m \times U_\ell)} \\ &\qquad \simeq \sum_{E,F} \left(\rho_n^E \otimes \rho_n^F\right)^{\Delta U_n} \otimes (\rho_m^E)^{U_m} \otimes (\rho_\ell^F)^{U_\ell}. \end{aligned} \tag{32}$$

From the above, we see that the homogeneous components of

$$P(M_{n(m+\ell)})^{\Delta U_n \times (U_m \times U_\ell)}$$

simultaneously compute the multiplicities of irreducible representations of $GL_m \times GL_\ell$ in restrictions of representations of $GL_{m+\ell}$, and the multiplicities of irreducible representations of GL_n in tensor products of such representations.

This construction can be iterated. Remarks similar to the above apply if we break up m into many summands: If we write $m = \sum_{i=1}^a m_i$, then we can think of the algebra

$$P(M_{nm})^{\Delta U_n \times (\prod_i U_{m_i})}$$

either as describing restrictions from GL_m to the block diagonal subgroup $\prod_i GL_{m_i}$, or as describing manyfold tensor products of representations of GL_n, of representations of depths limited by the m_i. Here ΔGL_n indicates the diagonal subgroup in $(GL_n)^a$, where a is the number of summands into which m is decomposed.

7 The Pieri algebra

We first apply the ideas of §6 to the case when $\ell = 1$. From the point of view of tensor products, this amounts to computing products of the form $\rho_n^D \otimes S^r(\mathbf{C}^n)$, a general representation with a symmetric power of the standard representation. This is the classical *Pieri Rule* [14].

From the point of view of branching, it amounts to describing restrictions of representations of GL_{m+1} to $GL_m \times GL_1$. This is the famous computation that formed the basis of the paper of Gelfand and Tsetlin, and led to Gelfand–Tsetlin patterns [16].

As is explained in [21] or [17], this situation is again multiplicity-free. Accordingly, the algebra

$$P(M_{n(m+1)})^{\Delta U_n \times U_m}$$

is a polynomial ring on a canonical set of generators. Again, according to [21], these generators are

$$\delta_k = \det \begin{bmatrix} x_{11} & x_{12} & x_{13} & \cdots & x_{1k} \\ x_{21} & x_{22} & x_{23} & \cdots & x_{2k} \\ x_{31} & x_{32} & x_{33} & \cdots & x_{3k} \\ & \cdot & \cdot & \cdot & \cdot & \cdot \\ & \cdot & \cdot & \cdot & \cdot & \cdot \\ & \cdot & \cdot & \cdot & \cdot & \cdot \\ x_{k1} & x_{k2} & x_{k3} & \cdots & x_{kk} \end{bmatrix} \quad \text{and}$$

$$\delta_{(k-1,1)} = \det \begin{bmatrix} x_{11} & x_{12} & x_{13} & \cdots & x_{1(k-1)} & y_{11} \\ x_{21} & x_{22} & x_{23} & \cdots & x_{2(k-1)} & y_{21} \\ x_{31} & x_{32} & x_{33} & \cdots & x_{3(k-1)} & y_{31} \\ & \cdot & \cdot & \cdot & \cdot & \cdot & \cdot \\ & \cdot & \cdot & \cdot & \cdot & \cdot & \cdot \\ & \cdot & \cdot & \cdot & \cdot & \cdot & \cdot \\ x_{k1} & x_{k2} & x_{k3} & \cdots & x_{k(k-1)} & y_{k1} \end{bmatrix}. \tag{33}$$

Here x_{ab} for $1 \leq a \leq n$, $1 \leq b \leq m$, are the entries of a typical matrix in M_{nm}, and y_{a1} are the entries of the $(m+1)$-th column of a matrix in $M_{n(m+1)}$.

8 SAGBI theory: Lattice cones and highest terms

To go beyond the Pieri Rule, we must interpret what it says in a robust way. This involves looking at the highest terms of monomials in the functions δ_k and $\delta_{(k-1,1)}$.

The general idea is to approximate general rings by rings with a more transparent structure. The traditional approach of describing rings in terms of generators and relations effectively amounts to comparing a general (affine) ring with polynomial rings. But since there is only one polynomial ring in each dimension, this is a somewhat Procrustean approach to dealing with rings. It might seem desirable to have a larger collection of "model algebras", with which to compare a general affine ring. Over the last several decades, it has become appreciated that there is a plausible collection of candidates for such model algebras.

A notable reason that we feel comfortable with polynomial rings is that they have a transparent multiplication. A polynomial ring in generators x_i has a basis of monomials

$$x^{\vec{a}} = x_1^{a_1} x_2^{a_2} x_3^{a_3} \dots x_n^{a_n}, \tag{34}$$

where $\vec{a} = \begin{bmatrix} a_1 \\ a_2 \\ a_3 \\ . \\ . \\ . \\ a_n \end{bmatrix}$ is an n-tuple of nonnegative integers (aka a *multi-index*). These monomials multiply according to the standard rule

$$x^{\vec{a}} \cdot x^{\vec{b}} = x^{\vec{a}+\vec{b}}. \tag{35}$$

In other words, the monomials form a semigroup under multiplication. Since the monomials form a basis, the polynomial ring $P(x_1, x_2, x_3, \dots, x_n)$ just amounts to the semigroup ring on the semigroup of monomials. This semigroup is of course $(\mathbf{Z}^+)^n$, the free abelian semigroup on n generators: $P(x_1, x_2, x_3, \dots, x_n) \simeq \mathbf{C}((\mathbf{Z}^+)^n)$. This suggests perhaps enlarging the set of "model algebras", with which we could compare more general algebras, to include semigroup rings.

Lattice cones: A fairly natural and general class of semigroups are the *lattice cones*. They consists of the integer points in a convex cone in cartesian n-space. More explicitly, for a linear function λ on $\mathbf{R}^n$, let the *positive half-space for* λ be the set

$$H_\lambda^+ = \{\vec{x} \in \mathbf{R}^n : \lambda(\vec{x}) \geq 0\}. \tag{36}$$

For a collection $L = \{\lambda_i\}$ of linear functions, define the *positive cone for* L by

$$C_L^+ = \cap_i H_{\lambda_i}^+. \tag{37}$$

This cone is closed under vector addition: it is a semigroup. It is also closed under multiplication by positive scalars, and it is convex: given two points in C_L^+, the whole line segment between them also lies in L.

Consider the intersection $C_L^+ \cap \mathbf{Z}^n$. It is clearly a semigroup. Moreover, when possible, it is finitely generated [3].

Theorem 8.1 (Gordan's Lemma). *If the functionals λ_i in L are rational, in the sense that they take rational values on $\mathbf{Z}^n$, then $C_L^+ \cap \mathbf{Z}^n$ is finitely generated as a semigroup.*

We call a semigroup $C_L^+ \cap \mathbf{Z}^n$ defined by rational functionals λ_i a *lattice cone*. (In [3], it is called a *normal affine semigroup*.)

Although lattice cones are finitely generated subsemigroups of $\mathbf{Z}^n$, they are obtained somewhat indirectly. The straightforward way to obtain a finitely generated subsemigroup of $\mathbf{Z}^n$ would be to select a finite set B of integral vectors $\vec{b}_i \in \mathbf{Z}^n$, and take the semigroup they generate. It turns out that such a semigroup is not too far from being a lattice cone. Let S_B be the semigroup generated by B. We can also describe S_B as the set of all linear combinations of elements of B with nonnegative integer coefficients. Let C_B be the set of all linear combinations of elements of B with nonnegative real coefficients. Standard duality results [3] show that C_B is a rational polyhedral cone. Clearly $S_B \subset C_B \cap \mathbf{Z}^n$. Moreover, it can be shown [3] that there is a whole number J such that $J \cdot (C_B \cap \mathbf{Z}^n) \subset S_B$. Thus S_B looks like a "lattice cone with holes". It is gotten by deleting some elements from $C_B \cap \mathbf{Z}^n$, but retains enough that one can recapture $C_B \cap \mathbf{Z}^n$ as a subsemigroup of the rational vector space spanned by B. (Note that $C_B \cap \mathbf{Z}^n$ is not canonically determined by S_B as an abstract semigroup, but depends on how B is embedded in $\mathbf{Z}^n$. For example, the subsemigroup of $(\mathbf{Z}^+)^2$ generated by $\vec{v}_1 = \begin{bmatrix} c_1 \\ 0 \end{bmatrix}$ and $\vec{v}_2 = \begin{bmatrix} d \\ c_2 \end{bmatrix}$ will be isomorphic to $(\mathbf{Z}^+)^2$, for any positive integers c_1, c_2 and d. However, the lattice cones $C_{\{\vec{v}_1, \vec{v}_2\}} \cap \mathbf{Z}^2$ run through infinitely many isomorphism classes of semigroup, and for a given isomorphism class, there are infinitely many inequivalent embeddings $(\mathbf{Z}^+)^2 \to C_B$.)

It should be clear from the above that the affine semigroup rings, among which the lattice cones form a distinguished subset, provide a rich class of potential model algebras. The next question would be, given a general affine ring, how might we find a semigroup ring to model it?

A fairly general answer to this question is provided by SAGBI theory [50]. This adapts the tools developed to allow computations with ideals (Gröbner basis theory) to directly describe rings. In particular, it makes use of the idea of *term order* on the monomials in the polynomial ring $P(x_1, x_2, x_3, \ldots, x_n)$. A term order is a total ordering $\leq$ on $(\mathbf{Z}^+)^n$ satisfying the following conditions.

(i) $\vec{0}$ is the minimal element in $(\mathbf{Z}^+)^n$.

(ii) Any decreasing sequence is finite. (Noetherian condition.) (38)

(iii) If $\vec{a} \leq \vec{b}$, then $\vec{a} + \vec{c} \leq \vec{b} + \vec{c}$, for any $\vec{a}$, $\vec{b}$, $\vec{c}$ in $(\mathbf{Z}^+)^n$.

(Compatibility with addition.)

Given a term order $\leq$ on $(\mathbf{Z}^+)^n$, and any polynomial p in n variables, we can define $LT(p) = x^{\vec{a}}$, where $\vec{a}$ is the largest exponent among all monomials appearing with nonzero coefficients in the monomial expansion of p. The compatibility condition on term orders implies that for any pair p, q of polynomials, we have

$$LT(pq) = LT(p)LT(q). \tag{39}$$

Now let $R \subset P(x_1, x_2, x_3, \ldots, x_n)$ be a subalgebra of the polynomials. Define $LT(R) = \{LT(p) : p \in R\}$ to be the collection of leading terms of all elements of R. The condition (39) says that $LT(R)$ is a semigroup. It may or may not be finitely generated, but if it is, a general result [6] says that the associated semigroup ring is a good approximation to R, in a suitable technical sense.

Theorem 8.2. *For a subring $R \subset P(x_1, x_2, x_3, \ldots, x_n)$, if the semigroup $LT(R)$ is finitely generated, then the associated semigroup ring $\mathbf{C}(LT(R))$ is a flat deformation of R.*

We refer to [6] for the definition of flat deformation, as well as the proof of Theorem 8.2. We note that the flat deformations constructed by Theorem 8.2 are not canonical, and the nature of their relation to the original ring R is not completely clear. The deformed ring $\mathbf{C}(LT(R))$ may require many more generators than does R itself. For our purposes of understanding the (G, H) branching algebras $R(G/U_G)^{U_H}$, whose main features of interest are the $\widehat{A}_G \times \widehat{A}_H$ multigrading, and the associated homogeneous components, it will be enough to know that the term order $\geq$ is compatible with the multigrading, in the sense that $\mathbf{C}(LT(R))$ also possesses a $\widehat{A}_G \times \widehat{A}_H$ multigrading, and the homogeneous components are taken to homogeneous components, so that we can compute the dimensions of these components in $\mathbf{C}(LT(R)$. This will be more or less clear from the constructions given below.

Thus, term orders give us a method of looking for semigroup rings that model a general ring. We should observe that term orders exist in abundance, and it is quite easy to construct them [7]. One method uses linear functionals. Let $\{\lambda_i\}$ be a collection of linear functions on $\mathbf{R}^n$ that are positive, in the sense that their values on the positive orthant $(\mathbf{R}^+)^n$ are nonnegative. To a multi-index $\vec{a}$, we can associate the sequence

$$\vec{a} \to (\lambda_1(\vec{a}), \lambda_2(\vec{a}), \lambda_3(\vec{a}), \ldots$$

of real numbers. We can order $(\mathbf{Z}^+)^n$, by the lexicographic order on this sequence: $\vec{a} > \vec{b}$ if

$$\begin{aligned} &\text{(i) } \lambda_1(\vec{a}) > \lambda_1(\vec{b}), \text{ or if} \\ &\text{(ii) } \lambda_1(\vec{a}) = \lambda_1(\vec{b}), \text{ and } \lambda_2(\vec{a}) > \lambda_2(\vec{b}), \text{ or if} \\ &\text{(iii) } \lambda_1(\vec{a}) = \lambda_2(\vec{b}), \text{ and } \lambda_2(\vec{a}) = \lambda_2(\vec{b}), \text{ and } \lambda_3(\vec{a}) > \lambda_3(\vec{b}), \text{ or if} \end{aligned} \tag{40}$$

etc.

If the λ_i span $(\mathbf{R}^n)^*$, then this will define a total order on $(\mathbf{Z}^+)^n$. Indeed, if λ_1 is *totally irrational*, in the sense that its kernel does not contain any rational vectors (equivalently, the entries of λ_1 should be linearly independent over the rationals), then the values of λ_1 alone will define a term order on $(\mathbf{Z}^+)^n$.

Among this type of term order, we single out a particular collection that is important for our purposes. For this, we first impose a total ordering on the variables of the polynomial ring. Essentially, this amounts to choosing a permutation σ of the indices $\{1, 2, 3, \dots, n\}$, and declaring that

$$x_{\sigma(1)} > x_{\sigma(2)} > x_{\sigma(3)} > \dots > x_{\sigma(n)}.$$

(If our variables are not named x_i in the standard way, we should somehow specify the desired total order.) Then we use the functionals $\lambda_1 = \sum_i a_i, \lambda_2 = a_{\sigma_1}, \lambda_3 = a_{\sigma_2}, \dots$ To determine which of two monomials $x^{\vec{a}}$ or $x^{\vec{b}}$ is larger with respect to this order, we first look at the total degree of the monomials, and choose the one with larger degree if these are unequal. If they are the same, then we choose the one with the larger exponent for $x_{\sigma(1)}$. If these also are the same, we choose the one with the larger exponent for $x_{\sigma(2)}$; and so forth. This term order is often called the *graded lexicographic order* (with respect to the given ordering of the coordinates).

To summarize, given a ring R of polynomials and a term ordering $\leq$ on monomials, if we can determine the semigroup $LT(R)$, and show in particular that it is finitely generated, then the semigroup ring $\mathbf{C}(LT(R))$ forms a "good approximation" to R, in the sense of Theorem 8.2.

SAGBI theory turns out to be very effective for studying the rings of interest here. Before going into the details, we should point out that, in allowing general semigroup rings as model algebras, one is in some sense going to the opposite extreme from allowing only the polynomial ring as the single model algebra in a given dimension. There is a huge zoo of semigroup rings, and even of lattice cones. Already in dimension 2, there is a large variety of lattice cones, and their associated semigroups can require arbitrarily many generators. Dimension 3 represents a large step up in complexity from dimension 2, and dimension 4 is still wilder.

Therefore, it is worthwhile to notice that the lattice cones that appear in the calculations below are among the nicest imaginable. They are from the very pleasant class of lattice cones studied by Hibi [18]. We will call them *Hibi cones*.

A Hibi cone is attached to a partially ordered set (poset). Briefly, given a partially ordered set Γ, with order relation $\succeq$, the Hibi cone attached to Γ is the cone

$$(\mathbf{Z}^+)^{\Gamma,\succeq} \tag{41}$$

of nonnegative, integer-valued, order preserving functions on Γ. Degeneration to Hibi rings allows your algebra to be described in terms of "patterns" of some sort, analogous to Gelfand–Tsetlin patterns.

In contrast to the plethora of lattice cones, there are obviously only finitely many Hibi cones in each dimension. Further, these cones have very nice properties.

These are discussed for example in [22], see also [18] and [49]. Generators and relations for Hibi cones can be given explicitly in terms of Γ. Also, there is a canonical decomposition of a Hibi cone into integral simplicial subcones that provides an abstract version of the "standard monomial theory" of Hodge [19]. Thus, although we will not emphasize it, the algebras analyzed below will all have a standard monomial theory. (We note that standard monomial theory has been studied extensively by Lakshmibai and various coworkers [39, 40, 41], etc.)

9 Highest terms for the Pieri algebra

We apply the ideas of §8 to the Pieri algebra. We use variables x_{ab} and y_{a1} for entries of a matrix in $M_{n(m+1)}$, as in formulas (33). We want to define a term order on monomials in these variables. First, we totally order the variables by the rule:

$$\begin{aligned} x_{11} > x_{21} > x_{31} > \ldots > x_{n1} > x_{12} > \\ x_{22} > \ldots > x_{nm} > y_{11} > y_{21} > \ldots > y_{n1}. \end{aligned} \tag{42}$$

In other words, we order the x_{ab} according to the lexicographic order on their indices, and the y_{c1} according to the value of c, and we say that any of the x_{ab} is greater than any of the y_{c1}.

We will use the graded lexicographic order on the monomials in these variables. It is easy to check that, with respect to this term order, the leading terms of the generators (33) are

$$LT(\delta_k) = \prod_{a=1}^{k} x_{aa}, \qquad \text{and} \qquad LT(\delta_{(k-1,1)}) = (\prod_{a=1}^{k-1} x_{aa}) y_{k1}. \tag{43}$$

We assume for convenience that $m < n$. The case of $m \geq n$ can be dealt with similarly. From inspection of these generators, we see that the following relations hold.

(i) x_{ab} with $a \neq b$ does not appear in any highest term;
(ii) If x_{aa} appears in a highest term, then so does $x_{(a-1)(a-1)}$.
(iii) If y_{a1} appears in a highest term, then so does $x_{(a-1)(a-1)}$.

From these conditions, it is easy to check that the highest term of a monomial in the δ_k and the $\delta_{(k-1,1)}$ has the form

$$\left(\prod_{a=1}^{m} x_{aa}^{\ell_{ao}}\right)\left(\prod_{b=1}^{m+1} y_{b1}^{d_{b1}}\right), \tag{44}$$

where the exponents ℓ_{ao} and d_{b1} satisfy

$$\ell_{ao} \geq \ell_{(a+1)o} + d_{(a+1)1}.$$

Since we automatically have $d_{b1} \geq 0$, we know that $\ell_{ao} + d_{a1} \geq \ell_{ao}$. Hence the full collection of inequalities between the exponents of highest terms can be captured by

$$\ell_{ao} \geq \ell_{(a+1)o} + d_{(a+1)1} \geq \ell_{(a+1)o}. \tag{45}$$

It is also not hard to argue that, conversely, any collection of exponents satisfying the inequalities (45) comes from the highest term of a monomial in the δ_k and $\delta_{(k-1,1)}$.

If we think of this algebra as computing tensor products of a general representation $\rho_n^{D_o}$ of GL_n with a symmetric power $S^d(\mathbf{C}^n)$ of the standard representation, then these inequalities can be interpreted nicely in terms of diagrams. The exponents ℓ_{ao} are the lengths of the rows of the diagram D_o, and the sums $\ell_{ao} + d_{a1}$ are the lengths of the rows of a constituent of the tensor product. The inequality $\ell_{ao} \geq \ell_{(a+1)o} + d_{(a+1)1}$ says that the length of a given row in the tensor product constituent is not longer than the next higher row of the original representation $\rho_n^{D_o}$. This is frequently called the "interleaving condition". This gives a nice way to present the possible constituents, and it also provides a good basis for inductive arguments, as we will see in the next section.

One can regard this situation as defining a nested pair of diagrams: the diagram D_o of the original ρ_n^D, and the diagram D_1 of the constituent of $\rho_n^D \otimes S^d(\mathbf{C}^n)$. One can distinguish the original D_o as a subdiagram of D_1 by leaving the boxes of D_o blank, while filling the boxes of $D_1 - D_o$ with a 1. In row a, the boxes of D_o are the leftmost ℓ_{ao} boxes, and the boxes with a 1 are the rightmost d_{a1} boxes. In other words, the a-th row of D_1 is obtained from the a-th row of D_o by adding d_{a1} boxes at the end, labeled with 1s. This is an example of a *skew tableau*. In terms of the tableau, the interleaving conditions can be interpreted as saying that the boxes with 1 in them form a *skew row*, meaning that no two are in the same column.

The lengths of the rows of D_o are ℓ_{ao}, and the lengths of the rows of D_1 are $\ell_{a1} = \ell_{ao} + d_{a1}$. The interleaving conditions are then expressed by the inequalities

$$\ell_{ao} \geq \ell_{(a+1)1} \geq \ell_{(a+1)o}. \tag{46}$$

Finally, also for later discussion, we note that, if we consider the set of points

$$\Gamma_{(n,m,1)} = \left\{ \begin{bmatrix} -a \\ -a \end{bmatrix} : 1 \leq a \leq m \right\} \quad \cup \quad \left\{ \begin{bmatrix} -a+1 \\ -a \end{bmatrix} : 1 \leq a \leq m+1 \right\}, \tag{47}$$

and if we define

$$\gamma\left(\begin{bmatrix} -a \\ -a \end{bmatrix} \right) = \ell_{ao} \qquad \text{and} \qquad \gamma\left(\begin{bmatrix} -a+1 \\ -a \end{bmatrix} \right) = \ell_{a1} = \ell_{ao} + d_{a1}, \tag{48}$$

then $\Gamma_{(n,m,1)}$ is a totally ordered subset of $\mathbf{Z}^2$ with its standard (coördinatewise) partial order, and γ is an increasing function on this set. Moreover, the map that takes the highest terms to the collection of such increasing functions on Γ is a bijective map of semigroups.

10 Iterated Pieri algebras

We now consider the algebra $P(M_{n(m+p)})$. (Note that we have abandoned ℓ as the index denoting the columns beyond the first m, because we need to use ℓ for indicating lengths of rows of diagrams.) We think of this as

$$P(M_{n(m+p)}) \simeq P(M_{nm}) \otimes P(M_{np}) \simeq P(M_{nm}) \otimes P(\mathbf{C}^n)^{\otimes p}. \tag{49}$$

That is, we are thinking of this as a GL_n-module obtained by starting with $P(M_{nm})$, and tensoring p times in succession with $P(\mathbf{C}^n)$.

We consider the GL_m highest weight vectors in this algebra, thinking of GL_m as a subgroup of GL_{m+p}. Note that this is the same as considering the highest weight vectors for the full product block diagonal subgroup $GL_m \times (GL_1)^p \subset GL_{m+p}$, since the unipotent radical of $(GL_1)^p$ is trivial. We have

$$P(M_{n(m+p)})^{U_m} \simeq P(M_{nm})^{U_m} \otimes P(\mathbf{C}^n)^{\otimes p}. \tag{50}$$

We want to describe the GL_n-module structure of this algebra. In other words, we want to describe the algebra

$$P(M_{n(m+p)})^{U_m \times U_n} \simeq \left(P(M_{nm})^{U_m} \otimes P(\mathbf{C}^n)^{\otimes p}\right)^{U_n}. \tag{51}$$

Because this algebra results from repeating the process that gave the Pieri rule, we call it an *iterated Pieri algebra*.

Each stage in the process leading to the iterated Pieri algebra is essentially the same as the creation of the original Pieri algebra: to pass from $P(M_{nm})^{U_m} \otimes P(\mathbf{C}^n)^{\otimes j}$ to $P(M_{nm})^{U_m} \otimes P(\mathbf{C}^n)^{\otimes (j+1)}$ we start with a representation of GL_n, corresponding to some diagram D_j, and we add boxes at the end of each row of D_j, observing the interleaving condition, to obtain a new diagram D_{j+1}. In more detail, if the a-th row of D_j has length ℓ_{aj}, then we add $d_{a(j+1)}$ boxes at the end to obtain a row of length $\ell_{a(j+1)} = \ell_{aj} + d_{a(j+1)}$. For the same reasons as in the case of the ordinary Pieri algebra, the interleaving conditions should hold: each row of D_{j+1} should not be longer than the previous row of D_j:

$$\ell_{aj} + d_{a(j+1)} \leq \ell_{(a-1)j}. \tag{52}$$

If for each j, we mark the boxes added in passing from D_j to D_{j+1} with the label $j+1$, then we will obtain a diagram with boxes labeled with nothing, or with a number from 1 to p. The labels will increase weakly as you move from left to right in a row. The boxes with a given label b will form a skew row. This is equivalent to saying that the labels in the (labeled) boxes increase strictly as you move down a column. Such a diagram with boxes labeled in this way is called a *semistandard tableau*.

We can explicitly exhibit GL_n highest weight vectors that will represent all these highest weights. These will be products of highest weight vectors of copies of the exterior powers $\Lambda^k(\mathbf{C}^n)$. These will correspond to column tableaux whose diagram consists of a single column, i.e., with all rows having length one. Denote the column tableau by T. Suppose that c rows of T are unlabeled, and that there are d rows with labels, and that the $r = (c+b)$-th row of T is labeled with j_b. Then we have the corresponding GL_n highest weight vector

$$\delta_T = \det \begin{bmatrix} x_{11} & x_{12} & \dots & x_{1c} & y_{1j_1} & y_{1j_2} & \dots & y_{1j_d} \\ x_{21} & x_{22} & \dots & x_{2c} & y_{2j_1} & y_{2j_2} & \dots & y_{2j_d} \\ x_{31} & x_{32} & \dots & x_{3c} & y_{3j_1} & y_{3j_2} & \dots & y_{3j_d} \\ \cdot & \cdot & \cdots & \cdot & \cdot & \cdot & \cdots & \cdot \\ \cdot & \cdot & \cdots & \cdot & \cdot & \cdot & \cdots & \cdot \\ \cdot & \cdot & \cdots & \cdot & \cdot & \cdot & \cdots & \cdot \\ \cdot & \cdot & \cdots & \cdot & \cdot & \cdot & \cdots & \cdot \\ x_{s1} & x_{s2} & \dots & x_{sc} & y_{sj_1} & y_{sj_2} & \dots & y_{sj_d} \end{bmatrix}. \tag{53}$$

Here $s = c + d$. If we extend the order of the variables described in §9 by ordering the y_{ab} in the same way as we ordered the x_b—by lexicographic ordering of the indices—and if we declare all x_{ab} to precede all y_{cd}, then the highest term of δ_T is easily determined to be just the diagonal term of δ_T:

$$LT(\delta_T) = \left(\prod_{a=1}^{c} x_{aa}\right)\left(\prod_{b=1}^{d} y_{(c+b)j_b}\right). \tag{54}$$

Now consider any semistandard skew tableau $\mathbf{T}$, and write it as a union of column tableaux T_c. with T_c being the c-th column of $\mathbf{T}$. Then set

$$\delta_{\mathbf{T}} = \prod_{c\geq 1} \delta_{T_c}. \tag{55}$$

Then we can check that

$$LT(\delta_{\mathbf{T}}) = \left(\prod_{a} x_{aa}^{\ell_{ao}}\right)\left(\prod_{b,c} y_{bc}^{d_{bc}}\right), \tag{56}$$

where ℓ_{ao} is the number of unlabeled boxes in the a-th row, and d_{bc} is the number of boxes in the b-th row labeled with a c. These exponents fill out the lattice cone specified by the interlacing conditions at each stage:

$$\begin{aligned} \ell_{a(k-1)} &= \ell_{ao} + \sum_{c=1}^{k-1} d_{ac} \leq \ell_{ao} + \sum_{c=1}^{k} d_{ac} \\ &= \ell_{ak} \leq \ell_{(a-1)(k-1)} = \ell_{(a-1)o} + \sum_{c=1}^{k-1} d_{(a-1)c}. \end{aligned} \tag{57}$$

As in the case of $p = 1$, there is a nice way to describe the interlacing conditions in terms of order-preserving functions on a partially ordered set, namely the subset of $\mathbf{Z}^2$ defined by the conditions

$$\Gamma_{(n,m,p)} = \left\{ \begin{bmatrix} -a+k \\ -a \end{bmatrix} : 1 \leq a \leq n;\ 0 \leq k \leq p;\ k \geq a-m \right\}. \tag{58}$$

If we define

$$\gamma_{\mathbf{T}}\left(\begin{bmatrix} -a+k \\ -a \end{bmatrix} \right) = \ell_{a,k}, \tag{59}$$

then the interlacing conditions simply translate to the statement that $\gamma_{\mathbf{T}}$ is an increasing function on $\Gamma_{(n,m,p)}$.

To summarize the argument of this section:

Theorem 10.1. *For the term order described above, the semigroup*

$$LT(P(M_{n(m+p)})^{U_m \times U_n})$$

is a lattice cone isomorphic to the cone $(\mathbf{Z}^+)^{\Gamma_{(n,m,p)},\succeq}$ *of order preserving, nonnegative integer-valued functions on the poset* $\Gamma_{(n,m,p)}$.

Remarks 10.2.

(a) The results above are a (perhaps slightly more general and flexible) combination of Hodge's standard monomial theory [19] combined with the Gelfand–Tsetlin [16] results on torus eigenbases for representations of GL_n, based on branching from GL_n to GL_{n-1}. The two points of view were finally combined in [15], which introduced the idea of toric deformation in this context (see also [53] for the Grassmannian case; this used highest terms), and these ideas were further refined in [38,45] and [32]. Extensions of toric deformation arguments to other flag algebras and more general multiplicity-free actions are given in [1,4,5,32].

(b) We would like to highlight the extent to which the logic of this account depends on the highest terms. First, using the skew tableau/GT pattern technology, one finds a combinatorial parametrization of the constituents of the tensor products that counts the multiplicity of any given representation in the indicated tensor products. Then one encodes the parameters as a collection of exponents, and observes that these exponents exactly constitute a lattice cone, isomorphic to the Hibi cone $(\mathbf{Z}^+)^{\Gamma_{(n,m,p)},\succeq}$. Finally, one defines a term order and produces highest weight vectors for GL_n whose highest terms have the previously described exponents. These highest weight vectors are monomials in an explicit set of elements, the δ_T for column tableaux T. The fact that these monomials have distinct highest terms guarantees that they are linearly independent, and therefore, the monomials for a given highest weight span a space of the required dimension. These monomials therefore exhaust the highest weight vectors of a given type, and therefore span the algebra being studied. It is the upper bound given by the combinatorial parameters, together with the lower bound given by the distinctness of the highest terms, and the matching of highest terms with combinatorial parameters, that guarantees everything is found.

In this reasoning, we have paid no direct attention to the algebra structure of the iterated Pieri algebra $\left(P(M_{nm} \otimes P(\mathbf{C}^n)^{\otimes p}\right)^{U_m \times U_n}$. We find at the end that the δ_T must be generators for the algebra, but we have not given any consideration to the relations between them. Thanks to general SAGBI theory, we have exhibited $\left(P(M_{nm}) \otimes P(\mathbf{C}^n)^{\otimes p}\right)^{U_m \times U_n}$ as a flat deformation of the Hibi ring on the poset $\Gamma_{(n,m,p)}$, and we note that the relations for a Hibi ring are easily described [18,22]. Although this does not determine the relations between the δ_T, it does allow one to make some conclusions about their qualitative nature. Detailed exploration of this topic is beyond the scope of this article.

(c) The above discussion was built around the case when $m < n$. If $m \geq n$, the set $\Gamma_{(n,m,p)}$ is somewhat simpler. Indeed, then the condition $k \geq a - m$ never applies, and can be omitted. Then the points of $\Gamma_{(n,m,p)}$ form a parallelogram with two horizontal sides and two diagonal sides. When $m < n$, the region also has a vertical side on the lower left, and when $m + p > n$, it is a pentagon.

11 Double Pieri algebras

We have been concentrating on describing $P(M_{n(m+p)})^{U_m}$ as a module for GL_n. However, the starting situation of $P(M_{nm})$ is symmetric in m and n, and we could clearly tensor $P(M_{nm})$ with copies of $P(\mathbf{C}^m)$ and get an analogous decomposition of the resulting GL_m module.

In fact, we can do both at the same time. Since the original ring is a sum of tensor products $\rho_n^D \otimes \rho_m^D$, and since the tensoring with $P(\mathbf{C}^n)$ (respectively, with $P(\mathbf{C}^m)$) has no effect on the module structure for the other member of the pair (GL_n, GL_m), we may conclude that the $GL_n \times GL_m$ constituents of

$$P(M_{nm}) \otimes \left(P(\mathbf{C}^n)^{\otimes p}\right) \otimes \left(P(\mathbf{C}^m)^{\otimes q}\right) = P(M_{nm}) \otimes P(M_{np}) \otimes P(M_{qm})$$

will be parametrized by pairs of semistandard skew tableaux (T, T'), one for GL_n (with up to n rows), and one for GL_m (with up to m rows). The tableaux are independent of each other, except that they must have the same initial diagram: $D_o = D_o'$.

We will refer to the algebras

$$\mathbf{L}_{(n,p)(m,q)} = (P(M_{nm}) \otimes P(M_{np}) \otimes P(M_{qm}))^{U_n \times U_m} \tag{60}$$

as *double (iterated) Pieri algebras*. It turns out that we can analyze them by a direct extension of the ideas used on the (single) iterated Pieri algebras

$$(P(M_{nm}) \otimes P(M_{np}))^{U_n \times U_m}.$$

In particular, we can write down a set of functions whose highest terms will encode a general pair (T, T') of the skew tableaux of the sort that arise.

Indeed, consider the functions

$$\delta_{(s;I;J)} = \det \begin{bmatrix} x_{11} & x_{12} & x_{13} & \cdots & x_{1(s+h)} & y_{1c_1} & y_{1c_2} & \cdots & y_{1c_g} \\ x_{21} & x_{22} & x_{23} & \cdots & x_{2(s+h)} & y_{2c_1} & y_{2c_2} & \cdots & y_{2c_g} \\ x_{31} & x_{32} & x_{33} & \cdots & x_{3(s+h)} & y_{3c_1} & y_{3c_2} & \cdots & y_{3c_g} \\ & \cdot & \cdot & \cdot & \cdot & \cdot & \cdot & \cdot & \\ & \cdot & \cdot & \cdot & \cdot & \cdot & \cdot & \cdot & \\ & \cdot & \cdot & \cdot & \cdot & \cdot & \cdot & \cdot & \\ x_{k1} & x_{k2} & x_{k3} & \cdots & x_{k(s+h)} & y_{kc_1} & y_{kc_2} & \cdots & y_{kc_g} \\ z_{d_11} & z_{d_12} & z_{d_13} & \cdots & z_{d_1(s+h)} & 0 & 0 & \cdots & 0 \\ z_{d_21} & z_{d_22} & z_{d_23} & \cdots & z_{d_2(s+h)} & 0 & 0 & \cdots & 0 \\ z_{d_31} & z_{d_32} & z_{d_33} & \cdots & z_{d_3(s+h)} & 0 & 0 & \cdots & 0 \\ & \cdot & \cdot & \cdot & \cdot & \cdot & \cdot & \cdot & \\ & \cdot & \cdot & \cdot & \cdot & \cdot & \cdot & \cdot & \\ & \cdot & \cdot & \cdot & \cdot & \cdot & \cdot & \cdot & \\ z_{d_h1} & z_{d_h2} & z_{d_h3} & \cdots & z_{d_h(s+h)} & 0 & 0 & \cdots & 0 \end{bmatrix}, \tag{61}$$

where $I = \{c_1 < c_2 < c_3 < \ldots < c_g\}$, and $J = \{d_1 < d_2 < d_3 < \ldots < d_h\}$ are subsets of the whole numbers from 1 to p, and from 1 to q, respectively. (Also, $k = s + g$, but we did not want to use such a cumbersome subscript for the row index of the matrix.)

Here the x_{ab} are the entries in a matrix in M_{nm}, the y_{ac} are the coordinates of the copies of $\mathbf{C}^n$, and the z_{db} are the coordinates of the copies of $\mathbf{C}^m$ (thought of as row vectors). We order these coordinates in similar fashion to earlier sections. We give each set of coordinates a total order defined by the lexicographic order on their subscripts (except, for the z_{db}, we look at the second index first). We then declare all the x_{ab} to dominate all the y_{ac}, which in their turn dominate all the z_{db}. (In fact, it is irrelevant which have precedence between the y_{ac} and the z_{db}, since these variables play essentially independent roles, as we will see.) For term order, we will use the graded lexicographic order with respect to this ordering of the variables.

With respect to this order, it is not difficult to check that

$$LT(\delta_{(s;I;J)}) = \left(\prod_{a=1}^{s} x_{aa}\right)\left(\prod_{i=1}^{g} y_{(s+i)c_i}\right)\left(\prod_{j=1}^{h} z_{d_j(s+j)}\right). \tag{62}$$

We see that this highest term records the data from one column tableau for GL_n, and a second column tableau for GL_m. The two tableaux have the same number of blank entries, namely s, corresponding to the factors x_{aa}, then they have arbitrary (and independent) increasing entries in the non-blank boxes. Thus, products of these elements can be used to create a highest term that encodes any pair of semistandard skew tableaux with the same diagram of blank boxes. The exponents of these leading terms will fill out a lattice cone defined by the interleaving conditions for each semistandard tableau.

This result can also be formulated cleanly in terms of Hibi rings. We have seen (Theorem 10.1) that, for the algebra $\left(P(M_{nm})^{U_m} \otimes P(\mathbf{C}^n)^{\otimes p}\right)^{U_n}$, the corresponding semigroup $LT(\left(P(M_{nm})^{U_m} \otimes P(\mathbf{C}^m)^{\otimes p}\right)^{U_n})$ of highest terms is isomorphic to the poset lattice cone $(\mathbf{Z}^+)^{(\Gamma_{(n,m,p)},\succeq)}$. A similar result of course holds for the iterated Pieri algebra $\left(P(M_{nm})^{U_n} \otimes P(\mathbf{C}^n)^{\otimes q}\right)^{U_m}$ for GL_m.

Now consider the reflection

$$\tau : \begin{bmatrix} a \\ b \end{bmatrix} \to \begin{bmatrix} b \\ a \end{bmatrix}$$

of $\mathbf{R}^2$. This of course preserves $\mathbf{Z}^2$, and defines an order preserving isomorphism of $\mathbf{Z}^2$. Thus, we could as well formulate Theorem 10.1 as saying that

$$\left(P(M_{nm})^{U_n} \otimes P(\mathbf{C}^m)^{\otimes q}\right)^{U_m} \simeq (\mathbf{Z}^+)^{(\tau(\Gamma_{(m,n,q)}),\succeq)}.$$

We further note that $\tau(\Gamma_{(m,n,q)})$ intersects $\Gamma_{(n,m,p)}$ in the set of their (common) intersection with the diagonal, i.e., the points

$$\left\{ \begin{bmatrix} -a \\ -a \end{bmatrix} : 1 \leq a \leq \min(n, m) \right\}.$$

Also, given a pair (T, T') of semistandard tableaux arising from the double Pieri algebra, the values of the functions γ_T in $(\mathbf{Z}^+)^{(\Gamma_{(n,m,p)},\succeq)}$ and of $\gamma_{T'}$ in $(\mathbf{Z}^+)^{(\tau(\Gamma_{(m,n,q)}),\succeq)}$ corresponding to T and to T' respectively, agree on the intersection $\tau(\Gamma_{(m,n,q)}) \cap \Gamma_{(n,m,p)}$, since these values just record the lengths of the rows of the original diagram from which the skew tableaux T and T' are built.

Therefore, the two functions γ_T and $\gamma_{T'}$ fit together to define a single function $\gamma_{T,T'}$ on

$$\Gamma_{(n,p)(m,q)} = \Gamma_{(n,m,p)} \cup \tau(\Gamma_{(m,n,q)}). \tag{63}$$

We claim that the function $\gamma_{T,T'}$ is an increasing function on $\Gamma_{(n,p)(m,q)}$. This follows because the intersection $\Gamma_{(n,m,p)} \cap \tau(\Gamma_{(m,n,q)})$ is "solid", in the sense that, if $\vec{\mu}$ is in $\Gamma_{(n,m,p)}$, and $\vec{\nu}$ is in $\tau(\Gamma_{(m,n,q)})$, and $\vec{\mu} \succeq \vec{\nu}$ with respect to the standard order on $\mathbf{Z}^2$, then there is a point $\vec{\eta}$ in $\Gamma_{(n,m,p)} \cap \tau(\Gamma_{(m,n,q)})$ for which $\vec{\mu} \succeq \vec{\eta} \succeq \vec{\nu}$. Thus,

$$\gamma_{T,T'}(\vec{\mu}) = \gamma_T(\vec{\mu}) \geq \gamma_T(\vec{\eta}) = \gamma_{T'}(\vec{\eta}) \geq \gamma_{T'}(\vec{\nu}) = \gamma_{T,T'}(\vec{\nu}).$$

Essentially the same argument applies when $\vec{\nu} \succeq \vec{\mu}$.

Thus, pairs (T, T') of semi-standard tableaux with equal unlabeled subdiagrams give rise to order preserving functions on $\Gamma_{(n,p)(m,q)}$. The converse may be shown by similar arguments. We can summarize this discussion with the following statement.

Theorem 11.1. *The semigroup of leading terms $LT(\mathbf{L}_{(n,p)(m,q)})$ of the double Pieri algebra $\mathbf{L}_{(n,p)(m,q)}$, with respect to the term order described above, is isomorphic to the poset lattice cone $(\mathbf{Z}^+)^{(\Gamma_{(n,p)(m,q)},\succeq)}$.*

Remarks 11.2.

(a) Although the theory of diagrams associated to the iterated Pieri algebras has a long history and a substantial literature, going back to [16], diagrams for objects like the double Pieri algebra have received much less attention. The paper [10] discusses double tableaux in connection with describing a $GL_n \times GL_m$-compatible basis for $P(M_{nm})$; this work and [11] and [8] use double tableaux to establish the first and second fundamental theorems of classical invariant theory, in the sense of Weyl [56], in a characteristic free manner.
(b) The iterated Pieri algebras of §10 appear as the "boundary values" $\mathbf{L}_{(n,p)(m,0)}$ and $\mathbf{L}_{(n,0)(m,q)}$ of the four-parameter family $\mathbf{L}_{(n,p)(m,q)}$ of double Pieri algebras.

12 Applications: Pieri rules for classical groups

Because of the reciprocity discussed in §6, between tensor products and branching, the description of the iterated Pieri algebras $P(M_{n(m+p)})^{U_m \times U_n}$ given above implies the Hodge–Gelfand–Tsetlin description of the flag algebra for GL_n [15, 16, 19, 32, 38, 53], and indeed, it was in this context that the technology of tableaux and patterns was developed. That connection in itself should be sufficient impetus for studying the iterated Pieri algebras.

However, Pieri algebras and closely related Hibi rings play a much wider role in representation theory. The description of the flag algebra for Sp_{2n} [32], and the branching rules from Sp_{2n} to Sp_{2k} for $k < n$ [33] is quite parallel to GL_n, even though there is no multiplicity-one branching rule to get things started. (Some would argue that there is such a multiplicity-one branching rule [47], but it requires leaving the family of reductive groups.) Also, analogs of Gelfand–Tsetlin patterns for symplectic groups go back to Zhelobenko [37, 46, 58].) Indeed, the branching algebras for Sp_{2n} are isomorphic to appropriate branching algebras for GL_{2n} [36]. Hibi rings also describe many branching rules for O_n, but only with some restrictions, known as the *stable range*. It seems likely that describing the full flag algebra for SO_n will require a mild generalization of Hibi rings.

Moreover, Pieri algebras provide a context for understanding a wide range of phenomena in the representation theory of the classical groups. It is explained in [26] how to use the iterated Pieri algebras in a proof of the Littlewood–Richardson rule, from which a large family of branching rules for classical symmetric pairs can be understood [28].

With Pieri algebras and related Hibi rings being so useful in representation theory, the extension to the double Pieri algebras is then perhaps justified by its simplicity. However, these algebras also appear naturally in some calculations for the classical groups. In particular, they provide a description of the analog for other classical groups of the iterated Pieri rule.

As noted in §6, the Pieri rule for GL_n has two interpretations, related by a reciprocity law: it describes either tensoring a representation with a representation corresponding to a diagram with only one row, or restricting a representation from GL_{n+1} to GL_n. For the orthogonal group O_n or the symplectic group Sp_{2m}, branching to smaller subgroups of the same type bears strong similarities to the case of GL_n, as mentioned above. However, tensoring with representations whose associated diagram has only one row – symmetric powers of the standard representation of Sp_{2n}, or spherical harmonics for O_n, is more complicated than for GL_n; in particular, the resulting tensor products are not always multiplicity-free. In [54], a description of the tensoring with one-rowed representations is given in terms of diagrams, but it involves erasing boxes as well as adding boxes, making iteration not straightforward.

In this section, we will show how to combine the results above with some facts from classical invariant theory (in the formulation of [20]) to describe the analogs for symplectic groups of the Pieri rule and the iterated Pieri rule. An essentially parallel discussion can be given for orthogonal groups, and also a similar treatment of a more general Pieri rule for GL_n, that considers tensoring not only with the symmetric powers $S^d(\mathbf{C}^n)$ of the standard representation, but also with their duals. These are discussed in more detail in [25].

We recall some facts from the dual pair formulation of classical invariant theory [20]. Let $Sp_{2n} \subset GL_{2n}$ be the subgroup that preserves the symplectic form

$$< \vec{u}, \vec{v} >= \sum_{i=1}^{n} u_i v_{(2n+1-i)} - u_{(2n+1-i)} v_i, \tag{64}$$

for vectors

$$\vec{u} = \begin{bmatrix} u_1 \\ u_2 \\ u_3 \\ . \\ . \\ . \\ u_{2n} \end{bmatrix} \quad \text{and} \quad \vec{v} = \begin{bmatrix} v_1 \\ v_2 \\ v_3 \\ . \\ . \\ . \\ v_{2n} \end{bmatrix} \quad \text{in } \mathbf{C}^{2n}.$$

Let Sp_{2n} act on $M_{2nm} \simeq (\mathbf{C}^{2n})^m$ by the restriction of the action (20)) of GL_{2n}. Then the symplectic pairings

$$r_{cd} = \sum_{i \le n} x_{ic} x_{(2n+1-i)d} - x_{(2n+1-i)c} x_{id}, \tag{65}$$

of the columns of X in M_{2nm} give polynomials that are invariant under the action of Sp_{2n}. They span a space that is invariant under the action of GL_m on the right, and under this action, their span is irreducible and isomorphic to $\Lambda^2(\mathbf{C}^p)$, the second exterior power of $\mathbf{C}^p$ (aka, the skew-symmetric $p \times p$ matrices). The First Fundamental Theorem (FFT) of classical invariant theory [17, 56] says that these quadratic polynomials generate the full algebra $P(M_{2nm})^{Sp} = I_{Sp}$ of Sp_{2n}-invariant polynomials.

Let

$$\Delta_{cd} = \sum_{i \leq n} \frac{\partial^2}{\partial x_{ic} \partial x_{(2n+1-i)d}} - \frac{\partial^2}{\partial_{x_{(2n+1-i)c}} \partial_{x_{id}}} \tag{66}$$

be the *partial Laplacians*. They are second order, constant coefficient differential operators, and they are also invariant under (that is, they commute with) the action of Sp_{2n}. Moreover, they generate the full algebra of Sp_{2n}-invariant constant coefficient differential operators. This fact is more or less the Fourier transform of the FFT.

By use of the partial Laplacians, the FFT can be extended to describe the full isotypic decomposition of Sp_{2n} acting on $P(M_{2nm})$ [20]. Define

$$H(M_{2nm}, Sp) = \cap_{c,d} \ker \Delta_{cd} = \cap_{D \in \Lambda^2(\mathbf{C}^m)} \ker D. \tag{67}$$

These are the Sp_{2n}*-harmonics* in $P(M_{2nm})$.

Then:

(i) The multiplication map

$$H(M_{2nm}, Sp) \otimes I_{Sp} \rightarrow P(M_{2nm}) \tag{68}$$

is onto.

(ii)

$$H(M_{2nm}, Sp) \simeq \sum_D \sigma_G^D \otimes \rho_m^D \tag{69}$$

as $Sp_{2n} \times GL_m$-module, where σ^D is the irreducible Sp_{2n} representation generated by the GL_{2n} highest weight vector in ρ_{2n}^D, for D with at most n rows. (As was the case for (GL_{2n}, GL_m)-duality, the diagram D here should also have at most m rows.)

Moreover, when m is small compared to n (precisely, when $m < n$; this condition is referred to as the *stable range*), we have the following additional properties.

(i) $I_{Sp} \simeq P(\Lambda^2(\mathbf{C}^m))$ is the full polynomial algebra on the generators r_{cd}; that is, the r_{cd} are algebraically independent.

(ii) $H(M_{2nm})^{U_{Sp_{2n}}} = P(M_{2nm})^{U_{GL_{2n}}}$ is a subalgebra of $P(M_{2nm})$. (70)

(iii) $H(M_{2nm}, Sp_{2n}) \otimes I_{Sp} \simeq P(M_{2nm})$. That is, the multiplication

mapping is a linear isomorphism.

(Indeed, an isomorphism of $(Sp_{2n} \times GL_m)$-modules.)

In the stable range, we can describe the analog for Sp_{2n} of the iterated Pieri algebra, using a double Pieri algebra for GL_{2n}. We want to consider tensor products of a general representation σ^D of Sp_{2n} with the symmetric powers $S^d(\mathbf{C}^{2n})$ of the basic representation of Sp_{2n}. (The restriction of $S^d(\mathbf{C}^{2n})$ from GL_{2n} to Sp_{2n} remains irreducible.)

As a model for the representations of Sp_{2n}, we will use $H(M_{2nm}, Sp)^{U_{GL_m}}$, the GL_m highest weight vectors in the Sp_{2n} harmonics. According to equation (69), this contains one copy of each irreducible representation $\sigma^D_{Sp_{2n}}$ for diagrams D having up to m rows (assuming that $m < n$, which is the stable range). According to equation (68), we can write

$$H(M_{2nm}, Sp) \simeq P(M_{2nm})/J, \tag{71}$$

where J is the ideal generated by the space $\Lambda^2(\mathbf{C}^m)$ of quadratic Sp invariants (cf. formula (65)).

Thus, for the iterated Pieri Rule, we want to compute

$$\left((P(M_{2nm})/J)^{U_{GL_m}} \otimes P(M_{2nq})\right)^{U_{Sp_{2n}}}.$$

We proceed as follows. We abbreviate

$$U_{GL_m} = U'_m, U_{Sp_{2n}} = U_{Sp}, \text{ and } U_{GL_{2n}} = U_{2n}.$$

We calculate

$$\begin{aligned}\Big((P(M_{2nm})/J)^{U'_m} \otimes P(M_{2nq})\Big)^{U_{Sp}} &\simeq \left(P(M_{2nm})/J \otimes P(M_{2nq})\right)^{U'_m \times U_{Sp}} \\ &\simeq \left(P(M_{2n(m+q)})/J_1)\right)^{U'_m \times U_{Sp}}\end{aligned} \tag{72}$$

where J_1 denotes the ideal in $P(M_{2n(m+q)})$ generated by $\Lambda^2(\mathbf{C}^m)$. Continuing, this equals

$$\begin{aligned}&\left((H(M_{2n(m+q)}) \otimes P(\Lambda^2(\mathbf{C}^{m+q})))/J_1\right)^{U'_m \times U_{Sp}} \\ &\simeq \left((H(M_{2n(m+q)}) \otimes P(\Lambda^2(\mathbf{C}^m) \oplus \Lambda^2(\mathbf{C}^q) \oplus (\mathbf{C}^m \otimes \mathbf{C}^q)))/J_1\right)^{U'_m \times U_{Sp}} \\ &\simeq \left(H(M_{2n(m+q)}) \otimes P(\Lambda^2(\mathbf{C}^q) \oplus (\mathbf{C}^m \otimes \mathbf{C}^q))\right)^{U'_m \times U_{Sp}}\end{aligned} \tag{73}$$

$$
\begin{aligned}
&\simeq \left(H(M_{2n(m+q)})^{U_{Sp}} \otimes P(\Lambda^2(\mathbf{C}^q)) \otimes P(M_{mq})\right)^{U'_m} \\
&\simeq \left(H(M_{2n(m+q)})^{U_{Sp}} \otimes P(M_{mq})\right)^{U'_m} \otimes P(\Lambda^2(\mathbf{C}^q)) \\
&= \left((P(M_{2n(m+q)})^{U_{2n}} \otimes P(M_{mq})\right)^{U'_m} \otimes P(\Lambda^2(\mathbf{C}^q)) \\
&\simeq \left((P(M_{2nm}) \otimes P(M_{2nq}))^{U_{2n}} \otimes P(M_{mq})\right)^{U'_m} \otimes P(\Lambda^2(\mathbf{C}^q)) \\
&\simeq \left(P(M_{2nm}) \otimes P(M_{2nq}) \otimes P(M_{mq})\right)^{U_{2n}\times U'_m} \otimes P(\Lambda^2(\mathbf{C}^q)) \\
&= \mathbf{L}_{(2n,q)(m,q)} \otimes P(\Lambda^2(\mathbf{C}^q)).
\end{aligned}
$$

Here $\mathbf{L}_{(2n,q)(m,q)}$ is as in formula (60).

References

1. V. Alexeev and M. Brion, Toric degenerations of spherical varieties, *Selecta Math. (N.S.)* **10** (2004), 453–478.
2. A. Borel, *Linear Algebraic Groups*. Second edition. Graduate Texts in Mathematics, Vol. 126. Springer-Verlag, New York, 1991.
3. W. Bruns and J. Herzog, *Cohen-Macaulay Rings*, Cambridge Univ. Press, 1998.
4. P. Caldero, Toric degenerations of Schubert varieties, *Trans. Groups* **7** (2002), 51–60.
5. Chirivi, R. LS algebras and application to Schubert varieties. *Trans. Groups* **5** (2000), 245–264.
6. A. Conca, J. Herzog and G. Valla, SAGBI basis with applications to blow-up algebras, *J. Reine Angew. Math.* **474** (1996), 113–138.
7. D. Cox, J. Little and D. O'Shea, *Ideals, Varieties, and Algorithms. An Introduction to Computational Algebraic Geometry and Commutative Algebra*, Second edition, Undergraduate Texts in Mathematics, Springer-Verlag, New York, 1997.
8. C. DeConcini and C. Procesi, A characteristic-free approach to invariant theory, *Adv. Math.* **21** (1976), 330–354.
9. R. Dehy and R. W. T. Yu, Degeneration of Schubert varieties of SL_n/B to toric varieties, *Ann. Inst. Fourier (Grenoble)* **51** (2001), 1525–1538.
10. J. Désarménien, J. Kung, and G.-C. Rota, Invariant theory, Young bitableaux, and combinatorics, *Adv. Math.* **27** (1978), 63–92.
11. P. Doubilet, G.-C. Rota and J. Stein, *Foundations of Combinatorial Theory, IX*, Stud. in App. Math. **53** (1974), 185–216.
12. W. Fulton, *Introduction to Toric Varieties*, Ann. Math. Studies, Princeton University Press, Princeton, New Jersey, 1993.
13. W. Fulton, *Young Tableaux*, Cambridge University Press, Cambridge, UK, 1997.
14. W. Fulton, Eigenvalues, invariant factors, highest weights, and Schubert calculus, *Bull. Amer. Math. Soc.* **37** (2000), 209–249.
15. N. Gonciulea and V. Lakshmibai, Degenerations of flag and Schubert varieties to toric varieties, *Trans. Groups* **1** (1996), 215–248.
16. I. M. Gelfand and M. L. Tsetlin, Finite-dimensional representations of the group of unimodular matrices. (Russian) *Doklady Akad. Nauk SSSR* (N.S.) **71**, (1950). 825–828. English translation: I. M. Gelfand, *Collected Papers. Vol II*. Edited by S. G. Gindikin, V. W. Guillemin, A. A. Kirillov, B. Kostant, and S. Sternberg. Springer-Verlag, Berlin, 1988, 653–656.
17. R. Goodman and N. Wallach, *Representations and Invariants of the Classical Groups*, Cambridge Univ. Press, 1998.

18. T. Hibi, Distributive lattices, affine semigroup rings and algebras with straightening laws, in: Commutative Algebra and Combinatorics, Adv. Stud. Pure Math., **11**, 1987, 93–109.
19. W. Hodge, Some enumerative results in the theory of forms, *Proc. Cam. Phi. Soc.* **39** (1943), 22–30.
20. R. Howe, Remarks on classical invariant theory, *Trans. Amer. Math. Soc.* **313** (1989), 539–570.
21. R. Howe, Perspectives on invariant theory, *The Schur Lectures, I.* Piatetski-Shapiro and S. Gelbart (eds.), Israel Mathematical Conference Proceedings, 1995, 1–182.
22. R. Howe, Weyl Chambers and Standard Monomial Theory for Poset Lattice Cones, *Q. J. Pure Appl. Math.* **1** (2005) 227–239.
23. R. Howe, *A Century of Lie theory*. American Mathematical Society centennial publications, Vol. II (Providence, RI, 1988), Amer. Math. Soc., Providence, RI, 1992, 101–320.
24. R. Howe, S. Jackson, S. T. Lee, E-C Tan and J. Willenbring, Toric degeneration of branching algebras, *Adv. Math.* **220** (2009), 1809–1841.
25. R. Howe, S. Kim and S.T. Lee, Iterated Pieri algebras and Hibi algebras, preprint.
26. R. Howe and S. T. Lee, Why should the Littlewood–Richardson rule be true?, *Bull. Amer. Math. Soc.* (N.S.) **49** (2012), 187–236.
27. R. Howe, E-C. Tan and J. Willenbring, Reciprocity algebras and branching for classical symmetric pairs, in *Groups and Analysis - the Legacy of Hermann Weyl*, London Mathematical Society Lecture Note Series No. 354, Cambridge University Press, 2008, 191–231.
28. R. Howe, E-C. Tan and J. Willenbring, Stable branching rules for classical symmetric pairs, *Trans. Amer. Math. Soc.* **357** (2005) 1601–1626.
29. R. Howe, E-C. Tan and J. Willenbring, A basis for GL_n tensor product algebra, *Adv. in Math.* **196** (2005), 531–564.
30. M. Kashiwara, Crystallizing the q-analogue of universal enveloping algebras, *Comm. Math. Phys.* **133** (1990), 249–260.
31. K. Kaveh, SAGBI bases and degeneration of spherical varieties to toric varieties, *Michigan Math. J.*, **53** (2005), 109–121.
32. S. Kim, Standard monomial theory for flag algebras of $GL(n)$ and $Sp(2n)$, *J. Algebra* **320** (2008), 534–568.
33. S. Kim, Standard monomial bases and degenerations of $SO_m(\mathbf{C})$ representations, *J. Algebra* **322** (2009), 3896–3911.
34. S. Kim, Distributive lattices, affine semigroups, and branching rules of the classical groups, preprint arXiv:1004.0054
35. S. Kim and S. T. Lee, Pieri algebras for the orthogonal and symplectic groups, Israel J. Math., **195** (2013), 215–245.
36. S. Kim and O. Yacobi, A basis for the symplectic group branching algebra, J. Algebraic Combin., **35** (2012), 269–290.
37. A. A. Kirillov, A remark on the Gelfand–Tsetlin patterns for symplectic groups. *J. Geom. Phys.* **5** (1988), 473–482.
38. M. Kogan and E. Miller, Toric degeneration of Schubert varieties and Gelfand–Tsetlin polytopes, *Adv. Math.* **193** (2005), 1–17.
39. V. Lakshmibai, The development of standard monomial theory, II. *A tribute to C. S. Seshadri*, (Chennai, 2002), Birkhäuser, Basel, 2003, 283–309.
40. V. Lakshmibai and K. Raghavan, *Standard Monomial Theory. Invariant theoretic approach.* Encyclopaedia of Mathematical Sciences, **137**. Invariant Theory and Algebraic Transformation Groups, 8. Springer-Verlag, Berlin, 2008.
41. V. Lakshmibai, P. Littelmann, P. Magyar, Standard monomial theory and applications. Notes by Rupert W. T. Yu. NATO Adv. Sci. Inst. Ser. C Math. Phys. Sci., **514**; *Representation theories and algebraic geometry* (Montreal, PQ, 1997), Kluwer Acad. Publ., Dordrecht, 1998, 319–364.
42. P. Littelmann, Paths and root operators in representation theory, Ann. of Math. **142** (1995), 499–525.
43. G. Lusztig, Canonical bases arising from quantized enveloping algebras, *J. Amer. Math. Soc.* **3** (1990), 447–498.
44. I. Macdonald, *Symmetric Functions and Hall Polynomials*, Oxford University Press, 1995.

45. E. Miller and B. Sturmfels, *Combinatorial Commutative Algebra*, Graduate Texts in Mathematics **227**, Springer-Verlag, New York, 2005.
46. A. Molev, A basis for representations of symplectic Lie algebras, *Comm. Math. Phys.* **201** (1999), 591–618.
47. R. Proctor, Odd symplectic groups. *Invent. Math.* **92** (1988), 307–332.
48. R. Proctor, Young tableaux, Gelfand patterns, and branching rules for classical groups. J. Algebra **164** (1994), 299–360.
49. V. Reiner, Signed posets. J. Combin. Theory Ser. A **62** (1993), 324–360.
50. L. Robbiano and M. Sweedler, Subalgebra bases, Proceedings Salvador 1988 Eds. W. Bruns, A. Simis, Lecture Notes Math **1430** (1990), 61–87.
51. M. Schützenberger, La correspondance de Robinson, in *Combinatoire et représentations du groupe symétrique* (D. Foata, ed.), Lecture Notes in Mathematics **579**, 1976, 59–113.
52. R. Stanley, *Enumerative Combinatorics*, Volume 2, Cambridge University Press, 1999.
53. B. Sturmfels, Gröbner Bases and Convex Polytopes, Univ. Lecture Series, Vol. 8, Amer. Math. Soc., Providence, RI, 1996.
54. S. Sundaram, Orthogonal tableaux and an insertion algorithm for SO(2n + 1). J. Combin. Theory Ser. A **53** (1990), 239–256.
55. N. Wallach, Quantum computing and entanglement for mathematicians, in *Representation theory and complex analysis*, Lecture Notes in Math. **1931**, Springer, Berlin, 2008, 345–376.
56. H. Weyl, *The Classical Groups*, Princeton University Press, Princeton, N.J., 1946.
57. D. Zhelobenko, *Compact Lie Groups and Their Representations*, Translations of Mathematical Monographs **40**, American Mathematical Society, Providence, R.I., 1973.
58. D. Zhelobenko, On Gelfand–Zetlin bases for classical Lie algebras, in: Kirillov, A. A. (ed.) *Representations of Lie groups and Lie algebras*, Budapest: Akademiai Kiado: 1985, 79–106.

Action of the conformal group on steady state solutions to Maxwell's equations and background radiation

Bertram Kostant and Nolan R. Wallach

The first author extends birthday greetings to the second author Nolan Wallach. Nolan has been a long-time highly-valued friend and has been a brilliant, inspiring, and creative collaborator.

— From the first author

Abstract The representation of the conformal group ($PSU(2,2)$) on the space of solutions to Maxwell's equations on the conformal compactification of Minkowski space is shown to break up into four irreducible unitarizable smooth Fréchet representations of moderate growth. An explicit inner product is defined on each representation. The energy spectrum of each of these representations is studied and related to plane wave solutions. The steady state solutions whose luminosity (energy) satisfies Planck's Black Body Radiation Law are described in terms of this analysis. The unitary representations have notable properties. In particular they have positive or negative energy, they are of type $A_{\mathfrak{q}}(\lambda)$ and are quaternionic. Physical implications of the results are explained.

Keywords: Maxwell's equations • Conformal compactification • Conformal group • Unitary representations • Background radiation

Mathematics Subject Classification: 78A25, 58Z05, 22E70, 22E45

Research partially supported by NSF grant DMS 0963035.

B. Kostant
Department of Mathematics, MIT, Cambridge, MA 02139, USA
e-mail: kostant@math.mit.edu

N.R. Wallach
Department of Mathematics, University of California at San Diego,
La Jolla, CA 92093-0112, USA
e-mail: nwallach@ucsd.edu

R. Howe et al. (eds.), *Symmetry: Representation Theory and Its Applications*,
Progress in Mathematics 257, DOI 10.1007/978-1-4939-1590-3_14

1 Introduction

The purpose of this paper is to analyze the steady state solutions of Maxwell's equations in a vacuum using the tools of representation theory. By steady state we mean those solutions that extend to the conformal compactification of Minkowski space (cf. [D]). We will now explain this. We look upon the solutions of Maxwell's equations as tensor valued on $\mathbb{R}^4$ with the flat Lorentzian metric given by

$$-dx_1^2 - dx_2^2 - dx_3^2 + dt^2,$$

where x_1, x_2, x_3 and $x_4 = t$ yield the standard coordinates of $\mathbb{R}^4$. The space $\mathbb{R}^4$ with this metric will be denoted by $\mathbb{R}^{1,3}$. The conformal compactification is the space $S^3 \times_{\{\pm 1\}} S^1$ (modulo the product action of ± 1) with (up to a positive scalar multiple on both factors) the product metric with the negative of the constant curvature 1 metric on S^3 and the usual metric on S^1. The injection f of $\mathbb{R}^4$ into $S^3 \times_{\{\pm 1\}} S^1$ is the inverse of a variant of stereographic projection (see Section 2). This embedding is not an isometry, but it is conformal. Our approach to Maxwell's equations uses the equivalent formulation in terms of differential 2-forms on Minkowski space. More generally, if (M, g) is an oriented Lorentzian four manifold (signature $(-,-,-,+)$ or $(1,3)$) and if " $*$ " denotes the Hodge star operator on 2-forms relative to the volume form γ, with $g(\gamma,\gamma) = -1$, then there is a version of Maxwell's equations on $\Omega^2(M)$ (differential two forms) given by

$$d\omega = d * \omega = 0.$$

Let M and N be four-dimensional Lorentzian manifolds, let $F : M \to N$ be a conformal transformation, and let ω be a solution to Maxwell's equations on N. Then $F^*\omega$ is a solution to Maxwell's equations on M. Since f is conformal we see that the pullback of solutions to Maxwell's equations on $S^3 \times_{\{\pm 1\}} S^1$ yields solutions to the usual Maxwell equations.

The group of conformal transformations of $S^3 \times_{\{\pm 1\}} S^1$ is locally the group $SO(4,2)$ and thus the solutions to Maxwell's equations on $S^3 \times_{\{\pm 1\}} S^1$ form a representation of this group. We can interpret this as follows: We first note that we can replace $S^3 \times_{\{\pm 1\}} S^1$ with the group $U(2)$. If on Lie$(U(2))$ we put the Lorentzian form that corresponds to the quadratic form $-\det X$, then the corresponding bi-invariant metric on a $U(2)$ is isometric, up to positive scalar multiple, with $S^3 \times_{\{\pm 1\}} S^1$. We interpret this space as the Shilov boundary of the Hermitian symmetric space that corresponds to $G = SU(2,2)$ (which is locally isomorphic with $SO(4,2)$).

We denote by Maxw the space of complex solutions in $\Omega^2(U(2))$ to Maxwell's equations. We show that there is a canonical nondegenerate G-invariant Hermitian form on Maxw. Further, we show that as a smooth Fréchet representation of G, Maxw splits into the direct sum of four irreducible (Fréchet) representations (of moderate growth) that are mutually orthogonal relative to the form. This form is positive definite on two of the irreducible pieces and negative definite on the other two. Since the $K = S(U(2) \times U(2))$ isotypic components of Maxw are all finite dimensional we see that this yields four unitary irreducible representations of G. Two of the representations are holomorphic (negative energy in the physics literature) and two are anti-holomorphic (positive energy). We also describe them in terms of the $A_{\mathfrak{q}}(\lambda)$ that yield second continuous cohomology (the four theta stable parabolics $\mathfrak{q}$ involved relate these representations to twister theory) and in terms of quaternionic representations ($SU(2,2)$ is the quaternionic real form of $SL(4,\mathbb{C})$). These representations are actually representations of $PSU(2,2)$. In this group there is a dual pair $PSU(1,1), SO(3)$ establishing an analogue of Howe duality in each of the four representations. The realization of these representations is intimately related to the work in [K]. We use the decomposition of the restriction of these representations to $PSU(1,1)$ to analyze the frequency distribution of the solutions in each of the $PSU(2,2)$ representations.

We interpret the plane wave solutions as generalized Whittaker vectors on Maxw and the solutions as wave packets of the Whittaker vectors. We study those wave packets with frequency spectrum fitting Planck's black body radiation law for intensity and arbitrary temperature. This means that we can fit our solutions to the measured background radiation on a steady state universe. We give a short discussion of how Segal's equilibrium solution [S1] might be related to our analysis. We make no assertions as to how such a steady state universe might physically exist. There are many suggestions in the literature (e.g., the work of Hoyle et al. [HBN]). All seem complicated. However, we will content ourselves with the assertion that the big bang is not necessarily the only possible interpretation of background radiation.

There is also an interpretation of the red shift that can be gleaned from this work involving the relationship between the measurement of time from the proposed "big bang" and the steady state "time" which is periodic but with a large period appearing to move faster as we look backwards or forwards in terms of "standard" time. (See [S2].)

We are aware that many of the aspects of representation theory in this paper could have been done in more generality. We have constrained our attention to the four representations at hand since the main thrust of this paper is to show how representation theory can be used to study well-known equations in physics.

Parts of this work should be considered expository. Related work has been done by [HSS] on the action of the conformal group on solutions of the wave equation and [EW] relating positive energy representations to generalized Dirac equations.

We would like to thank the referee for the careful review of this paper, for the explanation of Segal's work on the background radiation and for pointing out many references that we should have included.

2 Conformal compactification of Minkowski space

Let $\mathbb{R}^{1,3}$ denote $\mathbb{R}^4$ with the pseudo-Riemannian (Lorentzian) structure given by $(x,y) = -x_1y_1 - x_2y_2 - x_3y_3 + x_4y_4$. Here $x_i, i = 1,2,3,4$, are the standard coordinates on $\mathbb{R}^4$ and we identify the tangent space at every point with $\mathbb{R}^4$.

We can realize this space as the space of 2×2 Hermitian matrices (a 4-dimensional vector space over $\mathbb{R}$), V, with the Lorentzian structure corresponding to the quadratic form given by the determinant. Note that

$$\det \begin{bmatrix} x_4 + x_3 & x_1 + i x_2 \\ x_1 - i x_2 & x_4 - x_3 \end{bmatrix} = (x, x).$$

We can also realize the space in terms of skew-Hermitian matrices $\mathfrak{u}(2) = iV$ and noting that the form becomes $-\det$. With this interpretation, and realizing that $\mathfrak{u}(2) = \mathrm{Lie}(U(2))$ we have an induced Lorentzian structure, $\langle \ldots, \ldots \rangle$ on $U(2)$. We also have a transitive action of $K = U(2) \times U(2)$ on $U(2)$ by right and left translation

$$(g_1, g_2)u = g_1 u g_2^{-1}.$$

Since the isotropy group at I is $M = \mathrm{diag}(U(2)) = \{(g, g) | g \in U(2)\}$, we see that K acts by isometries on $U(2)$ with this structure.

We will now consider a much bigger group that acts. We first consider the indefinite unitary group $G \cong U(2,2)$ given by the elements, $g \in M_4(\mathbb{C})$ such that

$$g \begin{bmatrix} 0 & iI_2 \\ -iI_2 & 0 \end{bmatrix} g^* = \begin{bmatrix} 0 & iI_2 \\ -iI_2 & 0 \end{bmatrix}.$$

Here as usual g^* means conjugate transpose. $G \cap U(4)$ is the group of all matrices of the 2×2 block form

$$\begin{bmatrix} A & B \\ -B & A \end{bmatrix}$$

satisfying $AB^* = BA^*$ and $AA^* + BB^* = I$. These equations are equivalent to the condition

$$A \pm iB \in U(2).$$

It is easy to see that the map

$$\Psi : G \cap U(4) \to U(2) \times U(2)$$

given by

$$\begin{bmatrix} A & B \\ -B & A \end{bmatrix} \mapsto (A - iB, A + iB)$$

defines a Lie group isomorphism. This leads to the action of $G \cap U(4)$ on $U(2)$ given by

$$\begin{bmatrix} A & B \\ -B & A \end{bmatrix} \cdot x = (A - iB)x(A + iB)^{-1}.$$

Note that the stabilizer of I_2 is the subgroup isomorphic with M given by the elements

$$\begin{bmatrix} A & 0 \\ 0 & A \end{bmatrix}, \ A \in U(2).$$

We extend this to a map of G to $U(2)$ given by

$$\Phi : \begin{bmatrix} A & B \\ C & D \end{bmatrix} \mapsto (A + iC)(A - iC)^{-1}.$$

This makes sense since $A - iC$ is invertible if $\begin{bmatrix} A & B \\ C & D \end{bmatrix} \in G$. We consider the subgroup P of G that consists of the matrices

$$\begin{bmatrix} g & gX \\ 0 & (g^*)^{-1} \end{bmatrix}$$

with $g \in GL(2, \mathbb{C})$ and $X \in H$ (in other words, $X^* = X$). Then every element of G can be written in the form kp with $k \in G \cap U(4)$ and $p \in P$. We note that $\Phi(kp) = \Phi(k) = k \cdot I$. Now, $U(4) \cap K$ acts transitively on G/P and the stabilizer of the identity coset is the group $\begin{bmatrix} A & 0 \\ 0 & A \end{bmatrix}$, $A \in U(2)$. We will identify K with $U(4) \cap G$ (under Ψ) and M with the stabilizer of the identity. Thus $G/P = K/M$.

We consider the subgroup $\overline{N}$:

$$\begin{bmatrix} I & 0 \\ Y & I \end{bmatrix}, Y^* = Y.$$

If we write

$$\begin{bmatrix} I & 0 \\ Y & I \end{bmatrix} = kp$$

with

$$k = \begin{bmatrix} A & B \\ -B & A \end{bmatrix}$$

as above, then $A \in GL(2,\mathbb{C})$ and

$$-BA^{-1} = Y.$$

One can see that if we set

$$k(Y) = \begin{bmatrix} \frac{I}{\sqrt{I+Y^2}} & \frac{-Y}{\sqrt{I+Y^2}} \\ \frac{Y}{\sqrt{I+Y^2}} & \frac{I}{\sqrt{I+Y^2}} \end{bmatrix},$$

then

$$k(Y)P = \begin{bmatrix} I & 0 \\ Y & I \end{bmatrix} P.$$

This gives an embedding of H into $U(2)$

$$Y \longmapsto (I + iY)(I - iY)^{-1},$$

the Cayley transform. We next explain how this is related to the Cayley transform in the sense of bounded symmetric domains.

We note that it is more usual to look upon G (in its more usual incarnation) as the group of all elements $g \in GL(4,\mathbb{C})$ such that

$$g \begin{bmatrix} I & 0 \\ 0 & -I \end{bmatrix} g^* = \begin{bmatrix} I & 0 \\ 0 & -I \end{bmatrix}.$$

Let us set G_1 equal to this group. The relationship between the two groups is given as follows. Set

$$L = \frac{1}{\sqrt{2}} \begin{bmatrix} I & iI \\ I & -iI \end{bmatrix}$$

(a unitary matrix) if

$$\sigma(g) = LgL^*;$$

then σ defines an isomorphism of G onto G_1. G_1 has an action by linear fractional transformations on the bounded domain $\mathbf{D}$, given as the set of all $Z \in M_2(\mathbb{C})$ such that $ZZ^* < I$ (here $<$ is the order defined by the cone of positive definite Hermitian matrices. If $g = \begin{bmatrix} A & B \\ C & D \end{bmatrix} \in G_1$, then

$$g \cdot Z = (AZ + B)(CZ + D)^{-1}.$$

We note that if $Y^* = Y$, then

$$\sigma\left(\begin{bmatrix} I & 0 \\ Y & I \end{bmatrix}\right) \cdot I = (I + iY)(I - iY)^{-1}$$

and

$$\sigma\left(\begin{bmatrix} I & Y \\ 0 & I \end{bmatrix}\right) \cdot I = I.$$

The embedding F of Minkowski space $\mathbb{R}^{1,3}$ into $U(2)$ given by

$$(x_1, x_2, x_3, x_4) \mapsto \begin{bmatrix} x_4 + x_3 & x_1 + ix_2 \\ x_1 - ix_2 & x_4 - x_3 \end{bmatrix} = X \longmapsto (I + iX)(I - iX)^{-1}$$

embeds it as a dense open subset. However, it is only a conformal embedding. Indeed

Lemma 2.1. *The embedding F is conformal with*

$$(F^* \langle \ldots, \ldots \rangle)_x = 4\left(1 + 2\sum x_i^2 + (x, x)^2\right)^{-1} (\ldots, \ldots)_x.$$

Proof. We note that $T_u(U(2)) = \{uX | X \in \mathfrak{u}(2)\}$. Furthermore, $\langle uX, uX\rangle_u = -\det(X)$. Now let $Y \in M_2(\mathbb{C})$ be such that $Y^* = Y$, that is $Y \in H$. Let $Q(Y) = (I + iY)(I - iY)^{-1}$. We calculate $\langle dQ_Y(v), dQ_Y(v)\rangle_{Q(Y)}$ for $v \in H$ thought of as being an element of $T_Y(H)$. We get

$$\begin{aligned} dQ_Y(v) &= iv(I - iY)^{-1} + (I + iY)(I - iY)^{-1} iv(I - iY)^{-1} \\ &= i(I + iY)(I - iY)^{-1}((I - iY)(I + iY)^{-1} + I)v(1 - iY)^{-1} \\ &= 2iQ(Y)(I + iY)^{-1} v(1 - iY)^{-1}. \end{aligned}$$

Thus

$$\langle dQ_Y(v), dQ_Y(v)\rangle_{Q(Y)} = 4\frac{\det(v)}{\det(I + Y^2)} = 4\frac{(v, v)_Y}{\det(I + Y^2)}.$$

Now calculate $\det(I + Y^2)$ in terms of the x_i. □

More generally, we have

Lemma 2.2. *The action of G (or G_1) on $U(2)$ given by the linear fractional transformations is conformal relative to the pseudo-Riemannian metric $\langle \ldots, \ldots \rangle$ on $U(2)$.*

Proof. Let

$$\phi(Z) = g \cdot Z = (AZ + B)(CZ + D)^{-1}$$

with $g \in G_1$. Then if $X \in \mathfrak{u}(2)$ we have

$$\begin{aligned} d\phi_Z(ZX) &= (AZX - \phi(Z)CZX)(CZ + D)^{-1} \\ &= \phi(Z)(\phi(Z)^{-1}AZX - CZX)(CZ + D)^{-1}. \end{aligned}$$

Thus since $-\det X = \langle ZX, ZX \rangle_Z$ we have

$$\begin{aligned} &\langle d\phi_Z(ZX), d\phi_Z(ZX) \rangle_{\phi(Z)} \\ &\qquad = \det\left((\phi(Z)^{-1}AZ - CZ)(CZ + D)^{-1}\right) \langle ZX, ZX \rangle_Z . \end{aligned}$$

This proves the conformality. □

We note that

$$\begin{aligned} &(\phi(Z)^{-1}AZX - CZX)(CZ + D)^{-1} \\ &\qquad = (CZ + D)((AZ + B)^{-1}AZ - (CZ + D)^{-1}CZ)(CZ + D)^{-1}. \end{aligned}$$

Thus the conformal factor is

$$\begin{aligned} &\det((AZ + B)^{-1}AZ - (CZ + D)^{-1}CZ) \\ &\qquad = \det((AZ + B)^{-1}B - (CZ + D)^{-1}D). \end{aligned}$$

3 Maxwell's equations on compactified Minkowski space

We will first recall Maxwell's equations in Lorentzian form. For this we need some notation. If M is a smooth manifold, then $\Omega^k(M)$ will denote the space of smooth k-forms on M. We note that if (M, g) is an n-dimensional pseudo-Riemannian manifold, then g induces nondegenerate forms on each fiber $\wedge^k T(M)_x^*$ which we will also denote as g_x. If M is oriented, then there is a unique element $\gamma \in \Omega^n(M)$ such that if $x \in M$ and $v_1, \ldots, v_n$ is an oriented pseudo-orthonormal basis of $T(M)_x$ (i.e., $|g_x(v_i, v_j)| = \delta_{ij}$), then $\gamma_x(v_1, \ldots, v_n) = 1$. Using γ we can define the Hodge $*$ operator on M as follows: If $\omega \in \wedge^k T(M)_x^*$, then $*\omega$ is defined to be the unique element of $\wedge^{n-k} T(M)_x^*$ such that $\eta \wedge *\omega = g_x(\eta, \omega)\gamma_x$ for all $\eta \in \wedge^k T(M)_x^*$.

The next result is standard.

Lemma 3.1. *Let $F : M \to N$ be a conformal orientation-preserving mapping of oriented pseudo-Riemannian manifolds. If* $\dim M = \dim N = 2k$, *then* $F^* * \omega = *F^*\omega$ *for* $\omega \in \Omega^k$.

With this notation in place we can set up Maxwell's equations. Take $t = x_4$ in $\mathbb{R}^{1,3}$ and let $\omega \longmapsto *\omega$ denote the Hodge star operator on differential forms with respect to the Lorentzian structure $(\ldots,\ldots)$ and the orientation corresponding to $\gamma = dx_1 \wedge dx_2 \wedge dx_3 \wedge dt$. Then Maxwell's equations in an area free of current (simple media e.g., in a vacuum, with the dielectric constant, the permeability and thus the speed of light normalized to 1) can be expressed in terms of 2-forms as

$$d\omega = d * \omega = 0 \tag{3.1}$$

with d the exterior derivative. We note that in this formulation if $\mathbf{E} = (e_1, e_2, e_3)$ and $\mathbf{B} = (b_1, b_2, b_3)$ are respectively the electric field intensity and the magnetic field intensity vectors, then

$$\begin{aligned}\omega = {} & b_1 dx_2 \wedge dx_3 - b_2 dx_1 \wedge dx_3 + b_3 dx_1 \wedge dx_2 \\ & -e_1 dx_1 \wedge dt - e_2 dx_2 \wedge dt - e_3 dx_3 \wedge dt.\end{aligned}$$

The equations (3.1) are then the same as

$$\nabla \cdot \mathbf{E} = \nabla \cdot \mathbf{B} = 0$$

and

$$\frac{\partial}{\partial t}\mathbf{E} = -\nabla \times \mathbf{B}, \quad \frac{\partial}{\partial t}\mathbf{B} = \nabla \times \mathbf{E}.$$

Here the $*$ operation is just $\mathbf{E} \to \mathbf{B}$ and $\mathbf{B} \to -\mathbf{E}$, which is the duality between electricity and magnetism in the physics literature.

We note that if $M = \mathbb{R}^{1,3}$ and $N = U(2)$ with the Lorentzian structures described in the previous section, if F is the map described above, and if $\omega \in \Omega^2(U(2))$ satisfies the equations (3.1), then $F^*\omega$ satisfies Maxwell's equations on $\mathbb{R}^{1,3}$. We will thus call the equations (3.1) Maxwell's equations on compactified Minkowski space.

The group G (or G_1) acts on $U(2)$ by conformal diffeomorphisms. Thus we see that the space of solutions to Maxwell's equations defines a representation of G (which acts by pullback). Most of the rest of this article will be devoted to that analysis of this representation.

Denote by $\Omega^k(U(2))_{\mathbb{C}}$ the complex-valued k-forms. Endow it with the C^∞-topology which is a Fréchet space structure and the corresponding action of G on $\Omega^2(U(2))$ defines it as a smooth Fréchet representation of G moderate growth. To see this, we note that as a G-homogeneous space $U(2) \cong G/P$. Let μ denote the isotropy action of P on $V = T_{1P}(G/P)_{\mathbb{C}}$ (i.e., the action of P on $\mathrm{Lie}(G)/\mathrm{Lie}(P) \otimes \mathbb{C}$). Then the space $\Omega^k(U(2))_{\mathbb{C}}$ with the C^∞ topology and G action by pullback is just the C^∞ induced representation

$$\mathrm{Ind}_P^G(\wedge^k V^*)^\infty.$$

Furthermore, since it is as a K-representation

$$\mathrm{Ind}_M^K(\wedge^k V^*)^\infty,$$

Frobenius reciprocity implies that the representation is admissible (that is, the multiplicities of the K-types is finite). The maps d and $d*$ are continuous maps in this topology to $\Omega^3(U(2))_\mathbb{C}$; thus the solutions of Maxwell's equations on $U(2)$ define an admissible, smooth Fréchet representation of moderate growth.

4 The K-isotypic components of the space of solutions to Maxwell's equations on compactified Minkowski space: Step 1

In this section we will begin determination of the K-isotypic components of the space of solutions to Maxwell's equations. We will proceed by first determining the isotypic components of $\ker d$ on $\Omega^2(U(2))_\mathbb{C}$. We will then use explicit calculations for the case at hand of d and the Hodge star operator to complete the picture. We will now begin the first step.

We note that $U(2)$ is diffeomorphic with $SU(2) \times S^1$ under the map

$$u, z \mapsto u \begin{bmatrix} z & 0 \\ 0 & 1 \end{bmatrix}$$

with $u \in SU(2)$ and $z \in S^1 = \{z \in \mathbb{C} | |z| = 1\}$. We note that $SU(2)$ is diffeomorphic with S^3 which implies that we have

$$H^1(U(2), \mathbb{C}) = \mathbb{C},$$

$$H^2(U(2), \mathbb{C}) = 0,$$

and

$$H^3(U(2), \mathbb{C}) = \mathbb{C}.$$

So de Rham's theorem implies that we have the following short exact sequences

$$0 \to \mathbb{C}1 \to C^\infty(U(2), \mathbb{C}) \to \ker d_{|\Omega^1(U(2))_\mathbb{C}} \to \mathbb{C}\mu \to 0, \tag{4.1}$$

$$0 \to \ker d_{|\Omega^1(U(2))} \to \Omega^1(U(2))_\mathbb{C} \to \ker d_{|\Omega^2(U(2))_\mathbb{C}} \to 0; \tag{4.2}$$

in both sequences the map to the kernel is given by d. Also μ is the image of $\det^* \frac{dz}{z}$ in the quotient space. These are all morphisms of smooth Fréchet representations of G of moderate growth.

We also note that the center of G consists of the multiples of the identity and so acts trivially on $U(2)$. Now $K = U(2) \times U(2)$ and the multiples of the identity correspond under this identification with the diagonal elements $C = \{(zI, zI) | |z| = 1\}$ and M is the diagonal $U(2)$ in K. The actual groups acting on $U(2)$ are K/C and M/C. We define $K_1 = SU(2) \times U(2)$ and $M_1 = \{(u, u) | u \in SU(2)\}$. Then under the natural map $K/M = K_1/M_1$ we still have a redundancy of $\mu_2 = \{\pm(I, I)\}$. We will use the notation $(\tau_{p,q,r}, F^{(p,q,r)})$ for the representation of K_1 on $S^p(\mathbb{C}^2) \otimes S^q(\mathbb{C}^2)$ ($S^p(\mathbb{C}^2)$ the pth symmetric power) given by $\tau_{p,q,r}(u, vz) = z^r S^p(u) \otimes S^q(u)$ where $p, q \in \mathbb{Z}_{\geq 0}$, $r \in \mathbb{Z}$ and $r \equiv q \bmod 2$. If V is a closed K-invariant subspace of $\Omega^k(U(2))_{\mathbb{C}}$, then we denote by $V_{p,q,r}$ its $\tau_{p,q,r}$ isotypic component.

Lemma 4.1. *As a representation of K_1, the space of K_1-finite vectors of $\ker d_{|\Omega^2(U(2))_{\mathbb{C}}}$ splits into a direct sum*

$$\bigoplus_{\substack{k \geq 0 \\ r \equiv k \bmod 2}} (F^{k+2,k,r} \oplus F^{k,k+2,r} \oplus F^{k+1,k+1,r}).$$

Furthermore, if $p - q \neq 0$, then

$$d : \Omega^1(U(2))_{p,q,r} \to (\ker d_{|\Omega^2(U(2))_{\mathbb{C}}})_{p,q,r}$$

is a bijective K-intertwining operator.

Proof. The Peter–Weyl theorem implies that $L^2(U(2))$ is a Hilbert space direct sum

$$\bigoplus_{\tau \in \widehat{U(2)}} V^\tau \otimes (V^\tau)^*,$$

where $\widehat{U(2)}$ is the set of equivalence classes of irreducible finite-dimensional representations of $U(2)$ and V^τ is a choice of representative of τ. We have the exact sequence

$$1 \to \{\pm(I, 1)\} \to SU(2) \times S^1 \to U(2) \to 1$$

with the last map $u, z \longmapsto zu$. This implies (as above) that if we define $V^{p,r}$ to be the representation, $\tau_{p,r}$ of $SU(2) \times S^1$on $S^p(\mathbb{C}^2)$ with $\tau_{p,r}(u, z)v = z^r S^p(u)v$, then $\tau_{p,r}$ is the lift of an irreducible representation of $U(2)$ if and only if $r \equiv p \bmod 2$. These representations give a complete set of representatives for $\widehat{U(2)}$. We note that the dual representation of $\tau_{p,r}$ is equivalent with $\tau_{p,-r}$. We therefore see that the space of K-finite vectors in $C^\infty(U(2))_{\mathbb{C}}$ is isomorphic with the direct sum

$$\bigoplus_{\substack{p \in \mathbb{Z}_{\geq 0} \\ r \equiv p \bmod 2}} F^{p,p,r}.$$

We now apply Frobenius reciprocity to analyze the isotypic components of $\Omega^1(U(2))_{\mathbb{C}}$. As we have noted as a representation of K it is just the smooth induced representation of M to K where

$$M = \Delta(U(2)) = \{(u,u)|u \in U(2)\}$$

is acting on $\mathrm{Lie}(U(2))_{\mathbb{C}}$ under $\mathrm{Ad}(u)$. Thus in terms of the parameters above (identifying M with $U(2)$) we have

$$\mathrm{Lie}(U(2))_{\mathbb{C}} \cong F^{0,0} \oplus F^{2,0}.$$

Now Frobenius reciprocity implies that

$$\dim \mathrm{Hom}_K(F^{p,q,r}, \Omega^1(U(2))_{\mathbb{C}}) \tag{4.3}$$

$$= \dim \mathrm{Hom}_M(F^{p,q,r}, F^{0,0}) + \dim \mathrm{Hom}_M(F^{p,q,r}, F^{2,0}). \tag{4.4}$$

The argument above says that $\dim \mathrm{Hom}_M(F^{p,q,r}, F^{0,0}) = 0$ unless $p = q$ and $r \equiv p \bmod 2$. Now the Clebsch–Gordan formula implies that

$$F^{p,q,r}_{|M} \cong \bigoplus_{j=0}^{\min(p,q)} F^{p+q-2j,r}.$$

This implies that $\dim \mathrm{Hom}\, M(F^{p,q,r}, F^{2,0}) = 0$ unless $p = q$ or $|p-q| = 2$ and in either of these cases it is 1. Now the exact sequences (4.1) and (4.2) above imply the theorem. □

Remark 4.2. In the physics literature if ω is a solution to Maxwell's equations (as in the beginning of Section 3), then a one-form β such that $d\beta = \omega$ yields in the **E**, **B** formulation a potential **A**. In our formulation if we pull back to Minkowski space and we write

$$\beta = \sum_{i=1}^{4} a_i dx_i,$$

then considering $t = x_4$ and writing $\phi = a_4$ and $\mathbf{A} = (a_1, a_2, a_3)$, we then have

$$\mathbf{E} = \nabla \times \mathbf{A}, \quad \mathbf{B} = -\frac{\partial \mathbf{A}}{\partial t} + \nabla \phi.$$

This is the dual of what one normally finds in the physics literature. There it is pointed out that this potential has the ambiguity of a gradient field. We will see that the only isotypic components of Maxwell's equations are $\tau_{p,q,r}$ with $|p-q| = 2$ and $r = \pm(\max(p,q))$. Thus using only those Peter–Weyl coefficients yields a unique potential.

We will use the above lemma and some direct calculations to describe the K-isotypic components of Maxwell's equations in the next section.

5 The K-isotypic components of the space of solutions to Maxwell's equations on compactified Minkowski space: Step 2

We retain the notation of the previous section. Let $x_4 = iI$ and

$$x_1 = \begin{bmatrix} i & 0 \\ 0 & -i \end{bmatrix}, \quad x_2 = \begin{bmatrix} 0 & 1 \\ -1 & 0 \end{bmatrix}, \quad x_3 = \begin{bmatrix} 0 & i \\ i & 0 \end{bmatrix}.$$

We will use the usual identification of Lie($U(2)$)(left invariant vector fields) with skew-Hermitian 2×2 matrices (which we denote, as usual, by $\mathfrak{u}(2)$). Thus if $x \in \mathfrak{u}(2)$ then x_u is the tangent vector at 0 to the curve $t \longmapsto ue^{tx}$. We note that $((x_j)_u, (x_k)_u)_u = \varepsilon_j \delta_{j,k}$ with $\varepsilon_j = -(-1)^{\delta_{j4}}$. Thus x_1, x_2, x_3, x_4 define a pseudo-orthonormal frame on $U(2)$. We use this frame to define γ. Since there will be many uses of the star operator and pullbacks we will use the notation $J\omega = *\omega$ for $\omega \in \wedge^2 T^*(U(2))_u$ for all $u \in U(2)$. We define α_j to be the left invariant one-form on $U(2)$ defined by $\alpha_j(x_k) = \delta_{jk}$. We note that

$$J\alpha_1 \wedge \alpha_2 = \alpha_3 \wedge \alpha_4, J\alpha_1 \wedge \alpha_3 = -\alpha_2 \wedge \alpha_4, J\alpha_2 \wedge \alpha_3 = \alpha_1 \wedge \alpha_4$$

and

$$J\alpha_1 \wedge \alpha_4 = -\alpha_2 \wedge \alpha_3, J\alpha_2 \wedge \alpha_4 = \alpha_1 \wedge \alpha_3, J\alpha_3 \wedge \alpha_4 = -\alpha_1 \wedge \alpha_2.$$

From this we note

Lemma 5.1. *We have $J^2 = -I$ on each space $\wedge^2 T^*(U(2))_u$. Furthermore, a basis of the eigenspace for i in $\wedge^2 T^*(U(2))_u \otimes \mathbb{C}$ is*

$$\mathcal{B}_i = \{\alpha_1 \wedge \alpha_4 + i\alpha_2 \wedge \alpha_3, \alpha_2 \wedge \alpha_4 - i\alpha_1 \wedge \alpha_3, \alpha_3 \wedge \alpha_4 + i\alpha_1 \wedge \alpha_2\},$$

a basis for the eigenspace $-i$ is

$$\mathcal{B}_{-i} = \{\alpha_1 \wedge \alpha_4 - i\alpha_2 \wedge \alpha_3, \alpha_2 \wedge \alpha_4 + i\alpha_1 \wedge \alpha_3, \alpha_3 \wedge \alpha_4 - i\alpha_1 \wedge \alpha_2\}.$$

If $\mu \in \mathcal{B}_i$ and $\nu \in \mathcal{B}_{-i}$, then $\mu \wedge \nu = 0$.

We look upon J as an operator on $\Omega^2(U(2))_{\mathbb{C}}$. Since J preserves the real vector space $\Omega^2(U(2))$ we get a decomposition

$$\Omega^2(U(2))_{\mathbb{C}} = \Omega^2(U(2))_i \oplus \Omega^2(U(2))_{-i}$$

with $J_{|\Omega^2(U(2))_{\pm i}} = \pm iI$. If $\omega \in \Omega^2(U(2))_{\mathbb{C}}$, then we denote by $\overline{\omega}$ the complex conjugate of ω relative to the real space $\Omega^2(U(2))$. We note

Lemma 5.2. *With the notation above, we have*

$$\Omega^2(U(2))_{\pm i} = \bigoplus_{\mu \in \mathcal{B}_{\pm i}} C^\infty(U(2), \mathbb{C})\mu.$$

We now calculate the exterior derivatives of the α_j. We observe that if α is a left-invariant element of $\Omega^1(U(2))$, then

$$d\alpha(x, y) = -\alpha([x, y])$$

for $x, y \in \mathrm{Lie}(U(2))$. This implies that

$$d\alpha_1 = -2\alpha_2 \wedge \alpha_3, d\alpha_2 = 2\alpha_1 \wedge \alpha_3, d\alpha_3 = -2\alpha_1 \wedge \alpha_2. \tag{5.1}$$

We also note that $d\alpha_4 = 0$.

We denote by χ_k the character of $SU(2) \times S^1$ given by $\chi_k(u, z) = z^k$. We denote by π the covering map $\pi : SU(2) \times S^1 \to U(2)$ given by $\pi(u, z) = uz$. Then we have

Lemma 5.3. *We have*

$$(\ker d_{|\Omega^2(U(2))_{\mathbb{C}}})_{k,k,r} = \chi_r dC^\infty(U(2), \mathbb{C})_{k,k,0} \wedge \alpha_4$$

which is defined on $U(2)$ *if* $r \equiv k \bmod 2$.

Proof. We note that if $f \in C^\infty(U(2), \mathbb{C})$, then $df = (x_4 f)\alpha_4 + \nu$ with $\nu = \sum_j (x_j f)\alpha_j$. Thus if $df \wedge \alpha_4 = 0$, then $\nu = 0$. If $\nu = 0$, then $\pi^* f(u, z) = \pi^* f(I, z)$. A function in $C^\infty(U(2), \mathbb{C})_{k,k,0}$ with this property exists if and only if $k = 0$. It is also clear that $\chi_r dC^\infty(U(2), \mathbb{C})_{k,k,0} \wedge \alpha_4$ is contained in $\ker d$. Thus, since each of the isotypic components of $\ker d_{|\Omega^2(U(2))_{\mathbb{C}}}$ is irreducible and we have accounted for all of them by Lemma 4.1, the result follows. □

We denote by Maxw the space of complex solutions to the Maxwell equations (as described in the previous section). Then Maxw is a closed subspace of $\Omega^2(U(2))_{\mathbb{C}}$ yielding a smooth Fréchet representation of G of moderate growth under the action $\mu(g)\omega = (g^{-1})^*\omega$. We also note that J preserves Maxw and commutes with the action of G. This implies that $\mathrm{Maxw} = \mathrm{Maxw}_i \oplus \mathrm{Maxw}_{-i}$ (corresponding to the i and $-i$ eigenspaces of J on Maxw). We also note

Lemma 5.4. $\mathrm{Maxw}_{\pm i} = \{\omega \in \Omega^2(U(2))_{\mathbb{C}} | J\omega = \pm i\omega$ *and* $d\omega = 0\}$.

We can now eliminate some isotypic components of Maxw.

Lemma 5.5. $\mathrm{Maxw}_{k,k,r} = 0$ *for all* $k \in \mathbb{Z}_{\geq 0}$ *and all* $r \in \mathbb{Z}$.

Proof. If $\omega \in \mathrm{Maxw}_{k,k,r}$, then since

$$\mathrm{Maxw}_{k,k,r} = \mathrm{Maxw}_{k,k,r} \cap \mathrm{Maxw}_i \oplus \mathrm{Maxw}_{k,k,r} \cap \mathrm{Maxw}_{-i}$$

with each of the summands K-invariant and since each isotypic component is irreducible, we see that $J\omega = i\omega$ of $J\omega = -i\omega$. In either case Lemma 5.1 implies that ω is not an element of $\Omega^1(U(2))_{\mathbb{C}} \wedge \alpha_4$. But Lemma 5.2 implies that it must be of that form. This implies that $\omega = 0$. $\square$

We are finally ready to give the isotypic components of Maxw.

Theorem 5.6. $\mathrm{Maxw}_{p,q,r}$ *is nonzero if and only if* (p, q, r) *is in one the following forms:*

$$(k+2, k, k+2), (k+2, k, -(k+2)), (k, k+2, k+2), (k, k+2, -(k+2)).$$

If it is nonzero it is irreducible. Moreover, the spaces $\mathrm{Maxw}_{k+2,k,k+2}$ *and* $\mathrm{Maxw}_{k,k+2,-k-2}$ *are contained in* Maxw_i *and the spaces* $\mathrm{Maxw}_{k+2,k,-k-2}$ *and* $\mathrm{Maxw}_{k,k+2,k+2}$ *are contained in* Maxw_{-i}.

The proof will occupy the rest of the section. If $x \in \mathfrak{sl}(2, \mathbb{C})$ with $x = v + iw, v, w \in \mathfrak{su}(2)$, then we define the left-invariant vector field $x^L f(u) = \frac{d}{dt}(f(ue^{tv}) + if(ue^{tw}))_{|t=0}$ and $x^R f(u) = \frac{d}{dt}(f(e^{-tv}u) + if(e^{-tw}v))_{|t=0}$. We will think of these vector fields as being on $SU(2)$ or $SU(2)/\{\pm I\} = U(2)/S^1 I = PSU(2)$. We set $e = \frac{1}{2}(x_2 - ix_3)$, $f = -\frac{1}{2}(x_2 + ix_3)$ and $h = -ix_1$. Then e, f, h form the standard basis of $\mathfrak{sl}(2, \mathbb{C})$.

Let $\xi_e^L, \xi_f^L, \xi_h^L$ (respectively, $\xi_e^R, \xi_f^R, \xi_h^R$) be the left (resp. right) invariant one-forms on $PSU(2)$ that form a dual basis to e^L, f^L, h^L (respectively, e^R, f^R, h^R). Let $p : U(2) \to PSU(2)$ be the obvious quotient homomorphism and let $\alpha_b^a = p^* \xi_b^a$ for $a = L$ or R and $b = e, f$ or h. Now $((u, v)^{-1})^* \alpha_b^L(x^L) = \alpha_b^L(\mathrm{Ad}(v)^{-1} x^L)$ and $((u, v)^{-1})^* \alpha_b^R(x^R) = \alpha_b^R(\mathrm{Ad}(u)^{-1} x^R)$. Thus, if

$$g = \left(\begin{bmatrix} z & 0 \\ 0 & z^{-1} \end{bmatrix}, \begin{bmatrix} w & 0 \\ 0 & w^{-1} \end{bmatrix}\right)$$

with $z, w \in S^1$, then

$$(g^{-1})^* \alpha_e^L = w^{-2} \alpha_e^L, (g^{-1})^* \alpha_f^L = w^2 \alpha_f^L, (g^{-1})^* \alpha_h^L = \alpha_h^L$$

and

$$(g^{-1})^* \alpha_e^R = u^{-2} \alpha_e^R, (g^{-1})^* \alpha_f^R = u^2 \alpha_f^R, (g^{-1})^* \alpha_h^R = \alpha_h^R.$$

This implies that

$$\mathrm{Span}_{\mathbb{C}}(\alpha_e^L, \alpha_f^L, \alpha_h^L) = \Omega^1(U(2))_{0,2,0}$$

and

$$\mathrm{Span}_{\mathbb{C}}(\alpha_e^R, \alpha_f^R, \alpha_h^R) = \Omega^1(U(2))_{2,0,0}.$$

Also relative to the positive root system $g \to z^2, g \to w^2$ the highest weight space of $\Omega^1(U(2))_{0,2,0}$ in $\mathbb{C}\alpha_f^L$ and that of $\Omega^1(U(2))_{2,0,0}$ is $\mathbb{C}\alpha_f^R$.

We can now describe highest weight vectors for the isotypic components $\Omega^1(U(2))_{k,k+2,l}$ and $\Omega^1(U(2))_{k+2,k,l}$. Let e_1 and e_2 be the standard basis of $\mathbb{C}^2$. Fix a $U(2)$ invariant inner product $\langle \ldots, \ldots \rangle$ on each space $S^k(\mathbb{C}^2)$. Define

$$\phi_k(u) = \langle S^k(u)e_1^k, e_2^k \rangle.$$

Then ϕ_k is a highest weight vector for $C^\infty(U(2))_{k,k,k} = \Omega^0(U(2))_{k,k,k}$. Also we define $\chi_l(u,z) = z^l$ for $u \in SU(2)$ and $z \in S^1$. We note that if $l \equiv k \bmod 2$, then

$$\psi_{k,l}(uz) = z^{l-k}\phi_k(uz) = \chi_{l-k}(u,z)\phi_k(uz)$$

is defined and is a highest weight vector for $\Omega^0(U(2))_{k,k,l}$. This implies that $\psi_{k,l}\alpha_f^L$ is a highest weight vector for $\Omega^1(U(2))_{k,k+2,l}$ and $\psi_{k,l}\alpha_f^R$ is a highest weight vector for $\Omega^1(U(2))_{k+2,k,l}$. We have shown that Maxw is multiplicity free and Maxw $=$ Maxw$_i$ $\oplus$ Maxw$_{-i}$. We have also proved that $\mathrm{Maxw}_{k,k+2,l} = d\Omega^1(U(2))_{k,k+2,l}$ and $\mathrm{Maxw}_{k+2,k,l} = d\Omega^1(U(2))_{k+2,k,l}$. We have proved that

(1) $\mathrm{Maxw}_{k,k+2,l} \neq 0$ if and only if $Jd\psi_{k,l}\alpha_f^L = \lambda J\psi_{k,l}\alpha_f^L$ with $\lambda \in \{\pm i\}$ and $\mathrm{Maxw}_{k+2,k,l} \neq 0$ if and only if $Jd\psi_{k,l}\alpha_f^R = \lambda J\psi_{k,l}\alpha_f^R$ with $\lambda \in \{\pm i\}$.

We are now left with a computation. One checks as above (using that $d\alpha_f^L(X^L, Y^L) = -\alpha_f^L([X,Y]^L))$

$$d\alpha_f^L = 2\alpha_h^L \wedge \alpha_f^L;$$

we therefore have

$$d(\psi_{k,l}\alpha_f^L) = il\psi_{k,l}\alpha_4 \wedge \alpha_f^L + (k+2)\psi_{k,l}\alpha_h^L \wedge \alpha_f^L. \tag{5.2}$$

We consider $il\alpha_4 \wedge \alpha_f^L + (k+2)\alpha_h^L \wedge \alpha_f^L$. We observe that $\alpha_h^L = i\alpha_1$ and $\alpha_f^L = -(\alpha_2 - i\alpha_3)$. Thus (using the calculations leading to Lemma 5.1) the right-hand side of the equation above is equal to

$$li\alpha_4 \wedge \alpha_2 - l\alpha_4 \wedge \alpha_3 + i(k+2)\alpha_1 \wedge (\alpha_2 - i\alpha_3)$$

which equals

$$(-li\alpha_2 \wedge \alpha_4 + (k+2)J\alpha_2 \wedge \alpha_4) + (-l\alpha_1 \wedge \alpha_4 - i(k+2)J\alpha_1 \wedge \alpha_4.$$

We therefore see that if $l > 0$, then this expression is an element of $\Omega^2(U(2))_{-i} + (l-k-2)(-i\alpha_2 \wedge \alpha_4 - (l-k-2)\alpha_1 \wedge \alpha_4)$. If $l \leq 0$, then it is an element of $\Omega^2(U(2))_i + (l+k+2)(-i\alpha_2 \wedge \alpha_4 - (l+k+2)\alpha_1 \wedge \alpha_4)$. Thus we have

(2) $\mathrm{Maxw}_{k,k+2,l} \neq 0$ only if $l = k+2$ or $l = -(k+2)$. Furthermore, $\mathrm{Maxw}_{k,k+2,k+2} \subset \mathrm{Maxw}_{-i}$ and $\mathrm{Maxw}_{k,k+2,-(k+2)} \subset \mathrm{Maxw}_i$.

We note that everything that we have done could have been done with right-invariant vector fields to complete the proof of the theorem. However we will proceed in a different way. Let $\eta : U(2) \to U(2)$ be defined by $\eta(u) = u^{-1}$. Then for $x \in \mathfrak{u}(2)$, we have $d\eta_u(x_u^L) = x_{u^{-1}}^R$. This implies that

$$\langle d\eta_u(x_u^L), d\eta_u(x_u^L)\rangle_{u^{-1}} = \langle x_{u^{-1}}^R, x_{u^{-1}}^R\rangle_{u^{-1}} = -\det x$$

since

$$\langle x_{u^{-1}}^R, x_{u^{-1}}^R\rangle_{u^{-1}} = \langle dR(u^{-1})_I(x_I^R), dR(u^{-1})_I(x_I^R)\rangle_{u^{-1}} = \langle x_I^R, x_I^R\rangle_I .$$

This proves that η is an isometry. It also implies that

$$(\eta^*\alpha_f^R)_u(x_u^L) = (\alpha_f^R)_{u^{-1}}(d\eta_u(x_u^L)) = (\alpha_f^R)_{u^{-1}}(x_{u^{-1}}^R).$$

Hence $\eta^*\alpha_f^R = \alpha_f^L$. Now $\eta^* d(\psi_{k,l}\alpha_f^R) \in \Omega^2(U(2))_{k,k+2,-l}$ since $\eta^*\phi_k$ is a highest weight vector for $C^\infty(U(2))_{k,k,-k}$. We also note that $\eta^*\gamma = \gamma$. Thus $\eta^*\mathrm{Maxw} = \mathrm{Maxw}$. Hence, if $d(\psi_{k,l}\alpha_f^R) \in \mathrm{Maxw}_{k+2,k,l}$, then $\eta^* d(\psi_{k,l}\alpha_f^R) \in \mathrm{Maxw}_{k,k+2,-l}$. So, if $l > 0$ then, we must have $l = k+2$ and if $l \le 0$, then $l = -k-2$. Since η^* commutes with J the last assertion also follows.

Remark 5.7. We have

$$\eta^*\mathrm{Maxw}_{k+2,k,k+2} = \mathrm{Maxw}_{k,k+2,-k-2}, \tag{5.3}$$

$$\eta^*\mathrm{Maxw}_{k+2,k,-k-2} = \mathrm{Maxw}_{k,k+2,k+2} \text{ and } (\eta^*)^2 = I. \tag{5.4}$$

6 The Hermitian form

We retain the notation of the previous sections.

We note that

$$H_3(U(2), \mathbb{R}) = \mathbb{R}.$$

The form $\nu = \alpha_1 \wedge \alpha_2 \wedge \alpha_3$ restricted to $SU(2)$ satisfies

$$\int_{SU(2)} \nu = 2\pi^2$$

and $d\nu = 0$. This implies that the class of $SU(2)$ in the third homology over $\mathbb{R}$ is a basis. (In fact it is well known that this is true over $\mathbb{Z}$). We note that this implies, in particular, that if $\omega \in \Omega^3(U(3))_{\mathbb{C}}$ satisfies $d\omega = 0$, then if M is a compact submanifold such that there exists a smooth family of diffeomorphisms of $U(2)$, Φ_t such that $\Phi_0 = I$ and $\Phi_1(SU(2)) = M$, then

$$\int_{SU(2)} \omega = \int_M \omega.$$

In particular we have (since $U(2,2)$ is connected),

Lemma 6.1. *If $g \in U(2,2)$ and if $\omega \in \Omega^3(U(3))_{\mathbb{C}}$ satisfies $d\omega = 0$, then $\int_{SU(2)} \omega = \int_{gSU(2)} \omega = \int_{SU(2)} g^*\omega$.*

We will apply this observation to define $U(2,2)$ invariant sesquilinear forms on the spaces $\mathrm{Maxw}_{\pm i}$.

Lemma 6.2. *Let $\alpha \in \Omega^1(U(2))_{\mathbb{C}}$ and $\omega \in \Omega^2(U(2))_{\mathbb{C}}$ be such that $\omega, d\alpha \in$ Maxw_i (resp. Maxw_{-i}). Then $d(\alpha \wedge \overline{\omega}) = 0$.*

Proof. If $d\alpha \in \mathrm{Maxw}_i$, then $\overline{\omega} \in \mathrm{Maxw}_{-i}$. We note that Maxwell's equations imply that $d\overline{\omega} = 0$. Hence $d(\alpha \wedge \overline{\omega}) = d\alpha \wedge \overline{\omega}$ and Lemma 5.1 implies that $\mathrm{Maxw}_i \wedge \mathrm{Maxw}_{-i} = 0$. Obviously the same argument works for $-i$. □

Proposition 6.3. *If $\mu, \omega \in \mathrm{Maxw}_i$ (or Maxw_{-i}), there exists $\alpha \in \Omega^1(U(2))_{\mathbb{C}}$ such that $d\alpha = \omega$. The expression*

$$\int_{SU(2)} \alpha \wedge \overline{\mu}$$

depends only on ω and μ (and not on the choice of α). Furthermore, the integral defines a Hermitian form $\langle \omega, \mu \rangle$ on Maxw_i (or Maxw_{-i}) that satisfies $\langle g^*\omega, g^*\mu \rangle = \langle \omega, \mu \rangle$ for all $g \in U(2,2)$.

Proof. Suppose that $\beta \in \Omega^1(U(2))_{\mathbb{C}}$ is such that $d\beta = 0$ and $\int_{S^1 I} \beta = 0$. Then since $H_1(U(2), \mathbb{C})$ is spanned by the class of $S^1 I$, de Rham's theorem implies that there exists $f \in C^\infty(U(2), \mathbb{C})$ such that $df = \beta$. Set $\iota(\beta) = \int_{S^1 I} \beta$. We note that

$$\iota(\alpha_1) = 2\pi.$$

We also note that if $\nu \in \Omega^2(U(2))$, then $(\alpha_1 \wedge \nu)_{|SU(2)} = 0$.

We observed that if $\omega \in \mathrm{Maxw}_{\pm i}$, then there exists $\alpha \in \Omega^1(U(2))_{\mathbb{C}}$ such that $d\alpha = \omega$. If $d\beta = \omega$ then $d(\beta - \alpha) = 0$ and $\int_{S^1 I}(\beta - \alpha - \frac{\iota(\beta-\alpha)}{2\pi}\alpha_1) = 0$ so $\beta - \alpha - \frac{\iota(\beta-\alpha)}{2\pi}\alpha_1 = -df$ with $f \in C^\infty(U(2), \mathbb{C})$. This implies that

$$\int_{SU(2)} \alpha \wedge \overline{\mu} - \int_{SU(2)} \beta \wedge \overline{\mu} = \int_{SU(2)} (df - \frac{\iota(\beta - \alpha)}{2\pi}\alpha_1) \wedge \overline{\mu}$$

$$= \int_{SU(2)} d(f\overline{\mu}) - \frac{\iota(\beta - \alpha)}{2\pi} \int_{SU(2)} \alpha_1 \wedge \overline{\mu} = 0.$$

Both of the integrals are 0. We next observe that $\mu = d\xi$. We have

$$\overline{\langle \mu, \omega \rangle} - \langle \omega, \mu \rangle = \int_{SU(2)} \left(\overline{\xi} \wedge d\alpha - \alpha \wedge d\overline{\xi} \right)$$
$$= \int_{SU(2)} \left(\overline{\xi} \wedge d\alpha - d\overline{\xi} \wedge \alpha \right) = - \int_{SU(2)} d\left(\overline{\xi} \wedge \alpha \right) = 0.$$

We are left with the proof of $U(2,2)$ invariance. We will concentrate on Maxw_i. We note that

$$\int_{S^1 I} \alpha_1 = 2\pi.$$

Let $g \in U(2,2)$. If $\omega, \mu \in \mathrm{Maxw}_i$ and $\alpha \in \Omega^1(U(2))_{\mathbb{C}}$ satisfies $d\alpha = \omega$, then $dg^*\alpha = g^*\omega$. Thus

$$\langle g^*\omega, g^*\mu \rangle = \int_{SU(2)} g^*\alpha \wedge g^*\overline{\mu} = \int_{SU(2)} \alpha \wedge \overline{\mu} = \langle \omega, \mu \rangle$$

by Lemma 6.1. □

We will now calculate $\langle \ldots, \ldots \rangle$ on each of the isotypic components of Maxw. We set $\alpha_{k,k+2,l} = \psi_{k,l}\alpha_f^L$ as in the previous section. If $l = k+2$ or $-(k+2)$, then $\omega_{k,k+2,l} = d\alpha_{k,k+2,l}$ is a highest weight vector for $\mathrm{Maxw}_{k,k+2,l}$.

Lemma 6.4. *If* $l = \pm(k+2)$, *then* $\langle \omega_{k,k+2,l}, \omega_{k,k+2,l} \rangle = -\frac{4k+8}{k+1}\pi^2$.

Proof. We have seen in formula (5.2) in the previous section that (in the notation therein)

$$\omega_{k,k+2,l} = il\psi_{k,l}\alpha_4 \wedge \alpha_f^L + (k+2)\psi_{k,l}\alpha_h^L \wedge \alpha_f^L.$$

Thus since $\alpha_{4|SU(2)} = 0$ we have

$$\alpha_{k,k+2,l} \wedge \overline{\omega_{k,k+2,l}}_{|SU(2)} = (k+2)|\psi_{k,l}|^2 \alpha_f^L \wedge \overline{\alpha_h^L \wedge \alpha_f^L}.$$

Using the expressions for α_h^L and α_f^L one sees that

$$\alpha_f^L \wedge \overline{\alpha_h^L \wedge \alpha_f^L}_{|SU(2)} = -2\alpha_1 \wedge \alpha_2 \wedge \alpha_{3|SU(2)}.$$

Normalized invariant measure on SU(2) is $\mu = \frac{1}{2\pi^2}\alpha_1 \wedge \alpha_2 \wedge \alpha_{3|SU(2)}$. On $S^k(\mathbb{C}^2)$ we put the inner product defined by the tensor product. Thus e_1^k and e_2^k are unit vectors. Also

$$|\psi_{k,l}(u)|^2 = \left\langle S^k(u)e_1^k, e_2^k \right\rangle \overline{\left\langle S^k(u)e_1^k, e_2^k \right\rangle}.$$

The Schur orthogonality relations imply that

$$\int_{SU(2)} |\psi_{k,l}(u)|^2 \mu = \frac{1}{k+1}.$$

□

7 Four unitary ladder representations of $SU(2,2)$

We will consider $U(2,2)$ in the usual form, that is, if $I_{2,2} = \begin{bmatrix} I & 0 \\ 0 & -I \end{bmatrix}$, then (as in Section 2)

$$G_1 = \{g \in GL(4,\mathbb{C}) | g I_{2,2} g^* = I_{2,2}\}.$$

In this form K is the subgroup of block diagonal matrices. $\mathfrak{g} = \text{Lie}(U(2,2))_{\mathbb{C}} = M_4(\mathbb{C})$. We set

$$\mathfrak{p}^+ = \left\{ \begin{bmatrix} 0 & X \\ 0 & 0 \end{bmatrix} \mid X \in M_2(\mathbb{C}) \right\}$$

and

$$\mathfrak{p}^- = \left\{ \begin{bmatrix} 0 & 0 \\ Y & 0 \end{bmatrix} \mid Y \in M_2(\mathbb{C}) \right\}.$$

Then $\mathfrak{g} = \text{Lie}(K)_{\mathbb{C}} \oplus \mathfrak{p}$ (here $\mathfrak{p}$ is the orthogonal complement of $\text{Lie}(K)_{\mathbb{C}}$ relative to the trace form) and as a K-module $\mathfrak{p} = \mathfrak{p}^+ \oplus \mathfrak{p}^-$. We leave it to the reader to check.

Lemma 7.1. *As a representation of K (under the adjoint action) $\mathfrak{p}^+ \cong F^{1,1,-1}$ and $\mathfrak{p}^- \cong F^{1,1,1}$ and*

$$\wedge^2(\mathfrak{p}^+)^* \cong F^{2,0,2} \oplus F^{0,2,2}$$

and

$$\wedge^2(\mathfrak{p}^-)^* \cong F^{2,0,-2} \oplus F^{0,2,-2}.$$

Let $\pi(g)\omega = (g^{-1})^*\omega$ for $\omega \in \text{Maxw}$. Then we already observed that with the C^∞-topology, (π, Maxw) is an admissible smooth Fréchet representation of moderate growth. We set Maxw_K equal to the space of K-finite vectors of Maxw and we will use module notation for the action of $\mathfrak{g} = \text{Lie}(U(2,2))_{\mathbb{C}} = M_4(\mathbb{C})$ on Maxw_K, thereby having an admissible finitely generated $(\mathfrak{g}, K)$-module.

Lemma 7.2. *$\mathfrak{p}^+$ annihilates* $\mathrm{Maxw}_{2,0,2}$ *and* $\mathrm{Maxw}_{0,2,2}$*, whereas $\mathfrak{p}^-$ annihilates* $\mathrm{Maxw}_{2,0,-2}$ *and* $\mathrm{Maxw}_{0,2,-2}$.

Proof. Using the Clebsch–Gordon formula we have

$$F^{1,1,-1} \otimes F^{2,0,2} \cong F^{3,1,1} \oplus F^{1,1,1},$$

$$F^{1,1,-1} \otimes F^{0,2,2} \cong F^{1,3,1} \oplus F^{1,1,1},$$

$$F^{1,1,1} \otimes F^{2,0,-2} \cong F^{3,1,-1} \oplus F^{1,1,-1},$$

$$F^{1,1,1} \otimes F^{0,2,-2} \cong F^{1,3,-1} \oplus F^{1,1,-1}.$$

Theorem 5.6 implies that none of the K-types on the right of these equations occurs in Maxw. □

We set

$$(\mathrm{Maxw}_{2,0}^+)_K = \sum_{k\geq 0} \mathrm{Maxw}_{k+2,k,k+2}, \ (\mathrm{Maxw}_{0,2}^+)_K = \sum_{k\geq 0} \mathrm{Maxw}_{k,k+2,k+2},$$

$$(\mathrm{Maxw}_{2,0}^-)_K = \sum_{k\geq 0} \mathrm{Maxw}_{k+2,k,-k-2}, \ (\mathrm{Maxw}_{0,2}^-)_K = \sum_{k\geq 0} \mathrm{Maxw}_{k,k+2,-k-2}.$$

We will drop the sub-K for the completions of these spaces. We note Theorem 5.6 implies that each of these spaces is totally contained in Maxw_i or Maxw_{-i}. This implies that the Hermitian form, $\langle \ldots, \ldots \rangle$ from the previous section is defined on each of these spaces.

We have

Theorem 7.3. *Each of the spaces* $\mathrm{Maxw}_{2,0}^+$, $\mathrm{Maxw}_{2,0}^-$, $\mathrm{Maxw}_{0,2}^+$, $\mathrm{Maxw}_{0,2}^-$ *is an invariant irreducible subspace for π. Furthermore, the form $\langle \ldots, \ldots \rangle$ is positive definite on* $\mathrm{Maxw}_{2,0}^+$, $\mathrm{Maxw}_{2,0}^-$ *and negative definite on* $\mathrm{Maxw}_{0,2}^+$, $\mathrm{Maxw}_{0,2}^-$.

Proof. We consider $\mathrm{Maxw}_{2,0}^+$. We note that using the Clebsch–Gordan formula as above we have

$$\begin{aligned}\mathfrak{p}^-\mathrm{Maxw}_{k+2,k,k+2} \subset\ & \mathrm{Maxw}_{k+3,k+1,k+3} + \mathrm{Maxw}_{k+3,k-1,k+3} \\ & + \mathrm{Maxw}_{k+1,k+2,k+3} + \mathrm{Maxw}_{k+1,k-1,k+3}\end{aligned}$$

and

$$\begin{aligned}\mathfrak{p}^+\mathrm{Maxw}_{k+2,k,k+2} \subset\ & \mathrm{Maxw}_{k+3,k+1,k+1} + \mathrm{Maxw}_{k+3,k-1,k+1} \\ & + \mathrm{Maxw}_{k+1,k+1,k+1} + \mathrm{Maxw}_{k+1,k-1,k+1}.\end{aligned}$$

Using the form of the K-types of Maxw we see that

$$\mathfrak{p}^- \mathrm{Maxw}_{k+2,k,k+2} \subset \mathrm{Maxw}_{k+3,k+1,k+3}$$

and

$$\mathfrak{p}^+ \mathrm{Maxw}_{k+2,k,k+2} \subset \mathrm{Maxw}_{k+1,k-1,k+1}.$$

This implies that $(\mathrm{Maxw}^+_{2,0})_K$ is a $(\mathfrak{g},K)$-submodule of Maxw. Hence $\mathrm{Maxw}^+_{2,0}$ is G-invariant. The same argument proves

$$\mathfrak{p}^- \mathrm{Maxw}_{k,k+2,k+2} \subset \mathrm{Maxw}_{k+1,k+3,k+3},$$

$$\mathfrak{p}^+ \mathrm{Maxw}_{k,k+2,k+2} \subset \mathrm{Maxw}_{k-1,k+1,k+1},$$

$$\mathfrak{p}^+ \mathrm{Maxw}_{k+2,k,-(k+2)} \subset \mathrm{Maxw}_{k+3,k+1,-(k+3)},$$

$$\mathfrak{p}^- \mathrm{Maxw}_{k+2,k,-(k+2)} \subset \mathrm{Maxw}_{k+1,k-1,-(k+1)},$$

and

$$\mathfrak{p}^+ \mathrm{Maxw}_{k,k+2,-(k+2)} \subset \mathrm{Maxw}_{k+1,k+2,-(k+3)},$$

$$\mathfrak{p}^- \mathrm{Maxw}_{k,k+2,-(k+2)} \subset \mathrm{Maxw}_{k-1,k+1,-(k+1)}.$$

This proves the G-invariance of the indicated spaces. We next note that since the multiplicities of the K-types of Maxw are all one, the Hermitian form $\langle \ldots, \ldots \rangle$ restricted to each space $\mathrm{Maxw}_{k,l,m}$ is 0, positive definite or negative definite. Thus Lemma 6.4 implies that the form is negative definite on $(\mathrm{Maxw}^+_{0,2})_K$ and $(\mathrm{Maxw}^-_{0,2})_K$. Recall that $\eta^*(\mathrm{Maxw}^+_{0,2})_K = (\mathrm{Maxw}^-_{2,0})_K$ and $\eta^* \mathrm{Maxw}^-_{0,2})_K = (\mathrm{Maxw}^+_{2,0})_K$. Furthermore η is orientation-reversing on $SU(2)$, thus the form on $(\mathrm{Maxw}^+_{2,0})_K$ and $(\mathrm{Maxw}^-_{2,0})_K$ is positive definite. If one of the representations were not irreducible, then it would have a finite-dimensional invariant subspace (by the above formulae). Since these representations are all unitarizable and do not contain one-dimensional invariant subspaces the representations are all irreducible. □

Remark 7.4. The proof above implies that the last six inclusions above are all equalities.

We set $\pi^\varepsilon_{i,j}$ equal to the action of G on $\mathrm{Maxw}^\varepsilon_{i,j}$ for $\varepsilon = +,-$ and $i, j = 0,2$ or $2,0$.

Theorem 7.5. *The representations $\pi^\varepsilon_{i,j}$ for $\varepsilon = +,-$ and $i, j = 0,2$ or $2,0$ all have trivial infinitesimal character (that is equal to the restriction to the center of $U(\mathfrak{g})$ of the augmentation homomorphism to its center). Furthermore*

$$H^2(\mathfrak{g},K,\mathrm{Maxw}^\varepsilon_{i,j}) = \mathbb{C}, \varepsilon = +,- \textit{ and } i,j = 0,2 \textit{ or } 2,0.$$

Proof. By the above, all four of the representations are the spaces of C^∞ vectors of an irreducible unitary representation. Since the center of $\mathfrak{g}$ acts trivially and

$$\dim \mathrm{Hom}_K(\wedge^2\mathfrak{p}, M) = 1$$

for each M as in the statement, it is enough to prove that the Casimir operator corresponding to the trace form on $\mathfrak{g}$ acts by 0 on each of the representations (c.f. [BW]). We will show that the center of the enveloping algebra acts correctly. We look at the case $\mathrm{Maxw}^+_{2,0}$ and leave the other cases to the reader. By Lemma 7.2, $(\mathrm{Maxw}^+_{2,0})_K$ is a highest weight module relative to the Weyl chamber (in ε notation) $\varepsilon_1 > \varepsilon_2 > \varepsilon_3 > \varepsilon_4$. We calculate the highest weight $\lambda = a\varepsilon_1 + b\varepsilon_2 + c\varepsilon_3 + d\varepsilon_4$. The representation factors through the adjoint group so $a+b+c+d = 0$. By definition of $F^{2,0,2}$ $a - b = 2, c - d = 0, c + d = 2$. Solving the four equations yields $-2\varepsilon_2 + \varepsilon_3 + \varepsilon_4$. ρ for the chamber that we are studying is $\frac{3}{2}\varepsilon_1 + \frac{1}{2}\varepsilon_2 - \frac{1}{2}\varepsilon_3 - \frac{3}{2}\varepsilon_4$. Thus $\lambda + \rho = \frac{3}{2}\varepsilon_1 - \frac{3}{2}\varepsilon_2 + \frac{1}{2}\varepsilon_3 - \frac{1}{2}\varepsilon_4 = \sigma\rho$ for σ the cyclic permutation (243). □

Let θ be the Cartan involution of G_1 corresponding to K. In light of the Vogan–Zuckerman theorem [VZ] this implies that there are four θ stable parabolic subalgebras $\mathfrak{q}^\varepsilon_{i,j}$ $\varepsilon = +,-$ and $i,j = 0,2$ or $2,0$, such that $(\mathrm{Maxw}^\varepsilon_{i,j})_K$ is isomorphic with $A_{\mathfrak{q}^\varepsilon_{i,j}}(0)$ (c.f. [BW]).

Theorem 7.6. *One has* $\mathfrak{q}^+_{2,0}$ *is the parabolic subalgebra*

$$\{[x_{ij}] \in M_4(\mathbb{C}) \mid x_{21} = x_{31} = x_{41} = 0\},$$

$$\mathfrak{q}^+_{0,2} = \{[x_{ij}] \in M_4(\mathbb{C}) \mid x_{41} = x_{42} = x_{43} = 0\},$$

and $\mathfrak{q}^-_{0,2} = (\mathfrak{q}^+_{0,2})^T, \mathfrak{q}^-_{2,0} = (\mathfrak{q}^+_{2,0})^T$. *Thus* $(\mathrm{Maxw}^\varepsilon_{i,j})_K$ *is isomorphic with* $A_{\mathfrak{q}^\varepsilon_{i,j}}(0)$.

Proof. If $\mathfrak{q}$ is a θ-stable parabolic subalgebra with nilpotent radical $\mathfrak{u}$ and if $\mathfrak{u}_n = \mathfrak{u} \cap \mathfrak{p}$, we set $2\rho_n(\mathfrak{q})(h) = \mathrm{tr}\,\mathrm{ad}(h)_{|\mathfrak{u}_n}$ for $h \in \mathfrak{h}$ the Cartan subalgebra of diagonal matrices. One checks that if the parabolics are given as in the theorem, then $2\rho_n(\mathfrak{q}^+_{2,0})$ is the highest weight of $F^{2,0,2}$, $2\rho_n(\mathfrak{q}^+_{0,2})$ is that of $F^{0,2,2}$, $2\rho_n(\mathfrak{q}^-_{2,0})$ is the lowest weight of $F^{2,0,-2}$, and $2\rho_n(\mathfrak{q}^-_{0,2})$ is the lowest weight of $F^{0,2,-2}$. Now the result follows from the main theorem (c.f. [W1]). □

Remark 7.7. This result implies that if we consider the open orbit X of $U(2,2)$ in $\mathbb{P}^3(\mathbb{C})$ that has no non-constant holomorphic functions, then there are two holomorphic line bundles and two anti-holomorphic line bundles such that their degree 1 smooth sheaf cohomology yields the four versions of solutions of Maxwell's equations. The relation between the two realizations is related to the Penrose–Twistor transform for solutions of the wave equation.

Another realization of these representations is in [GW]. The observation here is that $SU(2,2)$ is the quaternionic real form of $SL(4,\mathbb{C})$. Since $K \cap SU(2,2) = S(U(2) \times U(2))$ there are two invariant quaternionic structures on $SU(2,2)/K \cap$

$SU(2,2)$ and on each there are two line bundles on which the methods of [GW] apply. Each yields an element of the "analytic continuation" of the corresponding quaternionic discrete series which is our third realization. We will allow the interested reader to check these assertions.

Finally, these representations appear in Howe duality between $U(2,2)$ and the $U(1)$ in its center. The details are worked out in [BW], VIII, 2.10–2.12 in the notation therein, and the pertinent representations are V_2 and V_{-2} yielding $(\mathrm{Maxw}^+_{2,0})_K$ and $(\mathrm{Maxw}^+_{0,2})_K$. The other two constituents are obtained by duality.

8 A dual pair in $PSU(2,2)$

We continue to consider $U(2,2)$ as the group G_1 in Section 2 and $PSU(2,2)$ the quotient by the center. Let S be the image of the subgroup of all matrices of the form

$$\begin{bmatrix} aI & bI \\ \overline{b}I & \overline{a}I \end{bmatrix}$$

with $|a|^2 - |b|^2 = 1$. Then S is isomorphic with $PSU(1,1)$. The image of the subgroup of $PSU(2,2)$ that centralizes every element of S is the image of the group of all block 2×2 diagonal elements of G_1, C which is isomorphic with $SO(3)$. These two groups form a reductive dual pair (the commutant of C is S). Before we analyze this pair, we will indicate its relationship with time in Minkowski space. If we write coordinates in Section 2 as $x_1, x_2.x_3$ and $x_4 = t$, then we have according to our rules

$$(0,0,0,t) \longmapsto \frac{(1+it)}{1-it}I.$$

In terms of the linear fractional action of $U(2,2)$ on $U(2)$ this corresponds to

$$\begin{bmatrix} \frac{1+it}{\sqrt{1+t^2}}I & 0 \\ 0 & \frac{1-it}{\sqrt{1+t^2}}I \end{bmatrix}.I.$$

That is, the time axis is the orbit of $K \cap S$ with $-I$ deleted.

As we observed, the center of $U(2,2)$ acts trivially on Maxw and thus we can consider the action of the group $C \times S$ through CS on Maxw. Let $\mathcal{H}^k$ denote the representation of $SO(3)$ on the spherical harmonics of degree k. We denote by $D_{2k}, k \in \mathbb{Z} - \{0\}$, the discrete series of S (D_{2k} has K-types $2k + 2\mathrm{sgn}(k)\mathbb{Z}_{\geq 0}$).

Theorem 8.1. *As a representation of* $C \times S$, $(\mathrm{Maxw}^+_{2,0})_K$ *and* $(\mathrm{Maxw}^+_{0,2})_K$ *are equivalent with*

$$\bigoplus_{k\geq 1} \mathcal{H}^k \otimes D_{-(2k+2)}$$

and $(\mathrm{Maxw}^-_{2,0})_K$ *and* $(\mathrm{Maxw}^-_{0,2})_K$ *are equivalent with*

$$\bigoplus_{k\geq 1} \mathcal{H}^k \otimes D_{2k+2}.$$

Proof. By considering the K-types we see that each of the modules has finite $K \cap CS$-multiplicities. We will give details for $M = (\mathrm{Maxw}^+_{2,0})_K$. Noting that $S \cap K$ is

$$\left\{ \begin{bmatrix} aI & \\ & a^{-1}I \end{bmatrix} \mid |a| = 1 \right\}$$

we see that the characters that appear for $S \cap K$ on M are $a^{-4}, a^{-6}, \ldots$ each with finite multiplicity. Thus, as a representation of S, M is a direct sum of highest weight modules. It is therefore enough to check that the character of M as a $CS \cap K$ module is correct. We note that in the formulation of the K-types, we can look upon $CS \cap K$ as the image of the group of all matrices

$$\begin{bmatrix} u(y) & 0 \\ 0 & u(y)x^{-2} \end{bmatrix}$$

with $u(y) = \begin{bmatrix} y & 0 \\ 0 & y^{-1} \end{bmatrix}$ and $|x|$ and $|y| = 1$. Set $t(x, y)$ equal to this element. Then if $\mathrm{Ch}(V)$ denotes the character of a $CS \cap K$ and χ_k denotes the character of $SU(2)$ on $S^k(\mathbb{C}^2)$, we have as a formal sum

$$\mathrm{Ch}(M)(t(x, y)) = \sum_{k=0}^{\infty} \chi_{k+2}(u(y))\chi_k(u(y))x^{-2k-4}$$

replacing x with x^{-1}, and plugging in $\chi_k(u(y)) = y^k(1 + y^{-2} + \ldots + y^{-2k})$ and summing we have

$$\frac{x^4(y^4 - x^2y^2 + y^2 + 1)}{(1 - x^2)(y^2 - x^2)(1 - x^2y^2)}.$$

If we multiply by $1 - x^2$ and expand in powers of x, we see that the series is

$$x^4 \sum_{k=0}^{\infty} \chi_{2k+2}(u(y))x^{2k}.$$

This implies the result in this case since the character of D_{-2k} is

$$\frac{x^{-2k}}{1 - x^{-2}} \left(\text{resp. } \frac{x^{2k}}{1 + x^2}\right)$$

if $k > 0$ (resp. $k < 0$). The other highest weight case is exactly the same. The cases $(\mathrm{Maxw}^-_{\overline{2},0})_K$ and $(\mathrm{Maxw}^-_{\overline{0},2})_K$ are handled in the same way but with the powers of x inverted. □

9 Plane wave solutions and (degenerate) Whittaker vectors

Recall that a plane wave solution to Maxwell's are those in the following form:

$$\mathbf{E} = e^{i(z,x)}\mathbf{E}_o$$

and

$$\mathbf{B} = e^{i(z,x)}\mathbf{B}_o$$

with $\mathbf{E}_o$ and $\mathbf{B}_o$ constant vectors, $z = (u_1, u_2, u_3, \omega) \in \mathbb{R}^4$, $\mathbf{u} = (u_1, u_2, u_3)$, $x = (x_1, x_2, x_3, t) \in \mathbb{R}^4$, and as before, $(z, x) = -\sum u_i x_i + \omega t$. To satisfy Maxwell's equations we must have $\mathbf{u} \cdot \mathbf{E}_o = 0, \mathbf{u} \cdot \mathbf{B}_o = 0$ and

$$\nabla \times \mathbf{E} = \frac{\partial}{\partial t}\mathbf{B}, \quad \nabla \times \mathbf{B} = -\frac{\partial}{\partial t}\mathbf{E}.$$

The first implies that $\mathbf{u} \times \mathbf{E}_o = -\omega\mathbf{B}_o$ and the second $\mathbf{u} \times \mathbf{B}_o = \omega\mathbf{E}_o$. Thus, if the solution is non-constant and $\omega > 0$ (resp. $\omega < 0$) $\frac{\mathbf{u}}{\|\mathbf{u}\|}, \frac{\mathbf{B}_o}{\|\mathbf{B}_o\|}, \frac{\mathbf{E}_o}{\|E_o\|}$ (resp. $\frac{\mathbf{u}}{\|\mathbf{u}\|}, \frac{\mathbf{B}_o}{\|\mathbf{B}_o\|}, \frac{-\mathbf{E}_o}{\|E_o\|}$) is an orthonormal basis of $\mathbb{R}^3$ obeying the "right-hand screw law". Putting these equations together yields $\omega^2 = \|\mathbf{u}\|^2$, that is, z is isotropic in $\mathbb{R}^{1,3}$, that is on the null light cone. This fits with the fact that if $\mathbf{E}, \mathbf{B}$ form a solution to Maxwell's equations, then their individual components satisfy the wave equation.

The purpose of this section is to interpret the plane wave solutions in the context of the four representations we have been studying. In the last few sections we have been emphasizing the realization of $U(2,2)$ which we have denoted by G_1. We now revert to the form G since the embedding of Minkowski space into $U(2)$ is clearer for that realization. Recall that the embedding is implemented in two stages; we map (x_1, x_2, x_3, x_4) to

$$\begin{bmatrix} I & 0 \\ Y & I \end{bmatrix}, Y = \begin{bmatrix} x_4 + x_3 & x_1 + ix_2 \\ x_1 - ix_2 & x_4 - x_3 \end{bmatrix}.$$

We denote the image group by $\overline{N}$ (note $Y^* = Y$). Then we consider the image of I under the action of G on $U(2)$ by linear fractional transformations. In this context to simplify the form subgroup S we must apply a Cayley transform. If we set

$$L = \frac{1}{\sqrt{2}}\begin{bmatrix} I & iI \\ I & -iI \end{bmatrix},$$

then the transform is given by

$$\sigma^{-1}(g) = L^{-1}gL.$$

Thus, in this incarnation, we take S to be the set of elements of $PSU(2,2)$ as $G/\operatorname{Center}(G)$ in the form

$$\begin{bmatrix} aI & bI \\ cI & dI \end{bmatrix}$$

with $\operatorname{Im} a\overline{b} = 0, a\overline{b} - b\overline{c} = 1, \operatorname{Im} c\overline{d} = 0$. This implies that the image of $S \cap \overline{N}$ is the set of matrices

$$\begin{bmatrix} I & 0 \\ cI & I \end{bmatrix}$$

with $c \in \mathbb{R}$. Let $z = (k_1, k_2, k_3, \omega)$ be, as above, an element in the light cone. Then considering z, as giving a linear functional on $\operatorname{Lie}(\overline{N})$, this functional restricted to $\begin{bmatrix} 0 & 0 \\ xI & 0 \end{bmatrix} \in \operatorname{Lie}(S \cap \overline{N})$ is given by the coefficient of q in the quadratic polynomial

$$\frac{1}{2}\det \begin{bmatrix} \omega + k_3 + qx & k_1 + ik_2 \\ k_1 - ik_2 & \omega - k_3 + qx \end{bmatrix}$$

which is ωx. We will record this as a lemma since we will need to apply it later in this section.

Lemma 9.1. *If $z = (k_1, k_2, k_3, \omega)$ is in the null light cone and if we consider z as a linear functional on* $\operatorname{Lie}(\overline{N})$ *(as above), then its value on*

$$\begin{bmatrix} 0 & 0 \\ xI & 0 \end{bmatrix}$$

is ωx.

We now relate these plane wave solutions to the representations $\mathrm{Maxw}^{\pm}_{\alpha,\beta}$ with $(\alpha, \beta) \in \{(2,0), \{0,2)\}$. Let $H^{\pm}_{\alpha,\beta}$ be the corresponding Hilbert space completions of these smooth Fréchet representations of moderate growth. Then the $\mathrm{Maxw}^{\pm}_{\alpha,b}$ are the spaces of C^{∞} vectors. Fix one of these representations and denote it by (π, H). Let C_K denote the Casimir operator of K. Let H_K be the space of K-finite vectors and let $\langle v, w\rangle_k = \left\langle (I + C_K)^{kd} v, w \right\rangle$ for $v, w \in H_K$, $\langle \ldots, \ldots \rangle$ the unitary structure and $d = \dim K$. Let H^k be the Hilbert space completion of H_K with respect to $\langle \ldots, \ldots \rangle_k$. Then $H^0 = H$ and the maps $T_{k,l} : H^l \to H^k$ for $k > l \geq 0$ that are the identity on H_K are nuclear (see [GV] for the definition).

Thus we have $\cap_{k\geq 0}H^k = H^\infty$. We set H^{-k} equal to the dual Hilbert space to H^k. Thus $(H^\infty)' = \cup_{k\geq 0}H^{-k}$. We thus have a rigged Hilbert space in the sense of [GV], Chapter 1. We now consider (π, H) as a unitary representation of $\overline{N}$. We have (c.f. [W1], Theorem 14.10.3) a direct integral decomposition of (π, H) as a representation of $\overline{N}$ given as follows:

$$\int_{\text{supp}(\pi)\subset\widehat{\overline{N}}} \mathbb{C}_\chi \otimes W_\chi d\tau(\chi)$$

with $\text{supp}(\pi)$ defined as in [W1, Volume 2, p. 337], τ a Borel measure on $\text{supp}(\pi)$, and $\chi \to W_\chi$ a Borel measurable Hilbert vector bundle over $\text{supp}(\pi)$. In this case $\text{supp}(\pi)$ is just the support of τ as a distribution. The theorem of Gelfand–Kostyuchenko (c.f. [GV], p. 117 Theorem 1′ and the remarks at the end of Subsection 1.4.4) implies that there exists $k > 0$ and a family of nuclear operators $T_\chi : H^k \to W_\chi$ such that if $u \in H^k$, then $u(\chi) = T_\chi(u)$ for τ-almost all χ. This implies that for all χ and all $u \in H^k$,

$$T_\chi(\pi(\overline{n})u) = \chi(\overline{n})T_\chi(u).$$

If T_χ is always 0, then $H = 0$. We see that there is a subset A of $\text{supp}(\pi)$ of full measure with $T_\chi \neq 0$ for $\chi \in A$. By definition of the H^k we see that if $\chi \in A$, then there exists $\lambda \in W_\chi'$ such that $\lambda \circ T_\chi \neq 0$. We have for this choice

$$\lambda \circ T_\chi(\pi(\overline{n})u) = \chi(\overline{n})\lambda \circ T_\chi(u)$$

for all $\overline{n} \in \overline{N}$ and all $u \in H^\infty$. If $\chi \in \widehat{\overline{N}}$, then set

$$\text{Wh}_\chi(\pi) = \{\mu \in (H^\infty)' | \mu \circ \pi(\overline{n}) = \chi(\overline{n})\mu\}.$$

We have with this notation

Theorem 9.2. *Let H^∞ be one of* $\text{Maxw}^\pm_{2,0}$ *or* $\text{Maxw}^\pm_{0,2}$. *Then*

1. $\dim \text{Wh}_\chi(\pi) \leq 3$.
2. *Writing*

$$\chi_Z\left(\begin{bmatrix} I & 0 \\ Y & I \end{bmatrix}\right) = e^{i(Z,Y)}$$

with $Z^* = Z$ (this describes all possible χ). Then $\text{Wh}_{\chi_Z}(\pi) = 0$ if $\det(Z) \neq 0$ (i.e., the support of π is contained in the null light cone). Furthermore, the support of π is the closure of a single orbit of the action of $GL(2, \mathbb{C})$ on $\{Z \in M_2(\mathbb{C}) | Z^* = Z\}$ given by $g \cdot Z = gZg^*$.

3. *Both* $\pi^+_{2,0}, \pi^+_{0,2}$ *have support equal to the set of all* χ_Z *with* $\det Z = 0$ *and* $\operatorname{tr} Z \leq 0$ *(negative light cone) and both* $\pi^-_{2,0}, \pi^-_{0,2}$ *have support equal to the set of all* χ_Z *with* $\det Z = 0$ *and* $\operatorname{tr} Z \geq 0$ *(positive light cone).*

Note: The discussion in the next section indicates that $\dim \mathrm{Wh}_\chi(\pi) = 3$. The proof of the theorem will use earlier work of the authors [K, W2, W3].

Proof. We will do all the details for $\mathrm{Maxw}^+_{2,0}$. The case $\mathrm{Maxw}^+_{0,2}$ is essentially the same. The cases $\mathrm{Maxw}^-_{2,0}$ and $\mathrm{Maxw}^-_{0,2}$ are done with an interchange of $\mathfrak{p}^\pm$ with $\mathfrak{p}^\mp$. Set $\mathfrak{q} = \mathfrak{k} \oplus \mathfrak{p}^+$ where $\mathfrak{k} = \mathrm{Lie}(K)_\mathbb{C}$. We consider $F^{2,0,2}$ to be the $\mathfrak{q}$-module with $\mathfrak{k}$ acting through the differential of the K action and $\mathfrak{p}^+$ acting by 0. Set

$$N(F^{2,0,2}) = U(\mathfrak{g}) \otimes_{U(\mathfrak{q})} F^{2,0,2}.$$

Then Lemma 7.2 implies that we have a surjective $(\mathfrak{g}, K)$-module homomorphism

$$p : N(F^{2,0,2}) \to (\mathrm{Maxw}^+_{2,0})_K.$$

Thus $p^* \mathrm{Wh}_\chi(\pi^+_{2,0})$ is contained in

$$\mathcal{W}_\chi = \{\lambda \in \mathrm{Hom}_\mathbb{C}(N(F^{2,0,2}), \mathbb{C}) | \lambda(Yv) = d\chi(Y)\lambda(v), Y \in \mathrm{Lie}(\overline{N})\}.$$

Since $p^* : \mathrm{Wh}_\chi(\pi^+_{2,0}) \to \mathcal{W}_\chi$ is injective ($(\mathrm{Maxw}^+_{2,0})_K$ is dense in $\mathrm{Maxw}^+_{2,0}$) the dimension estimate will follow if we prove that $\dim \mathcal{W}_\chi = 3$. For this we use the observation in [W3] that if

$$L = \left\{ \begin{bmatrix} g & 0 \\ 0 & (g^{-1})^* \end{bmatrix} | g \in GL(2, \mathbb{C}) \right\},$$

then $\mathrm{Lie}(L\overline{N})_\mathbb{C}$ and $\mathfrak{q}$ are opposite parabolic subalgebras (i.e., $\mathrm{Lie}(L\overline{N})_\mathbb{C} \cap \mathfrak{q}$ is a Levi factor of both parabolic subalgebras). This implies that $N(F^{2,0,2})$ is a free $\mathrm{Lie}(\overline{N})_\mathbb{C}$ module on $\dim F^{2,0,2} = 3$ generators. This clearly implies that $\dim \mathcal{W}_\chi = 3$.

We now note if $m = \begin{bmatrix} g & 0 \\ 0 & (g^{-1})^* \end{bmatrix}$ then $m \begin{bmatrix} I & 0 \\ Y & I \end{bmatrix} m^{-1} = \begin{bmatrix} I & 0 \\ g \cdot Y & I \end{bmatrix}$. Set $\overline{n} = \overline{n}(Y) = \begin{bmatrix} I & 0 \\ Y & I \end{bmatrix}$. If $\lambda \in \mathrm{Wh}_{\chi_Z}(\pi^+_{2,0})$, then

$$\begin{aligned} \lambda \circ \pi^+_{2,0}(m)(\pi^+_{2,0}(\overline{n})u) &= \lambda(\pi^+_{2,0}(m)\pi^+_{2,0}(\overline{n})u) \\ &= \lambda(\pi^+_{2,0}(m\overline{n}m^{-1})\pi^+_{2,0}(m)u) \\ &= \chi_Z(m\overline{n}m^{-1})\lambda \circ \pi^+_{2,0}(m)(u) \\ &= \chi_{g^{-1} \cdot Z}(\overline{n})\lambda \circ \pi^+_{2,0}(m)(u). \end{aligned}$$

Thus $\lambda \circ \pi^+_{2,0}(m) \in \mathrm{Wh}_{\chi_{g^{-1}Z}}(\pi^+_{2,0})$. This implies that $\mathrm{supp}(\pi^+_{2,0})$ is a union of orbits.

Using results in [W3] we can show that if $\mathrm{Wh}_{\chi_Z}(\pi^+_{2,0}) \neq 0$ with $\det Z \neq 0$, then the GK-dimension of $(\mathrm{Maxw}^+_{2,0})_K$ is at least $\dim \overline{N}$=4. Indeed, this condition implies that the dual module to $(\mathrm{Maxw}^+_{2,0})_K$ contains an irreducible submodule of GK-dimension 4. However as a $\mathfrak{p}^-$-module $(\mathrm{Maxw}^+_{2,0})_K$ is graded with the k-th level of the grad being isomorphic with the K-module $F^{k+2,k,k+2}$ whose dimension is $(k+3)(k+1)$. Thus the GK-dimension of $(\mathrm{Maxw}^+_{2,0})_K$ is 3. Thus we must have $\det Z = 0$. To complete the proof, it is enough to prove 3. Since there are three orbits of L in set of all Z with $\det Z = 0$:

$$\mathcal{O}^+ = \{Z | \det Z = 0, \operatorname{tr} Z > 0\}, \mathcal{O}^+ = \{Z | \det Z = 0, \operatorname{tr} Z < 0\}, \{0\}.$$

Thus we must prove that if $\mathrm{Wh}_{\chi_Z}(\pi^+_{2,0}) \neq 0$ and $Z \neq 0$ then $\operatorname{tr} Z < 0$. This follows from Theorem 17 (p. 306) in [W2] (in this reference the roles of χ and χ^{-1} are reversed). We can also prove the result directly using Lemma 9.1. Let $\lambda \neq 0$ be an element of $\mathrm{Wh}_{\chi_Z}(\pi^+_{2,0})$ with $Z \neq 0$. Then set $\eta = d\chi_Z$. If

$$Z = \begin{bmatrix} \omega + k_3 & k_1 + ik_2 \\ k_1 - ik_2 & \omega - k_3 x \end{bmatrix},$$

then $\operatorname{tr} Z = 2\omega$. In Lemma 9.1 we saw that

$$\eta \begin{bmatrix} 0 & 0 \\ xI & 0 \end{bmatrix} = \omega x.$$

Thus if S is as in the previous section and $\nu = \chi_{|N \cap S}$, then in the notation of [W2] Theorem 2, $r_\nu = \omega$. Thus, that theorem (also see [K]) implies that (taking into account the reversal of signs mentioned above) $r_\nu < 0$. This completes the proof.

□

Remark 9.3. The above results imply that in the steady state solutions to Maxwell's equations the plane wave solutions should be looked upon as 2-currents. That is, using γ, they are distributions on Maxw.

10 Planck's black body radiation law

We will now use the results of the previous section to study the solutions of Maxwell's equations that have intensities that follow Planck's black body radiation law. We first note that we can take the measure on the (dual) null light cone $\{(u_1, u_2, u_3, \omega) | u_1^2 + u_2^2 + u_3^2 = \omega^2\}$ to be the $SO(3,1)$ invariant measure

$$d\chi = \frac{du_1 du_2 du_3}{(u_1^2 + u_2^2 + u_3^2)^{\frac{1}{2}}},$$

which in polar coordinates is given by

$$rdrd\mu$$

with $d\mu$ the normalized $SO(3)$ invariant measure on the 2-sphere. We will study one of the representations supported on the positive null light cone. Theorem 5.6 implies that representation $\pi_{0,2}^-$ has that property and is contained in Maxw_i and its contragradient $\pi_{0,2}^+$ is in Maxw_{-i}. Using the material on plane wave solutions we have the Hilbert space that corresponds to $\pi_{0,2}^-$ pulled back to Minkowski space is given by the functions

$$(x,t) \mapsto e^{-i(\sum u_i x_i - \omega t)}(\mathbf{E}(u,\omega) - i\mathbf{B}(u,\omega))$$

with $\mathbf{E}(u,\omega)$ and $\mathbf{B}(u,\omega)$ corresponding to a plane wave solution for the given character (see the beginning of Section 9). Thus the real solutions are given by

$$\cos(\sum u_i x_i - \omega t)\mathbf{E}(u,\omega) + \sin(\sum u_i x_i - \omega t)\mathbf{B}(u,\omega).$$

The intensity per unit surface area, S, is given by ($u = \omega\mu$ since $\omega^2 = r^2$)

$$\begin{aligned} I_\omega &= \int_S \Big(\cos\big(\omega(\sum \mu_i x_i - t)\big)^2 \|\mathbf{E}(\omega\mu,\omega)\|^2 \\ &\qquad + \sin\big(\omega(\sum \mu_i x_i - t)\big)^2 \|\mathbf{B}(\omega\mu,\omega)\|^2 \Big) d\mu \\ &= \int_S \|\mathbf{E}(u,\omega)\|^2 d\omega = \int_S \|\mathbf{B}(u,\omega)\|^2 d\omega. \end{aligned}$$

(see the beginning of section 9 to see that $\|\mathbf{E}(\mathbf{u},\omega)\| = \|\mathbf{B}(\mathbf{u},\omega)\|$). Thus to satisfy the Planck law (after normalizing all constants that can be normalized to 1) the above integral must be equal to

$$I_\omega = \frac{\omega^3}{e^{\frac{\omega}{T}} - 1}.$$

Since

$$\int_0^\infty \frac{\omega^4}{e^{\frac{\omega}{T}} - 1} d\omega < \infty$$

and on the positive null light cone $\omega = r$, steady state solutions satisfying Planck's law exist in abundance.

In [S1], Segal analyzes the background radiation in what he calls Einstein's universe (essentially the universal covering space of $U(2)$). Thus the time parameter becomes "repetitious" rather than periodic (this is also the case in [P]). From the discussion in the beginning of Section 8, it follows that the energy levels studied by Segal correspond to the eigenvalues of

$$L = \frac{1}{2}\begin{bmatrix} I & 0 \\ 0 & -I \end{bmatrix}$$

which corresponds to $-i$ times the infinitesimal generator of the image of the group of all elements

$$e^{i\theta L} = \begin{bmatrix} e^{\frac{i\theta}{2}} I & 0 \\ 0 & e^{-\frac{i\theta}{2}} I \end{bmatrix}$$

in $PSU(2,2)$, which under the action on $U(2)$ described in Section 2, we have $e^{i\theta L} \cdot I = e^{-i\theta} I$. In our notation (using the same notation for the differential and the representation of K)

$$\tau_{k,k+2,\pm(k+2)}(L) = \mp(k+2)I$$

and also

$$\tau_{k+2,k,\pm(k+2)}(L) = \mp(k+2)I.$$

Thus the positive eigenvalues of L on Maxw appear in $\mathrm{Maxw}_{0,2}^{-}$ and $\mathrm{Maxw}_{2,0}^{-}$. This implies that the multiplicity of the eigenvalue $n = k + 2$ with $k \geq 0$ is $2(k+1)(k+3) = 2(n+1(n-1) = 2(n^2-1)$. These dimensions of the positive eigenvalues of H in the space of solutions to Maxwell's equations with the spectrum bounded below by 2 is stated in [S1] with references to [SJOPS] and [PS]. Segal then argues that this implies that the energy levels of the "equilibrium state" differ "by an unobservably small amount" from Planck's Law. The discussion in the beginning of Section 8 combined with the harmonic analysis of Section 9 indicates that Segal's "equilibrium solution" is very close to one of the solutions above satisfying Planck's Law.

11 Conclusion

In this paper we have shown (Theorem 7.3) that the solutions to Maxwell's equations that extend to the conformal compactification of Minkowski space break up into four irreducible unitary representations: 2 positive energy and 2 negative energy (in the sense of lowest weight or highest weight respectively) as a representation of the conformal group of the wave equation. The support (see Section 9) of each of the representations is either the forward or backward null light cone. The ones with positive energy (Theorem 9.2) yield only positive frequencies and thus according to Planck's formula $E = h\nu$ have positive energy in the sense of field theory. We observe that solutions that extend to the full compactification (i.e., steady state) can have their frequency spectrum constrained (see Section 10) in such a way that one has background radiation that follows Planck's Black Body Law with a temperature of about 2.7 degrees Kelvin (or any other temperature for that matter). This says that although the actual steady state models to the universe that fit the astronomical observations are complicated this work indicates that there can be a background radiation that fits the measurements that is not the outgrowth of an initial very high temperature source. We would also like to point out that the big bang models for the universe have had difficulty fitting the observations also, leading to theories involving inflation and the return of the cosmological constant. An interesting alternative which is not unrelated to this paper can be found in [P] where the conformal structure is emphasized and time does almost cycle. There is also the chronometric theory of [S1] which we related to our work at the end of the last section.

We have also shown that these representations fit in larger contexts. However, although many of the results in this paper generalize to $SO(n,2)$ (resp. $U(n,2)$), the beautiful geometric structures that appear in the case $n = 4$ (resp. $n = 2$) do not.

References

[BW] A. Borel and N. Wallach, *Continuous Cohomology, Discrete Subgroups, and Representations of Reductive Groups,* Second edition. Mathematical Surveys and Monographs, 67. American Mathematical Society, Providence, RI, 2000.

[D] P. A. M. Dirac, Wave equations in conformal space, *Ann. of Math.* **37** (1936), 429–442.

[EW] T. Enright and N. Wallach, Embeddings of unitary highest weight representations and generalized Dirac operators, *Math. Ann.* **307** (1997), 627–646.

[GV] I. M. Gelfand and N. Y. Vilenkin, *Generalized Functions Vol. 4: Applications of Harmonic Analysis*, Academic Press, 1964.

[HBN] F. Hoyle, G. Burbidge, J. V. Narlikar, A quasi-steady state cosmological model with creation of matter, *The Astrophysical Journal* **410** (1993), 437–457.

[GW] B. Gross and N. Wallach, On quaternionic discrete series representations, and their continuations, *J. Reine Angew. Math.*, **481**(1996), 73–123.

[HSS] M. Hunziker, M. Sepanski, and R. Stanke, The minimal representation of the conformal group and classical solutions to the wave equation, *J. Lie Theory* **22** (2012), 301–360.

[K] B. Kostant, On Laguerre polynomials, Bessel functions, Hankel transform and a series in the unitary dual of the simply-connected covering group of $SL(2,\mathbb{R})$, *Represent. Theory* **4** (2000), 181–224.

[PS] Stephen M. Paneitz, Irving E. Segal, Analysis in space-time bundles. I. General considerations and the scalar bundle. *J. Funct. Anal.* **47** (1982), no. 1, 78–142.

[P] Roger Penrose, *Cycles of Time*, Alfred Knopf, New York, 2011.

[SJOPS] I. E. Segal, H. P. Jakobsen, B. Ørsted, S. M. Paneitz, B. Speh, Covariant chronogeometry and extreme distances: elementary particles. *Proc. Nat. Acad. Sci. U.S.A.* **78** (1981), no. 9, part 1, 5261–5265.

[S1] Irving Segal, Radiation in the Einstein universe and the cosmic background, *Physical Review D* **28**(1983), 2393–2402.

[S2] Irving Segal, Geometric derivation of the chronometric red shift, *Proc. Nat. Acad. Sci.* **90** (1993), 11114–11116.

[V] M. Vergne, Some comments on the work of I. E. Segal in group representations, *J. Funct. Anal.* **190** (2002), 29–37.

[VZ] D. Vogan and G. Zuckerman, Unitary representations and continuous cohomology, *Comp/ Math.* **53** (1984), 51–90.

[W1] Nolan R. Wallach, *Real Reductive Groups I,II*, Academic Press, New York, 1988, 1992.

[W2] N. Wallach, Generalized Whittaker vectors for holomorphic and quaternionic representations, *Comment. Math. Helv.* **78** (2003), 266–307.

[W3] N. R. Wallach, Lie algebra cohomology and holomorphic continuation of generalized Jacquet integrals, *Advanced Studies in Pure Math.* **14** (1988), 123–151.

Representations with a reduced null cone

Hanspeter Kraft and Gerald W. Schwarz

In honor of Nolan Wallach

Abstract Let G be a complex reductive group and V a G-module. Let $\pi\colon V \to V/\!\!/G$ be the quotient morphism defined by the invariants and set $\mathcal{N}(V) := \pi^{-1}(\pi(0))$. We consider the following question. Is the null cone $\mathcal{N}(V)$ reduced, i.e., is the ideal of $\mathcal{N}(V)$ generated by G-invariant polynomials? We have complete results when G is SL_2, SL_3 or a simple group of adjoint type, and also when G is semisimple of adjoint type and the G-module V is irreducible.

Keywords: Null cone • Null fiber • Quotient morphism • Semisimple groups • Representations

Mathematics Subject Classification: 20G20, 22E46

1 Introduction

Let G be a reductive complex algebraic group and let V be a finite-dimensional G-module. Let $\pi\colon V \to V/\!\!/G$ be the categorical quotient morphism given by the G-invariant functions on V, and let

$$\mathcal{N} := \pi^{-1}(\pi(0)) = \{v \in V \mid \overline{Gv} \ni 0\} \subset V$$

be the *null cone*. We say that V is *coreduced* if $\mathcal{N}$ is reduced. This means that the ideal $I(\mathcal{N}) \subset \mathcal{O}(V)$ of the set $\mathcal{N}$ is generated by the invariant functions

H. Kraft
Mathematisches Institut der Universität Basel, Rheinsprung 21, CH-4051 Basel, Switzerland
e-mail: Hanspeter.Kraft@unibas.ch

G.W. Schwarz
Department of Mathematics, Brandeis University, Waltham, MA 02454-9110, USA
e-mail: schwarz@brandeis.edu

R. Howe et al. (eds.), *Symmetry: Representation Theory and Its Applications*,
Progress in Mathematics 257, DOI 10.1007/978-1-4939-1590-3_15

$\mathfrak{m}_0 := \mathcal{O}(V)^G \cap I(\mathcal{N})$, the homogeneous maximal ideal of $\mathcal{O}(V)^G$, and so $I(\mathcal{N}) = I_G(\mathcal{N}) := \mathfrak{m}_0\,\mathcal{O}(V)$, where we use $\mathcal{O}(X)$ to denote the regular functions on a variety X. If it is important to specify the group or representation involved, we will use notation such as (V, G), $\mathcal{N}(V)$, $\mathcal{N}_G(V)$, etc.

We say that V is *strongly coreduced* if every fiber of π is reduced. We can reformulate this in terms of slice representations. Let Gv be a closed orbit. Then G_v is reductive and we have a splitting of G_v-modules,

$$V = T_v(Gv) \oplus N_v.$$

Then (N_v, G_v) is the *slice representation of* G_v *at* v. We show that the fiber $\pi^{-1}(\pi(v))$ is reduced if and only if (N_v, G_v) is coreduced (Remark 4.1). Hence V is strongly coreduced if and only if every slice representation of V is coreduced.

A main difficulty in our work is that, a priori, V may be coreduced but π may have a nonreduced fiber $F \neq \mathcal{N}$. We conjecture that this is never the case:

Conjecture 1. A coreduced G-module is strongly coreduced.

Recall that V is *cofree* if $\mathcal{O}(V)$ is a free module over $\mathcal{O}(V)^G$. Equivalently, $\pi\colon V \to V/\!\!/G$ is flat. In the cofree case the associated cone to any fiber F (see [BK79] or [Kra84, II.4.2]) is the null cone. From this one can immediately see that $\mathcal{N}$ reduced implies that F is reduced, so the conjecture holds when V is cofree. There is another case in which the associated cone of F is $\mathcal{N}$: the case in which the isotropy group H of the closed orbit $Gv \subset F$ has the same rank as G, i.e., contains a maximal torus T of G. This was first noticed by RICHARDSON [Ric89, Proposition 5.5]. If the slice representation of H at v is not coreduced, then neither is (V, G) (Proposition 5.1). For an irreducible representation V of G, having $V^T \neq 0$ means that the weights of V are in the root lattice; equivalently, the center of G acts trivially on V. Hence we have a representation of the adjoint group $G/Z(G)$. This explains why many of our results require that the group be adjoint, or that at least one of the irreducible components of our representation contains a zero weight vector.

For SL_2 and SL_3 we have a complete classification of the coreduced modules. In the case of SL_3, the coreduced modules are either cofree (for which there is a finite list of possibilities) or one has the direct sum of arbitrarily many copies of $\mathbb{C}^3$ and $(\mathbb{C}^3)^*$ (§11). We would guess that something similar holds for representations of any simple algebraic group G. Besides the infinite series of coreduced representations involving the representations of the smallest dimension for SL_n and Sp_n (see §9), any coreduced module should be cofree (and for this there is a finite list of possibilities). If one replaces coreduced by *strongly* coreduced, then our techniques should be sufficient to establish this result. For irreducible representations we show that strongly coreduced and cofree are equivalent (§4), which is the first step. If G is adjoint, our classification of the coreduced representations produces a finite list of cofree representations. In case G is semisimple adjoint and not simple and the representation is faithful and irreducible, our classification shows that there are finitely many coreduced representations, all of which are again cofree. In these cases the group has two simple factors (Theorem 8.3). We would also guess that one can add at most finitely many representations to these and still remain coreduced, but we have not carried out the details. In general, to carry the classification any further, new techniques are needed.

Here is a summary of the contents of this paper. In §2 we present elementary results and determine the coreduced representations of tori (Proposition 2.13). In Section 3 we show how to use covariants to prove that a null cone is not reduced and as an application we determine the coreduced representations of SL_2 (Theorem 3.7). In §4 we show that every cofree irreducible representation of a simple algebraic group is coreduced (Theorem 4.10) and that, sort of conversely, every irreducible representation of a simple group which is strongly coreduced is cofree (Theorem 4.12). In §5 we consider modules V with $V^T \neq 0$, T a maximal torus of G. We develop methods (based on weight multiplicities) to show that a slice representation at a zero weight vector is not coreduced (we say that V has a *bad toral slice*). We then show that V has a bad toral slice if all the roots of G appear in V with multiplicity two or more (Proposition 5.17).

In §6 we apply our techniques to find the maximal coreduced representations of the adjoint exceptional groups (Theorem 6.9). The case of F_4 is rather complicated and needs some heavy computations (see Appendix A). In §7 we do the same thing for the classical adjoint groups (Theorem 7.1), and in §8 we determine the irreducible coreduced representations of semisimple adjoint groups (Theorem 8.3). This is not straightforward, e.g., the representation $(\mathbb{C}^7 \otimes \mathbb{C}^7, \mathsf{G}_2 \times \mathsf{G}_2)$ is not coreduced, but showing this is difficult (see Appendix B).

In §9 we show that, essentially, the classical representations of the classical groups are coreduced (with a restriction for SO_n). This is a bit surprising, since these representations are often far from cofree. In §11 we classify the coreduced representations of SL_3 (not just PSL_3). To do this, we need to develop some techniques for finding irreducible components of null cones (see §10). These techniques should be useful in other contexts.

Acknowledgments. We thank Michel Brion and John Stembridge for helpful discussions and remarks, and Jan Draisma for his computations. We thank the referee for helpful comments.

2 Preliminaries and elementary results

We begin with some positive results. Let G be a reductive group. We do not assume that G is connected, but when we say that G is semisimple this includes connectivity. Let V be a finite-dimensional G-module.

Proposition 2.1. *Suppose that G is semisimple and that V satisfies one of the following conditions.*

(1) $\dim V/\!\!/G \leq 1$;
(2) $V = \operatorname{Ad} G$.

Then V is coreduced.

Proof. If $\dim V/\!\!/G = 0$, then $\mathcal{N} = V$ is reduced. If $\dim V/\!\!/G = 1$, then $\mathcal{O}(V)^G$ is generated by a homogeneous irreducible polynomial f and its zero set $\mathcal{N}$ is reduced. If V is the adjoint representation of G, then $\mathcal{N}$ is irreducible of codimension equal to the rank ℓ of G. Since $\mathcal{N}$ is defined by ℓ homogeneous invariants and the rank of $d\pi$ is ℓ on an open dense subset of $\mathcal{N}$, it follows that $\mathcal{N}$ is reduced and even normal ([Kos63]). □

Example 2.2. Suppose that G is finite and acts nontrivially on V. Then $\mathcal{N} = \{0\}$ (as set) is not reduced since not all the coordinate functions can be G-invariant.

Let V be a G-module. Then $V/\!\!/G$ parameterizes the closed G-orbits in V. Let Gv be a closed orbit and let N_v be the slice representation of G_v. We say that Gv is a *principal orbit* and that G_v is a *principal isotropy group* if $\mathcal{O}(N_v)^{G_v} = \mathcal{O}(N_v^{G_v})$. In other words, N_v is the sum of a trivial representation and a representation N_v' with $\mathcal{O}(N_v')^{G_v} = \mathbb{C}$. The principal isotropy groups form a single conjugacy class of G and the image of the principal orbits in $V/\!\!/G$ is open and dense. We say that V is *stable* if $N_v' = (0)$; equivalently, there is an open dense subset of V consisting of closed orbits. If G is semisimple and there is a nonempty open subset of points with reductive isotropy group, then V is stable. In particular, if the general point in V has a finite isotropy group, then the representation is stable with finite principal isotropy groups.

Our example above generalizes to the following.

Remark 2.3. Let V be a G-module where $G/G^0 \neq \{e\}$. If G/G^0 acts nontrivially on the quotient $V/\!\!/G^0$, then V is not coreduced. Note that, for example, G/G^0 acts nontrivially if the principal isotropy group of (V, G) is trivial.

Proposition 2.4. *Assume that (V, G^0) is not coreduced. Then (V, G) is not coreduced.*

Proof. The null cones for G^0 and G are the same (as sets). There is an $f \in I(\mathcal{N})$ which is not in $I_{G^0}(\mathcal{N})$. Hence $f \notin I_G(\mathcal{N})$ and (V, G) is not coreduced. □

For the next three more technical results we have to generalize the definition of coreducedness to pointed G-varieties.

Definition 2.5. A *pointed G-variety* is a pair (Y, y_0) where Y is an affine G-variety and y_0 is a fixed point. A pointed G-variety (Y, y_0) is *coreduced* if the fiber $\pi^{-1}(\pi(y_0))$ is reduced where $\pi\colon Y \to Y/\!\!/G$ is the quotient morphism.

Lemma 2.6. *Let (X, x_0) be a pointed G-variety and $Y \subset X$ a closed G-stable subvariety containing x_0. Assume that the ideal $I(Y)$ of Y is generated by G-invariant functions. Then (X, x_0) is coreduced if and only if (Y, x_0) is coreduced.*

Proof. Let $\mathfrak{m} \subset \mathcal{O}(X)^G$ be the maximal ideal of $\pi(x_0)$ where $\pi: X \to X/\!\!/G$ is the quotient morphism, and let $\mathfrak{n} \subset \mathcal{O}(Y)^G$ denote the image of $\mathfrak{m}$. By assumption, the ideal $\mathfrak{m}\,\mathcal{O}(X)$ contains $I(Y)$ and so $\mathcal{O}(Y)/(\mathfrak{n}\,\mathcal{O}(Y)) \simeq \mathcal{O}(X)/(I(Y)+\mathfrak{m}\,\mathcal{O}(X)) = \mathcal{O}(X)/(\mathfrak{m}\,\mathcal{O}(X))$. □

Example 2.7. Let V be a G-module. Denote by θ_n the n-dimensional trivial representation and let $F \subset V \oplus \theta_n$ be a G-stable hypersurface containing 0. If G is semisimple, then F is defined by a G-invariant polynomial. Hence $(F, 0)$ is coreduced if and only if V is coreduced.

Lemma 2.8. *Let (X, x_0) be a pointed G-variety. Let H be a reductive group acting on X such that G sends H-orbits to H-orbits. Assume that every G-invariant function on X is H-invariant. If (X, x_0) is coreduced with respect to G, then so is $(X/\!\!/H, \pi(x_0))$.*

Proof. Put $Y := X/\!\!/H$. Then G acts on Y, because G preserves the H-invariant functions $\mathcal{O}(X)^H \subset \mathcal{O}(X)$. Suppose that f is an element of $\mathcal{O}(Y)$ which vanishes on the null fiber $\mathcal{N}_G(Y, \pi_H(x_0))$. By assumption, $\mathcal{N}_G(X, x_0) = \pi_H^{-1}(\mathcal{N}_G(Y, \pi_H(x_0)))$ and so $f \circ \pi_H$ vanishes on $\mathcal{N}_G(X, x_0)$. Hence $f \circ \pi_H = \sum_i a_i b_i$ where the a_i are G-invariant and vanish at x_0. Since $f \circ \pi_H$ is H-invariant, we may average the b_i over H and assume that they are in $\mathcal{O}(X)^H$. But then $a_i = \bar{a}_i \circ \pi_H$ and $b_i = \bar{b}_i \circ \pi_H$ for unique $\bar{a}_i \in \mathcal{O}(Y)^G$ and $\bar{b}_i \in \mathcal{O}(Y)$. Thus $f = \sum_i \bar{a}_i \bar{b}_i$ and so $(Y, \pi_H(x_0))$ is coreduced. □

Example 2.9. If (V, G) is a coreduced representation and $H \subset G$ a closed normal subgroup, then $(V/\!\!/H, \pi_H(0))$ is coreduced (with respect to G/H).

Example 2.10. Let (X, x_0) be a pointed G-variety, let W be a G-module of dimension n and let $H = \mathrm{SO}_n$ acting as usual on $\mathbb{C}^n$. Consider the pointed $(G \times H)$-variety $(Y, y_0) := (X \times (W \otimes \mathbb{C}^n), (x_0, 0))$. Assume that $G \to \mathrm{GL}(W)$ has image in $\mathrm{SL}(W)$. We claim that if (Y, y_0) is a coreduced $(G \times H)$-variety, then $(X \times S^2(W^*), (x_0, 0))$ is a coreduced G-variety.

By Classical Invariant Theory (see [Pro07]), the generators of the invariants of $(n\mathbb{C}^n, \mathrm{SO}_n)$ are the inner product invariants f_{ij} of the copies of $\mathbb{C}^n$ together with the determinant d. The relations are generated by the equality $d^2 = \det(f_{ij})$. Identifying $n\mathbb{C}^n$ with $W \otimes \mathbb{C}^n$ we see that the quadratic invariants transform by the representation $S^2(W^*)$ of G, the determinant d transforms by $\bigwedge^n(W^*) = \theta_1$, and the relation is G-invariant. Now applying Lemmas 2.8 and 2.6 gives the claim.

Lemma 2.11. *Let (Y, y_0) be a pointed G-variety and $Z \subset Y$ a closed G-stable subvariety containing y_0. Suppose that there is a G-equivariant retraction $p: (Y, y_0) \to (Z, y_0)$. If (Y, y_0) is coreduced, then so is (Z, y_0).*

Proof. Clearly, if y is in the null cone of Y, then $p(y)$ is in the null cone of Z. Thus if $f \in \mathcal{O}(Z)$ vanishes on the null cone of Z, then $\tilde{f} := p^* f \in \mathcal{O}(Y)$ vanishes on the null cone of Y. By hypothesis we have that $\widetilde{f} = \sum_i a_i b_i$ where the a_i are invariants vanishing at y_0. Restricting to Z we get a similar sum for f. Hence Z is coreduced. □

Example 2.12. *(1)* If $(V_1 \oplus V_2, G)$ is a coreduced representation, then so is (V_i, G), $i = 1, 2$.
(2) If (V, G) is a coreduced representation and $H \subset G$ a closed normal subgroup, then (V^H, G) is also coreduced.
(3) Let V_i be a G_i module, $i = 1, 2$. Then $(V_1 \oplus V_2, G_1 \times G_2)$ is coreduced if and only if both (V_1, G_1) and (V_2, G_2) are coreduced. Here we use that $\mathcal{N}(V_1 \oplus V_2) = \mathcal{N}(V_1) \times \mathcal{N}(V_2)$.

We finish this section with the case of tori which is quite easy. We will then see in Section 5 that this case can be applied to representations containing zero weights.

Proposition 2.13. *Let V be a T-module where T is a torus. Let $\alpha_1, \dots, \alpha_n$ be the nonzero weights of V. Then V is coreduced if and only if the solutions of*

$$\sum_i m_i \alpha_i = 0, \quad m_i \in \mathbb{N},$$

are generated by solutions where the m_i are zero or one.

It is well-known that the monoid of relations $\sum_i m_i \alpha_i = 0$, $m_i \in \mathbb{N}$ is generated by the *indecomposable* relations, i.e., by relations which cannot be written as a sum of two nontrivial relations. *So a necessary and sufficient condition for coreducedness is that the indecomposable relations $\sum_i n_i \alpha_i = 0$, $n_i \in \mathbb{N}$, satisfy $n_i = 0$ or 1.*

Proof. We may assume that $V^T = 0$. Let $x_1, \dots, x_n$ be a weight basis of V^* corresponding to the α_i. Suppose that there is an indecomposable relation where, say, $m_1 \geq 2$. Then the monomial $x_1 x_2^{m_2} \dots x_n^{m_n}$ vanishes on the null cone, but it is not in the ideal of the invariants. Hence our condition is necessary.

On the other hand, suppose that the indecomposable relations are of the desired form. Now any polynomial vanishing on $\mathcal{N}(V)$ is a sum of monomials with this property, and a monomial p vanishing on $\mathcal{N}(V)$ has a power which is divisible by an invariant monomial q without multiple factors. But then p is divisible by q and so $\mathcal{N}(V)$ is reduced. □

Corollary 2.14. *Let $T = \mathbb{C}^*$. Then $\mathcal{N}(V)$ is reduced if and only if $\mathcal{O}(V)^T = \mathbb{C}$ or the nonzero weights of V are $\pm k$ for a fixed $k \in \mathbb{N}$.*

3 The method of covariants

In this section we explain how covariants can be used to show that a representation is not coreduced. As a first application we classify the coreduced representations of SL_2.

Let G be a reductive group and V a representation of G. A G-equivariant morphism $\varphi: V \to W$ where W is an irreducible representation of G is called a *covariant of V of type W*. Clearly, covariants of type W can be added and multiplied with invariants and thus form an $\mathcal{O}(V)^G$-module $\operatorname{Cov}(V, W)$ which is known to be finitely generated (see [Kra84, II.3.2 Zusatz]).

A nontrivial covariant $\varphi: V \to W$ defines a G-submodule $\varphi^*(W^*) \subset \mathcal{O}(V)$ isomorphic to the dual W^*, and every irreducible G-submodule of $\mathcal{O}(V)$ isomorphic to W^* is of the form $\varphi^*(W^*)$ for a suitable covariant φ. Moreover, φ vanishes on the null cone $\mathcal{N}$ if and only if $\varphi^*(W^*) \subset I(\mathcal{N})$. We say that φ is a *generating covariant* if φ is not contained in $\mathfrak{m}_0 \operatorname{Cov}(V, W)$, or equivalently, if $\varphi^*(W^*)$ is not contained in $I_G(\mathcal{N})$. Thus we obtain the following useful criterion for non-coreducedness.

Proposition 3.1. *If φ is a generating covariant which vanishes on $\mathcal{N}$, then V is not coreduced.*

Remark 3.2. Let $f \in \mathcal{O}(V)^G$ be a generating homogeneous invariant of positive degree, i.e., $f \in \mathfrak{m}_0 \setminus \mathfrak{m}_0^2$. Then the differential $df: V \to V^*$ is a generating covariant. In fact, using the contraction $(df, \mathrm{Id}) = \deg f \cdot f$, we see that if $df = \sum_i f_i \varphi_i$ where the f_i are homogeneous nonconstant invariants, then $f = \frac{1}{\deg f} \sum_i f_i(\varphi_i, \mathrm{Id}) \in \mathfrak{m}_0^2$.

Example 3.3. Let G be SL_2 and $V = \mathfrak{sl}_2 \oplus \mathfrak{sl}_2$ where $\mathfrak{sl}_2 = \operatorname{Lie} \mathrm{SL}_2$ is the Lie algebra of SL_2. Then the null cone $\mathcal{N}(V)$ consists of commuting pairs of nilpotent matrices, and so the covariant

$$\varphi: \mathfrak{sl}_2 \oplus \mathfrak{sl}_2 \to \mathfrak{sl}_2, (A, B) \mapsto AB - \frac{1}{2} \operatorname{tr}(AB) \begin{bmatrix} 1 & 0 \\ 0 & 1 \end{bmatrix}$$

vanishes on $\mathcal{N}(V)$, i.e., $\varphi^*(\mathfrak{sl}_2) \subset I(\mathcal{N})$. But $\varphi^*(\mathfrak{sl}_2)$ is bihomogeneous of degree $(1, 1)$ and therefore is not contained in $I_G(\mathcal{N})$ because there are no invariants of degree 1.

Example 3.4. Let G be SO_4 and $V = \mathbb{C}^4 \oplus \mathbb{C}^4 \oplus \mathbb{C}^4$. The weights of $\mathbb{C}^4$ relative to the maximal torus $T = \mathrm{SO}_2 \times \mathrm{SO}_2$ are $\pm\varepsilon_1$, $\pm\varepsilon_2$, and the degree 2 invariants (dot products) $q_{ij}: (v_1, v_2, v_3) \mapsto v_i \cdot v_j$, $1 \le i \le j \le 3$, generate the invariant ring. Let V_{++} denote the span of the positive weight vectors and let V_{+-} denote the span of the weight vectors corresponding to ε_1 and $-\varepsilon_2$. Then $\mathcal{N} = GV_{++} \cup GV_{+-}$, and an easy calculation shows that every homogeneous covariant $V \to \mathbb{C}^4$ of degree 3 vanishes on the null cone $\mathcal{N}$. Now, using LiE (see [vLCL92]), one finds that there are 19 linearly independent homogeneous covariants of type $\mathbb{C}^4$ of degree 3, whereas there are 6 linearly independent invariants in degree 2, and obviously 3 linear covariants of type $\mathbb{C}^4$. Therefore, there is at least one generating covariant of type $\mathbb{C}^4$ in degree 3 and so V is not coreduced. (See Theorem 9.1(4) for a more general statement.)

We now use our method to classify the cofree SL_2-modules. Starting with the natural representation on $\mathbb{C}^2$ we get a linear action on the coordinate ring $\mathcal{O}(\mathbb{C}^2) = \mathbb{C}[x, y]$ where x has weight 1 with respect to the standard torus $T = \mathbb{C}^* \subset \mathrm{SL}_2$.

The homogeneous components $R_m := \mathbb{C}[x, y]_m$ of degree m give all irreducible representations of SL_2 up to isomorphism. A binary form $f \in R_m$ will be written as

$$f = \sum_{i=0}^{m} a_i \binom{m}{i} x^i y^{m-i}$$

so that the corresponding coordinate functions x_i are weight vectors of weight $m - 2i$. The null cone of R_m consists of those forms f that have a linear factor of multiplicity strictly greater than $\frac{m}{2}$. More generally, for any representation V of SL_2, we have

$$\mathcal{N} = \mathrm{SL}_2 V_+,$$

where V_+ is the sum of all weight spaces of strictly positive weight. In particular, $\mathcal{N}$ is always irreducible.

Example 3.5. The binary forms of degree 4 have the following invariants:

$$A := x_0x_4 - 4x_1x_3 + 3x_2^2 \quad \text{and} \quad H := \det \begin{bmatrix} x_0 & x_1 & x_2 \\ x_1 & x_2 & x_3 \\ x_2 & x_3 & x_4 \end{bmatrix}$$

classically called "Apolare" and "Hankelsche Determinante" which generate the invariant ring (see [Sch68]). It is easy to see that the null cone $\mathcal{N}(R_4) = \mathrm{SL}_2(\mathbb{C}x^3y \oplus \mathbb{C}x^4)$ is the closure of the 3-dimensional orbit of x^3y and thus has codimension 2. A simple calculation shows that the Jacobian $\mathrm{Jac}(H, A)$ has rank 2 at x^3y. It follows that $\mathcal{N}(R_4)$ is a reduced complete intersection. (One can deduce from $\operatorname{rank}\mathrm{Jac}(H, A) = 2$ that A, H generate the invariants.)

Example 3.6. The representation $kR_1 = R_1 \oplus R_1 \oplus \cdots \oplus R_1$ (k copies) can be identified with the space $M_{2\times k}(\mathbb{C})$ of $2 \times k$-matrices where SL_2 acts by left multiplication. Then the null cone $\mathcal{N}$ is the closed subset of matrices of rank ≤ 1, which is the determinantal variety defined by the vanishing of the 2×2-minors $M_{ij} = x_{1i}x_{2j} - x_{2i}x_{1j}$, $1 \leq i < j \leq k$. It is known that the ideal of $\mathcal{N}$ is in fact generated by the minors M_{ij}. This is an instance of the so-called Second Fundamental Theorem, see [Pro07, Chap. 11, Section 5.1]. Thus $\mathcal{N}$ is reduced, and the minors M_{ij} generate the invariants.

Theorem 3.7. *Let V be a nontrivial coreduced representation of* SL_2 *where* $V^{\mathrm{SL}_2} = (0)$. *Then V is isomorphic to R_2, R_3, R_4 or kR_1, $k \geq 1$.*

It is shown in [Dix81] that the representations R_n are not coreduced for $n \geq 5$. Our proof is based on the following results.

Lemma 3.8. *Let V be a representation of* SL_2 *and $\varphi: V \to R_m$ a homogeneous covariant of degree d.*

(1) *If* $d > m$, *then* $\varphi(\mathcal{N}) = (0)$.
(2) *If* $\pm\,\mathrm{Id}$ *acts trivially on* V *and* $2d > m$, *then* $\varphi(\mathcal{N}) = (0)$.

Proof. Let $V_+ \subset V$ be the sum of the positive weight spaces. Since $\mathcal{N} = \mathrm{SL}_2\, V_+$ it suffices to show that φ vanishes on V_+. Choose coordinates $x_1, \dots, x_n$ on V consisting of weight vectors and let $z = x_1^{k_1} x_2^{k_2} \cdots x_n^{k_n}$ be a monomial occurring in a component of φ. Then $\sum_i k_i = d$ and the weight of z occurs in R_m. Since $m < d$ the monomial z must contain a variable x_i with a weight ≤ 0, and so, z vanishes on V_+. This proves (1).

For (2) we remark that V contains only even weights and so a variable x_i with nonpositive weight has to appear in z as soon as $2d > m$. □

Lemma 3.9. *Let* V *be a nontrivial representation of* SL_2 *not isomorphic to* R_1, R_2, R_3 *or* R_4. *Then the principal isotropy group is either trivial or equal to* $\{\pm\,\mathrm{Id}\}$.

Proof. This is well known for the irreducible representations R_j, $j \geq 5$. Let T and U denote the usual maximal torus and maximal unipotent subgroup of SL_2. Denote by H_i the generic stabilizer of R_i, $i = 1, 2, 3$ and 4. Then we have $H_1 = U$, $H_2 = T$, $H_3 = \{\begin{bmatrix} \zeta & 0 \\ 0 & \zeta^2 \end{bmatrix} \mid \zeta^3 = 1\} \simeq \mathbb{Z}/3$ and

$$H_4 = \{\begin{bmatrix} \zeta & 0 \\ 0 & \zeta^3 \end{bmatrix} \mid \zeta^4 = 1\} \cup \{\begin{bmatrix} 0 & \zeta \\ -\zeta^3 & 0 \end{bmatrix} \mid \zeta^4 = 1\} \simeq \widetilde{Q}_8,$$

the group of quaternions of order 8. It is easy to see that the generic stabilizer of H_1 and H_3 on any nontrivial representations of SL_2 is trivial, and that the generic stabilizer of H_2 and H_4 on the R_{2j}, $j > 0$, is $\{\pm\,\mathrm{Id}\}$. □

Proof (of Theorem 3.7). For $V = R_2$ or R_3 the quotient $V /\!\!/ \mathrm{SL}_2$ is one-dimensional and so both are coreduced. In Examples 3.5 and 3.6 we have seen that R_4 and kR_1 are coreduced. So it remains to show that any other representation V of SL_2 is not coreduced.

By Lemma 3.9 we can assume that the principal isotropy group is trivial or $\{\pm\,\mathrm{Id}\}$. In the first case, V contains a closed orbit isomorphic to SL_2. By Frobenius reciprocity, we know that the multiplicity of R_m in $\mathcal{O}(\mathrm{SL}_2)$ is equal to $\dim R_m = m + 1$. This implies that the rank of the $\mathcal{O}(V)^G$-module $\mathrm{Cov}(V, R_m)$ is at least $m + 1$. Since we may assume that V is not kR_1, removing all but one copy of R_1 gives a module V' which still has trivial principal isotropy groups. By Example 2.12(1) it is sufficient to show that V' is not coreduced. Now there has to be a generating homogeneous covariant $\varphi\colon V' \to R_1$ of degree > 1. By Lemma 3.8(1) and Proposition 3.1 it follows that V' is not coreduced, hence neither is V.

Now assume that the principal isotropy group is $\{\pm\,\mathrm{Id}\}$. As above this implies that the rank of $\mathrm{Cov}(V, R_2)$ is at least 3. Since $R_2 \oplus R_2$ is not coreduced (Example 3.3) we can assume that V contains at most one summand isomorphic to R_2. It follows that there is a generating homogeneous covariant $\varphi\colon V \to R_2$ of degree > 1, and the claim follows by Lemma 3.8(2) and Proposition 3.1. □

4 Cofree representations

Let G be a (connected) reductive group, V a G-module and $\pi\colon V \to V/\!\!/G$ the quotient morphism.

Remark 4.1. Let Gv be a closed orbit with slice representation (N_v, G_v). Then, by LUNA's slice theorem, the fiber $F := \pi^{-1}(\pi(v))$ is isomorphic to $G \times^{G_v} \mathcal{N}(N_v)$ which is a bundle over $G/G_v \simeq Gv$ with fiber $\mathcal{N}(N_v)$. Hence F is reduced if and only if N_v is coreduced.

If the fiber $F = \pi^{-1}(z)$ is reduced, then F is smooth in a dense open set $U \subset F$ which means that the rank of the differential $d\pi_u$ equals $\dim V - \dim_u F$ for $u \in U$. Thus we get the following criterion for non-coreducedness.

Lemma 4.2. *If X is an irreducible component of $\mathcal{N}(V)$ and the rank of $d\pi$ on X is less than the codimension of X in V, then V is not coreduced.*

Recall that a G-module V is said to be *cofree* if $\mathcal{O}(V)$ is a free $\mathcal{O}(V)^G$-module. Equivalently, $\mathcal{O}(V)^G$ is a polynomial ring (V is *coregular*) and the codimension of $\mathcal{N}(V)$ is $\dim V/\!\!/G$. See [Sch79] for more details and a classification of cofree representations of simple groups.

Proposition 4.3. *Let V be a cofree G-module. If the null cone is reduced, then so is every fiber of $\pi\colon V \to V/\!\!/G$, and every slice representation of V is coreduced.*

Proof. Since V is cofree, the map π is flat. By [Gro67, 12.1.7], note that the set $\{v \in V \mid \pi^{-1}(\pi(v)) \text{ is reduced at } v\}$ is open in V. But this set is a cone. Thus if the null cone is reduced, then so is any fiber of π, and every slice representation is coreduced. □

For a cofree representation V the (schematic) null cone $\mathcal{N}(V)$ is a complete intersection. Using SERRE's criterion [Mat89, Ch. 8] one can characterize quite precisely when $\mathcal{N}(V)$ is reduced.

Proposition 4.4. *Let V be a cofree G-module. Then V is coreduced if and only if rank $d\pi = \operatorname{codim} \mathcal{N}(V)$ on an open dense subset of $\mathcal{N}(V)$.*

Example 4.5. Let $G = \mathrm{SL}_n$ and $V := S^2(\mathbb{C}^n)^* \oplus \mathbb{C}^n$. The quotient map $\pi\colon V \to \mathbb{C}^2$ is given by the two invariants $f := \det(q)$ and $h := q(v)$ of bidegrees $(n, 0)$ and $(1, 2)$ where $(q, v) \in V$. An easy calculation shows that for $n = 2$ the differential $d\pi$ has rank ≤ 1 on the null cone. Hence

$$(S^2(\mathbb{C}^2)^* \oplus \mathbb{C}^2, \mathrm{SL}_2)$$

is not coreduced, which we already know from Theorem 3.7.

We claim that for $n \geq 3$ the null cone is irreducible and reduced. Set $q_k := x_k^2 + x_{k+1}^2 + \cdots + x_n^2$,

$$X_k := \{q_k\} \times \{v \in \mathbb{C}^n \mid q_k(v) = 0\} \subset V \text{ and } X_{n+1} := \{0\} \times \mathbb{C}^n.$$

Then $\mathcal{N}(V) = \bigcup_{k=2}^{n+1} \mathrm{SL}_n \cdot X_k$. Since $\dim \mathrm{SL}_n \cdot X_k = \dim \mathrm{SL}_n q_k + n - 1$ for $2 \le k \le n$ we get $\operatorname{codim} \mathrm{SL}_n \cdot X_2 = 2 < \operatorname{codim} \mathrm{SL}_n \cdot X_k$ for all $k > 2$, and so $\mathcal{N}(V) = \overline{\mathrm{SL}_n \cdot X_2}$ is irreducible. Moreover,

$$df_{(q_2,v)}(q,w) = a_{11} \quad \text{and} \quad dh_{(q_2,v)}(q,w) = \sum_{i=2}^{n} 2v_i y_i + q(v),$$

where $v = (v_1, \dots, v_n)$, $q = \sum a_{ij} x_i x_j$ and $w = (y_1, \dots, y_n)$. It follows that the two linear maps $df_{(q_2,v)}$ and $dh_{(q_2,v)}$ are linearly independent on a dense open set of X_2, hence the claim.

In order to see that the null cone is reduced in a dense set we can use the following result due to KNOP [Kno86]. Recall that the *regular sheet* $\mathcal{S}_V$ of a representation (V, G) of an algebraic group is the union of the G-orbits of maximal dimension.

Proposition 4.6. *Let (V, G) be a representation of a semisimple group and let $\pi : V \to V/\!\!/G$ be the quotient map. Assume that $x \in V$ belongs to the regular sheet and that $\pi(x)$ is a smooth point of the quotient. Then π is smooth at x.*

Corollary 4.7. *Let (V, G) be a cofree representation of a semisimple group. Assume that the regular sheet $\mathcal{S}_V$ of V meets the null cone $\mathcal{N}(V)$ in a dense set. Equivalently, assume that every irreducible component of the null cone contains a dense orbit. Then (V, G) is coreduced.*

Let θ be an automorphism of finite order of a semisimple group H and let G denote the identity component of the fixed points H^θ. Given any eigenspace V of θ on the Lie algebra $\mathfrak{g}$ of G, we have a natural representation of G on V. These representations (V, G) are called θ-representations. They have been introduced and studied by VINBERG, see [Vin76]. Among other things he proved that θ-representations are cofree and that every fiber of the quotient map contains only finitely many orbits. As a consequence of Corollary 4.7 above we get the following result.

Corollary 4.8. *Every θ-representation (V, G) where G is semisimple is coreduced.*

Remark 4.9. The corollary above was first established by PANYUSHEV [Pan85].

Finally we can prove the main result of this section.

Theorem 4.10. *An irreducible cofree representation V of a simple group G is coreduced.*

Proof. It follows from the classification (see [Pop76, KPV76, Sch79]) that the only irreducible cofree representations of the simple groups that are not θ-representations (or have one-dimensional quotient) are the spin representation of Spin_{13} and the half-spin representations of Spin_{14}. For these representations, GATTI–VINIBERGHI [GV78] show that every irreducible component of the null cone has a dense orbit. □

Example 4.11. Here is an example of a *cofree but not coreduced* representation. Let $G = \mathrm{SL}_2$ and let $V = \mathfrak{sl}_2 \oplus \mathfrak{sl}_2 \simeq 2R_2$. Example 3.3 shows that V is not coreduced. Here is a different proof. Each copy of R_2 has a weight basis $\{x^2, xy, y^2\}$ relative to the action of the maximal torus $T = \mathbb{C}^*$. The null cone is $U^-(\mathbb{C}x^2 \oplus \mathbb{C}x^2)$ where U^- is the maximal unipotent subgroup of G opposite to the usual Borel subgroup. One can easily see that $d\pi_{(\alpha x^2, \beta x^2)}$ is nontrivial only on the vectors $(\gamma y^2, \delta y^2)$, giving a rank of two. But V is cofree with $\operatorname{codim} \mathcal{N}(V) = 3$. Thus V is not coreduced.

We can prove a sort of converse to the theorem above. Recall that V is *strongly coreduced* if every fiber of π is reduced; equivalently, every slice representation of V is coreduced.

Theorem 4.12. *A strongly coreduced irreducible representation of a simple group G is cofree.*

If G is simple, then we use the ordering of Bourbaki [Bou68] for the simple roots α_j of G and we let φ_j denote the corresponding fundamental representations. We use the notation ν_j to denote the 1-dimensional representation of $\mathbb{C}^*$ with weight j.

Proof. We use the techniques of [KPV76] (but we follow the appendix of [Sch78]). Let V be non-coregular (which is the same as V not being cofree by the classifications [Pop76, KPV76]). If V is $\varphi_1^3(\mathsf{A}_3)$ or $\varphi_2^3(\mathsf{A}_3)$, then there is a closed orbit with finite stabilizer whose slice representation is not coregular. Thus the slice representation is certainly not trivial, hence V is not strongly coreduced. Otherwise, there is a copy $T = \mathbb{C}^* \subset \mathrm{SL}_2 \subset G$ and a closed orbit Gv, $v \in V^T$, such that the identity component G_v^0 of the stabilizer G_v of v has rank 1. Moreover, one of the following occurs:

(1) $G_v^0 = T$ and the slice representation of G_v, restricted to T, has at least 3 pairs of nonzero weights $\pm a$, $\pm b$, $\pm c$ (where we could have $a = b = c$).
(2) The module is $\varphi_1\varphi_2(\mathsf{A}_3)$ or $\varphi_2\varphi_3(\mathsf{A}_3)$, G_v centralizes $G_v^0 = T$ and the slice representation is $\theta_2 + \nu_1 + \nu_{-1} + \nu_2 + \nu_{-2}$ where θ_n denotes the n-dimensional trivial representation.
(3) $G_v^0 = \mathrm{SL}_2$ and the slice representation of G_v, restricted to T, contains at least 4 pairs of weights $\pm a$, $\pm b$, ...

If, in case (1) above, the weights are not of the form $\pm a$ for a fixed a, then the G-module V is not strongly coreduced by Corollary 2.14. The same remark holds in case (3). Of course, in case (2), the module is not strongly coreduced. We went through the computations again and saw in which cases the weights of the slice representations were of the form $\pm a$ for a fixed a. One gets a list of representations as follows. (The list is complete up to automorphisms of the group.)

(4) $\varphi_i(\mathsf{A}_n)$, $5 \le i$, $2i \le n+1$.
(5) $\varphi_n(\mathsf{B}_n)$, $n \ge 7$.
(6) $\varphi_n(\mathsf{D}_n)$, $n \ge 9$.
(7) $\varphi_i(\mathsf{C}_n)$, $3 \le i \le n$, $n \ge 5$.

For the groups of type A and C, consider $\mathrm{SL}_2 \subset G$ such that the fundamental representation restricted to SL_2 is $2R_1 + \theta_1$. For the groups of type B and D consider $\mathrm{SL}_2 \subset G$ such that the fundamental representation restricted to SL_2 is $4R_1 + \theta$. Then using the techniques of [KPV76] one sees that there is a closed orbit in V^{SL_2} whose stabilizer has identity component SL_2 such that the slice representation restricted to SL_2 contains at least two copies of R_2. Hence the slice representation is not coreduced. □

We know from the [KPV76] and [Pop76] that an irreducible representation of a simple group is cofree if and only if it is coregular. Thus we get the following corollary.

Corollary 4.13. *Let (V, G) be an irreducible representation of a simple group. Then the following are equivalent:*

$$(V, G) \textit{ cofree} \iff (V, G) \textit{ coregular} \iff (V, G) \textit{ strongly coreduced.}$$

5 The method of slices and multiplicity of weights

Let G be a reductive group and $T \subset G$ a maximal torus. It is well known that the orbit Gv is closed for any zero weight vector $v \in V^T$. We say that V *is a G-module with a zero weight* if $V^T \neq (0)$. The basic result for such modules is the following.

Proposition 5.1. *Let V be a G-module with a zero weight. If the slice representation at a zero weight vector is not coreduced, then neither is V.*

For the proof we use the following result originally due to RICHARDSON [Ric89, Proposition 5.5]. If $X \subset V$ is a closed subset of a vector space V, then the *associated cone $\mathcal{C}X$* of X is defined to be the zero set of $\{\operatorname{gr} f \mid f \in I(X)\}$ where $\operatorname{gr} f$ denotes the (nonzero) homogeneous term of f of highest degree. If V is a G-module and X a closed subset of a fiber $F \neq \mathcal{N}(V)$ of the quotient map, then $\mathcal{C}X = \overline{\mathbb{C}^* X} \setminus \mathbb{C}^* X$ (cf. [BK79, §3]).

Lemma 5.2. *Suppose that G_v has the same rank as G. Then the associated cone of $F := \pi^{-1}(\pi(v))$ is equal to $\mathcal{N}(V)$.*

Proof. We know that the associated cone of every fiber of π is contained in $\mathcal{N}(V)$. For the reverse inclusion we can assume that $T \subset G_v$. Let $v_0 \in \mathcal{N}(V)$. Then $\overline{Tgv_0} \ni 0$ for a suitable $g \in G$. This implies that $\overline{T(gv_0 + v)} \ni v$ and so $\mathbb{C}gv_0 + v \subset F$. The lemma follows since $gv_0 \in \overline{\mathbb{C}^*(\mathbb{C}gv_0 + v)}$. □

Proof (of Proposition 5.1). Suppose that $\mathcal{N}(V)$ is reduced, and let $0 \neq f \in I(F)$ where F is as in the lemma above. Then the leading term $\operatorname{gr} f$ lies in the ideal of $\mathcal{N}(V)$, so that there are homogeneous $f_i \in \mathfrak{m}_0$ and homogeneous $h_i \in \mathcal{O}(V)$ such that $\operatorname{gr} f = \sum_i f_i h_i$ where $\deg f_i + \deg h_i = \deg \operatorname{gr} f$ for all i.

Then $\widetilde{f} := \sum_i (f_i - f_i(v))h_i$ lies in $I_G(F)$ and $\operatorname{gr} f = \operatorname{gr} \widetilde{f}$. Replacing f by $f - \widetilde{f}$ we are able to reduce the degree of f. Hence by induction we can show that $f \in I_G(F)$. Thus F is reduced. □

Remark 5.3. The proof above shows that the conclusion of Proposition 5.1 holds in a more general situation. *If the slice representation at $v \in V$ is not coreduced and the associated cone of $\pi^{-1}(\pi(v))$ is equal to $\mathcal{N}(V)$, then V is not coreduced.*

Example 5.4. We use the proposition above to give another proof that the irreducible representations R_{2m} of SL_2 are not coreduced for $m \geq 3$ (see Theorem 3.7). We have $R_{2m}^T = \mathbb{C}x^m y^m$, and a zero weight vector v has stabilizer $T \simeq \mathbb{C}^*$ if m is odd and $N(T)$ if m is even. The slice representation of T at v has the weights $\pm 4, \dots, \pm 2m$ (each with multiplicity one), and so, for $m \geq 3$, we have at least the weights ± 4 and ± 6. But then the slice representation restricted to T is not coreduced by Corollary 2.14, hence neither are the representations R_{2m} of SL_2 for $m \geq 3$.

Let G be semisimple with Lie algebra $\mathfrak{g}$. If μ is a dominant weight of G, let $V(\mu)$ denote the corresponding simple G-module. Recall that the following are equivalent:

(i) $V(\mu)$ has a zero weight;
(ii) All weights of $V(\mu)$ are in the root lattice;
(iii) μ is in the root lattice;
(iv) The center of G acts trivially on $V(\mu)$.

Remark 5.5. Let V be a nontrivial simple G-module with a zero weight. Then the short roots are weights of $V(\mu)$ and the highest short root is the smallest nontrivial dominant weight. This follows from the following result due to STEINBERG, see [Ste98]. (We thank John Stembridge for informing us of this result.)

Lemma 5.6. *Let $\nu \prec \mu$ be dominant weights. Then there are positive roots β_i, $i = 1, \dots, n$, such that*

(1) $\mu - \nu = \beta_1 + \beta_2 + \cdots + \beta_n$*, and*
(2) $\mu - \beta_1 - \cdots - \beta_j$ *is dominant for all* $j = 1, \dots, n$.

Example 5.7. Let G be a semisimple group and $\mathfrak{g}$ its Lie algebra. If V is a G-module with a zero weight, then the representation of G on $\mathfrak{g} \oplus V$ is not coreduced. This is a special case of a result of PANYUSHEV, see [Pan99, Theorem 4.5].

Proof. Let $T \subset G$ be a maximal torus and α a (short) root. Put $T_\alpha := (\ker \alpha)^0$. Then, for a generic $x \in \operatorname{Lie} T_\alpha \subset \mathfrak{g}$ we have $G_x = \operatorname{Cent}_G T_\alpha = G_\alpha \cdot T_\alpha$ where $G_\alpha \simeq \mathrm{SL}_2$ or $\simeq \mathrm{PSL}_2$, and so the slice representation at $x \in \mathfrak{g}$ is $\operatorname{Lie}(G_x) \simeq \mathfrak{sl}_2 + \theta_{\ell-1}$ where $\ell = \operatorname{rank} G$. Now consider the slice representation (S, G_x) at $(x, 0) \in \mathfrak{g} \oplus V$. Since V has a zero weight all short roots are weights of V (Remark 5.5) and thus the fixed points S^{T_α} are of the form $(\mathfrak{sl}_2 \oplus W, G_\alpha)$ where W is a nontrivial SL_2-module. Using Example 2.12(2) the claim follows from Theorem 3.7. □

If there is a $v \in V^T$ such that $G_v^0 = T$, then we can use Proposition 2.13 to show that the slice representation at v is not coreduced by giving an indecomposable relation of the weights of the slice representation which involves coefficients ≥ 2. We will see that this is a very efficient method to prove non-coreducedness in many cases.

The lemma below follows from [Ric89, Proposition 3.3]. For completeness we give a proof.

Lemma 5.8. *Let G be a semisimple group and let V be a G-module. Then all the roots of $\mathfrak{g}$ are weights of V if and only if there is a zero weight vector $v \in V^T$ whose isotropy group is a finite extension of the maximal torus T of G.*

Proof. Clearly if $(G_v)^0 = T$, then the roots of $\mathfrak{g}$ are weights of V. Conversely, assume all the roots appear and let α be a root of $\mathfrak{g}$. The weight spaces, with weight a multiple of α, form a submodule of V for the action of the corresponding copy of SL_2. Since α occurs as a weight of V, this module is not the trivial module. Hence there is a $v \in V^T$ such that $x_\alpha(v) \neq 0$ where $x_\alpha \in \mathfrak{g}$ is a root vector of α. Thus the kernel of x_α is a proper linear subspace of V^T and there is a $v \in V^T$ which is not annihilated by any x_α. Then the isotropy subalgebra of v is $\mathfrak{t}$. □

Definition 5.9. We say that a representation V of G has *a toral slice* if there is a $v \in V^T$ such that $G_v^0 = T$. We say that V has *a bad slice* if there is a $v \in V^T$ such that the slice representation at v restricted to G_v^0 is not coreduced, and that V has *a bad toral slice* if, in addition, $G_v^0 = T$.

Now Proposition 5.1 can be paraphrased by saying that *a representation with a bad slice is not coreduced.*

Example 5.10. Consider the representation $(S^{3k}(\mathbb{C}^3), \mathrm{SL}_3)$, $k \geq 2$. Then the isotropy group of the zero weight vector is a finite extension of the maximal torus T of SL_3, and the slice representation W of the torus contains the highest weight $2k\alpha + k\beta$ as well as the weights $-k\alpha$ and $-k\beta$. Thus there is the "bad" relation

$$(2k\alpha + k\beta) + 2(-k\alpha) + (-k\beta) = 0,$$

and so V has a bad toral slice.

Example 5.11. The following representations are not coreduced.

(1) $G = \mathrm{SL}_2 \times \mathrm{SL}_2$ on $(\mathfrak{sl}_2 \otimes \mathfrak{sl}_2) \oplus (R_i \otimes R_j)$, $i + j \geq 1$;
(2) $G = \mathrm{SL}_2 \times \mathrm{SL}_2 \times \mathrm{SL}_2$ on $(\mathfrak{sl}_2 \otimes \mathfrak{sl}_2 \otimes \mathbb{C}) \oplus (\mathbb{C} \otimes \mathfrak{sl}_2 \otimes \mathfrak{sl}_2)$.

Proof. (1) Let $t \in \mathfrak{sl}_2$ be a nonzero diagonal matrix. The stabilizer of

$$v = t \otimes t \in \mathfrak{sl}_2 \otimes \mathfrak{sl}_2$$

is $\mathbb{C}^* \times \mathbb{C}^*$. If i is odd, then the slice representation contains the weights $(\pm 2, \pm 2)$ and $(\pm 1, 0)$ or $(\pm 1, \pm 1)$, and so we find the bad relations

$$(2,2) + (2,-2) + 4(-1,0) = 0 \quad \text{or} \quad (2,2) + 2(-1,-1) = 0.$$

The same argument applies if j is odd. If i and j are both even and $i > 0$, then the slice representation contains the weights $(\pm 2, \pm 2)$ and $(\pm 2, 0)$, and so we find the bad relation

$$(2,2) + (2,-2) + 2(-2,0) = 0.$$

(2) The stabilizer of $v = t \otimes t \otimes x + x \otimes t \otimes t$ where $0 \neq x \in \mathbb{C}$ is $\mathbb{C}^* \times \mathbb{C}^* \times \mathbb{C}^*$. The slice representation contains the weights $(\pm 2, \pm 2, 0)$ and $(0, \pm 2, 0)$, and we can proceed as in (1).

□

Example 5.12. Consider the second fundamental representation of Sp_6: $\varphi_2(\mathsf{C}_3) = \bigwedge_0^2 \mathbb{C}^6 := \bigwedge^2(\mathbb{C}^6)/\mathbb{C}\beta$ where $\beta \in \bigwedge^2(\mathbb{C}^6)$ is the invariant form. It has the isotropy group $\mathrm{Sp}_2 \times \mathrm{Sp}_4$ with slice representation $\bigwedge_0^2 \mathbb{C}^4 + \theta_1$. We claim that $(2\bigwedge_0^2 \mathbb{C}^6, \mathrm{Sp}_6)$ is not coreduced, although it is cofree ([Sch79]). In fact, the slice representation is $(2\bigwedge_0^2 \mathbb{C}^4 + \mathbb{C}^2 \otimes \mathbb{C}^4 + \theta_2, \mathrm{Sp}_2 \times \mathrm{Sp}_4)$. Quotienting by Sp_2 we get a hypersurface $F \subset 3\bigwedge_0^2 \mathbb{C}^4 + \theta_3$ defined by an Sp_4-invariant function. Now the claim follows from Example 2.7, because $(3\bigwedge_0^2 \mathbb{C}^4, \mathrm{Sp}_4) = (3\mathbb{C}^5, \mathrm{SO}_5)$ is not coreduced as we will see in Theorem 9.1(4).

Next we want to show that a representation V is not coreduced if the weights contain all roots with multiplicity at least 2. This needs some preparation.

Lemma 5.13. *Let (V,G) and (W,H) be two representations. Let $v \in V$ and $w \in W$ be nonzero zero weight vectors with slice representations $(N_V \oplus \theta_n, G_v)$ and $(N_W \oplus \theta_m, H_w)$ where $N_V^{G_v} = 0$ and $N_W^{H_w} = 0$. Then the slice representation $N_{V\otimes W}$ of $G_v \times H_w$ at $v \otimes w$ contains*

$$(V^{\oplus(m-1)} \oplus N_V, G_v) \oplus (W^{\oplus(n-1)} \oplus N_W, H_w)$$
$$\oplus ((\mathfrak{g}/\mathfrak{g}_v \oplus N_V) \otimes (\mathfrak{h}/\mathfrak{h}_w \oplus N_W), G_v \times H_w).$$

Proof. The lemma follows from the decomposition $(V, G_v) = (\mathfrak{g}/\mathfrak{g}_v \oplus N \oplus \theta_n)$ and similarly for (W, H_w), and the fact that

$$T_{v\otimes w}((G \times H)v \otimes w) = \mathfrak{g}(v) \otimes w + v \otimes \mathfrak{h}(w) \subset \mathfrak{g}/\mathfrak{g}_v \otimes \theta_m + \theta_n \otimes \mathfrak{h}/\mathfrak{h}_w. \quad \square$$

Corollary 5.14. (1) *The two slice representations (N_V, G_v) and (N_W, H_w) occur as subrepresentations of the slice representation at $v \otimes w$. In particular, if (V,G) has a bad slice, then so does $(V \otimes W, G \times H)$.*

(2) *The slice representation at $v \otimes w$ contains $N_V \otimes N_W$, $\mathfrak{g}/\mathfrak{g}_v \otimes \mathfrak{h}/\mathfrak{h}_w$, $\mathfrak{g}/\mathfrak{g}_v \otimes N_W$ and $N_V \otimes \mathfrak{h}/\mathfrak{h}_w$.*

(3) *If $n > 1$ (resp. $m > 1$), then the slice representation contains a copy of W (resp. V).*

Remark 5.15. Since G_v and H_w have maximal rank, the isotropy group of $v \otimes w$ can at most be a finite extension of $G_v \times H_w$. Note also that the corollary generalizes in an obvious way to a representation of the form

$$(V_1 \otimes V_2 \otimes \cdots \otimes V_k, G_1 \times G_2 \times \cdots \times G_k),$$

where each (V_i, G_i) is a representation with a zero weight.

Proposition 5.16. *Let $G = G_1 \times \cdots \times G_k$ be a product of simple groups and $V = V_1 \otimes \cdots \otimes V_k$ a simple G-module where $k > 1$. Assume that all roots of G occur in V. Then V is coreduced if and only if G is of type $\mathsf{A}_1 \times \mathsf{A}_1$ and $V = \mathfrak{sl}_2 \otimes \mathfrak{sl}_2$.*

Proof. By Lemma 5.8 the product $T = T_1 \times \cdots \times T_k$ of the maximal tori appears as the connected component of the isotropy group of an element $v_1 \otimes \cdots \otimes v_k \in V^T$ where v_i is a generic element in $V_i^{T_i}$. Denote by $W_i := N_{V_i}$ the slice representation at v_i. Then the tensor products $W_{i_1} \otimes \cdots \otimes W_{i_m}$ where $i_1 < \cdots < i_m$ appear as subrepresentations of the slice representation at v (see Remark 5.15 above).

First, assume that $k > 2$. Choose simple roots α, β, γ of G_1, G_2, G_3, respectively. Then

$$(\alpha + \beta) + (\beta + \gamma) + (\alpha + \gamma) + 2(-\alpha - \beta - \gamma) = 0$$

is an indecomposable relation with a coefficient > 1.

Now assume that $k = 2$ and that $\operatorname{rank} G_1 > 1$ and choose two adjacent simple roots α, β of G_1 so that $\alpha + \beta$ is again a root. Let γ be a simple root of G_2. Then the relation

$$(\alpha + \gamma) + (\alpha - \gamma) + 2(\beta - \gamma) + 2(-(\alpha + \beta) + \gamma) = 0$$

is indecomposable, but contains coefficients > 1.

As a consequence, G is of type

$$\mathsf{A}_1 \times \mathsf{A}_1,$$

and $V = R_{2r} \otimes R_{2s}$. Calculating the representation of the maximal torus of $\mathsf{A}_1 \times \mathsf{A}_1$ at the zero weight vector shows that V is coreduced only for $r = s = 1$. □

Proposition 5.17. *Let G be a semisimple group and let V be a G-module. Assume that all roots of G are weights of V with multiplicity at least 2. Then V admits a bad toral slice.*

Proof. Choose a generic element v of the zero weight space V^T of V. Then $(G_v)^0 = T$ by Lemma 5.8, and all roots occur in the slice representation W of T at v as well as the highest weights of V. We will show that there is a bad relation.

If not all simple factors of G are of type A, then there is always a root α which expressed in terms of simple roots has some coefficient ≥ 2: $\alpha = \sum_i n_i \alpha_i$ where

$\{\alpha_1, \ldots, \alpha_r\}$ is a set of simple roots of $\mathfrak{g}$, $n_i \in \mathbb{N}$, and $n_j > 1$ for some j. But then $\alpha + \sum_i n_i(-\alpha_i) = 0$ is a bad relation and thus N is not coreduced.

We may thus assume that G is of type

$$\mathsf{A}_{n_1} \times \mathsf{A}_{n_2} \times \cdots \times \mathsf{A}_{n_k},$$

for $n_1 \geq n_2 \geq \cdots \geq n_k \geq 1$. Let $\{\alpha_1, \ldots, \alpha_n\}$ be a set of simple roots, $n := n_1 + n_2 + \cdots + n_k$. We can assume that the highest weights of the irreducible components of V are all of the form $\lambda = \sum_i n_i \alpha_i$ where $n_i \in \{0, 1\}$. It is easy to see that such a weight is dominant if and only if λ is a sum of highest roots. Thus each irreducible component V_k of V is a tensor product of certain $\mathfrak{sl}_{n_j}$'s. Now it follows from the previous proposition that either V_k is isomorphic to $\mathfrak{sl}_j$ or isomorphic to $\mathfrak{sl}_2 \otimes \mathfrak{sl}_2$. If $n_1 > 1$, then $\mathfrak{sl}_{n_1} \oplus \mathfrak{sl}_{n_1}$ must occur and so V is not coreduced (Example 5.7). The remaining cases where G is of type $\mathsf{A}_1 \times \mathsf{A}_1 \times \cdots \times \mathsf{A}_1$ follow immediately from Example 5.11. □

We finish this section with a criterion for the non-coreducedness of an irreducible representation of a simple group. We begin with a lemma about weights and multiplicities. Let U denote a maximal unipotent subgroup of G.

Lemma 5.18. *Let μ, ν be nonzero dominant weights of $\mathfrak{g}$.*

(1) *If there is a weight of $V(\mu)$ of multiplicity m, then there are nonzero weights in $V(\mu + \nu)$ with multiplicity $\geq m$.*
(2) *Suppose that zero is a weight of $V(\mu)$. Then the multiplicities of the nonzero weights of $V(\mu)$ are bounded below by the multiplicities of the (short) roots.*

Proof. Let $v_\mu \in V(\mu), v_\nu \in V(\nu)$ be highest weight vectors. Recall that the coordinate ring $\mathcal{O}(G/U)$ is a domain and contains every irreducible representation of G exactly once. Therefore, the multiplication with v_ν is injective and sends $V(\mu)$ into $V(\mu + \nu)$, i.e., $V(\mu) \otimes \mathbb{C}v_\nu \hookrightarrow V(\mu + \nu)$ as a T-submodule, and we have (1).

For (2), recall that the highest short root is the smallest dominant weight (Remark 5.5). Looking at root strings (see Remark 5.5 and Lemma 5.6) we see that the multiplicity of the highest short root has to be at least that of an arbitrary nonzero weight. □

The following criterion will be constantly used for the classifications in the following sections. Let G be a simple group. We use the notation $\varphi, \psi, \ldots$ for irreducible representations of G and denote by $\varphi\psi$ the Cartan product of φ and ψ.

Criterion 5.19. *Let φ, ψ be irreducible representations of G with a zero weight. Then $\varphi\psi$ has a bad toral slice in the following cases.*

(i) *φ has a bad toral slice.*
(ii) *$\varphi\psi$ contains a nonzero weight of multiplicity > 1.*
(iii) *The zero weight of φ has multiplicity > 1.*

Proof. As in the proof above, every nonzero weight vector $w \in \psi$ defines an embedding $\varphi \hookrightarrow \varphi\psi$ which shows that $\varphi\psi$ contains all sums of two (short) roots and therefore all roots. Thus $\varphi\psi$ has a toral slice. In case (i) we choose for $w \in \psi$ a weight vector of weight 0 and obtain a T-equivariant embedding $\varphi \hookrightarrow \varphi\psi$ which shows that $\varphi\psi$ has a bad toral slice.

In case (ii) the (short) roots occur in $\varphi\psi$ with multiplicity at least 2. Now let α be a short root. Then 2α and α occur in a toral slice representation, and we have the bad relation $(2\alpha) + 2(-\alpha) = 0$.

Finally, (iii) implies (ii) by Lemma 5.18(1). □

Remark 5.20. Let G be of type A, D or E so that all roots have the same length. If $\varphi = \varphi_{i_1}\varphi_{i_2}\cdots\varphi_{i_k}$ is a coreduced representation with a zero weight, then either $k = 1$ or all φ_{i_j} are multiplicity-free. In all other cases, φ has a bad toral slice. (If $k > 1$ and if one of the φ_{i_j} has a weight space of multiplicity ≥ 2, then the roots occur in φ with multiplicity ≥ 2, by Lemma 5.18, and thus φ is not coreduced, by Proposition 5.17.)

6 Coreduced representations of the exceptional groups

Let G be an exceptional simple group. In this section we classify the coreduced representations V of G which contain a zero weight. We know that each irreducible summand of V is coreduced (Example 2.12(1)). We show that all coreduced representations with a zero weight are contained in *maximal* coreduced representations all of which we determine. The types E_n and G_2 are easy consequences from what we have done so far, but the type F_4 turns out to be quite involved.

Proposition 6.1. *Let G be a simple group of type E and let V be a G-module with a zero weight. If V is coreduced, then V is the adjoint representation of G. Any other V with a zero weight has a bad toral slice.*

Proof. Since the groups of type E are simply laced, every irreducible representation φ with a zero weight contains all roots and thus has a toral slice. Now it follows from Lemma 6.2 below that every representation of the form $\varphi \oplus V$ where V is nontrivial has a bad toral slice. Hence a coreduced representation with a zero weight is irreducible.

(a) Let $G = \mathsf{E}_8$. One can check with LiE that the fundamental representations of G, except for the adjoint representation $\varphi_1(\mathsf{E}_8)$, contain the roots with multiplicity ≥ 2. Since the zero weight of $\varphi_1(\mathsf{E}_8)$ has multiplicity ≥ 2, it follows from Criterion 5.19 that every irreducible representation except for the adjoint representation has a bad toral slice.

(b) Let $G = \mathsf{E}_7$. Of the fundamental representations only $\varphi_1 = \mathfrak{g} = \operatorname{Ad} G$, φ_3, φ_4 and φ_6 are representations of the adjoint group. Using LiE one shows that every fundamental representation, except for φ_1 and the 56-dimensional representation φ_7, has a nonzero weight of multiplicity at least 6. Hence, by Remark 5.20, the only other candidates for a coreduced representation besides

φ_1 are φ_7^{2k}, $k \geq 1$. But φ_7^2, contains the roots with multiplicity 5. Thus every irreducible representation except for the adjoint representation has a bad toral slice.

(c) Let $G = \mathsf{E}_6$. From the fundamental representations only $\varphi_2 = \mathfrak{g}$ and φ_4 are representations of the adjoint group. By LiE, φ_3, φ_4 and φ_5 have nonzero weights of multiplicity at least 5, and $\varphi_1^2, \varphi_1\varphi_6$ and φ_6^2 have nonzero weights of multiplicities at least 4. Thus all irreducible representations with a zero weight, except for the adjoint representation φ_2, have a bad toral slice. □

Lemma 6.2. *Let G be a simple group of type E, V a G-module and $T \subset G$ a maximal torus. Then V, considered as a representation of T, is not coreduced.*

Proof. We have to show that the weights $\Lambda = \{\lambda_i\}$ of V admit a "bad relation," i.e., an indecomposable relation $\sum_i n_i \lambda_i = 0$ where $n_i \geq 0$ and at least one $n_j \geq 2$ (Proposition 2.13). This is clear if Λ contains the roots, in particular for all representations of E_8.

For E_7 we first remark that $\omega_1, \omega_3, \omega_4, \omega_6$ are in the root lattice and $\omega_7 \prec \omega_2, \omega_5$ in the usual partial order. This implies that for every dominant weight λ we have either $\omega_1 \prec \lambda$ or $\omega_7 \prec \lambda$. Thus the weights of V either contain the roots or the Weyl orbit of ω_7. Using LiE one calculates the Weyl orbit of ω_7 and shows that there is a "bad relation" among these weights.

Similarly, for E_6 one shows that for a dominant weight λ not in the root lattice, one has either $\omega_1 \prec \lambda$ or $\omega_6 \prec \lambda$. Then, using LiE, one calculates the Weyl orbit of ω_1 and shows that there is a "bad relation" among these weights. Since ω_6 is dual to ω_1 its weights also have a "bad relation." □

We prepare to consider F_4. The following result will be used several times in connection with slice representations at zero weight vectors.

Lemma 6.3. *Let G be semisimple and let V be a G-module where $V^G = 0$. Let $H \subset G$ be a maximal connected reductive subgroup which fixes a nonzero point $v \in V$. Then $Gv \subset V$ is closed with stabilizer a finite extension of H.*

Proof. Since H is maximal, $N_G(H)/H$ is finite so that Gv is closed [Lun75]. Similarly, G_v can only be a finite extension of H. □

For the maximal subgroups of the simple Lie groups see the works of Dynkin [Dyn52b, Dyn52a].

Example 6.4. Let $V = \varphi_2(\mathsf{C}_n)$, $n \geq 3$. Then $H := \mathsf{C}_1 \times \mathsf{C}_{n-1}$ is a maximal subgroup of C_n where $(\varphi_1(\mathsf{C}_n), \mathsf{C}_1 \times \mathsf{C}_{n-1}) = \varphi_1(\mathsf{C}_1) \oplus \varphi_1(\mathsf{C}_{n-1})$. Now H fixes a line in V. Thus a finite extension of H (actually H itself) is the stabilizer of a closed orbit, and one easily sees that the slice representation is $\theta_1 + \varphi_2(\mathsf{C}_{n-1})$. By induction one sees that the principal isotropy group of $\varphi_2(\mathsf{C}_n)$ is a product of n copies of SL_2.

Example 6.5. Let $G = \mathsf{F}_4$ which is an adjoint group. Now $\varphi_1 = \mathfrak{g}$ and φ_4 is the irreducible 26-dimensional representation whose nonzero weights are the short roots. The representations φ_2 and φ_3 contain the roots with multiplicities at least two.

Moreover, $\varphi_1^2, \varphi_1\varphi_4$ and φ_4^2 contain the roots with multiplicities at least 3. Hence every irreducible representation of G except for φ_1 and φ_4 has a bad toral slice.

Proposition 6.6. *The representations $\varphi_1(\mathsf{F}_4)$ and $2\varphi_4(\mathsf{F}_4)$ are the maximal coreduced representations of F_4. Moreover, the representation $2\varphi_4(\mathsf{F}_4)$ contains a dense orbit in the null cone.*

Proof. The sum $\varphi_1 + \varphi_4$ is not coreduced because the slice representation of the maximal torus is not coreduced (the nonzero weights are the short roots and these contain a bad relation). This leaves us to consider copies of φ_4. We know that $2\varphi_4$ is cofree ([Sch79]). So it suffices to show that $3\varphi_4$ is not coreduced and that $2\varphi_4$ contains a dense orbit in the null cone. For both statements we use some heavy calculations which are given in Appendix A, see Proposition A.1. □

Example 6.7. Let $G = \mathsf{G}_2$ which is an adjoint group. The fundamental representation φ_1 of dimension 7 and φ_2 (adjoint representation) are the only coreduced irreducible representations. This follows from Criterion 5.19, because φ_1^2 contains a nonzero weight of multiplicity ≥ 2.

Proposition 6.8. *Let $G = \mathsf{G}_2$. Then $2\varphi_1$ and the adjoint representation φ_2 are the maximal coreduced representations of G.*

Proof. See [Sch88] for the invariant theory of G_2. The invariants of $2\varphi_1$ are just the SO_7-invariants, so this representation is coreduced (see Theorem 9.1(4)). Now $\bigwedge^3(\varphi_1)$ contains a copy of φ_1, and it is easy to see that the corresponding covariant vanishes on the null cone of $3\varphi_1$. In fact, this holds for any covariant of type φ_1 of degree ≥ 3. Since the covariant is alternating of degree 3, it cannot be in the ideal of the quadratic invariants. More precisely, we have $S^2(\varphi_1 \otimes \mathbb{C}^3)^G = \theta_1 \otimes S^2\mathbb{C}^3$, and so in $S^3(\varphi_1 \otimes \mathbb{C}^3)$ we have

$$S^2(\varphi_1 \otimes \mathbb{C}^3)^G \cdot (\varphi_1 \otimes \mathbb{C}^3) = \varphi_1 \otimes (S^2\mathbb{C}^3 \cdot \mathbb{C}^3),$$

and this space does not contain $\varphi_1 \otimes \bigwedge^3 \mathbb{C}^3$. Thus $3\varphi_1$ is not coreduced.

To see that $\varphi_1 + \varphi_2$ is not coreduced we choose a nontrivial zero weight vector in $\varphi_2 = \mathfrak{g}$ which is annihilated by a short root α. Then the isotropy group has rank 2 and semisimple rank 1, and the slice representation contains two copies of (R_2, A_1), hence is not coreduced (Theorem 3.7). □

Let us summarize our results.

Theorem 6.9. *The following are the maximal coreduced representations of the exceptional groups containing a zero weight.*

(1) *For E_n: the adjoint representations $\varphi_2(\mathsf{E}_6), \varphi_1(\mathsf{E}_7), \varphi_1(\mathsf{E}_8)$.*
(2) *For F_4: $\mathrm{Ad}\,\mathsf{F}_4 = \varphi_1(\mathsf{F}_4)$ and $2\varphi_4(\mathsf{F}_4)$.*
(3) *For G_2: $\mathrm{Ad}\,\mathsf{G}_2 = \varphi_2(\mathsf{G}_2)$ and $2\varphi_1(\mathsf{G}_2)$.*

Remark 6.10. The proofs above and in Appendix A show that if an irreducible representation (V, G) of an adjoint exceptional group G is not coreduced, then V has a bad slice.

7 Coreduced representations of the classical groups

In this section we classify the coreduced representations V of the *simple adjoint groups* of classical type. If G is adjoint and simply laced, i.e., of type A or D, then a reducible representation V is not coreduced by Proposition 5.17, and so the maximal coreduced representations are all irreducible. We will see that this is also true for G of type C_n, $n \geq 3$, but not for type B_n.

The case of SL_2 has been settled in Theorem 3.7 even without assuming that the center acts trivially. So we may assume that the rank of G is at least 2.

Theorem 7.1. *Let G be a simple classical group of rank at least 2. Then, up to automorphisms, the following representations are the maximal coreduced representations of $G/Z(G)$.*

(1) $G = \mathsf{A}_n$, $n \geq 2$: $\operatorname{Ad}\mathsf{A}_n = \varphi_1\varphi_n$, $\varphi_2^2(\mathsf{A}_3)$, $\varphi_1^3(\mathsf{A}_2)$;
(2) $G = \mathsf{B}_n$, $n \geq 2$: $\operatorname{Ad}\mathsf{B}_n = \varphi_2(\mathsf{B}_n)$ (φ_2^2 *if* $n{=}2$), $\varphi_1^2(\mathsf{B}_n)$, $n\varphi_1(\mathsf{B}_n)$;
(3) $G = \mathsf{C}_n$, $n \geq 3$: $\operatorname{Ad}\mathsf{C}_n = \varphi_1^2(\mathsf{C}_n)$, $\varphi_2(\mathsf{C}_n)$, $\varphi_4(\mathsf{C}_4)$;
(4) $G = \mathsf{D}_n$, $n \geq 4$: $\operatorname{Ad}\mathsf{D}_n = \varphi_2(\mathsf{D}_n)$, $\varphi_1^2(\mathsf{D}_n)$.

In Section 4 we showed that every irreducible cofree representation of a simple group is coreduced. Looking at the list above and the one in Theorem 6.9 we see that we have the following partial converse.

Corollary 7.2. *Let G be a simple adjoint group and V an irreducible representation of G. Then V is coreduced if and only if V is cofree.*

We start with type A_n, $n \geq 2$. Recall that $\varphi_p := \bigwedge^p \mathbb{C}^{n+1}$, $p = 1, \ldots, n$.

Lemma 7.3. *Consider the representations φ_p and φ_q of SL_{n+1} where $1 \leq p \leq q \leq n$ and $n \geq 2$. Then there is a nonzero weight of $\varphi_p\varphi_q$ of multiplicity ≥ 2 except in the cases*

(1) φ_1^2 *or* φ_n^2,
(2) $\varphi_1\varphi_n$,
(3) $\varphi_2^2(\mathrm{SL}_4)$,

where the zero weight has multiplicity greater than one in (2) and (3).

Proof. It is easy to calculate that the weight $2\varepsilon_1 + \cdots + 2\varepsilon_{p-1} + \varepsilon_p + \cdots + \varepsilon_{q+1}$ occurs in $\varphi_p \otimes \varphi_q$ with multiplicity $q - p + 2$ and that it occurs in $\varphi_{p-1} \otimes \varphi_{q+1}$ once. Since $\varphi_p \otimes \varphi_q = \varphi_p\varphi_q + \varphi_{p-1} \otimes \varphi_{q+1}$ we see that our weight occurs with multiplicity $q - p + 1$ in $\varphi_p\varphi_q$. This gives us a nonzero weight of multiplicity at least two, except in the following two cases:

(1) $\varphi_1\varphi_n$ where the above weight is the zero weight, and
(2) φ_p^2 where $1 \leq p \leq n$.

However, in the second case, we can suppose, by duality, that $2p \leq n+1$. If $2p \leq n$, then one sees as above that $\varepsilon_1 + \cdots + \varepsilon_{2p}$ occurs with multiplicity $\frac{1}{p}\binom{2p}{p-1}$ which is at least 2 as long as $p > 1$. If $2p = n+1$, then one computes that $\varepsilon_1 - \varepsilon_2 = 2\varepsilon_1 + \varepsilon_3 + \cdots + \varepsilon_{2p}$ occurs with multiplicity $\frac{1}{p-1}\binom{2p-2}{p-2}$ which is ≥ 2 as long as $p > 2$. Thus the only possibilities are φ_1^2 and $\varphi_2^2(\mathrm{SL}_4)$. □

The next lemma was proved by STEMBRIDGE. We give a slightly different version of his proof.

Lemma 7.4. *Let φ be an irreducible representation of PSL_{n+1}, $n \geq 2$. Then the roots of G occur with multiplicity at least two in φ, except in the following cases.*

(1) *The adjoint representation $\varphi_1\varphi_n$;*
(2) *$\varphi_1^{k(n+1)}(\mathrm{SL}_{n+1})$ or its dual, $k = 1, 2, \ldots$;*
(3) *$\varphi_2^2(\mathrm{SL}_4) = \varphi_1^2(\mathsf{D}_3)$.*

Proof. The representation φ has highest weight $\lambda = \sum_i \lambda_i\omega_i$ where the ω_i are the fundamental dominant weights and $\sum_i i\lambda_i$ is a multiple of n. Now, Lemma 5.18 together with Lemma 7.3 above implies that the only irreducible representations of PSL_{n+1} containing the roots with multiplicity one are those listed. □

Proposition 7.5. *Let $n \geq 2$. The nontrivial irreducible coreduced representations of PSL_{n+1} are the adjoint representation $\varphi_1\varphi_n$, $\varphi_2^2(\mathrm{SL}_4)$, $\varphi_1^3(\mathrm{SL}_3)$ and $\varphi_2^3(\mathrm{SL}_3)$. All other irreducible representations admit a bad toral slice.*

Proof. By Proposition 5.17 we know that the only candidates for coreduced irreducible representations of PSL_n are those listed in Lemma 7.4 above. So it remains to show that $S^{km}(\mathbb{C}^m)$ is not coreduced for $m > 3$ and for $m = 3, k > 2$. For $m \geq 4$ the slice representation at a generic fixed point of the maximal torus T contains the weights $\beta_i := km\varepsilon_i$ and the weight $\alpha := -k(2\varepsilon_1 + 2\varepsilon_2 + \varepsilon_3 + \cdots + \varepsilon_{m-2})$ of the monomial $(x_3 \cdots x_{m-2}x_{m-1}^2x_m^2)^k$ which satisfy the indecomposable relation $m\alpha + 2\beta_1 + 2\beta_2 + \beta_3 + \cdots + \beta_{m-2} = 0$, and so the slice representation is not coreduced.

For $m = 3$ and $k > 1$ we have the weights $\beta_i := 3k\varepsilon_i$ and the weight $\alpha := -3(k-1)\varepsilon_1 - 3(k-2)\varepsilon_2$ of the monomial $x_2^3x_3^{3(k-1)}$ which satisfy the indecomposable relation $k\alpha + (k-1)\beta_1 + (k-2)\beta_2 = 0$. Again it follows that the slice representation is not coreduced. □

Now we look at type B_n.

Proposition 7.6. *Let $G = \mathrm{SO}_{2n+1}$ be the adjoint group of type B_n, $n \geq 2$. Then the only nontrivial irreducible coreduced representations are the adjoint representation φ_2, the standard representation φ_1 and φ_1^2. All other irreducible representations admit a bad toral slice.*

The representations φ_2 and φ_1^2 are maximal coreduced, whereas $k\varphi_1$ is coreduced if and only if $k \leq n$.

Proof. The highest weights of irreducible representations of G are just sums of the highest weights $\omega_1, \dots, \omega_{n-1}, 2\omega_n$ of the representations $W_\ell := \bigwedge^\ell(\mathbb{C}^{2n+1})$ for $1 \le \ell \le n$. For $\ell = 2m+1$ or $2m$, $m \ge 2$, one can compute that the weights of W_ℓ contain the roots of G with multiplicity $\binom{n-2}{m-1}$ which is at least 2. Thus W_ℓ admits a non-coreduced slice representation of a maximal torus and is therefore not coreduced for $\ell \ge 4$. For $\ell = 3$, hence $n \ge 3$, we have the weights $\pm\varepsilon_i \pm \varepsilon_j \pm \varepsilon_k$ where $1 \le i < j < k \le n$ as well as the weights $\pm\varepsilon_i$, $1 \le i \le n$ where the latter have multiplicity ≥ 2. Now the relation

$$(\varepsilon_1 + \varepsilon_2 + \varepsilon_3) + (-\varepsilon_1 + \varepsilon_2 + \varepsilon_3) + 2(-\varepsilon_2) + 2(-\varepsilon_3) = 0 \tag{1}$$

is indecomposable and so the slice representation of the maximal torus is not coreduced.

Now let V be an irreducible representation of G with highest weight $\lambda = \sum_i m_i\omega_i$. If $m_i > 0$ for some $i \ge 3$ then, by Criterion 5.19, V has a non-coreduced slice representation of a maximal torus, and thus is not coreduced.

Hence we are left with $\lambda = r\omega_1 + s\omega_2$ where s is even in case $n = 2$. Let us first assume that $n > 2$. Since φ_2^2 contains the roots with multiplicity $\ge \dim W_2^T = n \ge 3$ and since the weights of $\varphi_1\varphi_2$ contain the indecomposable weight relation $(2\varepsilon_1+\varepsilon_2)+2(-\varepsilon_1)+(-\varepsilon_2) = 0$ and the short roots occur with multiplicity > 1, we are reduced to the highest weights $r\omega_1$. If $r \ge 3$, we have the roots and the weights $3\varepsilon_1$ and $-2\varepsilon_1$ which lead one to see that the slice representation is not coreduced.

The arguments in the case $n = 2$ are the same (one has to replace φ_2 by φ_2^2 everywhere).

Finally, we have to look at direct sums of φ_1, φ_2 and φ_1^2. We will see in Theorem 9.1(2) that $k\varphi_1$ is coreduced if and only if $k \le n$. Since φ_2 and φ_1^2 contain all roots it remains to show that $\varphi_1^2 + \varphi_1$ and $\varphi_2 + \varphi_1$ are not coreduced. First, consider φ_1^2, $n \ge 4$. The subgroup $\mathrm{SO}_3 \times \mathrm{SO}_{2n-2}$ is maximal in SO_{2n+1}, it has rank n and has slice representation $\varphi_1^4(\mathsf{A}_1) \oplus \varphi_1^2(\mathsf{D}_{n-1}) + \theta_1$. If we add a copy of $\varphi_1(\mathsf{B}_n)$, then we have a subrepresentation $(\varphi_1^4 + \varphi_1^2, \mathsf{A}_1)$ which is not coreduced. The details work out similarly for $n = 2$ and $n = 3$. We are left with $\operatorname{Ad} G + \varphi_1$. The slice representation of the group $\mathrm{SO}_3 \times (\mathrm{SO}_2)^{n-1}$ contains two copies of the standard representation of SO_3 on $\mathbb{C}^3$ which is not coreduced (Theorem 3.7). Hence $\operatorname{Ad} G + \varphi_1$ is not coreduced. □

For type D_n we get the following result. Recall that only irreducible representations of PSO_{2n} can be coreduced (Proposition 5.17).

Proposition 7.7. *Let $G = \mathrm{PSO}_{2n}$ be the adjoint group of type D_n, $n \ge 4$. Then the only nontrivial coreduced representations are the adjoint representation φ_2, φ_1^2, $\varphi_3^2(\mathsf{D}_4)$ and $\varphi_4^2(\mathsf{D}_4)$, and these are maximal coreduced. All other irreducible representations admit a bad toral slice.*

Proof. The highest weights of representations of SO_{2n} are just sums of the highest weights $\omega_1, \dots, \omega_{n-2}, \omega_{n-1}+\omega_n$ of the representations $W_\ell := \bigwedge^\ell(\mathbb{C}^{2n})$ for $1 \le \ell \le n-1$ and twice the highest weights ω_{n-1} and ω_n of the two half-spin representations. Moreover, $\varphi_{n-1}^2 \oplus \varphi_n^2 \simeq W_n := \bigwedge^n(\mathbb{C}^{2n})$.

The representations W_{2m} for $m > 1$ contain the roots of G with multiplicity $\binom{n-2}{m-1} \geq 2$. The representations W_k for k odd, $k > 1$, have no zero weights, but they contain the weights of φ_1 with multiplicity greater than one. Hence the Cartan products $W_k W_\ell$ for $k, \ell \leq n-1$ odd, $k + \ell \geq 4$, contain the adjoint representation more than once, so that the representations are not coreduced. We already know that φ_1^2 is coreduced and by Criterion 5.19 no power φ_1^{2k} is coreduced for $k \geq 2$.

It remains to consider those representations φ of PSO_{2n} that are Cartan products with φ_{n-1}^2 or φ_n^2. If $n \geq 6$ is even, then both contain the roots at least three times, hence φ is not coreduced. If $n = 4$, then φ_3^2 and φ_4^2 are outer isomorphic to φ_1^2 which is coreduced. If φ is not exactly one of these representations, then it is not coreduced by Criterion 5.19. If n is odd, then φ_{n-1}^2 and φ_n^2 both contain the weights of W_1 at least three times, and so φ contains the roots with multiplicity at least 3 and is not coreduced. □

For type C_n we will use the following lemma.

Lemma 7.8. *Let $H_1, \ldots, H_4$ be copies of SL_2 and let $V_i \simeq \mathbb{C}^2$ have the standard action of H_i. Let $H = \prod_i H_i$ and $V = \bigoplus_{i<j} V_{ij}$ where $V_{ij} = V_i \otimes V_j$. Then (V, H) is not coreduced.*

Proof. Consider the subrepresentation $V' := V_{12} \oplus V_{14} \oplus V_{23} \oplus V_{34} \oplus V_{24}$. We have the quotient mapping (by H_1) from $V_{12} \oplus V_{14}$ to $V_{24}' \oplus \theta_2$ where V_{24}' is another copy of V_{24}. The image is a hypersurface F defined by an equation saying that the invariant of $(V_{24}', H_2 \times H_4)$ is the product of the coordinate functions on θ_2. By Lemmas 2.6 and 2.8 (see Examples 2.7 and 2.9) the representation $V_{24}' \oplus V_{23} \oplus V_{34} \oplus V_{24} \oplus \theta_2$ of $H_2 \times H_3 \times H_4$ is coreduced if V' is coreduced. Quotienting by the action of H_3, we similarly obtain a representation $(V_{24}' \oplus V_{24}'' \oplus V_{24} \oplus \theta_4, H_2 \times H_4) \simeq (3\mathbb{C}^4 \oplus \theta_4, \mathrm{SO}_4)$ which is not coreduced (Example 3.4). Hence (V', H) and (V, H) are not coreduced. □

The fundamental representations φ_i of C_n are given by $\varphi_1 = \mathbb{C}^{2n}$, $\varphi_2 = \bigwedge^2 \mathbb{C}^{2n}/\mathbb{C}\beta$, and $\varphi_i = \bigwedge^i \mathbb{C}^{2n}/\beta \wedge \bigwedge^{i-2} \mathbb{C}^{2n}$ for $i = 3, \ldots, n$ where $\beta \in \bigwedge^2 \mathbb{C}^{2n}$ is the invariant form. They can be realized as the irreducible subspaces $\bigwedge_0^i(\mathbb{C}^{2n}) \subset \bigwedge^i(\mathbb{C}^{2n})$ of highest weight $\omega_i := \varepsilon_1 + \cdots + \varepsilon_i$. The generators of the representations of the adjoint group $G = \mathrm{PSp}_{2n}$ are the φ_i for i even and the $\varphi_i\varphi_j$ for i and j odd.

Proposition 7.9. *Let $G = \mathrm{PSp}_{2n}$ be the adjoint group of type C_n, $n \geq 3$. Then the nontrivial irreducible coreduced representations of G are the adjoint representation φ_1^2, φ_2 and $\varphi_4(\mathsf{C}_4)$, and these are all maximal. Moreover, all other irreducible representations admit a bad slice.*

Proof. (a) First consider the case of $\varphi_i\varphi_j$ where i and j are odd. We may suppose that $j \geq 3$. Then φ_j contains the weight $\varepsilon_1 + \varepsilon_2 + \varepsilon_3$ (it is a dominant weight which is the highest weight of φ_j minus a sum of positive roots). By the action of the Weyl group we have all the weights $\pm\varepsilon_1 \pm \varepsilon_2 \pm \varepsilon_3$. In φ_i (and φ_j) we similarly have all the weights $\pm\varepsilon_k$. Thus $\varphi_i\varphi_j$ contains the roots $2\varepsilon_1$ and $\varepsilon_1 - \varepsilon_2$, hence all the roots. Moreover, we have the following indecomposable relation of weights in $\varphi_i\varphi_j$ (none of which are roots):

$$(2\varepsilon_1+\varepsilon_2+\varepsilon_3)+(2\varepsilon_1-\varepsilon_2-\varepsilon_3) \\ +2(-\varepsilon_1+2\varepsilon_2+\varepsilon_3)+2(-\varepsilon_1-2\varepsilon_2-\varepsilon_3)=0. \tag{2}$$

Hence $\varphi_i\varphi_j$ has a bad toral slice and is therefore not coreduced. The same holds for every Cartan product of $\varphi_i\varphi_j$ with any other representation of G.

Now φ_1^4 is a representation of G, but since φ_1^2 contains the trivial representation n times, φ_1^4 contains the adjoint representation at least n times, hence has a bad toral slice and is not coreduced. Therefore, the adjoint representation φ_1^2 is the only coreduced irreducible representation $\varphi=\varphi_{i_1}\varphi_{i_2}\cdots\varphi_{i_m}$ of G where at least one i_k is odd.

(b) Now we consider representations φ_{2i}, $2i\le n$. These representations, one can show as above, contain the short roots of G. But the long roots do not occur. Hence the connected component of the isotropy group at a generic zero weight vector is covered by a product $H:=\prod_{j=1}^n H_j$, where each H_j is the copy of SL_2 in G corresponding to the positive long root $2\varepsilon_j$. If $n\ge 5$ and $2i\ge 4$, then the slice representation contains the subrepresentation

$$\bigoplus_{1\le j<k\le n} V_{jk} \quad \text{where} \quad V_{jk}:=(\mathbb{C}^2\otimes\mathbb{C}^2, H_j\times H_k),$$

which is not coreduced (Lemma 7.8). Finally, one easily sees that any product $\varphi_{2i}\varphi_{2j}$ contains all the roots as well as the zero sum of weights given above in equation (2). This includes the case where a factor is φ_2 or φ_4. Hence the irreducible coreduced representations of G are as claimed.

(c) It remains to show that the coreduced representations of G are all irreducible. As seen above, the connected component of the isotropy group at a generic zero weight vector of φ_2 is covered by a product $H:=\prod_{j=1}^n H_j$, where each H_j is the copy of SL_2 in G corresponding to the positive long root $2\varepsilon_j$. If we add another copy of φ_2 or the adjoint representation φ_1^2, then the slice representation contains $\bigoplus_{1\le j<k\le n} V_{jk}$ where $V_{jk}:=(\mathbb{C}^2\otimes\mathbb{C}^2, H_j\times H_k)$, which is not coreduced for $n\ge 4$. The same holds if $n=4$ and we add a copy of $\varphi_4(\mathsf{C}_4)$. This proves the claim for $n\ge 4$, because φ_1^2 and $\varphi_4(\mathsf{C}_4)$ contain all roots. For $\varphi_2(\mathsf{C}_3)+\varphi_1^2(\mathsf{C}_3)$ we have the slice representation of $H=H_1\times H_2\times H_3$ on $V_{12}\oplus V_{13}\oplus V_{23}\oplus\mathfrak{h}_1\oplus\mathfrak{h}_2\oplus\mathfrak{h}_3\oplus\theta_2$ where the H_i are copies of SL_2 and the V_{ij} are as above. Consider the subrepresentation $(V', H_1\times H_2):=(V_{12}\oplus\mathfrak{h}_1\oplus\mathfrak{h}_2, H_1\times H_2)$. The principal isotropy group of $\mathfrak{h}_1$ is $\mathbb{C}^*\times H_2$ where $\mathbb{C}^*$ acts on V_{12} with weights ±1. Let $\mathfrak{h}_2'$ denote a second copy of $\mathfrak{h}_2$. Then the quotient of V_{12} by $\mathbb{C}^*$ is a quadratic hypersurface in $\mathfrak{h}_2'+\theta_1$ which equates the quadratic invariant of $\mathfrak{h}_2'$ and the square of the coordinate function on θ_1. Thus, as in Lemma 7.8, the fact that the representation $(\mathfrak{h}_2+\mathfrak{h}_2'+\theta_1, H_2)$ is not coreduced (Example 5.12) implies that V' is not coreduced, hence neither is $\varphi_2(\mathsf{C}_3)+\varphi_1^2(\mathsf{C}_3)$. Finally, $2\varphi_2(\mathsf{C}_3)$ is not coreduced as we have seen in Example 5.12. □

Remark 7.10. The proofs above show that if an irreducible representation (V, G) of an adjoint classical group G is not coreduced, then V has a bad slice. We have already seen in the previous section that the same holds for the exceptional groups (Remark 6.10).

8 Irreducible coreduced representations of semisimple groups

In this section we classify the irreducible coreduced representations of adjoint semisimple groups.

Example 8.1. The representation $(\mathbb{C}^{2n+1} \otimes \mathbb{C}^{2m+1}, \mathrm{SO}_{2n+1} \times \mathrm{SO}_{2m+1})$ is the isotropy representation of a symmetric space. (Consider the automorphism θ of $\mathrm{SO}_{2(n+m+1)}$ given by conjugation with $\begin{bmatrix} \mathrm{Id}_{2n+1} & \\ & -\mathrm{Id}_{2m+1} \end{bmatrix}$.) It now follows from [KR71, Theorem 14, p. 758] that this representation is coreduced for all $n, m \geq 1$. Of course, this is also an example of a θ-representation, hence coreduced by Corollary 4.8.

Example 8.2. The representation $(V, G \times H) = (\mathbb{C}^3 \otimes \varphi_1(\mathsf{G}_2), \mathrm{SO}_3 \times \mathsf{G}_2)$ is coreduced. In fact, (V, H) is cofree and the quotient $V/\!\!/H$ is the SO_3-module $\varphi_1^4 \oplus \theta_2$ which is cofree and coreduced. Hence $(V, G \times H)$ is cofree too. Now the proper nontrivial slice representations of $(3\varphi_1, \mathsf{G}_2)$ are $(2\mathbb{C}^3 + 2(\mathbb{C}^3)^* + \theta_3, \mathrm{SL}_3)$ (coreduced by Theorem 9.1) and $(2\mathbb{C}^2 + \theta_6, \mathrm{SL}_2)$ (coreduced by Theorem 3.7). Thus every fiber of $\pi\colon V \to V/\!\!/H$ is reduced, except for the zero fiber, which has codimension 7. Thus the null cone of $(V, G \times H)$, which has codimension 4, is reduced off of a subset of V of codimension 7, hence $(V, G \times H)$ is coreduced.

Surprisingly, these two examples are the only irreducible coreduced representations besides those where G is simple.

Theorem 8.3. *The coreduced irreducible representations of a semisimple non-simple adjoint group are*

$$(\varphi_1(\mathsf{B}_n) \otimes \varphi_1(\mathsf{B}_m), \mathsf{B}_n \times \mathsf{B}_m) \quad \textit{and} \quad (\varphi_1^2(\mathsf{A}_1) \otimes \varphi_1(\mathsf{G}_2), \mathsf{A}_1 \times \mathsf{G}_2).$$

The proof needs some preparation. We first construct a list of non-coreduced representations which will help to rule out most candidates.

Example 8.4. Let $(V, G) = (\mathbb{C}^n \otimes \mathbb{C}^m + \mathbb{C}^n, \mathrm{SO}_n \times \mathrm{SO}_m)$ where $m, n \geq 2$. We show that V is not coreduced. There are three cases. Recall that

$$(S^2(\mathbb{C}^n) \oplus \mathbb{C}^n, \mathrm{SO}_n)$$

is not coreduced even for $n = 2$.

(1) $n < m$. Quotienting by the action of SO_m we obtain $S^2(\mathbb{C}^n) \oplus \mathbb{C}^n$ which is not coreduced, hence neither is V (Example 2.9).
(2) $n = m$. By Example 2.10 the representation V is not coreduced since quotienting by O_m we obtain the non-coreduced representation $S^2(\mathbb{C}^n) \oplus \mathbb{C}^n$.
(3) $n > m$. We have at most n copies of $\mathbb{C}^n$, so by Example 2.10 we may quotient by the action of O_n to arrive at the representation

$$S^2(\mathbb{C}^m) \oplus \mathbb{C}^m$$

which is not coreduced. Hence V is not coreduced.

We have seen in Corollary 5.14 that for two representations (V, G) and (W, H) with a zero weight, if (V, G) has a bad slice, then so does $(V \otimes W, G \times H)$. Together with Remarks 6.10 and 7.10 this implies that we need only consider tensor products of the irreducible coreduced representations (V, G) of the simple adjoint groups. They fall into five types.

(1) $(V, G) = \varphi_1^2(\mathsf{A}_1) = (\mathbb{C}^3, \mathrm{SO}_3)$.
(2) $(V, G) = \varphi_1^4(\mathsf{A}_1)$ or there is a slice representation (W, H) where $H^0 = T$ is a maximal torus of G (of rank at least 2) and W contains weight spaces of roots α, β and $-(\alpha + \beta)$ or W contains θ_2 and weight spaces $\pm\alpha$.
(3) $(V, G) = \varphi_1(\mathsf{B}_n)$, $n \geq 2$.
(4) $(V, G) = \varphi_1(\mathsf{G}_2)$.
(5) $(V, G) = \varphi_4(\mathsf{F}_4)$, $\varphi_4(\mathsf{C}_4)$, or $\varphi_2(\mathsf{C}_n)$, $n \geq 3$.

Note that the representations $\varphi_1^2(\mathsf{D}_n)$, $n \geq 3$ and $\varphi_1^2(\mathsf{B}_n)$, $n \geq 2$, are of type (2) as is the representation $\varphi_1^3(\mathsf{A}_2)$. We consider tensor products of the various types of representations.

Lemma 8.5. *Let (V_1, G_1) be of type (2) and let (V_2, G_2) be of arbitrary type. Then $(V_1 \otimes V_2, G_1 \times G_2)$ has a bad slice.*

Proof. We leave the case that (V_1, G_1) or (V_2, G_2) is $\varphi_1^4(\mathsf{A}_1)$ to the reader. It will be clear from our techniques what to do in this case. Let T_1 be a maximal torus of G_1 fixing v_1. First, assume that the weights of the slice representation at v_1 contain roots α, β and $-\alpha - \beta$. Let T_2 be a maximal torus of G_2. Suppose first that $(V_2, G_2) = (\mathbb{C}^3, \mathrm{SO}_3)$, Let $v_2 \in V_2$ be a zero weight vector and let γ be a nonzero weight of (V_2, T_2). Then by Corollary 5.14 the slice representation of $T_1 \times T_2$ at $v_1 \otimes v_2$ contains the weights

$$-\gamma + \alpha,\ \gamma + \beta, \gamma - \alpha - \beta, \text{ and } -\gamma - \alpha - \beta.$$

We have the minimal zero sum

$$2(-\gamma + \alpha) + 2(\gamma + \beta) + (\gamma - \alpha - \beta) + (-\gamma - \alpha - \beta) = 0,$$

hence the slice representation of $T_1 \times T_2$ is not coreduced. The same argument works in case (V_2, G_2) is of type (2). Now suppose that $(V_2, G_2) = \varphi_1(\mathsf{B}_n)$, $n \geq 2$. Then we have a slice representation of $\mathrm{SO}_{2n} \times T_1$ containing the irreducible components $\mathbb{C}^{2n} \otimes \mathbb{C}_\alpha$, $\mathbb{C}^{2n} \otimes \mathbb{C}_\beta$ and $\mathbb{C}^{2n} \otimes \mathbb{C}_{-\alpha-\beta}$. Quotienting by SO_{2n} we obtain a representation of T_1 with weights

$$2\alpha,\ 2\beta,\ \alpha+\beta,\ -\alpha,\ -\beta \text{ and } -2\alpha-2\beta.$$

Hence the slice representation is not coreduced. The same argument works in case (V_2, G_2) is of type (5). For type (4) we get a slice representation of $\mathrm{SL}_3 \times T_1$ containing

$$\mathbb{C}^3 \otimes \mathbb{C}_\alpha,\ \mathbb{C}^3 \otimes \mathbb{C}_\beta, (\mathbb{C}^3)^* \otimes \mathbb{C}_{-\alpha-\beta}, \text{ and } (\mathbb{C}^3)^* \otimes \mathbb{C}_\alpha,$$

and quotienting by SL_3 we obtain a T_1-representation with weights $-\beta$, $-\alpha$, 2α and $\alpha+\beta$. Hence we have a non-coreduced slice representation.

Finally, assume that the slice representation at v_1 contains θ_2 and weights $\pm\alpha$ and that (V_2, G_2) is of arbitrary type. Let $\pm\gamma$ be nonzero weights of V_2. Because of the θ_2, the slice representation at $v_1 \otimes v_2$ contains the weights of V_2 (Corollary 5.14). Hence we have weights $\pm\alpha \pm \gamma$ and $\pm\gamma$. and the minimal bad relation

$$(\alpha+\gamma) + (-\alpha+\gamma) - 2(\gamma) = 0.$$

Thus $(V_1 \otimes V_2, G_1 \times G_2)$ is not coreduced. □

We are left with type (1) and types (3–5).

Lemma 8.6. *Suppose that (V_1, G_1) is of type (1) or (3) or (5) and that (V_2, G_2) is of type (5). Then $(V_1 \otimes V_2, G_1 \times G_2)$ has a bad slice.*

Proof. First assume that (V_1, G_1) is $\varphi_1(\mathsf{B}_n)$, $n \geq 1$ (type (1) or type (3)). If (V_2, G_2) is $\varphi_4(\mathsf{F}_4)$, then there is a (principal) slice representation of D_4 on θ_2 where $(\varphi_4(\mathsf{F}_4), \mathsf{D}_4) = (\varphi_1 + \varphi_3 + \varphi_4 + \theta_2)$ while (V_1, G_1) has a slice representation of SO_{2n} on θ_1 where $(V_1, \mathrm{SO}_{2n}) = (\mathbb{C}^{2n} + \theta_1, \mathrm{SO}_{2n})$. By Corollary 5.14 there is a subrepresentation of a slice representation of $(V_1 \otimes V_2, G_1 \times G_2)$ which is of the form $(\mathbb{C}^{2n} \otimes \mathbb{C}^8 \oplus \mathbb{C}^{2n}, \mathrm{SO}_{2n} \times \mathrm{SO}_8)$. It follows from Example 8.4 that the slice representation is not coreduced.

If (V_2, G_2) is $\varphi_2(\mathsf{C}_m)$, $m \geq 4$, then there is a slice representation $(W, H) = (\theta_2 + \varphi_2, \mathrm{SL}_2 \times \mathrm{SL}_2 \times \mathsf{C}_{m-2})$ where $(\varphi_2(\mathsf{C}_m), H)$ contains

$$(\mathbb{C}^2 \otimes \mathbb{C}^2, \mathrm{SL}_2 \times \mathrm{SL}_2) \simeq (\mathbb{C}^4, \mathrm{SO}_4).$$

There is a non-coreduced subrepresentation of the slice representation of $(V_1 \otimes V_2, G_1 \times G_2)$ of the form $(\mathbb{C}^{2n} \otimes \mathbb{C}^4 \oplus \mathbb{C}^{2n}, \mathrm{SO}_{2n} \times \mathrm{SO}_4)$. The case of $\varphi_2(\mathsf{C}_3)$ is only notationally different and the case of $\varphi_4(\mathsf{C}_4)$ is similar. Finally, if (V_1, G_1) is of type (5), then the same techniques produce a non-coreduced slice representation at a zero weight vector. □

We leave the proof of the following to the reader.

Lemma 8.7. *A tensor product* $(V_1 \otimes V_2, G_1 \times G_2)$ *has a bad slice if* (V_1, G_1) *is* $\varphi_1(\mathsf{G}_2)$ *(type (4)) and* (V_2, G_2) *is of type (5).*

We are now left with the problem of tensor products of representations of types (1), (3) and (4). First, we handle types (1) and (3).

Proposition 8.8. *Let* $3 \leq 2k+1 \leq 2m+1 \leq 2n+1$ *and*

$$(V, G) = (\mathbb{C}^{2n+1} \otimes \mathbb{C}^{2m+1} \otimes \mathbb{C}^{2k+1}, \mathrm{SO}_{2n+1} \times \mathrm{SO}_{2m+1} \times \mathrm{SO}_{2k+1}).$$

Then the slice representation at the zero weight vector is not coreduced.

Proof. The slice representation at the zero weight vector is

$$(W, H) = (\mathbb{C}^{2n} \otimes \mathbb{C}^{2m} \otimes \mathbb{C}^{2k} + \mathbb{C}^{2n} \otimes \mathbb{C}^{2m} + \mathbb{C}^{2m} \otimes \mathbb{C}^{2k} + \mathbb{C}^{2n} \otimes \mathbb{C}^{2k}, \mathrm{SO}_{2n} \times \mathrm{SO}_{2m} \times \mathrm{SO}_{2k}).$$

If $k > 1$, consider the subrepresentation $\mathbb{C}^{2m} \otimes \mathbb{C}^{2k} \oplus \mathbb{C}^{2n} \otimes \mathbb{C}^{2k}$. Quotienting by $\mathrm{SO}_{2m} \times \mathrm{SO}_{2n}$ we get $(2S^2(\mathbb{C}^{2k}), \mathrm{SO}_{2k})$ which is not coreduced. Using Example 2.9 we see that (V, G) is not coreduced.

Now assume that $k = 1$ but $m > 1$. We have a subrepresentation

$$\mathbb{C}^{2n} \otimes \mathbb{C}^{2m} \oplus \mathbb{C}^{2m} \otimes \mathbb{C}_\nu \oplus \mathbb{C}^{2m} \otimes \mathbb{C}_{-\nu},$$

where the $\mathbb{C}_{\pm\nu}$ are irreducible representations of $\mathrm{SO}_{2k} \simeq \mathbb{C}^*$ of weight ± 1. Quotienting by O_{2n} we obtain the representation

$$(S^2(\mathbb{C}^{2m}) \oplus \mathbb{C}^{2m} \otimes \mathbb{C}_\nu \oplus \mathbb{C}^{2m} \otimes \mathbb{C}_{-\nu}, \mathrm{SO}_{2m} \times \mathrm{SO}_2).$$

Let $\pm\varepsilon_1, \dots, \pm\varepsilon_m$ be the weights of $\mathbb{C}^{2m}$ for the action of the maximal torus T of SO_{2m}. Then the slice representation of $S^2(\mathbb{C}^{2m})$ at a generic zero weight vector is, up to trivial factors, the sum of the $\mathbb{C}_{\pm 2\varepsilon_i}$. Hence we have a slice representation of $T \times \mathrm{SO}_2$ containing

$$\mathbb{C}_{-2\varepsilon_1} \oplus \mathbb{C}_{-2\varepsilon_2} \oplus (\mathbb{C}_{\varepsilon_1} \otimes \mathbb{C}_\nu) \oplus (\mathbb{C}_{\varepsilon_2} \otimes \mathbb{C}_{-\nu}).$$

This last representation is not coreduced.

Now assume that $n \geq m = k = 1$. We rename the weight ε_1 of $\mathrm{SO}_{2m} = \mathrm{SO}_2$ to be just ε. Then we have the subrepresentation

$$(\mathbb{C}^{2n} \otimes \mathbb{C}_\varepsilon) \oplus (\mathbb{C}^{2n} \otimes \mathbb{C}_\nu) \oplus (\mathbb{C}_{-\varepsilon} \otimes \mathbb{C}_{-\nu}).$$

Quotienting by O_{2n} we get a representation

$$\mathbb{C}_{2\varepsilon} \oplus \mathbb{C}_{2\nu} \oplus (\mathbb{C}_\varepsilon \otimes \mathbb{C}_\nu) \oplus (\mathbb{C}_{-\varepsilon} \otimes \mathbb{C}_{-\nu})$$

of $\mathbb{C}^* \times \mathbb{C}^*$ which is not coreduced. □

Proposition 8.9. *Let* $(V, G) = (\mathbb{C}^{2n+1} \otimes \mathbb{C}^7, \mathrm{SO}_{2n+1} \times \mathsf{G}_2)$, $n \geq 2$, *or*

$$(\mathbb{C}^3 \otimes \mathbb{C}^3 \otimes \mathbb{C}^7, \mathrm{SO}_3 \times \mathrm{SO}_3 \times \mathsf{G}_2).$$

Then (V, G) *has a bad slice.*

Proof. We leave the latter case to the reader. In the former case we have the slice representation (minus the trivial factor)

$$(W, H) = (\mathbb{C}^{2n} \otimes (\mathbb{C}^3 \oplus (\mathbb{C}^3)^*), \mathrm{SO}_{2n} \times \mathrm{SL}_3).$$

If $n \geq 3$, then quotienting by O_{2n} we obtain the representation

$$(S^2(\mathbb{C}^3) \oplus S^2(\mathbb{C}^{3*}) \oplus \mathbb{C}^3 \otimes \mathbb{C}^{3*}, \mathrm{SL}_3)$$

which is not coreduced.

We are left with the case $(W, H) = (\mathbb{C}^4 \otimes (\mathbb{C}^3 \oplus \mathbb{C}^{3*}), \mathrm{SO}_4 \times \mathrm{SL}_3)$. Consider a 1-parameter subgroup ρ of $\mathrm{SO}_4 \times \mathrm{SL}_3$ whose action on $\mathbb{C}^4$ has weights ± 1 and on $\mathbb{C}^3$ has weights $2, 0, -2$. Then Z_ρ, the span of the positive weight vectors, has dimension 12 (which is not surprising since (W, H) is self-dual of dimension 24). Note that Z_ρ is in the null cone and is stable under a Borel subgroup B of H. Now one can show that the dimension of $U^- Z_\rho$ is $17 = 12 + \dim U^-$, the maximal possible, where U^- is the maximal unipotent subgroup of H opposite B. Hence HZ_ρ is a component of the null cone (see section 10 for more details).

The positive weights of ρ on W are 1 and 3 and the negative weights are -1 and -3. This implies that the differential of an invariant of degree > 4 vanishes on Z_ρ, hence on HZ_ρ. But we have only 4 generating invariants in degree at most 4, and so the null cone is not reduced along HZ_ρ, because $\operatorname{codim} HZ_\rho = 7$. □

We are left with the case $\mathsf{G}_2 \times \mathsf{G}_2$ acting on $\mathbb{C}^7 \otimes \mathbb{C}^7$.

Proposition 8.10. *The representation* $(\mathbb{C}^7 \otimes \mathbb{C}^7, \mathsf{G}_2 \times \mathsf{G}_2)$ *is not coreduced.*

We have two proofs of this, and both need some computations. They are given in Appendix B.

9 Classical invariants

Classical Invariant Theory describes the invariants of copies of the standard representations of the classical groups, e.g., the $\mathrm{GL}(V)$- or $\mathrm{SL}(V)$-invariants of $mV \oplus nV^*$ or the $\mathrm{Sp}(V)$-invariants of mV where $mV := V^{\oplus m}$ denotes the direct sum of m copies of V. In this context we will prove the following theorem.

Theorem 9.1. (1) *The representation* $(pV \oplus qV^*, \mathrm{GL}(V))$ *is coreduced for all* $p, q \geq 0$. *The null cone is irreducible if and only if* $p + q \leq n$.

(2) *The representation* $(pV \oplus qV^*, \mathrm{SL}(V))$ *is coreduced for all* $p, q \geq 0$. *The null cone is irreducible in the following cases:* $p + q \leq n$ *or* $(p, q) = (n, 1)$ *or* $(p, q) = (1, n)$.
(3) *The representations* $(mV, \mathrm{Sp}(V))$ *are coreduced for all* $m \geq 0$, *and the null cone is irreducible and normal.*
(4) *The representations* $(mV, \mathrm{O}(V))$, $(mV, \mathrm{SO}(V))$ *are coreduced if and only if* $2m \leq \dim V$. *The null cone is irreducible and normal for* $2m < \dim V$.

The basic reference for this section is [Sch87]. Denote by T_m, B_m and U_m the subgroups of GL_m consisting of diagonal, upper triangular, and upper triangular unipotent matrices. If λ is a dominant weight, i.e., $\lambda = \sum_{i=1}^m \lambda_i \varepsilon_i \in X(T_n) = \bigoplus_{i=1}^m \mathbb{Z}\varepsilon_i$ and $\lambda_1 \geq \lambda_2 \geq \cdots \geq \lambda_m$, we denote by ψ_λ or $\psi_\lambda(m)$ the corresponding irreducible representation of GL_m. In the following, we will only deal with *polynomial* representations of GL_m, so that $\lambda_i \geq 0$ for all i. Set $|\lambda| := \sum \lambda_i$ and define the *height* of a dominant weight by $\mathrm{ht}(\lambda) := \max\{i \mid \lambda_i > 0\}$.

The famous CAUCHY formula describes the decomposition of the symmetric powers of a tensor product where we consider $\psi_\lambda(m) \otimes \psi_\mu(k)$ as a representation of $\mathrm{GL}_m \times \mathrm{GL}_k$ (see [Sch87, (1.9) Theorem]).

Proposition 9.2.

$$S^d(\mathbb{C}^m \otimes \mathbb{C}^k) = \bigoplus_{|\lambda|=d,\, \mathrm{ht}(\lambda) \leq \min\{m,k\}} \psi_\lambda(m) \otimes \psi_\lambda(k).$$

If λ is a dominant weight of height r, then $\psi_\lambda(m)$ makes sense for any $m \geq r$. In fact, ψ_λ is a functor and $\psi_\lambda(V)$ is a well-defined $\mathrm{GL}(V)$-module for every vector space V of dimension $\geq r$. In particular, if $\rho\colon G \to \mathrm{GL}(V)$ is a representation of a reductive group G, then all $\psi_\lambda(V)$ for $\mathrm{ht}(\lambda) \leq \dim V$ are representations of G as well. From the CAUCHY formula we thus get

$$\mathcal{O}(mV)_d = S^d(\mathbb{C}^m \otimes V^*) = \bigoplus_{|\lambda|=d,\, \mathrm{ht}(\lambda) \leq \min(m, \dim V)} \psi_\lambda(m) \otimes \psi_\lambda(V^*)$$

as a representation of $\mathrm{GL}_m \times G$. Taking U_m-invariants we find

$$\mathcal{O}(mV)_d^{U_m} = S^d(\mathbb{C}^m \otimes V^*)^{U_m} = \bigoplus_{|\lambda|=d,\, \mathrm{ht}(\lambda) \leq \min(m, \dim V)} \psi_\lambda(V^*), \qquad (*)$$

where the torus $T_m \subset \mathrm{GL}_m$ acts on $\psi_\lambda(V^*)$ with weight λ. Thus the algebra $\mathcal{O}(mV)^{U_m}$ is $\mathbb{Z}^m$-graded, and the homogeneous component of weight λ is the G-module $\psi_\lambda(V^*)$. In particular, $\mathcal{O}(mV)^{U_m}$ is multiplicity-free as a $\mathrm{GL}(V)$-module. It follows that the product $\psi_\lambda(V^*) \cdot \psi_\mu(V^*)$ in $\mathcal{O}(mV)$ is equal to $\psi_{\lambda+\mu}(V^*)$. This leads to the following definition.

Definition 9.3. Let G be a connected reductive group and let A be a G-algebra, i.e., a commutative $\mathbb{C}$-algebra with a locally finite and rational action of G by algebra automorphisms. Two simple submodules $U, V \subset A$ are called *orthogonal* if the product $U \cdot V \subset A$ is either zero or simple and isomorphic to the Cartan (highest weight) component of $U \otimes V$.

The result above can therefore be expressed by saying that all irreducible $\mathrm{GL}(V)$-submodules of $\mathcal{O}(mV)^{U_m}$ are orthogonal to each other. The following crucial result is due to BRION [Bri85, Lemme 4.1].

Proposition 9.4. *Let A be a G-algebra and let $V_1, V_2, W \subset A$ be simple submodules. Assume that V_1, V_2 are both orthogonal to W. Then any simple factor of $V_1 \cdot V_2$ is orthogonal to W.*

We will also need the following result about U-invariants (see [Kra84, III.3.3]).

Proposition 9.5. *Let G be a connected reductive group, $U \subset G$ a maximal unipotent subgroup, and let A be a finitely generated G-algebra. Then A is reduced, resp. a domain, resp. normal if and only if A^U is reduced, resp. a domain, resp. normal.*

Another consequence of formula $(*)$ is that $\mathcal{O}(mV)^{U_m} = \mathcal{O}(nV)^{U_n}$ for all $m \geq n = \dim V$.

We start with the groups $\mathrm{GL}(V)$ and $\mathrm{SL}(V)$ acting on $W := pV \oplus qV^*$. It is known that the $\mathrm{GL}(V)$-invariants are generated by the bilinear forms

$$f_{ij}\colon (v_1, \ldots, v_p, \xi_1, \ldots, \xi_q) \mapsto \xi_j(v_i).$$

If V_i^* is the ith copy of V^* in $W^* \subset \mathcal{O}(W)$ and V_j the jth copy of V, then $V_i^* \cdot V_j = \mathfrak{sl}(V) \oplus \mathbb{C} f_{ij}$ in $\mathcal{O}(W)$, and so V_i^* and V_j are orthogonal in $\mathcal{O}(W)/I$ where I is the ideal generated by the invariants f_{ij}. It follows from Proposition 9.4 above that all simple submodules of $\mathcal{O}(pV)$ are orthogonal to all simple submodules in $\mathcal{O}(qV^*)$ modulo I. Thus the $\mathrm{GL}(V)$-homomorphism

$$\mathcal{O}(pV)^{U_p} \otimes \mathcal{O}(qV^*)^{U_q} \to (\mathcal{O}(pV \oplus qV^*)/I)^{U_p \times U_q}$$

is surjective, and the same holds if we take invariants under $U := U_p \times U_V \times U_q \subset \mathrm{GL}_p \times \mathrm{GL}(V) \times \mathrm{GL}_q$ where $U_V \subset \mathrm{GL}(V)$ is a maximal unipotent subgroup. This also shows that the $(U_p \times U_q)$-invariants do not change once $p \geq n$ or $q \geq n$, so that we can assume that $p, q \leq n$.

Now we have $\mathcal{O}(pV)^U = \mathbb{C}[x_1, \ldots, x_p]$ where $x_i \in \bigwedge^i V^*$ is a highest weight vector. Similarly, $\mathcal{O}(qV^*)^U = \mathbb{C}[y_1, \ldots, y_q]$, and thus we get a surjective homomorphism

$$\varphi\colon \mathbb{C}[x_1, \ldots, x_p, y_1, \ldots, y_q] \to (\mathcal{O}(pV \oplus qV^*)/I)^U. \qquad (**)$$

Proof (of both Theorem 9.1(1) and (2)). We claim that the kernel of φ is generated by the products $x_r y_s$ where $r + s > n$. This implies that we have an isomorphism

$$(\mathcal{O}(pV \oplus qV^*)/I)^U \simeq \mathbb{C}[x_1, \ldots, x_p, y_1, \ldots, y_q]/(x_i y_j \mid i + j > n),$$

and so $\mathcal{N}_{p,q} := \mathcal{N}(pV \oplus qV^*)$ is reduced, by Proposition 9.5. We also see that the ideal $(x_i y_j \mid i + j > n)$ is prime if and only if it is (0), i.e., when $p + q \leq n$. This proves the theorem for $\mathrm{GL}(V)$.

To prove the claim we first remark that the kernel of φ is spanned by monomials, because φ is equivariant under the action of the maximal torus $T_p \times T_q$. Moreover, it is not difficult to see that $\varphi(x_r y_s) = 0$ if $r + s > n$; see [Sch87, Remark 1.23(2)].

Now let $f := x_{i_1} \cdots x_{i_p} y_{j_1} \cdots y_{j_q}$ be a monomial which is not in the ideal $(x_i y_j \mid i + j > n)$. Then $r + s \leq n$ where $r := \max(p_i)$ and $s := \max(q_j)$. If $(e_1, \dots, e_n)$ is a basis of V and $(e_1^*, \dots, e_n^*)$ the dual basis of V^*, then we can assume that $x_i = e_{n-i+1}^* \wedge e_{n-i+2}^* \wedge \cdots \wedge e_n^*$ and $y_j = e_1 \wedge_2 \wedge \cdots \wedge e_j$. Now it is clear that the monomial f does not vanish at the point $w := (0, \dots, 0, e_{n-r+1}, \dots, e_n, e_1^*, \dots, e_s^*, 0, \dots, 0)$ which is in the null cone $\mathcal{N}_{p,q}$.

For the group $\mathrm{SL}(V)$ there are more invariants, namely the determinants $d_{i_1 \cdots i_n} := \det \begin{bmatrix} v_{i_1} \\ \vdots \\ v_{i_n} \end{bmatrix}$ where $i_1 < i_2 < \cdots < i_n$, and $d^*_{j_1 \cdots j_n} := \det \begin{bmatrix} \xi_{j_1} \\ \vdots \\ \xi_{j_n} \end{bmatrix}$ where $j_1 < j_2 < \cdots < j_n$. These invariants only appear if $p \geq n$, resp. $q \geq n$. In particular, we have the same invariants and the same null cone in case $p, q < n$. From the surjectivity of the map φ in $(**)$ above we see that there remain only the cases where either $p = n$ and $q \leq n$, or $q = n$ and $p \leq n$. Let J denote the ideal generated by the $\mathrm{SL}(V)$-invariants. Then $J^U = I^U + (x_n)$ if $p = n > q$, $J^U = I^U + (y_n)$ if $p < n = q$, and $J^U = I^U + (x_n, y_n)$ if $p = n = q$. Hence

$$(\mathcal{O}(pV \oplus qV^*)/J)^U \simeq \mathbb{C}[x_1, \dots, x_{p'}, y_1, \dots, y_{q'}]/(x_i y_j \mid i + j > n)$$

where $p' := \min(p, n-1)$ and $q' := \min(q, n-1)$. The rest of the proof is as above. $\square$

Next we study the case where V is a symplectic space, i.e., V is equipped with a nondegenerate skew form β, $\dim V = 2n$. We have the group $G := \mathrm{Sp}(V) \subset \mathrm{GL}(V)$ which preserves β. Then the invariants of mV are generated by the bilinear maps

$$\beta_{ij} : (v_1, \dots, v_n) \mapsto \beta(v_i, v_j),\ 1 \leq i < j \leq m.$$

We denote by $\psi_k := \bigwedge_0^k V^* \subset \bigwedge^k V^*$ $(k = 1, \dots, n)$ the fundamental representations of $\mathrm{Sp}(V)$ where $\bigwedge^k V^* = \bigwedge_0^k V^* \oplus \beta \wedge \bigwedge^{k-2} V^*$. We know from equation $(*)$ that $\mathcal{O}(mV)_k^{U_m}$ contains a unique copy of $\bigwedge^k V^*$ for $k \leq \min(m, n)$.

Lemma 9.6. *Let $I \subset \mathcal{O}(mV)$ be the ideal generated by the invariants β_{ij}. Then in $\mathcal{O}(mV)^{U_m}$ we have*

(1) $\bigwedge^k V^* = \psi_k \pmod I$ *for* $k = 1, \dots, \min(m, n)$;
(2) $\bigwedge^k V^* \cdot \bigwedge^\ell V^* = \psi_k \psi_\ell \pmod I$ *for* $1 \leq k \leq \ell \leq \min(m, n)$.

Proof. Part (1) is clear since $\psi_{k+2} = \bigwedge^{k+2} V^*/\beta \bigwedge^k V^*$. For part (2) let $x_1, \dots, x_n \in V^*$ correspond to the positive weights $\varepsilon_1, \dots, \varepsilon_n$ and let $y_1, \dots, y_n$ correspond to the $-\varepsilon_j$. A simple submodule occurring in $\psi_k \cdot \psi_\ell$ has a highest weight vector containing a unique term $\gamma := x_1 \wedge \cdots \wedge x_k \cdot \alpha$ where α is an ℓ-fold wedge product of a certain number of x_i and y_j. But the only possibility for obtaining a highest weight of $\mathrm{Sp}(V)$ is $\alpha = x_1 \wedge \cdots \wedge x_q \wedge y_{k-r+1} \wedge \cdots \wedge y_k$ where $q \leq \ell$ and $r = \ell - q$. This gives the highest weight of a (unique) copy of $\psi_p \psi_q$ where $p = k - r$.

Suppose that $r > 0$. We have an element β_r in $(\bigwedge^r(V^*) \otimes \bigwedge^r(V^*))^{\mathrm{Sp}(V)}$ where $\beta_r(v_1 \wedge \cdots \wedge v_r, w_1 \wedge \cdots \wedge w_r) = \det(\beta(v_i, w_j))$. Here the v_i and w_j are elements of V. It is easy to see that β_r projects to a nontrivial invariant element β_r' of $\psi_r \psi_r$, and that $\beta_r' \in I^r$. Then the product of β_r' with $\psi_p \psi_q \subset \psi_p \cdot \psi_q$ is a copy of $\psi_p \psi_q$ in $\psi_k \psi_\ell$, and we have (2). □

Proof (of Theorem 9.1(3)). It follows from the lemma above and Proposition 9.4 that all simple submodules in $\mathcal{O}(mV)^{U_m}$ are orthogonal and the covariants are generated by $\psi_1, \dots, \psi_{m'}$ where $m' := \min(m, n)$. Let $U_V \subset \mathrm{Sp}(V)$ be a maximal unipotent subgroup and let $x_k \in \bigwedge_0^k V^* \subset \mathcal{O}(mV)_k^{U_m}$ be a highest weight vector. Then we have a surjective homomorphism

$$\varphi \colon \mathbb{C}[x_1, \dots, x_{m'}] \to (\mathcal{O}(mV)/I)^{U_m \times U_V}.$$

If $W \subset V$ is a maximal isotropic subspace, then $W^{\oplus m}$ is contained in the null cone of mV, and for a suitable choice of W, the function x_k does not vanish on $W^{\oplus m}$ for $k \leq m'$. This implies that φ is an isomorphism, because the action of T_m on $\mathbb{C}[x_1, \dots, x_{m'}]$ has one-dimensional weight spaces, and so the kernel of φ is linearly spanned by monomials. Now the theorem for Sp_n follows from Proposition 9.5. □

Finally, let V be a quadratic space, i.e., an n-dimensional vector space with a nondegenerate quadratic form q. The $\mathrm{O}(V)$-invariants of mV are generated by the bilinear maps

$$q_{ij} \colon (v_1, \dots, v_m) \mapsto q(v_i, v_j), \; 1 \leq i \leq j \leq m.$$

The $\mathrm{SO}(V)$-modules $\psi_k := \bigwedge^k V^*$ are simple if $2k < n$. For $n = 2m$ $\psi_m := \bigwedge^m V^*$ is simple as an $\mathrm{O}(V)$-module, but decomposes as $\psi_m = \psi_m^+ \oplus \psi_m^-$ as an $\mathrm{SO}(V)$-module.

Lemma 9.7. *Let $2m \leq n$ and let $I \subset \mathcal{O}(mV)$ be the ideal generated by the invariants q_{ij}. Then in $\mathcal{O}(mV)^{U_m}$ we have*

(1) $\psi_k \cdot \psi_\ell = \psi_k \psi_\ell \pmod{I}$ *for* $1 \leq k \leq \ell \leq \min(m, \frac{n-1}{2})$;
(2) *If $n = 2m$, then* $\psi_m^+ \cdot \psi_m^- = 0 \pmod{I}$.

Proof. Let $n = 2s$ or $2s + 1$ so that $m \leq s$. We consider a weight basis $x_1, \dots, x_s$ and $y_1, \dots, y_s$ (and a zero weight element z if n is odd). First suppose that n is even. For (1) we can then proceed as in the symplectic case. The only difference is that we use the invariant bilinear form q to generate an element q_r' lying in $(\psi_r\psi_r)^{\mathrm{SO}(V)}$ and in I^r. As for (2), the highest weight vectors are $x_1\wedge\cdots\wedge x_m$ and $x_1\wedge\cdots\wedge x_{m-1}\wedge y_m$. Their product is the image of $q_1'\otimes\psi_{m-1}\psi_{m-1}$ in $\psi_m^+\psi_m^-$. The argument of (1) shows that any other irreducible occurring in $\psi_m^+\cdot\psi_m^-$ also lies in I.

Now suppose that n is odd. Then the argument for (1) above goes through except when the zero weight vector appears in the expression for α. So suppose that $\alpha = x_1\wedge\cdots\wedge x_{\ell-1}\wedge z$. Then

$$x_1\wedge\cdots\wedge x_k\cdot\alpha + (x_1\wedge\cdots\wedge x_{\ell-1}\wedge x_{\ell+1}\cdots\wedge x_k\wedge z)\cdot(x_1\wedge\cdots\wedge x_\ell)$$

is a vector in $\psi_k\cdot\psi_\ell$. It is obtained from $(x_1\wedge\cdots\wedge x_k)\cdot(x_1\wedge\cdots\wedge x_\ell)$ by applying elements of U^-. Hence we don't have a new irreducible component of $\psi_k\cdot\psi_\ell$. □

Proof (of Theorem 9.1(4)). Choose highest weight vectors $x_k\in\bigwedge^k V^*$ for $2k<n$ and $x_m^+\in\psi_m^+$, $x_m^-\in\psi_m^-$ for $n=2m$. The preceding lemma and Proposition 9.4 show that the induced maps

$$\mathbb{C}[x_1,\dots,x_m]\to(\mathcal{O}(mV)/I)^{U_m\times U_V}\ \text{for } 2m<n,\ \text{and}$$
$$\mathbb{C}[x_1,\dots,x_m^+,x_m^-]/(x_m^+x_m^-)\to(\mathcal{O}(mV)/I)^{U_m\times U_V}\ \text{for } 2m=n$$

are surjective. The weights $\nu(x_k)$ of the highest weight vectors (with respect to $T_m\times T_V$, T_V a maximal torus of $\mathrm{SO}(V)$) are linearly independent, except that in case $n=2m$ we have $\nu(x_m^+)+\nu(x_m^-)=2\nu(x_{m-1})$. It follows that the algebras on the left-hand side are multiplicity free, and so the kernels of the two maps are spanned by monomials. But it is easy to see that none of the x_k, $x_m^\pm$ vanish on the null cone, and so the two maps are isomorphisms. Again using Proposition 9.5 we obtain the theorem for the groups $\mathrm{O}(V)$ and $\mathrm{SO}(V)$ in the case where $2m\le n$.

It remains to show that the null cone is not reduced for $2m>n$. Let $n=2k$. (The case $n=2k-1$ is similar and will be left to the reader.) We may take $m=k+1$. Then in degree $k+1$ we find the submodule $M:=\bigwedge^{k+1}\mathbb{C}^{k+1}\otimes\bigwedge^{k+1}V^*$, by CAUCHY's formula (Proposition 9.2). The $\mathrm{SO}(V)$-module $\bigwedge^{k+1}V^*$ is simple and isomorphic to $\psi_{k-1}=\bigwedge^{k-1}V$. We claim that M vanishes on the null cone $\mathcal{N}$, but is not contained in the ideal I generated by the invariants.

The first part is clear, because $\mathcal{N}=\mathrm{O}(V)\cdot(k+1)W$ where $W\subset V$ is a maximal isotropic subspace, and every function $f_1\wedge\cdots\wedge f_{k+1}$ vanishes on $(k+1)W$ because $\dim W=k$.

For the second part, we remark that the module ψ_{k-1} appears the first time in degree $k-1$, in the form $\bigwedge^{k-1}\mathbb{C}^{k+1}\otimes\bigwedge^{k-1}V^*$. If $M\subset I$, then M must belong to the product

$$\mathcal{O}((k+1)V)_2^{\mathrm{SO}(V)}\cdot(\textstyle\bigwedge^{k-1}\mathbb{C}^{k+1}\otimes\bigwedge^{k-1}V^*)$$

which is a quotient of $(S^2(\mathbb{C}^{k+1})\otimes\bigwedge^{k-1}\mathbb{C}^{k+1})\otimes\bigwedge^{k-1}V^*$. But the tensor product $S^2(\mathbb{C}^{k+1})\otimes\bigwedge^{k-1}\mathbb{C}^{k+1}$ does not contain the "determinant" $\bigwedge^{k+1}\mathbb{C}^{k+1}$ as a GL_{k+1}-module. □

10 Non-reduced components of the null cone

We need some information about null cones (see [Hes80, KW06, Ric89] for more details). Let G be a connected reductive complex group, $T \subset G$ a maximal torus and V a G-module. Let $X(T) = \mathrm{Hom}(T, \mathbb{C}^*)$ denote the character group of T and let $Y(T) = \mathrm{Hom}(\mathbb{C}^*, T)$ denote the group of 1-parameter subgroups of T. Then $Y(T)$ and $X(T)$ are dually paired: $\langle \rho, \mu \rangle = n$ if $\mu(\rho(t)) = t^n$. For any $\rho \in Y(T)$ we set

$$Z_\rho := \{v \in V \mid \lim_{t\to 0} \rho(t)v = 0\} = \bigoplus_{\mu\in X(T), \langle\rho,\mu\rangle>0} V_\mu$$

where $V_\mu \subset V$ denotes the weight space of weight μ. These Z_ρ are called *positive weight spaces.* Then the Hilbert–Mumford theorem says that $\mathcal{N}$ is the union of the sets GZ_ρ, $\rho \in Y(T)$. In fact, one needs only a finite number of elements of $Y(T)$. Pick a system of simple roots for G. Then using the action of the Weyl group, we can assume that any given ρ is positive when paired with the simple roots $\alpha_1, \ldots, \alpha_\ell \in X(T)$, $\ell = \dim T$. In fact, we can always assume that the pairings are strictly positive and that ρ only takes the value 0 on the zero weight. We call such elements of $Y(T)$ *generic*. Now Z_ρ is stable under the action of the Borel B, thus GZ_ρ is closed in the Zariski topology, and GZ_ρ is irreducible. Thus there are finitely many generic ρ_i such that the sets GZ_{ρ_i} are the irreducible components of $\mathcal{N}$.

Remark 10.1. We will use this description of the null cone to show that a given homogeneous covariant $\tau: V \to W$ of degree d vanishes on the null cone, generalizing Lemma 3.8. It suffices to show that τ vanishes on Z_ρ for the relevant generic ρ's. Denote by $\mu_1, \ldots, \mu_m$ the weights of Z_ρ. *If $\tau \neq 0$ on Z_ρ, then the highest weight μ of W is of the form $\sum_i d_i\mu_i$ where $\sum_i d_i = d$.* (This follows from the B-equivariance of τ.) Hence τ vanishes if μ cannot be expressed as such a sum.

Let $\Lambda(V)$ denote the set of weights of V. For $\rho \in Y(T)$, let Λ_ρ denote the subset of $\Lambda(V)$ of elements which pair strictly positively with ρ. A subset $\Lambda \subset \Lambda(V)$ is called *admissible* if $\Lambda = \Lambda_\rho$ for a generic ρ. In this case set $Z_\Lambda := Z_\rho$. We will often switch between looking at generic elements of $Y(T)$ (or $Y(T) \otimes \mathbb{Q}$) and corresponding subsets $\Lambda \subset \Lambda(V)$. We say that an admissible Λ is *dominant* if GZ_Λ is a component of the null cone.

Here is a way to show that the null cone $\mathcal{N}$ is not reduced.

Proposition 10.2. *Let $\Lambda \subset \Lambda(V)$ be dominant and let $W \subset V$ be a T-stable complement of Z_Λ. Assume that for any $z \in Z_\Lambda$ the differential $d\pi_z$ restricted to W has rank $<$ $\operatorname{codim}_V GZ_\Lambda$, or, equivalently, there is a subspace $W' \subset W$ of dimension $>$ $\operatorname{codim}_{GZ_\Lambda} Z_\Lambda$ such that the differential of any invariant vanishes on W'. Then no point of $GZ_\Lambda \subset \mathcal{N}$ is reduced.*

Proof. Either condition implies that the rank of $d\pi_z$ is less than the codimension of GZ_Λ for any $z \in Z_\Lambda$. □

Remark 10.3. Let $(v_1, \dots, v_n)$ be a basis of V consisting of weight vectors of weight $\mu_1, \dots, \mu_n$, and let $(x_1, \dots, x_n)$ be the dual basis. If $f = x_{i_1} \cdots x_{i_d}$ is a monomial of weight zero, then an x_i such that $\mu_i \notin \Lambda$ has to appear. If two such x_i appear in f, then clearly $(df)|_{Z_\Lambda} = 0$. This gives our first method to show that $\mathcal{N}$ is not reduced.

(1) Let Λ' be the complement of Λ in $\Lambda(V)$. Let $d \in \mathbb{N}$ be minimal such that every zero weight monomial f containing exactly one factor x_i corresponding to a weight from Λ' has degree $\leq d$.
(2) Show that there are not enough invariants of degree $\leq d$, i.e., show that the number of invariants of degree $\leq d$ is strictly less than the codimension of GZ_Λ.

If W is irreducible of highest weight λ, we denote the highest weight of the dual representation, W^* by λ^*. The next result will give us another way to see if the null cone is not reduced. It uses the method of covariants introduced in section 3 (see Proposition 3.1).

Proposition 10.4. *Let $\varphi\colon V \to W$ be a covariant, where W is irreducible of highest weight λ. Let $\Lambda \subset \Lambda(V)$ be admissible and assume that φ does not vanish on GZ_Λ. Then $\lambda^* \in \mathbb{N}\Lambda$.*

Proof. Let W^* be the subspace of $\mathcal{O}(V)$ corresponding to φ. Let f be a highest weight vector of W^*. Then f has weight λ^* and f does not vanish on GZ_Λ by assumption. It follows that f contains a monomial $m = x_{i_1} x_{i_2} \cdots x_{i_d}$ where the corresponding v_{i_k} all belong to Z_Λ, i.e., $\lambda^* = \mu_{i_1} + \mu_{i_2} + \cdots + \mu_{i_d} \in \mathbb{N}\Lambda$. □

Remark 10.5. This proposition will be used in the following way.

(1) Find a suitable highest weight λ and an integer d such that λ^* cannot be written as a sum of more than d weights from Λ.
(2) Show that there are generating covariants of type W_λ in degree $> d$.

By the proposition above this implies that the generating covariants from (2) vanish on GZ_Λ. In order to apply Proposition 3.1 one has to fix d and check (1) for any admissible Λ.

We finish this section by giving some criteria to find the dominant Λ among the admissible ones. Let Λ_1 and Λ_2 be admissible subsets of $\Lambda(V)$. Set $Z_i := Z_{\Lambda_i}$, $i = 1, 2$. We say that Λ_2 *dominates* Λ_1, and we write $\Lambda_1 \leq \Lambda_2$, if $GZ_1 \subset GZ_2$. Given $\sigma \in \mathcal{W}$, the Weyl group of G, let $\Lambda_1^{(\sigma)} := \{\lambda \in \Lambda_1 \mid \sigma(\lambda) \in \Lambda_2\}$ and let $Z_1^{(\sigma)}$ denote the sum of the weight spaces with weights in $\Lambda_1^{(\sigma)}$.

Lemma 10.6. *Let Λ_1 and Λ_2 be admissible. Then Λ_2 dominates Λ_1 if and only if there is a $\sigma \in \mathcal{W}$ such that $BZ_1^{(\sigma)}$ is dense in Z_1.*

Proof. Suppose that $\Lambda_1 \leq \Lambda_2$. Let $z \in Z_1$. Then there is a $g \in G$ such that $gz \in Z_2$. Write $g = u\sigma b$ where $b \in B$, $u \in U$ and $\sigma \in \mathcal{W}$ (Bruhat decomposition). Since b and u preserve the Z_i, we see that $bz \in Z_1^{(\sigma)}$. Thus Z_1 is the union of the constructible subsets $BZ_1^{(\sigma)}$, $\sigma \in \mathcal{W}$, and one of them must be dense.

Conversely, suppose that some $BZ_1^{(\sigma)}$ is dense in Z_1. Since $\sigma(BZ_1^{(\sigma)})$ lies in GZ_2 and GZ_2 is closed, we see that $GZ_1 \subset GZ_2$. □

The condition that $BZ_1^{(\sigma)}$ is dense in Z_1 has some consequences for the weights of $Z_1^{(\sigma)}$. Denote by Φ^+ the set of positive roots, i.e., the weights of $\mathfrak{b} := \operatorname{Lie} B$.

Lemma 10.7. *Let Z be a B-module and $Z' \subset Z$ a T-stable subspace. If BZ' is dense in Z, then*

$$\Lambda(Z) \subset \Lambda(Z') + (\Phi^+ \cup \{0\}).$$

In particular, $\Lambda(Z')$ contains the set $\Omega := \{\lambda \in \Lambda(Z) \mid \lambda \notin \Lambda(Z) + \Phi^+\}$.

Proof. The tangent map of $B \times Z' \to Z$ at a point (e, z_0) has the form $(X, v) \mapsto Xz_0 + v$, and so $\mathfrak{b}\, Z' + Z' = Z$. If $z \in Z'$ is a weight vector of weight λ, then $\mathfrak{b}\, z \subset \bigoplus_{\omega \in \Phi_+ \cup \{0\}} Z_{\lambda+\omega}$, hence $\Lambda(Z)$ is as claimed. □

Proposition 10.8. *Let $\Lambda_1, \Lambda_2 \subset \Lambda(V)$ be admissible subsets. Define $\Omega_1 := \{\lambda \in \Lambda_1 \mid \lambda \notin \Lambda_1 + \Phi^+\}$ and suppose that $\mathbb{Q}_{\geq 0}\Omega_1$ contains the simple roots. Then $\Lambda_1 \leq \Lambda_2$ implies that $\Lambda_1 \subset \Lambda_2$.*

Proof. Let σ be as in Lemma 10.6. Then $\Lambda_1^{(\sigma)}$ contains Ω_1 by Lemma 10.7. This in turn implies that Λ_2 is positive on $\sigma(\alpha_j)$, $j = 1, \ldots, \ell$. Thus each $\sigma(\alpha_j)$ is a positive root and so σ is the identity. Hence $\Omega_1 \subset Z_2$ and thus $\Lambda_1 \subset \Lambda_2$. □

Corollary 10.9. *Suppose that $G = \mathrm{SL}_3$ with simple roots α and β. Let $\Lambda = \Lambda_\rho \subset \Lambda(V)$ be admissible and maximal with respect to set inclusion. Suppose that Λ contains nonzero weights of the form $\lambda_1 := -a\alpha + b\beta$ and $\lambda_2 := c\alpha - d\beta$ where the coefficients a, b, c and d are nonnegative rational numbers. Then Λ is dominant.*

Proof. Let $\Omega \subset \Lambda$ be the minimal elements. We may assume that λ_1 and λ_2 are in Ω. Clearly $b, c \neq 0$. If $a = 0$ or $d = 0$, then the hypotheses of Proposition 10.8 are satisfied. If $a, d \neq 0$, then $\langle \rho, \lambda_1 \rangle > 0$ and $\langle \rho, \lambda_2 \rangle > 0$ forces $bc - ad > 0$. Thus the inverse of the matrix $\left(\begin{smallmatrix} c & -a \\ -d & b \end{smallmatrix}\right)$ has positive entries, so that the hypotheses of Proposition 10.8 are satisfied and Λ is dominant. □

See Example 11.2 below for a calculation of components of a null cone.

11 Coreduced representations of SL_3

In this section we classify the coreduced representations of $G = \mathrm{SL}_3$ (Theorems 11.10 and 11.12).

We denote the representation $V := \varphi_1^r \varphi_2^s$ by $V[r,s]$, $r, s \in \mathbb{N}$, and its highest weight by $[r,s]$. Let α and β be the simple roots of G. We denote a weight $p\alpha + q\beta$ of a representation of G by (p,q) where p, $q \in (1/3)\mathbb{Z}$ and $p + q \in \mathbb{Z}$. Hence $\alpha = (1,0) = [2,-1]$, $\beta = (0,1) = [-1,2]$, and so

$$[r,s] = (\frac{2r+s}{3}, \frac{r+2s}{3}) \text{ and } (p,q) = [2p-q, 2q-p].$$

Moreover, $[r,s]$ is in the root lattice if and only if $r \equiv s \mod 3$.

We leave the following lemma to the reader (see Lemma 5.6).

Lemma 11.1. *Let $V := V[r,s]$ be an irreducible representation of G and set $(p,q) = [r,s]$.*

(1) *The dominant weights of $V[r,s]$ are the weights $[r',s']$ obtained starting with $[r,s]$ and using the following inductive process: $[r',s']$ gives rise to $[r'-2,s'+1]$ if $r' \geq 2$ and to $[r'+1,s'-2]$ if $s' \geq 2$. Finally, $[1,1]$ gives rise to $[0,0]$. Equivalently, the dominant weights of $V[r,s]$ are those of the form $(k,l) := (p-a, q-b)$ where a, $b \in \mathbb{N}$, $0 \leq k \leq 2l$ and $0 \leq l \leq 2k$.*
(2) *The Weyl group orbit of the dominant weight (k,l) is*

   a. (k,l), $(l-k,l)$, $(k,k-l)$, $(l-k,-k)$, $(-l,k-l)$, $(-l,-k)$ if $k \neq 2l$ and $l \neq 2k$,
   b. $(2l,l)$, $(-l,l)$, $(-l,-2l)$ if $k = 2l$ and
   c. $(k,2k)$, $(k,-k)$ and $(-2k,-k)$ if $l = 2k$.

(3) *Let (p,q) be dominant, $p \neq q$, and let $\mathcal{W} \cdot (p,q)$ be the Weyl group orbit of (p,q). Then*

$$\max\left\{\frac{-k}{\ell} \mid (k,\ell) \in \mathcal{W} \cdot (p,q)\right\} = \frac{\min(p,q)}{|p-q|},$$

$$\min\left\{\frac{-k}{\ell} \mid (k,\ell) \in \mathcal{W} \cdot (p,q),\ \frac{-k}{\ell} > 0\right\} = \frac{|p-q|}{\min(p,q)}.$$

Suppose that $\Lambda(V)$ is not contained in the root lattice. Then let Λ_α denote the weights (p,q) of V where $p > 0$. We define Λ_β similarly. Note that Λ_α is stable under the simple reflection σ_β and that Λ_β is stable under the simple reflection σ_α.

Example 11.2. Consider the module $V = V[3,1]$. Then the dominant weights are $[3,1]$, $[1,2]$, $[2,0]$ and $[0,1]$. Thus the weights of V are

(1) $(7/3, 5/3)$, $(-2/3, 5/3)$, $(7/3, 2/3)$, $(-2/3, -7/3)$, $(-5/3, 2/3)$, $(-5/3, -7/3)$ (the $\mathcal{W}$-orbit of $[3,1]$);
(2) $(4/3, 5/3)$, $(1/3, 5/3)$, $(4/3, -1/3)$, $(1/3, -4/3)$, $(-5/3, -1/3)$, $(-5/3, -4/3)$ (the $\mathcal{W}$-orbit of $[1,2]$);
(3) $(4/3, 2/3)$, $(-2/3, 2/3)$, $(-2/3, -4/3)$ (the $\mathcal{W}$-orbit of $[2,0]$);
(4) $(1/3, 2/3)$, $(1/3, -1/3)$, $(-2/3, -1/3)$ (the $\mathcal{W}$-orbit of $[0,1]$).

Let $\rho \in Y(T) \otimes \mathbb{Q}$ be generic. We may assume that $\rho(\alpha) = 1$ and, of course, we have $\rho(\beta) > 0$. Then $\rho(\beta)$ has to avoid the values $2/5$, $5/2$, 4, $1/4$ and 1, so there are six cases to consider.

Case 1: Let Λ_1 correspond to $2/5 < \rho(\beta) < 1$. It is easy to see that Λ_1 is maximal. Then Λ_1 is dominant by Corollary 10.9 since $(-2/3, 5/3)$ and $(1/3, -1/3)$ are ρ-positive.

Case 2: Let Λ correspond to $0 < \rho(\beta) < 1/4$. Then $\Lambda = \Lambda_\alpha$ is $\sigma := \sigma_\beta$-stable so that $\sigma(\Lambda^{(\sigma)}) = \sigma(\Lambda \cap \Lambda_1)$. Now $\Lambda \cap \Lambda_1$ is $\Lambda \setminus \{(1/3, -4/3)\}$, hence $\Lambda^{(\sigma)}$ is $\Lambda \setminus \{(1/3, 5/3)\}$ where $(1/3, 5/3)$ has multiplicity one. Thus $UZ_\Lambda^{(\sigma)}$ is dense in Z_Λ so that $\Lambda < \Lambda_1$. (One can also see directly that $U^- Z_1$ has Z_Λ in its closure.) Now it is easy to calculate that $\dim GZ_\Lambda < \dim GZ_1$, so that $\Lambda = \Lambda_\alpha$ is not dominant.

Case 3: Let Λ correspond to $1/4 < \rho(\beta) < 2/5$. Then $\Lambda \subset \Lambda_1$.

Case 4: Let Λ_2 correspond to $5/2 < \rho(\beta) < 4$. Then Λ_2 is maximal and $(-5/3, 2/3)$ and $(4/3, -1/3)$ are ρ-positive, so that Λ_2 is dominant by Corollary 10.9.

Case 5: Let Λ correspond to $1 < \rho(\beta) < 5/2$. Then $\Lambda \subset \Lambda_1$.

Case 6: Let Λ correspond to $\rho(\beta) > 4$. Then $\Lambda = \Lambda_\beta$ and as in Case 2 we see that $\Lambda < \Lambda_1$ and that Λ is not dominant.

Thus there are only two components of the null cone, GZ_{Λ_1} and GZ_{Λ_2} corresponding to cases 1 and 4. Note that neither Λ_α nor Λ_β is dominant.

Lemma 11.1 does not tell us anything about multiplicities of weights, but the following result gives us some lower bounds, which suffice for our uses. If $[r, s]$ is a weight of V, then we denote by $V_{[r,s]} \subset V$ the corresponding weight space.

Lemma 11.3. *Let $r = r_0 + r'$ and $s = s_0 + s'$ where $r' \equiv s' \mod 3$. Then every weight of $V[r_0, s_0]$ occurs in $V[r, s]$ with multiplicity at least the dimension of the zero weight space $V[r', s']_{[0,0]}$.*

Proof. As in Lemma 5.18 this follows from the fact that $\mathcal{O}(G/U)$ is a domain and that the product of the copies of $V[r_0, s_0]$ and $V[r', s']$ in $\mathcal{O}(G/U)$ is just the copy of $V[r, s]$ □

Example 11.4. Consider $V[3, 2]$. Then the multiplicity of $[1, 0]$ is at least the multiplicity of the zero weight in $V[2, 2]$, which is 3. The multiplicity of $[2, 1]$ is similarly seen to be at least 2. Thus the multiplicities of the dominant weights of $V[3, 2]$ are at least as follows: $[3, 2]$, $[4, 0]$, $[1, 3]$ and $[0, 2]$ with multiplicity one, $[2, 1]$ with multiplicity two and $[1, 0]$ with multiplicity three. In fact, these multiplicities are correct, except that $[0, 2]$ actually has multiplicity two.

In Example 11.2 we have seen that neither Λ_α nor Λ_β is dominant. But this is an exception as shown by the following result.

Lemma 11.5. *Let $V = V[r, s]$ where $r \geq s$.*

(1) *If $r - s \equiv 1 \mod 3$, then Λ_β is dominant.*
(2) *If $r - s \equiv 2 \mod 3$ and $[r, s] \neq [3, 1]$ or $[5, 0]$, then Λ_α is dominant.*

Proof. For $t > 0$ define $\rho_t \in Y(T)$ by $\rho_t(\alpha) = 1$ and $\rho_t(\beta) = t$, and set $\Lambda_t := \Lambda_{\rho_t}$. Define

$$\mathcal{T} := \{t > 0 \mid \rho_t(\lambda) = 0 \text{ for some } \lambda \in \Lambda(V[r,s]), \lambda \neq 0\}.$$

We have $\mathcal{T} = \{t_1, t_2, \ldots, t_m\}$ where $0 < t_1 < t_2 < \cdots < t_m$, so there are $m+1$ admissible subsets $\Lambda^{(i)}$, $i = 0, \ldots, m$, defined by $\Lambda^{(i)} := \Lambda_t$ for $t_i < t < t_{i+1}$, where $t_0 = 0, t_{m+1} = \infty$. Clearly, $\Lambda^{(0)} = \Lambda_\alpha$, $\Lambda^{(m)} = \Lambda_\beta$, and Λ_α (resp. Λ_β) is not maximal if and only if $\Lambda_\alpha \subset \Lambda^{(1)}$ (resp. $\Lambda_\beta \subset \Lambda^{(m-1)}$). Note that if $\rho_t((k,l)) = 0$, then $t = -k/l$.

(1) First suppose that $r - s \equiv 1 \mod 3$ and let $(p,q) = [r,s]$. Then $[1,0] = (2/3, 1/3)$ is a weight of V, and the α-string through $[1,0]$ has the form

$$\Sigma = ((-q, 1/3), (-q+1, 1/3), \ldots, (2/3, 1/3), \ldots, (q+1/3, 1/3))$$

where $(-q, 1/3)$ is in the $\mathcal{W}$-orbit of $(q+1/3, q)$. Note that $\#\Sigma = 2q + 4/3$. Since the case $V = V[1,0]$ is obvious we can assume that $q \geq 4/3$, hence $\#\Sigma \geq 4$.

Claim 1: We have $t_m = 3q$ and $t_{m-1} = 3q - 3$, and these values are attained at the first two weights $(-q, 1/3)$ and $(-q+1, 1/3)$ of the α-string Σ. In particular, $\Lambda_\beta \supset \Lambda^{(m-1)}$ and $\#(\Sigma \cap \Lambda^{(m')}) \leq \#\Sigma - 2$ for $m' \leq m-2$.

This implies that Λ_β is dominant. In fact, suppose that $\Lambda_\beta < \Lambda$ for some admissible Λ. Set $Z_\beta := Z_{\Lambda_\beta}$. There is a $\sigma \in \mathcal{W}$ such that $BZ_\beta^{(\sigma)}$ is dense in Z_β and $\sigma(\Lambda_\beta^{(\sigma)}) \subset \Lambda$ (Lemma 10.6). Clearly $\Lambda_\beta^{(\sigma)}$ has to contain a subset Σ' of the α-string Σ which omits at most one element and contains $(-q, 1/3)$ (see Lemma 10.7). Since Σ' contains at least 3 elements it is easy to see that $\sigma = e$ and $\sigma = \sigma_\alpha$ are the only elements from $\mathcal{W}$ which send Σ' to elements which have at least one positive α or β coefficient. Thus $\sigma(\Sigma') \subset \Lambda \cap \Sigma$. By the claim above, this implies that $\Lambda = \Lambda_\beta$ or $\Lambda = \Lambda^{(m-1)}$ and so $\Lambda \subset \Lambda_\beta$.

(2) Now suppose that $r - s \equiv 2 \mod 3$. Then $[0,1] = (1/3, 2/3)$ is a weight of V, and the β-string through $[0,1]$ has the form

$$\Sigma = ((1/3, -q+1/3), (1/3, -q+4/3), \ldots, (1/3, 2/3), \ldots, (1/3, q))$$

where $(1/3, -q+1/3)$ is in the $\mathcal{W}$-orbit of $(q-1/3, q)$. Note that $\#\Sigma = 2q + 4/3$.

Claim 2: If $\#\Sigma \geq 6$ (i.e., $q \geq 8/3$), then $t_1 = 1/(3q-1)$ and $t_2 = 1/(3q-4)$, and these values are attained at the first two weights $(1/3, -q+1/3)$ and $(1/3, -q+4/3)$ of the β-string Σ. Moreover, $\Lambda_\alpha \supset \Lambda^{(1)}$ and $\#(\Sigma \cap \Lambda^{(m')}) \leq \#\Sigma - 2$ for $m' \geq 2$.

Now the same argument as above implies that Λ_α is dominant. Note that the condition $q \geq 8/3$ is satisfied for $[r,s] \neq [2,0]$, $[3,1]$ or $[5,0]$. For $V[2,0]$ there are only two admissible sets, Λ_α and Λ_β, both are dominant and $\mathcal{N} = GZ_\alpha = GZ_\beta$.

(3) It remains to prove the two claims. Let $r - s \equiv 1 \mod 3$. We use the first formula given in Lemma 11.1(3) for a dominant (p', q'):

$$\mu_{(p',q')} := \max\left\{\frac{-k}{\ell} \mid (k,\ell) \in \mathcal{W}\cdot(p',q')\right\} = \frac{\min(p',q')}{|p'-q'|}$$

By assumption we have $q \geq 4/3$. If $(p', q') \leq (p, q)$ is dominant, then $|p' - q'| \geq 1/3$. Thus

$$t_m = \max(\mu_{(p',q')} \mid (p',q') \text{ dominant}, (p',q') \leq (p,q)) = \mu_{(q+1/3,q)} = 3q,$$

and this value is attained at a single weight of V, namely at the weight $(-q, 1/3) \in \mathcal{W}\cdot(q+1/3, q)$. It follows that t_{m-1} is either equal to $\mu_{(q-2/3,q-1)} = 3(q-1)$ or equal to $\mu_{(p',q)}$ for a suitable $p' \leq p$, $p' \neq q+1/3$. But then $p' = q-2/3$ or $p' = q+4/3$ and in both cases we get $\mu_{(p',q)} \leq 3(q-1)$, because $q \geq 4/3$. Hence $t_{m-1} = 3(q-1)$ and this value is attained at the weight $(-q+1, 1/3) \in \mathcal{W}\cdot(q-2/3, q-1)$. As a consequence, $\Lambda_\beta \supset \Lambda^{m-1} = \Lambda_\beta \setminus \{(-q, 1/3)\}$, and $(-q, 1/3), (-q+1, 1/3) \notin \Lambda^{m'}$ for $m' \leq m-2$. This proves Claim 1.

For $r - s \equiv 2 \mod 3$ we use the second formula in Lemma 11.1(3) for a dominant (p', q'):

$$\nu_{(p'q')} := \min\left\{\frac{-k}{\ell} \mid (k,\ell) \in \mathcal{W}\cdot(p',q'),\ \frac{-k}{\ell} > 0\right\} = \frac{|p'-q'|}{\min(p',q')}.$$

The minimal values of $|p'-q'|$ are $1/3$ and $2/3$ and they are attained at $(q'-1/3, q')$ and $(q'+2/3, q')$. Thus, for a fixed q' the minimal values of $\nu_{(p',q')}$ are $1/(3q'-1)$ and $2/(3q')$. Since $q \geq 8/3 > 4/3$ we get

$$t_1 = \min\big(\nu_{(p',q')} \mid (p',q') \leq (p,q) \text{ dominant}\big) = \nu_{(q-1/3,q)} = 1/(3q-1),$$

and this value is attained at a single weight, namely at $(1/3, -q+1/3) \in \mathcal{W}\cdot(q-1/3, q)$. It follows that t_2 is either equal to $\nu(q+2/3, q) = 2/(3q)$ or equal to $\nu(q-4/3, q-1) = 1/(3q-4)$. Since $q \geq 8/3$ we get $3q-4 = (3/2)q + ((3/2)q - 4) \geq (3/2)q$. Hence $t_2 = 1/(3q-4)$ and this value is attained at $(1/3, -q+4/3) \in \mathcal{W}\cdot(q-4/3, q-1)$. Now Claim 2 follows as above. □

Remark 11.6. Let $\Lambda = \Lambda_\alpha$ or Λ_β. Then Z_Λ is stabilized by a parabolic subgroup of codimension 2, hence $\operatorname{codim}_{GZ_\Lambda} Z_\Lambda \leq 2$.

We need the following estimate on the dimension of $S^3(V)^G$:

Proposition 11.7. *Let $r \geq s \geq 0$. Then*

(1) *The multiplicity of $[r-s, 0]$ in $V[r,0] \otimes V[0,s]$ is $\binom{s+2}{2}$.*
(2) *The multiplicity of $[r-s, 0]$ in $V[r,s]$ is $s+1$.*
(3) *The multiplicity of $V[s,r]$ in $V[r,s] \otimes V[r,s]$ is at most $s+1$.*
(4) *The dimension of $S^3(V[r,s])^G$ is at most $s+1$, hence there are at most $s+1$ linearly independent cubic invariants of $V[r,s]$.*

Proof. Let e_1, e_2 and e_3 be the usual basis of $\mathbb{C}^3$ and let f_1, f_2, f_3 be the dual basis. Then the weight vectors of weight $[r-s,0]$ in $V[r,0]\otimes V[0,s]$ have basis the vectors $e_1^{r-t}m \otimes f_1^{s-t}m^*$, where $0 \le t \le s$ and m is a monomial of degree t in e_2 and e_3 and m^* is the same monomial in f_2 and f_3. Thus the dimension of this weight space is $1+\cdots+(s+1)$, giving (1).

Part (2) follows from the fact that

$$V[r,0]\otimes V[0,s] = V[r,s] \oplus V[r-1,0]\otimes V[0,s-1].$$

This is an immediate consequence of Pieri's formula (see [Pro07, formula (10.2.2) in 9.10.2]).

The multiplicity of $V[s,r]$ in $V[r,s]\otimes V[r,s]$ is bounded by the multiplicity of the weight $[r,s]-(r,s)$ in $V[r,s]$ since $[r,s]+([r,s]-(r,s)) = [s,r]$. Now $[r,s]-(r,s) = 1/3(-r+s, r-s)$ which is in the $\mathcal{W}$-orbit of $1/3(2r-2s, r-s) = [r-s,0]$. Thus (2) implies (3). Clearly (3) implies (4). □

Example 11.8. Assume that $r \ge s \ge 1$ and that $r-s \equiv 2 \mod 3$. Then the multiplicities of the weights of $V[0,1]$ and $V[3,1]$ in $V[r,s]$ are $\ge s$, and the multiplicities of the weights of $V[2,0]$ are $\ge s+1$ in case $r \ge 5$.

(In fact, for $V[3,1]$ the multiplicities are $\ge \dim V[r-3,s-1]_{[0,0]}$ by Lemma 11.3 and $\dim V[r-3,s-1]_{[0,0]} \ge \dim V[r-3,s-1]_{[r-s-2,0]} = s$ by Proposition 11.7(2). The other cases follow by similar arguments.)

Proposition 11.9. *Let $V = V[r,s]$ where $r+s \ge 4$ or $(r,s) = (2,1)$ or $(r,s) = (1,2)$. Then there is an irreducible component $\mathcal{N}_1$ of $\mathcal{N}$ such that the rank of $d\pi$ is less than the codimension of $\mathcal{N}_1$ in V on $\mathcal{N}_1$. In particular, $\mathcal{N}$ is not reduced.*

An immediate consequence is

Theorem 11.10. *Let V be an irreducible representation of $G = \mathrm{SL}_3$. Then V is coreduced if and only if V is on the following list:*

(1) $V[1,0]$, $V[2,0]$, $V[3,0]$;
(2) $V[0,1]$, $V[0,2]$, $V[0,3]$;
(3) $V[1,1]$.

Equivalently, V is coreduced if and only if it is cofree.

Proof (of Proposition 11.9). We may assume that $V = V[r,s]$ where $r \ge s$ and $V[r,s]$ does not appear in (1), (2) or (3) of the theorem. Let $(p,q) = [r,s]$. We are constantly applying Remarks 10.3 and 10.5.

Case 1: Assume that $r-s \equiv 1 \mod 3$ and consider $\Lambda = \Lambda_\beta$ which is dominant by Lemma 11.5. Recall that $\operatorname{codim}_{GZ_\Lambda} Z_\Lambda \le 2$. First suppose that $s \ge 1$ and $r > 2$. Then $[1,3]$ and $[0,2]$ are weights of V. Let $\rho \in Y(T)$ correspond to Λ_β. Then ρ is negative on the weights $(2/3,-2/3)$ and $(-4/3,-2/3)$ in the $\mathcal{W}$-orbit of $[0,2]$, on the weights $(-7/3,-2/3)$ and $(5/3,-2/3)$ in the $\mathcal{W}$-orbit of $[1,3]$ and on the weight $(-1/3,-2/3)$ in the $\mathcal{W}$-orbit of $[1,0]$ which occurs with multiplicity at least $s + 1$ since $[r - s, 0]$ has multiplicity $s + 1$ by Proposition 11.7(2). These negative weights can be paired with at most quadratic expressions in the positive weights (just look at the coefficients of β). Now there are at most $s + 1$ cubic invariants (and no quadratic invariants), hence $\mathcal{N}$ is not reduced if $s \geq 1, r > 2$.

If $(r, s) = (2, 1)$, then we have the following negative weights: $(2/3, -2/3)$, $(-4/3, -2/3)$ and $(-1/3, -2/3)$ (with multiplicity 2). There is only a one-dimensional space of degree 3 invariants, and so $\mathcal{N}$ is not reduced.

If $s = 0$, then the cases to consider are $V[4, 0]$, $V[7, 0]$, etc. If $r \geq 7$, then we have a dominant weight $[1, 3]$ whose $\mathcal{W}$-orbit contains $(-7/3, -2/3)$ and $(5/3, -2/3)$. We still have $(2/3, -2/3)$, $(-4/3, -2/3)$ and $(-1/3, -2/3)$. Since there is at most one degree 3 invariant, $\mathcal{N}$ is not reduced.

We are left with the case of $V[4, 0]$. Here we have negative weights $(2/3, -2/3)$, $(-4/3, -2/3)$ and $(-1/3, -2/3)$ as well as $(-1/3, -5/3)$ and $(-4/3, -5/3)$ in the $\mathcal{W}$-orbit of $[2, 1]$. Thus $\mathcal{N}$ is not reduced since there are only two irreducible invariants of degree ≤ 6 (the Poincaré series of $\mathcal{O}(V)^G$ is $1 + t^3 + 2t^6 + \dots$).

Case 2: Assume that $r - s \equiv 2 \mod 3$. For the cases $[r, s] = [3, 1]$ or $[5, 0]$ see Example 11.11 below. So we may assume that $\Lambda = \Lambda_\alpha$ is dominant. If $s \geq 1$ (and so $r \geq 5$), then among the dominant weights we have $[3, 1]$ with multiplicity at least s, $[2, 0]$ with multiplicity at least $s + 1$ and $[0, 1]$ with multiplicity at least s (see Example 11.8). The $\mathcal{W}$-orbit of $[3, 1]$ contains the weights $(-2/3, 5/3)$ and $(-2/3, -7/3)$ with negative α-coefficient, the $\mathcal{W}$-orbit of $[2, 0]$ contains $(-2/3, 2/3)$ and $(-2/3, -7/3)$ and the $\mathcal{W}$-orbit of $[0, 1]$ contains $(-2/3, -1/3)$. Since there is at most an $(s + 1)$-dimensional space of degree 3 invariants, $\mathcal{N}$ is not reduced. If $s = 0$ (and so $r \geq 5$), then we have the weights $[3, 1]$, $[2, 0]$ and $[0, 1]$ with multiplicity one, and $\mathcal{N}$ is not reduced because $\dim S^3(V)^G \leq 1$.

Case 3: If $r - s \equiv 0 \mod 3$, then we are in the adjoint case and the claim follows from Proposition 7.5. □

Example 11.11. Let $V = V[3, 1]$. Then from Example 11.2 we see that there are two dominant Λ, one corresponding to $\rho(\alpha) = 1$ and $2/5 < \rho(\beta) < 1$ (choose $\rho(\beta) = 1/2$) and the other to $\rho(\alpha) = 1$ and $5/2 < \rho(\beta) < 4$ (choose $\rho(\beta) = 3$). Neither Λ_α nor Λ_β is dominant. Consider the case where $\rho(\beta) = 1/2$. Then the minimal positive weights (in terms of their ρ-value) are $(1/3, -1/3)$ and $(-2/3, 5/3)$, both having ρ-value $1/6$. Now consider the covariants of type $V[1, 0]$. The highest weight is $(2/3, 1/3)$ where $\rho(2/3, 1/3) = 5/6$. Thus the highest degree in which the covariant could occur in $S^*(V)$ and not vanish on GZ_ρ is 5. One gets the same bound in case $\rho(\beta) = 3$. The Poincaré series of the invariants is $1+t^3+\dots$ and for the $V[1, 0]$ covariants it is $4t^5 + 44t^8 + \dots$. Thus there are generating covariants in degree 8, which vanish on $\mathcal{N}$, so that $\mathcal{N}$ is not reduced.

If $V = V[5, 0]$, then the calculations of Example 11.2 show that the dominant Λ again correspond to $\rho(\beta) = 1/2$ or 3. (The only new weights are (10/3,5/3), (−5/3,5/3) and (−5/3, −10/3) and they give rise to no new ratios.) Hence the highest degree in which the covariant $V[1, 0]$ could occur in $S^*(V)$ and not vanish on GZ_ρ is again 5. The covariant $V[1, 0]$ first occurs in degree 5, with multiplicity one. But since the principal isotropy group of V is trivial, the $V[1,0]$ covariants have to have generators in higher degree, and these necessarily vanish on $\mathcal{N}$. Thus $\mathcal{N}$ is not reduced.

We now have the following result, which uses Theorem 11.10.

Theorem 11.12. *Let $G = \mathrm{SL}_3$ and V be a nontrivial reducible coreduced G-module with $V^G = 0$. Then, up to isomorphism and taking duals, we have the following list:*

(1) $kV[1,0] + \ell V[0,1]$, $k + \ell \geq 2$.
(2) $V[2,0] + V[0,1]$.

Proof. We already know that the representations in (1) and (2) are coreduced by Theorem 9.1 and Example 4.5. We have to show that combinations not on the list are not coreduced.

Consider $V[1,1]$ together with another irreducible. For $V[1,1]$ there is a closed orbit with isotropy group double covered by $\mathrm{SL}_2 \times \mathbb{C}^*$. The slice representation (as representation of the double cover) is $\theta_1 + R_2$. If we add $V[1,0]$ we get an additional copy of $R_1 \otimes \nu_1 + \nu_{-2}$ in the slice representation. Quotienting by $\mathbb{C}^*$ we get the hypersurface in $\theta_1 + R_2 + R_2$ where the quadratic invariant of the second copy of R_2 vanishes. The hypersurface is not coreduced (see Example 3.3), hence $V[1,1] + V[1,0]$ is not coreduced. For $V[1,1] + V[2,0]$ one can easily see that the slice representation of the maximal torus is not coreduced, and for $V[3,0]$ one uses the slice representation at the zero weight vector to rule out a coreduced sum involving $V[3,0]$.

Next consider $2V[2,0]$ which is cofree with generating invariants in bidegrees $(3,0), (2,1), (1,2), (0,3)$, and choose the 1-parameter subgroup ρ with weights $(1,1,-2)$. Then one can easily see that the codimensions of GZ_ρ is $4 = \operatorname{codim} \mathcal{N}$ and that the rank of the differentials of the invariants is 2 on GZ_ρ. Hence the representation is not coreduced.

The representation $V[2,0] + V[0,2]$ is again cofree with generating invariants in degrees (3,0), (1,1), (2,2), (0,3). For the same ρ one computes that the rank is 3, while the codimension of GZ_ρ is $4 = \operatorname{codim} \mathcal{N}$, so this possibility is ruled out. We cannot add $V[1,0]$ to $V[2,0]$ since the rank of the two generating invariants is only 1 on the null cone.

Finally, we have to show that $V := V[2,0] + 2V[0,1]$ (not cofree but coregular) is not coreduced. Consider the 1-parameter subgroup with weights $(2,-1,-1)$. This clearly gives a maximal dimensional component of the null cone and it has codimension $3 = \dim V/\!\!/G - 1$. The 1-parameter subgroup with weights $(1,1,-2)$ gives something of codimension 5, which is too small to be an irreducible component of $\mathcal{N}$ since it is cut out by 4 functions. Hence $\mathcal{N}$ is irreducible. But V has the slice representation $2R_2 + \theta_1$ of SL_2 whose null cone is not reduced but also has codimension three. Thus the associated cone to the fiber $F \simeq G \times^{\mathrm{SL}_2} \mathcal{N}(2R_2)$ is $\mathcal{N}(V)$. We know that F is not reduced (Example 3.3), hence by Remark 5.3, V is not coreduced. □

Appendix A Computations for F_4

Let G be a simple group of type F_4 and let $V = \varphi_4$ be the 26-dimensional representation of G. The main result of this appendix is the following proposition.

Proposition A.1. *The representation $(V^{\oplus n}, G)$ is coreduced if and only if $n \leq 2$. Moreover, V and $V \oplus V$ are both cofree and contain a dense orbit in the null cone.*

We will use the notation introduced in Section 10. The nonzero weights of V are the short roots of F_4. Hence Z_ρ is the span of the positive short root spaces for any generic $\rho \in Y(T)$ which implies that the null cone $\mathcal{N}(V^{\oplus n})$ is irreducible for any n. We also know that $V \oplus V$ is cofree with $\dim(V \oplus V)/\!\!/G = 8$ [Sch79], hence $\dim \mathcal{N}(V \oplus V) = 44$. Let us look at the following statements which imply the proposition.

(a) *$V \oplus V \oplus V$ is not coreduced.*
(b) *$V \oplus V$ is coreduced.*
(c) *There is a dense orbit in the null cone of $V \oplus V$.*

Although we know that (c) implies (b) (Corollary 4.7) we will present direct proofs of all three claims. They are based on some explicit computations.

Proof (of statement (a)). There is a maximal subgroup of type B_4 of F_4 where $(\varphi_4(\mathsf{F}_4), \mathsf{B}_4) = \varphi_1 + \varphi_4 + \theta_1$. (For more see the discussion after Lemma A.2.) The slice representation of B_4 on $\varphi_4(\mathsf{F}_4)$ is $\varphi_1(\mathsf{B}_4) + \theta_1$. To prove that $3\varphi_4(\mathsf{F}_4)$ is not coreduced, it suffices to prove that $V_1 := 3\varphi_1(\mathsf{B}_4) + 2\varphi_4(\mathsf{B}_4)$ is not coreduced. Now D_4 is a maximal subgroup of B_4 and $V' := 2\varphi_1(\mathsf{D}_4) + 2\varphi_3(\mathsf{D}_4) + 2\varphi_4(\mathsf{D}_4)$ is a slice representation of V_1 at a zero weight vector. So we have to show that V' is not coreduced. Since our representations are self-dual, we will deal with the symmetric algebra $S(V')$ in place of $\mathcal{O}(V')$.

Recall that G is now D_4 and V' is as above. We have $\bigwedge^2 \varphi_1 = \bigwedge^2 \varphi_3 = \bigwedge^2 \varphi_4 = \varphi_2$, the adjoint representation. In the tensor product of three copies of φ_2 we have 7 copies of φ_2, but only five of them are in the ideal generated by the invariants. (This can be checked using LiE). We will show now that every covariant of type φ_2 in $\bigwedge^2 \varphi_1 \otimes \bigwedge^2 \varphi_3 \otimes \bigwedge^2 \varphi_4 \subset S(V')_{(2,2,2)}$ vanishes on the null cone, i.e., vanishes on Z_ρ for every generic $\rho \in Y(T)$.

Recall that the weights of φ_1 are $\pm\varepsilon_i$, those of φ_3 are $1/2(\pm\varepsilon_1 \pm \varepsilon_2 \pm \varepsilon_3 \pm \varepsilon_4)$ where the number of minus signs is even. The weights of φ_4 look similar, but have an odd number of minus signs. We use the notation $(\pm \pm \pm\pm)$ for these weights.

There is an outer automorphism τ of D_4 of order 2 (coming from the Weyl group of B_4) which normalizes the maximal torus, fixes $\varepsilon_1, \varepsilon_2, \varepsilon_3$ and sends ε_4 to $-\varepsilon_4$. If G is of type D_4 and if $\rho_i : G \to \mathrm{GL}(V_i)$ denotes the ith fundamental representation φ_i, then $\rho_1 \circ \tau \simeq \rho_1$, $\rho_2 \circ \tau \simeq \rho_2$, and $\rho_3 \circ \tau \simeq \rho_4$. Thus there is a linear automorphism $\mu\colon V' \xrightarrow{\sim} V'$ which is τ-equivariant, i.e., $\mu(gv) = \tau(g)\mu(v)$. It follows that μ has the following properties:

(1) μ sends G-orbits to G-orbits. In particular, $\mu(\mathcal{N}) = \mathcal{N}$.
(2) $\mu(V'_\lambda) = V'_{\tau(\lambda)}$.
(3) If $\psi: V' \to V''$ is a covariant of type φ_2 in $\bigwedge^2 \varphi_1 \otimes \bigwedge^2 \varphi_3 \otimes \bigwedge^2 \varphi_4$, then so is $\psi \circ \mu: V' \to V''$.

This implies that for every 1-PSG ρ we have $\mu(Z_\rho) = Z_{\tau(\rho)}$ and that if all $\psi: V' \to V''$ as in (3) vanish on Z_ρ, then they vanish on $Z_{\tau(\rho)}$ too.

As a consequence, we can assume that $\varepsilon_1 > \varepsilon_2 > \varepsilon_3 > \varepsilon_4 > 0$. This implies that the following weights are positive:

$$\{\varepsilon_1, \varepsilon_2, \varepsilon_3, \varepsilon_4\} \subset \Lambda(\varphi_1),$$
$$\{(++++), (+-+-), (++--)\} \subset \Lambda(\varphi_3),$$
$$\{(+++-), (++-+), (+-++)\} \subset \Lambda(\varphi_4).$$

Since $(-++-) < (-+++)$ we see that there are only three cases of maximal positive weight spaces to be considered.

(1) $\{\varepsilon_1, \varepsilon_2, \varepsilon_3, \varepsilon_4\}$, $\{(++++), (+-+-), (++--), (-++-)\}$ and $\{(+++-), (+-++), (++-+), (-+++)\}$;
(2) $\{\varepsilon_1, \varepsilon_2, \varepsilon_3, \varepsilon_4\}$, $\{(++++), (+-+-), (++--), (+--+)\}$ and $\{(+++-), (+-++), (++-+), (+---)\}$;
(3) $\{\varepsilon_1, \varepsilon_2, \varepsilon_3, \varepsilon_4\}$, $\{(++++), (+-+-), (++--), (+--+)\}$ and $\{(+++-), (+-++), (++-+), (-+++)\}$.

Now we have to calculate the positive weights in $\bigwedge^2 \varphi_1$, $\bigwedge^2 \varphi_3$ and $\bigwedge^2 \varphi_4$ which come from the positive weights in the two copies of φ_1, φ_3, and φ_4. For $\bigwedge^2 \varphi_1$ we get $\{\varepsilon_i + \varepsilon_j \mid i < j\}$ in all three cases. For the two others we find the following sets.

(1) $\bigwedge^2 \varphi_3$: $\{\varepsilon_i + \varepsilon_j \mid i < j < 4\} \cup \{\varepsilon_i - \varepsilon_4 \mid i < 4\}$; $\bigwedge^2 \varphi_4$: $\{\varepsilon_i + \varepsilon_j \mid i < j\}$.
(2) $\bigwedge^2 \varphi_3$: $\{\varepsilon_1 \pm \varepsilon_j \mid j > 1\}$; $\bigwedge^2 \varphi_4$: $\{\varepsilon_1 \pm \varepsilon_j \mid j > 1\}$.
(3) $\bigwedge^2 \varphi_3$: $\{\varepsilon_1 \pm \varepsilon_j \mid j > 1\}$; $\bigwedge^2 \varphi_4$: $\{\varepsilon_i + \varepsilon_j \mid i < j\}$.

Now it is easy to see that in all three cases there is no way to write the highest weight $\varepsilon_1 + \varepsilon_2$ of φ_2 as a sum of three positive weights, one from each $\bigwedge^2$. Hence (V', D_4) is not coreduced and we have proved (a). □

Proof (of statement (b)). Since $V \oplus V$ is cofree the null cone is (schematically) a complete intersection. Therefore it suffices to find an element $v \in \mathcal{N}(V \oplus V)$ such that the differential $d\pi_v$ of the quotient morphism $\pi: V \oplus V \to Y$ at v has maximal rank $8 = \dim Y$.

The nonzero weights of $\varphi_4(\mathsf{F}_4)$ are $\pm\varepsilon_i$, $i = 1, \dots, 4$ (the nonzero weights of $\varphi_1(\mathsf{B}_4)$) and $(1/2)(\pm\varepsilon_1 \pm \varepsilon_2 \pm \varepsilon_3 \pm \varepsilon_4)$ (the weights of $\varphi_4(\mathsf{B}_4)$). We will abbreviate the latter weights as $(\pm\pm\pm\pm)$ from now on. The positive weights are the ε_i and the weights of $\varphi_4(\mathsf{B}_4)$ where the coefficient of ε_1 is positive. Let $v_{\pm i}$ denote a nonzero vector in the weight space of $\pm\varepsilon_i$, let v_0 denote a zero weight vector and let v_{++++} denote a nonzero vector in the weight space $(++++)$ and similarly for v_{+++-}, etc. We claim that $d\pi$ has rank 8 at the point $v = (v_2 + v_3 + v_{+--+} + v_{+---}) \in 2\varphi_1(\mathsf{B}_4) + 2\varphi_4(\mathsf{B}_4)$.

The invariants of $2\varphi_4(\mathsf{F}_4)$ are the polarizations of the degree 2 invariant and the degree 3 invariant of one copy of $\varphi_4(\mathsf{F}_4)$ together with an invariant of degree $(2, 2)$ [Sch78]. The restriction of the degree 2 invariant to $\varphi_1(\mathsf{B}_4) + \varphi_4(\mathsf{B}_4)$ is the sum of the degree two invariants there (see [Sch78] for descriptions of the invariants of $(2\varphi_1 + 2\varphi_4, \mathsf{B}_4)$.). Clearly the differentials of the degree 2 invariants of $2\varphi_4(\mathsf{F}_4)$ at v have rank 3 when applied to the subspace spanned by the vectors v_{-2}, v_{-3} in the two copies of $\varphi_1(\mathsf{B}_4)$. There is only one degree 3 generator in $(\varphi_1 + \varphi_4, \mathsf{B}_4)$ and it is the contraction of φ_1 with the copy of φ_1 in $S^2(\varphi_4)$. Another way to think of the invariant is as the contraction of φ_4 with the copy of φ_4 in $\varphi_1 \otimes \varphi_4$. Now the highest weight vector of the copy of φ_4 in $\varphi_1 \otimes \varphi_4$ is (up to some nonzero coefficients)

$$v_1 \otimes v_{-+++} + v_2 \otimes v_{+-++} + v_3 \otimes v_{++-+} + v_4 \otimes v_{+++-} + v_0 \otimes v_{++++}.$$

From this one derives the form of the other weight vectors of $\varphi_4 \subset \varphi_1 \otimes \varphi_4$ and restricting to v one gets contributions to the weights $(++-+)$, $(++--)$, $(+-++)$ and $(+-+-)$. Thus the differential of the degree 3 invariant of $\varphi_1(\mathsf{F}_4)$ at v vanishes on φ_4 except on v_{--+-}, v_{--++}, v_{-+--} and v_{-+-+}. Now polarizing it is easy to see that the four generators of degree 3 have differential of rank 4 at v when applied to vectors in $2\varphi_4(\mathsf{B}_4)$.

There remains the generator of degree 4. Restricted to B_4 one easily sees that the invariant is a sum of two generators (modulo products of the generators of degree 2), one of which is the invariant which contracts the copy of $\bigwedge^2(\varphi_1) \subset S^2(2\varphi_1)$ with the copy in $\bigwedge^2(\varphi_4) \subset S^2(2\varphi_4)$ and the other which is of degree 4 in $2\varphi_4(\mathsf{B}_4)$ (and doesn't involve $2\varphi_1$). Now the highest weight vector of $\bigwedge^2(\varphi_1) \subset \bigwedge^2(\varphi_4)$ is (up to nonzero scalars)

$$v_{++++} \wedge v_{++--} + v_{+++-} \wedge v_{++-+}$$

from which it follows that the weight vector of weight $-\varepsilon_2 - \varepsilon_3$ does not vanish on $v_{+---} + v_{---+}$. Now in $\bigwedge^2(\varphi_1) \subset S^2(2\varphi_1)$ we have $v_2 \wedge v_3$ of weight $\varepsilon_2 + \varepsilon_3$. Hence the differential of the degree 4 invariant evaluated at v does not vanish on v_{---+} and the rank of the differentials of the 8 invariants is indeed 8. Thus $2\varphi_4(\mathsf{F}_4)$ is coreduced. □

Proof (of statement (c)). Recall that the root system R of G has the following 3 parts A, B and C:

$$A = \{\pm\varepsilon_i\}, \quad B = \{\pm\varepsilon_i \pm \varepsilon_j \mid i < j\}, \quad C = \{\frac{1}{2}(\pm\varepsilon_1 \pm \varepsilon_2 \pm \varepsilon_3 \pm \varepsilon_4)\}$$

with cardinality $\#A = 8$, $\#B = 24$ and $\#C = 16$. Thus

$$\mathfrak{g} = \operatorname{Lie} G = \mathfrak{h} \oplus \bigoplus_{\beta \in A \cup B \cup C} \mathfrak{g}_\beta$$

where $\mathfrak{h} = \operatorname{Lie} T$ is the Cartan subalgebra and $T \subset G$ a maximal torus. The weights $\Lambda = \Lambda_V$ of the representation V are the short roots $A \cup C$ together with the zero weight 0. The nonzero weight spaces V_λ are 1-dimensional, and the zero weight space $V_0 = V^T$ has dimension 2. This implies the following.

Lemma A.2. *Let $\beta \in R$ be a root and $\lambda \in \Lambda$ a weight of V. If $\beta + \lambda$ is a weight of V, then $\mathfrak{g}_\beta V_\lambda$ is a nontrivial subspace of $V_{\beta+\lambda}$.*

Note that $\mathfrak{g}_\beta V_0 = V_\beta$ and $\mathfrak{g}_\beta V_{-\beta} \subset V_0$ is 1-dimensional for every short root $\beta \in A \cup C$.

The subspace $\mathfrak{g}' := \mathfrak{h} \oplus \bigoplus_{\beta \in A \cup B} \mathfrak{g}_\beta \subset \mathfrak{g}$ is the Lie algebra of a maximal subgroup $G' \subset G$ of type B_4, and the representation V decomposes under G' into $V = \theta_1 \oplus \varphi_1(\mathsf{B}_4) \oplus \varphi_4(\mathsf{B}_4)$ where $\varphi_4(\mathsf{B}_4) = \bigoplus_{\gamma \in C} V_\gamma$, $\varphi_1(\mathsf{B}_4) = V_A \oplus \bigoplus_{\alpha \in A} V_\alpha$, and $V_A := \varphi_1(\mathsf{B}_4)^T \subset V_0$. It follows that $\mathfrak{g}_\alpha V_{-\alpha} = V_A$ for $\alpha \in A$, but $\mathfrak{g}_\gamma V_{-\gamma} \not\subseteq V_A$ for $\gamma \in C$, so that $\mathfrak{g}_\alpha V_{-\alpha} + \mathfrak{g}_\gamma V_{-\gamma} = V_0$.

The basic idea for the calculations is the following. To every vector $v \in V$ we define its *weight support* $\omega(v) \subset A \cup C \cup \{0_A, 0_C\}$ in the following way. Write v as a sum of weight vectors, $v = \sum_A v_\alpha + \sum_C v_\gamma + v_0$. Then

$$\omega(v) := \{\alpha \in A \mid v_\alpha \neq 0\} \cup \{\gamma \in C \mid v_\gamma \neq 0\} \cup \begin{cases} \emptyset & \text{if } v_0 = 0, \\ \{0_A\} & \text{if } v_0 \in V_A \setminus \{0\}, \\ \{0_C\} & \text{if } v_0 \in V_0 \setminus V_A. \end{cases}$$

This extends in an obvious way to the weight support of elements from $V \oplus V$. Now we look at a pair $v = (v', v'') = (v_{\alpha'} + v_{\gamma'}, v_{\alpha''} + v_{\gamma''}) \in V \oplus V$ where $\alpha', \alpha'' \in A$ and $\gamma', \gamma'' \in C$ are distinct weights. Define

$$\Omega(v) := \{\omega(x_\beta v) \mid \beta \in A \cup B \cup C\} \cup \{\alpha', \alpha'', \gamma', \gamma''\},$$

where $x_\beta \in \mathfrak{g}_\beta$ is a (nonzero) root vector. This is the set of weight supports of generators of $\mathfrak{g}v$ where we use that $\mathfrak{h}v = \mathbb{C}v_{\alpha'} \oplus \mathbb{C}v_{\gamma'} \oplus \mathbb{C}v_{\alpha''} \oplus \mathbb{C}v_{\gamma''}$.

Our problem can now be understood in the following way. We are given a matrix M of column vectors from which we want to calculate the rank. We replace M by the "support matrix" $\Omega(M)$ which is obtained from M by replacing each nonzero entry by 1. How can one find a lower bound for the rank of M from $\Omega(M)$?

There is an obvious procedure. We first look for a column of $\Omega(M)$ which contains a single 1, let us say in row i. Then we remove all other 1's in row i and repeat this procedure as often as possible to obtain a matrix $\Omega(M)'$. It is clear that this "reduced" matrix $\Omega(M)'$ is again the support matrix of a matrix M' which is obtained by column reduction from M. This first step is called "column reduction."

Now we apply row reduction to M' which amounts to looking at rows of $\Omega(M)'$ which contain a single 1. Then we delete all other 1's in the corresponding column. Again we repeat this procedure as often as possible and obtain a matrix $\Omega(M)''$. We call this procedure "row reduction." It is clear now that a lower bound for the rank of M is given by the number of columns of $\Omega(M)''$ containing a single 1.

Now we choose $v \in V \oplus V$ as above with weights

$$(\alpha', \gamma', \alpha'', \gamma'') = (\varepsilon_3, 1/2(\varepsilon_1 - \varepsilon_2 - \varepsilon_3 + \varepsilon_4), \varepsilon_2, 1/2(\varepsilon_1 - \varepsilon_2 - \varepsilon_3 - \varepsilon_4)).$$

We obtain a set $\Omega(v)$ with 45 elements where each element's weight support has cardinality at most two. (We use Mathematica® to perform these and the following calculations.) After applying the "column reduction" we obtain a new set $\Omega(v)'$ that contains 44 elements where 34 of them contain a single weight. For the remaining 10 elements, the "row" reduction produces 10 sets with a single weight. Thus we get $\dim Gv = \dim \mathfrak{g}v = 44 = 2\dim V - \dim V/\!\!/G$, and we are done. □

Remark A.3. We are grateful to Jan Draisma who did some independent calculations (using GAP) to show that there is a dense orbit in the null cone of $V \oplus V$.

Appendix B Computations for $\mathsf{G}_2 \times \mathsf{G}_2$

The main result of this appendix is the following proposition. We will give two proofs.

Proposition B.1. *The representation* $(\mathbb{C}^7 \otimes \mathbb{C}^7, \mathsf{G}_2 \times \mathsf{G}_2)$ *is not coreduced.*

Proof (First Proof of Proposition B.1). The nontrivial part of the slice representation at the zero weight vector is $G := \mathrm{SL}_3 \times \overline{\mathrm{SL}_3}$ on the four possible versions of ($W := \mathbb{C}^3$ or W^*) tensored with ($\overline{W} := \overline{\mathbb{C}^3}$ or $\overline{W}^*$). Set $V_1 := W \otimes \overline{W}$, $V_2 := W \otimes \overline{W}^*$, $V_3 := W^* \otimes \overline{W}$ and $V_4 := W^* \otimes \overline{W}^*$.

Lemma B.2. *The G-module*

$$V := V_1 + V_2 + V_3 + V_4$$

is not coreduced.

We have a group N of order 8 that acts on V by interchanging W and W^*, $\overline{W}$ and $\overline{W}^*$ as well as interchanging W, W^* with $\overline{W}$, $\overline{W}^*$. Then N normalizes the action of G. Here are the steps in the proof of the proposition above.

(1) We show that there is a minimal generator f of the invariants of (V, G) which is multihomogeneous of degree (3,3,3,3) in the four irreducible subspaces of V.
(2) We show that, up to the action of the Weyl group and N, there are eight 1-parameter subgroups ρ of G such that the union of the GZ_ρ is $\mathcal{N}(V)$.
(3) We show that for each such ρ, the differential of f vanishes on Z_ρ.

It then follows from Remark 3.2 that V is not coreduced.

Let $R = \mathbb{C}[a_1, \dots, a_n]$ be a finitely generated $\mathbb{N}^d$-graded ring where the a_i are homogeneous. Recall that $a_1, \dots, a_m$ are a *regular sequence* in R if a_1 is not a zero divisor and a_{j+1} is not a zero divisor in $R/(Ra_1 + \cdots + Ra_j)$, $1 \le j < m$.

We may write R as a quotient $R = \mathbb{C}[x_1, \ldots, x_n]/I$ where the image of x_i in R is a_i, $i = 1, \ldots, n$. Let I_s denote the elements of I homogeneous of degree $s := (s_1, \ldots, s_d)$, $s_j \in \mathbb{N}$ and let $\bar{I}$ denote $I/(x_1, \ldots, x_m)$. We leave the proof of the following lemma to the reader.

Lemma B.3. *Let R, etc. be as above. Then $I_s \to \bar{I}_s$ is an isomorphism for all $s \in \mathbb{N}^d$.*

The lemma says that we can determine the dimension of the space of relations of the a_i in degree s by first setting $a_1, \ldots, a_m$ to zero.

Lemma B.4. *There is a generator f of $R := \mathcal{O}(V)^G$ of multidegree* $(3, 3, 3, 3)$.

We used LiE to compute a partial Poincaré series of R:

$$\begin{aligned} 1 + (ps + qr) + 2(p^2s^2 + q^2r^2) + (p^3 + q^3 + r^3 + s^3) \\ + 2(p^3q^3 + p^3r^3 + q^3s^3 + r^3s^3) + 3(p^3s^3 + q^3r^3) \\ + 2(q^3ps + r^3ps + p^3qr + s^3qr) \\ + 6(q^3p^2s^2 + r^3p^2s^2 + p^3q^2r^2 + s^3q^2r^2) \\ + 13(p^3q^3r^3 + p^3q^3s^3 + p^3r^3s^3 + q^3r^3s^3) \\ + 4pqrs + 10(pq^2r^2s + p^2qrs^2) + 18(pq^3r^3s + p^3qrs^3) \\ + 37p^2q^2r^2s^2 + 86(p^2q^3r^3s^2 + p^3q^2r^2s^3) + 265p^3q^3r^3s^3. \end{aligned}$$

If there were no relations among the generators of R of degree at most $(3, 3, 3, 3)$, then the Poincaré series would indicate that we have generators in degree (a, b, c, d) of a certain multiplicity which we denote by $\text{gen}(a, b, c, d)$. We list the relevant $\text{gen}(a, b, c, d)$, modulo symmetries given by the group N.

(1) $\text{gen}(0110) = 1$
(2) $\text{gen}(0220) = 1$
(3) $\text{gen}(3000) = 1$
(4) $\text{gen}(3300) = 1$
(5) $\text{gen}(0330) = 0$
(6) $\text{gen}(1111) = 3$
(7) $\text{gen}(3110) = 1$
(8) $\text{gen}(3220) = 3$
(9) $\text{gen}(3330) = 3$
(10) $\text{gen}(1221) = 5$
(11) $\text{gen}(1331) = 2$
(12) $\text{gen}(2222) = 14$
(13) $\text{gen}(3223) = 13$
(14) $\text{gen}(3333) = 11$

It is easy to see that the representations $V_i + V_j$, $1 \leq i < j \leq 4$ are cofree. Now $V_2 + V_3$ has generators in degrees (3,0), (0,3), (1,1) and (2,2) while $V_1 + V_2$ has generators in degrees (3,0), (0,3) and (3,3). Thus it is easy to see that we have a regular sequence in R consisting of the (determinant) invariants of degree 3 and those of degree (0110), (0220), (1001) and (2002).

Lemma B.5. *Suppose that we are in one of the cases above, except for* (2222), (3223) *and* (3333). *Then R has* $\mathrm{gen}(abcd)$ *generators in degree* $(abcd)$.

Proof. We set the invariants of our regular sequence equal to zero and see if we have any relations. But then there are no nonlinear polynomials in the remaining generators in the degrees we are worried about. □

Proof (of Lemma B.4). As usual, we set the elements of our regular sequence equal to zero. This does not change the number of minimal generators of degree (3333). Now how can we have fewer generators than gen(3333) in degree (3333)? This can only occur if there is a degree $(abcd)$ with an "unexpected" relation such that $(3333) - (abcd)$ is the degree of a generator not in our regular sequence. Thus the only problem could occur because of relations in degree (2222) multiplied by the 3 generators f_1, f_2 and f_3 in degree (1111). Moreover, modulo our regular sequence, the unexpected relations in degree (2222) have the form $(r = \sum_{ij} c_{ij} f_i f_j) = 0$. Thus there are unexpected relations $r_1, \dots r_d$, $d \leq 6$. For each relation r_k we add an additional generator y_k in degree (2222) and to get the correct count of non-generators in degree (3333) we have to adjust our formal count by adding $3d$ (from the product of the y_k by the f_i) and subtracting the dimension of the span of the $f_i r_k$ in the polynomial ring $\mathbb{C}[f_1, f_2, f_3]$. But the correction is by less than 11:

Case 1. $d \leq 5$. Then we have a correction of at most $3d - d \leq 10$.

Case 2. $d = 6$. Then the correction is $18 - \dim S^3(\mathbb{C}^3) = 8$. □

We now have our generator f of degree (3333). Next we need to calculate the irreducible components of the null cone, up to the action of N.

Let ρ be a 1-parameter subgroup of $G := \mathrm{SL}_3 \times \overline{\mathrm{SL}}_3$ whose weights for $\mathbb{C}^3$ are a, b and c and whose weights for $\overline{\mathbb{C}}^3$ are $\overline{a}$, $\overline{b}$ and $\overline{c}$. We have that $a \geq b \geq c$ and similarly for $\overline{a}$, etc. We also can assume that no weight of V is zero. Of course, many choices of a, etc. will give the same subset Z_ρ in V. We say that a particular choice of a, etc. is a *model* if it gives the correct Z_ρ.

The action of our group N does not change the weights that occur, just in which of the four components they occur. Thus to show that df vanishes on $\mathcal{N}(V)$, we can always reduce to the case that $a > \overline{a}$ and that the other numbers are negative (or zero). For every possibility we will give a model such that df vanishes on Z_ρ.

Lemma B.6. *We have that* $c - \overline{b} \leq c - \overline{c} \leq c + \overline{a} \leq b + \overline{a} \geq b - \overline{c} \geq b - \overline{b}$. *Moreover,* $c - \overline{b} < 0$ *and not both* $c - \overline{c}$ *and* $b - \overline{b}$ *are positive.*

Proof. The string of inequalities is obvious. If $c - \overline{b} > 0$, then $b - \overline{c} > 0$ and adding we get that $-a + \overline{a} > 0$ which is a contradiction. Similarly, not both $c - \overline{c}$ and $b - \overline{b}$ can be positive. □

Given the lemma, there are eight possibilities for the signs of $c-\overline{b}, c-\overline{c}, \dots, b-\overline{b}$ which we present in matrix form. In another matrix, we present the values a, b, c, $\overline{a}$, $\overline{b}$ and $\overline{c}$ of a 1-parameter subgroup ρ which is a model for the signs. Note that the signs tell you exactly which vectors in V are in the positive weight space of ρ.

$$\begin{pmatrix} -1 & -1 & -1 & -1 & -1 & -1 \\ -1 & -1 & -1 & 1 & -1 & -1 \\ -1 & -1 & -1 & 1 & 1 & -1 \\ -1 & -1 & -1 & 1 & 1 & 1 \\ -1 & -1 & 1 & 1 & -1 & -1 \\ -1 & -1 & 1 & 1 & 1 & -1 \\ -1 & -1 & 1 & 1 & 1 & 1 \\ -1 & 1 & 1 & 1 & 1 & -1 \end{pmatrix} \quad \begin{pmatrix} 4 & -2 & -2 & 1 & 0 & -1 \\ 8 & -3 & -5 & 4 & -2 & -2 \\ 4 & -1 & -3 & 2 & 0 & -2 \\ 3 & 0 & -3 & 2 & -1 & -1 \\ 6 & -3 & -3 & 4 & -2 & -2 \\ 8 & -3 & -5 & 6 & -2 & -4 \\ 7 & -2 & -5 & 6 & -3 & -3 \\ 4 & -2 & -2 & 3 & 0 & -3 \end{pmatrix}$$

Proof of Lemma B.2. Consider signs which have the 1-parameter subgroup ρ with weights $\begin{pmatrix} 8 & -3 & -5 & 6 & -2 & -4 \end{pmatrix}$ as model. Then the largest negative weight occurring in V is -14 while the positive weights occurring in $V_1, \dots, V_4$ are

$$(1,3,4,6,14),\ (1,2,10,12),\ (1,1,3,9,11),\ (7,7,5,9).$$

Now consider a monomial m in the weight vectors which occurs in f. If df does not vanish on Z_ρ then there is a monomial with only one negative weight vector. But the sum of the positive weights occurring in m is at least $3+3+3+2*5 = 19$ which is greater than 14. Hence this is impossible and df vanishes on GZ_ρ. One similarly (and more easily) sees that df vanishes on GZ_ρ in the other 7 cases.

This finishes the first proof of Proposition B.1. □

Second Proof of Proposition B.1. The weights of $V = \mathbb{C}^7$ are the short roots of $G := \mathsf{G}_2$ together with 0, and all weight spaces are 1-dimensional. We use the notation $\Lambda := \{\pm\alpha, \pm\beta, \pm(\alpha+\beta), 0\}$ where $\alpha+\beta$ is the highest weight. Thus the weight spaces of $V \otimes V$ are given by the tensor products $V_\mu \otimes V_\nu$, $(\mu,\nu) \in \Lambda \times \Lambda$.

We first determine the maximal positive subspaces of $W := V \otimes V$, up to the action of the Weyl group. If ρ is a one-parameter subgroup of $G \times G$ we denote by W_ρ the sum of the ρ-positive weight spaces, i.e.,

$$V_\rho := \bigoplus_{\rho(\mu,\nu)>0} W_{(\mu,\nu)}.$$

The 1-PSG ρ is defined by the values $a := (\rho,(\alpha,0))$, $b := (\rho,(\beta,0))$, $a' := (\rho,(0,\alpha))$, $b' := (\rho,(0,\beta))$. Using the action of the Weyl group, we can assume that

$$a,b,a',b' > 0, \quad a \geq b, \quad a' \geq b', \quad \{a,b,a+b\} \cap \{a',b',a'+b'\} = \emptyset.$$

We can also assume that $a > a'$; we will then get the other maximal positive subspaces by the symmetry $(\mu,\nu) \mapsto (\nu,\mu)$. Now V_ρ depends only on the relative position of the values $a + b > a \geq b$ and the values $a' + b' > a' \geq b'$. It is not difficult to see that there are eight cases.

(1) $a + b > a > b > a' + b' > a' \geq b'$ represented by $\rho = (5, 4, 2, 1)$;
(2) $a + b > a > a' + b' > b > a' \geq b'$ represented by $\rho = (6, 4, 3, 2)$;
(3) $a + b > a' + b' > a \geq b > a' \geq b'$ represented by $\rho = (6, 5, 4, 3)$;
(4) $a + b > a > a' + b' > a' > b > b'$ represented by $\rho = (5, 2, 3, 1)$;
(5) $a + b > a' + b' > a > a' > b > b'$ represented by $\rho = (6, 4, 5, 3)$;
(6) $a + b > a > a' + b' \geq a' > b' \geq b$ represented by $\rho = (7, 1, 4, 2)$;
(7) $a + b > a' + b' > a > a' > b > b'$ represented by $\rho = (6, 2, 4, 3)$;
(8) $a' + b' > a + b > a > a' \geq b' > b$ represented by $\rho = (6, 2, 5, 4)$.

To get the full set of maximal positive subspaces we have to add the 8 ρ's obtained from the list above by replacing (a, b, a', b') with (a', b', a, b).

Now we used LiE to look at the covariants of type $\theta_1 \otimes V$. The multiplicities of this covariant in degrees 1 to 9 are $(0, 0, 1, 1, 3, 5, 12, 18, 41)$, and the dimensions of the invariants in these degrees are $(0, 1, 1, 3, 2, 8, 7, 17, 19)$. It follows that at most $37 = 1 \cdot 12 + 1 \cdot 5 + 3 \cdot 3 + 2 \cdot 1 + 8 \cdot 1$ covariants of degree 9 are in the ideal generated by the invariants, hence there are generating covariants of this type in degree 9. Now we have to show that for every positive weight space V_ρ the highest weight $(0, \alpha + \beta)$ of $\theta \otimes V$ cannot be expressed as a sum of 9 weights from V_ρ. Because of duality, each V_ρ has dimension $24 = (7 * 7 - 1)/2$. If we denote by Λ_ρ the set of weights of V_ρ, this amounts to prove that the system

$$\sum_{\lambda \in \Lambda_\rho} x_\lambda \lambda = (0, \alpha + \beta), \quad \sum_{\lambda \in \Lambda_\rho} x_\lambda = 9$$

has no solution in nonnegative integers x_λ. Note that the first condition consists in 4 linear equations in 24 variables. Now we used Mathematica to show that there are no solutions for each one of the sixteen maximal positive weight spaces Z_ρ given above.

References

[BK79] Walter Borho and Hanspeter Kraft, *Über Bahnen und deren Deformationen bei linearen Aktionen reduktiver Gruppen*, Comment. Math. Helv. **54** (1979), no. 1, 61–104.

[Bou68] N. Bourbaki, *Eléments de mathématique. Fasc. XXXIV. Groupes et algèbres de Lie. Chapitre IV: Groupes de Coxeter et systèmes de Tits. Chapitre V: Groupes engendrés par des réflexions. Chapitre VI: systèmes de racines*, Actualités Scientifiques et Industrielles, No. 1337, Hermann, Paris, 1968.

[Bri85] Michel Brion, *Représentations exceptionnelles des groupes semi-simples*, Ann. Sci. École Norm. Sup. (4) **18** (1985), no. 2, 345–387.

[Dix81] Jacques Dixmier, *Sur les invariants des formes binaires*, C. R. Acad. Sci. Paris Sér. I Math. **292** (1981), no. 23, 987–990.

[Dyn52a] E. B. Dynkin, *Maximal subgroups of the classical groups*, Trudy Moskov. Mat. Obšč. **1** (1952), 39–166.

[Dyn52b] E. B. Dynkin, *Semisimple subalgebras of semisimple Lie algebras*, Mat. Sbornik N.S. **30(72)** (1952), 349–462 (3 plates).

[GV78] V. Gatti and E. Viniberghi, *Spinors of* 13*-dimensional space*, Adv. in Math. **30** (1978), no. 2, 137–155.

[Gro67] A. Grothendieck, *Éléments de géométrie algébrique. IV. Étude locale des schémas et des morphismes de schémas IV*, Inst. Hautes Études Sci. Publ. Math. (1967), no. 32, 361.

[Hes80] Wim H. Hesselink, *Characters of the nullcone*, Math. Ann. **252** (1980), no. 3, 179–182.

[KPV76] Victor G. Kac, Vladimir L. Popov, and Ernest B. Vinberg, *Sur les groupes linéaires algébriques dont l'algèbre des invariants est libre*, C. R. Acad. Sci. Paris Sér. A-B **283** (1976), no. 12, Ai, A875–A878.

[Kno86] Friedrich Knop, *Über die Glattheit von Quotientenabbildungen*, Manuscripta Math. **56** (1986), no. 4, 419–427.

[Kos63] Bertram Kostant, *Lie group representations on polynomial rings*, Amer. J. Math. **85** (1963), 327–404.

[KR71] Bertram Kostant and Stephen Rallis, *Orbits and representations associated with symmetric spaces*, Amer. J. Math. **93** (1971), 753–809.

[Kra84] Hanspeter Kraft, *Geometrische Methoden in der Invariantentheorie*, Aspects of Mathematics, D1, Friedr. Vieweg & Sohn, Braunschweig, 1984.

[KW06] Hanspeter Kraft and Nolan R. Wallach, *On the null cone of representations of reductive groups*, Pacific J. Math. **224** (2006), no. 1, 119–139.

[Lun75] D. Luna, *Adhérences d'orbite et invariants*, Invent. Math. **29** (1975), no. 3, 231–238.

[Mat89] Hideyuki Matsumura, *Commutative Ring Theory*, second ed., Cambridge Studies in Advanced Mathematics, Vol. 8, Cambridge University Press, Cambridge, 1989, Translated from the Japanese by M. Reid.

[Pan85] D. I. Panyushev, *Regular elements in spaces of linear representations. II*, Izv. Akad. Nauk SSSR Ser. Mat. **49** (1985), no. 5, 979–985, 1120.

[Pan99] Dmitri I. Panyushev, *Actions of "nilpotent tori" on G-varieties*, Indag. Math. (N.S.) **10** (1999), no. 4, 565–579.

[Pop76] V. L. Popov, *Representations with a free module of covariants*, Funkcional. Anal. i Priložen. **10** (1976), no. 3, 91–92.

[Pro07] Claudio Procesi, *Lie Groups*, Universitext, Springer, New York, 2007, An approach through invariants and representations.

[Ric89] R. W. Richardson, *Irreducible components of the nullcone*, Invariant theory (Denton, TX, 1986), Contemp. Math., Vol. 88, Amer. Math. Soc., Providence, RI, 1989, pp. 409–434.

[Sch68] Issai Schur, *Vorlesungen über Invariantentheorie*, Bearbeitet und herausgegeben von Helmut Grunsky. Die Grundlehren der mathematischen Wissenschaften, Band 143, Springer-Verlag, Berlin, 1968.

[Sch78] Gerald W. Schwarz, *Representations of simple Lie groups with regular rings of invariants*, Invent. Math. **49** (1978), no. 2, 167–191.

[Sch87] Gerald W. Schwarz, *On classical invariant theory and binary cubics*, Ann. Inst. Fourier (Grenoble) **37** (1987), no. 3, 191–216.

[Sch88] Gerald W. Schwarz, *Invariant theory of G_2 and* Spin_7, Comment. Math. Helv. **63** (1988), no. 4, 624–663.

[Sch79] Gerald W. Schwarz, *Representations of simple Lie groups with a free module of covariants*, Invent. Math. **50** (1978/79), no. 1, 1–12.

[Ste98] John R. Stembridge, *The partial order of dominant weights*, Adv. Math. **136** (1998), no. 2, 340–364.

[vLCL92] M. A. A. van Leeuwen, A. M. Cohen, and B. Lisser, *LiE: A package for Lie Group Computations*, CAN, Computer Algebra Netherland, Amsterdam, 1992.

[Vin76] È. B. Vinberg, *The Weyl group of a graded Lie algebra*, Izv. Akad. Nauk SSSR Ser. Mat. **40** (1976), no. 3, 488–526, 709.

M-series and Kloosterman–Selberg zeta functions for $\mathbb{R}$-rank one groups

Roberto J. Miatello and Nolan R. Wallach

The first author wishes to congratulate Nolan Wallach on this occasion and to thank him for his most generous teaching and support.

R.J. Miatello

Abstract For an arbitrary Lie group G of real rank one, we give a formula for the Fourier coefficient $D^{\chi}_{\chi'}(\xi, \nu)$ of the **M**-series (a type of Poincaré series) defined in [17], in terms of Kloosterman–Selberg zeta functions $\zeta_{\chi,\chi',\xi}(\mu)$. As a consequence we show that the meromorphic continuation of $\zeta_{\chi,\chi',\xi}(\nu)$ to $\mathbb{C}$ follows from the meromorphic continuation of the **M**-series. We also give a description of the pole set in the region $\mathrm{Re}\,\nu \geq 0$.

Keywords: Kloosterman sum • Kloosterman–Selberg zeta function • Fourier coefficient • Whittaker vector • rank one Lie group

Mathematics Subject Classification: 11F30, 11F70, 11L05, 11M36

Partially supported by CONICET, SecytUNC, MINCyT (Argentina) and NSF grant DMS-0963036.

R.J. Miatello
FaMAF-CIEM, Facultad de Matemática, Astronomia y Fisica,
Universidad Nacional de Córdoba, Córdoba 5000, Argentina
e-mail: miatello@famaf.unc.edu.ar

N.R. Wallach
Department of Mathematics, University of California at San Diego,
La Jolla, CA 92093-0112, USA
e-mail: nwallach@ucsd.edu

R. Howe et al. (eds.), *Symmetry: Representation Theory and Its Applications*, Progress in Mathematics 257, DOI 10.1007/978-1-4939-1590-3_16

1 Introduction

The usual Kloosterman sum for $c \in \mathbb{Z}_{>0}, n, m \in \mathbb{Z}$ is given by the formula

$$S(m,n,c) = \sum_{\substack{a,d \,\in\, \mathbb{Z}/c\mathbb{Z} \\ ad \,=\, 1}} e^{\frac{2\pi i(am+dn)}{c}}.$$

In 1965 Selberg defined a Dirichlet series

$$\zeta_{m,n}(s) = \sum_{c \geq 1} \frac{S(m,n,c)}{c^{2s}}.$$

Using Weil's estimates on the Kloosterman sums, Selberg proved that the series converges absolutely for $\operatorname{Re} s > \frac{3}{4}$ and also proved a meromorphic continuation to $\operatorname{Re} s \geq \frac{1}{2}$. One can show that if Γ is a discrete subgroup of $\mathrm{SL}(2,\mathbb{R})$ such that $\operatorname{vol}(\Gamma \backslash \mathrm{SL}(2,\mathbb{R})) < \infty$ and if λ is an eigenvalue of the (positive standardly normalized) Laplacian on $L^2(\Gamma \backslash \mathrm{SL}(2,\mathbb{R})/\mathrm{SO}(2))$, then $\lambda = s(1-s)$ for some s with $\operatorname{Re} s \geq \frac{1}{2}$. Selberg proved that if $0 \neq \lambda = s(1-s)$ is such an eigenvalue with Γ a Hecke congruence subgroup, then there exist m, n such that s is a pole of $\zeta_{m,n}(s)$ with $\operatorname{Re} s \geq \frac{1}{2}$. Thus s must, in fact, satisfy $\frac{1}{2} \leq \operatorname{Re} s \leq \frac{3}{4}$ which implies Selberg's famous spectral gap for congruence subgroups, that is, there are no eigenvalues of the Laplacian with $0 < \lambda < \frac{3}{16}$.

To prove his results Selberg introduced the suitable Poincaré series $P_m(z,s)$ and computed the inner product of two such series in two different ways, namely, first by using the spectral expansion and secondly by using the Bruhat decomposition. It is in this second way where Kloosterman sums come into play.

In [17] we studied a type of Poincaré series, we call them **M**-series, that give Γ-invariant eigenfunctions of the Casimir operator for any $\mathbb{R}$-rank one Lie group G, Γ a lattice in G. Let $P = MAN$ be a cuspidal parabolic subgroup of G with MA the standard Levi factor and A the standard split component. The Poincaré series, denoted by $\mathbf{M}^\chi(\xi,\nu)$ for $\xi \in \widehat{M}$, $\nu \in \mathfrak{a}^*$, with $\mathfrak{a}$ the complexified Lie algebra of A, forms a one-parameter family of Γ-invariant eigenfunctions similar to the Eisenstein series, but they do not define automorphic forms since they do not satisfy the condition of moderate growth. They are constructed from Whittaker vectors associated to a generic character χ of $\Gamma \cap N \backslash N$. They are initially given by a convergent series for $\operatorname{Re} \nu > \rho$ and they can be meromorphically continued to all of $\mathfrak{a}^*$. An important property is that although they are exponentially increasing, their residues at the poles in the half-plane $\operatorname{Re} \nu \geq 0$ do define square-integrable automorphic forms and, under some conditions, they generate all of $L^2_{\mathrm{disc}}(\Gamma \backslash G)$.

The goal of this paper is the study of the χ'-Fourier coefficients $D^{\chi}_{\chi'}(\xi,\nu)$ of $\mathbf{M}^\chi(\xi,\nu)$ in connection with the Kloosterman–Selberg zeta function $\zeta_{\chi,\chi',\xi}(\nu)$ for an arbitrary real rank one group G. The coefficient $D^{\chi}_{\chi'}(\xi,\nu)$ is naturally connected to $\zeta_{\chi,\chi',\xi}(\nu)$, since the expansion at infinity of $\mathbf{M}^{\chi}(\xi,\nu)$ yields a similar "expansion" for $D^{\chi}_{\chi'}(\xi,\nu)$ and the first term of this expansion is a multiple of $\zeta_{\chi,\chi',\xi}(\nu)$. Using this fact, we will prove that the meromorphic continuation of $\zeta_{\chi,\chi',\xi}(\nu)$ to $\mathfrak{a}^*$ follows from the continuation of $D^{\chi}_{\chi'}(\xi,\nu)$ (see Theorem 4.2) and we will give a result on the location of the poles of $\zeta_{\chi,\chi',\xi}(\nu)$ in the closed right half-plane (see Theorem 5.2). The pole set of $D^{\chi}_{\chi'}(\xi,\nu)$ in the right half-plane is essentially the set of spectral parameters associated to the M-type ξ, while the pole set of $\zeta_{\chi,\chi',\xi}(\nu)$ will also involve translates of spectral parameters associated to many other M-types $\sigma \in \widehat{M}$. The more regular behavior of $D^{\chi}_{\chi'}(\xi,\nu)$ can be attributed to the fact that the $\mathbf{M}^{\chi}(\xi,\nu)$ are Casimir eigenfunctions, while the Selberg-type Poincaré series satisfies a shift equation. On the other hand, $D^{\chi}_{\chi'}(\xi,\nu)$ is a more complicated function than $\zeta_{\chi,\chi',\xi}(\nu)$, since it is an infinite sum of translates of Kloosterman–Selberg type zeta functions (see (6)).

In the last section, we give the explicit expression of $D^{\chi}_{\chi'}(\xi,\nu)$ in terms of Kloosterman–Selberg zeta functions in the cases of $G = \mathrm{SL}(2,\mathbb{R})$, $G = \mathrm{SO}(n+1,1)$, $n \geq 3$ and $G = \mathrm{SU}(2,1)$.

The Kloosterman–Selberg zeta function has been studied by several authors. Goldfeld and Sarnak ([7]) reobtained Kuznetsov's estimate for averages of Kloosterman sums and Cogdell and Piatetski-Shapiro ([4]) carried out by spectral methods, the meromorphic continuation to $\mathbb{C}$ in the case of $G = \mathrm{SL}(2,\mathbb{R})$. Elstrodt–Grunewald–Menicke ([5]), Cogdell–Li–Piatetski-Shapiro–Sarnak ([2]) and Li ([14]) meromorphically continued the zeta function for discrete subgroups acting on real hyperbolic n-space $\mathbb{R}H^n$ and complex hyperbolic n-space $\mathbb{C}H^n$ respectively. These authors use a generalization of Selberg's Poincaré series and, by first proving adequate estimates for the associated Kloosterman sums, obtain an analog of Selberg's lower bound for λ_1. We note that in the case of [14] this bound is valid only for generic cusp forms, that is, forms having at least one nonzero Fourier coefficient.

2 The definition

Let G be a connected simple Lie group of real rank one, K a maximal compact subgroup, θ the corresponding Cartan involution, and let Γ be a discrete subgroup of finite covolume. Let $P = MAN$ be a cuspidal parabolic subgroup with given Langlands decomposition. Here MA is the standard Levi factor (i.e., $M = K \cap P$) and A is the standard split component. Let $s^* \in K$ such that $s^* a (s^*)^{-1} = a^{-1}$. The Bruhat lemma implies that

$$G = Ps^*N \cup P.$$

This union is disjoint and if $g \in Ps^*N$, then $g = n_1(g)m(g)a_g s^* n_2(g)$ with $n_i(g) \in N, i = 1,2$, $m(g) \in M$, and $a_g \in A$ uniquely determined. We will use the notation $m(g)$ and m_g interchangeably. Let $\chi, \chi' : N \to S^1$ be unitary characters of N, trivial on $\Gamma_N := \Gamma \cap N$. We set $\Gamma_P := \Gamma \cap P$.

Let $\mathfrak{g}$, $\mathfrak{p}$, $\mathfrak{m}$, $\mathfrak{a}$, $\mathfrak{n}$ be respectively the *complexified* Lie algebras of G, P, M, A, N. Relative to A, $\mathfrak{n} = \mathfrak{n}_\lambda \oplus \mathfrak{n}_{2\lambda}$ with λ and 2λ the restricted roots with $[\mathfrak{n}, \mathfrak{n}] = \mathfrak{n}_{2\lambda}$ (if G is locally $\mathrm{SO}(n, 1)$ one has $\mathfrak{n}_{2\lambda} = \{0\}$). We set $\rho = \frac{1}{2}(\dim \mathfrak{n}_\lambda + 2 \dim \mathfrak{n}_{2\lambda})\lambda$. If $\nu \in \mathfrak{a}^*$, then $\nu = z\lambda$ with $z \in \mathbb{C}$. We write $\operatorname{Re} \nu = (\operatorname{Re} z)\lambda$ and $t\lambda > u\lambda$ if $t, u \in \mathbb{R}$ and $t > u$.

One can give a Lie-theoretic definition of Kloosterman sums and of the zeta function introduced by Selberg.

Let (σ, H_σ) be a unitary representation of M. Fix $\mathcal{S}$ to be a set of representatives of the double cosets in $\Gamma_N \backslash \Gamma \cap Ps^* N / \Gamma_N$.

The *Kloosterman–Selberg zeta function* is defined as the operator-valued series for $\nu \in \mathfrak{a}^*$

$$\begin{aligned} \zeta_{\chi,\chi',\sigma}(\nu) &= \sum_{\gamma \in \Gamma_N \backslash (\Gamma - \Gamma_P)/\Gamma_N} \chi(n_1(\gamma))\chi'(n_2(\gamma))a_\gamma^{\nu+\rho}\sigma(m(\gamma)) \\ &= \sum_{m_\gamma a_\gamma \in \mathcal{S}_{MA}} S(\chi, \chi'; m_\gamma a_\gamma)a_\gamma^{\nu+\rho}\sigma(m_\gamma). \end{aligned} \tag{1}$$

Here, S_{MA} is the set of $m_\gamma a_\gamma$ with γ running through $\Gamma \cap Ps^* N$ and

$$S(\chi, \chi'; ma) = \sum_{\gamma} \chi(n_1(\gamma))\, \chi'(n_2(\gamma)) \tag{2}$$

is a *generalized Kloosterman sum*, where, for each $ma \in \mathcal{S}_{MA}$, γ runs over the $\gamma \in \mathcal{S}$ such that $m_\gamma a_\gamma = ma$. In Example 6.1 we show that if $G = SL(2, \mathbb{R})$, Γ is a Hecke congruence subgroup, σ is the trivial representation, P is the upper triangular parabolic subgroup of G, and χ and χ' are appropriate characters of N, then (2) defines the usual Kloosterman sum.

The series in (1) makes sense as a formal series since

$$\chi(n_1(\mu\gamma\eta))\chi'(n_2(\mu\gamma\eta))a_{\mu\gamma\eta}^{\nu+\rho}\sigma(m(\mu\gamma\eta)) = \chi(n_1(\gamma))\chi'(n_2(\gamma))a_\gamma^{\nu+\rho}\sigma(m(\gamma))$$

for $\mu, \eta \in \Gamma_N$ and $\gamma \in \Gamma \setminus \Gamma_P$.

Lemma 2.1. *Let ω be a compact subset of $\{\nu \in \mathfrak{a}^* | \operatorname{Re} \nu > \rho\}$. Then the series in* (1) *converges absolutely and uniformly on ω. Furthermore, there exists a constant $C_\omega < \infty$, depending only on Γ and ω, such that*

$$\left\| \zeta_{\sigma,\chi,\chi'}(\nu) \right\| \leq \zeta_{1,1,1}(\operatorname{Re} \nu) \leq C_\omega$$

for $\nu \in \omega$.

Proof. We note that

$$\begin{aligned} \left\| \zeta_{\sigma,\chi,\chi'}(\nu) \right\| &\leq \sum_{\gamma \in \Gamma_N \backslash (\Gamma - \Gamma_P)/(\Gamma_N)} \left\| \chi(n_1(\gamma))\chi'(n_2(\gamma))a_\gamma^{\nu+\rho}\sigma(m(\gamma)) \right\| \\ &= \zeta_{1,1,1}(\operatorname{Re} \nu). \end{aligned}$$

Let E_Γ be the standard spherical Eisenstein series

$$E_\Gamma(P,\nu,g)=\sum_{\gamma\in\Gamma_N\backslash\Gamma}a(\gamma g)^{\nu+\rho}$$

with $g=n(g)a(g)k(g)$ for $g\in G, n(g)\in N, a(g)\in A$ and $k(g)\in K$. In the proof of [17] Proposition 2.7 (iii) we observed that the constant term of $E_\Gamma(P,\nu,g)$ at $g=1$ is (up to normalization of measures)

$$|\Gamma_P/\Gamma_N|+\sum_{\gamma\in\Gamma_N\backslash(\Gamma-\Gamma_P)/\Gamma_N}\int_N a(\gamma n)^{\nu+\rho}dn$$

with the latter an absolutely convergent series of absolutely convergent integrals for $\operatorname{Re}\nu>\rho$. We note that if $\gamma\in\Gamma_N\backslash(\Gamma-\Gamma_P)/\Gamma_N$, then

$$\int_N a(\gamma n)^{\nu+\rho}dn=a_\gamma^{\nu+\rho}c(\nu).$$

with $c(\nu)$, the Harish-Chandra C-function. □

Our next task is to carry out the meromorphic continuation when σ is finite dimensional and χ is generic. To do this we need to recall several results from [17].

3 The χ-constant term of the M-series

We retain the notation of the preceding section. Let (ξ,H_ξ) be an irreducible continuous representation of M. Let for $\nu\in\mathfrak{a}_\mathbb{C}^*$, $\pi_{\xi,\nu}$ be the principal series action of G on $\operatorname{Ind}_M^K(\xi)$. For each generic χ such that $\chi_{|\Gamma_N}=\{1\}$, a meromorphic family of operators, $\mathbf{M}^\chi(\xi,\nu)$, from the space of K-analytic vectors in $\operatorname{Ind}_M^K(\xi)$, H_ω^ξ, to H_ξ was introduced, such that $\mathbf{M}^\chi(\xi,\nu)\circ\pi_{\xi,\nu}(\gamma)=\mathbf{M}^\chi(\xi,\nu)$ for all $\gamma\in\Gamma$. These operators have the property that if $u\in H_K^\xi$ (the K-finite vectors) and $\eta\in H_\xi^*$, then $g\mapsto\eta(\mathbf{M}^\chi(\xi,\nu)\pi_{\xi,\nu}(g)u)$ satisfies all the conditions of an automorphic function except for moderate growth. We will discuss the spectral properties of this meromorphic family in $L^2(\Gamma\backslash G)$ later in this paper. Here we will confine our attention to one result that is pertinent to the meromorphic continuation of the Kloosterman–Selberg zeta function. We will now introduce the notation needed to describe the result in [17] that we need.

We recall some material from the first appendix in [17]. Let $\bar{\mathfrak{n}}=\theta\mathfrak{n}$. Let B be the Killing form of Lie(G) and let $x\longmapsto\bar{x}$ be complex conjugation of $\mathfrak{g}$ relative to Lie(G). Set $\langle x,y\rangle=-B(\theta x,\bar{y})$ for $x,y\in\mathfrak{g}$. Then $\langle\,,\,\rangle$ is a positive definite K-invariant inner product on $\mathfrak{g}$. Thus $\langle\,,\,\rangle_{|\bar{\mathfrak{n}}}$ is a positive definite M-invariant inner product on $\bar{\mathfrak{n}}$. Let $Y_1,\ldots,Y_p$ be an orthonormal basis of $\bar{\mathfrak{n}}_{-\lambda}$. Extend the orthonormal basis with $Y_{p+1},\ldots,Y_{p+q}$ an orthonormal basis of $\bar{\mathfrak{n}}_{-2\lambda}$

(here $q = 0, 1, 3, 7$). Let $Y^{I,J} = Y_1^{i_1} \cdots Y_p^{i_p} Y_{p+1}^{j_1} \cdots Y_{p+q}^{j_q}$, in the standard multi-index notation. On $U(\overline{\mathfrak{n}})$ we put a Hermitian inner product also denoted $\langle\, ,\, \rangle$ such that $\langle Y^{I,J}, Y^{L,M} \rangle = \frac{1}{I!J!}\delta_{I,L}\delta_{J,M}$. This form is $\mathrm{Ad}(M)$-invariant. In [17], A.1 we defined entire functions $a_{I,J}^{\xi,\chi}(\nu)$ on $\mathfrak{a}^*$ with values in $\mathrm{End}_{\mathbb{C}}(H_\xi)$ such that if ω is a compact subset of $\mathfrak{a}^*$, then

$$\left\| a_{I,J}^{\xi,\chi}(\nu) \right\| \le \frac{C_{\xi,\chi,\omega}^{|I|+|J|}}{(|I|+|J|)!} \tag{3}$$

with $C_{\xi,\omega}$ a positive constant. These functions were part of the following expansion:

$$\varpi^{\chi}(\xi,\nu)(u) = \sum_{I,J} \frac{a_{I,J}^{\xi,\chi}(\nu)}{\sqrt{I!J!}} (\pi_{\xi,\nu}(Y^{I,J})u)(1) \tag{4}$$

of a Whittaker vector in the dual of H_K^ξ relative to χ and the action $\pi_{\xi,\nu}$. If $\gamma \in P$, then we set $\chi^\gamma = \chi \circ \mathrm{Ad}(\gamma)^{-1}$.

The following result is a combination of Proposition 2.7 (ii) in [17] and its proof.

Proposition 3.1. *If χ, χ' are nontrivial characters of N, $u \in H_K^\xi$, and if $\mu \in H_\xi^*$, then*

$$\int_{\Gamma_N \backslash N} \chi'(n)^{-1} \mu(\mathbf{M}^{\chi}(\xi,\nu)\pi_{\xi,\nu}(n)u)\, dn = \mu(D_{\chi'}^{\chi}(P,\xi,\nu) J_{\xi,\nu}^{\chi'}(u))$$
$$+\mu\left(\left(\sum_{\gamma \in \Gamma_N \backslash \Gamma_P} \delta_{\chi',\chi^\gamma} \xi(m(\gamma))\chi(n(\gamma))\right) \varpi^{\chi'}(\xi,\nu)u\right),$$

where $D_{\chi'}^{\chi}(P,\xi,\nu)$ is meromorphic for $\nu \in \mathfrak{a}^*$ with values in $\mathrm{End}_{\mathbb{C}}(H_\xi)$ *and, for* $\mathrm{Re}\,\nu > \rho$, *is given by the formula*

$$D_{\chi'}^{\chi}(P,\xi,\nu)$$
$$= \sum_{\gamma \in \Gamma_N \backslash (\Gamma - \Gamma \cap P)/\Gamma_N} \chi(n_1(\gamma))\chi'(n_2(\gamma)) a_\gamma^{\nu+\rho} \xi(m(\gamma)) \tau(\chi',\chi,m(\gamma)a_\gamma,\nu)$$

with

$$\tau(\chi',\chi,ma,\nu) = \sum_{I,J} \frac{a_{I,J}^{\xi,\chi}(\nu)}{\sqrt{I!J!}} d\chi'(\mathrm{Ad}(mas^*)^{-1} Y^{I,J}).$$

Furthermore, $J_{\xi,\nu}^{\chi'}$ is the Jacquet integral corresponding to $\pi_{\xi,\nu}$ and χ'.

We note that since $d\chi(\mathfrak{n}_{2\lambda}) = 0$ and $\mathrm{Ad}(a^{-1})Y^{I,J} = a^{|I|\lambda+2|J|\lambda}Y^{I,J}$ we have

$$\tau(\chi', \chi, ma, \nu) = \sum_I \frac{a_{I,0}^{\xi,\chi}(\nu)}{\sqrt{I!}} a^{|I|\lambda} d\chi'(\mathrm{Ad}(ms^*)^{-1}Y^{I,0}). \tag{5}$$

We need the following fact about the Jacquet integral (c.f. [22]).

Theorem 3.2. *If χ is nontrivial, then $\nu \mapsto J_{\xi,\nu}^{\chi}$ extends to a weakly holomorphic family of maps, from the C^∞-vectors of H^ξ to H_ξ, that is never* 0. *Thus, for each $\nu \in \mathfrak{a}^*$ we can choose a compact neighborhood of ν and $u \in H_K^\xi$ such that*

$$\left\| J_{\xi,\nu}^{\chi}(u) \right\| \geq C_u > 0.$$

4 The meromorphic continuation

We retain the notation of the previous section. If σ is a finite-dimensional representation of M that is equivalent with the direct sum $\xi_1 \oplus \cdots \oplus \xi_r$ with ξ_i irreducible, then we define

$$D_{\chi'}^{\chi}(P, \sigma, \nu) = D_{\chi'}^{\chi}(P, \xi_1, \nu) \oplus \cdots \oplus D_{\chi'}^{\chi}(P, \xi_r, \nu),$$

and

$$a_{I,J}^{\sigma,\chi}(\nu) = a_{I,J}^{\xi_1,\chi}(\nu) \oplus \cdots \oplus a_{I,J}^{\xi_r,\chi}(\nu).$$

The formula in Prop. 2.7 [17] remains true for $D_{\chi'}^{\chi}(P, \sigma, \nu)$ with ξ replaced by σ. For the purpose of the meromorphic continuation we will define some unitary representations of M. Let, for $j \in \mathbb{Z}_{\geq 0}$, (σ_j, V_j), denote the representation of M on the $-j\lambda$ weight space of A in $U(\bar{\mathfrak{n}})/U(\bar{\mathfrak{n}})\bar{\mathfrak{n}}_{-2\lambda}$. We can think of the elements $\frac{Y^{I,0}}{\sqrt{I!}}$ with $|I| = j$ as an orthonormal basis of this space. We can also think of these elements as an orthonormal basis of V_j^*.

For ω a compact subset of the half space $\mathrm{Re}\,\nu > \rho$ we have

$$\begin{aligned}
&D_{\chi'}^{\chi}(P, \sigma, \nu) \\
&= \sum_{\gamma \in \Gamma_N \backslash (\Gamma \smallsetminus \Gamma_P)/\Gamma_N} \chi(n_1(\gamma))\chi'(n_2(\gamma))a_\gamma^{\nu+\rho}\sigma(m(\gamma))\tau(\chi', \chi, m(\gamma)a_\gamma, \nu) \\
&= \sum_{\gamma \in \Gamma_N \backslash (\Gamma \smallsetminus \Gamma_P)/\Gamma_N} \chi(n_1(\gamma))\chi'(n_2(\gamma))a_\gamma^{\nu+\rho}\sigma(m(\gamma)) \\
&\qquad \times \sum_I \frac{a_{I,0}^{\sigma,\chi}(\nu)}{\sqrt{I!}} a^{|I|\lambda} d\chi'(\mathrm{Ad}(m(\gamma)s^*)^{-1}Y^{I,0}).
\end{aligned}$$

The estimate in (3) in the previous section implies that if we take the term by term norm of the individual terms (for γ and I) we have

$$\left\| \chi(n_1(\gamma))\chi'(n_2(\gamma))a_\gamma^{\nu+\rho}\sigma(m(\gamma))\frac{a_{I,0}^{\sigma,\chi}(\nu)}{\sqrt{I!}}a^{|I|\lambda}d\chi'\big(\mathrm{Ad}(m(\gamma)s^*)^{-1}Y^{I,0}\big) \right\| \leq \frac{a_\gamma^{\nu+j\lambda+\rho}\,\|d\chi'\|^j\,(\max(C_{\sigma_j,\chi,\omega}))^j}{j!}.$$

Now Lemma in Appendix A in [17] implies that the series defining $D_{\chi'}^{\chi}(P,\sigma,\nu)$ for $\mathrm{Re}\,\nu > \rho$ converges absolutely; we can thus interchange the order of summation. Define η_j to be the element of V_j^* given by $d\chi' \circ \mathrm{Ad}(s^*)^{-1}$ restricted to V_j. We have, after the obvious manipulations, for $z \in H_\xi$,

$$D_{\chi'}^{\chi}(P,\sigma,\nu)(z) = \sum_{j=0}^{\infty}\sum_{|I|=j} a_{I,0}^{\sigma,\chi}(\nu)(\mathrm{Id}\otimes\eta_j)\zeta_{\chi,\chi',\sigma\otimes\sigma_j^*}(\nu+j\lambda)(z\otimes Y^{I,0}). \tag{6}$$

For each $r \in \mathbb{N}_0 = \mathbb{N}\cup\{0\}$ we define

$$D_{\chi'}^{\chi}(P,\sigma,\nu)_{\geq r}(z) = \sum_{j\geq r}^{\infty}\sum_{|I|=j} a_{I,0}^{\sigma,\chi}(\nu)(\mathrm{Id}\otimes\eta_j)\zeta_{\chi,\chi',\sigma\otimes\sigma_j^*}(\nu+j\lambda)(z\otimes Y^{I,0}).$$

Then the same estimates as above imply

Lemma 4.1. *$D_{\chi'}^{\chi}(P,\sigma,\nu)_{\geq r}$ is holomorphic in ν for* $\mathrm{Re}\,\nu > \rho - r\lambda$.

We now relate this result to the Kloosterman–Selberg zeta functions.

Since we are only interested in the range $\mathrm{Re}\,\nu \geq 0$ we can replace $D_{\chi}^{\chi}(P,\xi,\nu)(z)$ with

$$I_\xi(\nu)\zeta_{\xi,\chi,\chi}(\nu)(z) + \sum_{j=1}^{r}\sum_{|I|=j} a_{I,0}^{\xi,\chi}(\nu)(\mathrm{Id}\otimes\eta_j)\zeta_{\chi,\chi,\xi\otimes\sigma_j^*}(\nu+j\lambda)(z\otimes Y^{I,0}) \tag{7}$$

where r is the smallest integer such that $\rho < r\lambda$ (i.e., $r = \frac{p+2q}{2}+1$ if p is even and $\frac{p+2q}{2}+\frac{1}{2}$ if p is odd) and $I_\xi(\nu)$ is an entire function.

Theorem 4.2. *The Kloosterman–Selberg zeta function, $\zeta_{\sigma,\chi,\chi'}(\nu)$, has a meromorphic continuation to all of $\mathfrak{a}^*$ for every unitary finite-dimensional representation σ of M.*

Proof. We set

$$I_\sigma(\nu) = I_{\xi_1}(\nu)\oplus\ldots\oplus I_{\xi_r}(\nu).$$

Suppose that we have shown that $D^{\chi}_{\chi'}(P,\mu,\nu)$ has a meromorphic continuation to the set $\operatorname{Re}\nu > \rho - r\lambda$ for all finite-dimensional unitary representations μ of M. Then if σ is a unitary irreducible representation of M, we have for $z \in H_{\xi}$ if $\operatorname{Re}\nu > \rho - j|\lambda|$

$$D^{\chi}_{\chi'}(P,\sigma,\nu)(z) = I_{\sigma}(\nu)\zeta_{\sigma,\chi,\chi'}(\nu)(z) \tag{8}$$

$$+\sum_{j=1}^{r}\sum_{|I|=j} a^{\sigma,\chi}_{I,0}(\nu)(\mathrm{Id}\otimes\eta_j)\zeta_{\chi,\chi',\sigma\otimes\sigma_j^*}(\nu+j\lambda)(z\otimes Y^{I,0}) + D^{\chi}_{\chi'}(P,\sigma,\nu)_{\geq r+1}(z).$$

The left-hand side of the equation is meromorphic for all ν and every term other than the first one is meromorphic for $\operatorname{Re}\nu > \rho-(j+1)\lambda$, by our assumption. Thus the first term is also meromorphic for $\operatorname{Re}\nu > \rho-(j+1)\lambda$. □

5 Spectral properties

In this section we will study the relationship between the poles of the $\mathbf{M}^{\chi}$-series, those of the $D^{\chi}_{\chi'}$ and those of the $\zeta_{\chi,\chi'}$ with the discrete spectrum of $L^2(\Gamma\backslash G)$. We retain the notation of the previous sections. We will study these meromorphic functions in the half-plane $\operatorname{Re}\nu \geq 0$. We will recall some material from Section 3 of [17].

Let $(\xi.H_{\xi})$ be an irreducible unitary representation of M. Let ξ^* be the contragredient representation. Then for $\operatorname{Re}\nu > \rho$ we have

$$\mathbf{M}^{\chi}(\xi,\nu)(u) = \sum_{\gamma\in\Gamma_N\backslash\Gamma} \varpi^{\chi}(\xi,\nu)(\pi_{\xi,\nu}(\gamma)u)$$

with (in the notation of (4))

$$\varpi^{\chi}(\xi,\nu)(u) = \sum_{I,J}\frac{a^{\xi,\chi}_{I,J}(\nu)}{\sqrt{I!J!}}(\pi_{\xi,\nu}(Y^{I,J})u)(1).$$

In [17, A.1] the coefficients $a^{\xi,\chi}_{I,J}(\nu)$ are given as a product

$$a^{\xi,\chi}_{I,J}(\nu) = I_{\xi}(\nu)b^{\xi,\chi}_{I,J}(\nu)$$

with $b^{\xi,\chi}_{I,J}$ rational in ν with values in $\operatorname{End}_{\mathbb{C}}(H_{\xi})$ and $I_{\xi}(\nu)$ an entire function such that $a^{\xi,\chi}_{I,J}(\nu)$ is entire. We recall the definition of $b^{\xi,\chi}_{I,J}(\nu)$. Let $M^{\xi^*,-\nu}$ be the Verma module

$$M^{\xi^*,-\nu} = U(\mathfrak{g})\otimes_{U(\mathfrak{p})}H_{\xi^*,-\nu}$$

with $H_{\xi^*,-\nu}$, the $\mathfrak{p} = \mathfrak{m} \oplus \mathfrak{a} \oplus \mathfrak{n}$-module acting as follows: $\mathfrak{m}$ acts by $d\xi^*$, $h \in \mathfrak{a}$ acts by $(-\nu - \rho)(h)I$ and $\mathfrak{n}$ acts by 0. Let $\overline{\mathfrak{p}} = \mathfrak{m} \oplus \mathfrak{a} \oplus \overline{\mathfrak{n}}$; then as a $\overline{\mathfrak{p}}$-module $M^{\xi^*,-\nu} = U(\overline{\mathfrak{n}}) \otimes H_{\xi^*,-\nu}$ (here $\overline{\mathfrak{n}}$ acts by multiplication in the first factor and $\mathfrak{m} \oplus \mathfrak{a}$ acts by ad on $U(\overline{\mathfrak{n}})$). Thus $M^{\xi^*,-\nu} = \bigoplus_{I,J} Y^{I,J} \otimes H_{\xi^*,-\nu}$. In [17, A.1] we observed that if $M^{\xi^*,-\nu}$ is irreducible, then the formal completion of $M^{\xi^*,-\nu}$ relative to $\overline{\mathfrak{n}}$ (that is, the formal series in the $Y^{I,J}$) is equivalent with

$$U(\mathfrak{n})^* \otimes H_{\xi^*,-\nu}$$

as an $\mathfrak{n}$-module with $\mathfrak{n}$ acting only on the first factor by $xf(n) = f(nx)$. In particular, if $\phi : \mathfrak{n} \to \mathbb{C}$ is a Lie algebra homomorphism, if $\mu \in H_{\xi^*} = H_\xi^*$ and if $M^{\xi^*,-\nu}$ is irreducible, then there is a unique element of the completion

$$T_{\xi,\nu} = \sum_{I,J} Y^{I,J} \otimes c_{I,J}^{\xi,\phi}(\nu)(\mu)$$

such that $c_{I,J}^{\xi,\chi}(\nu) \in \operatorname{End}(H_\xi^*)$ and

$$xT_{\xi,\nu}(u) = \phi(x)T_{\xi,\nu}(u) \text{ and } c_{0,0}^{\xi,\phi}(\nu)(\mu) = \mu.$$

We now consider the $\mathfrak{g}$-module map

$$S_{\xi,\nu} : M^{\xi^*,\nu} \to (H_K^\xi)^*$$

given by

$$S_{\xi,\nu}(x \otimes \mu)(u) = \mu((x\delta_{\xi,\nu})(u))$$

for $u \in (H_K^\xi)^*$. Here $\delta_{\xi,\nu}(u) = v(u)$. This leads us to a formal formula for $I_\xi(\nu)^{-1}\varpi^\chi(\xi,\nu)$ (with $d\chi = \phi$)

$$I_\xi(\nu)^{-1}\mu(\varpi^\chi(\xi,\nu)(u)) = \sum c_{I,J}^{\xi,\phi}(\nu)(\mu)(Y^{I,J}\delta_{\xi,\nu})(u).$$

If V is a finite-dimensional vector space and if $L \in \operatorname{End}(V^*)$, then we define $L^T \in \operatorname{End}(V)$ by $\mu(L^T(u)) = L(\mu)(u)$. ($\mu \in V^*$ and $u \in V$). Then

$$b_{I,J}^{\xi,\chi}(\nu) = (c_{I,J}^{\xi,\chi}(\nu))^T.$$

Arguing as in [17, A.1] (and also as in [8]) one can prove that $b_{I,J}^{\xi,\chi}(\nu)$ is rational in ν for fixed ξ with poles at the discrete set of rational multiples of λ such that $M^{\xi^*,-\nu}$ is reducible. We set $\mathcal{P}(\xi)$ equal to the (finite) set of ν with $\operatorname{Re}\nu \geq 0$ such that $M^{\xi^*,-\nu}$ is reducible. We replace $\mathbf{M}^\chi(\xi,\nu)$ with $I_\xi(\nu)^{-1}\mathbf{M}^\chi(\xi,\nu)$, $D_{\chi'}^\chi(P,\xi,\nu)$ by $I_\xi(\nu)^{-1}D_{\chi'}^\chi(P,\xi,\nu)$ and $\tau(\chi',\chi,ma,\nu)$ with $I_\xi(\nu)^{-1}\sum_I \frac{a_{I,0}^{\xi,\chi}(\nu)}{\sqrt{I!}} a^{|I|\lambda} d\chi'(\operatorname{Ad}(ms^*)^{-1}Y^{I,0})$.

With this notation in place we can rephrase a combination of Theorem 3.2 and Corollary 3.5 in [17] as follows:

Theorem 5.1. *Suppose that* $\nu_0 \in \mathrm{Spec}(\sigma, \chi)$*, that is,* $\mathrm{Re}\,\nu_0 \geq 0$ *and an irreducible quotient,* V*, of* π_{ξ,ν_0} *occurs in the discrete spectrum of* $L^2(\Gamma\backslash G)$ *and an automorphic form,* f*, corresponding to* V *is such that*

$$\int_{\Gamma_N\backslash N} f(ng)\chi(n)^{-1}dn \neq 0$$

with χ *nontrivial. Then* $\nu D_\chi^\chi(P,\xi,\nu)$ *has a pole at* ν_0*. If* $\nu_0 \notin \mathcal{P}(\sim)$ *(or* V *is spherical), then the converse is true.*

Remark 1. *In fact, we show (in the converse situation) that one can construct the image of* V *from residues of* $\nu M^\chi(\xi,\nu)$ *(here we identify* ν *with* z *if* $\nu = z\lambda$*).*

We can now obtain information on the pole set $\mathcal{P}(\boldsymbol{\zeta}(\sigma,\chi,\chi'))$ of $\zeta_{\sigma,\chi,\chi'}(\nu)$ by using relation (6) together with Theorem 5.1.

Theorem 5.2. *Let* σ *be a finite-dimensional representation of* M *and let* χ, χ' *be generic unitary characters of* N mod Γ_N*. Denote by* $\mathcal{P}(D(\sigma,\chi,\chi'))$*,* $\mathcal{P}(\boldsymbol{\zeta}(\sigma,\chi,\chi'))$ *and* $\mathcal{P}(\varpi(\sigma,\chi))$*, the pole sets of* $D_{\chi'}^\chi(P,\sigma,\nu)$*,* $\zeta_{\sigma,\chi,\chi'}(\nu)$ *and* $\varpi^\chi(\sigma,\nu)$*, respectively.*

Let $\mathrm{Spec}(\sigma,\chi)$ *be as defined in Theorem 5.1 and for any* $\Omega \subset \mathfrak{a}^*$*, let* $\Omega_+ = \Omega \cap \{\nu \in \mathfrak{a}^* : \mathrm{Re}\,\nu \geq 0\}$*. Let, for each* $j \in N$*,* (σ_j, V_j) *denote the representation of* M *on the* $-j\lambda$ *weight space of* A *on* $U(\overline{\mathfrak{n}})/U(\overline{\mathfrak{n}})\overline{\mathfrak{n}}_{-2\lambda}$*.*

Then we have

$$\mathcal{P}(\boldsymbol{\zeta}(\sigma,\chi,\chi'))_+ \subset \bigcup_{0\leq\sum k_j\leq[\rho]} \left\{\mathrm{Spec}(\sigma\otimes\sigma_{k_1}^*\otimes\ldots\otimes\sigma_{k_r}^*,\chi) - \sum_1^r k_j\right\}$$
$$\cup \bigcup_{0\leq\sum k_j\leq[\rho]} \left\{\mathcal{P}(\varpi(\sigma\otimes\sigma_{k_1}^*\otimes\ldots\otimes\sigma_{k_r}^*,\chi)) - \sum_1^r k_j\right\}.$$

Proof. We know from Theorem 5.1 (see also [17], Thm 2.5 and Thm 3.2) that for every σ, χ

$$\mathcal{P}(D(\sigma,\chi,\chi))_+ \subset \mathrm{Spec}(\sigma,\chi) \cup \mathcal{P}(\varpi(\sigma,\chi)). \tag{9}$$

Now if we apply (6) and (9) we see that

$$\mathcal{P}(\boldsymbol{\zeta}(\sigma,\chi,\chi')_+ \subset \mathrm{Spec}(\sigma,\chi) \cup \mathcal{P}(\varpi(\sigma,\chi))$$
$$\cup \bigcup_{k=1}^{[\rho]} \{\mathcal{P}(\boldsymbol{\zeta}(\sigma\otimes\sigma_k^*,\chi,\chi')) - k\}.$$

Similarly, for each $k \in \mathbb{N}$, $\mathcal{P}(\boldsymbol{\zeta}(\sigma \otimes \sigma_k^*, \chi, \chi'))_+$ is included in

$$\operatorname{Spec}(\sigma \otimes \sigma_k^*, \chi) \cup \mathcal{P}(\varpi(\sigma \otimes \sigma_k^*, \chi)) \cup \bigcup_{h=1}^{[\rho]} \{\mathcal{P}(\boldsymbol{\zeta}(\sigma \otimes \sigma_k^* \otimes \sigma_h^*, \chi, \chi')) - h\}.$$

Thus

$$\begin{aligned}\mathcal{P}(\boldsymbol{\zeta}(\sigma, \chi, \chi'))_+ \subset \operatorname{Spec}(\sigma, \chi) &\cup \bigcup_{k_1=1}^{[\rho]} \{\operatorname{Spec}(\sigma \otimes \sigma_{k_1}^*, \chi) - k_1\} \\ &\cup \mathcal{P}(\varpi(\sigma, \chi)) \cup \bigcup_{k_1=1}^{[\rho]} \{\mathcal{P}(\varpi(\sigma \otimes \sigma_{k_1}^*, \chi)) - k_1\} \\ &\cup \bigcup_{k_1+k_2 \le [\rho]} \{\mathcal{P}(\zeta(\sigma \otimes \sigma_{k_1}^* \otimes \sigma_{k_2}^*, \chi)) - (k_1 + k_2)\}.\end{aligned}$$

By iteration of this argument we arrive at

$$\begin{aligned}&\mathcal{P}(\boldsymbol{\zeta}(\sigma, \chi, \chi'))_+ \subset \\ &\bigcup_{\sum k_j \le [\rho]} \left\{\operatorname{Spec}(\sigma \otimes \sigma_{k_1}^* \otimes \ldots \otimes \sigma_{k_r}^*, \chi) - \sum_{j=1}^{r} k_j \,:\, k_j \in \mathbb{N}_0,\ r \ge 0\right\} \\ &\cup \bigcup_{\sum k_j \le [\rho]} \left\{\mathcal{P}(\varpi(\sigma \otimes \sigma_{k_1}^* \otimes \ldots \otimes \sigma_{k_r}^*, \chi)) - \sum_{j=1}^{r} k_j \,:\, k_j \in \mathbb{N}_0,\ r \ge 0\right\}\end{aligned}$$

as asserted. □

Note. In [2] the authors carry out the meromorphic continuation of $\zeta(\sigma, \chi, \chi')$ in the case of $G = \mathrm{SO}(n+1, 1)$ and give a precise description of its poles in all of $\mathfrak{a}$. In this paper, we have given a short uniform proof of the meromorphic continuation of the zeta function for arbitrary rank one groups, based on the meromorphic continuation of the **M**-series, by connecting the zeta function with the Fourier coefficient $D_\chi^\chi(P, \xi, \nu)$ defined in [17][1].

[1]The contents of this paper were described by the second named author in a lecture in Marseille-Luminy, in 2002, in honor of Jacques Carmona.

6 Examples

In this section we exhibit explicitly the function $D_{\chi}^{\chi}(1,\nu)$ for $\mathrm{Re}(\nu) > \rho$, for P the standard parabolic and 1 the trivial representation of M, together with its expression in terms of Kloosterman–Selberg zeta functions for the groups $G = \mathrm{SL}(2,\mathbb{R})$, $G = \mathrm{SO}(n+1,1)$, $n \geq 3$ and $G = \mathrm{SU}(2,1)$. The starting point is the expression

$$\begin{aligned}
D_{\chi'}^{\chi}(1,\nu) &= \sum_{\delta \in S(N,N)} \chi(n_1(\delta))\, \chi'(n_2(\delta)) a_\delta^{\nu+\rho} \tau(\chi',\chi, m_\delta a_\delta, \nu) \\
&= \sum_{\delta \in S(N,N)} \chi(n_1(\delta))\, \chi'(n_2(\delta)) a_\delta^{\nu+\rho} \\
&\quad \times \sum_{i \in \mathbb{N}^n} a_I(1,\nu) a_\delta^{n\alpha}(d\chi)\left(\mathrm{Ad}(m_\delta s^*)^{-1} Y(I)^t\right) \\
&= \sum_{n \in \mathbb{N}} \sum_{I \in \mathbb{N}^n} a_I(1,\nu) \\
&\quad \times \sum_{\delta \in S(N,N)} \chi(n_1(\delta))\, \chi'(n_2(\delta)) a_\delta^{\nu+\rho+n\alpha}(d\chi)\left(\mathrm{Ad}(m_\delta s^*)^{-1} Y(I)^t\right).
\end{aligned}$$

6.1 $G = \mathrm{SL}(2,\mathbb{R})$

Let first $G = \mathrm{SL}_2(\mathbb{R})$ and $\Gamma = \mathrm{SL}_2(\mathbb{Z})$. Then $G = NAK$ with $A = \left\{a_t = \begin{pmatrix} t & 0 \\ 0 & 1/t \end{pmatrix} : t > 0\right\}$, $K = \mathrm{SO}(2)$, $N = \left\{\begin{pmatrix} 1 & * \\ 0 & 1 \end{pmatrix}\right\}$. Let $P = MAN = \left\{\begin{pmatrix} * & * \\ 0 & * \end{pmatrix}\right\}$, $M = \{\pm I\}$.

We have in this case $a_t^\lambda = \begin{pmatrix} t & 0 \\ 0 & 1/t \end{pmatrix}^\lambda = t^2$ and $\rho = \frac{1}{2}\lambda$.

We fix unitary characters on N, $\chi\begin{pmatrix} 1 & x \\ 0 & 1 \end{pmatrix} = e^{2\pi i m_1 x}$ and $\chi'\begin{pmatrix} 1 & x \\ 0 & 1 \end{pmatrix} = e^{2\pi i m_2 x}$ with $m_1, m_2 \in \mathbb{Z} \smallsetminus \{0\}$.

For $\gamma = \begin{pmatrix} a & b \\ c & d \end{pmatrix} \in \Gamma \cap P \begin{pmatrix} 0 & -1 \\ 1 & 0 \end{pmatrix} N$ one computes that

$$n_1(\gamma) = \begin{pmatrix} 1 & a/c \\ 0 & 1 \end{pmatrix}, n_2(\gamma) = \begin{pmatrix} 1 & d/c \\ 0 & 1 \end{pmatrix}, m_\gamma = \mathrm{sign}(c)\begin{pmatrix} 1 & 0 \\ 0 & 1 \end{pmatrix}, a_\gamma = \begin{pmatrix} 1/|c| & 0 \\ 0 & |c| \end{pmatrix}. \tag{10}$$

Thus, if $m_\gamma a_\gamma = \pm\begin{pmatrix} 1/k & 0 \\ 0 & k \end{pmatrix}$ with $k \in \mathbb{N}$, using (2) and (10), it is not hard to check that $S(\chi_{m_1}, \chi_{m_2}, m_\gamma a_\gamma) = S(m_1, m_2; k)$, a classical Kloosterman sum.

We have that $\zeta_{\chi_{m_1}, \chi_{m_2}, 1}(\nu) = 2\sum_{c \geq 1} S(m_1, m_2; c) c^{-2\nu-1}$ is a multiple of the Selberg zeta function with $\nu = s - \frac{1}{2}$ and $\sigma = 1$.

Note. We give the Kloosterman sums in another standard example, that of $G = \mathrm{SL}_2(\mathbb{C})$, $\Gamma = \mathrm{SL}_2(\mathcal{O})$ where $\mathcal{O}$ is the ring of integers in some imaginary quadratic field extension F of $\mathbb{Q}$. Here $N = \left\{\left(\begin{smallmatrix} 1 & * \\ 0 & 1 \end{smallmatrix}\right)\right\}$, $\chi\left(\begin{smallmatrix} 1 & x \\ 0 & 1 \end{smallmatrix}\right) = e^{2\pi i\,\mathrm{Tr}(m_1 x)}$, $\chi'\left(\begin{smallmatrix} 1 & x \\ 0 & 1 \end{smallmatrix}\right) = e^{2\pi i\,\mathrm{Tr}(m_2 x)}$ where Tr is the trace of F over $\mathbb{Q}$ and where $m_1, m_2 \in F \setminus \{0\}$ are such that $\mathrm{Tr}(m_j x) \in \mathbb{Z}$ for all $x \in \mathcal{O}$. With a similar reasoning as before we find that, for $\xi = \left(\begin{smallmatrix} 1/u & 0 \\ 0 & u \end{smallmatrix}\right) \in MA$ with $u \in \mathcal{O} \setminus \{0\}$, $S(\xi) = \sum_v e^{2\pi i(\mathrm{Tr}((m_1\widetilde{v}+m_2 v)/u))}$, where v runs over representatives of $\mathcal{O}/u\mathcal{O}$ for which $\widetilde{v} \in \mathcal{O}$ can be found such that $\widetilde{v}v \in 1 + u\mathcal{O}$.

Going back to $G = \mathrm{SL}_2(\mathbb{R})$ let $\chi = \chi_r$, $\chi' = \chi_{r'}$, with $r, r' \in \mathbb{Z}$, in the notation of [17, Section 4]. By using the explicit expression of $\tau(\chi', \chi, ma, \nu)$, we get

$$
\begin{aligned}
D^{\chi}_{\chi'}(\xi,\nu) &= \sum_{\delta\in S(N,N)} \chi(n_1(\delta))\,\chi(n_2(\delta))\xi(m_\delta)a_\delta^{\rho}\tau(\chi',\chi,m_\delta a_\delta,\nu) \\
&= \sum_{n=0}^{\infty} \frac{(-1)^n(4\pi r r'/h)^n}{n!\prod_{s=1}^{n}(2\nu+s)} \sum_{\delta\in S(N,N)} \chi(n_1(\delta))\,\chi'(n_2(\delta))a_\delta^{\nu+\rho+n\alpha}\xi(m_\delta) \\
&= \sum_{n=0}^{\infty} \frac{(-1)^n(4\pi r r'/h)^n}{n!\prod_{s=1}^{n}(2\nu+s)} \zeta_{\chi,\chi',\xi}(\nu+n\alpha).
\end{aligned}
$$

6.2 $G = \mathrm{SO}(n+1,1)$, $n \geq 3$

From [18, Cor.A.2] we see that if $\mathrm{Re}\,\nu > \rho$, then

$$
\begin{aligned}
D^{\chi}_{\chi}(1,\nu) &= \sum_{\delta\in S(N,N)} \chi(n_1(\delta))\,\chi(n_2(\delta))a_\delta^{\nu+\rho} \\
&\quad\times \sum_{j,k\geq 0} \frac{\lambda^{2j+6k}a_\delta^{(j+4k)\alpha}\langle Y_1, \mathrm{Ad}(m_\delta)Y_1\rangle^j}{2^k j!k!\prod_{r=0}^{2k+j-1}(\nu+n/2+r)\prod_{r=1}^{k}(\nu+1/2+r)}
\end{aligned}
$$

$$
= \sum_{j,k\geq 0} \frac{\lambda^{2j+6k}\sum_{\delta\in S(N,N)}\chi(n_1(\delta))\,\chi(n_2(\delta))a_\delta^{\nu+\rho+(j+4k)\alpha}\langle Y_1, \mathrm{Ad}(m_\delta)Y_1\rangle^j}{2^k j!k!\prod_{r=0}^{2k+j-1}(\nu+n/2+r)\prod_{r=1}^{k}(\nu+1/2+r)}.
$$

6.3 $G = \mathrm{SU}(2, 1)$

In the notation of Proposition A.4 and Corollary A.5 in [18] we see that if $\mathrm{Re}\,\nu > \rho$, then

$$D_\chi^\chi(1, \nu) = \sum_{\delta \in S(N,N)} \chi(n_1(\delta))\,\chi(n_2(\delta)) a_\delta^{\nu+\rho} \times \sum_{j,k \geq 0} \frac{\prod_{r=1}^{j+k}(\nu + r) a_\delta^{(j+k)\alpha} e^{3i(j-k)\theta_{m_\delta}}}{j!k!\prod_{r=1}^{j}(\frac{\nu}{2} + r)(\nu + r)\prod_{r=1}^{k}(\frac{\nu}{2} + r)(\nu + r)}$$

$$= \sum_{j,k \geq 0} \frac{\prod_{r=k+1}^{j+k}(\nu + r)\xi_\chi^{2j}\bar{\xi}_\chi^{2k}}{j!k!\prod_{r=1}^{j}(\frac{\nu}{2} + r)(\nu + r)\prod_{r=1}^{k}(\frac{\nu}{2} + r)} \times \sum_{\delta \in S(N,N)} \chi(n_1(\delta))\,\chi(n_2(\delta)) a_\delta^{\nu+\rho+(j+k)\alpha} e^{3i(j-k)\theta_{m_\delta}}.$$

References

1. R.W. Bruggeman, R.J. Miatello: *Estimates of Kloosterman sums for groups of real rank one,* Duke Math. J. **80** (1995) 105–137.
2. J. Cogdell, J.-S. Li, I. Piatetski-Shapiro, P. Sarnak: *Poincaré series for* SO$(n, 1)$, Acta Math. **167** (1991), 229–285.
3. J.W. Cogdell, I.I. Piatetski-Shapiro: *The meromorphic continuation of Kloosterman–Selberg zeta functions,* Complex Analysis and Geometry (Pisa 1988), pp. 23–35, Lecture Notes in Math. **1422**, Springer Verlag, 1990.
4. J.W. Cogdell, I.I. Piatetski-Shapiro: *The Arithmetic and Spectral Analysis of Poincaré Series,* Perspectives in Mathematics **13**, Academic Press, 1990.
5. J. Elstrodt, F. Grunewald, J. Mennicke: *Kloosterman sums for Clifford algebras and a lower bound for the positive eigenvalues of the Laplacian for congruence subgroups acting on hyperbolic spaces,* Invent. math. **101** (1990), 641–685.
6. S. Gelbart, D. Soudry: *On Whittaker models and the vanishing of Fourier coefficients of cusp forms,* Proc. Indian Acad. Sci. Math. Sci. 97, Ramanujan Birth Centenary Volume, pp. 67–74.
7. D. Goldfeld, P. Sarnak: *Sums of Kloosterman sums,* Invent. Math. **71** (1983), 243–250.
8. R. Goodman, N. Wallach: *Whittaker vectors and conical vectors,* Jour. Functional Analysis. Math. **39** (1980) ,199–279.
9. S. Helgason: *Groups and Geometric Analysis,* Academic Press, 1984.
10. H. Kim, F. Shahidi: *Functorial products for* $GL(2) \times GL(3)$ *and functorial symmetric cube for* $GL(2)$, C.R. Acad. Sci. Paris **331** (2001), 599–604.
11. N.V. Kuznetsov: *The Petersson hypothesis for forms of weight zero and the Linnik hypothesis (Russian),* Preprint Khabarovsk, 1977.
12. N. V. Kuznetsov: *The Petersson hypothesis for parabolic forms of weight zero and the Linnik hypothesis. Sums of Kloosterman sums (Russian),* Mat. Sbornik **111** (153) (1980), 334–383.
13. R. P. Langlands: *On the Functional Equations Satisfied by Eisenstein Series,* Lecture Notes in Math. **544**, Springer Verlag, 1976.

14. J.-S. Li: *Kloosterman–Selberg zeta functions on complex hyperbolic spaces,* Amer. J. of Math. **113** (1991), 653–731.
15. Ju. V. Linnik: *Additive problems and eigenvalues of the modular operators,* Proc. Int. Congr. Math., 1962, Stockholm, pp. 270–284, Inst. Mittag-Leffler, 1963.
16. W. Luo, Z. Rudnick, P. Sarnak: *On Selberg's eigenvalue conjecture,* Geometric and Functional Analysis **5** (1995), 387–401.
17. R.J. Miatello, N. R. Wallach: *Automorphic forms constructed from Whittaker vectors,* J. Funct. Analysis **86** (1989), 411–487.
18. R.J. Miatello, N. R. Wallach: *Kuznetsov formulas for rank one groups,* J. Funct. Anal. **93** (1990), 171–207.
19. P. Sarnak: *Additive number theory and Maass forms,* Number Theory (New York 1982), pp. 286–309, Lecture Notes in Math. **1052**, Springer Verlag 1984.
20. A. Selberg: *On the estimation of Fourier coefficients of modular forms,* Number Theory, Proc. Symp. Pure Math. VIII, A.M.S. (1965), 1–15 .
21. N.R. Wallach: *Real Reductive Groups II,* Academic Press, 1989..
22. N.R. Wallach: *Lie Algebra Cohomology and Holomorphic Continuation of Generalized Jacquet Integrals,* Advanced Studies in Math **14** (1988), pp. 123–151.
23. A. Weil: *On some exponential sums,* Proc. Nat. Acad. Sci USA **34** (1948) 204–207.

Ricci flow and manifolds with positive curvature

Lei Ni

Dedicated to Nolan Wallach on his 70th birthday

Abstract This is an expository article based on the author's lecture delivered at the conference *Lie Theory and Its Applications* in March 2011, UCSD. We discuss various notions of positivity and their relations with the study of the Ricci flow, including a proof of the assertion, due to Wolfson and the author, that the Ricci flow preserves the positivity of the complex sectional curvature. We discuss the examples of Wallach of the manifolds with positive pinched sectional curvature and the behavior of Ricci flow on some examples. Finally we discuss the recent joint work with Wilking on the manifolds with pinched flag curvature and some open problems.

Keywords: Positivity of the Curvature • Ricci flow • Flag curvature pinching

Mathematics Subject Classification: 53C44

1 Introduction

This article is based on the author's lecture delivered at the conference *Lie Theory and Its Applications* in March 2011, UCSD.

Gauss curvature was defined for surfaces in three-dimensional Euclidean space $\mathbb{R}^3$ by the determinant of the second fundamental form of the embedding with respect to the first fundamental form, namely the induced metric. The *Theorem*

The research of the author was partially supported by NSF grant DMS-1105549.

L. Ni
Department of Mathematics, University of California at San Diego, La Jolla, CA 92093, USA
e-mail: lni@math.ucsd.edu

R. Howe et al. (eds.), *Symmetry: Representation Theory and Its Applications*, Progress in Mathematics 257, DOI 10.1007/978-1-4939-1590-3_17

Egregium of Gauss [11] asserts that it is in fact an invariant depending only on the first fundamental form, namely the metric of the given surface. Let (M, g) be a Riemannian manifold with metric $g = g_{ij} dx^i \otimes dx^j$. For any given point $p \in M$, let T_pM be the tangent space at p and let $\exp_p : T_pM \to M$ be the exponential map at p. The concept of the sectional curvature was introduced by Riemann [26], which can be described via the Gauss curvature in the following way. For any two-dimensional subspace σ, say spanned by e_1, e_2 with $\{e_i\}$ being an orthonormal frame of T_pM, take an open neighborhood (of the origin) $U \subset \sigma$, the sectional curvature $K(\sigma)$ is defined by the Gauss curvature of the surface $\exp_p U$ at p. It is the same as $R(e_1, e_2, e_1, e_2)$, where $R(\cdot, \cdot, \cdot, \cdot)$ is the curvature tensor defined by

$$R(X, Y, Z, W) = \langle -\nabla_X \nabla_Y Z + \nabla_Y \nabla_X Z + \nabla_{[X,Y]} Z, W \rangle,$$

which measures the commutability of the second-order covariant differentiations.

Understanding the topology/differential topology of manifolds with positive sectional curvature has been one of the central problems in the study of Riemannian geometry. In this article we shall illustrate how Hamilton's Ricci flow can be applied to study manifolds with positive sectional curvature. In this regard we shall focus on (1) Ricci flow and various notions of positivity; (2) Wilking's general result on the invariance of various positive cones; (3) examples of manifolds with positive sectional curvature, particularly by Wallach and Aloff–Wallach, and on which the Ricci flow does not preserve the positivity of the sectional curvature by the author and by Böhm and Wilking; (4) the most recent classification result by Wilking and the author on manifolds with so-called pinched flag curvature. The selection of the topics is of course completely subjective. One should consult the excellent survey articles [29, 32] on the subject of the manifolds with positive sectional curvature, particularly on more comprehensive overviews about recent progress via other techniques, e.g., the actions of the isometry groups. These articles also contain many more open problems, some of which ambitious readers may find interesting.

Acknowledgment. The author would like to thank Ann Kostant for the careful editing of the article.

2 Ricci flow and preserved positivities

Let $\mathrm{Ric} = r_{ij} dx^i \otimes dx^j$ be the Ricci curvature tensor which is defined as $r_{ij} = g^{kl} R_{ikjl}$. The Ricci flow is a parabolic PDE which deforms a Riemannian metric by its Ricci curvature:

$$\frac{\partial}{\partial t} g_{ij} = -2 r_{ij}. \tag{1}$$

It is parabolic since under a "good coordinate" (precisely the harmonic coordinate),

$$r_{ij} = -\frac{1}{2} g^{st} \frac{\partial^2}{\partial x^s \partial x^t} g_{ij} + o(1).$$

Here $o(1)$ means terms involving at most the first-order derivatives. This also explains the number "2" in the equation (1).

Since the "good coordinates" are not invariant under the flow, to prove the short time existence, the most economic approach is via the De-Turck trick:

First solve the Ricci–DeTurck equation

$$\begin{cases} \frac{\partial g_{ij}(x,t)}{\partial t} &= -2\,\mathrm{Ric}_{ij}(g)(x,t) + \nabla_j W_i + \nabla_i W_j, \\ g(x,0) &= g_0(x). \end{cases} \tag{2}$$

Here $W_i = g_{ir} g^{st} \left(\Gamma^r_{st} - \widetilde{\Gamma}^r_{st} \right)$ with $\Gamma^r_{st}, \widetilde{\Gamma}^r_{st}$ being the Christoffel symbols for the metric $g_{ij}(x,t)$ and a fixed background metric $\widetilde{g}_{ij}$ respectively. Computation under the local coordinates shows that the Ricci–DeTurck equation is a quasilinear strictly parabolic system, whose short time existence can be proved via, say, a modified standard implicit function theorem argument. Denote its solution by $\bar{g}(x,t)$. Now let W be the vector field given by $W = W^i \frac{\partial}{\partial x^i}$ where $W^j = \bar{g}^{st} \left(\bar{\Gamma}^j_{st} - \widetilde{\Gamma}^j_{st} \right)$. Let Φ_t be the diffeomorphism generated by the vector field $-W(x,t)$. Define $g(x,t) = \Phi_t^*(\bar{g}(x,t))$. Direct calculation shows that

$$\begin{aligned} \frac{\partial}{\partial t} g(x,t) &= \Phi_t^*(-2\,\mathrm{Ric}(\bar{g}) + \bar{\nabla}_i W_j + \bar{\nabla}_j W_i) \\ &\quad + \frac{\partial}{\partial s} \Phi_{t+s}^*(\bar{g}(x,t)) \Big|_{s=0} \\ &= -2\,\mathrm{Ric}(g)(x,t). \end{aligned}$$

This approach avoids appealing to the Nash–Moser inverse function theorem which is the original method adapted by Hamilton in his groundbreaking paper [15].

Tedious, but straightforward calculations show that the curvature tensor of $g(x,t)$ satisfies

$$\begin{aligned} \frac{\partial}{\partial t} R_{ijkl} &= \Delta R_{ijkl} + g^{pq} g^{st} R_{ijpt} R_{klqs} + 2\left(B_{ikjl} - B_{iljk} \right) \\ &\quad - g^{pq} \left(R_{pjkl} r_{qi} + R_{ipkl} r_{qj} + R_{ijpl} r_{qk} + R_{ijkp} r_{ql} \right). \end{aligned}$$

Here $B_{ijkl} = g^{ps} g^{qt} R_{piqj} R_{sktl}$, r_{ij} being the Ricci curvature.

To make the computation easier, Hamilton in [16] introduced the gauge fixing trick (due to Karen Uhlebeck) to get rid of the last four terms. Let E denote a vector bundle which is isomorphic to TM. Then consider the map $u : E \to TM$ satisfying $\frac{\partial u}{\partial t} = \mathrm{Ric}\, u$. Here, by abusing notation, $\mathrm{Ric}_i^j = r_i^j = r_{ik} g^{jk}$ is viewed as a symmetric transformation of TM.

If we pull back the changing metric on TM by u and call it h, it is easy to see that

$$\begin{aligned}\frac{\partial}{\partial t}h(X,Y) &= \frac{\partial}{\partial t}g(u(X),u(Y))\\ &= -2\operatorname{Ric}(u(X),u(Y)) + g(\operatorname{Ric} u(X),u(Y))\\ &\quad +g(u(X),\operatorname{Ric} u(Y))\\ &= 0.\end{aligned}$$

As long as the flow exists, u is an isometry between the fixed metric h on E and the changing metric $g(t)$ on TM. Again by possibly abusing notation, we pull back the curvature tensor R at time t, and denote by $\widetilde{R}$,

$$\widetilde{R}(e_a,e_b,e_c,e_d) = R(u(e_a),u(e_b),u(e_c),u(e_d)).$$

Using the previous convention we simply abbreviate it as $\widetilde{R}(a,b,c,d)$ or $\widetilde{R}_{abcd}$.

The connection (which shall be denoted by D) can also be pulled back through $u(D_i a) = \nabla_i u(a)$. Hence there exists a time-dependent metric connection D on the vector bundle E. It is easy to see that u is invariant, namely $Du = 0$.

Direct calculation shows that

$$D_i\widetilde{R}(a,b,c,d) = (\nabla_i R)(u(a),u(b),u(c),u(d)).$$

On the other hand,

$$\begin{aligned}\frac{\partial}{\partial t}\widetilde{R}_{abcd} &= \frac{\partial}{\partial t}R_{u(a)u(b)u(c)u(d)} + R(\operatorname{Ric} u(a),u(b),u(c),u(d))\\ &\quad +R(u(a),\operatorname{Ric} u(b),c,d)\\ &\quad +R(u(a),u(b),\operatorname{Ric} u(c)),u(d))\\ &\quad +R(u(a),u(b),u(c),\operatorname{Ric} u(d)).\end{aligned}$$

Hence

$$\frac{\partial}{\partial t}\widetilde{R}_{abcd} = \Delta\widetilde{R}_{abcd} + 2\widetilde{\mathrm{Rm}}^2_{abcd} + 2\widetilde{\mathrm{Rm}}^{\#}_{abcd}. \tag{3}$$

Here $\widetilde{\mathrm{Rm}}^2$ and $\widetilde{\mathrm{Rm}}^{\#}$ are the corresponding quadratic operations on $\widetilde{R}$ with

$$\begin{aligned}\widetilde{\mathrm{Rm}}^2_{ijkl} &= g^{pq}g^{st}\widetilde{R}_{ijpt}\widetilde{R}_{klqs}\\ \widetilde{\mathrm{Rm}}^{\#}_{ijkl} &= 2\left(B_{ikjl} - B_{iljk}\right).\end{aligned}$$

In [16], Hamilton also observed that there is a Lie algebraic interpretation on the second reaction term in the diffusion reaction equation (3) satisfied by the curvature tensor. First, there exists a natural identification between $\wedge^2\mathbb{R}^n$ and

$\mathfrak{so}(n)$, the Lie algebra of $\mathsf{SO}(n)$. The identification can be done by first defining $X \otimes Y(Z) = \langle Y, Z\rangle X$. Then $e_i \wedge e_j$ can be identified with $E_{ij} - E_{ji}$, where E_{ij} is the matrix with 0 components, except 1 at the (i, j)-th position. The product on $\mathfrak{so}(n)$ is taken to be $\langle v, w\rangle = \frac{1}{2}\operatorname{trace}(v^{\mathrm{tr}} w)$, so that the identification is an isometry.

The curvature tensor can be viewed as a symmetric transformation between $\wedge^2\mathbb{R}^n$ via the equation

$$\langle \mathrm{Rm}(X \wedge Y), Z \wedge W\rangle = R(X, Y, Z, W).$$

We denote all such transformations by $S^2_B(\wedge^2(\mathbb{R}^n))$, where B stands for the first Bianchi identity. For any Rm_1 and $\mathrm{Rm}_2 \in S^2(\wedge^2(\mathbb{R}^n))$ we define $\langle \mathrm{Rm}_1, \mathrm{Rm}_2\rangle = \sum \langle \mathrm{Rm}_1(b^\alpha), \mathrm{Rm}_2(b^\alpha)\rangle$. Here $\{b^\alpha\}$, with $1 \le \alpha \le \frac{n(n-1)}{2}$, is an orthonormal basis of $\mathfrak{so}(n)$.

Lemma 2.1 (Hamilton). *With the above notation,* $\mathrm{Rm}^\#$ *is given via the following equation:*

$$\langle \mathrm{Rm}^\#(v), w\rangle = \frac{1}{2}\sum_{\alpha,\beta} \langle [\mathrm{Rm}(b^\alpha), \mathrm{Rm}(b^\beta)], v\rangle \langle [b^\alpha, b^\beta], w\rangle. \tag{4}$$

This, together with Hamilton's tensor maximum principle which, roughly put, asserts that the "nonnegativity" condition is preserved by the diffusion reaction equation as long as it is preserved by the ODE with the reaction term as the vector fields. This fact immediately implies that the Ricci flow preserves the nonnegativity of Rm, namely the nonnegativity of *the curvature operator*, since clearly $\mathrm{Rm}^2 \ge 0$, and if $\mathrm{Rm} \ge 0$, the above lemma asserts that $\mathrm{Rm}^\# \ge 0$. Thus the reaction term

$$\mathrm{Rm}^2 + \mathrm{Rm}^\# \ge 0$$

as long as $\mathrm{Rm} \ge 0$. This was first obtained in [16].

The second preserved positivity is on the *complex sectional curvature*. To define the terms we need to complexify the tangent bundle at any given point p and denote it as $T^{\mathbb{C}}_p M = T_p M \otimes \mathbb{C}$. Now extend linearly the curvature tensor to $\otimes^4 T^{\mathbb{C}}_p M$. Then we say that Rm has nonnegative complex sectional curvature if for any $X, Y \in T^{\mathbb{C}}_p M$,

$$\langle \mathrm{Rm}(X \wedge Y), \overline{X \wedge Y}\rangle = R(X, Y, \bar{X}, \bar{Y}) \ge 0. \tag{5}$$

It seems that the *nonpositivity of complex sectional curvature* was first introduced in [24] (1985) for Riemannian manifolds. For a Kähler manifold, given any nonzero $X \in T'_p M$ (holomorphic), $Y \in T''_p M$ (anti-holomorphic), then

$$\langle \mathrm{Rm}(X \wedge Y), \overline{X \wedge Y}\rangle < 0$$

is equivalent to *Siu's strong negativity* [25] (the condition introduced a few years earlier, under which Siu proved the holomorphicity of harmonic maps between Kähler manifolds). The following proof first appeared in [22].

Proposition 2.2. *Let (M, g_0) be a compact Riemannian manifold. Assume that $g(t)$ is a solution to (RF) on $M \times [0, T]$ with $g(0) = g_0$. Suppose that g_0 has nonnegative complex sectional curvature. Then $g(t), 0 \le t \le T$, has nonnegative complex sectional curvature.*

Proof. View $\langle \mathrm{Rm}(U \wedge V), \overline{U} \wedge \overline{V}\rangle$ as a linear functional $\ell_{U\wedge V}(\cdot)$ on $\mathrm{Rm} \in \mathbb{R}^N$ with N being the dimension of $S^2_B(\wedge^2\mathbb{R}^n)$. The cone $\mathcal{C}_{PCS}$ is defined as the set $\{\mathrm{Rm} \in \mathbb{R}^N \,|\, \ell_{U\wedge V}(\mathrm{Rm}) \ge 0, \text{ for all } U \wedge V\}$. By Hamilton's tensor maximum principle, it suffices to check that the ODE

$$\frac{d\,\mathrm{Rm}}{dt} = Q(\mathrm{Rm}) := \mathrm{Rm}^2 + \mathrm{Rm}^\# \tag{6}$$

preserves the cone. It is then sufficient to show the following. If $\mathrm{Rm}_0 \in \partial\mathcal{C}_{PCS}$, which amounts to $\ell_{U_0\wedge V_0}(\mathrm{Rm}_0) = 0$ for some $U_0 \wedge V_0$ and $\ell_{U\wedge V}(\mathrm{Rm}_0) \ge 0$ for all $U \wedge V$, then we need to check that $Q(\mathrm{Rm}_0) \in T_{\mathrm{Rm}_0}\mathcal{C}_{PCS}$. Let K be the collection of all $U_0 \wedge V_0$ satisfying $\ell_{U_0\wedge V_0}(\mathrm{Rm}_0) = 0$. Then at Rm_0, the tangent cone is given by the intersection of halfplanes $\ell_{U\wedge V}(\mathrm{Rm} - \mathrm{Rm}_0) \ge 0$ for all $U \wedge V \in K$. Hence in order to show that the ODE (6) preserves $\mathcal{C}_{PCS}$ it suffices to verify the *null vector condition*: If, for some $\mathrm{Rm} \in \mathcal{C}_{PCS}$, there exists $U \wedge V$ satisfying $\langle \mathrm{Rm}(U \wedge V), \overline{U} \wedge \overline{V}\rangle = 0$, then $\langle Q(\mathrm{Rm})(U \wedge V), \overline{U} \wedge \overline{V}\rangle \ge 0$. Since $\langle \mathrm{Rm}^2(U \wedge V), \overline{U} \wedge \overline{V}\rangle = \langle \mathrm{Rm}(U \wedge V), \overline{\mathrm{Rm}(U \wedge V)}\rangle \ge 0$ always, it suffices to show that $\langle \mathrm{Rm}^\#(U \wedge V), \overline{U} \wedge \overline{V}\rangle \ge 0$, which, via the definition, amounts to

$$R_{Up\overline{U}q}R_{Vp\overline{V}q} - R_{Up\overline{V}q}R_{Vp\overline{U}q} \ge 0, \tag{7}$$

where $\{e_p\}$ is a orthonormal basis of T_pM (which is identified to $\mathbb{R}^n$). Now for any U_1 and V_1, consider the function

$$I(z) := \langle \mathrm{Rm}((U + zU_1) \wedge (V + zV_1)), \overline{(U + zU_1) \wedge (V + zV_1)}\rangle$$

which satisfies that $I(z) \ge 0$ and $I(0) = 0$. Hence $\frac{\partial^2}{\partial z \partial \bar{z}} I(z)|_0 \ge 0$, which implies that

$$\begin{aligned}\langle \mathrm{Rm}(U \wedge V_1), \overline{U \wedge V_1}\rangle + 2\mathcal{R}e(\langle \mathrm{Rm}(U \wedge V_1), \overline{U_1 \wedge V}\rangle)& \\ + \langle \mathrm{Rm}(U_1 \wedge V), \overline{U_1 \wedge V}\rangle \ge 0.&\end{aligned} \tag{8}$$

Let $A_{ij} = R_{iVj\overline{V}} = R_{Vi\overline{V}j}$, $B_{ik} = R_{iV\overline{U}k}$, $C_{kl} = R_{Uk\overline{U}l}$ and $A = (A_{ij})$, $B = (B_{ik})$, $C = (C_{kl})$; then (7) asserts that

$$\mathcal{M}_1 := \begin{pmatrix} A & B \\ \overline{B}^{\mathrm{tr}} & C \end{pmatrix} \geq 0.$$

It is easy to check that (8) is equivalent to trace$(A\overline{C} - B\overline{B}) \geq 0$, since $\mathcal{M}_1 \geq 0$ implies that

$$\mathcal{M}_2 := \begin{pmatrix} 0 & -I \\ I & 0 \end{pmatrix} \begin{pmatrix} A & B \\ \overline{B}^{\mathrm{tr}} & C \end{pmatrix} \begin{pmatrix} 0 & I \\ -I & 0 \end{pmatrix} = \begin{pmatrix} C & -\overline{B}^{\mathrm{tr}} \\ B & A \end{pmatrix} \geq 0.$$

The theorem follows from $2\,\mathrm{trace}(A\overline{C} - B\overline{B}) = \mathrm{trace}(\mathcal{M}_1\overline{\mathcal{M}_2}) \geq 0$, a simple fact from the linear algebra. □

This proof was discovered shortly after the proof on the invariance of nonnegativity of isotropic curvature in [6] and [19], which seemed a bit mysterious at the time. We were led to such a notion since at that time it was the only condition left in a table of [14, page 18], whose invariance was not yet clear at the point before the above proof (in 2007) and further development afterwards.

Recall that $X \wedge Y$ is called an isotropic plane if for any $W \in \sigma$ where σ is the plane span$\{X, Y\}$, $\langle W, W\rangle = 0$. The curvature operator is said to have nonnegative isotropic curvature if (5) holds for any isotropic plane $X \wedge Y$. In [6], it has been observed that $M \times \mathbb{R}^2$ has nonnegative isotropic curvature and is also preserved under the Ricci flow. After we discovered the above presented proof, we suspected that (M, g) having nonnegative complex sectional curvature is equivalent to $M \times \mathbb{R}^2$ nonnegative isotropic curvature. Our speculation was also motivated by an observation of Brendle and Schoen at that time that (M, g) having nonnegative complex sectional curvature is equivalent to $M \times \mathbb{R}^4$ has nonnegative isotropic curvature. When I discussed our speculation with Nolan, I got the confirmed answer the same day! Interested readers are referred to [22] for Nolan's simple proof of this equivalence. In view of this equivalence, the first proof to the proposition was obtained in [6] via the more involved isotropic curvature invariance. The above proof provides a simple alternative.

The *complex sectional curvature* not only has a long root in the study of geometry as pointed out above, but also motivated (according to [30]) the formulation of the following general invariant cone result of Wilking, which provides so far the most general result on invariant conditions, while with the simplest proof (at the same time illuminating the possible previous mystery related to the isotropic curvature).

First we set up some notation. The complexified Lie algebra $\mathfrak{so}(n, \mathbb{C})$ can be identified with $\wedge^2(\mathbb{C}^n)$. Its associated Lie group is $\mathsf{SO}(n, \mathbb{C})$, namely all complex matrices A satisfying $A \cdot A^{\mathrm{tr}} = A^{\mathrm{tr}} \cdot A = \mathrm{id}$. Recall that there exists the natural action of $\mathsf{SO}(n, \mathbb{C})$ on $\wedge^2(\mathbb{C}^n)$ by extending the adjoint action $g \in \mathsf{SO}(n)$ on $x \otimes y$ ($g(x \otimes y) = gx \otimes gy$). For any $a \in \mathbb{R}$, let $\Sigma_a \subset \wedge^2(\mathbb{C}^n)$ be a subset which is invariant under the adjoint action of $\mathsf{SO}(n, \mathbb{C})$. Let $\widetilde{\mathcal{C}}_{\Sigma_a}$ be the cone of curvature operators satisfying that $\langle \mathrm{Rm}(v), \bar{v}\rangle \geq a$ for any $v \in \Sigma_a$. Here we view the space of algebraic curvature operators as a subspace of $S^2(\wedge^2(\mathbb{R}^n))$ satisfying the first Bianchi identity. In [30], the following result is proved.

Theorem 2.3 (Wilking). *Assume that $(M, g(t))$, for $0 \leq t \leq T$, is a solution of Ricci flow on a compact manifold. Assume that* $\mathrm{Rm}(g(0)) \in \widetilde{\mathcal{C}}_{\Sigma_a}$. *Then* $\mathrm{Rm}(g(t)) \in \widetilde{\mathcal{C}}_{\Sigma_a}$ *for all $t \in [0, T]$.*

3 Manifolds with positive and nonnegative sectional curvature

Unfortunately, the Ricci flow does not preserve the nonnegativity of the sectional curvature when the dimension is greater than three. This fact may have been known for the ODE (6) long before the concrete geometric example illustrated in [20]. But nothing was written down explicitly before [20]. Moreover, the geometric example says more than that the ODE (6) does not preserve such a condition. Compact examples were constructed later in [7]. But before we present these examples we recall the examples of Wallach [27] and Aloff–Wallach [2] on manifolds with positive sectional curvature since this, together with the above mentioned examples (about Ricci flow non-invariance), shows the subtlety of the sectional curvature.

We say that (M, g) is δ-pinched if $K(\sigma) > 0$ for all σ such two planes and if

$$\frac{\inf_\sigma(K(\sigma))}{\sup_\sigma(K(\sigma))} = r > \delta.$$

By compactness, it is easy to see that if (M, g) has positive sectional curvature, there must be some $\delta > 0$ such that (M, g) is δ-pinched.

Until the work of Marcel Berger ([4], 1961) the only known simply connected manifolds that admitted a $\delta > 0$ pinched structure were the spheres and projective spaces over $\mathbb{C}$ and $\mathbb{H}$ (the quaternions) and the projective plane over the octonions $\mathbb{O}$. Berger proved that two new examples have this property. One is of dimension 7 and another of dimension 13.

In 1969, Wallach set out to classify the homogeneous, simply connected, examples of positive pinching. In 1970, in the Bulletin of AMS he announced a partial result, which, in particular, asserted that in even dimensions the spaces had to be diffeomorphic with spheres and projective spaces over $\mathbb{C}, \mathbb{H}$ and the projective plane over the octonions or the full flag variety in $\mathbb{C}^3$ or $\mathbb{H}^3$. A breakthrough came when Wallach realized that he had overlooked one possible example: $F_4/\mathrm{Spin}(8)$, the manifold of flags in the 2-dimensional octonion projective plane.

Theorem 3.1 (Wallach). *The flag varieties in the 2-dimensional projective plane over $\mathbb{C}, \mathbb{H}$ and the octonions (dimensions 6, 12 and 24), namely* $\mathsf{SU}(3)/T^2, \mathsf{Sp}(3)/(\mathsf{Sp}(1) \times \mathsf{Sp}(1) \times \mathsf{Sp}(1)), \mathsf{F}_4/\mathsf{Spin}(8)$ *admit a homogeneous positive pinching metric.*

He also considered $\mathsf{SU}(3)/T$ with T a circle group embedded in $\mathsf{SU}(3)$. Up to conjugacy these are of the form

$$T_{k,l} = \left\{ \sigma_{k,l}(z) = \begin{pmatrix} z^k & 0 & 0 \\ 0 & z^l & 0 \\ 0 & 0 & z^{-(k+l)} \end{pmatrix}, |z| = 1 \right\}$$

where $k, l \in \mathbb{Z}$, gives rise to the spaces $W^7_{k,l} = \mathsf{SU}(3)/T_{k,l}$. The following was the main result of [2].

Theorem 3.2 (Aloff–Wallach). *For each k, l such that $k, l, k+l$ are not 0, there exists a one parameter family of positively pinched metrics $\langle\cdot,\cdot\rangle$ with $0 < t < 1$, yielding $W^7_{k,l,t}$. Moreover*

$$H_4(W^7_{k,l}, \mathbb{Z}) = \mathbb{Z}/(k^2 + l^2 + kl)\mathbb{Z}.$$

This result asserting the infinite topological type of 7-dimensional manifolds with positive sectional curvature shows that the subject is quite intricate since Gromov [13] showed that there exists $C(n)$ such that the Betti numbers of any compact Riemannian manifold with *nonnegative* sectional curvature is bounded by $C(n)$.

Now we explain the examples on Ricci flow invariance. After we told Nolan about our noncompact example [20] and pointed out the question on possible compact examples, he immediately suggested that we study some of the metrics on $\mathsf{SU}(3)/T^2$, which admit nonnegative sectional curvature and share a very similar Lie algebraic structure as the compact examples in [7], which we state below.

Theorem 3.3 (Böhm–Wilking). *On the 12-dimensional flag manifold*

$$M = \mathsf{Sp}(3)/(\mathsf{Sp}(1) \times \mathsf{Sp}(1) \times \mathsf{Sp}(1))$$

there exists an $\mathsf{Sp}(3)$-adjoint homogenous metric g with which, as the initial data shows, the Ricci flow cannot preserve the positivity of the sectional curvature.

The metrics in [20], where the Ricci flow does not preserve the nonnegativity of the sectional curvature, reside on noncompact manifolds. Precisely, they are complete metrics on the total space of the tangent bundle over spheres. The fact that the Ricci flow solution with bounded curvature does not preserve the nonnegativity follows from the following structure result proved in [20].

Theorem 3.4. *Let $(M, g_{ij}(x,t))$ be a solution to the Ricci flow with nonnegative sectional curvature. Assume also that M is simply-connected. Then M splits isometrically as $M = N \times M_1$, where N is a compact manifold with nonnegative sectional curvature. M_1 is diffeomorphic to $\mathbb{R}^k$ and for the restriction of metric $g_{ij}(x,t)$ on M_1 with $t > 0$, there is a strictly convex exhaustion function on M_1. Moreover, the soul of M_1 is a point and the soul of M is $N \times \{o\}$ if o is the soul of M_1.*

It was remarked in [7] that on the 6-dimensional manifold $\mathsf{SU}(3)/T^2$, there exists a metric of positive sectional curvature, which is not preserved by the Ricci flow. It would be interesting to find out whether or not such a four-dimensional compact example exists. Due to a general convergence theorem in the next section, one cannot expect that the ODE (6) preserves the nonnegativity of the sectional curvature. The intricacy of the problem in dimension 4 is of course also related to the celebrated Hopf conjecture on the existence of a positively curved metric on $\mathbb{S}^2 \times \mathbb{S}^2$. It is also interesting to find out if such a metric exists on the seven-dimensional examples of [2].

It has been computed that the pinching constant δ on the nonsymmetric examples with positive sectional curvature is rather small (considerably smaller than $1/4$ for example).

Recently, Cheung and Wallach [10] gave a detailed study on how the sectional curvature evolves under the Ricci flow of homogenous metrics on flag varieties.

4 Flag curvature pinching

First, we start with a general Ricci flow convergence theorem, which first appeared in [29], since this result and the above examples of Berger, Wallach and Aloff–Wallach also illuminate the reason why the Ricci flow does not preserve the sectional curvature.

Theorem 4.1 (Böhm–Wilking). *Let C be an $\mathsf{O}(n)$-invariant convex cone of full dimension in the vector space of algebraic curvature operators $S^2_B(\mathfrak{so}(n))$ with the following properties:*

(i) C is invariant under the ODE $\frac{d\,\mathrm{Rm}}{dt} = \mathrm{Rm}^2 + \mathrm{Rm}^\#$.
(ii) C contains the cone of nonnegative curvature operators, or slightly weaker all nonnegative curvature operators of rank 1.
(iii) C is contained in the cone of curvature operators with nonnegative sectional curvature.

Then for any compact manifold (M, g) whose curvature operator is contained in the interior of C at every point $p \in M$, the normalized Ricci flow evolves g to a limit metric of constant sectional curvature.

Assume that the nonnegativity of the sectional curvature is preserved (in the sense of ODE); then the above result would conclude that any manifold with positive sectional curvature is a space form.

We should remark that the above result was first proved in [8] for C being the cone of nonnegative curvature operators. Then, it was observed in [6] that the proof of [8] for the case of C being the nonnegative curvature operator cone can be transplanted, verbatim, to cover the case where C is the cone of nonnegative complex sectional curvatures. It appeared first in [29] with the above generality. In [5], a slightly different argument was adapted to prove the above result for the case of C being the cone of the nonnegative complex sectional curvatures.

Flag curvature pinching was first introduced by Andrews–Nguyen [3], who proved a $1/4$-flag pinching condition is invariant under the Ricci flow in dimension four and obtained a classification result for such manifolds in dimension four. First we introduce the definition.

Assume that (M, g) has nonnegative sectional curvature. Fixing a point $x \in M$, for any nonzero vector $e \in T_xM$, we define the flag curvature in the direction e by the symmetric bilinear form $R_e(X, X) = R(e, X, e, X)$. Restricting $R_e(\cdot, \cdot)$ to the subspace orthogonal to e, it is semi-positive definite. We say that (M, g) *has λ-pinched flag curvature ($1 > \lambda \geq 0$) if the eigenvalues of the symmetric bilinear form $R_e(\cdot, \cdot)$, restricted to the subspace orthogonal to e, are λ-pinched for all nonzero vectors e*, namely

$$R_e(X, X) \geq \lambda(x) R_e(Y, Y) \tag{9}$$

for any X, Y in the subspace orthogonal to e, with $|X| = |Y|$.

The λ-pinched flag curvature condition is equivalent to saying that $K(\sigma_1) \geq \lambda K(\sigma_2)$ for a pair of planes σ_1 and σ_2 such that $\sigma_1 \cap \sigma_2 \neq \{0\}$.

It is easy to see that if an algebraic curvature operator has λ-pinched flag curvature, then its sectional curvature is λ^2-pinched. This estimate is indeed sharp. Precisely, in [21] there exists an example of an algebraic curvature operator, such that its $1/4$-flag pinched and its sectional curvature are no better than $1/16$-pinched.

The first result of [21] is a classification result.

Theorem 4.2. *Let (M^n, g) be a compact nonnegatively curved Riemannian manifold with $1/4$-pinched flag curvature and the scalar curvature* $\mathrm{Scal}(x) > 0$ *for some $x \in M$. Then (M, g) is diffeomorphic to a spherical space form or isometric to a finite quotient of a rank-one symmetric space.*

In view of the convergence result Theorem 4.1, the key towards the above result is the following.

Theorem 4.3. *Let (M^n, g) be a nonnegatively curved Riemannian manifold. If (M, g) has a quarter pinched flag curvature, then (M, g) has nonnegative complex sectional curvature.*

If we assume the stronger assumption that the sectional curvature is $1/4$-pinched, the nonnegativity of the complex sectional curvature was essentially proved earlier by Hernández [17] and Yau–Zheng [31] in the 1990s. (What was proved there is that if a curvature operator has negative sectional curvature and $1/4$-pinched sectional curvature, then it must have nonpositive complex sectional curvature. By flipping the sign, the argument can be transplanted to the case of nonnegative sectional curvature.) An immediate consequence of this fact, together with Proposition 2.2, Theorem 4.1, is Brendle–Schoen's sectional curvature $1/4$-pinching sphere theorem [6].

For the proof of Theorem 4.3, first observe the following lemma.

Lemma 4.4. *Given any complex plane $\sigma \subset \mathbb{C}^n = \mathbb{R}^n \otimes \mathbb{C}$, where $\mathbb{R}^n$ is equipped with an inner product $\langle \cdot, \cdot \rangle$ which is extended bilinearly to $\mathbb{C}^n$, there must exist unit vectors $U, V \in \sigma$ such that*

$$\langle U, U\rangle, \langle V, V\rangle \in \mathbb{R} \text{ with } 1 \geq \langle U, U\rangle \geq \langle V, V\rangle \geq 0,\ \langle U, V\rangle = \langle U, \bar{V}\rangle = 0.$$

Particularly, if $U = X + \sqrt{-1}Y$, $V = Z + \sqrt{-1}W$, *it implies that*

$$|X| \geq |Y|, |Z| \geq |W| \text{ and } \{X, Y, Z, W\}$$

are mutually orthogonal.

Proof. Let $f(\widetilde{U}) \doteqdot Re\left(\langle \widetilde{U}, \widetilde{U}\rangle\right)$ be the functional defined on the unit sphere (with respect to the norm $|\widetilde{U}| = \sqrt{\langle \widetilde{U}, \bar{\widetilde{U}}\rangle}$) inside σ. Let U be the maximizing vector, at which f attains the maximum $\bar{\lambda}$, with $\bar{\lambda} \in [0, 1]$. Clearly for such U, $f(U) = |\langle U, U\rangle|$. Let V be a unit vector such that it is perpendicular to U (namely $\langle U, \bar{V}\rangle = 0$). By the maximizing property of U, from the first variation, it is easy to see that $\langle U, V\rangle = 0$ for any $V \in \sigma$ with $\langle U, \bar{V}\rangle = 0$ and $|V| = 1$. To see this let $h(\theta) = f(\cos\theta U + \sin\theta V)$. Since $h(0) = \bar{\lambda} \geq h(\theta)$, we have $h'(0) = 0$, which, together with the same conclusion with V replaced by $Ve^{\sqrt{-1}\pi/2}$, implies the claim. Among all possible choices of such V, which can be parametrized by S^1, there clearly exists one with $\langle V, V\rangle \geq 0$. □

It is clear from the proof that σ is isotropic if and only if $\bar{\lambda}(\sigma) = 0$. It also makes sense to define $\underline{\lambda}(\sigma)$ to be the minimum of $|\langle U, U\rangle|$ for any $U \in \sigma$ with unit length. Since the inner product induces one on the space of 2-planes $\sigma = U \wedge V$, similarly one may define the $\mu(\sigma)$ as $|\langle U \wedge V, U \wedge V\rangle|$ among all $\sigma = U \wedge V$ of unit length. We may call σ *weakly isotropic* if $\underline{\lambda}(\sigma) = 0$. Clearly both σ being isotropic and σ being weakly isotropic are invariant under the adjoint action of $\mathsf{SO}(n, \mathbb{C})$. Hence Theorem 2.3 implies the Ricci flow invariance on the complex sectional curvature nonnegativity for all such 2-planes.

We may define for any $a, b \in [0, 1]$, $\Sigma_a^{\bar{\lambda}} = \{\sigma\,|\bar{\lambda}(\sigma) \leq a\}$, $\Sigma_b^{\mu} = \{\sigma\,|\mu(\sigma) \leq b\}$ and $\Sigma_{a,b} = \{\sigma\,|\bar{\lambda}(\sigma) \leq a, \mu(\sigma) \leq b\}$. It is a natural question to ask if the nonnegativity on $\Sigma_a^{\bar{\lambda}}$, Σ_b^{μ}, or $\Sigma_{a,b}$ is invariant for any $(a, b) \in [0, 1] \times [0, 1]$ since Theorem 2.3 implies that it is the case when $(a, b) = (0, 0)$ and $(a, b) = (1, 1)$. Related to this, it is also interesting to ask whether or not the condition

$$\langle \mathrm{Rm}(U \wedge V), \overline{U \wedge V}\rangle + \underline{\lambda}(\sigma)\|U \wedge V\|^2 \geq 0 \tag{10}$$

is preserved under the Ricci flow. Here σ is the plane spanned by $\{U, V\}$.

The key to Theorem 4.3 is the following result generalizing a useful lemma of Berger.

Proposition 4.5. *Assume that* (M, g) *has* λ*-pinched flag curvature with dimension* $n \geq 4$. *Assume that the sectional curvature is nonnegative at* x *and* $X, Y, Z, W \in T_xM$ *are four vectors mutually orthogonal. Then*

$$6\frac{1+\lambda}{1-\lambda}|R(X, Y, Z, W)| \leq k(X, Z) + k(Y, Z) + k(X, W) + k(Y, W)$$
$$+2k(X, Y) + 2k(Z, W).$$

If equality holds and $\mathrm{Rm}(x) \neq 0$, *then vectors* X, Y, Z, W *have the same norm.*

In [21], results were obtained for manifolds with flag-pinching constant below $1/4$ (note that flag curvature pinching is always *pointwise*).

Theorem 4.6. *For any dimension $n \geq 4$ and $C > 0$, there is an $\epsilon > 0$ such that the following holds. Let (M^n, g) be a nonnegatively curved Riemannian orbifold of dimension n with $\frac{1-\epsilon}{4}$ pinched-flag curvature and scalar curvature satisfying $1 \leq \mathrm{Scal} \leq C$. Then the following holds.*

(i) When $n = 2m + 1$, M admits a metric of constant curvature;
(ii) When $n = 2m$, either M is diffeomorphic to the quotient of rank one symmetric space by a finite isometric group action or it is diffeomorphic to the quotient of a weighted complex projective space by a finite group action.

If one replaces the flag pinching (pointwise) condition by a global sectional curvature pinching, a similar result was obtained by Petersen and Tao [23] earlier.

Since here ϵ is depending on n, we would like to point out a related result and some open problems. A theorem of Abresch and Meyer [1] asserts that *any simply connected odd-dimensional manifold with sectional curvature K satisfying $\frac{1}{4(1+10^{-6})^2} \leq K \leq 1$ is homeomorphic to a sphere.* Note that here a *global* instead of *pointwise* pinching is assumed. An obvious question arises whether or not one can weaken the assumption to a pointwise one and improve the conclusion from the homeomorphism to the diffeomorphism.

Since Micallef–Moore [18] proved (using harmonic spheres) that *any simply-connected manifold with positive isotropic curvature is a homotopy sphere (hence homeomorphic)*, it is natural to ask if this can be improved to diffeomorphic.

In [28] Wilking obtained homotopic classification result for manifolds with positive curvature and "large" enough symmetry. Can the method of using the isometry group and the method of the Ricci flow be combined to get a better result?

Grove–Shiohama [12] (see also [9, Theorem 6.13]) proved *a sphere theorem by assuming that the sectional curvature is bounded from below by one (namely $K \geq 1$) and* $\mathrm{diam}(M) > \frac{\pi}{2}$. Can this be upgraded to a "diffeomorphism"?

References

1. Abresch, U.; Meyer, W.- T., A sphere theorem with a pinching constant below $\frac{1}{4}$. (English summary) *J. Differential Geom.* **44** (1996), no. 2, 214–261.
2. Aloff, S.; Wallach, N. R., An infinite family of distinct 7-manifolds admitting positively curved Riemannian structures. *Bull. Amer. Math. Soc.* **81** (1975), 93–97.
3. Andrews, B.; Nguyen, H., Four-manifolds with 1/4-pinched flag curvatures. (English summary) *Asian J. Math.* **13** (2009), no. 2, 251–270.
4. Berger, M., Les variétés riemanniennes homogènes normales simplement connexes à courbure strictement positive. (French) *Ann. Scuola Norm. Sup. Pisa (3)* **15**(1961), 179–246.
5. Brendle, S., Ricci flow and the sphere theorem. *GSM* **111**, AMS 2010.
6. Brendle, S.; Schoen, R., Manifolds with 1/4-pinched curvature are space forms. *J. Amer. Math. Soc.* **22** (2009), no. 1, 287–307.

7. Böhm, C.; Wilking, B., Nonnegatively curved manifolds with finite fundamental groups admit metrics with positive Ricci curvature. *Geom. Funct. Anal.* **17** (2007), no. 3, 665–681.
8. Böhm, C.; Wilking, B., Manifolds with positive curvature operators are space forms. *Ann. of Math.* (2) 167 (2008), no. 3, 1079–1097.
9. Cheeger, J.; Ebin, D., *Comparison theorems in Riemannian geometry.* North-Holland Mathematical Library, Vol. 9. North-Holland Publishing Co., New York, 1975.
10. Cheung, M.; Wallach, N., *Ricci flow and curvature on the variety of flags on the two dimensional projective space over the complexes, quaternions and the octonions.* Preprint, 2012. To appear in proceedings of AMS
11. Do Carmo, M., *Differential Curves and Surfaces.* Prentice Hall, 1976.
12. Grove, K.; Shiohama, K., A generalized sphere theorem. *Ann. of Math.* **106**(1977), 201–211.
13. Gromov, M. Curvature, diameter and Betti numbers. *Comment. Math. Helv.* **56** (1981), no. 2, 179–195.
14. Gromov, M., Positive curvature, macroscopic dimension, spectral gaps and higher signatures. (English summary) *Functional analysis on the eve of the 21st century*, Vol. *II* (New Brunswick, NJ, 1993), 1–213, Progr. Math., Vol. 132, Birkhäuser Boston, 1996.
15. Hamilton, R. S., Three-manifolds with positive Ricci curvature. *J. Differential Geom.* **17** (1982), no. 2, 255–306.
16. Hamilton, R. S., Four-manifolds with positive curvature operator. *J. Differential Geom.* **24** (1986), no. 2, 153–179.
17. Hernández, L., Kähler manifolds and 1/4-pinching. *Duke Math. J.* **62** (1991), no. 3, 601–611.
18. Micallef, M. J.; Moore, J. D., Minimal two-spheres and the topology of manifolds with positive curvature on totally isotropic two-planes. *Ann. of Math.* (2) **127** (1988), no. 1, 199–227.
19. Nguyen, H. T., Isotropic curvature and the Ricci flow. *Int. Math. Res. Not.*, 2010, no. 3, 536–558.
20. Ni, L., Ricci flow and nonnegativity of sectional curvature. (English summary) *Math. Res. Lett.* **11** (2004), no. 5–6, 883–904.
21. Ni, L.; Wilking, B., Manifolds with 1/4-pinched flag curvature. (English summary) *Geom. Funct. Anal.* **20** (2010), no. 2, 571–591.
22. Ni, L.; Wolfson, J., Positive complex sectional curvature, Ricci flow and the differential sphere theorem. ArXiv:0706.0332, 2007.
23. Petersen, P.; Tao, T., Classification of almost quarter-pinched manifolds. (English summary) *Proc. Amer. Math. Soc.* **137** (2009), no. 7, 2437–2440.
24. Sampson, J. H., Harmonic maps in Kähler geometry. Harmonic mappings and minimal immersions (Montecatini, 1984), 193–205, *Lecture Notes in Math.*, Vol. 1161, Springer, Berlin, 1985.
25. Siu, Y. T., The complex-analyticity of harmonic maps and the strong rigidity of compact Kähler manifolds. *Ann. of Math.* (2) 112 (1980), no. 1, 73–111.
26. Spivak, M., *Differential Geometry.* 2nd Vol. Publish or Perish, 1979.
27. Wallach, N. R., Compact homogeneous Riemannian manifolds with strictly positive curvature. *Ann. of Math.* (2) 96 (1972), 277–295.
28. Wilking, B., Positively curved manifolds with symmetry. *Ann. of Math.* (2) **163** (2006), no. 2, 607–668.
29. Wilking, B., Nonnegatively and positively curved manifolds. *Surveys in differential geometry.* Vol. XI, 25–62, Surv. Differ. Geom., 11, Int. Press, Somerville, MA, 2007.
30. Wilking, B., A Lie algebraic approach to Ricci flow invariant curvature condition and Harnack inequalities. *J. Reine Angew. Math.* **679**(2013), 223–247.
31. Yau, S.-T.; Zheng, F., Negatively $\frac{1}{4}$-pinched Riemannian metric on a compact Kähler manifold. *Invent. Math.* **103** (1991), no. 3, 527–535.
32. Ziller, W., Examples of Riemannian manifolds with non-negative sectional curvature. *Surveys in differential geometry.* Vol. XI, 63–102, Int. Press, Somerville, MA, 2007.

Remainder formula and zeta expression for extremal CFT partition functions

Floyd L. Williams

Dedicated to Nolan Wallach. Thank you Nolan for your mentorship and your inspiration that have markedly shaped my mathematical life.

Abstract We derive a remainder formula for the coefficients of modular invariant partition functions of extremal conformal field theories of central charge $c = 24k$, where k is a positive integer. The formula encodes, in particular, asymptotics of these coefficients and it provides for additional corrections to Bekenstein–Hawking black hole entropy. We also relate these partition functions to a Patterson–Selberg zeta function. More generally, when c is divisible by 8 we relate this zeta function to vacuum characters of affine E_8 and $E_8 \times E_8$.

Keywords: Affine E_8 • Extremal conformal field theory • Modular j-invariant • q-expansion • Black hole entropy • Kloosterman sum • Zeta function

Mathematics Subject Classification: 11FO3, 11F25, 11F30, 11M36, 81T40, 81R10, 83C57

1 Introduction

In the paper [1] G. Höhn considers holomorphic conformal field theories of central charge $c = 24k$, for $k = 1, 2, 3, \ldots$. Such a theory is called an *extremal conformal field theory* (ECFT) and was proposed by E. Witten [2] to be dual to 3-dimensional pure gravity with a negative cosmological constant; also see [3–7]. Its construction in general remains an open question. For $k = 1$ one has of course the classical construction of I. Frenkel, J. Lepowsky, and A. Meurman (FLM) [8].

F.L. Williams
Department of Mathematics and Statistics, University of Massachusetts at Amherst, Amherst, MA 01003, USA
e-mail: williams@math.umass.edu

R. Howe et al. (eds.), *Symmetry: Representation Theory and Its Applications*, Progress in Mathematics 257, DOI 10.1007/978-1-4939-1590-3_18

Their theory, moreover, possesses the deep property of *monstrous symmetry*: its states transform as a representation of the Fischer–Griess group M of order $|M| = 2^{46} \cdot 3^{20} \cdot 5^9 \cdot 7^6 \cdot 11^2 \cdot 13^2 \cdot 17 \cdot 19 \cdot 23 \cdot 29 \cdot 31 \cdot 41 \cdot 47 \cdot 59 \cdot 71 \simeq 10^{54}$, M (called the *monster* or *friendly giant*) being "secretly" the symmetry group of 3-dimensional quantum gravity.

One can construct a well-defined partition function $Z_k(\tau)$ on the upper 1/2-plane Π^+, ($\mathfrak{I}(\tau) > 0$, τ being a "temperature" variable) for any ECFT. Moreover, $Z_k(\tau)$ is *modular invariant*: it is invariant under the standard linear fractional action of the modular group $SL(2, \mathbb{Z})$ on Π^+; here $\mathbb{Z}$ denotes the set of integers. $Z_k(\tau)$ has a q-expansion in powers q^n for $q \stackrel{\text{def}}{=} \exp(2\pi i \tau)$, $-k \leq n < \infty$. Our interest here is in the coefficients $b_{k,n}$ of q^n for $n \geq 1$, since the subleading terms of their asymptotics, as $n \to \infty$, correspond to black hole entropy corrections.

Using some important, relatively new estimates of N. Brisebarre and G. Philibert [9] we present in Section 3 a *remainder* formula for each $b_{k,n}$, from which the asymptotic behavior of $b_{k,n}$ (for every fixed k) is immediately read off. In particular, one need not assume that k and n are large with n/k fixed as is assumed in [2]. We point out that apart from the remainder formula, the asymptotic behavior of $b_{k,n}$ can actually be deduced, immediately, from a 50-year old result of M. Knopp [10]; here we consider $Z_k(\tau)$ *abstractly* as a modular form of weight 0.

Similarly, from this abstract point of view, an *exact* formula for each $b_{k,n}$ (a formula stated in[3]) immediately follows from a general result in [9]. An alternate version of this exact formula is also presented in Section 5. Its proof is based on an exact formula of H. Petersson and (independently) H. Rademacher [11, 12] applied to the ECFT of FLM in case $k = 1$. In this alternate version slightly simpler Kloosterman sums (than those in [3, 13]) are incorporated.

The $b_{k,n}$, remarkably, are known to be positive integers. At the physical level they are the number of states with Virasoro energy $L_0 = n$ and one obtains (as is known) the Bekenstein–Cardy–Hawking black hole entropy $S = 4\pi\sqrt{kn} = 2\pi\sqrt{\frac{cL_0}{6}}$ [14] by way of the logarithm of $b_{k,n}$. In addition some careful (small) corrections (especially for large n) are obtained by way of the remainder formula.

Section 2 is given to a brief review of the construction of $Z_k(\tau)$ for the convenience of the reader, especially as some definitions such as Eisenstein series, Hecke operators, or the j-invariant might be less familiar to physicists. In fact, we shall need some of the ingredients involved in that construction. A partition function, zeta function connection is discussed in the final section, Section 6.

We conclude the paper with a personal reflection concerning Nolan Wallach.

2 Review of a construction of $Z_k(\tau)$

The ECFT partition function $Z_k(\tau)$ of level $k \in \mathbb{Z}$, $k \geq 1$, can be constructed from the FLM partition function $Z_1(\tau)$ of level 1 and the holomorphic sector

$$Z_{\text{hol}}(\tau) \stackrel{\text{def}}{=} \prod_{n=2}^{\infty} \frac{1}{1-q^n} \tag{1}$$

of the one-loop gravity partition function [15, 16]

$$Z_{\text{gravity}}^{1-\text{loop}}(\tau) = Z_{\text{hol}}(\tau)\overline{Z}_{\text{hol}}(\tau) \tag{2}$$

with the help of *Hecke operators* $T(r)$, $1 \le r \le k$. Again $q \stackrel{\text{def}}{=} \exp(2\pi i \tau)$ for $\tau \in \Pi^+$. The bar "—" in the factorization (2) denotes complex conjugation. $Z_1(\tau)$ is given by

$$Z_1(\tau) \stackrel{\text{def}}{=} J(\tau) \stackrel{\text{def}}{=} j(\tau) - 744 \tag{3}$$

for

$$j(\tau) \stackrel{\text{def}}{=} \frac{1728(60G_4(\tau))^3}{(60G_4(\tau))^3 - 27(140G_6(\tau))^2} \tag{4}$$

where

$$G_w(\tau) \stackrel{\text{def}}{=} \sum_{(m,n)\in\mathbb{Z}\times\mathbb{Z}-\{(0,0)\}} \frac{1}{(m+n\tau)^w} \tag{5}$$

is a holomorphic Eisenstein series of weight w, $w = 4, 6, 8, 10, 12, \ldots$. The action of $T(r)$ on $Z_1(\tau)$ is given in terms of a sum over positive divisors d of r:

$$(T(r)Z_1)(\tau) \stackrel{\text{def}}{=} \frac{1}{r} \sum_{d>0,\ d|r} \sum_{m=1}^{d-1} Z_1\left(\frac{r\tau + dm}{d^2}\right). \tag{6}$$

$j(\tau)$ has the Fourier q-expansion

$$j(\tau) = q^{-1} + \sum_{n=0}^{\infty} c_n q^n \tag{7}$$

where, remarkably, every coefficient c_n is an integer. For example,

$$\begin{aligned} &c_0 = 744, \quad c_1 = 196,884, \quad c_2 = 21,493,760 \\ &c_3 = 864,299,970, \quad c_4 = 20,245,856,256 \\ &c_5 = 333,202,640,600, \quad c_6 = 4,252,023,300,096 \end{aligned} \tag{8}$$

so that $Z_1(\tau) = j(\tau) - c_0$ in (3). An expository discussion of the modular j-invariant $j(\tau)$, the Eisenstein series $G_w(\tau)$, the action of Hecke operators on more general modular forms than the action on $J(\tau)$ given in (6), and of other related matters is given, for example, in [17]. Finally, let $p(n)$ be the partition function on the set of positive integers $\mathbb{Z}^+$: $p(n)$ is the number of ways of writing $n \in \mathbb{Z}^+$ as a sum of elements in $\mathbb{Z}^+$. For convenience we also define $p(-1) = 0$.

All of the ingredients necessary for defining $Z_k(\tau)$ are now in place. In fact, define

$$Z_0(\tau) \stackrel{\text{def}}{=} q^{-k} Z_{\text{hol}}(\tau) = \sum_{r=-k}^{\infty} a_r(k) q^r,$$

$$a_r(k) \stackrel{\text{def}}{=} p(r+k) - p(r+k-1), \quad r \geq -k; \tag{9}$$

see (1). The q-expansion here follows from Euler's generating function for $p(n)$. Then we set

$$Z_k(\tau) \stackrel{\text{def}}{=} a_0(k) + \sum_{r=1}^{\infty} a_{-r}(k) r (T(r)J)(\tau); \tag{10}$$

see (6), [2]. For $k = 1$, the right-hand side of (10) reduces to $(T(1)J)(\tau) = J(\tau)$, which by (3) shows that (10) is an extension of the definition of the FLM partition function from level 1 to an arbitrary level k.

Using the q-expansion (7) and a corresponding q-expansion of $(T(r)J)(\tau)$ one obtains the q-expansion

$$Z_k(\tau) = a_{-k}(k) q^{-k} + \cdots + a_{-1}(k) q^{-1} + a_0(k) + \sum_{n=1}^{\infty} b_{k,n} q^n \tag{11}$$

of $Z_k(\tau)$, where (see (9))

$$a_{-k}(k) = 1, \quad b_{k,n} \stackrel{\text{def}}{=} \sum_{r=1}^{k} r a_{-r}(k) c_n^{(r)}, \; n \geq 1, \tag{12}$$

for

$$c_n^{(r)} = \sum_{d>0,\, d|r,\, d|n} \frac{c_{rn/d^2}}{d}. \tag{13}$$

Further details on the derivation of (11) are provided in [17], for example. Using (8) and Table 1 one can compute

Table 1. Values of $a_{-r}(k)$, $0 \leq r < k \leq 12$

k	$a_0(k)$	$a_{-1}(k)$	$a_{-2}(k)$	$a_{-3}(k)$	$a_{-4}(k)$	$a_{-5}(k)$	$a_{-6}(k)$	$a_{-7}(k)$	$a_{-8}(k)$	$a_{-9}(k)$	$a_{-10}(k)$	$a_{-11}(k)$
1	0											
2	1	0										
3	1	1	0									
4	2	1	1	0								
5	2	2	1	1	0							
6	4	2	2	1	1	0						
7	4	4	2	2	1	1	0					
8	7	4	4	2	2	1	1	0				
9	8	7	4	4	2	2	1	1	0			
10	12	8	7	4	4	2	2	1	1	0		
11	14	12	8	7	4	4	2	2	1	1	0	
12	21	14	12	8	7	4	4	2	2	1	1	0

$$Z_1(\tau) = q^{-1} + 196,884q + 21,493,760q^2 + 864,299,970q^3 + \cdots \tag{14}$$
$$Z_2(\tau) = q^{-2} + 1 + 42,987,520q + 40,491,909,396q^2 + \cdots$$
$$Z_3(\tau) = q^{-3} + q^{-1} + 1 + 2,593,096,794q + 12,756,091,394,048q^2 + \cdots$$

for example. $Z_k(\tau)$ automatically inherits the key property of modular invariance by (10) since $J(\tau)$ is modular invariant.

3 A remainder formula

The remainder formula that we present in this section (formula 19), which encapsulates a bit more than just the asymptotic behavior of the number of quantum states $b_{k,n}$ in (12) (for large n), is based on the following result for the coefficients c_n in the q-expansion (7). Namely, by Theorem 1.3 of [9], for $n, p \in \mathbb{Z}^+$

$$c_n = c_\infty(n)\left(\sum_{k=0}^{p-1} \frac{(-1)^k(1,k)}{(8\pi\sqrt{n})^k} + \frac{r_p(n)}{n^{p/2}}\right), \tag{15}$$

where

$$c_\infty(n) \overset{\text{def}}{=} \frac{e^{4\pi\sqrt{n}}}{\sqrt{2}n^{3/4}}, \quad (1,k) \overset{\text{def}}{=} \prod_{j=0}^{k-1} \frac{[4-(2j+1)^2]}{4^k k!},$$

$$|r_p(n)| \leq \frac{|(1,p)|}{\sqrt{2}(4\pi)^p} + 62\sqrt{2}e^{-2\pi\sqrt{n}}n^{p/2}. \tag{16}$$

$(1,k)$ is a *Hankel symbol* $(\alpha,k) \overset{\text{def}}{=} \Gamma(\frac{1}{2}+\alpha+k)/k!\Gamma(\frac{1}{2}+\alpha-k)$ (with $\alpha=1$, $(\alpha,0) \overset{\text{def}}{=} 1$) employed of course in the expression of the asymptotic behavior of Bessel functions. Also the upper bound result

$$c_n \leq c_\infty(n), \quad n \geq 1 \tag{17}$$

is established in [9]. The idea now is to write (13) as $c_n^{(r)} = c_{rn} + S(r,n)$, where $S(r,n)$ is the corresponding sum over divisors $d \geq 2$, apply (15) to c_{rn} use (17), estimate $S(r,n)$ in terms of the Riemann zeta function value $\zeta(3/2)$ and then to apply these various steps to $b_{k,n}$ in (12). In the end one reaches the following conclusion. For $k,n,p \in \mathbb{Z}^+$, let

$$b_{k,n}^\infty \overset{\text{def}}{=} kc_\infty(kn) \overset{\text{def(16)}}{=} \frac{ke^{4\pi\sqrt{kn}}}{\sqrt{2}(kn)^{3/4}}. \tag{18}$$

Then, in terms of some of the preceding notation,

$$\begin{aligned} b_{k,n} = b_{k,n}^\infty \Bigg[& 1 + \sum_{m=1}^{p-1} \frac{(-1)^m(1,m)}{(8\pi\sqrt{kn})^m} + \frac{r_p(kn)}{(kn)^{p/2}} + \frac{S(k,n)}{c_\infty(kn)} \\ & + \frac{1}{k^{1/4}} \sum_{1\leq r<k} \frac{r^{1/4}a_{-r}(k)}{e^{4\pi\sqrt{n}(\sqrt{k}-\sqrt{r})}} \left(1 + \sum_{m=1}^{p-1} \frac{(-1)^m(1,m)}{(8\pi\sqrt{kn})^m} \right. \\ & \left. + \frac{r_p(kn)}{(kn)^{p/2}} + \frac{S(k,n)}{c_\infty(kn)} \right) \Bigg] \end{aligned} \tag{19}$$

where for $1 \leq r \leq k$

$$\begin{aligned} \left| \frac{r_p(rn)}{(rn)^{p/2}} \right| &\leq \frac{|(1,p)|}{\sqrt{2}(4\pi)^p(rn)^{p/2}} + 62\sqrt{2}e^{-2\pi\sqrt{rn}}, \\ 0 < \frac{S(r,n)}{c_\infty(kn)} &\leq \frac{r^{3/2}\zeta(3/2)}{2e^{2\pi\sqrt{rn}}}, \frac{n^{3/2}\zeta(3/2)}{2e^{2\pi\sqrt{rn}}}. \end{aligned} \tag{20}$$

If $p = 1$, for example, the terms in (19) with the summation from $m = 1$ to $m = p-1$ are interpreted to have the value 0.

In particular, from (19) and (20) one can immediately read off the asymptotic result

$$b_{k,n} \sim \frac{ke^{4\pi\sqrt{kn}}}{\sqrt{2}(kn)^{3/4}}, \qquad n \to \infty \tag{21}$$

for every fixed $k \geq 1$, by definition (18). For $k = 1$ however, $b_{1,n} = c_n$ (by (12), (13)) and (21) reduces to the classical asymptotic formula

$$c_n \sim \frac{ke^{4\pi\sqrt{n}}}{\sqrt{2}n^{3/2}}, \qquad n \to \infty \tag{22}$$

in [11, 12]. Of course (22) also follows from (15), (16).

4 Quantum corrections to black hole entropy

As indicated in the introduction one can derive corrections to black hole entropy $S = 4\pi\sqrt{kn}$ by considering the logarithm of $b_{k,n}$, especially for large n. Off hand, (21) yields the logarithmic correction

$$\log b_{k,n} \simeq S + \frac{1}{4}\log k - \frac{3}{4}\log n - \frac{1}{2}\log 2 \tag{23}$$

to S, as in [3] of course. In the remainder formula (19) the sum over $r < k$ involves terms of exponential decay with regard to the growth of n, and the estimate for $S(k,n)/c_\infty(kn)$ in (20) involves a bound of exponential decay. If such terms are disregarded and if the estimate $\log(1+x) \simeq x$ for small x is invoked, the refinement

$$\log b_{k,n} \simeq S+\frac{1}{4}\log k-\frac{3}{4}\log n-\frac{1}{2}\log 2+\sum_{m=1}^{p-1}\frac{(-1)^m(1,m)}{(8\pi\sqrt{kn})^m}+\frac{r_p(kn)}{(kn)^{p/2}}, \tag{24}$$

for any $k, p \geq 1$, of (23) is obtained. Here a bound for the remainder term $r_p(kn)/(kn)^{p/2}$ is given by (20) (taking $r = k$ there). Also in (24) the following table provides values of the Hankel symbols $(1,m)$ (and thus of $|(1,p)|$ in (20)); see definition (16).

For the choice $p = 5$, for example, (24) with Table 2 provides for the following additional terms for (23):

$$\log b_{k,n} \simeq S + \frac{1}{4}\log k - \frac{3}{4}\log n - \frac{1}{2}\log 2 - \frac{0.75}{8\pi\sqrt{kn}} - \frac{0.4688}{(8\pi\sqrt{kn})^2}$$
$$-\frac{0.8203}{(8\pi\sqrt{kn})^3} - \frac{2.3071}{(8\pi\sqrt{kn})^4} + \frac{r_5(kn)}{(kn)^5}. \tag{25}$$

We have remarked that every coefficient c_n in equation (7) is an *integer*. One knows that every c_n is also *positive* – as is suggested, for example, by the values in (8). It follows that the $c_n^{(r)}$ in (13) are positive, and hence the $b_{k,n}$ in (12) are positive and they are integers. In particular the $\log b_{k,n}$ in (23)–(25) are well-defined real numbers.

Table 2. Hankel symbols

k	$(1, k)$
1	$3/4 = \quad 0.750000000$
2	$-15/32 = -0.468750000$
3	$105/128 = \quad 0.820312500$
4	$-4725/2048 = -2.307128906$
5	$72765/8192 = \quad 8.882446289$
6	$-2837835/65536 = -43.30192566$
7	$66891825/262144 = \quad 255.1720619$
8	$-14783093325/8388608 = -1762.282053$
9	$468131288625/33554432 = \quad 13951.39958$
10	$-50565381191325/268435456 = -188370.7240$

5 An exact formula for $b_{k,n}$

It is possible to give, in fact, an *exact* formula for the coefficients $b_{k,n}$. This formula is stated in [3]. We consider an alternate version. Its derivation is based on an exact formula [11, 12] for the coefficients c_n, $n \geq 1$, in (7):

$$c_n = \frac{2\pi}{\sqrt{n}} \sum_{m=1}^{\infty} \frac{A_m(n)}{m} I_1\left(\frac{4\pi\sqrt{n}}{m}\right) \tag{26}$$

where

$$I_1(x) = \sum_{j=0}^{\infty} \frac{\left(\frac{x}{2}\right)^{2j+1}}{j!(j+1)!} \tag{27}$$

is a modified Bessel function and where

$$A_m(n) = \sum_{h \in \mathbb{Z}/m\mathbb{Z}} e^{-\frac{2\pi i}{m}(nh+h')} \tag{28}$$

is a *Kloosterman sum*, with the sum taken over h that are relatively prime to m, and with h' chosen to satisfy $hh' \equiv -1 (\text{mod } m)$. Then by (13)

$$c_n^{(r)} = \frac{2\pi}{\sqrt{rn}} \sum_{\substack{d>0 \\ d|r,\, d|n}} \sum_{m=1}^{\infty} \frac{A_m\left(\frac{rn}{d^2}\right)}{m} I_1\left(\frac{4\pi\sqrt{rn}}{md}\right), \tag{29}$$

and hence

$$b_{k,n} = \frac{2\pi}{\sqrt{n}} \sum_{r=1}^{k} \sqrt{r} a_{-r}(k) \sum_{\substack{d>0 \\ d|r,\, d|n}} \sum_{m=1}^{\infty} \frac{A_m\left(\frac{rn}{d^2}\right)}{m} I_1\left(\frac{4\pi\sqrt{rn}}{md}\right) \tag{30}$$

by (12), which is the desired formula.

It is of interest perhaps to point out that the original version of (30) in (5.18) of [3] (where $a_{-r}(k)$ correspond to $F_{\Delta'}$ 0 there) also follows from a general, abstract result in [9], given that $Z_k(\tau)$ is a modular form of weight 0. For this we use an alternative version

$$rc_n^{(r)} = 2\pi \sqrt{\frac{r}{n}} \sum_{m=1}^{\infty} \frac{A_m(n,r)}{m} I_1\left(\frac{4\pi\sqrt{rn}}{m}\right) \tag{31}$$

of formula (29), where the Kloosterman sum

$$A_m(n,r) \stackrel{\text{def}}{=} \sum_{h\in\mathbb{Z}/m\mathbb{Z}} e^{-\frac{2\pi i}{m}(nh+rh')}$$

$$h \text{ relatively prime to } m$$

$$hh' \equiv -1 (\operatorname{mod} m) \tag{32}$$

slightly generalizes the sum $A_m(n)$ in (28). An application of Theorem 5.2 of [9] to equation (5.13) on page 345 of [17], for example, leads immediately to formula (31). Then by (12)

$$b_{k,n} = 2\pi \sum_{r=1}^{k} a_{-r}(k) \sqrt{\frac{r}{n}} \sum_{m=1}^{\infty} \frac{A_m(n,r)}{m} I_1\left(\frac{4\pi\sqrt{rn}}{md}\right) \tag{33}$$

for $k, n \geq 1$ where again $a_{-r}(k) \stackrel{\text{def}}{=} p(k-r) - p(k-r-1)$ by (9); see Table 1 above. One can also obtain formula (33) by an application of Theorem 5.2 of [9] to equation (11) directly. Note that formula (31), which is of independent interest, coincides with formula (26) when $r = 1$. The Kloosterman sums in (30) are simpler than those in (33), or in [13]. On the other hand, (30) also follows from (33) by a general Selberg identity on Kloosterman sums.

Rademacher type formulas such as formula (33) appear initially in black hole physics in the reference [13]. Compare also, for example, the reference [18] where a Rademacher formula is employed in the discussion of an entropy bound within the framework of two-dimensional conformal field theory.

6 A zeta function connection

In the reference [19] the author attached to the three-dimensional BTZ black hole with a conical singularity a natural zeta function, in terms of which one-loop quantum corrections of R. Mann–S. Solodukhin [20] to Bekenstein–Hawking entropy were expressed. That zeta function was constructed as a *conical deformation* of an appropriate *Patterson–Selberg zeta function* $Z_\Gamma(s)$. Details and further references for that work appear in [17], where it is shown, moreover, that the holomorphic sector $Z_\Gamma(s)$ of the one-loop gravity partition function (see definitions (1), (2)) is also expressed in terms of $Z_\Gamma(s)$.

On the other hand the partition function $Z_k(\tau)$ was constructed to satisfy the "axiomatic" condition

$$\begin{aligned} Z_k(\tau) &= \text{ Virasoro decendants } + \text{ BTZ black holes} \\ &= q^{-k} Z_{\text{hol}}(\tau) + \mathcal{O}(q) = q^{-k}\left[Z_{\text{hol}}(\tau) + \mathcal{O}(q^{k+1})\right]. \end{aligned} \tag{34}$$

Thus we have a zeta function connection (of $Z_k(\tau)$ with $Z_\Gamma(s)$) that (for the record) we express more succinctly in this section.

Given real numbers a, b with $a > 0$, let

$$\begin{aligned} \gamma = \gamma_{(a,b)} &\overset{\text{def}}{=} \begin{bmatrix} e^{a+ib} & 0 \\ 0 & e^{-(a+ib)} \end{bmatrix}, \\ \Gamma = \Gamma_{(a,b)} &\overset{\text{def}}{=} \{\gamma^n \mid n \in \mathbb{Z}\}, \\ Z_\Gamma(s) &\overset{\text{def}}{=} \prod_{0 \le k_1, k_2 \in \mathbb{Z}}^{\infty} \left[1 - \left(e^{i2b}\right)^{k_1} \left(e^{-i2b}\right)^{k_2} e^{-(k_1+k_2+s)2a}\right]. \end{aligned} \tag{35}$$

Γ is the subgroup of $SL(2,\mathbb{C})$ generated by γ, and $Z_\Gamma(s)$ (which is an entire function of $s \in \mathbb{C}$) is the Patterson–Selberg zeta function corresponding to Γ [21]. Since $a > 0$, $\tau_{(a,b)} \overset{\text{def}}{=} (b + ia)/\pi \in \Pi^+$, and by Theorem 3.26 of [17]

$$\begin{aligned} Z_{\text{hol}}\left(\tau_{(a,b)}\right) &= Z_\Gamma\left(3 - i\frac{b}{a}\right) / Z_\Gamma\left(2 - i\frac{2b}{a}\right), \\ Z^{1\text{-loop}}_{\text{gravity}}\left(\tau_{(a,b)}\right) &= \frac{Z_\Gamma\left(3 - i\frac{b}{a}\right) Z_\Gamma\left(3 + i\frac{b}{a}\right)}{Z_\Gamma\left(2 - i\frac{2b}{a}\right) Z_\Gamma\left(2 + i\frac{2b}{a}\right)}. \end{aligned} \tag{36}$$

By equations (9) and (11)

$$Z_k(\tau) = q^{-k} Z_{\text{hol}}(\tau) + \sum_{n=1}^{\infty} [b_{k,n} - a_n(k)] q^n, \tag{37}$$

which is (34). Hence, by the first equation in (36), we see that $Z_k(\tau)$ and $Z_\Gamma(s)$ are related by

$$Z_k\left(\frac{b+ia}{\pi}\right) = q^{-k}\frac{Z_\Gamma\left(3-i\frac{b}{a}\right)}{Z_\Gamma\left(2-i\frac{2b}{a}\right)} + \sum_{n=1}^{\infty}[b_{k,n} - a_n(k)]q^n, \tag{38}$$

for real numbers a, b with $a > 0$. Again $a_n(k), b_{k,n}$ are given by (9) and (12); $q^{-k} = exp(-2ki(b+ia))$.

A special choice, for example, is $a = \beta/2, b = \theta/2$; i.e., $\tau = (\theta + i\beta)/2\pi$ corresponds to the anti-de Sitter–Hawking temperature β^{-1} and angular potential θ.

ECFT partition functions have been constructed, more generally, for theories with central charge divisible by 8: $c = 8m$. Explicit formulas for such functions are derived in [6] for c up to 88, where the cases $c = 48, 72$ are already handled in [2]; also see [1]. Here (34) is replaced by the requirement

$$Z_{8m}(\tau) = q^{-m/3}\left[Z_{\text{hol}}(\tau) + \mathcal{O}\left(q^{[m/3]+1}\right)\right], \tag{39}$$

(see definition (1)), which does reduce to (34) in the special case $m = 3k$. Similar to (10), $Z_{8m}(\tau)$ is known to assume the form (see definition (4))

$$Z_{8m}(\tau) = j(\tau)^{m/3}\sum_{r=0}^{[m/3]} a_r\, j(\tau)^{-r}, \tag{40}$$

for computable coefficients a_r [1]. Namely, the a_r are found by comparing terms of order $q^{r-m/3}$ with terms in (39). Strictly speaking, we are considering only genus 1 (i.e., torus) partition functions in this paper. The authors in [6] consider also genus 2 partition functions.

In the simple cases $m = 1, 2$, for example, one has

$$Z_8(\tau) = j(\tau)^{1/3} = q^{-1/3}\left[1 + 248q + 4124q^2 + 34752q^3 + \cdots\right],$$
$$Z_{16}(\tau) = j(\tau)^{2/3} = q^{-2/3}\left[1 + 496q + 69752q^2 + 2115008q^3 + \cdots\right]. \tag{41}$$

Applying the first formula in (36) again, we obtain the corresponding generalization (for a, b real, $a > 0$)

$$Z_{8m}\left(\frac{b+ia}{\pi}\right) = q^{-m/3}Z_\Gamma\left(3-i\frac{b}{a}\right)/Z_\Gamma\left(2-i\frac{2b}{a}\right) + \mathcal{O}\left(q^{[m/3]-m/3+1}\right) \tag{42}$$

of (38), by (39). Since $j(\tau)^{1/3}, j(\tau)^{2/3}$ are vacuum characters of $\hat{E}_8, \hat{E}_8 \times \hat{E}_8$ for affine level 1 theories [22, 23], equations (41), (42) (in particular) provide for a connection between these characters and the zeta function $Z_\Gamma(s)$.

As is well known, due to initial observations of J. McKay, J. Conway, S. Norton, J. Thompson, and others, the initial Fourier coefficients c_n in (7), (8) are integral combinations of dimensions of irreducible representations of the monster group M mentioned in Section 1. Some historic remarks on this, with references, are presented in [8]. For example $c_3 = 2\cdot 1 + 2\cdot(196,833) + 21,296,876,842,609,326$, where $1, 196,833, 21,296,876,842,609,326$ are dimensions of irreducible representations, 1 being the dimension of the trivial representation. Of course this observation is part of a much more general "moonshine" phenomenon/conjecture that was eventually resolved/proved by R. Borcherds. The 1st bracket in (41) ($= q^{1/3}$ times an $\hat{E}_8$ character) is a McKay–Thompson series for M, of class $3C$.

It is a remarkable fact that the set of zeros of the Patterson–Selberg (P-S) zeta function coincides with the set of "nontrivial" poles (called *resonances*) of a scattering matrix defined by *reflection coefficients* in the asymptotics of a particular solution of the Schrödinger's equation for a Pöschel–Teller potential [24–26]. This fact can be viewed as a mathematical type of AdS/CFT (or bulk/boundary) correspondence, where the zeta zeros represent a bulk quantity and the scattering matrix a boundary quantity. The original AdS/CFT correspondence (due to J. Maldacena) referred to a (boldly) proposed duality between $(d+1)$-dimensional gravity on AdS space (i.e., on anti-deSitter space) and conformal field theory (CFT) on its d-dimensional boundary.

Connections between 3-dimensional AdS gravity (where a *negative* cosmological constant is present) and the P-S zeta function are explored in [17, 27], where in the review article [27] a bulk-boundary correspondence is set up even for a 2-dimensional black hole vacuum with a *parabolic* generator of its holonomy. The discussion in Section 6 here shows that a zeta function connection exists also on the CFT side, and thus connections have life on both sides of the correspondence.

7 Acknowledgements

The author extends special thanks to Prof. Alex Maloney for his helpful remarks that clarified a key point in the papers [2, 3].

A Personal Reflection

I first met Nolan Wallach in 1962 when I was a first year graduate student at Washington University. Nolan was an advanced student at the time who was well into the study of Lie groups and differential geometry, with Professors like William Boothby and Jun-Ichi Hano. Prof. Hano later served as Nolan's thesis advisor. The title of the 1966 thesis was "A classification of real simple Lie algebras". I eventually developed an interest in a book by L. Pontrjagin entitled *Topological Groups*, a book that became an initial part of the Princeton Mathematical Series, along with other classics such as those of H. Weyl and C. Chevalley. At some point I came across a quotation of E. Cartan: "If an n-sphere is a topological group, then $n = 0, 1$, or 3" – a statement that I found a bit fascinating and which stirred my mathematical curiosity. With some help from Prof. Hano, I struggled through the first four chapters of the book, the fourth chapter being my initial exposure to representation theory (of compact groups). Evidently, Nolan was not so impressed with my little success and he felt that it might be more important if I tried to learn something about Lie groups, from a more modern text, namely that I should study from Chevalley's Princeton book *Theory of Lie Groups*. As a young student with a somewhat modest background, I found the suggestion, and the book, rather intimidating, although Chevalley's writing seemed especially elegant, careful, and scholarly. It was a hard book for me, indeed, but I kept in mind the above mentioned quotation of E. Cartan which continued to stimulate me, and most importantly it was Nolan's kind assistance with the reading that made all the difference in the end. I began to realize through that experience that he was teaching me to reach toward higher standards, which was the value of having a wonderful mentor, and friend as he was, and has always been. Thank you Nolan for your mentorship and your inspiration that have markedly shaped my mathematical life. And congratulations to you for your manifold, stellar mathematical achievements that tower substantially over those of your former, fellow Washington University students, and in fact over many of the great mathematical masters of the day. Always with fondest recollections, Floyd.

References

1. Höhn, G., *Selbstduale Vertexoperatorsuperalgebren und das Babymonster*. Bonner Mathematische Schriften, 286, Universität Bonn Mathematisches Institut, Bonn, 1996, Dissertation, Rheinische Friedrich-Wilhelms Universitat Bonn, 1995.
2. Witten, E., *Three-dimensional gravity revisited*, 2007 preprint, arXiv: 0706.3359.
3. Maloney, A.; Witten, E., *Quantum gravity partition function in three dimensions*. J. High Energy Phys. 2010, No. 2, 1–58; arXiv: 0712.0155, 2007.
4. Maloney, A., *Physics and the monster-quantum gravity, modular forms, and Faber polynomials*. Concordia University lecture, 2007.
5. Osorio, A.V., *Modular invariance and 3d gravity*, 2008 thesis supervised by Vandoren, A.; pdf available online.
6. Avramis, S.; Kehagias, A.; Mattheopoulou, C., *Three-dimensional AdS gravity and extremal CFTs at* $c = 8m$. arXiv: 0708.3386, 2007.
7. Gaiotto, D., *Monster symmetry and extremal CFTs*, 2008 preprint, arXiv: 0801.0988.
8. Frenkel, I.; Lepowsky, J.; Meurman. A., *A moonshine module for the monster*, in *Vertex Operators in Mathematics and Physics*. Edited by Lepowsky, J., Mandelstam, S., and Singer, I. (Berkeley, CA., 1983). Math. Sci. Research Inst. Publ. Vol. 3, Springer, New York, 1985, 231–273.
9. Brisebarre, N.; Philibert, G., *Effective lower and upper bounds for the Fourier coefficients of powers of the modular invariant* j. Journal of the Ramanujan Math. Soc. 2005, 20, 255–282.
10. Knopp, M., *Automorphic forms of non-negative dimension and exponential sums*. Michigan Math. J., 1960, 7, 257–281.

11. Petersson, H., *Über die Entwicklungskoeffizienten der automorphen formen.* Acta Math. 1932, 58, 501–512.
12. Rademacher, H., *The Fourier coefficients of the modular invariant* $J(\tau)$. Amer. J. Math. 1938, 60, 501–512.
13. Dijkgraaf, R. ; Maldacena, J.; Moore, G.; Verlinde, E., *A black hole Farey tail.* arXiv: hepth/0005003, 2007.
14. Cardy, J., *Operator content of two-dimensional conformally invariant theories.* Nuclear Phys. B, 1986, 270, 186–204.
15. Giombi, S.; Maloney, A.; Yin, X., *One-loop partition functions of 3d gravity.* J. High Energy Phys. 2008, 007, 1–24; arXiv:0804. 1173, 2008.
16. Giombi, S., *One-loop partition functions of 3d gravity.* Harvard University lecture, 2008.
17. Williams, F., *Lectures on zeta functions, L-functions and modular forms with some physical applications*, and also the lecture *The role of the Patterson–Selberg zeta function of a hyperbolic cylinder in three-dimensional gravity with a negative cosmological constant*, in A Window into Zeta and Modular Physics. Edited by Kirsten, K. and Williams, F. Math. Sci. Research Inst. Publ. Vol. 57, Cambridge Univ. Press, 2010, 7–100 and 329–351.
18. Birmingham, D.; Siddhartha, S., *Exact black hole entropy bound in conformal field theory.* Phys. Rev. D 2001, 63, 047501-1-047501-3, arXiv:hep-th/0008051, 2000.
19. Williams, F., *Conical defect zeta function for the BTZ black hole*, from One Hundred Years of Relativity: Proceedings of the Einstein Symposium (Iais, Romania, 2005). Scientific Annals of Alexandru Ioan Univ. 2005, TOM LI-LII, 54–58.
20. Mann, R.; Solodukhin, S., *Quantum scalar field on a three-dimensional (BTZ) black hole instanton; heat kernel, effective action, and thermodynamics.* Phys. Rev. D, 1997, 55, 3622–3632.
21. Patterson, S., *The Selberg zeta function of a Kleinian group*, in Number Theory, Trace Formulas, and Discrete Groups (Oslo, Sweden, 1987). Academic Press, Boston, MA. 1989, 409–441.
22. Kac, V. G., *An elucidation of "Infinite-dimensional algebras and the very strange formula", $E_8^{(1)}$ and the cube root of the modular invariant* j, Adv. in Math. 1980, 35, 264–273.
23. Lepowsky, J., *Euclidean Lie algebras and the modular function* j, from the Santa Cruz Conf. on Finite Groups, AMS Proc. Sympos. Pure Math, 1980, 37, 567–570.
24. Patterson, S. J.; Perry, P., *The divisor of Selberg's zeta function for Kleinian groups.* Duke Math. J. 2001, 106, 321–390 (with an appendix by Epstein, C.).
25. Perry, P.; Williams, F., *Selberg zeta function and trace formula for the BTZ black hole.* Internat. J. Pure and Appl. Math. 2003, 9, 1–21.
26. Guillopé, L.; Zworski, M., *Upper bounds on the number of resonances for non-compact Riemann surfaces.* J. Funct. Anal. 1995, 129, 364–389.
27. Williams, F., *Remarks on the Patterson–Selberg zeta function, black hole vacua and extreme CFT partition functions*, in a volume dedicated to the 75th birthday of Stuart Dowker, J. Phys. A, Math. Theor. 2012, 45 (19 pp).

Principal series representations of infinite-dimensional Lie groups, I: Minimal parabolic subgroups

Joseph A. Wolf

To Nolan Wallach on the occasion of his seventieth birthday

Abstract We study the structure of minimal parabolic subgroups of the classical infinite-dimensional real simple Lie groups, corresponding to the classical simple direct limit Lie algebras. This depends on the recently developed structure of parabolic subgroups and subalgebras that are not necessarily direct limits of finite-dimensional parabolics. We then discuss the use of that structure theory for the infinite-dimensional analog of the classical principal series representations. We look at the unitary representation theory of the classical lim-compact groups $U(\infty)$, $SO(\infty)$ and $Sp(\infty)$ in order to construct the inducing representations, and we indicate some of the analytic considerations in the actual construction of the induced representations.

Keywords: Principle series representation • Infinite-dimensional Lie group • Minimal parabolic subgroup

Mathematics Subject Classification: 32L25, 22E46, 32L10

1 Introduction

This paper discusses some recent developments in a program of extending aspects of real semisimple group representation theory to infinite-dimensional real Lie groups. The finite-dimensional theory is entwined with the structure of parabolic

Research partially supported by a Simons Foundation grant.

J.A. Wolf
Department of Mathematics, University of California, Berkeley, CA 94720-3840, USA
e-mail: jawolf@math.berkeley.edu

R. Howe et al. (eds.), *Symmetry: Representation Theory and Its Applications*, Progress in Mathematics 257, DOI 10.1007/978-1-4939-1590-3_19

subgroups, and that structure has recently been worked out for the classical direct limit groups such as $SL(\infty, \mathbb{R})$ and $Sp(\infty; \mathbb{R})$. Here we explore the consequences of that structure theory for the construction of the counterpart of various Harish-Chandra series of representations, specifically the principal series.

The representation theory of finite-dimensional real semisimple Lie groups is based on the now-classical constructions and Plancherel formula of Harish-Chandra. Let G be a real semisimple Lie group, e.g., $SL(n; \mathbb{R})$, $SU(p,q)$, $SO(p,q)$, ... Then one associates a series of representations to each conjugacy class of Cartan subgroups. Roughly speaking, this goes as follows. Let $\mathrm{Car}(G)$ denote the set of conjugacy classes $[H]$ of Cartan subgroups H of G. Choose $[H] \in \mathrm{Car}(G)$, $H \in [H]$, and an irreducible unitary representation χ of H. Then we have a "cuspidal" parabolic subgroup P of G constructed from H, and a unitary representation π_χ of G constructed from χ and P. Let Θ_{π_χ} denote the distribution character of π_χ. The Plancherel Formula: if $f \in \mathcal{C}(G)$, the Harish-Chandra Schwartz space, then

$$f(x) = \sum_{[H]\in \mathrm{Car}(G)} \int_{\hat{H}} \Theta_{\pi_\chi}(r_x f) d\mu_{[H]}(\chi) \tag{1.1}$$

where r_x is right translation and $\mu_{[H]}$ is a Plancherel measure on the unitary dual $\hat{H}$.

In order to consider any elements of this theory in the context of real semisimple direct limit groups, we have to look more closely at the construction of the Harish-Chandra series that enter into (1.1).

Let H be a Cartan subgroup of G. It is stable under a Cartan involution θ, an involutive automorphism of G whose fixed point set $K = G^\theta$ is a maximal compactly embedded[1] subgroup. Then H has a θ-stable decomposition $T \times A$ where $T = H \cap K$ is the compactly embedded part and (using lower case gothic letters for Lie algebras) $\exp : \mathfrak{a} \to A$ is a bijection. Then $\mathfrak{a}$ is commutative and acts diagonalizably on $\mathfrak{g}$. Any choice of a positive $\mathfrak{a}$-root system defines a parabolic subalgebra $\mathfrak{p} = \mathfrak{m} + \mathfrak{a} + \mathfrak{n}$ in $\mathfrak{g}$ and thus defines a parabolic subgroup $P = MAN$ in G. If τ is an irreducible unitary representation of M and $\sigma \in \mathfrak{a}^*$, then $\eta_{\tau,\sigma} : man \mapsto e^{i\sigma(\log a)}\tau(m)$ is a well defined irreducible unitary representation of P. The equivalence class of the unitarily induced representation $\pi_{\tau,\sigma} = \mathrm{Ind}_P^G(\eta_{\tau,\sigma})$ is independent of the choice of a positive $\mathfrak{a}$-root system. The group M has (relative) discrete series representations, and $\{\pi_{\tau,\sigma} \mid \tau \text{ is a discrete series rep of } M\}$ is the series of unitary representations associated to $\{\mathrm{Ad}(g)H \mid g \in G\}$.

One of the most difficult points here is dealing with the discrete series. In fact the possibilities of direct limit representations of direct limit groups are somewhat limited except in cases where one can pass cohomologies through direct limits without change of cohomology degree. See [14] for limits of holomorphic discrete series, [15] for Bott–Borel–Weil theory in the direct limit context, [11] for some

[1]A subgroup of G is *compactly embedded* if it has compact image under the adjoint representation of G.

nonholomorphic discrete series cases, and [24] for principal series of classical type. The principal series representations in (1.1) are those for which M is compactly embedded in G, equivalently the ones for which P is a minimal parabolic subgroup of G.

Here we work out the structure of the minimal parabolic subgroups of the finitary simple real Lie groups and discuss construction of the associated principal series representations. As in the finite-dimensional case, a minimal parabolic has structure $P = MAN$. Here $M = P \cap K$ is a (possibly infinite) direct product of torus groups, compact classical groups such as $Spin(n)$, $SU(n)$, $U(n)$ and $Sp(n)$, and their classical direct limits $Spin(\infty)$, $SU(\infty)$, $U(\infty)$ and $Sp(\infty)$ (modulo intersections and discrete central subgroups).

Since this setting is not standard we start by sketching the background. In Section 2 we recall the classical simple real direct limit Lie algebras and Lie groups. There are no surprises. Section 3 sketches their relatively recent theory of complex parabolic subalgebras. It is a little bit complicated and there are some surprises. Section 4 carries those results over to real parabolic subalgebras. There are no new surprises. Then in Sections 5 and 6 we deal with Levi components and Chevalley decompositions. That completes the background.

In Section 7 we examine the real group structure of Levi components of real parabolics. Then we specialize this to minimal self-normalizing parabolics in Section 8. There the Levi components are locally isomorphic to direct sums in an explicit way of subgroups that are either the compact classical groups $SU(n)$, $SO(n)$ or $Sp(n)$, or their limits $SU(\infty)$, $SO(\infty)$ or $Sp(\infty)$. The Chevalley (maximal reductive part) components are slightly more complicated, for example involving extensions $1 \to SU(*) \to U(*) \to T^1 \to 1$ as well as direct products with tori and vector groups. The main result is Theorem 8.3, which gives the structure of the minimal self-normalizing parabolics in terms similar to those of the finite dimensional case. Proposition 8.12 then gives an explicit construction for minimal parabolics with a given Levi factor.

In Section 9 we discuss the various possibilities for the inducing representation. There are many good choices, for example tame representations or more generally representations that are factors of type II. The theory is at such an early stage that the best choice is not yet clear.

Finally, in Section 10 we indicate construction of the induced representations in our infinite-dimensional setting. Smoothness conditions do not introduce surprises, but unitarity is a problem, and we defer details of that construction to [26] and applications to [27].

I thank Elizabeth Dan-Cohen and Ivan Penkov for many very helpful discussions on parabolic subalgebras and Levi components.

2 The classical simple real groups

In this section we recall the real simple countably infinite-dimensional locally finite ("finitary") Lie algebras and the corresponding Lie groups. This material follows from results in [1], [2] and [6].

We start with the three classical simple locally finite countable-dimensional Lie algebras $\mathfrak{g}_{\mathbb{C}} = \varinjlim \mathfrak{g}_{n,\mathbb{C}}$, and their real forms $\mathfrak{g}_{\mathbb{R}}$. The Lie algebras $\mathfrak{g}_{\mathbb{C}}$ are the classical direct limits,

$$\begin{aligned} \mathfrak{sl}(\infty, \mathbb{C}) &= \varinjlim \mathfrak{sl}(n; \mathbb{C}), \\ \mathfrak{so}(\infty, \mathbb{C}) &= \varinjlim \mathfrak{so}(2n; \mathbb{C}) = \varinjlim \mathfrak{so}(2n+1; \mathbb{C}), \\ \mathfrak{sp}(\infty, \mathbb{C}) &= \varinjlim \mathfrak{sp}(n; \mathbb{C}), \end{aligned} \tag{2.1}$$

where the direct systems are given by the inclusions of the form $A \mapsto \left(\begin{smallmatrix} A & 0 \\ 0 & 0 \end{smallmatrix}\right)$. We will also consider the locally reductive algebra $\mathfrak{gl}(\infty; \mathbb{C}) = \varinjlim \mathfrak{gl}(n; \mathbb{C})$ along with $\mathfrak{sl}(\infty; \mathbb{C})$. The direct limit process of (2.1) defines the universal enveloping algebras

$$\begin{aligned} &\mathcal{U}(\mathfrak{sl}(\infty, \mathbb{C})) = \varinjlim \mathcal{U}(\mathfrak{sl}(n; \mathbb{C})) \text{ and } \mathcal{U}(\mathfrak{gl}(\infty, \mathbb{C})) = \varinjlim \mathcal{U}(\mathfrak{gl}(n; \mathbb{C})), \\ &\mathcal{U}(\mathfrak{so}(\infty, \mathbb{C})) = \varinjlim \mathcal{U}(\mathfrak{so}(2n; \mathbb{C})) = \varinjlim \mathcal{U}(\mathfrak{so}(2n+1; \mathbb{C})), \text{ and} \\ &\mathcal{U}(\mathfrak{sp}(\infty, \mathbb{C})) = \varinjlim \mathcal{U}(\mathfrak{sp}(n; \mathbb{C})). \end{aligned} \tag{2.2}$$

Of course each of these Lie algebras $\mathfrak{g}_{\mathbb{C}}$ has the underlying structure of a real Lie algebra. Besides that, their real forms are as follows ([1, 2, 6]).

If $\mathfrak{g}_{\mathbb{C}} = \mathfrak{sl}(\infty; \mathbb{C})$, then $\mathfrak{g}_{\mathbb{R}}$ is one of $\mathfrak{sl}(\infty; \mathbb{R}) = \varinjlim \mathfrak{sl}(n; \mathbb{R})$, the real special linear Lie algebra; $\mathfrak{sl}(\infty; \mathbb{H}) = \varinjlim \mathfrak{sl}(n; \mathbb{H})$, the quaternionic special linear Lie algebra, given by $\mathfrak{sl}(n; \mathbb{H}) := \mathfrak{gl}(n; \mathbb{H}) \cap \mathfrak{sl}(2n; \mathbb{C})$; $\mathfrak{su}(p, \infty) = \varinjlim \mathfrak{su}(p, n)$, the complex special unitary Lie algebra of real rank p; or $\mathfrak{su}(\infty, \infty) = \varinjlim \mathfrak{su}(p, q)$, complex special unitary algebra of infinite real rank.

If $\mathfrak{g}_{\mathbb{C}} = \mathfrak{so}(\infty; \mathbb{C})$, then $\mathfrak{g}_{\mathbb{R}}$ is one of $\mathfrak{so}(p, \infty) = \varinjlim \mathfrak{so}(p, n)$, the real orthogonal Lie algebra of finite real rank p; $\mathfrak{so}(\infty, \infty) = \varinjlim \mathfrak{so}(p, q)$, the real orthogonal Lie algebra of infinite real rank; or $\mathfrak{so}^*(2\infty) = \varinjlim \mathfrak{so}^*(2n)$

If $\mathfrak{g}_{\mathbb{C}} = \mathfrak{sp}(\infty; \mathbb{C})$, then $\mathfrak{g}_{\mathbb{R}}$ is one of $\mathfrak{sp}(\infty; \mathbb{R}) = \varinjlim \mathfrak{sp}(n; \mathbb{R})$, the real symplectic Lie algebra; $\mathfrak{sp}(p, \infty) = \varinjlim \mathfrak{sp}(p, n)$, the quaternionic unitary Lie algebra of real rank p; or $\mathfrak{sp}(\infty, \infty) = \varinjlim \mathfrak{sp}(p, q)$, quaternionic unitary Lie algebra of infinite real rank.

If $\mathfrak{g}_{\mathbb{C}} = \mathfrak{gl}(\infty; \mathbb{C})$, then $\mathfrak{g}_{\mathbb{R}}$ is one $\mathfrak{gl}(\infty; \mathbb{R}) = \varinjlim \mathfrak{gl}(n; \mathbb{R})$, the real general linear Lie algebra; $\mathfrak{gl}(\infty; \mathbb{H}) = \varinjlim \mathfrak{gl}(n; \mathbb{H})$, the quaternionic general linear Lie algebra; $\mathfrak{u}(p, \infty) = \varinjlim \mathfrak{u}(p, n)$, the complex unitary Lie algebra of finite real rank p; or $\mathfrak{u}(\infty, \infty) = \varinjlim \mathfrak{u}(p, q)$, the complex unitary Lie algebra of infinite real rank.

As in (2.2), given one of these Lie algebras $\mathfrak{g}_{\mathbb{R}} = \varinjlim \mathfrak{g}_{n,\mathbb{R}}$ we have the universal enveloping algebra. We will need it for the induced representation process. As in the finite-dimensional case, we use the universal enveloping algebra of the complexification. Thus when we write $\mathcal{U}(\mathfrak{g}_{\mathbb{R}})$ it is understood that we mean $\mathcal{U}(\mathfrak{g}_{\mathbb{C}})$. The reason for this is that we will want our representations of real Lie groups to be representations on complex vector spaces.

The corresponding Lie groups are exactly what one expects. First, the complex groups, viewed either as complex groups or as real groups,

$$\begin{aligned}
&SL(\infty;\mathbb{C}) = \varinjlim SL(n;\mathbb{C}) \text{ and } GL(\infty;\mathbb{C}) = \varinjlim GL(n;\mathbb{C}),\\
&SO(\infty;\mathbb{C}) = \varinjlim SO(n;\mathbb{C}) = \varinjlim SO(2n;\mathbb{C}) = \varinjlim SO(2n+1;\mathbb{C}),\\
&Sp(\infty;\mathbb{C}) = \varinjlim Sp(n;\mathbb{C}).
\end{aligned} \tag{2.3}$$

The real forms of the complex special and general linear groups $SL(\infty;\mathbb{C})$ and $GL(\infty;\mathbb{C})$ are

$$\begin{aligned}
&SL(\infty;\mathbb{R}) \text{ and } GL(\infty;\mathbb{R}) : \text{ real special/general linear groups},\\
&SL(\infty;\mathbb{H}) : \text{ quaternionic special linear group},\\
&(S)U(p,\infty) : \text{ (special) unitary groups of real rank } p < \infty,\\
&(S)U(\infty,\infty) : \text{ (special) unitary groups of infinite real rank.}
\end{aligned} \tag{2.4}$$

The real forms of the complex orthogonal and spin groups $SO(\infty;\mathbb{C})$ and $Spin(\infty;\mathbb{C})$ are

$$\begin{aligned}
&SO(p,\infty),\ Spin(p;\infty) : \text{ real orth./spin groups of real rank } p < \infty,\\
&SO(\infty,\infty),\ Spin(\infty,\infty) : \text{ real orthog./spin groups of real rank } \infty,\\
&SO^*(2\infty) = \varinjlim SO^*(2n), \text{ which doesn't have a standard name.}
\end{aligned} \tag{2.5}$$

Here

$$SO^*(2n) = \{g \in SL(n;\mathbb{H}) \mid g \text{ preserves the form } \kappa(x,y) := \textstyle\sum x^\ell i \bar{y}^\ell = {}^t x i \bar{y}\}.$$

Alternatively, $SO^*(2n) = SO(2n;\mathbb{C}) \cap U(n,n)$ with

$$SO(2n;C) \text{ defined by } (u,v) = \textstyle\sum (u_j v_{n+jr} + v_{n+j} w_j).$$

Finally, the real forms of the complex symplectic group $Sp(\infty;\mathbb{C})$ are

$$Sp(\infty;\mathbb{R}): \text{ real symplectic group,}$$
$$Sp(p,\infty): \text{ quaternion unitary group of real rank } p < \infty, \text{ and} \qquad (2.6)$$
$$Sp(\infty,\infty): \text{ quaternion unitary group of infinite real rank.}$$

3 Complex parabolic subalgebras

In this section we recall the structure of parabolic subalgebras of $\mathfrak{gl}(\infty;\mathbb{C})$, $\mathfrak{sl}(\infty);\mathbb{C})$, $\mathfrak{so}(\infty;\mathbb{C})$ and $\mathfrak{sp}(\infty;\mathbb{C})$. We follow Dan-Cohen and Penkov ([3,4]).

We first describe $\mathfrak{g}_\mathbb{C}$ in terms of linear spaces. Let V and W be nondegenerately paired countably infinite-dimensional complex vector spaces. Then $\mathfrak{gl}(\infty,\mathbb{C}) = \mathfrak{gl}(V,W) := V \otimes W$ consists of all finite linear combinations of the rank 1 operators $v \otimes w : x \mapsto \langle w, x\rangle v$. In the usual ordered basis of $V = \mathbb{C}^\infty$, parameterized by the positive integers, and with the dual basis of $W = V^* = (\mathbb{C}^\infty)^*$, we can view $\mathfrak{gl}(\infty,\mathbb{C})$ as infinite matrices with only finitely many nonzero entries. However V has more exotic ordered bases, for example parameterized by the rational numbers, where the matrix picture is not intuitive.

The rank 1 operator $v \otimes w$ has a well-defined trace, so trace is well defined on $\mathfrak{gl}(\infty,\mathbb{C})$. Then $\mathfrak{sl}(\infty,\mathbb{C})$ is the traceless part, $\{g \in \mathfrak{gl}(\infty;\mathbb{C}) \mid \text{trace } g = 0\}$.

In the orthogonal case we can take $V = W$ using the symmetric bilinear form that defines $\mathfrak{so}(\infty;\mathbb{C})$. Then

$$\mathfrak{so}(\infty;\mathbb{C}) = \mathfrak{so}(V,V) = \Lambda\mathfrak{gl}(\infty;\mathbb{C}) \text{ where } \Lambda(v \otimes v') = v \otimes v' - v' \otimes v.$$

In other words, in an ordered orthonormal basis of $V = \mathbb{C}^\infty$ parameterized by the positive integers, $\mathfrak{so}(\infty;\mathbb{C})$ can be viewed as the infinite antisymmetric matrices with only finitely many nonzero entries.

Similarly, in the symplectic case we can take $V = W$ using the antisymmetric bilinear form that defines $\mathfrak{sp}(\infty;\mathbb{C})$, and then

$$\mathfrak{sp}(\infty;\mathbb{C}) = \mathfrak{sp}(V,V) = S\mathfrak{gl}(\infty;\mathbb{C}) \text{ where } S(v \otimes v') = v \otimes v' + v' \otimes v.$$

In an appropriate ordered basis of $V = \mathbb{C}^\infty$ parameterized by the positive integers, $\mathfrak{sp}(\infty;\mathbb{C})$ can be viewed as the infinite symmetric matrices with only finitely many nonzero entries.

In the finite-dimensional setting, Borel subalgebra means a maximal solvable subalgebra, and parabolic subalgebra means one that contains a Borel. It is the same here except that one must use *locally solvable* to avoid the prospect of an infinite derived series.

Definition 3.1. A *Borel subalgebra* of $\mathfrak{g}_\mathbb{C}$ is a maximal locally solvable subalgebra. A *parabolic subalgebra* of $\mathfrak{g}_\mathbb{C}$ is a subalgebra that contains a Borel subalgebra. ◇

In the finite-dimensional setting a parabolic subalgebra is the stabilizer of an appropriate nested sequence of subspaces (possibly with an orientation condition in the orthogonal group case). In the infinite-dimensional setting here, one must be very careful as to which nested sequences of subspaces are appropriate. If F is a subspace of V, then $F^{\perp}$ denotes its annihilator in W. Similarly if $'F$ is a subspace of W, then $'F^{\perp}$ denotes its annihilator in V. We say that F (resp. $'F$) is *closed* if $F = F^{\perp\perp}$ (resp. $'F = {}'F^{\perp\perp}$). This is the closure relation in the Mackey topology [13], i.e., the weak topology for the functionals on V from W and on W from V.

In order to avoid repeating the following definitions later on, we make them in somewhat greater generality than we need just now.

Definition 3.2. Let V and W be countable dimensional vector spaces over a real division ring $\mathbb{D} = \mathbb{R}, \mathbb{C}$ or $\mathbb{H}$, with a nondegenerate bilinear pairing $\langle \cdot, \cdot \rangle : V \times W \to \mathbb{D}$. A *chain* or $\mathbb{D}$*-chain* in V (resp. W) is a set of $\mathbb{D}$-subspaces totally ordered by inclusion. An *generalized* $\mathbb{D}$*-flag* in V (resp. W) is a $\mathbb{D}$-chain such that each subspace has an immediate predecessor or an immediate successor in the inclusion ordering, and every nonzero vector of V (or W) is caught between an immediate predecessor successor (IPS) pair. An generalized $\mathbb{D}$-flag $\mathcal{F}$ in V (resp. $'\mathcal{F}$ in W) is *semiclosed* if $F \in \mathcal{F}$ with $F \neq F^{\perp\perp}$ implies $\{F, F^{\perp\perp}\}$ is an IPS pair (resp. $'F \in {}'\mathcal{F}$ with $'F \neq' F^{\perp\perp}$ implies $\{'F,' F^{\perp\perp}\}$ is an IPS pair). ◇

Definition 3.3. Let $\mathbb{D}$, V and W be as above. Generalized $\mathbb{D}$-flags $\mathcal{F}$ in V and $'\mathcal{F}$ in W form a *taut couple* when (i) if $F \in \mathcal{F}$ then $F^{\perp}$ is invariant by the $\mathfrak{gl}$-stabilizer of $'\mathcal{F}$ and (ii) if $'F \in {}'\mathcal{F}$, then its annihilator $'F^{\perp}$ is invariant by the $\mathfrak{gl}$-stabilizer of $\mathcal{F}$. ◇

In the $\mathfrak{so}$ and $\mathfrak{sp}$ cases one can use the associated bilinear form to identify V with W and $\mathcal{F}$ with $'\mathcal{F}$. Then we speak of a generalized flag $\mathcal{F}$ in V as *self-taut*. If $\mathcal{F}$ is a self-taut generalized flag in V, then [6] every $F \in \mathcal{F}$ is either isotropic or coisotropic.

Example 3.4. Here is a quick peek at an obvious phenomenon introduced by infinite dimensionality. Enumerate bases of $V = \mathbb{C}^{\infty}$ and $W = \mathbb{C}^{\infty}$ by $(\mathbb{Z}^{+})^{2}$, say $\{v_i = v_{i_1,i_2}\}$ and $\{w_j = w_{j_1,j_2}\}$, with $\langle v_i, w_j \rangle = 1$ if both $i_1 = j_1$ and $i_2 = j_2$ and $\langle v_i, w_j \rangle = 0$ otherwise. Define $\mathcal{F} = \{F_i\}$ ordered by inclusion where one builds up bases of the F_i first with the $v_{i_1,1}$, $i_1 \geqq 1$ and then the $v_{i_1,2}$, $i_1 \geqq 1$ and then the $v_{i_1,3}$, $i_1 \geqq 1$, and so on. One does the same for $'\mathcal{F}$ using the $\{w_j\}$. Now these form a taut couple of semiclosed generalized flags whose ordering involves an infinite number of limit ordinals. That makes it hard to use matrix methods. ◇

Theorem 3.5. *The self-normalizing parabolic subalgebras of the Lie algebras* $\mathfrak{sl}(V, W)$ *and* $\mathfrak{gl}(V, W)$ *are the normalizers of taut couples of semiclosed generalized flags in* V *and* W*, and this is a one-to-one correspondence. The self-normalizing parabolic subalgebras of* $\mathfrak{sp}(V)$ *are the normalizers of self-taut semiclosed generalized flags in* V*, and this too is a one-to-one correspondence.*

Theorem 3.6. *The self-normalizing parabolic subalgebras of* $\mathfrak{so}(V)$ *are the normalizers of self-taut semiclosed generalized flags* $\mathcal{F}$ *in* V*, and there are two possibilities:*

(1) *the flag* $\mathcal{F}$ *is uniquely determined by the parabolic, or*
(2) *there are exactly three self-taut generalized flags with the same stabilizer as* $\mathcal{F}$*.*

The latter case occurs precisely when there exists an isotropic subspace $L \in \mathcal{F}$ *with* $\dim_{\mathbb{C}} L^{\perp}/L = 2$*. The three flags with the same stabilizer are then*

$$\begin{aligned}&\{F \in \mathcal{F} \mid F \subset L \text{ or } L^{\perp} \subset F\},\\&\{F \in \mathcal{F} \mid F \subset L \text{ or } L^{\perp} \subset F\} \cup M_1,\\&\{F \in \mathcal{F} \mid F \subset L \text{ or } L^{\perp} \subset F\} \cup M_2,\end{aligned}$$

where M_1 *and* M_2 *are the two maximal isotropic subspaces containing* L*.*

Example 3.7. Before proceeding we indicate an example which shows that not all parabolics are equal to their normalizers. Enumerate bases of $V = \mathbb{C}^{\infty}$ and $W = \mathbb{C}^{\infty}$ by rational numbers with pairing

$$\langle v_q, w_r\rangle = 1 \text{ if } q > r, \quad = 0 \text{ if } q \leqq r.$$

Then $\mathrm{Span}\{v_q \otimes w_r \mid q \leqq r\}$ is a Borel subalgebra of $\mathfrak{gl}(\infty)$ contained in $\mathfrak{sl}(\infty)$. This shows that $\mathfrak{sl}(\infty)$ is parabolic in $\mathfrak{gl}(\infty)$. ◇

One pinpoints this situation as follows. If $\mathfrak{p}$ is a (real or complex) subalgebra of $\mathfrak{g}_{\mathbb{C}}$ and $\mathfrak{q}$ is a quotient algebra isomorphic to $\mathfrak{gl}(\infty;\mathbb{C})$, say with quotient map $f : \mathfrak{p} \to \mathfrak{q}$, then we refer to the composition $trace \circ f : \mathfrak{p} \to \mathbb{C}$ as an *infinite trace* on $\mathfrak{g}_{\mathbb{C}}$. If $\{f_i\}$ is a finite set of infinite traces on $\mathfrak{g}_{\mathbb{C}}$ and $\{c_i\}$ are complex numbers, then we refer to the condition $\sum c_i f_i = 0$ as an *infinite trace condition* on $\mathfrak{p}$.

These quotients can exist. In Example 3.4 we can take V_a to be the span of the $v_{i_1,a}$ and W_a the span of the dual $w_{i_1,a}$ for $a = 1, 2, \ldots$ and then the normalizer of the taut couple $(\mathcal{F}, {}'\mathcal{F})$ has infinitely many quotients $\mathfrak{gl}(V_a, W_a)$.

Theorem 3.8. *The parabolic subalgebras* $\mathfrak{p}$ *in* $\mathfrak{g}_{\mathbb{C}}$ *are the algebras obtained from self-normalizing parabolics* $\widetilde{\mathfrak{p}}$ *by imposing infinite trace conditions.*

As a general principle one tries to be explicit by constructing representations that are as close to irreducible as feasible. For this reason we will construct principal series representations by inducing from parabolic subgroups that are minimal among the self-normalizing parabolic subgroups. Still, one should be aware of the phenomenon of Example 3.7 and Theorem 3.8.

4 Real parabolic subalgebras and subgroups

In this section we discuss the structure of parabolic subalgebras of real forms of the classical $\mathfrak{sl}(\infty,\mathbb{C})$, $\mathfrak{so}(\infty,\mathbb{C})$, $\mathfrak{sp}(\infty,\mathbb{C})$ and $\mathfrak{gl}(\infty,\mathbb{C})$. In this section $\mathfrak{g}_{\mathbb{C}}$ will always be one of them and $G_{\mathbb{C}}$ will be the corresponding connected complex Lie group.

Also, $\mathfrak{g}_{\mathbb{R}}$ will be a real form of $\mathfrak{g}_{\mathbb{C}}$, and $G_{\mathbb{R}}$ will be the corresponding connected real subgroup of $G_{\mathbb{C}}$.

Definition 4.1. Let $\mathfrak{g}_{\mathbb{R}}$ be a real form of $\mathfrak{g}_{\mathbb{C}}$. Then a subalgebra $\mathfrak{p}_{\mathbb{R}} \subset \mathfrak{g}_{\mathbb{R}}$ is a *parabolic subalgebra* if its complexification $\mathfrak{p}_{\mathbb{C}}$ is a parabolic subalgebra of $\mathfrak{g}_{\mathbb{C}}$. ◇

When $\mathfrak{g}_{\mathbb{R}}$ has two inequivalent defining representations, in other words when

$$\mathfrak{g}_{\mathbb{R}} = \mathfrak{sl}(\infty;\mathbb{R}),\ \mathfrak{gl}(\infty;\mathbb{R}),\ \mathfrak{su}(*,\infty),\ \mathfrak{u}(*,\infty),\ \text{ or }\ \mathfrak{sl}(\infty;\mathbb{H}),$$

we denote them by $V_{\mathbb{R}}$ and $W_{\mathbb{R}}$, and when $\mathfrak{g}_{\mathbb{R}}$ has only one defining representation, in other words when

$$\mathfrak{g}_{\mathbb{R}} = \mathfrak{so}(*,\infty),\ \mathfrak{sp}(*,\infty),\ \mathfrak{sp}(\infty;\mathbb{R}),\ \text{ or }\ \mathfrak{so}^*(2\infty) \text{ as quaternion matrices},$$

we denote it by $V_{\mathbb{R}}$. The commuting algebra of $\mathfrak{g}_{\mathbb{R}}$ on $V_{\mathbb{R}}$ is a real division algebra $\mathbb{D}$. The main result of [6] is

Theorem 4.2. *Suppose that $\mathfrak{g}_{\mathbb{R}}$ has two inequivalent defining representations. Then a subalgebra of $\mathfrak{g}_{\mathbb{R}}$ (resp. subgroup of $G_{\mathbb{R}}$) is parabolic if and only if it is defined by infinite trace conditions (resp. infinite determinant conditions) on the $\mathfrak{g}_{\mathbb{R}}$-stabilizer (resp. $G_{\mathbb{R}}$-stabilizer) of a taut couple of generalized $\mathbb{D}$-flags $\mathcal{F}$ in $V_{\mathbb{R}}$ and $'\mathcal{F}$ in $W_{\mathbb{R}}$.*

Suppose that $\mathfrak{g}_{\mathbb{R}}$ has only one defining representation. A subalgebra of $\mathfrak{g}_{\mathbb{R}}$ (resp. subgroup) of $G_{\mathbb{R}}$ is parabolic if and only if it is defined by infinite trace conditions (resp. infinite determinant conditions) on the $\mathfrak{g}_{\mathbb{R}}$-stabilizer (resp. $G_{\mathbb{R}}$-stabilizer) of a self-taut generalized $\mathbb{D}$-flag $\mathcal{F}$ in $V_{\mathbb{R}}$.

5 Levi components of complex parabolics

In this section we discuss Levi components of complex parabolic subalgebras, recalling results from [4, 5, 8–10] and [25]. We start with the definition.

Definition 5.1. Let $\mathfrak{p}$ be a locally finite Lie algebra and $\mathfrak{r}$ its locally solvable radical. A subalgebra $\mathfrak{l} \subset \mathfrak{p}$ is a *Levi component* if $[\mathfrak{p},\mathfrak{p}]$ is the semidirect sum $(\mathfrak{r} \cap [\mathfrak{p},\mathfrak{p}]) \subsetplus \mathfrak{l}$. ◇

Every finitary Lie algebra has a Levi component. Evidently, Levi components are maximal semisimple subalgebras, but the converse fails for finitary Lie algebras. In any case, parabolic subalgebras of our classical Lie algebras $\mathfrak{g}_{\mathbb{C}}$ have maximal semisimple subalgebras, and those are their Levi components.

Definition 5.2. Let $X \subset V$ and $Y \subset W$ be paired subspaces, isotropic in the orthogonal and symplectic cases. The subalgebras

$$\mathfrak{gl}(X,Y) \subset \mathfrak{gl}(V,W) \quad \text{and } \mathfrak{sl}(X,Y) \subset \mathfrak{sl}(V,W),$$
$$\Lambda\mathfrak{gl}(X,Y) \subset \Lambda\mathfrak{gl}(V,V) \text{ and } S\mathfrak{gl}(X,Y) \subset S\mathfrak{gl}(V,V)$$

are called *standard*. ◇

Proposition 5.3. *A subalgebra $\mathfrak{l}_{\mathbb{C}} \subset \mathfrak{g}_{\mathbb{C}}$ is the Levi component of a parabolic subalgebra of $\mathfrak{g}_{\mathbb{C}}$ if and only if it is the direct sum of standard special linear subalgebras and at most one subalgebra $\Lambda\mathfrak{gl}(X,Y)$ in the orthogonal case, and at most one subalgebra $S\mathfrak{gl}(X,Y)$ in the symplectic case.*

The occurrence of "at most one subalgebra" in Proposition 5.3 is analogous to the finite-dimensional case, where it is seen by deleting some simple root nodes from a Dynkin diagram.

Let $\mathfrak{p}$ be the parabolic subalgebra of $\mathfrak{sl}(V,W)$ or $\mathfrak{gl}(V,W)$ defined by the taut couple $(\mathcal{F},{}'\mathcal{F})$ of semiclosed generalized flags.

Definition 5.4. Define two sets J and ${}'J$ by

$$J = \{(F',F'') \text{ IPS pair in } \mathcal{F} \mid F' = (F')^{\perp\perp} \text{ and } \dim F''/F' > 1\},$$
$${}'J = \{({}'F',{}'F'') \text{ IPS pair in } {}'\mathcal{F} \mid {}'F' = ({}'F')^{\perp\perp}, \dim {}'F''/{}'F' > 1\}.$$

Since $V \times W \to \mathbb{C}$ is nondegenerate, the sets J and ${}'J$ are in one-to-one correspondence by $(F''/F') \times ({}'F''/{}'F') \to \mathbb{C}$ is nondegenerate. We use this to identify J with J', and we write (F'_j, F''_j) and $({}'F'_j, {}'F''_j)$ treating J as an index set.

Theorem 5.5. *Let $\mathfrak{p}$ be the parabolic subalgebra of $\mathfrak{sl}(V,W)$ or $\mathfrak{gl}(V,W)$ defined by the taut couple $\mathcal{F}$ and ${}'\mathcal{F}$ of semiclosed generalized flags. For each $j \in J$ choose a subspace $X_j \subset V$ and a subspace $Y_j \subset W$ such that $F''_j = X_j + F'_j$ and ${}'F''_j = Y_j + {}'F_j{}'$ Then $\bigoplus_{j\in J} \mathfrak{sl}(X_j,Y_j)$ is a Levi component of $\mathfrak{p}$. The inclusion relations of $\mathcal{F}$ and ${}'\mathcal{F}$ induce a total order on J.*

Conversely, if $\mathfrak{l}$ is a Levi component of $\mathfrak{p}$ then there exist subspaces $X_j \subset V$ and $Y_j \subset W$ such that $\mathfrak{l} = \bigoplus_{j\in J} \mathfrak{sl}(X_j,Y_j)$.

Now the idea of finite matrices with blocks down the diagonal suggests the construction of $\mathfrak{p}$ from the totally ordered set J and the direct sum $\mathfrak{l} = \bigoplus_{j\in J} \mathfrak{sl}(X_j,Y_j)$ of standard special linear algebras. We outline the idea of the construction; see [5]. First, $\langle X_j, Y_{j'}\rangle = 0$ for $j \neq j'$ because the $\mathfrak{s}_j = \mathfrak{sl}(X_j,Y_j)$ commute with each other. Define $U_j := ((\bigoplus_{k\leqq j} X_k)^\perp \oplus Y_j)^\perp$. Then one proves $U_j = ((U_j \oplus X_j)^\perp \oplus Y_j)^\perp$. From that, one shows that there is a unique semiclosed generalized flag $\mathcal{F}_{\min}$ in V with the same stabilizer as the set $\{U_j, U_j \oplus X_j \mid j \in J\}$. One constructs similar subspaces ${}'U_j \subset W$ and shows that there is a unique semiclosed generalized flag ${}'\mathcal{F}_{\min}$ in W with the same stabilizer as the set $\{{}'U_j, {}'U_j \oplus Y_j \mid j \in J\}$. In fact $(\mathcal{F}_{\min}, {}'\mathcal{F}_{\min})$ is the minimal taut couple with IPS pairs $U_j \subset (U_j \oplus X_j)$ in $\mathcal{F}_0$ and $(U_j \oplus X_j)^\perp \subset ((U_j \oplus X_j)^\perp \oplus Y_j)$ in $'\mathcal{F}_0$ for $j \in J$. If $(\mathcal{F}_{\max}, '\mathcal{F}_{\max})$ is maximal among the taut couples of semiclosed generalized flags with IPS pairs $U_j \subset (U_j \oplus X_j)$ in $\mathcal{F}_{\max}$ and $(U_j \oplus X_j)^\perp \subset ((U_j \oplus X_j)^\perp \oplus Y_j)$ in $'\mathcal{F}_{\max}$, then the corresponding parabolic $\mathfrak{p}$ has Levi component $\mathfrak{l}$.

The situation is essentially the same for Levi components of parabolic subalgebras of $\mathfrak{g}_\mathbb{C} = \mathfrak{so}(\infty;\mathbb{C})$ or $\mathfrak{sp}(\infty;\mathbb{C})$, except that we modify Definition 5.4 of J to add the condition that F'' be isotropic, and we add the orientation aspect of the $\mathfrak{so}$ case.

Theorem 5.6. *Let $\mathfrak{p}$ be the parabolic subalgebra of $\mathfrak{g}_\mathbb{C} = \mathfrak{so}(V)$ or $\mathfrak{sp}(V)$, defined by the self-taut semiclosed generalized flag $\mathcal{F}$. Let $\widetilde{F}$ be the union of all subspaces F'' in IPS pairs (F', F'') of $\mathcal{F}$ for which F'' is isotropic. Let $\widetilde{'F}$ be the intersection of all subspaces F' in IPS pairs for which F' is closed ($F' = (F')^{\perp\perp}$) and coisotropic. Then $\mathfrak{l}$ is a Levi component of $\mathfrak{p}$ if and only if there are isotropic subspaces X_j, Y_j in V such that*

$$F_j'' = F_j' + X_j \text{ and } 'F_j'' = 'F_j + Y_j \text{ for every } j \in J$$

and a subspace Z in V such that $\widetilde{F} = Z + \widetilde{'F}$, where $Z = 0$ in case $\mathfrak{g}_\mathbb{C} = \mathfrak{so}(V)$ and $\dim \widetilde{F}/\widetilde{'F} \leqq 2$, such that

$$\mathfrak{l} = \mathfrak{sp}(Z) \oplus \bigoplus\nolimits_{j \in J} \mathfrak{sl}(X_j, Y_j) \text{ if } \mathfrak{g}_\mathbb{C} = \mathfrak{sp}(V),$$

$$\mathfrak{l} = \mathfrak{so}(Z) \oplus \bigoplus\nolimits_{j \in J} \mathfrak{sl}(X_j, Y_j) \text{ if } \mathfrak{g}_\mathbb{C} = \mathfrak{so}(V).$$

Further, the inclusion relations of $\mathcal{F}$ induce a total order on J which leads to a construction of $\mathfrak{p}$ from $\mathfrak{l}$.

6 Chevalley decomposition

In this section we apply the extension [4] to our parabolic subalgebras, of the Chevalley decomposition for a (finite-dimensional) algebraic Lie algebra.

Let $\mathfrak{p}$ be a locally finite linear Lie algebra, in our case a subalgebra of $\mathfrak{gl}(\infty)$. Every element $\xi \in \mathfrak{p}$ has a Jordan canonical form, yielding a decomposition $\xi = \xi_{ss} + \xi_{nil}$ into semisimple and nilpotent parts. The algebra $\mathfrak{p}$ is *splittable* if it contains the semisimple and the nilpotent parts of each of its elements. Note that ξ_{ss} and ξ_{nil} are polynomials in ξ; this follows from the finite-dimensional fact. In particular, if X is any ξ-invariant subspace of V, then it is invariant under both ξ_{ss} and ξ_{nil}.

Conversely, parabolic subalgebras (and many others) of our classical Lie algebras $\mathfrak{g}$ are splittable.

The *linear nilradical* of a subalgebra $\mathfrak{p} \subset \mathfrak{g}$ is the set $\mathfrak{p}_{\mathrm{nil}}$ of all nilpotent elements of the locally solvable radical $\mathfrak{r}$ of $\mathfrak{p}$. It is a locally nilpotent ideal in $\mathfrak{p}$ and satisfies $\mathfrak{p}_{\mathrm{nil}} \cap [\mathfrak{p}, \mathfrak{p}] = \mathfrak{r} \cap [\mathfrak{p}, \mathfrak{p}]$.

If $\mathfrak{p}$ is splittable, then it has a well-defined maximal locally reductive subalgebra $\mathfrak{p}_{\mathrm{red}}$. This means that $\mathfrak{p}_{\mathrm{red}}$ is an increasing union of finite-dimensional reductive Lie algebras, each reductive in the next. In particular $\mathfrak{p}_{\mathrm{red}}$ maps isomorphically under the projection $\mathfrak{p} \to \mathfrak{p}/\mathfrak{p}_{\mathrm{nil}}$. That gives a semidirect sum decomposition $\mathfrak{p} = \mathfrak{p}_{\mathrm{nil}} \oplus\!\!\!\!\supset \mathfrak{p}_{\mathrm{red}}$ analogous to the Chevalley decomposition mentioned above. Also, here,

$$\mathfrak{p}_{\mathrm{red}} = \mathfrak{l} \oplus\!\!\!\!\supset \mathfrak{t} \quad \text{and} \quad [\mathfrak{p}_{\mathrm{red}}, \mathfrak{p}_{\mathrm{red}}] = \mathfrak{l}$$

where $\mathfrak{t}$ is a toral subalgebra and $\mathfrak{l}$ is the Levi component of $\mathfrak{p}$. A glance at $\mathfrak{u}(\infty)$ or $\mathfrak{gl}(\infty; \mathbb{C})$ shows that the semidirect sum decomposition of $\mathfrak{p}_{\mathrm{red}}$ need not be direct.

7 Levi and Chevalley components of real parabolics

Now we adapt the material of Sections 5 and 6 to study Levi and Chevalley components of real parabolic subalgebras in the real classical Lie algebras.

Let $\mathfrak{g}_{\mathbb{R}}$ be a real form of a classical locally finite complex simple Lie algebra $\mathfrak{g}_{\mathbb{C}}$. Consider a real parabolic subalgebra $\mathfrak{p}_{\mathbb{R}}$. It has form $\mathfrak{p}_{\mathbb{R}} = \mathfrak{p}_{\mathbb{C}} \cap \mathfrak{g}_{\mathbb{R}}$ where its complexification $\mathfrak{p}_{\mathbb{C}}$ is parabolic in $\mathfrak{g}_{\mathbb{C}}$. Let τ denote complex conjugation of $\mathfrak{g}_{\mathbb{C}}$ over $\mathfrak{g}_{\mathbb{R}}$. Then the locally solvable radical $\mathfrak{r}_{\mathbb{C}}$ of $\mathfrak{p}_{\mathbb{C}}$ is τ-stable because $\mathfrak{r}_{\mathbb{C}} + \tau\mathfrak{r}_{\mathbb{C}}$ is a locally solvable ideal, so the locally solvable radical $\mathfrak{r}_{\mathbb{R}}$ of $\mathfrak{p}_{\mathbb{R}}$ is a real form of $\mathfrak{r}_{\mathbb{C}}$.

Let $\mathfrak{l}_{\mathbb{R}}$ be a maximal semisimple subalgebra of $\mathfrak{p}_{\mathbb{R}}$. Its complexification $\mathfrak{l}_{\mathbb{C}}$ is a maximal semisimple subalgebra, hence a Levi component, of $\mathfrak{p}_{\mathbb{C}}$. Thus $[\mathfrak{p}_{\mathbb{C}}, \mathfrak{p}_{\mathbb{C}}]$ is the semidirect sum $(\mathfrak{r}_{\mathbb{C}} \cap [\mathfrak{p}_{\mathbb{C}}, \mathfrak{p}_{\mathbb{C}}]) \oplus\!\!\!\!\supset \mathfrak{l}_{\mathbb{C}}$. The elements of this formula all are τ-stable, so we have proved

Lemma 7.1. *The Levi components of $\mathfrak{p}_{\mathbb{R}}$ are real forms of the Levi components of $\mathfrak{p}_{\mathbb{C}}$.*

If $\mathfrak{g}_{\mathbb{C}}$ is $\mathfrak{sl}(V, W)$ or $\mathfrak{gl}(V, W)$ as in Theorem 5.5, then $\mathfrak{l}_{\mathbb{C}} = \bigoplus_{j \in J} \mathfrak{sl}(X_j, Y_j)$ as indicated there. Initially the possibilities for the action of τ are

- τ preserves $\mathfrak{sl}(X_j, Y_j)$ with fixed point set $\mathfrak{sl}(X_{j,\mathbb{R}}, Y_{j,\mathbb{R}}) \cong \mathfrak{sl}(*; \mathbb{R})$,
- τ preserves $\mathfrak{sl}(X_j, Y_j)$ with fixed point set $\mathfrak{sl}(X_{j,\mathbb{H}}, Y_{j,\mathbb{H}}) \cong \mathfrak{sl}(*; \mathbb{H})$,
- τ preserves $\mathfrak{sl}(X_j, Y_j)$ with f.p. set $\mathfrak{su}(X'_j, X''_j) \cong \mathfrak{su}(*, *)$, $X_j = X'_j + X''_j$, and
- τ interchanges two summands $\mathfrak{sl}(X_j, Y_j)$ and $\mathfrak{sl}(X_{j'}, Y_{j'})$ of $\mathfrak{l}_{\mathbb{C}}$, with fixed point set the diagonal ($\cong \mathfrak{sl}(X_j, Y_j)$) of their direct sum.

If $\mathfrak{g}_\mathbb{C} = \mathfrak{so}(V)$ as in Theorem 5.6, $\mathfrak{l}_\mathbb{C}$ can also have a summand $\mathfrak{so}(Z)$, or if $\mathfrak{g}_\mathbb{C} = \mathfrak{sp}(V)$ it can also have a summand $\mathfrak{sp}(Z)$. Except when $A_4 = D_3$ occurs, these additional summands must be τ-stable, resulting in fixed point sets

- when $\mathfrak{g}_\mathbb{C} = \mathfrak{so}(V)$: $\mathfrak{so}(Z)^\tau$ is $\mathfrak{so}(*,*)$ or $\mathfrak{so}^*(2\infty)$,
- when $\mathfrak{g}_\mathbb{C} = \mathfrak{sp}(V)$: $\mathfrak{sp}(Z)^\tau$ is $\mathfrak{sp}(*,*)$ or $\mathfrak{sp}(*;\mathbb{R})$.

8 Minimal parabolic subgroups

We describe the structure of minimal parabolic subgroups of the classical real simple Lie groups $G_\mathbb{R}$.

Proposition 8.1. *Let $\mathfrak{p}_\mathbb{R}$ be a parabolic subalgebra of $\mathfrak{g}_\mathbb{R}$ and $\mathfrak{l}_\mathbb{R}$ a Levi component of $\mathfrak{p}_\mathbb{R}$. If $\mathfrak{p}_\mathbb{R}$ is a minimal parabolic subalgebra, then $\mathfrak{l}_\mathbb{R}$ is a direct sum of finite-dimensional compact algebras $\mathfrak{su}(p)$, $\mathfrak{so}(p)$ and $\mathfrak{sp}(p)$, and their infinite-dimensional limits $\mathfrak{su}(\infty)$, $\mathfrak{so}(\infty)$ and $\mathfrak{sp}(\infty)$. If $\mathfrak{l}_\mathbb{R}$ is a direct sum of finite-dimensional compact algebras $\mathfrak{su}(p)$, $\mathfrak{so}(p)$ and $\mathfrak{sp}(p)$ and their limits $\mathfrak{su}(\infty)$, $\mathfrak{so}(\infty)$ and $\mathfrak{sp}(\infty)$, then $\mathfrak{p}_\mathbb{R}$ contains a minimal parabolic subalgebra of $\mathfrak{g}_\mathbb{R}$ with the same Levi component $\mathfrak{l}_\mathbb{R}$.*

Proof. Suppose that $\mathfrak{p}_\mathbb{R}$ is a minimal parabolic subalgebra of $\mathfrak{g}_\mathbb{R}$. If a direct summand $\mathfrak{l}'_\mathbb{R}$ of $\mathfrak{l}_\mathbb{R}$ has a proper parabolic subalgebra $\mathfrak{q}_\mathbb{R}$, we replace $\mathfrak{l}'_\mathbb{R}$ by $\mathfrak{q}_\mathbb{R}$ in $\mathfrak{l}_\mathbb{R}$ and $\mathfrak{p}_\mathbb{R}$. In other words we refine the flag(s) that define $\mathfrak{p}_\mathbb{R}$. The refined flag defines a parabolic $\mathfrak{q}_\mathbb{R} \subsetneqq \mathfrak{p}_\mathbb{R}$. This contradicts minimality. Thus no summand of $\mathfrak{l}_\mathbb{R}$ has a proper parabolic subalgebra. Theorems 5.5 and 5.6 show that $\mathfrak{su}(p)$, $\mathfrak{so}(p)$ and $\mathfrak{sp}(p)$, and their limits $\mathfrak{su}(\infty)$, $\mathfrak{so}(\infty)$ and $\mathfrak{sp}(\infty)$, are the only possibilities for the simple summands of $\mathfrak{l}_\mathbb{R}$.

Conversely suppose that the summands of $\mathfrak{l}_\mathbb{R}$ are $\mathfrak{su}(p)$, $\mathfrak{so}(p)$ and $\mathfrak{sp}(p)$ or their limits $\mathfrak{su}(\infty)$, $\mathfrak{so}(\infty)$ and $\mathfrak{sp}(\infty)$. Let $(\mathcal{F}, '\mathcal{F})$ or $\mathcal{F}$ be the flag(s) that define $\mathfrak{p}_\mathbb{R}$. In the discussion between Theorems 5.5 and 5.6 we described a minimal taut couple $(\mathcal{F}_{\min}, '\mathcal{F}_{\min})$ and a maximal taut couple $(\mathcal{F}_{\max}, '\mathcal{F}_{\max})$ (in the $\mathfrak{sl}$ and $\mathfrak{gl}$ cases) of semiclosed generalized flags which define parabolics that have the same Levi component $\mathfrak{l}_\mathbb{C}$ as $\mathfrak{p}_\mathbb{C}$. By construction $(\mathcal{F}, '\mathcal{F})$ refines $(\mathcal{F}_{\min}, '\mathcal{F}_{\min})$ and $(\mathcal{F}_{\max}, '\mathcal{F}_{\max})$ refines $(\mathcal{F}, '\mathcal{F})$. As $(\mathcal{F}_{\min}, '\mathcal{F}_{\min})$ is uniquely defined from $(\mathcal{F}, '\mathcal{F})$ it is τ-stable. Now the maximal τ-stable taut couple $(\mathcal{F}^*_{\max}, '\mathcal{F}^*_{\max})$ of semiclosed generalized flags defines a τ-stable parabolic $\mathfrak{q}_\mathbb{C}$ with the same Levi component $\mathfrak{l}_\mathbb{C}$ as $\mathfrak{p}_\mathbb{C}$, and $\mathfrak{q}_\mathbb{R} := \mathfrak{q}_\mathbb{C} \cap \mathfrak{g}_\mathbb{R}$ is a minimal parabolic subalgebra of $\mathfrak{g}_\mathbb{R}$ with Levi component $\mathfrak{l}_\mathbb{R}$.

The argument is the same when $\mathfrak{g}_\mathbb{C}$ is $\mathfrak{so}$ or $\mathfrak{sp}$. □

Proposition 8.1 says that the Levi components of the minimal parabolics are the compact real forms, in the sense of [21], of the complex $\mathfrak{sl}$, $\mathfrak{so}$ and $\mathfrak{sp}$. We extend this notion.

The group $G_\mathbb{R}$ has the natural *Cartan involution* θ such that $d\theta((\mathfrak{p}_\mathbb{R})_{\text{red}}) = (\mathfrak{p}_\mathbb{R})_{\text{red}}$, defined as follows. Every element of $\mathfrak{l}_\mathbb{R}$ is elliptic, and $(\mathfrak{p}_\mathbb{R})_{\text{red}} = \mathfrak{l}_\mathbb{R} \oplus \mathfrak{t}_\mathbb{R}$

where $\mathfrak{t}_{\mathbb{R}}$ is toral, so every element of $(\mathfrak{p}_{\mathbb{R}})_{\text{red}}$ is semisimple. (This is where we use minimality of the parabolic $\mathfrak{p}_{\mathbb{R}}$.) Thus $(\mathfrak{p}_{\mathbb{R}})_{\text{red}} \cap \mathfrak{g}_{n,\mathbb{R}}$ is reductive in $\mathfrak{g}_{m,\mathbb{R}}$ for every $m \geqq n$. Consequently we have Cartan involutions θ_n of the groups $G_{n,\mathbb{R}}$ such that $\theta_{n+1}|_{G_{n,\mathbb{R}}} = \theta_n$ and $d\theta_n((\mathfrak{p}_{\mathbb{R}})_{\text{red}} \cap \mathfrak{g}_{n,\mathbb{R}}) = (\mathfrak{p}_{\mathbb{R}})_{\text{red}} \cap \mathfrak{g}_{n,\mathbb{R}}$. Now $\theta = \varinjlim \theta_n$ (in other words $\theta|_{G_{n,\mathbb{R}}} = \theta_n$) is the desired Cartan involution of $\mathfrak{g}_{\mathbb{R}}$. Note that $\mathfrak{l}_{\mathbb{R}}$ is contained in the fixed point set of $d\theta$.

The Lie algebra $\mathfrak{g}_{\mathbb{R}} = \mathfrak{k}_{\mathbb{R}} + \mathfrak{s}_{\mathbb{R}}$ where $\mathfrak{k}_{\mathbb{R}}$ is the $(+1)$-eigenspace of $d\theta$ and $\mathfrak{s}_{\mathbb{R}}$ is the (-1)-eigenspace. The fixed point set $K_{\mathbb{R}} = G_{\mathbb{R}}^{\theta}$ is the direct limit of the maximal compact subgroups $K_{n,\mathbb{R}} = G_{n,\mathbb{R}}^{\theta_n}$. We will refer to $K_{\mathbb{R}}$ as a *maximal lim-compact subgroup* of $G_{\mathbb{R}}$ and to $\mathfrak{k}_{\mathbb{R}}$ as a maximal *lim-compact subalgebra* of $\mathfrak{g}_{\mathbb{R}}$. By construction $\mathfrak{l}_{\mathbb{R}} \subset \mathfrak{k}_{\mathbb{R}}$, as in the case of finite-dimensional minimal parabolics. Also as in the finite-dimensional case (and using the same proof), $[\mathfrak{k}_{\mathbb{R}}, \mathfrak{k}_{\mathbb{R}}] \subset \mathfrak{k}_{\mathbb{R}}$, $[\mathfrak{k}_{\mathbb{R}}, \mathfrak{s}_{\mathbb{R}}] \subset \mathfrak{s}_{\mathbb{R}}$ and $[\mathfrak{s}_{\mathbb{R}}, \mathfrak{s}_{\mathbb{R}}] \subset \mathfrak{k}_{\mathbb{R}}$.

Lemma 8.1. *Decompose* $(\mathfrak{p}_{\mathbb{R}})_{\text{red}} = \mathfrak{m}_{\mathbb{R}} + \mathfrak{a}_{\mathbb{R}}$ *where* $\mathfrak{m}_{\mathbb{R}} = (\mathfrak{p}_{\mathbb{R}})_{\text{red}} \cap \mathfrak{k}_{\mathbb{R}}$ *and* $\mathfrak{a}_{\mathbb{R}} = (\mathfrak{p}_{\mathbb{R}})_{\text{red}} \cap \mathfrak{s}_{\mathbb{R}}$. *Then* $\mathfrak{m}_{\mathbb{R}}$ *and* $\mathfrak{a}_{\mathbb{R}}$ *are ideals in* $(\mathfrak{p}_{\mathbb{R}})_{\text{red}}$ *with* $\mathfrak{a}_{\mathbb{R}}$ *commutative. In particular* $(\mathfrak{p}_{\mathbb{R}})_{\text{red}} = \mathfrak{m}_{\mathbb{R}} \oplus \mathfrak{a}_{\mathbb{R}}$, *direct sum of ideals.*

Proof. Since $\mathfrak{l}_{\mathbb{R}} = [(\mathfrak{p}_{\mathbb{R}})_{\text{red}}, (\mathfrak{p}_{\mathbb{R}})_{\text{red}}]$ we compute $[\mathfrak{m}_{\mathbb{R}}, \mathfrak{a}_{\mathbb{R}}] \subset \mathfrak{l}_{\mathbb{R}} \cap \mathfrak{a}_{\mathbb{R}} = 0$. In particular $[[\mathfrak{a}_{\mathbb{R}}, \mathfrak{a}_{\mathbb{R}}], \mathfrak{a}_{\mathbb{R}}] = 0$. So $[\mathfrak{a}_{\mathbb{R}}, \mathfrak{a}_{\mathbb{R}}]$ is a commutative ideal in the semisimple algebra $\mathfrak{l}_{\mathbb{R}}$, in other words $\mathfrak{a}_{\mathbb{R}}$ is commutative. □

The main result of this section is the following generalization of the standard decomposition of a finite-dimensional real parabolic. We have formulated it to emphasize the parallel with the finite-dimensional case. However some details of the construction are rather different; see Proposition 8.12 and the discussion leading up to it.

Theorem 8.3. *The minimal parabolic subalgebra* $\mathfrak{p}_{\mathbb{R}}$ *of* $\mathfrak{g}_{\mathbb{R}}$ *decomposes as* $\mathfrak{p}_{\mathbb{R}} = \mathfrak{m}_{\mathbb{R}} + \mathfrak{a}_{\mathbb{R}} + \mathfrak{n}_{\mathbb{R}} = \mathfrak{n}_{\mathbb{R}} \inplus (\mathfrak{m}_{\mathbb{R}} \oplus \mathfrak{a}_{\mathbb{R}})$, *where* $\mathfrak{a}_{\mathbb{R}}$ *is commutative, the Levi component* $\mathfrak{l}_{\mathbb{R}}$ *is an ideal in* $\mathfrak{m}_{\mathbb{R}}$, *and* $\mathfrak{n}_{\mathbb{R}}$ *is the linear nilradical* $(\mathfrak{p}_{\mathbb{R}})_{\text{nil}}$. *On the group level,* $P_{\mathbb{R}} = M_{\mathbb{R}} A_{\mathbb{R}} N_{\mathbb{R}} = N_{\mathbb{R}} \rtimes (M_{\mathbb{R}} \times A_{\mathbb{R}})$ *where* $N_{\mathbb{R}} = \exp(\mathfrak{n}_{\mathbb{R}})$ *is the linear unipotent radical of* $P_{\mathbb{R}}$, $A_{\mathbb{R}} = \exp(\mathfrak{a}_{\mathbb{R}})$ *is isomorphic to a vector group, and* $M_{\mathbb{R}} = P_{\mathbb{R}} \cap K_{\mathbb{R}}$ *is limit–compact with Lie algebra* $\mathfrak{m}_{\mathbb{R}}$.

Proof. The algebra level statements come out of Lemma 8.1 and the semidirect sum decomposition $\mathfrak{p}_{\mathbb{R}} = (\mathfrak{p}_{\mathbb{R}})_{\text{nil}} \inplus (\mathfrak{p}_{\mathbb{R}})_{\text{red}}$.

For the group level statements, we need only check that $K_{\mathbb{R}}$ meets every topological component of $P_{\mathbb{R}}$. Even though $P_{\mathbb{R}} \cap G_{n,\mathbb{R}}$ need not be parabolic in $G_{n,\mathbb{R}}$, the group $P_{\mathbb{R}} \cap \theta P_{\mathbb{R}} \cap G_{n,\mathbb{R}}$ is reductive in $G_{n,\mathbb{R}}$ and θ_n-stable, so $K_{n,\mathbb{R}}$ meets each of its components. Now $K_{\mathbb{R}}$ meets every component of $P_{\mathbb{R}} \cap \theta P_{\mathbb{R}}$. The linear unipotent radical of $P_{\mathbb{R}}$ has Lie algebra $\mathfrak{n}_{\mathbb{R}}$ and thus must be equal to $\exp(\mathfrak{n}_{\mathbb{R}})$, so it does not effect components. Thus every component of P_{red} is represented by an element of $K_{\mathbb{R}} \cap P_{\mathbb{R}} \cap \theta P_{\mathbb{R}} = K_{\mathbb{R}} \cap P_{\mathbb{R}} = M_{\mathbb{R}}$. That derives $P_{\mathbb{R}} = M_{\mathbb{R}} A_{\mathbb{R}} N_{\mathbb{R}} = N_{\mathbb{R}} \rtimes (M_{\mathbb{R}} \times A_{\mathbb{R}})$ from $\mathfrak{p}_{\mathbb{R}} = \mathfrak{m}_{\mathbb{R}} + \mathfrak{a}_{\mathbb{R}} + \mathfrak{n}_{\mathbb{R}} = \mathfrak{n}_{\mathbb{R}} \inplus (\mathfrak{m}_{\mathbb{R}} \oplus \mathfrak{a}_{\mathbb{R}})$. □

The reductive part of the algebra $\mathfrak{p}_{\mathbb{R}}$ can be constructed explicitly. We do this for the cases where $\mathfrak{g}_{\mathbb{R}}$ is defined by a hermitian form $f : V_{\mathbb{F}} \times V_{\mathbb{F}} \to \mathbb{F}$ where $\mathbb{F}$ is $\mathbb{R}$, $\mathbb{C}$ or $\mathbb{H}$. The idea is the same for the other cases. See Proposition 8.12 below.

Write $V_{\mathbb{F}}$ for $V_{\mathbb{R}}$, $V_{\mathbb{C}}$ or $V_{\mathbb{H}}$, as appropriate, and similarly for $W_{\mathbb{F}}$. We use f for an $\mathbb{F}$-conjugate-linear identification of $V_{\mathbb{F}}$ and $W_{\mathbb{F}}$. We are dealing with a minimal Levi component $\mathfrak{l}_{\mathbb{R}} = \bigoplus_{j\in J} \mathfrak{l}_{j,\mathbb{R}}$ where the $\mathfrak{l}_{j,\mathbb{R}}$ are simple. Let $X_{\mathbb{F}}$ denote the sum of the corresponding subspaces $(X_j)_{\mathbb{F}} \subset V_{\mathbb{F}}$ and $Y_{\mathbb{F}}$ the analogous sum of the $(Y_j)_{\mathbb{F}} \subset W_{\mathbb{F}}$. Then $X_{\mathbb{F}}$ and $Y_{\mathbb{F}}$ are nondegenerately paired. Of course they may be small, even zero. In any case,

$$\begin{aligned} &V_{\mathbb{F}} = X_{\mathbb{F}} \oplus Y_{\mathbb{F}}^{\perp}, W_{\mathbb{F}} = Y_{\mathbb{F}} \oplus X_{\mathbb{F}}^{\perp}, \text{ and} \\ &X_{\mathbb{F}}^{\perp} \text{ and } Y_{\mathbb{F}}^{\perp} \text{ are nondegenerately paired.} \end{aligned} \tag{8.4}$$

These direct sum decompositions (8.4) now become

$$V_{\mathbb{F}} = X_{\mathbb{F}} \oplus X_{\mathbb{F}}^{\perp} \quad \text{and} \quad f \text{ is nondegenerate on each summand.} \tag{8.5}$$

Let X' and X'' be paired maximal isotropic subspaces of $X_{\mathbb{F}}^{\perp}$. Then

$$V_{\mathbb{F}} = X_{\mathbb{F}} \oplus (X'_{\mathbb{F}} \oplus X''_{\mathbb{F}}) \oplus Q_{\mathbb{F}} \text{ where } Q_{\mathbb{F}} := (X_{\mathbb{F}} \oplus (X'_{\mathbb{F}} \oplus X''_{\mathbb{F}}))^{\perp}. \tag{8.6}$$

The subalgebra $\{\xi \in \mathfrak{g}_{\mathbb{R}} \mid \xi(X_{\mathbb{F}} \oplus Q_{\mathbb{F}}) = 0\}$ of $\mathfrak{g}_{\mathbb{R}}$ has a maximal toral subalgebra $\mathfrak{a}^{\dagger}_{\mathbb{R}}$, contained in $\mathfrak{s}_{\mathbb{R}}$, in which every element has all eigenvalues real. One example, which is diagonalizable (in fact diagonal) over $\mathbb{R}$, is

$$\begin{aligned} \mathfrak{a}^{\dagger}_{\mathbb{R}} = \bigoplus\nolimits_{\ell\in C} \mathfrak{gl}(x'_\ell\mathbb{R}, x''_\ell\mathbb{R}) \text{ where} & \\ \{x'_\ell \mid \ell \in C\} \text{ is a basis of } X'_{\mathbb{F}} \text{ and} & \\ \{x''_\ell \mid \ell \in C\} \text{ is the dual basis of } X''_{\mathbb{F}}. & \end{aligned} \tag{8.7}$$

We interpolate the self-taut semiclosed generalized flag $\mathcal{F}$ defining $\mathfrak{p}$ with the subspaces $x'_\ell\mathbb{R} \oplus x''_\ell\mathbb{R}$. Any such interpolation (and usually there will be infinitely many) gives a self-taut semiclosed generalized flag $\mathcal{F}^{\dagger}$ and defines a minimal self-normalizing parabolic subalgebra $\mathfrak{p}^{\dagger}_{\mathbb{R}}$ of $\mathfrak{g}_{\mathbb{R}}$ with the same Levi component as $\mathfrak{p}_{\mathbb{R}}$. The decompositions corresponding to (8.4), (8.5) and (8.6) are given by $X^{\dagger}_{\mathbb{F}} = X_{\mathbb{F}} \oplus (X'_{\mathbb{F}} \oplus X''_{\mathbb{F}})$ and $Q^{\dagger}_{\mathbb{F}} = Q_{\mathbb{F}}$.

In addition, the subalgebra $\{\xi \in \mathfrak{p}_{\mathbb{R}} \mid \xi(X_{\mathbb{F}} \oplus (X'_{\mathbb{F}} \oplus X''_{\mathbb{F}})) = 0\}$ has a maximal toral subalgebra $\mathfrak{t}'_{\mathbb{R}}$ in which every eigenvalue is pure imaginary, because f is definite on $Q_{\mathbb{F}}$. It is unique because it has derived algebra zero and is given by the action of the $\mathfrak{p}_{\mathbb{R}}$-stabilizer of $Q_{\mathbb{F}}$ on the definite subspace $Q_{\mathbb{F}}$. This uniqueness tell us that $\mathfrak{t}'_{\mathbb{R}}$ is the same for $\mathfrak{p}_{\mathbb{R}}$ and $\mathfrak{p}^{\dagger}_{\mathbb{R}}$.

Let $\mathfrak{t}''_{\mathbb{R}}$ denote the maximal toral subalgebra in $\{\xi \in \mathfrak{p}_{\mathbb{R}} \mid \xi(X_{\mathbb{F}} \oplus Q_{\mathbb{F}})) = 0\}$. It stabilizes each $\mathrm{Span}(x'_\ell, x''_\ell)$ in (8.7) and centralizes $\mathfrak{a}^\dagger_{\mathbb{R}}$, so it vanishes if $\mathbb{F} \neq \mathbb{C}$. The $\mathfrak{p}^\dagger_{\mathbb{R}}$ analog of $\mathfrak{t}''_{\mathbb{R}}$ is 0 because $X^\dagger_{\mathbb{F}} \oplus Q_{\mathbb{F}} = 0$. In any case we have

$$\mathfrak{t}_{\mathbb{R}} = \mathfrak{t}^\dagger_{\mathbb{R}} := \mathfrak{t}'_{\mathbb{R}} \oplus \mathfrak{t}''_{\mathbb{R}}. \tag{8.8}$$

For each $j \in J$ we define an algebra that contains $\mathfrak{l}_{j,\mathbb{R}}$ and acts on $(X_j)_{\mathbb{F}}$ by: if $\mathfrak{l}_{j,\mathbb{R}} = \mathfrak{su}(*)$, then $\widetilde{\mathfrak{l}_{j,\mathbb{R}}} = \mathfrak{u}(*)$ (acting on $(X_j)_{\mathbb{C}}$); otherwise $\widetilde{\mathfrak{l}_{j,\mathbb{R}}} = \mathfrak{l}_{j,\mathbb{R}}$. Define

$$\widetilde{\mathfrak{l}_{\mathbb{R}}} = \bigoplus_{j \in J} \widetilde{\mathfrak{l}_{j,\mathbb{R}}} \quad \text{and} \quad \mathfrak{m}^\dagger_{\mathbb{R}} = \widetilde{\mathfrak{l}_{\mathbb{R}}} + \mathfrak{t}_{\mathbb{R}}. \tag{8.9}$$

Then, by construction, $\mathfrak{m}^\dagger_{\mathbb{R}} = \mathfrak{m}_{\mathbb{R}}$. Thus $\mathfrak{p}^\dagger_{\mathbb{R}}$ satisfies

$$\mathfrak{p}^\dagger_{\mathbb{R}} := \mathfrak{m}_{\mathbb{R}} + \mathfrak{a}^\dagger_{\mathbb{R}} + \mathfrak{n}^\dagger_{\mathbb{R}} = \mathfrak{n}^\dagger_{\mathbb{R}} \Subset (\mathfrak{m}_{\mathbb{R}} \oplus \mathfrak{a}^\dagger_{\mathbb{R}}). \tag{8.10}$$

Let $\mathfrak{z}_{\mathbb{R}}$ denote the centralizer of $\mathfrak{m}_{\mathbb{R}} \oplus \mathfrak{a}_{\mathbb{R}}$ in $\mathfrak{g}_{\mathbb{R}}$ and let $\mathfrak{z}^\dagger_{\mathbb{R}}$ denote the centralizer of $\mathfrak{m}_{\mathbb{R}} \oplus \mathfrak{a}^\dagger_{\mathbb{R}}$ in $\mathfrak{g}_{\mathbb{R}}$. We claim

$$\mathfrak{m}_{\mathbb{R}} + \mathfrak{a}_{\mathbb{R}} = \widetilde{\mathfrak{l}_{\mathbb{R}}} + \mathfrak{z}_{\mathbb{R}} \text{ and } \mathfrak{m}_{\mathbb{R}} + \mathfrak{a}^\dagger_{\mathbb{R}} = \widetilde{\mathfrak{l}_{\mathbb{R}}} + \mathfrak{z}^\dagger_{\mathbb{R}}, \tag{8.11}$$

for by construction $\mathfrak{m}_{\mathbb{R}} \oplus \mathfrak{a}_{\mathbb{R}} = \widetilde{\mathfrak{l}_{\mathbb{R}}} + \mathfrak{t}_{\mathbb{R}} + \mathfrak{a}_{\mathbb{R}} \subset \widetilde{\mathfrak{l}_{\mathbb{R}}} + \mathfrak{z}_{\mathbb{R}}$. Conversely, if $\xi \in \mathfrak{z}_{\mathbb{R}}$ it preserves each $X_{j,\mathbb{F}}$, each joint eigenspace of $\mathfrak{a}_{\mathbb{R}}$ on $X'_{\mathbb{F}} \oplus X''_{\mathbb{F}}$, and each joint eigenspace of $\mathfrak{t}_{\mathbb{R}}$, so $\xi \subset \widetilde{\mathfrak{l}_{\mathbb{R}}} + \mathfrak{t}_{\mathbb{R}} + \mathfrak{a}_{\mathbb{R}}$. Thus $\mathfrak{m}_{\mathbb{R}} + \mathfrak{a}_{\mathbb{R}} = \widetilde{\mathfrak{l}_{\mathbb{R}}} + \mathfrak{z}_{\mathbb{R}}$. The same argument shows that $\mathfrak{m}_{\mathbb{R}} + \mathfrak{a}^\dagger_{\mathbb{R}} = \widetilde{\mathfrak{l}_{\mathbb{R}}} + \mathfrak{z}^\dagger_{\mathbb{R}}$.

If $\mathfrak{a}_{\mathbb{R}}$ is diagonalizable as in the definition (8.7) of $\mathfrak{a}^\dagger_{\mathbb{R}}$, in other words if it is a sum of standard $\mathfrak{gl}(1;\mathbb{R})$'s, then we could choose $\mathfrak{a}^\dagger_{\mathbb{R}} = \mathfrak{a}_{\mathbb{R}}$, hence we could construct $\mathcal{F}^\dagger$ equal to $\mathcal{F}$, resulting in $\mathfrak{p}_{\mathbb{R}} = \mathfrak{p}^\dagger_{\mathbb{R}}$. In summary:

Proposition 8.12. *Let $\mathfrak{g}_{\mathbb{R}}$ be defined by a hermitian form and let $\mathfrak{p}_{\mathbb{R}}$ be a minimal self-normalizing parabolic subalgebra. In the notation above, $\mathfrak{p}^\dagger_{\mathbb{R}}$ is a minimal self-normalizing parabolic subalgebra of $\mathfrak{g}_{\mathbb{R}}$ with $\mathfrak{m}^\dagger_{\mathbb{R}} = \mathfrak{m}_{\mathbb{R}}$. In particular $\mathfrak{p}^\dagger_{\mathbb{R}}$ and $\mathfrak{p}_{\mathbb{R}}$ have the same Levi component. Further we can take $\mathfrak{p}_{\mathbb{R}} = \mathfrak{p}^\dagger_{\mathbb{R}}$ if and only if $\mathfrak{a}_{\mathbb{R}}$ is the sum of commuting standard $\mathfrak{gl}(1;\mathbb{R})$'s.*

Similar arguments give the construction behind Proposition 8.12 for the other real simple direct limit Lie algebras.

9 The inducing representation

In this section $P_{\mathbb{R}}$ is a self-normalizing minimal parabolic subgroup of $G_{\mathbb{R}}$. We discuss representations of $P_{\mathbb{R}}$ and the induced representations of $G_{\mathbb{R}}$. The latter are the *principal series* representations of $G_{\mathbb{R}}$ associated to $\mathfrak{p}_{\mathbb{R}}$, or more precisely to the pair $(\mathfrak{l}_{\mathbb{R}}, J)$ where $\mathfrak{l}_{\mathbb{R}}$ is the Levi component and J is the ordering on the simple summands of $\mathfrak{l}_{\mathbb{R}}$.

We must first choose a class $\mathcal{C}_{M_{\mathbb{R}}}$ of representations of $M_{\mathbb{R}}$. Reasonable choices include various classes of unitary representations (we will discuss this in a moment) and continuous representations on nuclear Fréchet spaces, but "tame" (essentially the same as II_1) may be the best with which to start. In any case, given a representation κ in our chosen class and a linear functional $\sigma : \mathfrak{a}_{\mathbb{R}} \to \mathbb{R}$ we have the representation $\kappa \otimes e^{i\sigma}$ of $M_{\mathbb{R}} \times A_{\mathbb{R}}$. Here $e^{i\sigma}(a)$ means $e^{i\sigma(\log a)}$ where $\log : A_{\mathbb{R}} \to \mathfrak{a}_{\mathbb{R}}$ inverts $\exp : \mathfrak{a}_{\mathbb{R}} \to A_{\mathbb{R}}$. We write E_κ for the representation space of κ.

We discuss some possibilities for $\mathcal{C}_{M_{\mathbb{R}}}$. Note that $\mathfrak{l}_{\mathbb{R}} = [(\mathfrak{p}_{\mathbb{R}})_{\text{red}}, (\mathfrak{p}_{\mathbb{R}})_{\text{red}}] = [\mathfrak{m}_{\mathbb{R}} + \mathfrak{a}_{\mathbb{R}}, \mathfrak{m}_{\mathbb{R}} + \mathfrak{a}_{\mathbb{R}}] = [\mathfrak{m}_{\mathbb{R}}, \mathfrak{m}_{\mathbb{R}}]$. Define

$$L_{\mathbb{R}} = [M_{\mathbb{R}}, M_{\mathbb{R}}] \text{ and } T_{\mathbb{R}} = M_{\mathbb{R}}/L_{\mathbb{R}} .$$

Then $T_{\mathbb{R}}$ is a real toral group with all eigenvalues pure imaginary, and $M_{\mathbb{R}}$ is an extension $1 \to L_{\mathbb{R}} \to M_{\mathbb{R}} \to T_{\mathbb{R}} \to 1$. Examples indicate that $M_{\mathbb{R}}$ is the product of a closed subgroup $T'_{\mathbb{R}}$ of $T_{\mathbb{R}}$ with factors of the group $L'_{\mathbb{R}}$ indicated in the previous section. That was where we replaced summands $\mathfrak{su}(*)$ of $\mathfrak{l}_{\mathbb{R}}$ by slightly larger algebras $\mathfrak{u}(*)$, hence subgroups $SU(*)$ of $L_{\mathbb{R}}$ by slightly larger groups $U(*)$. There is no need to discuss the representations of the classical finite-dimensional $U(n)$, $SO(n)$ or $Sp(n)$, where we have the Cartan highest weight theory and other classical combinatorial methods. So we look at $U(\infty)$.

Tensor Representations of $U(\infty)$. In the classical setting, one can use the action of the symmetric group $\mathfrak{S}_n$, permuting factors of $\otimes^n(\mathbb{C}^p)$. This gives a representation of $U(p) \times \mathfrak{S}_n$. Then we have the action of $U(p)$ on tensors picked out by an irreducible summand of that action of $\mathfrak{S}_n$. These summands occur with multiplicity 1. See Weyl's book [23]. Segal [17], Kirillov [12], and Strătilă and Voiculescu [18] developed and proved an analog of this for $U(\infty)$. However those "tensor representations" form a small class of the continuous unitary representations of $U(\infty)$. They are factor representations of type II_∞, but they are somewhat restricted in that they do not even extend to the class of unitary operators of the form $1 + (\text{compact})$. See [19, Section 2] for a summary of this topic. Because of this limitation one may also wish to consider other classes of factor representations of $U(\infty)$.

Type II_1 Representations of $U(\infty)$. Let π be a continuous unitary finite-factor representation of $U(\infty)$. It has a character $\chi_\pi(x) = \text{trace}\,\pi(x)$ (normalized trace). Voiculescu [22] worked out the parameter space for these finite-factor representations. It consists of all bilateral sequences $\{c_n\}_{-\infty<n<\infty}$ such that (i) $\det((c_{m_i+j-i})_{1\leqq i,j\leqq N} \geqq 0$ for $m_i \in \mathbb{Z}$ and $N \geqq 0$ and (ii) $\sum c_n = 1$. The character corresponding to $\{c_n\}$ and π is $\chi_\pi(x) = \prod_i p(z_i)$ where $\{z_i\}$ is the multiset of eigenvalues of x and $p(z) = \sum c_n z^n$. Here π extends to the group of all unitary operators X on the Hilbert space completion of $\mathbb{C}^\infty$ such that $X - 1$ is of trace class. See [19, Section 3] for a more detailed summary. This may be the best choice of class $\mathcal{C}_{M_\mathbb{R}}$. It is closely tied to the Olshanskii–Vershik notion (see [16]) of tame representation.

Other factor representations of $U(\infty)$. Let $\mathcal{H}$ be the Hilbert space completion of $\varinjlim \mathcal{H}_n$ where $\mathcal{H}_n$ is the natural representation space of $U(n)$. Fix a bounded hermitian operator B on $\mathcal{H}$ with $0 \leqq B \leqq I$. Then

$$\psi_B : U(\infty) \to \mathbb{C}\,, \text{ defined by } \psi_B(x) = \det((1 - B) + Bx)$$

is a continuous function of positive type on $U(\infty)$. Let π_B denote the associated cyclic representation of $U(\infty)$. Then ([20, Theorem 3.1], or see [19, Theorem 7.2]),

(1) ψ_B is of type I if and only if $B(I - B)$ is of trace class. In that case π_B is a direct sum of irreducible representations.
(2) ψ_B is factorial and type I if and only if B is a projection. In that case π_B is irreducible.
(3) ψ_B is factorial but not of type I if and only if $B(I - B)$ is not of trace class. In that case

    (i) ψ_B is of type II_1 if and only if $B - tI$ is Hilbert–Schmidt where $0 < t < 1$; then π_B is a factor representation of type II_1.
    (ii) ψ_B is of type II_∞ if and only if (a) $B(I - B)(B - pI)^2$ is trace class where $0 < t < 1$ and (b) the essential spectrum of B contains 0 or 1; then π_B is a factor representation of type II_∞.
    (iii) ψ_B is of type III if and only if $B(I - B)(B - pI)^2$ is not of trace class whenever $0 < t < 1$; then π_B is a factor representation of type III.

Similar considerations hold for $SU(\infty)$, $SO(\infty)$ and $Sp(\infty)$. This gives an indication of the delicacy in choice of type of representations of $M_\mathbb{R}$. Clearly factor representations of type I and II_1 will be the easiest to deal with.

It is worthwhile to consider the case where the inducing representation $\kappa \otimes e^{i\sigma}$ is trivial on $M_\mathbb{R}$, in other words it is a unitary character on $P_\mathbb{R}$. In the finite-dimensional case this leads to a $K_\mathbb{R}$-fixed vector, spherical functions on $G_\mathbb{R}$ and functions on the symmetric space $G_\mathbb{R}/K_\mathbb{R}$. In the infinite dimensional case it leads to open problems, but there are a few examples ([7, 24]) that may give accurate indications.

10 Parabolic induction

We view $\kappa \otimes e^{i\sigma}$ as a representation $man \mapsto e^{i\sigma}(a)\kappa(m)$ of $P_{\mathbb{R}} = M_{\mathbb{R}}A_{\mathbb{R}}N_{\mathbb{R}}$ on E_κ. It is well defined because $N_{\mathbb{R}}$ is a closed normal subgroup of $P_{\mathbb{R}}$. Let $\mathcal{U}(\mathfrak{g}_{\mathbb{C}})$ denote the universal enveloping algebra of $\mathfrak{g}_{\mathbb{C}}$. The *algebraically induced representation* is given on the Lie algebra level as the left multiplication action of $\mathfrak{g}_{\mathbb{C}}$ on $\mathcal{U}(\mathfrak{g}_{\mathbb{C}}) \otimes_{\mathfrak{p}_{\mathbb{R}}} E_\kappa$,

$$d\pi_{\kappa,\sigma,\mathrm{alg}}(\xi) : \mathcal{U}(\mathfrak{g}_{\mathbb{C}}) \otimes_{\mathfrak{p}_{\mathbb{R}}} E_\kappa \to \mathcal{U}(\mathfrak{g}_{\mathbb{C}}) \otimes_{\mathfrak{p}_{\mathbb{R}}} E_\kappa \text{ by } \eta \otimes e \mapsto (\xi\eta) \otimes e.$$

If $\xi \in \mathfrak{p}_{\mathbb{R}}$, then $d\pi_{\kappa,\sigma,\mathrm{alg}}(\xi)(\eta \otimes e) = \mathrm{Ad}(\xi)\eta \otimes e + \eta \otimes d(\kappa \otimes e^{i\sigma})(\xi)e$. To obtain the associated representation $\pi_{\kappa,\sigma}$ of $G_{\mathbb{R}}$ we need a $G_{\mathbb{R}}$-invariant completion of $\mathcal{U}(\mathfrak{g}_{\mathbb{C}}) \otimes_{\mathfrak{p}_{\mathbb{R}}} E_\kappa$ so that the $\pi_{\kappa,\sigma,alg}(\exp(\xi)) := \exp(d\pi_{\kappa,\sigma,\mathrm{alg}}(\xi))$ are well defined. For example we could use a C^k completion, $k \in \{0, 1, 2, \ldots, \infty, \omega\}$, representation of $G_{\mathbb{R}}$ on C^k sections of the vector bundle $\mathbb{e}_{\kappa\otimes e^{i\sigma}} \to G_{\mathbb{R}}/P_{\mathbb{R}}$ associated to the action $\kappa \otimes e^{i\sigma}$ of $P_{\mathbb{R}}$ on E_κ. The representation space is

$$\{\varphi : G_{\mathbb{R}} \to E_\kappa \mid \varphi \text{ is } C^k \text{ and } \varphi(xman) = e^{i\sigma}(a)^{-1}\kappa(m)^{-1}f(x)\}$$

where $m \in M_{\mathbb{R}}$, $a \in A_{\mathbb{R}}$ and $n \in N_{\mathbb{R}}$, and the action of $G_{\mathbb{R}}$ is

$$[\pi_{\kappa,\sigma,C^k}(x)(\varphi)](z) = \varphi(x^{-1}z).$$

In some cases one can unitarize $d\pi_{\kappa,\sigma,alg}$ by constructing a Hilbert space of sections of $\mathbb{e}_{\kappa\otimes e^{i\sigma}} \to G_{\mathbb{R}}/P_{\mathbb{R}}$. This has been worked out explicitly when $P_{\mathbb{R}}$ is a direct limit of minimal parabolic subgroups of the $G_{n,\mathbb{R}}$ [24], and more generally it comes down to transitivity of $K_{\mathbb{R}}$ on $G_{\mathbb{R}}/P_{\mathbb{R}}$ [26]. In any case the resulting representations of $G_{\mathbb{R}}$ depend on the choice of class $\mathcal{C}_{M_{\mathbb{R}}}$ of representations of $M_{\mathbb{R}}$.

References

1. A. A. Baranov, *Finitary simple Lie algebras*, J. Algebra **219** (1999), 299–329.
2. A. A. Baranov and H. Strade, *Finitary Lie algebras*, J. Algebra **254** (2002), 173–211.
3. E. Dan-Cohen, *Borel subalgebras of root-reductive Lie algebras*, J. Lie Theory **18** (2008), 215–241.
4. E. Dan-Cohen and I. Penkov, *Parabolic and Levi subalgebras of finitary Lie algebras*, Internat. Math. Res. Notices 2010, No. 6, 1062–1101.
5. E. Dan-Cohen and I. Penkov, *Levi components of parabolic subalgebras of finitary Lie algebras*, Contemporary Math. **557** (2011), 129–149.
6. E. Dan-Cohen, I. Penkov, and J. A. Wolf, *Parabolic subgroups of infinite-dimensional real Lie groups*, Contemporary Math. **499** (2009), 47–59.
7. M. Dawson, G. Ólafsson, and J. A. Wolf, *Direct systems of spherical functions and representations*, J. Lie Theory **23** (2013), 711–729.
8. I. Dimitrov and I. Penkov, *Weight modules of direct limit Lie algebras*, Internat. Math. Res. Notices 1999, No. 5, 223–249.

9. I. Dimitrov and I. Penkov, *Borel subalgebras of* $\mathfrak{l}(\infty)$, Resenhas IME-USP **6** (2004), 153–163.
10. I. Dimitrov and I. Penkov, *Locally semisimple and maximal subalgebras of the finitary Lie algebras* $gl(\infty)$, $sl(\infty)$, $so(\infty)$, *and* $sp(\infty)$, Journal of Algebra **322** (2009), 2069–2081.
11. A. Habib, *Direct limits of Zuckerman derived functor modules*, J. Lie Theory **11** (2001), 339–353.
12. A. A. Kirillov, *Representations of the infinite-dimensional unitary group*, Soviet Math. Dokl. **14** (1973), 1355–1358.
13. G. W. Mackey, *On infinite-dimensional linear spaces*, Trans. Amer. Math. Soc. **57** (1945), 155–207.
14. L. Natarajan, Unitary highest weight modules of inductive limit Lie algebras and groups, J. Algebra **167** (1994), 9–28.
15. L. Natarajan, E. Rodríguez-Carrington, and J. A. Wolf, *The Bott–Borel–Weil theorem for direct limit groups*, Trans. Amer. Math. Soc. **124** (2002), 955–998.
16. G. I. Olshanskii, *Unitary representations of infinite-dimensional pairs* (G, K) *and the formalism of R. Howe*, in Representations of Lie Groups and Related Topics, ed. A. Vershik and D. Zhelobenko, Advanced Studies in Contemporary Mathematics **7** (1990), 269–463.
17. I. E. Segal, *The structure of a class of representations of the unitary group on a Hilbert space*, Proc. Amer. Math. Soc. **8** (1957), 197–203.
18. S. Strătilă and D. Voiculescu, Representations of AF–algebras and of the Group $U(\infty)$, Lecture Notes Math. **486**, Springer–Verlag, 1975.
19. S. Strătilă and D. Voiculescu, *A survey of the representations of the unitary group* $U(\infty)$, in Spectral Theory, Banach Center Publ., **8**, Warsaw, 1982.
20. S. Strătilă and D. Voiculescu, *On a class of KMS states for the unitary group* $U(\infty)$, Math. Ann. **235** (1978), 87–110.
21. N. Stumme, *Automorphisms and conjugacy of compact real forms of the classical infinite-dimensional matrix Lie algebras*, Forum Math. **13** (2001), 817–851.
22. D. Voiculescu, *Sur les représentations factorielles finies du* $U(\infty)$ *et autres groupes semblables*, C. R. Acad. Sci. Paris **279** (1972), 321–323.
23. H. Weyl, *The Classical Groups, Their Invariants, and Representations*, Princeton Univ. Press, 1946.
24. J. A. Wolf, *Principal series representations of direct limit groups*, Compositio Mathematica, **141** (2005), 1504–1530.
25. J. A. Wolf, *Principal series representations of direct limit Lie groups*, in Mathematisches Forschungsinstitut Oberwolfach Report 51/210, Infinite-dimensional Lie Theory (2010), 2999–3003.
26. J. A. Wolf, *Principal series representations of infinite-dimensional Lie groups, II: Construction of induced representations*, Contemporary Mathematics, Vol. 598 (2013), 257–280.
27. J. A. Wolf, *Principal series representations of infinite-dimensional Lie groups, III: Function theory on symmetric spaces.* In preparation.

CPSIA information can be obtained at www.ICGtesting.com
Printed in the USA
BVOW10*2220070115

382405BV00001B/1/P

9 781493 915897